EUTOPIAS

Making Good Places
Ecologically & Culturally

Alan Wittbecker

40th Anniversary edition
2010

Books by Alan Wittbecker

Eutopias: A Poetic Commonwealth of Earth
Ordering Spaces and Living Places: Aesthetic & Ecological Dimensions of Place
Poetic Archaeology of the Flesh: Creative Language, Physics & the Ecology of Being
One Earth Many Worlds: The Role of Cosmologies on Ecological Impact & Accommodation
REviewing REthinking REturning: Essays of Life, Ecology, and Ecological Design (2 ed.)
Good Forestry from Good Theories & Good Practices: Essays on Ecosystem Medicine & Ecological Forestry (2 ed.)
Eutopian Essays: Towards Making Good Places with Thought Experiments
*[O]utopias Or [E]*utopias (Eutopias Part 1)
Topopoetics (Eutopias Part 2)
Global Emergency Actions (Eutopias Part 3)
Redesigning the Planet
Redesigning Places & Regions in the Planet

EUTOPIAS

Making Good Places
Ecologically & Culturally

Using Thought Experiments, Radical
Philosophy, Deep Ecology, Holeconomics, Real Politics,
Religious Sensitivity, Ecopoetics, Synergetic Design, & Common Sense

Alan Wittbecker

40th Anniversary edition
2010

Clio Press
Sarasota

For more information on sites and projects in text:
SynGeo ArchiGraph Co.: www.syngeo.org
Ecoforestry Institute: www.ecoforestry.net
G. P. Marsh Institute: www.marshinstitute.net
Pan Ecology: www.panecology.net
Rian Garcia Calusa: www.riangarciacalusa.com
Eutopian Ecologists: www.eutopias.net

Copyright © 1970, 1976, 1980, 1984, 1990, 1994, 2000, 2004, 2010 Alan Wittbecker

All rights reserved under International and Pan American Copyright Convention. No part of the book may be reproduced in any form or by any means, including information storage and retrieval systems, without the prior written consent of the Author or Publisher, excepting brief quotes used in reviews:
Clio Press, Mozart & Reason Wolf
SynGeo ArchiGraph, P.O. Box 370, Tallevast, Florida 34270

Publisher's Cataloging in Publication Data
Alan Wittbecker 1946-
Eutopias: Making Good Places Ecologically & Culturally
p. cm.
Includes Bibliographical References and Index
1. Human Ecology. 2. Radical Philosophy. 3. Ecological Design
I. Title.
GF75.W5851
ISBN 0-911385-43-6 (paper)

Book Design by Rian Garcia Calusa
Manufactured in the United States of America
10 9 8 7 6 5 4 3 2

40th Anniversary September 2010

Contents

Dedications

To Garrett Hardin, for his support and correspondence, as well as for his criticism, commiseration and bad Darwinian postcards. I have followed his lead on many things, from tackling unpopular topics to supporting the Hemlock Society, and I am grateful that he led the way.

Special thanks to John B. Cobb Jr., for his constant support and encouragement, and for inviting me to speak at the Center for Process Studies at Claremont (1996). All I really wanted was for him to write this book; I know that he would have done a better job.

Special thanks to Michael W. Fox, for showing me that being a professional pariah can be an honorable thing, and then for blazing that path. He offered me more help and inspiration than I can ever repay. I have tried to follow his lead in wolf studies, animal rights, renewable agriculture, and a vegan diet ever since.

Thanks to Arne Naess for hiring me to work on the Wolf Project and to lecture at the Center for Nature at the University of Oslo one spring semester (1987), between hiking, climbing and swimming expeditions. We met regularly over 25 years to discuss whether Deep Ecology was a form of Radical Ecology or Radical Ecology was a form of Deep Ecology. When he suggested a boxing match to resolve the issue, I found that my relative youth was not an advantage and he found that I had also boxed in college (I always gave him priority, anyway, since his article was published first).

Thanks to Norman Bowie for inviting me to work on this book at the Center for the Study of Values at the University of Delaware one summer (1986) as a Visiting Scholar. I got some new chapters finished, even with meeting friends and colleagues from 20 years earlier, when I dropped out to finish this book (and direct the local drug abuse clinic).

Thanks to Paolo Soleri for giving me the opportunity to conduct arcological thought experiments for two summers (1983 and 1992) at Arcosanti, as well as to clean toilets and kitchens and to work in the gardens. I still believe that is the best place on earth to live.

Thanks to Eugene Odum for his many comments after our conference presentations in 1983, 1986 and 1998. As a graduate student in ecology I learned from his textbook. As an instructor and ecosystem ecologist, I taught my first course using his textbook.

Thanks to Alan Drengson for publishing and promoting my articles, as well as for arranging for many of his students to take my first class in ecological forestry (1994). I am grateful for his help, friendship and correspondence.

Thanks to Emerson Wittbecker, for his patience and encouragement with all of my science projects, from fabric chemistry and interfacial polycondensation to potato proteins, fruit fly genetics, electricity generation, and stellar modeling. And, equally to Margaret Wittbecker, for her patience and encouragement on all of the wildlife collections, edible plants, forest explorations, and investigations in cultural ecology. Although she denied me the chance to drop out of grade school to live in the woods with the local hobo, Bill Carty, she has voluntarily read everything I ever wrote and then improved it.

And, of course, my continued gratitude to colleagues Michael Barnes, George Carroll, Nick Gier, Christyann Helm, Amy Ihrke, Twila Jacobsen, Neil Keefe, Nadya Kristoforova, Devorah Levi, Boyd Martin, Linda Martin-Schapf, William Odum, Nela Rachevitz, Karen Walter, and Precious Woulfe for their unselfish criticism, suggestions, support, or assistance. To my other tutors at International College: Neil Evernden, Buckminster Fuller, David Klein, Paul Shepard, and Henryk Skolimowski. Finally, if I had not been trapped in a mountain snowstorm for a month at the observatory on Mt. Lemon in early 1973, I might not have read books by Theodore Roszak, Ivan Illich, and Leopold Kohr, and become inspired and more determined to continue exploring these topics.

1.0. Making Good Places—The Idea Refined

This project was started in 1969, inspired by the works of Paolo Soleri, Garrett Hardin, Lewis Mumford, John B. Cobb, Jr., Leopold Kohr, Thomas Paine, Eugene Odum, and many others. I thought that their ideas could be combined into one synthetic project. The first draft resembled a series of footnotes to their works, held together by am ambitious outline. The book surrounded the idea of place with statistics and problems. The statistics on suffering and loss were disturbing and depressing—and are even more so now. I was working as a psychologist at a local hospital and taking graduate courses at night; in a geography course, we played a City Land-Use Game (CLUG), which led me to pay more attention to urban places. So, I tried to concentrate on the psychology of urban places. Driven by a sense of urgency, several of us at the University of Delaware started the Marsh Institute for Research and Education in Ecology; our first paying project was to clean up after Earth Day—there were two in 1970. Later that year, a partial draft was submitted for credit for linked graduate courses in Psychology, Philosophy, Communications, Art, Anthropology, and Geography. It was titled *Ordering Spaces and Living Places*, but was considered too rough and was rejected. Several months later the entire notebook was bound by Shamrock Press in Newark, Delaware as *Eutopias: A Poetic Commonwealth of Earth* and circulated to a limited readership.

1.1. History of the Book & Ideas

After courses in international communications and philosophy, I expanded the outline as a utopian project. Dizzy with enthusiasm, I came up with many brilliant and original ideas, from an explication of the good side, "eutopias," to the correlation of crowding and ugliness to sickness and crime. Then, I found that most all of these ideas had been described already— eutopias addressed by Buckminster Fuller and the ecological design charted by Ian McHarg. After reading Ian McHarg's *Design with Nature*, I played with his idea that the relative beauty of a place could be correlated with illness and crime. People seem to be unhappy for an indefinite number of reasons, such as lack of food or prestige, too few symbols, like money, or lack of meaning. I considered that living in good places would increase human happiness; good places could provide food, security, and meaning. So, I tried to locate good places everywhere and then describe them. I started to define what made them good. And then, I tried to figure out how to design good places, to define the kind of actions that would result in good places. By then I was teaching physics laboratories at a Canadian university.

At first, I wanted to show what good places looked like and then to distill their common features. The more I looked, the more I realized that it was almost impossible for good places to exist without a network of connected, surrounding good places. Even enclaves required a relatively stable environment and exchanges with an outside society. So, good places had to have good environments and good social contexts. As I traveled and began to describe places, I saw that they were already changing, because people like me wanted to live in good places, and if the places existed, then people would move there and enjoy them. Alas, when many people move to a good place, the place changes, and the reasons people move there are overwhelmed and disappear because we do not know how to respect limits. I moved to a place that I had identified as good and began teaching philosophy there, in Idaho, as a graduate fellow. I did not want to advertise good places only to have them be swamped and diminished by people who wanted to find a good place, rather than to stay in place and build good the qualities they desire (the irony of my own decision was not lost on me).

So, I rewrote the manuscript as a discussion of the kind of problems that existing communities had to respond to in order to survive. I sent out a group of chapters on the metaphysics, metaphors and metaecologies of place in 1976; it was accepted by Libra Press (New York), but published by Reason Wolfe (Wilmington) as *The Poetic Archaeology of the Flesh*. I was teaching ecology in Oregon, as a graduate teaching fellow, trying to finish my first degree, a doctorate. That same year, with several partners, I started a comprehensive design company, Nieman Ryan Community Designs (Portland, OR)—we designed everything from books and posters to wetlands and forests.

No longer comfortable listing examples of good places or emphasizing problems, I rewrote the book as an abstract discussion of the differences between utopian and eutopian thinking about societies and places. Then, I immediately rewrote it with an ecological perspective, expanding the discussion of good places with ecological definitions. This continued to the early 1980s, while I was teaching computing and mathematics and attempting to finish my formal education in Los Angeles, at a small institution dedicated to tutorial learning. Having learned from books, from teachers I had never met, I jumped at the chance to work with the brightest and wildest thinkers of the time. International let me select my own tutors: Buckminster Fuller, Paul Shepard, Joe Meeker, Michael W. Fox, John B, Cobb, Jr., Arne Naess, Paolo Soleri, David Klein, Henryk Skolimowski, and Neil Evernden. Fuller decided not to be an official tutor, but suggested new directions and ideas. Shepard did not like anything I had done, but that forced me to examine and criticize his ideas, before I incorporated them into my project. Meeker praised my integrative skills, but suggested I was more interested in collecting big names so I could bask in their reputations—perhaps that was some part of my choices, but the glow never paid the rent or got anything published. Fox took over as my chief advisor and introduced me to the whole dimension of inhumane farming practices and the misuse of science. Fox, Klein, and Naess encouraged me with wolf projects. Soleri and Skolimowski invited me to work at Arcosanti. Skolimowski and Cobb required numerous rewrites on ecological philosophy and process theory.

So, I thought of just describing a cosmological framework. But, the more I understood about human psychology, the more I realized that not all people would want good places. Like many mammals, some people preferred suboptimal environments, that is, they preferred danger and change to a safe comfortable environment. So, I realized that the framework had to be suboptimal and loose. It had to be culturally flexible, yet protective of each region. After a while, the manuscript incorporated many of the design ideas from our work in Oregon, Washington and Idaho; another version was circulated in 1983, and articles from it were published in several journals. Slowly, problems were replaced with cultural descriptions and ecological descriptions of good places, which fed back into our design work. To meet degree requirements, I focused only on a framework for human cosmologies; this excerpt was accepted as a dissertation and published in a limited edition in 1984 as *One Earth Many Worlds*.

Our design group realized that although we could influence local design, there was a whole dimension of untapped regional and global design possibilities, so we started an offshoot, SynGeo ArchiGraph, for that kind of design in Seattle in 1991. We approached global corporations, cities, and nations to offer ecological planning and design services that addressed populations and styles, as well as multiple scales and ecological principles. We finished several *pro bono* projects on regional design, including a design for the North Slope of Alaska, but were unable to interest anyone in using this approach or in hiring us for a design. Financially, the business was a failure. We took part-time jobs designing books, boulevards, forests, and streams in Massachusetts, New Hampshire, Washington, and Oregon. The book was expanded to include the designs of good places. To supplement my income, I taught courses in environmental science for a university with branches in Seattle and Vancouver (BC).

With another set of colleagues, I helped to set up education and training for the Ecoforestry Institute in Portland; we also assembled an Ecoforestry Management Team to manage forests. We cofounded the Forest Stewardship Council in Oaxaca, Mexico to try to set high standards for forest product certification. Many of the examples in this book are from our research, management and restoration experience with northwest coniferous forests. I taught workshops in forestry and courses in forest ecology and planning; finances improved. Modified further with global ecological designs, the manuscript was sent out again in 1992 and 1997. Individual articles were published during that time.

After brief participation in wolf projects in Alaska and Siberia, both truncated due to lost funding, I worked on a wolf survey in Bulgaria for two years with the Peace Corps (with six other grants). There, I was invited to submit regular essays on ecology as a newspaper columnist (in English) and a series of articles on ecology and forestry to a science magazine (in Bulgarian). I tore apart the book and rewrote it as a series of essays, mostly from previously published articles. My quiet interludes of writing were punctuated by the excitement of being lost in a blizzard, being caught in an avalanche, climbing mountains bare-handed, encountering fantastic wildlife, trying to teach English to children raised with television, walking the country, and marrying a naval engineer (who applied her project management skills to the wolf surveys).

In 2002, I sent out a series of essays from the book. One publisher suggested putting all the published reviews and essays together in one book, *REviewing REthinking REturning*, which they published that year, and which won a 2003 EPPIE award for best nonfiction book on the web. The following year, Cambridge Press and Ebooksonthenet published a second collection about forestry design and practices, *Good Forestry from Good Theories & Good Practices*. And, the next year, a third set of essays taken from this book, all previously published in journals, magazines, or other books, became *Eutopian Essays* (Urania Science Press 2004). Although I worried if these books would have any effect, I also wondered if my predecessors and professors would approve of my use of their works. Of course, I am not a professional writer, scientist or designer. I am an amateur in the real sense of the word: I love playing with ideas and forms. If I am not skilled enough or persuasive enough, that fault lies with me and not with the teachings or ideas.

I reintegrated some of the essays back into the matrix manuscript, concentrating more attention on the politics of place and international order. I kept working, planting trees and creating landscape designs. I started teaching courses in anthropology, ecology, and design at a small college in Florida. I continued teaching distance learning courses on forest ecology and design to students in Canada, Armenia, Chile, Germany, and Romania. These college and independent courses allowed me to try different approaches for dealing with culturally specific alternatives.

I kept working to involve people, not only in celebrations, such as earth day, but in lifestyle changes and joyous living. One of my arguments was that we had to treat the dismal effects of modern civilization, from deforestation to global disease patterns, as a catastrophic emergency that required immediate action, but we should not let this slow, global emergency overwhelm all joy and creativity. As I was writing about why it was necessary, I participated in animal surveys and planted thousands of trees to restore overused forests.

In 2004, I broke the book into three volumes: *[O]utopias or [E]utopias, Topopoetics*, and *Global Emergency Actions*. With the aid of several friends and the financial backing of another friend, we set up our own Urania Science Press and published these volumes, which are on the web and sold by CreateSpace and Amazon.

This work was written and rewritten over a period of forty years. It parallels the philosophical discussion of an early draft. In some places, I have left it the way it was originally, even though some of the ideas have been presented later by others. In other places, I have quoted their works, especially where they have presented things better or additional ideas that I did not consider. Very few arguments are used in this book. Many good arguments for cultural collapse and ecological damage have already been made. This book assumes those problems and offers proposals that might dissolve the problems. Arguments for or against these proposals might come later.

This project will never be finished to my satisfaction, but I want to present these ideas in raw outline form as a series of questions and suggestions. It remains an ambitious outline with many empty spaces. Of course, empty space may stimulate the creativity of others. The book was first written as an expanded outline. It was revised as a series of essays that made specific arguments. It was revised again as a series of answers to specific questions. The binding idea has always been the ecological and cultural creation of good places.

The form of this work reflects an effort to describe the ideas and creation of good places. It is not a linear path leading to a definite conclusion. It is a dialectical spiral, which gathers facts and ideas and reforms them in wider contexts, then tries to use repetition and charm to lead readers to agreement. But, it also requires more effort from you, the reader. The work presents its subjects in a loose form. If the metaphors and ideas stretch too far to hold, perhaps they are, as Ludwig Wittgenstein wrote of his own propositions, all nonsense, to be read and left behind.

1.2. Questioning Places

What are places? What are good places? How are they made or developed, located or designed? What makes something good? Can good places increase human productivity or happiness? Can good places provide opportunities for wildness and otherness, especially in the form of diverse habitats and unique species? Can we recognize good places, or describe their characteristics? Can we consciously design good places? These are some of the questions that I wanted to find answers for or to try to answer myself.

Like any "thing" or system, a place is extended in space and bounded, so it has an inside and outside. As part of a geographic approach, most places can be located on the earth. Like any finite thing, a place is a unique set embedded in a larger system; it is a local thing connected to other places by dynamic processes and cycles. Places have meaning for the species that have modified them by living there for generations. For humans, places have even more of a psychological dimension that reflects the personal investment of energy and emotion in a place.

Flatscapes, Christian Norberg-Schulz's appropriate term, have dulled landscapes, and places have become monotonous and boring, displaying fewer unique qualities and fostering weaker attachment. Places become homogenous and interchangeable, as people disperse everywhere their limited ideas and ways of relating to the nonhuman. David Brower urges that the human sense of place needs to be revived.

Places cannot be preserved, without cutting their vital connections and making them lifeless. Places cannot be restored to some Arcadian fantasy, without severely limiting their movement and development. Places cannot be created using a machine metaphor that boasts the replaceability or substitution of anything.

There have been attempts to define and design places. Edward Relph outlined the essence of place and disappearance of variety that results in placelessness. Christopher Alexander decomposed environmental objects and activities into their constitutive elements, to be reconstructed into designs that could fit local places. These formal solutions can improve strategies for design and can provide a matrix for the making of places. But, they should not assume that human variables can be manipulated to achieve a predicted response. And, they cannot ignore psychological or cultural patterns.

Ian McHarg showed that the relative beauty of a place could be correlated with illness and crime. People seem to be unhappy for an indefinite number of reasons, such as lack of food or prestige, too few symbols, like money, or lack of meaning. I considered that living in good places would increase human happiness; good places could provide food, security and meaning. So, I tried to locate good places everywhere and the describe them.

At first, I wanted to show what good places looked like and then to distill their common features. The more I looked, the more I realized that it was almost impossible for good places to exist without a network of connected, surrounding good places. Even isolated enclaves required a relatively stable environment and exchanges with an outside society. So, good places had to have good environments and good social contexts. I investigated the kind of problems that existing communities had to respond to in order to survive. But, first I tried to understand the philosophical and physical basis of place. Parts of this work were published as a metaphysical description of the depth of ecology into Being.

No longer comfortable listing examples of good places or emphasizing their problems, I rewrote the book as an abstract discussion of the differences between utopian and eutopian thinking about societies and places. Parts were published as philosophical essays on the thread of eutopias in utopian thought.

I thought of just describing a cosmological framework. The description of a holocosmology was published separately. But, the more I understood about human psychology, the more I realized that not all people would want good places. Slowly, problems were replaced with cultural descriptions and ecological descriptions of good places. I started paying more attention to very long-term trends, and defined a series of long-term catastrophes that worked against the survival of good places. Some of the solutions were published separately as a set of emergency actions.

Then, I started to define what made places good. And, I tried to figure out how to design good places, to define the kind of changes that would result in good places. Despite the advances in ecology, there seemed to be something missing, a description or process for making places, especially good places. Then I immediately rewrote it with an ecological perspective, expanding the discussion of good places with ecological definitions.

There must be a way to define places, using an ecological approach, and considering cultural modifications and adaptations to places. The approach has to be tradition-based and partially self-conscious. It has to be responsive to the genius of place, as well as to human meaning derived from the existential and phenomenological significance of the place. The concept of place incorporates physical, biological, and cultural dimensions. The solutions cannot, and do not, need to be precise; they can be unfinished and ambiguous. They can be fuzzy. It does not design or guarantee rootedness or workability, but it can identify limits. It can provide possibilities through a matrix that allows tradition and richness. It can provide direction from the understanding of the parts. It can understand how to make a fertile kind of soil, as a metaphor, where things can live in and develop. Living beings synthesize the parts and find meaning in living there, and in doing so revitalize a place.

This part of the whole utopian project, "Eutopias," is concerned with outlining a scientific basis for defining and designing good places. It is not a finished work, or even an eloquent presentation. It is, as Whitehead suggested of his work, an adventure of ideas. The adventure is open and partially formed. Please help and add to it.

1.3. Recognizing Emergencies

What emergencies? In the case of industrial nations, we have been embracing excess for many generations, so that we are crippled by stress and sickness. Perhaps all we need is a diet. The essence of a diet is to restore yourself to health, by restricting unhealthy consumption. As societies and cultures may also be guilty of this kind of behavior, so they need to put themselves on a diet. Archaic and agricultural nations have been strapped by historical inequities and unfair trading. Their challenge is to avoid simply repeating the same errors and consequences in the rush to acquire minimum standards and wealth. The solutions for all nations include trying new kinds of balance for self-reliance, paying attention to cultural and physical catastrophes, and striving for better equity. Because of the extent of our overuse and ecosystemic conversions, and their effects on natural sources, this situation is an ***emergency***.

The nature of an emergency requires everyone to drop their normal activities and normal behaviors and to respond to a catastrophe. The catastrophe is usually quite evident, a wall of fire or a massive surge of water that will destroy or has already destroyed homes and people, as well as insects and birds, plants and animals, and their habitats. But, we are finding that not all catastrophes are fast, human-scale or visible. The effects of those changes make us uneasy but not adrenaline-ready; the changes are reflected in starving children, hotter summers and stronger storms, failing food supplies, and collapsing infrastructures. We seem reluctant to give the causes of these catastrophes the status of real emergencies, partly because the catastrophes seem like natural events, such as a warming trend, and partly because they are related to our industrial habits, which provide us with necessities, as well as with comforts and luxuries.

We have to learn to recognize and respond to these slow catastrophes, these invisible catastrophes and these very large and long-term catastrophes. And, we have to do it now, before they crest and become overwhelming. We can do it. We have the evidence that things are taking a downward turn (the original meaning of the word catastrophe). We have acted on a large-scale before, in times of a world war. We were able to treat war as an emergency and to encourage or enforce remarkable changes, such as rationing or job-remolding. We were able to take these actions without destroying our citizens or our cultures.

Although nature, and our human nature, are not enemies to be vanquished, the current situation has similarities to war. Massive changes threaten our lifestyles. Resources are removed from our reach by thoughtless or inefficient use. Changing insect and animal populations seem to be attacking our food supplies. Species being forced to extinction. Habitats are collapsing; Dangerous chemical wastes are accumulating. Ozone holes are growing, extreme climates pressing, and the entire planet seems to be wobbling. Changes in climate and ocean balance, as well as renewed diseases and infiltrated toxic chemicals, threaten our lives. And, it is happening everywhere, at once.

We have been fooled by the fact that we cannot see an enemy. We have been misled by the slowness and subtlety of the penetration of our defenses. We have been betrayed by our own desire to continue our industrial dreaming at any cost. Some people have noticed changes and have been crying alarms, but they have not been loud enough or persuasive enough. Everybody needs to be awakened; everybody needs to participate, everybody needs to sacrifice and work towards peaceful solutions.

The big problems seem insurmountable, and simple actions will not save our civilization from catastrophes. Part-time participation will not be enough to reverse the degradation of ecosystems, and partial business greening will not stop the unraveling of global cycles.

What is needed is an immediate, comprehensive approach to this situation—wise actions in a eutopian framework. As a framework, ***Eutopias*** is a paratraditional nostrum for preserving what is good and useful in human cultures and sciences, and for reserving what is necessary for nature to keep regenerating itself, while addressing the cascading problems of the modern expansion and development with an emergency approach. Eutopias is a practical framework for allowing the creative anarchy of traditional-size cultures to be able to implement appropriate technology to deal with their resources and with other cultures through a revitalized and empowered international body that has the power of taxing global resources and properties for its own support, as well as the power to disarm and neutralize the unhealthy influences of large nations and corporations. Eutopias is a framework that limits human expansion to domestic and artificial areas, by specifying responsibilities and duties, while permitting the free operation of nature on the majority of the planet. It saves neopoetic areas and reserves wilderness. It encourages respect for natural and cultural capital. It recommends recognizing limits and planning for them using an ecological perspective and a metaphorical approach—it is metaphor-based as well as science-based, and limits-based as well as culture-based. Eutopias is concerned with saving human cultures and the environments that human cultures have come to fit in comfortably.

Why would Eutopias work? Because life has over three billion years experience with changing and adapting, because human life and cultures have over 50,0000 years of practical experience adapting and making changes, and because humans are immensely adaptable—if they can adjust to poverty and suffering, they can adjust to a few good changes. Perhaps it is already too late—limits have been passed and the catastrophes cannot all be reversed. We do not know, and may never know, but we can still act as if we were wise, as if doing the right thing makes a difference. And, we will have worked together to help others, to improve things and to make good places. If we act ***now,*** this month, this week, this day, this hour!

2.0. Places on Earth

Human consciousness of our effects on the planet and on each other has gradually increased, as shown by recycling programs and by celebrations such as earth day, which has been getting larger every year, as a celebration. But, celebrations did not seem to have lasting effects. Environmental deterioration has worsened in some places. Levels of consumption have increased; populations have increased.

This year's earth day celebration (1990) is over. What were we celebrating? That we *are going* to save the earth or maybe just still think about it? Perhaps we were celebrating our intention to go on a material diet or an opportunity to spend money on t-shirts and buttons. Perhaps earth day is a new spring-time variation of a new year's resolution—a temporary awareness, a limited intent, and a reason to party before business as usual. Or, perhaps it is a modern penance that allows us to buy a place in heaven by promising to save the earth with small tokens.

The token changes and vows do help, but are they enough? Will a little conservation avoid a great human disaster? Are these easy remedies reminiscent of medical cures for diseases and problems, such as smoking, overeating and stress, that could be avoided by simple denial. The implication is that a few small things, such as using less water or recycling bottles, will save the earth—that ozone depletion, rainforest destruction, population growth, and the polarities of wealth will somehow be corrected automatically, as governments and industries continue as before, adjusting their labels by using greener colors.

We have been told that saving the earth starts in the home. No wonder corporations give their blessings to this event—most pollution and waste is industrial and agricultural! Which issues have higher priority? Deadly local ones, such as toxic waste dumps or topsoil loss, or deadly global ones, such as greenhouse gases or chemical runoff? Is an alarm justified, or is caution enough? Should we listen to ecological Cassandras or to economic Neros? Are we too lazy to follow through with the effort that we already celebrated? Are we too cheap to deduct a required percentage from profits to pay the real environmental costs? Are we too crazy to stop our material and biological growth? Where is the will and the vision necessary to really make radical changes? We have, in fact, taken the easy way at every branch. We have assumed that corporations will choose the proper path of production and regulate their pollution. Yet, we know that they put profit first and only prevent pollution when forced to do so by public action or government regulations.

We have wasted thirty years attacking the symptoms and not their technological or social origins. We must acknowledge the failure of our remedial efforts, and our failure even to address the flaws of our designs and ideologies.

I do not want to participate in the wrong games. I do not want to drive fewer miles in a high-powered gas guzzler—I want to travel by train; I want my radio shipped by train and not truck; I do not want my vegetables shipped at all, but grown locally. I do not want farmers to do slightly less aerial spraying of fertilizers and biocides, I want organic produce. I do not want to recycle aluminum and plastic, I want returnable glass containers. I do not want safer coal-burning centralized power plants, I want local solar power. I do not want to give more money to the homeless, I want my tax money to help them build and keep their own homes. If industries cannot help me with what I want, then I want my government to encourage them and channel them, tax them and regulate them. And, if my government cannot do that, I want to encourage a change in government and support better candidates.

Leaders, corporate, religious or political, rarely decide the direction of development. People living in communities decide that. Decisions bubble up through consciousness. People get tired of consuming and seek more meaningful ways of living in nature. And, out of frustration, they participate in selecting leaders or run for office themselves, or, in my case, write a grand outline about it. The history of the book has ranged from statistics to the psychology of place, from the ecology of place to the politics of place.

This work is a collection of thoughts, but like Aristotle said about emotion: Thought by itself moves nothing. Our living together, and our ancient social tendencies, give rise to a shared morality that directs our behavior. I would want nothing more from this book than it direct a discussion and perhaps suggest good behaviors.

We need to propose and execute national policies to steer technology. I want my representatives to ban CFCs, to ban burning, to tax nonrecyclables— and if they do not, then I will run for office myself. We do not have time to look at all the information that we have collected, or to convert it to knowledge. We never have had time. We cannot connect with all the information flows. So, we will have to act as if the information we have is enough for wise decisions.

The earth does not need to be saved or healed, as if we could do either. The ways of life that we remember and prefer, the places that depend on other species and natural processes—these can be saved. Our own divided minds, that let the poor be enslaved by the wealthy, that let 'good' animals be domesticated and 'bad' animals be eradicated, can be healed. The sacrifices will have to be great; the changes will have to be radical. But, the celebrations will be meaningful only then.

The local environment, and that of the region and planet, offer opportunities and challenges to all living organisms. How humans respond to these things results in accomplishments or problems, in creations or deaths. Almost every place on earth has been claimed, and modified, by some living being. And, almost every living being has been influenced and changed by place. Places are patterns of things and webs of relations that can be understood by observing and participating.

2.1. Platopias: Facing Flatscapes from Losses

The local environment, and that of the region and planet, offers opportunities and challenges to all living organisms. How humans respond to these things results in accomplishments or problems, in creations or deaths. Every year numbers are collected in every area of human interest. Those numbers indicate the deterioration of water and air quality, the erosion of soil and land, the destruction of forests, the decline in health and longevity, the deaths of people, the fracture of cultures, and the wobble of planetary cycles. In the past few years, the numbers have worsened dramatically.

The numbers of human deaths are far more disturbing. There have been massive human die-offs in the past 106 years since the start of the Twentieth Century. These numbers tell a story of big death. From democides—the intentional killing of races, nations, tribes or communities—total deaths may range from a minimum of 150 million to a possible 350 million people, who were shot, knifed, burned, suffocated, poisoned, starved, crushed, drowned, hanged, bombed, or buried alive, in a plague of violence. This includes Josef Stalin's 1932-3 forced famine that killed 7 million people, as well as the Nazi holocaust (6 million) of Jews and Gypsies under Hitler by 1945, the current deaths in the Congo (by 2006, 4 million), Pol Pot's Democratic Kampuchea, Khmer Rouge 1975-79 (over 1.5 million), Sudan (1985-2004, 2 million), Ethiopia (1978, 2 million), North Korea (1948-87, perhaps 2 million murdered), and Pakistan (1940-50s, 1.5 million murdered). Rwanda (1994) and Bosnia-Herzegovina (1992-5), add 800,000 and 200,000 to the total. Is it possible to even imagine that number?

Deaths as the result of formal or declared wars are surprising less than democides, which are often internal to a nation, at about 61 million dead. The war between China and Japan, before and during the second world war (1932-1945), resulted in 14 million dead. The rest of the world may have experienced twenty one million dead during the six war years.

Famine, once the greatest producer of deaths in agricultural nations for thousands of years, may have killed 32 million in this century. One of the largest modern famines was the Great Famine in China, centered in the early 1950s; it killed 10.7 million people. There were two famines in India, in 1942 and 1965, that each killed 1.5 million people from drought. Two famines in Korea, in 1948-87 and 1995-8, each killed over 1 million people.

Disease, once the greatest producer of deaths during early phases of globalization, has killed many millions, especially at the end of the Roman empire, when the Justinian plague in 540-590, might have killed 100 million; the Black Death of the 1300s in Europe and Asia, and the beginning of the Spanish Empire, produced 46 million dead. In the Twentieth Century, the influenza pandemic in 1918 killed over 20 million people, and possibly as many as 50 million. The recent AIDS epidemic, from 1978 to 2001, counts for over 23 million.

Disasters, from drought, flooding, earthquakes and other regular planetary events, killed 21 million people. Accidents, from transportation from horses to space shuttles, killed half a million people. Murders and terrorism killed hundreds of thousands (perhaps 0.02 million). Nations, such as the U.S. or the United Kingdom, have killed thousands of innocents; the U.S. is guilty of indiscriminate bombings of Germany and Japan in 1945, then later in parts of Africa and the Middle East, and the U.K. caused thousands of deaths with its 1914-1919 food blockades of Germany and the Middle East. Such a list would also have to include Afghanistan, Angola, Albania, Ethiopia, Burundi, China (1917-1949), Croatia, Czechoslovakia, Indonesia, Iraq, Turkey (1919-1923), and Uganda.

Many of these overlap by category. For instance, ethnic cleansing can start from changes in the distribution of food, that lead to famine and disease. Poverty is never listed as a cause of death, but over 15 million, perhaps as many as 30 million people, die from a lack of clean water, food, medical services, or shelter, every year, including 2006.

The numbers on the living environment are disturbing and critical: The list of animal deaths in the United States in 1984 reads like a doomsday book of atrocities: 22,078 North Pacific fur seals clubbed to death; 17 million mammals trapped for fur in the U. S.—303 million throughout the world; 12 million unwanted pets put to death; 70 million laboratory animals used in experiments; 3.5 billion chickens killed for food; 700,000 cattle dead from transport-related injuries; 598,757 animals shot for sport on wildlife refuges. Although species are still being identified at the rate of 8,500 new insects species and 100 new fish species per year, probably 400 species are driven to a premature extinction and 1,000,000 species are threatened every year. Statistics for habitats are almost incomprehensible. Three billion cubic meters of wood are consumed annually. Twelve million hectares of forest are cleared annually and 10 million hectares are degraded. Marsh lands are filled in; coral reefs are mined; and grasslands are paved over.

Possibly 59 percent of arable land degraded; 50 percent of fresh water co-opted for human use; 50 percent of the planet's wetlands modified, drained or destroyed; 50 percent of the coral reefs damaged and perhaps 20 percent destroyed; extinctions that are uncounted, perhaps uncountable.

We tend to think of problems as unwanted 'side-effects' of the wanted main-effects, but all effects are equal, as Buckminster Fuller noted, and must be addressed as equal. A problem, from the Greek words 'to throw forward,' can be considered as a question proposed for solution. Most things identified as problems are embedded in a network. Nothing is simple; there is not one truth, there is not one way, there is not one goal. Problems could be considered also as challenges that we must respond to continuously, in the process of living, not as puzzles that have to be solved once for all time. A challenge is a calling into question or a demanding task (a challenge is 'a call to take part in'). It is about consciously choosing to see what can be done, rather than dismissing a conflict as terrible and unsolvable. When challenged by some situation, we react by habit, although this may be disconnected from other habits. Habits protect us from many problems. Addressing a problem often has to do with a power struggle, which becomes part of the problem of the problem. If problems are regarded as challenges that require a social response, then much of the conflict can be avoided.

The problems of cultures, of natural ecosystems, and of modern, industrial, corporate, urban civilization, have been documented quite thoroughly. We have identified most of the problems in the problematique, from erosion, pests, and fertility loss, to population migration and diseases, and we have addressed them separately, using technological innovations or political adjustments. But, we have not dealt with them in a whole pattern. We have not understood them as complex large dynamic systems.

Sometimes we forget, however, that the decisions of our ancestors can saddle us with losses, just as our losses will encumber our heirs with deforested landscapes on depleted soils, despoiled by exotic chemicals and hazardous wastes, in a network of impoverished habitats with an unstable climate, and of course, compounded by large intergenerational financial debts. The losses indicated by these numbers can be categorized simply.

The local environment, and those of the region and planet, offer opportunities and challenges to all living organisms. How humans respond to these things results in accomplishments or problems, in creations or deaths, in vital places or flat wastelands.

2.1.1. *Eight Terrible Losses*

This network of problems can be grouped under eight large categories, each of which contains a multitude of related problems. Each category extends numerous threads to the other categories.

First, there is the loss of nature and the wild, the loss of the whole and the pieces, ecosystems, habitat, species, and individuals. The loss of culture leads to the inability to adapt to unique environments due to unstable social environments. The loss of health results in weakness and sickness, the inability to maintain the self or to produce necessities. The loss of fitness leads to the inability to function under normal environmental conditions and to reproduce. The loss of equity results from the massive breakdown of the distribution of goods and luxuries. The loss of renewal reflects the inability of social systems to renew themselves or provide security and resources for their constituents. The loss of accord happens when people or cultures are unable to work together or control conflict. The loss of design is the inability to imagine, shape and build things that enhance life and safety; it is the inability to respond to changing circumstances.

All of these problems interact, so that it becomes difficult to isolate separate problems. These long-standing problems result from long-standing challenges to human health and happiness: Food shortages, disruptions in distribution, violence, stupidity, greed, forgetfulness—not just personal forgetfulness but larger forms of cultural amnesia—old diseases and new, lust for power, and the lust to consume. The challenges and our responses to them combine into many invisible, long-term trends, such as deforestation and the acquisition of property. These problems result in the death of many individual plants and animals, as well as entire species and whole habitats and ecosystems.

2.1.1.1. Loss of Nature (Habitats Species & Pieces)

The overuse of ecosystems results in deforestation, devegetation, and desertification, then in the depletion of raw materials and the depletion of agricultural productivity. Economic and political pressures, derived ultimately from population pressures, force farmers to intensify their efforts to increase crop production, instigating a dismal cycle of population expansion, environmental deterioration, and human poverty.

This overuse results in the attendant losses of many habitats, species, and individuals. Many species are not hunted to extinction; they are being lost as a result of the simplification of their habitats for human uses, from the expansion of agriculture to the pavement of cities and road systems. The rates of extinction are now far higher than the normal, unassisted rates. The loss of habitat also decreases any new speciation processes. Demands from a growing human population, as well as their violent conflicts, accelerate the losses. As the human world becomes more abstract and separated from the wild worlds, it becomes more isolated and human specific. As our habitats become more removed from natural processes, we tend to think they are all self-created and controlled, at least until a dramatic hurricane or earthquake overwhelms our abilities at control.

2.1.1.2. Loss of Culture

With the globalization of trade and the domination of national languages, smaller cultures are being absorbed into larger ones. Although people retain some of their traditions, the loss of their language means that the whole perspective of their culture, related to the uniqueness of place and human adaptation to place, is being lost. Knowledge of plants, animals, and processes is being lost at a time when modern science is half-heartedly trying to catalog knowledge of every place. A lack of knowledge is often related to illiteracy— or the lack of numeracy or ecolacy, to use Garrett Hardin's words, but this kind of loss is personal and cultural—it had been learned once and is lost as a result of a set of circumstances where it is devalued and ignored. Smaller, less-competitive cultures are being abandoned by new generations of people or subverted by global comparisons of luxuries and expenditures.

Population pressures, resource shortages, and manufacturing 'side effects' cause instability in many societies. Militarism, intolerance, crimes, and health problems are symptoms of the instability of cultures. Confusion and misinformation contribute further to the destruction of cultures. If cultures are lost and new forms cannot fit themselves to the patterns and uncertainty of natural systems, then people may not be able to adapt to the continued development of the earth and its ecosystems.

2.1.1.3. Loss of Health

The instability of cultures, as well as stress, insecurity, and insufficient diets, results in physical and psychological problems for people. Individual powerlessness and disillusion provoke further disintegration. With disruption of agricultural and manufacturing productivity, as a result of conflicts, people have less food and fewer necessities. Hunger affects over half the people in the world, in every culture, including many of the dominant cultures in China, Russia, and the United States. Extended hunger not only alters behavior and intelligence—that is the ability to evade hunger—but it allows diseases to run rampant. Many diseases have made dramatic comebacks where people live in large groups in unsanitary conditions with inadequate food.

Health can be reduced by long-term trends in lifestyles, from sedentary work to family size. The size of families is crucial for mental health, for instance. Basically, extended families allow more outlets for communication and spread the stress among members. Small families have to bear much more stress and members often become unhealthy. Loss of health leads to other losses.

2.1.1.4. Loss of Fitness

Fitness is the ability to function under normal environmental conditions. Fitness can be measured quantitatively by testing aerobic and anaerobic power, the capacity for sustained exercise, respiratory and heart rate markers; pulmonary gas exchange, mechanical power and strength, body composition; range of motion; and stamina. Stress, obesity, illness, erectile dysfunction, weak sperm, social conflict, and other kinds of dysfunctions, can reduce fitness. Some of this reduction, evidence of poor personal or reproductive health, is caused by chemicals, especially persistent chlorinated organic chemicals, such as polychlorinated biphenyts (PCBs) and dichlorodiphenyltrichloroethane (DDT).

Some of these chemicals also reduce our ability to reproduce, raise young, make good decisions about consumption and social conflict, and to persevere in general. The cost of fitness ranges from lower fertility and sperm viability to early death. However, a lifestyle dependent on physical and energy slaves, not to mention a diet riddled with addictions to cheap fats and sugars, decreases our fitness. All of the elements of a true addiction are present: We get a short term high, and we suffer long term health problems. The air pollution, the sedentary life styles, and the global warming are all injurious to health. Exposure to toxic chemicals and stressful city environments makes things worse, even as we acquire more information about the dangers and options.

Perhaps this fitness problem is a result of a long-term trend. As scavengers we had to gather seeds roots and fruits, as well as dead animals. As hunters we moved quite a bit; as farmers, we had to work harder and longer, planting, harvesting and storing food. Then we started using animals and better tools and fossil fuels; we invented labor-saving, and activity-saving, devices. So, now exercise is simply an activity outside the education or work day. As human populations expand and shift to cities to live, they are losing the ability to fit with natural ecosystems.

2.1.1.5. Loss of Equity

Although many resources are distributed unequally over the globe, as a result of different kinds of historical geological processes, trade can allow access to those resources. However, as a result of long-term processes of inequity, from keeping people enslaved to cultural hoarding, many people have far less than others. This has resulted in permanent overclasses and underclasses, which are maintained by physical force, as well as by the force of economic and religious myths. These myths tell all people that they participate in the "best possible economic system" regardless if it justifies the differences of inequity based on history or on perceived racial abilities. Most people are hungry; few people are fulfilled. Even low average levels of food and fulfillment can be maintained only through theft from other species and from future human generations, and through the degradation of billions of humans as well as the ecosystems on which they depend.

Other myths, such as the "market is free" or "growth is beneficial to all," suggest to people not to try other economic forms. But, the market is not free; it is orchestrated to benefit the rich. Continued growth is amoral and pathological, benefiting the elite of authoritarian regimes as much as the oligarchs of democratic ones. It refuses to recognize, much less to pay, all of its costs, such as depletion, loss of security—which may be most important—or extinction. The entire system perpetuates mass poverty and justifies it by blaming individuals, but the system itself fails to reduce inequity or poverty. This loss reduces effort; it is responsible for tiredness and low kinds of health, productivity, and esteem, things that are necessary for personal and systemic renewal.

2.1.1.6. Loss of Renewal

Self-renewing systems renew themselves through a process of renewal—yes, the definition is self-defining and circular. But, renewal is limited when too much of the system is lost through waste. Manufacturing for a large population in "free" economies results in the production of waste, and in the storage of solid wastes in landfills. The waste is unavoidable in the current system, which is unable to renew itself and has to force a linear flow and trust that natural recycling systems will be able to adjust to immensely higher volumes. Much loss in the current economic system is wrongly identified as waste. It is simply not economical to recover this waste in this system, due to the volume of "free" resources and goods from nature. The current political and economic systems are unable to use the unwanted or unexpected products of their systems.

Waste is a strangely inappropriate category; the waste of one system is almost always a potential resource for the next system. Even the waste of the sun, light, is a resource for plants on earth. People themselves are treated as a waste product of capital-producing system, although the system could be viewed as a waste-producing system that also produces capital.

Pollution is not so much waste as a kind of resource out of place, for example, in the acidification of rain, in salinization of waters, and in the eutrophication of water bodies. Many new chemical products, such as carbonated fluorocarbons (CFCs), pollute the atmosphere and water. Pollution interferes with natural cycles of renewal, from atmospheric cooling to purification of water by plants. Many manufacturing processes result in the production of new dangers, such as modified and uncontrolled genetic fragments, nuclear wastes from reactors and weapons, and new substances and products, which are not easily incorporated into natural cycles. The loss extends to human lives, as well.

2.1.1.7. Loss of Accord

As a result of the unequal distribution of natural resources, including unincorporated waste and pollution, and the unequal distribution of materials and wealth between people, economic conflicts arise, often becoming violent political conflicts. Accord, by definition, means agreement or the concurrence of will or action. It can mean harmony of mind, as well as harmony of sounds. As an agreement between the parties in a controversy, by which satisfaction for an injury is stipulated, accord allows people to reconcile their injuries and interests. Because many national boundaries were drawn as a result of colonial expansion and contraction, many cultures are artificially combined in large territories or stretched across several traditional territories. This has created the conditions for continuing cultural and political conflict. In addition to the normal conflict between different cultures with different ways and values, usually resolved by trade or distance, this new conflict resembles small permanent wars over large numbers of territories. This kind of hot conflict not only destroys habitats and resources, but it causes immense human suffering. As the rules change, and conflict includes noncombatants, as well as plants and animals, there is less accord between conflicting groups. Accord requires the ability to trust, which requires self-reliance and confidence. The lose of accord leads to other related losses.

2.1.1.8. Loss of Design

During the evolutionary development of human responses to environmental challenges, people tested many kinds of designs for their tools, houses, and living habits. These initial designs often worked well in situations of limited resources or limited power. With the application of more power and the acceleration of the interchange of dominant designs, we have been able to force standard designs to work under almost any conditions, often replacing adapted designs.

Many of our current problems are design problems, Victor Papenak judges, that is to say, bad design causes all varieties of suffering and waste; it causes death, from accidents or collapses, and immense waste by relying

on power to overcome its flaws. Many of these problems could be solved by better design, not just by more comprehensive industrial design or political design, but by the ecological design of entire processes and places. Of course some problems are caused by gravity, wildfires, or earthquakes, which are a natural part of the planetary system. While they may seem unsolvable, we can adapt to them with better designs, for instance, by reserving flood plains or landslide areas as parks without houses or buildings, and by adjusting the houses that are built to local conditions. Other problems are caused by human demands and the design response to them.

2.1.1.9. One Bad Combination of Losses

All of these problems interact, so that it becomes difficult to isolate separate problems. These longstanding problems result from longstanding challenges to human health and happiness: Food shortages, disruptions in distribution, violence, stupidity, greed, forgetfulness—not just personal forgetfulness but larger forms of cultural amnesia, old diseases and new, lust for power, and the lust to consume. These problems result in the death of many individual plants and animals, as well as entire species and whole habitats and ecosystems.

People die for many reasons: Environmental catastrophes, such as earthquakes, accidents, old age, lack of food, diseases, or personal violence. We know that people also die "indirectly" as a result of the destruction of natural ecosystems, specifically, as an example, the lack of clean, drinkable water. People die for economic reasons, such as not having a job to afford food, or not having a place to live, as a result of war or dislocation. It is almost impossible to calculate how many die in each category How many die directly from hunger, since hunger leads to disease, such as pneumonia, and disease gets the credit; how many die from unclean water, since the contaminant or the accident gets the credit? It is also difficult to calculate deaths from theft, economic dislocation, lack of planning, bad designs, or depression. Cause and effect make a complex dance of candidates in a network of things. Only specific diseases, accidents or bullets seem to be unambiguously fatal, and even these are often causally linked with many other contributing factors. We need to use better tools to understand and solve these problems, as well as to understand how a cosmology or economic system can promote the problems and then fail to solve them.

2.1.2. *Perceptual Challenges & Networks of Problems*

Many challenges to, or problems with, civilization are caused by our classical logic, which, in the western civilization, is deductive and bivalued; that is, we arrive at decisions from an overall image, and develop categories that are mutually exclusive. This binary approach results in an "either/or situation," rather than a "both/and" situation. For instance, the argument is tossed back and forth about whether our approach to resources should be anthropocentric or ecocentric, when in fact it should be neither. This argument occurs due to a misunderstanding of the concept of frame and focus.

Scale. Switching our attention from a focus to the larger frame usually involves a change in scale. Scale has several meanings. Basically, as it is used here, it has to do with the level of measurement in an ecosystem context.

Global & Local. Differences in scale are often called local or global differences. Concepts of scale apply to physical and biological fields. Local fields or events are those separated from other fields or events and exert minimal influence on the others. The internal connections inside a local field are stronger than the connections between local fields.

Maximum or Optimum. Western civilization has become enamored with the idea of the maximum; it is the goal for many kinds of planning and operations. Agriculture, for instance, is concerned with maximizing its products and profits, but it also narrows the ecological basis of food production and decreases human livelihood in the long run.

2.1.2.1. Focus/Frame

Everything that we focus our attention on can be seen to be of a larger framework. Focus and frame can be understood metaphorically—and, in fact, metaphor itself can be understood as consisting of two parts, according to Max Black: Focus and frame. The focus designates the figurative term signified through the process, and the frame refers to the subject or context.

Using this distinction, it can be seen that most of the fuss in ecosystem-based activities, such as forestry for illustration, has occurred at the focus level. Modern foresters have so long focused on trees that they forget that the forest is a frame that holds many foci, or points of view. Alan Drengson is fond of saying that we do not see the important operations of nature because we are looking through the wrong paradigm—a paradigm acts like a pair of glasses, focusing on what we want to see.

The elements of a forest are related psychologically, by foresters, as focus or frame, as contrast or uniformity, as dominant or recessive, or in a number of other pairs. For instance, forests can be considered by scientists as either matter systems or energy systems, but the focus on either frame permits subtle differences and limitations in interpretation. Some scientists describe organisms as being configured by energy through time, but, organisms are material patterns in space as well. Furthermore, it is not quite right to say that ecological forestry is ecocentric, because it is actually concerned with the frame and not the focus, the periphery and not the center; perhaps it should be called 'drymoperipheral' instead.

The anthropocentric needs of people can be contrasted with the needs of the forest itself. This dilemma can be answered by considering the focus/frame character of the situation. Both of the two basic views are merely aspects of one view of good forestry. Anthropocentric values that focus on commodities can only be considered in the context of the values that are contained in the whole; that is, they are derivative from the frame. Furthermore, it is not necessary to ascribe conscious purpose to the latter view. Massive disruption often results when a community falls out of balance with its local forest environment, and in fact industrial forestry only avoids the penalties for such disruption by trading advantageously with other communities in less powerful areas. Emphasizing balanced use, including natural forest requirements, rather than perpetual growth, as is done now, would permit the frame to stay pluralistic and multidimensional, and still provide resources to human populations.

2.1.2.2. Small & Large Scales

Switching our attention from a focus to the larger frame usually involves a change in scale. Scale has several meanings. Basically, as it is used here, it has to do with the level of measurement in a space/time energy/mass context. For instance, in forestry the measurements of leaf litter by foresters can be made at several scales: Single tree, stand, annual measurements, or stand life measurements. Processes that are unimportant at a small scale might be vital at a large scale. Too much litter in one stand in one year, for instance, might suppress a soil cycle; too little litter over a century might interfere with several regional or global atmospheric cycles.

Some patterns in forests are scale-dependent; for instance, hemlock trees may dominate small clusters, but be scattered all across the entire forested landscape. That is to say, the pattern changes with the scale. This is true of processes in forests as well. Canopies shade the understory annually, but fires in a lodgepole pine forest may increase dramatically the amount of light to forest floors once every 200-300 years.

The scale of the system defined, for example, wooded patch, ecosystem or temperate broadleaf forest, depends on the scale of the phenomenon being addressed. Mangrove forests are in phase with the frequency of hurricanes, although hurricanes may not influence the life histories of short-lived organisms as much as daily or seasonal cycles. Microbes, for instance, are affected more directly by short-term cycles of precipitation and temperature.

Problems in forestry arise where applications that work on a small physical scale are expanded to large scales, without thought for the difference or changes in patterns. For instance, it is well-known that a Douglas-fir tree is shade intolerant and grows best in openings that get light. Rather than simply remove single trees or small groups of trees, and release or plant the fir in the openings, industrial forestry applies the treatment to the entire landscape with large clearcuts, which alter the other conditions that firs require: Some shade, water, protection from browsing, and associated species. This is similar to the formal operation of Greek tragedy—applying a good idea to another situation or scale where it does not fit.

Other systems of cutting, such as high-grading or thinning from below, may also have unintended effects with changes in scale. High-grading is a form of group selection, in which the "best" and largest trees are removed. The practice was not overly destructive on a small scale, with many of the matriarch and patriarch trees remaining, but on a large scale, all old, large, heritage trees were removed. Thinning from below may be a good idea in a stand, but like high-grading, it is a form of biological selection—the gene pool is altered as the small and suppressed trees, some of which may be genetically superior to larger ones, are removed.

An inappropriate time scale, its shortness and urgency, is a cause of many other problems in industrial forestry. Although forests are considered renewable resources, they are slowly renewable, requiring hundreds or thousands of years to renew from catastrophic disturbance; this time is far longer than any economic plans, and really is nonrenewable on a human life scale. This has important implications on sustainability. For instance, we cannot keep cutting on short rotations without destroying the character of the forest; we need to plan for 500-year or 2000-year rotations.

Really large scales, bioregional or global, are not considered often at all. At large scales, other factors have to be considered. Many cosmic tendencies, such as disorder (entropy) and order (ektropy), as well as change, creativity and temporality, are relevant to forestry. But, managers tend to emphasize only those tendencies that seem positive, such as creativity, diversity, and order, and not those with negative connotations, such as entropy, death, and uniformity. The universe comes with the whole package, and we risk serious error by choosing only the tendencies we want.

2.1.2.3. Global & Local Fields

Differences in scale are often called local or global differences. Concepts of scale apply to physical and biological fields. Local fields or events are those separated from other fields or events and exert minimal influence on the others. That is, the internal connections inside a local field are stronger than the connections between local fields. Both quantum theory and relativity consider that local fields are generally noncausally and nonlocally connected—that is, no communication is instantaneous—through higher dimensional realities.

Local ecosystems are unique and original. In an ecosystem, each locality supports a segment of the total species population in a unique context, with a particular set of predators, competition, food, or physical habitat. The environment requires an enormous amount of minuscule local adaptations between the earth and its users. Production of pine cones, for example, is adapted to the local microsites; Loblolly pine planted fifty miles north or south of the seed source are less vigorous. The chemical DNA seems to reflect local conditions. Local ecosystems are separated from one another, not only in space, but by differences. For example, Dutch elm disease is a local problem, even if it seems ubiquitous in all forests with elms, produced by local actions and requiring local solutions.

Local systems, however, can affect global systems. Cutting down local forests may contribute to the discharge of greenhouse gases, such as carbon dioxide, into the local atmosphere. This may have the effect of increasing quantities in the global atmosphere, creating a global change, possibly a runaway increase in temperature.

Local systems emerge from a global system, which has characteristics that emerge from the interplay of local systems. Global systems have properties that no local systems have, such as an overall atmospheric temperature or global biogeochemical cycles. James Lovelock suggests that the interaction between hardwood forests and softwood forests may act as a global regulator of oxygen for the planet. Many things, such as human poverty and species extinctions, only seem global because they are happening in many local systems at the same time, and may effect global cycles.

2.1.2.4. Minimum & Maximum Limits

Our western civilization has become enamored with the idea of the maximum; it is the goal for many kinds of planning and operations. Agriculture, for instance, is concerned with maximizing its products and profits, but it also narrows the ecological basis of world food production and decreases human livelihood in the long run. We also strive to maximize good for ourselves, and the diversity or yield for nature, without a good grasp of what a maximum or minimum is.

In some senses, nature does try to maximize or minimize certain conditions. For instance a bubble minimizes surface tension by assuming a spherical shape that contains a maximum volume. Of course, we could also say that the bubble is an optimum form for the least possible surface area for a given interior volume. In principle, mathematicians reduce the question of maxima and minima to a geometric construct, often a three-dimensional surface with peaks and pits. In trying to apply the idea to a tree, however, the calculations become incredibly complex. A tree appears to create a maximum leaf area to collect radiation and a maximum number of seeds for reproduction, but it also tries to minimize evaporation and energy for its metabolism, both of which would set a smaller leaf area. So, leaf area reflects a compromise between gains and losses. The tree puts out an optimum or at least a satisfactory number of leaves. A forest is even more complex. In reality, the processes of trees and forests have many local maxima and minima; these can be represented graphically as surface potentials on a catastrophe field, as Rene Thom has done.

To survive, an ecosystem depends on the interactions and balance of many variables, most of which are not well understood. In agriculture or forestry we try to maximize one of those variables. When that happens, the balance or harmony is altered, and although it may take decades or centuries for the consequences to be known, the system is affected.

Many scientists and ecologists argue that we should maximize diversity. Diversity is an emergent property of ecosystems, arising from the activities of multitudinous beings learning to use the productivity of the system to augment their own flesh. The ecosystem is self-organizing, but diversity is not a goal of the system. Diversity is just a characteristic of mature ecosystems. But it is never an independent thing by itself; to think so is to be guilty of misplaced concreteness, according to A.N. Whitehead. In fact, nothing exists by itself, that is, not in relation to other characteristics, activities, or systems.

While it is meaningful to speak of an optimum diversity, as the result of limits and the interaction of many factors, a maximum diversity may never be reached in any system. As Paul Weiss noted, the patterns of organic nature are a combination of order and diversity; order involves constraint while diversity requires freedom for difference. Maximum order would result in a static universe, where a maximum diversity would create a nonordering chaos.

Few mammals—and humans are mammals—try for a maximum condition. Few mammals even try for an optimum; most strive for a satisficium, that is, a satisfactory amount, according to Francisco Varela. Some animals, individual rats, for instance, seek substandard, or more challenging, conditions. Varela analyzes the evolutionary process as satisficing rather than optimizing; that is, a suboptimal solution is adequate to continue living; striving for an optimum or maximum does not pay off in terms of investment of effort.

Individual people are concerned with having maximum freedom or producing maximum values, and in fact, the idea of unlimited value tends to encourage the effort. Values may be indefinite, but they may not be infinite. Like the principle of limited good, there may be a principle of limited value. Even aesthetic appreciation requires limits, to avoid having our over-appreciation overwhelm the values evident in nature.

2.2. Fashioning Tools for Challenges & Problems

A tool is a material object, or idea, that extends the human body or human thought. Technological things, such as tools or artifacts, are extensions of our physical being. The spoken language, using sound, engages and extends all the senses. The written language, using visual symbols or glyphs, can also engage and extend all senses, if the body resonates with them and recreates the full environment.

The users of tools had to have the ability to copy tools. Copying behavior may have shifting goals depending on context. Chimpanzees can learn to crack open oil-palm nuts using tools. This may start with the young chimp copying the mother's behavior, to be like her or to please her as part of the mother-infant bond. Young chimps may have to practice for three years to be able to coordinate hitting nuts with one stone as an anvil and another as a hammer. This nut-breaking is most often done at the end of the fruit season, when fruits are not available. The animals can get nine times the kilocalories that they put in as effort. Tool technology is a critical skill for eating and maintaining health. Stones are often kept at the nut "factory" although mothers may move the stones to familiarize infants with them. The copying has a lengthy investment by parent and child. The identity-based learning may shift to reward-based learning as chimps mature.

Tools may have first been used to acquire or store food, then to extend the efficiency of those processes. Tools have affects on human beings, from psychological distancing from things, as the tool is intermediary to a thing, to intimacy with the tool. There are physical effects as well: Loss of use of teeth, muscles, memory, loss of hand-eye coordination and perception of wild, but increase of the hand-eye coordination of tools. Each tool has an ideological bias, as skills or behaviors are amplified or ignored. To a man with a hammer, everything looks like a nail. To a woman with a grade sheet, everyone looks like a number, to a man with a computer everything is data, and to a woman with a camera everything is an image.

For tool-using cultures, tools had to do two things: First, they had to solve specific physical problems, and second, they had to serve the symbolic world of art in the construction of things, such as homes or monumental buildings. There are effects of tools on human cultures. The complexity of tools leads to rules for their use, and for who uses them, such as trade unions and clubs like masonry. Numbers become a tool to assess someone's behavior or worth. And, there are effects of tools on ecosystems, from the disturbance of soils to scale effects. Tools play a central role in the thought world of a culture, according to Neil Postman. Tools are not simply integrated into culture, they can reorder the culture. They may undermine the ideas in a culture or dominate them completely. At an extreme, tools may eliminate traditional world views and reduce the meaning of life to machinery. New technologies compete with old ones for dominance in a world view. The medium of technology contains an ideological bias. Tools attacks tools, according to Postman; printing attacks manuscripts, television attacks printing, painting attacks rock art, and photography attacks painting. Tool use has many effects, so care must be taken to use them consciously aware of their impacts and limits.

2.2.0.1. General Tools

Some tools, such as words or logic, we rarely think of as tools. Others, such as habits, stereotyping, myths, and even mathematics, we often use to hide or dismiss troublesome problems. Myths, for instance, are spread and remembered through stories told in every culture. Remembered stories become myths, accepted and not examined. Myths become part of the cultural heritage. The recital of myths is of great importance, though; myths provide a cognitive order for people. Myths are a major way of explaining the universe as it exists and communicating the cultural view of nature. Myths explain that which is otherwise incomprehensible, like death and disorder. To be understandable to members of a culture, the elements of a myth must be taken from the features and values of that culture. Neighborhoods share a common store of general myths of a larger field of culture or a nation, in which their local areas are embedded.

Myth sometimes starts as a cosmological account of the past. Any account of the past that is believed to be true is mythical. All interpretation and recounting of the past is a form of myth-making. That is why myths reflect the details of a culture. The experiences of many lives, in their daily succession, are coded in myths, along with the natural phenomena, features of the culture, supernatural beliefs, and moral values, in a social order in a physical place.

In the simplest human societies, mythology is the text of rites of passage. In Hindu, Chinese and Greek philosophy, it is the picture language of metaphysics. The first function is extended by the second; both harmoniously bind people to their world in its visible and transcendent aspects. Ludwig von Bertalanffy proposed that myth, magic and language are interconnected near the origin of symbolic activity, for instance, in the new stone age drawings at Altamira. Mythology can be said to be the womb of a person's initiation to life and death.

Levi-Strauss thought that mythology was explanatory, a product of observation and a reflection of reality. Myths often were created, however, out of inherited ideational pieces through fitting together. But thought proceeded through understanding, with the aid of distinctions and oppositions. Levi-Strauss argues that all myths conformed to a small number of structures. Most myths, for instance, have a built-in binary opposition. He cites the Zuni origin myth, where life and death are in opposition. He also contended that the oppositions were attempted to be resolved with an ambivalent mediating term; in the Zuni, this term was hunting, which was life-producing and death-producing. Levi-Strauss reasons that the human mind operated using oppositions, that were reflected in myths, and attempted to resolve them through myth.

Myth and rationality are polar; yet both exercise an ordering function in the world. One dimension may be analyzed by the other, but it is difficult. Mythology is a multidimensional expression, whereas rationality achieves clarity following a single idea in one dimension. One dramatic presentation in mythology could be transposed into rational thought, but it would require a series of analyses. Myth condenses a theme, which serves as a symbol for the interpretation of a situation in the world, thus ordering imagination. Philosophical thought orders reason, by dispersing a theme to trace the logical connection of its structure. The philosophical analysis of myth excavates implicit meanings, and tries to clarify them. It can bring to awareness ideas in the myth, but not replace them.

In Plato, myth and logical thought share authority. In general, myth attempts to explain natural phenomena by going beyond them to imprecisely determined causes, usually gods; reason, on the other hand, limits itself to a sufficient cause. Myth and logic are coextensive in many respects, but some aspects of myth are inaccessible to logic, and likewise some truths of logic are unformed in myth. Beginning with Socrates, and culminating with Aristotle, myth and logic became thoroughly separated; for Aristotle, myths provided material for drama and poetry, while experience promoted the rational sciences. By limiting the area of investigation, and curtailing the range of speech, logic achieved a greater degree of accuracy, but only at the cost of completeness of experience. Mythical thought, with its receptivity of images, has been eclipsed by the untiring activity of logical thought, which attempts to treat everything in its manner.

We have tried to produce a monolithic theory of the nature of myths, by describing them as inadequate attempts to describe nature, or as a protoscience explaining human origins, or as heuristic devices giving the whole vision to be completed by science.

In a scientific view, ethics and cosmology are separate. But in a good mythology, there is a coherence between areas. The individual feeds back into the cosmology in altered form what was received. It is almost like a closed loop between cosmology, culture and the individual. Archaic peoples translate the natural world into the language of myth. Being a narrative, a myth is aesthetic as well as intellectual. Myths develop in terms of their own internal logic, drawing together observations of the world. Claude Levi-Strauss described the process as bricolage, fitting the bits together, identifying impressions of life as sets and forming them into mythical systems; the world picture is a metaphorical puzzle. Bricolage is the mentality of synthesis, a technique for learning, creating, and expressing understanding, using whatever is available from the past and in the present to achieve an integrating form. This is what mythological thinking does, and what scientific thought might do. Mythologies are of major importance in the life of society and civilization.

Although ideology and mythology are systems of thought, they differ in at least one important respect. An ideology is a system of thought that aims to justify and preserve the status quo; its emphasis is on the present. A myth is characterized by traditional consciousness. Its emphasis is on the past and the sacred. Yet, it is a timeless understanding of the cosmos. Archaic myth makes all things into living beings, persons having life and movement. Myths sanctify the existence of a culture by underlining the continuity of human experience with recurring elements. Myths give places, even relatively featureless ones, special significance. Ascribing mythic significance to land strengthens human identification with it.

Joseph Campbell distinguishes three functions of mythology. The first function of a living mythology, the religious function, is to waken and maintain in the individual an experience of awe and respect in recognition of the ultimate mystery that transcends words and names, from which words turn back (the word *mythos* originally meant 'mute'). Myths are ways of teaching unobservable realities by way of observable symbols. This is so the individual has an identity. The second function of mythology is to provide a cosmology, an image of the universe. In fact, the world has to be recognized and assimilated by the mythopoetic imagination. Myth weaves human knowledge, skills and aspirations together in intersubjective realm of image that blends science and art. Mythic symbols can store and convey vast amounts of information concisely. Myth describes the range of behavior that ordinary people are capable of. Myths teach what it is important to know to live in a place. The third mythological function is the validation and maintenance of established order.

Malinowski argued that this function is primary, to justify human activities, in order to have those activities continued. Unless a cultural phenomenon satisfied a basic human need, it would not be repeated. But, myths can also go wrong. They can be constructed for other purposes, to protect interests or to deceive people about what is important.

In the U.S., for instance, the mentality of the frontier bloomed as the physical frontier was being closed. This mentality assumed the nature of a myth, that all people would prosper in the frontier way. Even though the 1890 census recognized that there was no more frontier, and warned people to use what they had, people have sustained the myth of limitlessness. The myth has become stronger than the logic of limits. Corporations and banks have used the myth to offset the dismal flavor of their economics. The victims, from miners to loggers, have been tricked into passionately defending their own exploitation and that of the environment.

The frontier of myth no longer exists. And we cannot understand new frontiers until we are divested of the old one. New myths are made all the time, from poetry and comic books to politics and advertising. Many of these new myths are self-consciously lies, for the purpose of manipulating people, justifying greedy behavior, or enforcing new behavior. We justify our temporarily successful behavior with myths that allow us to continue that behavior without being responsible. The myths of agriculture and forestry, as well as democracy and globalism, and progress and nationalism, are examined for their limits or flaws.

Myths change and evolve. Our old myths, that reality is hierarchical, that the earth is passively female, and that 'man' is lord over the earth, have proven to be dangerous. They should be retired to the scrolls of dead myths. The eighteenth century held that a golden age would appear if the priests and kings were deposed. In the nineteenth century, a central myth, founded at the 1851 exposition, was that industrialization would bring universal peace. A mythology can affect every element of our social, individual, spiritual, ecological and political life, all at once. Myths can be traps, if they are never renewed by every generation. Continued belief in myths about nature in the face of contradictory facts is a trap. Facts are always changing, as are myths, as is nature; and, myths and facts are based on nature and are interpretations.

Contemporary thinkers, such as McKinley, argue that we need a new mythology of humanity in nature. New myths can be developed, such as the myth of participating in organic beauty, where development, not growth, is without limit. The universe is a frontier; the mind is a frontier, but these frontiers are based on a whole and healthy planet and a nature of which humanity is a special part among many special parts. To achieve transformations of human culture, we must go beyond the authoritarian conspiracies and technocratic elitism, to create new myths. We will only learn to treat land as part of a community, and not as an exploited commodity, when we have new myths.

Mythology is a tool that can shape ideas and behaviors, good or bad. Many cultural or ecological problems can be examined using simple tools such as questioning or metaphors. Good words or good metaphors can illuminate a problem better than long explanations and tables of numbers.

2.2.0.2. Words as Tools

The etymological meanings of words sometimes describe the limits of their original uses. We should perhaps heed these limits when we use words in this kind of work. The history of a word sometimes yields clues about how it has been used, as well as nuances of thought.

The word 'nature,' for instance, means, literally, 'that which is born' or 'made again.' Nature is the self-making universe; it is the self-regulating power that moves itself. Nature is the wild environment of humanity, a feeling system that embodies the experience of its inhabitants, with mutual experiences and histories.

'Culture' is a word that refers to a form of human expression that is adaptive to the environment of nature. Culture is the complex whole—invisible, behavioral or material—that includes knowledge, belief, art, morals, law, custom, artifacts, things, and any other capabilities and habits developed or acquired by human beings as members of society in a specific environment. The world 'environment,' meaning surroundings, tends to imply a division between nature and culture. This division has been explored, widened and narrowed by many scholars. Perhaps a new word could be used to describe nature and culture as a whole, 'domiture.'

Domiture, a neologism made from the Greek word fragments for 'home' and 'again,' could do that. The system of culture is embedded in nature; domiture is a larger term to enclose the previous nature/culture dualism. It has to include human reason and human emotion, rather than simply putting them on opposite columns, as with order and chaos, higher and lower, linear and cyclic, as well as agriculture and wilderness. Nature needs a place to play, without controls. Culture needs a place to play, without being forced by nature. Domiture envelopes both concepts.

Beyond being fun, making new words can force us to reconsider certain things. Established words often are bound with connotations and cultural baggage. A new word can call attention to gaps in our knowledge. A new word can suggest new directions for conversation, yet also remind us—if the word is well-made—of its origin. In this sense the construction of words shares the operations of metaphors, by bringing together things from a different context.

2.2.0.2.1. Words like World Nation Place & Home

Other words, such as 'world' and 'nation,' describe whole things. A world is that part of the planet designated by one culture, and the perspective of the inhabitants (from the German word for 'man-image'). It can also mean the entire planet. A nation is a people or tribe or a people living in a territory united by a single government, a stable, historically-developed community of people with a distinct culture occupying a common territory, although it has also come to mean an artificial designation as a result of political violence.

People in a 'domiture' create plans, models and nostrums to improve their situations.

'Plan' means flat; that might imply that our plans are two-dimensional and that being two-dimensional might mean that some properties of three, four or eleven-dimensional reality drop out in two-dimensions. Therefore, we should consider things that might be lost by our planning. 'Blueprint' means a flat in blue ink; perhaps our blueprints need to be four-dimensional (or seven or eleven-dimensional) to be complete. The word 'model' means measure; we need to be sure not to ignore things that are not measurable, such as happiness, health, or completeness. The word *nostrum*, from the Latin word in the plural, means 'our,' and it originally referred to a medicine of secret composition. Often, we offer plans as nostrums to fix a set of problems or illnesses (parenthetically, *nost* in Greek meant to return home, as it is used in words like 'nostalgia').

Human situations unfold in place. The word 'place' comes from the Latin meaning 'open space' and originally from the Greek word for 'broad' as it was applied to a way or street. The use now should reflect the kind of space modified by human interactions, whether cities, homes or pathways. The word 'environment' is from the old French word for 'surroundings.' In normal use, it implies the immediate nonhuman surroundings. We need to make sure that it includes larger scales that incorporate the large cycles in nature, to understand that our local neighborhoods are linked in many ways by atmospheric and hydrological cycles, as well as by elemental cycles. These words are not used in some cryptic way, but just to illustrate what they originally described, in order to understand how people thought, and then to extend those thoughts.

The place is an ecosystem, the subject of ecology, often regarded as a home. The study of life in place is 'ecology' (from the Greek word *oikos*, a dwelling place or house. The word economy derives from *oikonomia*, household management.) Ecology is knowledge of the house, as economy is its management.

Insects and animals displayed a powerful attachment to places, as Adolf Portmann observed; this attachment was best understood as home. The fundamental ambiguity of existence is that humans have different capacities for feelings and awareness. Some feel strongly about a place or home; others never do. Several metaphors have been used to describe the human place on earth. The earth is a storehouse, property, or a spaceship. But the earth is not a spaceship or storehouse; it is home. Victor Ferkiss proclaimed that: "The world and humanity are one entity, one system in equilibrium. Earth is humanity's only home; humanity is one people in relationship to the earth."

There are a wide variety of meanings of home. It is a place of family residence, the family social unit, habitat, and place of origin. The word 'home' comes from the Middle English word *hom* and Old English *ham*, Old Norwegian *heimr*, Greek *kome*, and Sanskrit *kayati*, meaning village or home. The Old Norwegian word for home meant village or world. The word can be traced through the Greek to the Sanskrit, which meant "he is lying down." Its spectrum of reference is enormous. The word is used to describe house, village, city, bioregion, cultural world, and the earth. Its content is also ambiguous. Home can hold a single person, family, relatives, pets, domestic food animals, neighbors, or others.

Home living is simultaneous on different levels; the importance may shift from city to nation, or nation to state, or house to bioregion, or state to habitat. Each level can be a metaphor for the next. There are parallels between nature and a house, as the basis for home. Solar space is like the landscaping; wilderness is the foundation; conservation areas form the shell and provide services; and each bioregion is a unique room. The analogy cannot be carried too far, but it shows that a house is not, as Le Corbusier said, "a machine to live in." It is a matrix for home. Home is not just a house, either; it is a complex of significant events centered in place. It is the foundation of our individual identity on one level, and our role in the community, on another. What makes home different from house? Participation in its making, or commitment. People invest parts of themselves in a place, to make a home.

The concept of home has a mixed reputation. Paul Shepard finds humanitarians obsessed with the 'homelessness' of stray pets and wild animals. He points out how the fixation on shelter is taken over by the advertising of wood industries, who describe their meager reseeding efforts as 'creating a home for wildlife.' He sees protective organizations swaying to the tunes of propagandist lullabies. Perhaps, he is right. But, there is a misunderstanding of home. A home is not the house, not an undifferentiated place. Animals accustomed to rich woodlands are not at home in a replanted clearcut. The use of the word is a cheap advertising device, yet, it shows the importance of the meaning. A home is living thing, a house is not.

A home is a part of the environment claimed by feeling. Emotion creates an 'in-place.' A place must be found and made. Humans, like plants and animals, identify greatly with local environments. Maybe this is a function of the limbic system of the brain, a function we share with many territorial mammals. So far, no psychologists have studied what happens when a person sees her place, her very context, destroyed. These kind of catastrophes may be the basis for diseases, depression, or cancers, which also may be caused by the invasion of the home place by exotic substances.

The human ordering of the world makes places from wilderness. A place changes qualitatively; it becomes structured. Natural complexity decreases as the human complication increases, although the two are not mutually exclusive. Fitness is achieved after slow, progressive, reciprocal adaptations; it requires a stability of relationships between societies and place. Human places are complex integrations of nature and culture that develop in particular locations. The place precedes knowledge of it. The knowledge of place is one of the first links in a chain of knowledge. This knowing is essential to our existence. Being human is having and knowing a place. Only learning flowing from hospitable presence can promote life and enhance human existence.

Paul Shepard suggests that for each individual the organization of thinking and meaning is intimately related to specific places. Experience focuses on a place, which acts as the background for specific events. The features of the world are experienced meaningfully. The place is a matrix for ordering experience. The specificity of place is important. The earth extrudes itself into particular plants and animals; flexes mountains; and sweats weather. Places animate people (from the Latin word 'anima,' meaning 'inspire').

The inspiration of the sentiment of dependence is called impregnation. Animals and humans are imprinted early in life to particular places. Each difference in the landscape has meaning, that can be perceived, for instance, as when the aborigines of Northwest Australia perceive physical differences and even a symbolic landscape. In fact, they structure space according to myth, where Europeans use buildings and roads for structure. Every place has a unique identity as a result of combinations of factors.

Being apart from home can result in a disease, nostalgia. The word nostalgia was coined by Johannes Hofer a Swiss medical student in 1678 to describe an illness characterized by insomnia, palpitations, stupor, fever, and the persistent thought of home. The disease could result in death. For the Northern Aranda in Australia, as well as for émigré

Russians and other groups, it is not possible to stay away from home indefinitely and still live. Nostalgia can be a fatal disease. Thus far, the sense of place cannot be gleaned from an analysis of the human nervous system. Yet a place shapes the nervous system, somehow.

Permanence is important element in idea of home. In English the term for dwelling is 'to stay.' This is the symbolic opposite of moving or changing. It means to 'withstand time.' Dwelling resists and persists. The Royal Commission on Local Government in England and Wales found that people's attachment to 'home area' increased over time. Gaston Bachelard has written much about the significance of home: "For our home is our corner of the world. As has often been said, it is our first universe." The home is a springboard to understanding the universe.

The concept of home is proposed as a metaphor for the development of appropriate attitudes and participation in appropriate ways of living. The Pueblo Indians, for instance, see their American desert as a providential home, because of their attachment and knowledge.

2.2.0.2.2. Words like Ethics & Design

Our attitudes are often expressed in behavior or rules, as an ethics, that we use for making decisions about actions and about making in general. Making is a human activity that involves effort, using tools, in place. To 'make' means 'to bring into being' or produce something physically or mentally (from the English and Germans words, from the Greek meaning to 'knead,' press and stretch dough). It also means to build, fit, or create. The making of a home would be an 'ecopoetic' activity (from the Greek word fragments for 'house-making').

The word ethics is derived from the Greek word meaning 'custom' which itself came from the Sanskrit word for one's 'own doing' (Greek *ethos*, Sanskrit *svadha*). Since it was used in the plural, it meant 'doing together.' The word 'morality' comes from the Latin word for 'will of the people;' the singular meant the 'will' of a person (Latin *mores*, plural of *mos*, from *meare).* It was probably derived from the verb 'to measure,' as to measure one's way or to go one's way. Morals means the 'way of going together.' Ethics means 'doing together,' which of course one does living together. And, in an anthropometric universe, it is entirely appropriate. Ethics is a description of human conduct, or set of rules, determined by custom and culture. Ecologically, ethics is a limit on one's freedom of action in a culture.

Customs and actions combine with history as we redesign ecosystems to provide our needs. The word 'design' means 'marking off a pattern.' Our effort should be to make sure that we are using the whole patterns. Design is a human project in which, as Oliver Lucas says, "visual and physical parts are assembled in order to achieve a specific end result." Ecological design is the creative modification of ecosystems to repair or enhance their ability at self-organization and the maintenance of their complexity and diversity. The word design comes from Latin word meaning 'to mark.' Nature is self-making and self-designing, but we humans now influence every natural system, taking what we need from some ecosystems, enhancing a few, misusing others, and interfering with the rest. We need designs to restore the balance between human needs and natural processes.

Ecological designs focus on whole communities that work in the same self-sustaining and self-limiting ways as nature. By consciously creating meaningful ordered patterns, we can develop ways of producing widespread community wealth while positioning the community for a long, sustainable future in a healthy environment.

2.2.0.2.3. Words like Good

We judge the results of our efforts as good, or not good. Thinkers have puzzled over the term 'good' for centuries, constructing partial theories and contradictory systems. The word 'good' has an interesting and long history. The current version is derived from the old English, *god*, meaning 'suitable' or fitting, similar to the words meaning a 'suitable time' and to be 'suitable' or 'pleasing.' In an organic world, good things are defined by a free interplay of energies. Perhaps, as a working definition, we can just use 'harmony.' In Chinese medical tradition, the highest good is harmony, especially social harmony or good relations. A good person is one who creates and maintains harmony. Harmony is related to wholeness—indeed, the word 'whole' comes from the Indo-European root *kailo*, which is also the root for the words health and holy.

Science is a useful system for realizing new knowledge, but it has trouble dealing with concepts like good or bad. We can learn about ecological processes and try to manage them, but it is problematic to designate good theories or good practices. Normally we do not think of physical or biological events, such as gravity or cell division, as good or bad; they simply exist in a neutral way. But, the effects of these events are valued in the human realm by human intention, which is open to interpretation and ambiguity, and which is described by human symbols and signs.

Signs are arbitrary, as Ferdinand de Saussure noted; that is to say, they have no necessary connection to things. Signs can make many connections to physical and biological events, thus intensifying them; signs can also become dense with human meaning over time. It is the play of signs in human language and human culture that results in judging that something is good or bad; that is, the manipulation of signs, in a field of surprise, due to many levels of meaning or uncertainty, produces degrees of goodness.

Cultural structures provide rules that limit play and action, so that it stops short of death or destruction. All cultures work that way. Ethics and economics, for instance, are rules of behavior; and politics is the practice of changing the rules as society changes. Signal play without cultural structures can result in problems, especially when science or other devices claim to be independent of culture.

Goodness is thus intention and action in the context of the rules of a culture, using ambiguous signs. Goodness is a feature of the path of actions, as is badness, which is sometimes identified in extremes as evil. Of course, many cultural codes divide everything neatly into good or bad, but, everything is not that simple, and goodness cannot be grown and picked like a ripe tomato. By trying to focus on either extreme, of pure goodness or pure badness, we miss the ambiguity and uncertainty of everyday situations, most of which occur in a mixture. Our ethics and ideologies are not comprehensive enough to help us live in this mixture, with the inevitability of uncertainty, and with the possibility of enantiodromia, that is, the creation of the opposite of our intentions. Using new logic and metaphors can help us to adjust to such a reality.

The use of the word 'good' with human places or ecosystems is problematic. Good means different things to different people. Your standards or codes, personally or culturally, might be different from mine. Therefore, the meanings of the words will be different. The search for good is measured by personal criteria, personal judgment, and personal reflection. Furthermore, human beings cannot know, or even think of anything, according to Robert Zajonc, without some involvement of emotion, that is, having at least a vague feeling of good or bad. On the other hand, this sense is complicated by questions of what one 'ought' to do, that is, morality.

When we set goals, as for the goodness of ecological restoration, we base those goals on the meaning of the symbols, which can have many more than one meaning. A shift in context can change the meaning. Because there is no absolute good in our practices, it is difficult to determine the best practice to use, especially when applied to long-lived ecosystems, such as forests and deserts. Goodness is relative, but we must make decisions regardless. Throughout this book, the term goodness is related to overall harmony, but it cannot be considered independent of practice.

Any example of a practice in an ecosystem, such as forestry, can be good, if it follows its methods, regardless if it is totally external, human-centered, rule-bound, or simply pragmatic. Any application of management can be good, if the worker unconsciously follows certain codes or accidentally does the right thing for the ecosystem—the right thing being an absence of interference with the dynamic systems that shape and maintain ecosystem processes. Good management can happen for the most contradictory or trivial reasons: Self-limitation, true love for nature, accident, intention, techniques, or shame—this last may be why management gets altered with public interest and scrutiny.

2.2.0.2.4. Words like Reality & Thing

The word reality comes from the Latin *res*, from which we get the word 'reality,' and it means 'thing.' Being real was being a thing. Reality was thinghood in general. The word thing goes back to Old English words that mean 'object,' action, event, condition, or meeting. It may have originally meant 'something occurring at a certain time.' The word might have been used originally as a metaphor for some event under certain conditions. It might have been used for describing any form of existence, permanent or transitory, limited or determined by conditions. It is now a very general word indicating a form of existence limited by conditions. According to David Bohm, the Latin word *res* comes from the verb to think, such that things are what are thought about. Perhaps thought is a dance of things in the mind.

Occasionally, throughout this book, the verb forms of words is used to indicate an on-going process. In some languages, such as Hebrew of Hopi, the verb is taken as the primary part of a sentence. In other languages and cultures, the group identity, that is the larger subject, is taken as primary, over the separate individual.

Bohm proposes a rheomode, which uses a new grammatical construction, in which verbs are used in a new way; it becomes a new mode of language. Bohm says that words can be brought in to enrich discourse in the ordinary mode of language. The reconstruction of words is not essentially different from the construction of phrases and sentences. This approach is similar to the use of particles in a field in physics. The particles are only abstractions that may be convenient. In ordinary language, for instance, 'truth' is taken as a noun. But, truth is that process which works. We think it is established and stable, but we should recognize that it is constant (from the Latin *constare*, 'to stand together') only within limits of the flow.

2.2.0.3. Logic of Logics

As our words need to reflect a larger sense, so our logic needs to be aware of its scope and limits. Most Western science assumes a predicate logic and is constrained by that logic. This Aristotelian logic is deductive and binary, where categories are mutually exclusive, and substance and identity are permanent. In this logic, contradictions are false by definition. The difficulties with this logic are evident in many instances in physics and ecology, e.g., light cannot be a wave *and* a not-wave, or wolves cannot be a keystone species *and* a species not necessary to an ecosystem. Neither substance and identity seem to be permanent in any human sense of that word.

Truth itself is not an absolute something, or even a relative something. It is that which is lived in connection to the whole web of lives. It is not fixed, static, or a point. It cannot be abstracted by words or measured by numbers. There is no path to it; there are no tools that can isolate it.

Our science and theories are beset by logical problems. Would a more flexible logic solve the problems? A nonpredicative logic? There is another form of this logic that is multivalued, but it is also based on the same epistemology of a hierarchical universe in which values can be rank ordered. Magorah Maruyama calls these logics, and the epistemologies that contain them, homogenistic.

There is another independent-event form, with an inductive, statistical logic, which is used by scientists and existentialists. Independent-event is the reverse of homogenistic logic; both characterize 'Western' logic. The homogenistic logic can be characterized as classificational quantitative, competitive, uniform, nonreciprocally causal, and hierarchical.

Other logic systems exist, although they are not as predominant. Maruyama distinguishes two other epistemologies that have a typical logic based on them: Homeostatic, with a complementary logic, as exemplified by Chinese thought, and Morphogenetic, with a logic incorporating change, harmony, heterogeneity, and unrepeatable and irreversible processes, as demonstrated in Mandenka, Navajo, and Inuit thought.

Morphogenetic logic is quite different from homogenistic logic. The morphogenetic logic can be characterized as relational, qualitative, symbiotic, heterogenistic, reciprocally causal, and interactionist. Reciprocal causal processes can increase structure, differentiation and complexity in natural systems, according to Maruyama. Such a logic is more useful for addressing complex operations in ecosystems. As a many-valued logic, it can address values other than truth and falsehood. It can avoid certain fallacies embedded in two-valued logic.

No superlogic can combine categories, however, since they are based on different epistemologies and cosmologies. Any researcher, using any logic, will filter and distort data to some extent, as a result of epistemological 'selection.' A broader logic, however, such as the morphogenetic, could work at a metalevel and avoid many of the problems of polar thinking.

2.2.1. *Naming with Metaphors*

The concept of metaphor has been defined and used for over twenty centuries. Before that, metaphor seemed to characterize language itself. Metaphor is used in all advanced languages. For Aristotle, metaphor was a trope, the 'turning of phrase.' The word metaphor comes from the Greek word *metaphora*, which means 'carrying beyond.' Aristotle was one of the first to say what it represented in words: The process of transferring a word from one object of reference to another.

There are two principles governing the creation of a metaphor, according to Max Black: First, the just disclosed phenomenon is given the name of a previously identified one that resembles it, and second, this resemblance is discovered in the most 'essential' aspect of the new phenomenon, which calls forth a direct perceptual experience, already named. That is, a metaphor furnishes a label and emphasizes similarities between a known thing and an unknown one. It not only defines and extends new meanings, but also redescribes domains seen already through one metaphoric frame.

The metaphor itself consists of two parts: (1) the focus (or tenor or figure), which designates the figurative term signified through the process, and (2) the frame (or vehicle or ground), which refers to the subject or context. For example, in the sentence, "Light is a particle," the focus is 'light' and the frame is 'particle.' The focus of a metaphor is a primary system that refers to that which is to be understood, a brain or atom, for instance. The frame is a secondary system that provides a pool of names to be used for comparison: computer or solar system. The interactions of systems is mutual; the secondary system imposes a reorganization on the concepts of the primary, but the use of the metaphor alters perception of the secondary as well as the primary. Not only are the two concepts altered, but also the meaning of the concepts is altered. And this 'halo' meaning, in its unique reducibility, permits things to be said that could not be said otherwise. In the secondary and primary systems—for example Karl Popper's phrase "the brain is a computer"—opposites interpenetrate and are unified even though the metaphor is initially and perceived as absurd.

This interpenetration is a dialectic, where opposites are unified in a metaphorical synthesis that transcends the initial contradictory conjunction of the two systems. Thus, concept formation in art and science is a working out of contradictions, as in Bohr's metaphor, "the atom is a solar system." Logical opposites logically combined assert nothing in traditional logic. Metaphorical logic is not limited to traditional logic. It is more contextual and multivalued. The original meaning of logic in Greek was 'gathering up,' which is what metaphoric logic does. Logical consistency is no longer required.

The metaphor is a form of discourse where one of the terms is always denotative. The collocation of terms in a metaphor leads to a sort of intercourse, in the Platonic sense, a transaction or transference between the associated contexts. The focus is responded to as the frame, in addition to its own original, evoked responses.

Some distance between the focus and the frame of a metaphor is necessary; similarity must be accompanied by some disparity. If the two are too close together, then the perspective of double vision might be lost. Samuel Johnson claimed that it was useless to compare a tree with a bush. The power of the analogy is also related to the emotive potential of the matrices. It is equally useless to compare a sneeze to a cloud.

This is a form of reasoning by analogy, not by differences, which a digital computer uses to compare things. Metaphorical expression has an analogical unity in human existence. Its central organizing principle is the single person, with perception and imagination. The expression points the mind toward the broadest meaning. It points, not toward truth, but to an ever-enlarging relational field. It points to the frame, in fact. That which is the frame must be ambiguous. Frames themselves must be used as metaphors. The word 'nature,' for instance, has been used by Western philosophers in over thirty distinct senses. Under ambiguity many meanings may hide. Nature can be regarded as just a 'puzzle' or 'opponent' to be beaten, but both detract from a total experience of nature, in nature.

2.2.1.1. Metaphorical Basis of Science

The entire activity of science can be said to be guided by metaphors. Metaphors can allow scientists to deal with complex situations. For example: "The world is atomic," according to Leukippus and Demokritus; "atoms are billiard balls" for Dalton and Rutherford; "the world is mathematical," according to Pythagoras; "man is an animal" for Karl Pribram; "man is a system" for Ernst Laszlo; and "man is a computer" for Michael Arbib.

Bacon referred to true metaphor as "the footsteps of nature." Metaphor is the vital spirit of a scientific paradigm, according to Thomas Kuhn, and its best organizing relation. The notion of paradigm includes techniques, examples, community values, and a central metaphor. According to Kuhn, there is no methodological evolution of science; rather, normal science progresses by a succession of paradigms, which he described as noncompetitive and open-ended. He stated that paradigms are the traditions described by historical rubrics like: Ptolemaic astronomy or Copernican; Aristotelian dynamics or Newtonian. These examples include law, theory, application, and instrumentation together, and provide models from which the traditions of scientific research spring. In his view, science proceeds by working out problems uncovered by each current paradigm. Should problems occur that cannot be ignored, suppressed or resolved, then a revolution should occur to replace the paradigm. The new paradigm would have to include all the old data as well as the new problematical data. Metaphoric systems are the core of structural coherence of any paradigm.

The word 'individual' is a metaphor. As such it refers to aggregates (sand), functional units (populations), and autonomous, self-regulating beings (a wolf). The word 'whole' is a metaphor. A 'thing' is a metaphor, though it is so old it is considered a real term. A thing (or part) is that which can be separated from other things, thus intimately related to the idea of boundaries. The metaphor has a boundary. The meanings of a metaphor can be ontological (physical things) or structural (quantified or valued). Where we make boundaries usually depends on our past experiences. Interface as a synonym implies that the boundary is Janus-faced.

Metaphors in science are usually termed models and are used to bestow a more precise meaning to expressions. Science makes use of the metaphorical process to construct its models.

2.2.1.2. The Strength of Metaphors

The metaphor of a chain, as in the "great chain of being" described by A. O. Lovejoy, was used to order living beings in a single linear series. This implied that things were level and ordered one after the other. The use of a tree by Buffon to trace the genealogy of dogs in different climates added a dimension to understanding of the effect of the environment on animals. Metaphors are good to think with. The use of the tree metaphor by Ernest Haeckel, as the shape of life, is used in archaic cultures also. After all, the earth is a forest planet; eighty percent of the biomass on the planet is tied up in trees. And, trees are animated water. As a metaphor, a tree conveys strength, growth, and longevity. For the Kwakiutl people, for instance, the chief is the post of the world, the only one standing, the cedar that cannot be spanned, the post of heaven that holds up the sky and bridge it. The chief is the trunk of the cedar, the upper-class people are the branches, and the common people of the trunk. The chief's ancestors are the root of the tribe. If the chief and the heir die, "both tree and root are killed."

Darwin used the 'struggle for existence' as a metaphor: "I use this term in a large and metaphorical sense including dependence of one being on another, and including (which is more important) not only the life of the individual, but success in leaving progeny." The struggle is between individuals of a species, or different species, and with the physical conditions. Darwin's theory of evolution destroyed the tradition of natural economy with its essences ordered by reason, but he still used a rational economics for his metaphors of individualism. The metaphor seemingly arose from the socioeconomic view of the ruling classes of the time, that the suffering of the poor was inevitable. This metaphor formed part of the basis of a new worldview where nature was a theater of violent competition, especially after Tennyson made nature seem more poetic, as "red in tooth and claw." The frame where a metaphor originates often carries the conceptual baggage of that time. It can support the idea of superiority of 'favoured' races in the struggle for existence and emphasize the role of competition in biological and cultural situations, at the expense of other interactions.

'Natural selection' was also metaphoric. Darwin did regret later not calling the process 'natural preservation.' Wallace described evolution as a conservation process, similar to the centrifugal governor of a steam engine. Darwin used the tangled bank as another metaphor. The metaphor of the tangled bank evokes the mutuality and interdependence of the interactions of organisms in webs. Darwin said: "It is interesting to contemplate a tangled bank, clothed with many plants of many kinds, with birds singing on the bushes, with various insects flitting about, and with worms crawling through the damp earth, and to reflect that these elaborately constructed forms, so different from each other, and dependent upon each other in so complex a manner, have all been produced by laws acting around us."

2.2.1.3. Modeling Things

Metaphors emphasize likenesses between things, living things, languages, or human constructs. Metaphor in the broad sense includes similes, analogies, models, patterns, and paradigms. A visual representation of a metaphor can be a model. Modeling can be done with metaphors as well as with shapes and numbers.

2.2.1.3.1. Metaphorical Models

Darwin used 'natural selection' as a metaphor for evolution. Later, the philosopher Maurice Merleau-Ponty used an explosion as a metaphor for evolution. Merleau-Ponty considers that the direction of evolution relies on a pseudo-teleology that is the inverse of mechanism, on a great 'pattern mixed-upness.' Evolution is a radiation of life. Rather than relying on evolution as a basis for metaphysics, he called for a "phenomenal topology" of things as they loom upward bodily around us. Marjorie Grene noted that because the records take one every which way, evolutionary theory was a poor crutch for renewal of ontology.

It is possible to use metaphors for other complex assemblies that are difficult to understand. For instance, it could be said: The city is a sponge. The sponge cannot use the sun to get energy directly, but must use surrounding environment and the productivity of other organisms in the water. Sponges must circulate food brought in, using energy from that food. The city has own metabolism, that is a network of circulatory structures used for exchanges. Of course, the metaphor implies the city is a living organism. Perhaps another metaphor would permit the city to be understood as more self-creating, perhaps if the city were identified as a tree or a box.

Mario Bunge used black boxes as metaphors. The idea of the black box was originally conceptualized by electrical engineers to describe certain unknown systems devoid of structure. Black box theories include kinematics, thermodynamics, information theory, scattering-matrix theory, and circuit theory. The black box approach is useful for all theories whose variables are external and global, that are simple and have a high degree of generality. As theories are supported by observation, black boxes become translucent. Some empirical theories have root metaphors that range from black to almost transparent: 'man is an animal' according to Karl Pribram, or 'the brain is a hologram' for Michael Arbib. When we fill in the box, and make it transparent, we see the interactions of the components. In an ecosystem, if we make the box itself transparent, then we are left with a food web.

The box is a productive metaphor. The box can also be equated to a cage or trap. Karl Marx contended that we live in cages, part natural and part made, although human actions could modify them. The word cage is a metaphor; it implies being trapped. It is, however, a metaphor that can be expanded with a description in space as a four-dimensional box. Perhaps there is a better metaphor, since we depend on nature and society as a foundation for life, that of a trap. The word 'trap' is from the Old English 'to step.' A trap is a device for catching and holding animals or a stratagem for catching people. The idea of a trap can be related to the idea of closeness to limits or to the overconnection of links.

Taken in four dimensions, a trap can be a serial trap: The use of resources by a people, where the replenishment rate is constant ands the rate of use exceeds it. This trap results in ecosystem degradation that is less reversible. The industrial age mistakes the rate of discovery for the rate of recovery. Agriculture is an energy trap, because it allows the concentration of energy, that is, higher yields, but then it requires more energy be put into the system to maintain it. The system has to produce more energy than it uses to be useful. And be sustainable, with a surplus for trade. Instead of being free from economic want to develop their potential as creative human beings, people are trapped in a consumer cycle.

A filter is an interesting metaphor. For organisms, the environment is a filter that allows some characteristics to continue, by providing opportunities and challenges. Organisms put together structures based on historical patterns, and move through a filter of limits like minnows through a fish net. Perhaps, the idea of a filter is too passive. Perhaps a better model might be a kind of sorting filter, a mutual filtering of organisms and environments. Organisms and an environment co-order each other, sorting things out in a mutual process of activities and adjustments. In the next largest system, ecosystems are also filtered or interactively sorted.

Culture is also a filter or mutual sorter. The anthropologist Branislav Malinowski tried to show that almost all of a whole culture is can be seen as a mechanism to modify and satisfy the sexual needs of an individual. What a filter that would be. Of course, you could also argue that the mechanism of culture is to channel energy or distribute material goods. These mechanisms are only parts of a culture. There are some parallels with biological evolution, especially if we use memes as the unit of transmission. Memes are still filtered by the mental environment for fitness to that environment, although in this case the filters are partly due to memes. The scale of replication is related to positive feedback, as when a work of art fetches great sums of money or is written up from controversy. The information that is transmitted is in form. Culture is more than a simple filtering or sorting process. It is more than a formation process. Because it occurs with physical, living, conscious, and social systems, culture is more like a novation process, that is, it recombines things into new orders.

These models of box, trap, filter, and sorter will be used in subsequent discussions.

2.2.1.3.2. Conceptual or Mathematical Models

Models can be constructed with a series of verbal statements or with a combination of statements and mathematical equations, using data. There have been many such models of the world system recently. Each model gathered data, analyzed it, and made a set of predictions based on the data and analysis. Each of the models also recognized having very general problems. For instance, The *Limits to Growth* model had vastly more information than it could use in an orderly way, due to deficiencies in theories of structure. J. W. Forrester emphasized that modeling projects should be global or national, but not regional, and should draw heavily on mental databases. The Forrester model distributes values homogeneously over the globe, disregarding sociocultural and ecological systems. It tolerates an expansionist view of industrial human civilization. But, Dennis Meadows and others recognized the constraints on the planet with limited resources, and concluded that partial solutions could make things worse.

Mesarovic and Pestel established the necessity of using models to represent the objective aspects of world development, and defined a model as "a coherent and systematic set of descriptions of relevant relationships." The entire report used a strict scientific methodology to treat actual data, avoiding any hint of an abstract academic exercise. But, it reduced everything to measurements, perhaps invalid measurements. Although they carefully contrasted their model with the Forrester-Meadows thesis, there does not seem to be any essential difference at all. Their multilevel model is still a reductionistic, as is their attempt at regionalizing cultural areas of the world. They claim a complete set of descriptions, but the subsystems sets are grossly oversimplified, and the results are starkly primitive graphs, reduced to meaninglessness by uniform, noncontextual computer decision-making processes. In summary, their model assumes basic industrial values and loses most human values in the shuffle of facts.

The book, *Reshaping the International Order*, coordinated by Jan Tinbergen, recognizes a need for a new international order. The book identifies problem areas and examines the progress toward their solution. Tinbergen defines a difference between a forecast, as in *Limits to Growth*, and a plan, which is a human effort to control destiny. In fact, the plans described here are based on many forecasts. The second part of the book describes the architecture of the order and discusses strategies for steering change. One aim of development is towards an equitable social order, by reducing the differential between rich and poor. He explores the implications of his own plan, which accommodates the tension between developed and underdeveloped nations. Tinbergen recommends that poor nations negotiate with rich nations through collective bargaining to obtain a greater opportunities and rights. This tactic could bring some success.

Where the *Blueprint* group suggested more drastic steps, from disregard for the adaptive dynamics of supraeconomic and ecological factors, Ernst Laszlo's alternative is purposive policies that mobilize adaptive capacities in these dimensions. Focus should be directed toward dynamic equilibrium models achieved without undue trauma, and sustained without repression. The ethical dimension of order relates to attitudes about violence, satisfaction of needs, and sociopolitical conditions compatible with equality and worth of human beings. Laszlo proposes a model which accounts for biospheric and sociocultural inputs from the total environment; recognizes sociocultural systems as converter and subsystematic; relates outputs—material and social technics—to positive and negative feedback. This system is more comprehensive and consciously planned than most.

We have separate kinds of models related to plant productivity, human population, and world games. We can try to generate models dealing, not only with the more constant of natural and human characteristics, but with the limits of certainty and the vagaries of change.

2.2.1.3.3. Theoretical Models

A theory is an explanation or system of everything. It is an exposition of the abstract principles of a science, or it is a speculation. The dictionary definition states that a theory is a reasoned expectation, as opposed to practice. A scientific theory is a set of general explanatory statements about a natural process; the set is related as a model, which is a human linguistic construction, subject to human perspectives and limitations. Theories may incorporate guesses, critical observations, experimentation, and logical inference, as well as hypotheses and laws. A theory must be a specific enough statement, such that experimental results can be assessed as negative or positive. The results of an experiment can confirm or justify a theory, although other experiments could show that the theory is wrong. Good theories tend to have a limited number of generalizations. Theories can also be nested, so that, for instance, the theory of gravitation might be combined with other theories in a unified field theory.

Science works with theories, that is, specific directions of observation, based on previously identified facts, to lead to evidence, or proofs of hypotheses, about how things work, which can be used to predict other actions in the future. Theory is used to guide experimentation. But much experimentation these days occurs without theoretical guidance, or rather the theories used are not capable of addressing the context itself, so that the experiments tend to go out of control. Many experiments are simply conducted improperly, with no control group, no good theory, or no control of scale.

People who know a lot, but do not have theories that bind them to wise actions, tend to ignore ignorance, or to act in an arrogant manner; such behavior can be dangerous. Theory can reduce this kind of danger. Theory can describe the limits of itself, of knowledge and understanding, so that we do not try to do that which cannot be done, for instance, create perpetual motion machines. That is, theory deals with ignorance, the kinds of ignorance, and especially the kinds that cannot be removed. Theory also deals with ambiguity and uncertainty.

Theory has to account for those things that cannot be reduced or explained or understood. Some mysteries cannot be reduced, cannot be understood with human understanding. Some things are ineffable; even if they are understood, they cannot be reduced to words. Our theories and practices are human theories and practices; they are limited by real human limits, by the dimensions and terms of times that we humans experience.

Theories can help science to avoid claiming things that it cannot claim. Theory can distinguish between knowledge and theory, between theory and the unknowable. Theory needs to address the incompleteness and limits of science. There will always be limits and errors; these cannot be avoided. Theory needs to accommodate this truth. Fallibility is unavoidable; it is a part of certainty from a limited perspective with a limited knowledge; it is part of being incomplete and using incomplete knowledge. But, this is the limit of all being; this is the limit of finiteness.

What we know will always be incomplete, but we have to act under those circumstances, and therefore we have to act on partial knowledge in the face of partial ignorance. Theory is a guide to acting, under different kinds of unavoidable and irremovable ignorance, based on thinking about previous actions and projecting future actions. Theory is not necessary to get more information. But it is necessary to understand what can be predicted or understood. Theory can help distinguish between local and global problems. Theory can describe the fitness of our conduct within an ecological and cultural system.

Most theories in this century have been dominated by the impoverished philosophies of modern industrial cultures. These theories have no way of dealing with the profound changes of global warming and habitat destruction, extinctions, social decay, new diseases, or unrestrained technology. These changes are modifying the planet and may destroy us, as well as those things that we require for adventure or inspiration.

To be really effective, theories need to be rooted in conceptually rich philosophies, such as process philosophy, phenomenology, general systems theory, ecophilosophy, deep ecology, ecofeminism, or radical ecology, philosophies that are more fruitful and considerate of nonhuman beings and systems—and in fact consider such problems as what is good or what are human limits. These philosophies contribute to a new attitude that is more appropriate to the unity and interrelatedness of the earth. These philosophies provide a number of fundamental philosophical, historical, scientific, and cosmological principles that have been presented in other contexts by thinkers such as Cobb, Einstein, Fox, Naess, and Whitehead. Very few of these principles are absolute. Nevertheless, they are essential to the understanding of ecosystems and cultures. Principles, combined with common sense and good judgment, are necessary as guides in the absence of definite knowledge. They give us a deep foundation and a broad predictive ability that we need for good practices.

Good theories are just those that incorporate cultural values with ambiguity and uncertainty. Good theories are grounded in place and in communities. They must be used to place our science and technological capabilities in a culture that has goals and limits for our actions. In this sense, theories are tools for adaptation to place, within place, within specific places that we work towards making good.

2.2.2. *Questioning Things*

Questioning has a long history in human experience. The Greek philosopher Socrates was one of the more renowned questioners. He approached teaching through a disciplined, rigorous dialogue with people he met on the street, often by accident or by design. Socrates tried to get others to recognize the contradictions in their ideas; he assumed that incomplete or inaccurate ideas would be corrected during the process of questioning, and hence would lead to progressively greater truth. He never seemed to reach an end to questioning, however, perhaps because there was no end, that is, the process of questioning could refine any kind of knowledge or ignorance, indefinitely, or perhaps because by itself questioning could only do so much with definitions and concepts. His method was a common search, through conversation, for the goal of truth.

2.2.2.1. The Socratic Method

Socrates asked questions as part of a conversation with others. He seemed concerned with discovering what the opinions of others were based on, an invisible truth. The questioning forced the other participants in the conversation to try to agree on the truths beneath the opinions. Socrates professed ignorance of the truth himself, in fact or in pretense, ignorance being the first step in the pursuit of knowledge. He expressed skepticism that the other conversants actually had real knowledge. The process of questioning subjected opinions to real examples from real experiences—an empirical method—leading to a more general concept—this is the process of induction used in a scientific, homogenistic logic—and then the consequences of the definition were drawn out, through deduction. These definitions were refined by further questioning until all members of the conversation had a better grasp of the concepts. Through thorough questioning Socrates demonstrated that knowledge was quite often uncertain. There was no absolutely certain knowledge. Questions were also meant to examine life as well as belief and truth, and to show that often people were ignorant of their ignorance. Socrates held that disciplined questioning enabled the other to examine ideas logically and to be able to determine the validity of those ideas.

2.2.2.2. The Hardin Extension

For Socrates the goal of knowledge was the acquisition of concepts, such as justice, courage or wisdom. He thought that the truth could be contained in a correct definition. And, he was groping for more abstract definitions. This became a problem, as abstraction became removed from the specifics of living. Socrates was most concerned with examining concepts, but concepts are a small part of reality. Of course, constant questioning of concepts can expose the psychological basis of concepts, and perhaps that is what Socrates meant to do. But, questioning concepts often reaches limits fairly quickly. Socrates never had any answers, although he assumed that an absolute certain knowledge was possible to become established eventually.

The ecologist Garrett Hardin used questioning to illuminate partial knowledge and to track connections between things. Questions establish the limits of assumptions and perspectives. They can clarify the focus of a problem and test evidence related to any problem. Questioning can be used as a device to focus on a specific problem, not only the extent of the problem, but its aspects. Questioning can also be used to explore specific aspects of the dimensions of thought. Of course, questions can refine the process of critical thinking and can allow refocusing in a wider or narrower context. This type of questioning arrives at answers as workable hypotheses or guidelines for making decisions about operating in the world. Without certain knowledge, however, we can make adaptive decisions, based on partial knowledge. His questioning took on the form of asking what happened then, after an answer was arrived at. "And then what?" Hardin asks. Questioning works in a conversational way by weaving ideas. Conrad Lorenz decided that humanity would indeed have destroyed itself by its first inventions, were it not for the very wonderful fact that inventions and responsibility are both the achievements of the same specifically human faculty of asking questions.

More than an annoying part of a conversation, questions are legitimate ways to approach a known or unknown situation. More than just a way to turn around a conversation, questions are tools that allow you to surround a topic and define it more completely. More than simply an admission of ignorance, questions can form a phenomenological spiral that allows you to return to a subject from different perspectives with different levels of understanding.

2.2.2.3. Maslow's Questions about the Norms of a Society

Before even designing a good society, certain normative questions must be answered. These tentative answers are based on those questions first suggested by the psychologist Abraham

Maslow. The questions have been slightly modified before being answered briefly.

Is the norm to be universal, national, subcultural, familial, or individual? The norm could have universal elements, not only for humans but for all species impacted by humans. Maslow assumes that different norms must be on different levels, depending on the context. For example, there would be some universal human behavioral standards, but special local expectations, to conform with various cultures.

Should society be selective or unselective? The society should be unselected. It must account for all human variability; and accept it when possible and treat it when necessary. It would have to account for prisoners and misfits. The society should be pluralistic, and accept and use individual differences in constitution and character. Humans are not interchangeable; the insane and aged must be considered. It must integrate all people into a society or work in that direction.

Should society be pro-something or anti-something else? Society could be pro-industrial and pro-intellectual (or scientific), within the set limits of the society and the planet. But, industry must be properly scaled; and science must be cautious. The size of community cultures could be limited by function.

Should it be centralized or decentralized? The global unit could be centralized, electronically, at least, and socially planned; but individual cultures could be autarchistic, based on self-reliance and interdependence. Both should be flexible. Regions could be centralized for some functions.

Should society be tolerant? Society must tolerate all cultures in the nations. Each culture would determine styles and complexity for its individuals. It should aim for taoistic noninterference, but be available for help.

What should be done about injustices? Biological injustices exist and can be ameliorated; social injustices can be rectified. Not everything can be made equal or fair. But opportunities to use resources, for instance, can be granted equally to all groups in a region.

Should society determine family attitudes? Family or sexual attitudes can be institutionalized by the culture. All group adaptations would be determined by culture.

Should society be open to more than one religion? Society can tolerate any institutionalized religion, so long as it does not impinge on other groups. The spiritual life is a necessary part of a society.

How should leaders be chosen? Leaders within a culture could be chosen by traditional means, within the limits of human laws. Leaders of the international framework could be chosen by global referendum from the ranks of cultural leaders. The leaders would determine the relation of truth to people—who shall know how much about what and when.

For what is an individual responsible? The individual is responsible for the style and simplicity of their life and for its effects on nature and society. The individual is responsible for being tolerant of others, and is free to make many kinds of choices.

How these questions, and many others, are answered partly determines the shape of this project.

2.2.3. *Following Change with Gigatrends*

Things change, sometimes rapidly, sometimes slowly. We often refer to rapid change as a revolution or as a catastrophe; we refer to slower changes as trends or fate. A number of long-term ecological or cultural trends are evident now. A few of them can be fit into the megatrends identified by John Naisbitt over a decade ago.

Naisbitt identified ten larger patterns in society, including the move from an industrial society to an information society, the economic interdependence of the human world, and the restructuring of society from short-term considerations to long-term time frames, that is, he says, from two-year horizons to a "very long-term" time frame of "six to ten years." Some other counter-megatrends (negatrends?), such as the denigration of reason and science in the popular press and media, or the incredible explosion in information—without a corresponding increase or spread of wisdom—he ignored. Negative trends in general are left out of the popular picture. The things that are most popular to society, such as money, real estate, insurance, and politics, are the things that are treated as the most important. The things that will ultimately be most important, such as directing civilization or limiting human activities, are neglected.

The real long trends in ecosystems and cultures occupy the entire human calendar. We might call them gigatrends, since they are larger and more involving than megatrends (*giga* is from the Greek for very large or giant, whereas *mega* means large). Gigatrends are long or very-long-term trends, usually ignored by science and economics, such as atmospheric temperature increases or global deforestation. These gigatrends, on a global scale, include: Human populations increase exponentially; the impacts of a small percentage of people, the wealthy, increase exponentially; humanity takes over the habitats and functions of other animals, eliminating them and trying to take over their functions with chemicals, which is unsustainable for long-term; and ecosystems are simplified and degraded; deforestation, desertification, and exotic take-overs occur on a scale similar to the ice-age or a comet impact, but more immediately meaningful to humanity, since we depend on the ecosystems we are changing. There are, of course, positive gigatrends, although most of them seem to be negative.

Some of the negative gigatrends have been noticeable for thousands of years, but nothing has been done to halt them. Some of these negative gigatrends are hard to see, much less stop or reverse, because they are based on misunderstandings, fallacies, useless myths, and psychological blinders. For example, planners often treat exponential growth rates, such as human population or forest demand, the same as linear rates, such as rice production.

Environmental factors, such as climate or resource location, have shaped the course of human history to a greater extent than has been realized. The decline of Rome demonstrates that ignorance of forest ecology can have important consequences. There have been environmental catastrophes in the Tigris and Euphrates valley, Greece, Khmer, Maya, Midwest United States, and the Australian outback. These civilizations were very successful before they failed. Failure from success is tragic. For the Greeks, the operation of tragedy resulted from success taken to great lengths, that is, where successful behavior in one context is applied to all contexts, with the result that the opposite action occurs from the one desired. For example, humans in moderate numbers were able to take what they needed, such as wood, from natural ecosystems without interfering with the processes. Our dominance, once so successful because of our big brains and tool-using hands, has now become self-destructive. When human cultures adapted to ecosystems over long periods of time, the ecosystems also adapted to human cultures; when the human impact has been rapid and intense, as it has been in North America recently, the ecosystems collapsed or stabilized at a simpler state.

Although many gigatrends are interrelated, they can be discussed is several categories: human populations, human economics and technology, and ecosystems. Positive trends, often smaller and more recent, are discussed throughout.

2.2.3.1. Human Populations and Needs

With human success has come human influence on ecosystems. Humans have modified animal and plant associations in different ways, simplifying patterns of energy and chemical exchange, and solidifying themselves at the end of many food chains as a dominant species. Our domination is related to our large biomass, our large annual population increase, our high energy use, and our high structural organization of information and matter. None of these effects are exclusive to us as a species, but they are excessive, rapid, compounded, and very large-scale. Basic gigatrends include:

- Human populations increase exponentially; the impacts of a small percentage of people also increase exponentially.
- Humanity takes over the habitats and functions of other animals, but this is not sustainable in the long-term.
- There is a drawdown of nonrenewable resources to support our numbers and lifestyles; this can only be a temporary fix.
- Human interactions become more violent as a result of competition, inequity, and limits in the distribution of resources.
- Local planning ignores limits to carrying capacity, long-term deficits, other species, and ecosystems; regional and global planning are essentially nonexistent.
- The material goods and luxuries of human societies have been increasing. Few people can carry everything they own on their backs, a bicycle, or a bus.
- The social avoidance of hard decisions. No one wants to take responsibility for recognizing limits; no one can say no to their constituents or business partners.
- Increase in mind-altering substances, from fermented berries to coffee, tea, sugar, opium, and medical drugs.
- Shifts in our perception of humanity and the earth, from the center of the universe to central inhabitants of the earth to participants in a centerless arena. No culture is the center, or the evolutionary survivor of other cultures.

Eric Eckholm concludes that the United Nations must identify, analyze and marshal world resources against negative trends. A scientific method would take a long time, however, and poor countries cannot wait. They must attempt a rural regeneration of some kind, to stop urban drift and ecosystem destruction. Eckholm interprets negative trends as indicating the sinking of marginal peoples on marginal lands into a quiet helpless poverty, later leading to urban deterioration, which may be considerably less quiet.

2.2.3.2. Ecosystem Gigatrends

Ecosystems build up information. There are at least three different channels of information in an ecosystem: The genetic, in replicable individuals; an ecological based on interaction between cohabiting species, expressed in changes in their numbers; and the cultural, transmitted through individual learning based on experience. Feedback within the interaction of species is expensive memory with little storage capacity. Whenever succession starts again, after a volcanic eruption, for instance, old information in the form of interactions has not been saved. Genetic memory has a much larger capacity and is long-term. Cultural memory is further enlarged in complex vertebrates. The unconsidered use of information results in still more long-term trends:

- Ecological limits are ignored
- Ecosystems are simplified and degraded; deforestation, desertification, and exotic take-overs occur on a large scale
- Deforestation of the forest planet
- Stress on ecological systems
- Humans simplify ecosystems and keep them at early seral stages to harvest the increased productivity
- Much vegetation becomes a social artifact. In Scotland, for instance, forest cover was reduced from 55% of the total area to 5% by primitive stock-keeping and agriculture; the moors decreased by half, but meads increased eight-fold
- Biogeochemical cycles are disrupted, for instance, atmospheric carbon dioxide has increased since the 1700s. The atmosphere is heating up. Higher temperatures will reduce crop yield. Heat increase will lead to water shortage then food shortages and security problems
- Biological diversity decreases over the same time period. Human diversity increases by leaps or punctuations
- Extinction rates increasing
- Humans take over the productivity of nature. Mathis Wackernagel estimates all human demands on the environment surpassed the planet's regenerative capacity in 1980. By 1999, it surpassed it by twenty percent. What happens if we extrapolate this to 100 percent?

These trends contribute to a shifting planetary system that has less flexibility. Some of these negative gigatrends are hard to see, much less stop or reverse, because they are based on misunderstandings, fallacies, discredited myths, and psychological blinders. For example, planners often treat exponential growth rates the same as linear rates. Thus, if our forests were to last for 400 years at our current rate of use, before extinction, they would only last for 75 years with a demand growing at 3 percent per year, or 50 years at 6 percent—demand is growing now at over 3 percent per year. Dennis Meadows points out the speed with which surpluses disappear with increasing population and increasing per capita demand.

Many predictions about resources are based on fallacies. There is the fallacy of substitution, that states that a substitute can be found for any resource in short supply. This is not always true, especially with cultural preferences. Furthermore, there is a gross underestimation of the length of time for a substitute resource to attain traditional markets. For instance, the transition from wood to coal as an energy source took about 50 years, despite the fact that coal technology was established and attractive. Meadows identifies another fallacy: The expectation that people and institutions perceive problems and react to them rationally, e.g., that with the threat of wood shortages, prices would rise and consumers would value the resource more. Yet, the price of wood is still nowhere the real costs of production and wood is used for cheap, impermanent goods. Meadows suggests that the model of addiction might be more appropriate than adaptation for dealing with consumer demand.

Many of these gigatrends are based on myths, such as the economic myth of forestry, which as it is related by Chris Maser, is based on the rationale of "soil rent" theory, a classic economic theory that assumes, fallaciously, that all ecological variables are constant so that capital investment is easily calculated from the rate of growth of the crop species.

2.2.3.3. Economics and Technology Gigatrends

Eric Eckholm describes how economic and political pressures, which are derived ultimately from population pressures, force farmers to intensify their efforts to increase crop production. This instigates an "utterly dismal cycle" of population expansion, environmental deterioration, and poverty: As the population expands, forests are cleared for land, arable lands are used to capacity—and sometimes beyond; as the soil deteriorates, it requires more fertilizers that cause more hazardous conditions that decrease agricultural capacity; people starve, but the population increases, and marginal lands are used to meet increased demands—or food is imported from other lands, which are also experiencing stress.

Eckholm describes how the usage of such marginal lands could result in dust-bowls, when climactic conditions change. After the U.S. dust-bowl in the 1930s, national conservation programs were able to restore some of the mythical fecundity, through pasteurizing, strip cropping, terracing, and contour plowing. However, current production efforts are causing greater losses of topsoil, and farmers are abandoning some of the conservation methods for economic reasons. Eckholm concludes that free market conditions encourage dangerous trends. The lesson of the latest dust bowl may be forgotten until the next one. As forestry copies the agricultural model, so it contributes to soil losses and destabilizes ecosystems.

There are large long trends related to economics. Economics has been defined as the "way people make their living," although as a discipline it attempts to balance human wants with scarce resources. Over many cultural lifetimes, this balancing has resulted in definite long-term trends:

- Accumulation of goods in human houses.
- Shortages of timber. Most Old-World civilizations faced shortages of high-quality wood, from the Mesopotamians, Egyptians, Greeks, and Romans to the modern countries of Europe. Pressures on British woodlands in the 1300's forced people to turn to coal as a fuel source, a source then regarded as inferior to wood. The timber famine reached Europe in the 1700s. It had existed in China and India over a thousand years before. Countries that exhausted their wood supplies had to invade other countries or to find substitute such as coal or water power. Each substitute required more energy to produce. Eckholm warns of a serious firewood shortage over most of the earth due to population pressures on the remaining woodlands.
- The earth has been divided up; every part is claimed and owned—one would think that this gigatrend of ownership has ended with the finite territory of earth, but it shows signs of continuing through the solar system. With ownership comes management and control.
- Increase of technologies and information to the fewer wealthier countries. The flow of information accelerates into those countries already technologized better. The acceleration of change, as a result of intensity and scale of human exploitation.
- National and international goals (and trading regulations) have become more

important than local goals and considerations. According to industrial economic logic it is better to stockpile local forests and import cheaper wood from poorer nations. This puts stress on forests in poorer, or less greedy, nations. Management is becoming more intense. Its goal is always to raise productivity per hectare instead of questioning the demand or other options for getting wood.

- The polarization of countries by wealth, as a result of colonialization and advantageous trading. A small minority of nations controls both resources and productivity.
- Identifying a human minimum. Society seems to be working toward a minimum: a box for everyone, with sufficient heat and light, air and plumbing. Mere existence has become an acceptable option. We have gone from the green forest to the gray box. There is no place to hide, no place on earth that the air cannot stink and the rain not burn, that ugliness cannot reach and misery not touch.

Control has been the goal of science and technology. Scientists are obsessed with the treatment of weeds and vermin. The industrial cosmology has put humanity at war with the planet, which always misbehaves; weeds and vermin threaten us. Stability has been raised to sacred scientific state. Our culture has been distorted by the modern emphasis on the scientific method, with its devaluation of philosophical concepts. Rational values are exaggerated and spiritual events are ignored or suppressed. Most philosophies and technologies are not adequate to deal with nature and ecological relationships.

Even when beauty is considered, it is only cosmetic. Many landscapes and ecosystems are considered too impersonal or "ugly" to be valued. Joseph Meeker judged that a burned forest is ugly because it is truncated; but, a burned forest is as beautiful as a baby, and for the same reasons: it has potential, development, and being in the process of renewal. Our sense of beauty, of what is intrinsically meaningful, dominates our grasp of what is real. Until we understand that all phases of nature have beauty, we will not appreciate the meaning of the entire process.

Understanding beauty could even allow a deeper understanding of the utilitarian. For example, what is the use—or the beauty—of a burned forest, as related to the function of lightning or the planetary carbon cycle? Lightning is essential to life by breaking down amino acids into ammonia, methane, hydrogen, and water. Burned forests are necessary to maintain diversity and keep single species from taking over whole ecosystems. We have been persuaded that humans are able to transform any habitat on earth and benefit from the transformation, but we do not extend that power to nature.

2.2.3.4. Political

These trends are partly the result of our unconsciousness of large-scale, long-term events, partly the result of our cultural amnesia about things that make us unhappy, and partly the result of our cultivated indifference—doubtless from our remoteness from wild nature, remoteness as a result of our tools and the general abstraction of civilization. Many of these trends have to do with war and conflicts.

- Trends in war from amateurs to specialists, from special days to every day, even holy days, from soldiers to noncombatants, from fields of killing to any fields and any places, from destruction of soldiers to destruction of civilians, fields and ecosystems.
- Disarmament of civilian populations. In England recently (1995), murder rates were only a tenth what they were 800 years ago and half what they were 300 years ago. The

disarmament took place in many small steps from seizures to licensing, production controls, and fewer public exhibitions.
- Stress between countries polarized by wealth.
- Overmilitarization by wealthy countries. Overmilitarization by poor countries.
- The transition of states, from surplus distributors, to tribute-driven, to commercial exchanges.
- The political transitions from leaders, to chiefs, to tribute systems, to economic and political trade systems.
- The reach of the state into individual lives. From tribute to lack of arms, ownership, and licensing, to information-gathering.
- Trust and social cohesion decline.
- Crime rate, disorder, inequity, poverty and development of an underclass, and consumption.

Some of these trends have been reversed for short periods of time, in various places by different cultures, but the overall trend has been negative.

2.2.3.5. Positive trends

None of these negative gigatrends can be really reversed until human neutrality and remoteness is reeducated into participation and attentiveness. There are already a few trends flowing against the tide. Some positive gigatrends include:
- Adaptation of human cultures and ecosystems over time in Asia, Europe, and parts of Africa, resulting in stable domesticated landscapes.
- Setting aside of areas, such as preservation of ecosystem processes, reservation of archaic cultures, and conservation lands, from industrial interference.
- An increase in the scope of ethics, from family, tribe, nation, humanity, to include reverence for all living beings, first identified recently by Albert Schweitzer.
- An increase in the scope of ethics to include land and forests, identified by Aldo Leopold, and later by Ernst Laszlo to include all systems.
- An increase in the scope of law to include legal rights for ecosystems, such as forests, identified by Christopher Stone.
- Restoration of forests from abandoned fields, anthropogenic deserts, and ruined ecosystems.
- Practice of comprehensive approaches, such as ecological forestry and permaculture.
- The number of people investing in small businesses in forest products, businesses that are labor-intensive and practice rather than profit-oriented.

Combined with other new trends in housing—such as arcologies (Paolo Soleri) and bioshelters (John and Nancy Todd), and agriculture—agroforestry, permaculture (Bill Mollison), and tree crops (Russell Smith), these trends counter many of the negative ones.

The intent of describing large-scale trends or patterns is to have human patterns fit with observed patterns in nature; patterns have a form, sometimes repetition, and sometimes regularity, but each of these is caused by some limiting factor. Fitting the pattern can lead to both continuity and predictability, and both of these are needed to adapt human activities to natural limits.

Thinking we have conquered nature and are omnipotent, we have quit thinking. Satisfied with our comforts, we do not ask enough of ourselves. With these gigatrends possibly ending in tragedy for humanity, we must continue to ask questions. What kind of planet do we want? Wild or domestic? Managed or unmanaged? How shall we use the resources of nature? For products? To protect watersheds and maintain global biogeochemical processes? As a home for other beings? For recreation? As some kind of balance? How many different places, different ecosystems, or other cultures, do we need? What kinds, in what forms? How many should be wild? These questions lead to new strategies for living with the forests and wild ecosystems, strategies that we can test with thought experiments.

2.2.4. *Examining Possibilities with Thought Experiments*

Humanity is already engaged in a great experiment with the planet. We are replacing large, old, complex ecosystems with young, simple fragments, in which fires are suppressed, large predators are removed, large herbivore populations are encouraged, exotic species are introduced, soil is compacted, and excessive biomass is removed—all for the purposes of increasing the amount that can be harvested for human use. Our actions are experiments, whether we want them to be or not. Unfortunately the experiment is not only bad science—there is no control planet—it is ill-considered. This experimental course, which may be global and irreversible, cannot be unmade, not by planning or science, much less by our standard methods of ignorance, cupidity, or denial.

There must be a way to refine the experiments, to minimize our impacts, to be less reckless, and to anticipate the outcome of our experiments before we finish performing them. Not all experiments must be physically implemented. Albert Einstein and Leopold Infield suggest that knowledge of laws can be gained through the contemplation of idealized experiments created by thought, Gedanke-Experiment. For example, to address the equality of inertial and gravitational masses, that is, how the problem of general relativity is connected with gravitation, Einstein imagined an elevator at the top of an incredibly high building, and then imagined what research could be done in this local environment. Such experiments might seem "fantastic" in his words, but they might help us understand the universe.

Although ecosystems and political orders are orders of magnitude more complex than physical systems, perhaps we could imagine and use such experiments to help us understand what is happening with our complex planet that is composed of many interlocking ecological systems. One of the more comprehensive thought experiments conducted is "Daisy World" by James Lovelock, to show how the evolution of species would lead to the self-regulation of climate on an earth-like world. Freeman Dyson offers another example of a thought experiment from Isaac Asimov: Saturn's satellite Enceladus could be used to provide water and warmth to Mars; a rocket could be sent to the satellite, carrying self-reproducing automatons, which could make miniature solar sailboats to carry small blocks of ice from it— there would be enough ice to keep the Martian climate wet for about 10,000 years.

Thought experiments can give us clues about what can happen and what is the likelihood of that happening. "And then what?" asks Garrett Hardin again. Unlike medical doctors or scientists, we cannot either wait or directly experiment within a realistic time frame or scale. We cannot experiment at all in a traditional sense, where we hold most variables fixed, while changing one or two variables in experimental runs. Ecosystems operate over very long time spans; furthermore, their historical nature means that they cannot be restarted for tests.

Large-scale, long-term experiments are expensive and relatively few. Most experiments are short-range, small-scale, isolated, and detail dense. Most experiments do not present the hypotheses required for the management of ecosystems. Ecosystem management, because of uncertainties, lack of controls, age, and uniqueness, is an uncontrolled, large-scale experiment. Thought experiments can refine the design of our larger experiments by suggesting better hypotheses and considerate behaviors.

Thought experiments can help us avoid being overwhelmed by details. Thought experiments can help formulate goals and interpret information appropriate to scale. The idea of science is to manage our experiences with generalities. Once the thought experiments are started they can be refined with conceptual or mathematical models, which can simulate the changes and historical development of changes. Computer-based models can permit complex explorations, as well as suggest new patterns and further hypotheses. Through thought experiments, many dangers and expenses of activities can be avoided.

Thought experiments are vital to understanding the complexity of ecosystems. In practice, erring on the side of preservation—the prudent and conservative course—means minimizing the influence of human activities on the land. It means experimenting cautiously with new approaches to ecosystem management and being properly skeptical about any claims for sustainability. It means drastically reducing our demand for natural products, through conservation, reuse, recycling, and human population control, so that the greatest possible number of ecosystems are left wild and degraded lands can recover.

Thought experiments can also be used to examine possible scenarios of the future based on our actions. For example, if we continue the current trend of disequity, how might things play out? For example if the rich keep getting richer, how will they have to protect their wealth from the poor? Will laws be enough? Will they need ever larger armies of security personnel? With already four times the number of civic police, will they need even more? Will corporate police protect the wealth of their stockholders after the civic police give up? Will the poor collapse leaving rich enclaves that have to grow their own food? At what point will the gap be wide enough that the poor have to harvest the wealth of the rich by force? At what number of poor? Will the poor prey on each other first? At what point will the environment be used entirely for a few more years of life for rich or poor humans? At what point might the environment collapse?

Will regional groups, such as the northern hemisphere, form alliances to keep going, after writing off other regions or the southern hemisphere? Will this block be able to defend its resources? Will that extend the time of any collapse? Or accelerate it?

Will the United Nations be able to coordinate some kind of peaceful reorganization? Can a revitalized UN guarantee a rational economic and political strategy for all nations? Should this UN be dominated by a China or a United States, so that it may operate without as much discussion? Is it utopian to think of such reorganization or redistribution for equity? Is this less naïve than allowing the market to sort out entire cultures and regions and consign them to poverty and violence?

The thought experiment presented below is incomplete, but suggestive of the kinds that we could be creating and manipulating to guide our plans and models.

2.2.4.1. A Sample Thought Experiment with Arcologies

At present over three billion people live in cities, about half of the total world population of 6.3014 billion (revised and estimated for 1 September 2003). What if, as a thought experiment, essentially all people lived in arcologies, except for some small traditional communities living in wild ecosystems?

Historically, human hunting altered ecosystems, then converted forest, grassland,

and wetland ecosystems to agroecosystems that had to be managed. Now, the expansion of urban areas with roads, power grids, and other infrastructure, is interfering with the basic functioning of many ecosystems. Modification, conversion and destruction of ecosystems disrupts the complex interactions within and between ecosystems, the hydrology, soil structure, topography, and vegetation; it changes the complement of species, and it causes a loss of diversity. The new replacement systems are simpler, less mature, and less diverse.

We have achieved great horizontal growth, much like a fungus. However, if we want to be like a smarter fungus, slime molds for instance, we need to learn to cooperate to grow up and be more dense. The larger metropolitan regions are covering wild ecosystems and agricultural fields with single-family houses, malls, building, recreational areas, and roads— all of which are car-centric or auto-morphic. This means that energy and goods are also spread thin. Such systems of things are hard to control, hard to keep safe, and hard to remain interesting. It might be worthwhile to compare human systems to mature ecosystems; we are creating pioneer individuals that do not live well in concentrations. We are creating edge individuals and not those who can live in interiors and share resources, or can develop new resources with cleverness and intelligence. City designs do exist, however, which incorporate the properties of mature systems, as well as the characteristics of ecological thinking.

An arcology, as defined by Paolo Soleri, is a city which embodies the fusion of architecture with ecology. The arcology concept proposes a highly integrated and compact three-dimensional urban form that enables radical conservation of land, energy and resources. Arcology eliminates the automobile from within the city, and with it, the fifty percent of land devoted to automotive needs. The multi-use nature of arcology design would put living, working and public spaces within easy reach of each other and walking, supplemented by elevators and airport things, would become the main form of transportation within the city. An arcology would use passive solar architectural techniques such as the apse effect, greenhouse architecture and garment architecture to reduce the energy usage of the city, especially in terms of heating, lighting and cooling.

The small footprint of an arcology, combined with many built-in gardens, would allow rural space and agricultural fields to be closer to the city, and a part of the immediate urban environment. Wilderness, also, would be much closer to population centers in arcologies. Psychologically, the intelligent design would be more conducive to inspired living, the kind found in traditional culturally-significant cities at certain times. The proximity of agriculture and wilderness would allow people to participate more in them, with the full range of benefits that comes from growing and cultivating plants, as well as being able to immerse in the otherness of wild ecosystems.

The sizes of arcologies range from 250,000 to almost a million people, although smaller or larger ones are possible. For the sake of argument, assume that the average arcology is the size of Soleri's proposed Novanoah, at 400,000 people. At that size, it would take 15,734 arcologies to house the planetary population; this number is less than the number of cities in the United States in 2004, at 19,354. Assuming that the area under the arcology is about 5 square kilometers (almost two square miles), the surface area taken up by arcologies would only be 78,768 square kilometers, which is only 0.00054 percent of the land area of the planet—that is half of one thousandth of one percent—(149.45 billion square km, or roughly 0.0167 percent of the land area currently under concrete and asphalt, which is 4.71 million square kilometers or 3.15 percent of the land area of the planet).

The number of roads would be significantly reduced, from 676,750 square kilometers (roughly half of one percent of the total land surface in every country, according to the International Road Transport Union) to 230,000 km^2 (a generous number to be sure, but it could be as little as 50,000 km^2). Over 2 percent of the land is under road surfaces in the United States, with a smaller percentage in Europe, by comparison. Land under agricultural production would also be reduced, from 33 billion square kilometers (22 percent of the land area of the planet), partly due to improved practices, partly due to the integration of many kinds of agriculture into the city, and partly due to the carefully limited use of wild populations, without domestication or containment.

The shapes of arcologies would be as diverse as any. Many of Soleri's shapes are geometric. They could be large pyramids, filled with living and working spaces, connected by transits and illuminated by light wells. The traditional ziggurat modernized would offer a good ratio of sunlight and truck gardens to size. Arcologies could be empty tubular pyramids with modular dymaxion attachments, that could be moved between arcologies or new sites. They could fit the shape of the landscape, as does the Palouse Arcology (Wittbecker 1992). Arcologies could be built around small mountains, in bridges crossing canyons, or threading through coastal seas.

What would such a change mean to most people? Probably, few people would have the need for private cars. A typical day, for most workers, whether administrating, grading papers, policing, or making steel, would start with a walk to work, past local stores and businesses, playgrounds, and microfactories. Work would involve fewer layers of hierarchy; the pay range would only be from 1 to 7 times the minimum salary.

Since Soleri's heroic designs, arcologies have been confined to computer games and have become elements in science fiction and cyberpunk films. With more prototypes being designed, one may be built in the next twenty to thirty years. A proposed project for Tokyo Bay, the Shimizu TRY 2004 Mega-City Pyramid, if constructed, would become the largest artificial structure on the planet; it would be 2004 meters tall and house 750,000 people. The external structure of the pyramid will be an open network of megatrusses, supporting struts made from carbon nanotubes to allow the pyramid to stand against high winds, earthquakes, and tsunamis. The trusses will be coated with photovoltaic film to convert sunlight into electricity and help power the city. The building will be zoned into residential, commercial and leisure areas. Separate buildings for housing and offices would be suspended from the supporting structure with nanotube cables. Transportation would be provided by accelerating walkways, inclined elevators, and a Personal Rapid Transit system inside the trusses with individual driverless pods.

It seems that arcologies would be part of a whole package of changes, brought about by ecological planning on a global scale. The experiment would not require that arcologies replace archaic populations living in human-modified ecosystems, or even all low-density habitations or traditional cities. But, they could be new cities situated in infertile areas.

Many cultures could live in optimum configurations in their territories, as part of wild and domestic landscapes. But, we also need heroic architecture. Heroic design and extravagance in life is needed in general. It is not contradictory or antithetical to frugal lifestyles or to restoring a healthy environment. Life is exuberant; energy is used, lives are lived and used, not wasted or saved. Life is the accumulation of individual experiences that cannot be saved, stored, or owned. The heroic things in life are often those most admired or remembered by subsequent generations.

Thought experiments will be suggested throughout this work. The best response to a question about what would happen as a result of some actions under some circumstances may be a thought experiment. Through that, you can create explanations and discover answers in a dialogue with others.

2.2.4.2. Duality & Conversation

A thought experiment can be considered as a communication: Mind reaching to mind, intention to intention, or as a simple correspondence. Communication is a presentation of one's self to another, through a conversation, in a back and forth process. The word conversation is derived from the French word meaning 'to live with,' from the Latin, meaning 'to turn with.' Since all things are in some sense subjective and unique, the term conversation is appropriate.

For Gordon Pask, a theory of conversation is a theory of participants in conversation; all events are subjective. Conversation creates a domain, which is the appropriate frame of reference. The domain is the environment of a conversation. The conversation is a minimal situation for observing the psychological events of which the participants are conscious. Understanding between participants is pivotal. The conversation is relativistic and reflexive.

The biologist Francisco Varela uses conversation as a paradigm for interactions among autonomous systems. Conversation is direct. Each side has a perspective and this is the heart of the process. When conversation is considered as a totality, there is no distinction about what is contributed by whom. The process is a coherent event shared by the participants, not a simple information exchange. Both Pask and Varela apply their ideas to human systems. But, there is no reason why it could not apply to any interaction between living beings.

When different modes of description appear as opposites, it is more satisfactory to consider them complementary instead. This Varela says is necessary with tree/net duality and recursion/behavior duality. Complementarity solves the problem of opposites, such as focus/frame considered under an either/or logic.

2.2.4.2.1. Duality

Duality leads to the philosophical idea of trinity, with some similarity to the ideas of Charles S. Peirce, since the poles are related yet remain distinct; they are not one or two, but are really three. Varela offers the heuristic star, where the star is equivalent to the whole process. Thus:

* = it/process leading to it

Varela proposes that dualisms or dialectical contradictions such as mind/body or whole/part or being/becoming should be conceived of as stars, that consist of an it/becoming-it process. Both sides of the slash must be considered, and the process leading to it. Varela borrows Bateson's concept of minds jointly defined as conversational pattern (left of slash) and bodies as participants in pattern (right). Consider both sides of the slash. The slash is a compact indication of a transition between states. For instance:

* = whole/parts constituting whole

* = stability/approximation in time The predator/prey pair are not excluding opposites, but generate a whole unity, an autonomous domain where there is complementarity, stabilization and survival values for both.

* = ecosystem/species interaction

In general, any autonomous situation is on the left of slash, and corresponding process is on the other side: being/becoming or right brain/left. But that is incomplete; the network is the process that constitutes the trees. The duality is connected with processes in both directions. Varela sees the totality as emerging from part-by-part approximation of the trees (process leading to net). But sometimes the whole determines the parts. Mostly, the process mutually arises. Complementary elements mutually specify each other, so, in a sense there is no more duality.

Even Varela states that in natural systems, there is no real opposition, except where we put our values. Yet, the human brain, encultured, perceives primal dualities. For every apparent set of opposites there is a star that is the left hand of another equation.

Varela suggests using the idea of intersecting triangles, a star, as a solution. The star is an ancient alchemical symbol. Man reflects the macrocosm in a microscosm. The two sliding triangles represent Hermes Trismegistus aphorism: As above, so below.

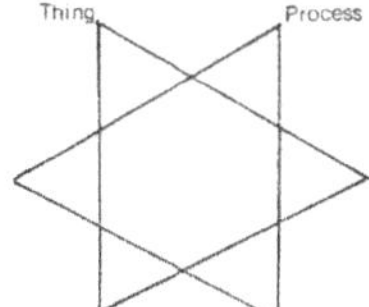

Figure 2-2421-1. The Star

Varela rotates them, however. By using two intersecting triangles in the shape of star, as two levels of logical type, they are noncontradictory and mutually-specifying, manifesting restrained complementarity. The star of Varela differs from dualism and Hegel's dialectics. There is no Hegelian synthesis because there is nothing new, only a more direct appraisal of how things are put together. This view of complementarity is a departure from the classical way of dialectics. Dualities are represented by imbrication of levels, where one term emerges from the other. Their basic form is asymmetry; both terms extend across levels. This dialectic is self-referential: it/becoming-it. The tree is rebalanced by constraints of the network.

The net is prior in relation, however. This explains problem of the space/time energy/mass field much better, where there is network from which space-time and energy-mass can be extracted. The network/tree duality can represent recursion/behavior duality by retaining interest in the connectivity of a system only. The nodes in tree or net represent elements in a system. Their links suggest interconnections. The reciprocal connectivity of a net suggests coordination; a tree structure suggests the sequential subordination of a system's parts.

But the image of the star is not quite comprehensive enough. The sides of a star are not inversely related. The sides are directly related; as one side increases, the other side increases. The relationship of sides cannot be described as conflict only. For example, '= whole / part' is not complete. There are parts in wholes, in motion, and in context. Similarly, for good/evil, good and evil resides in the knowledge of human actions in cultures in place.

2.2.4.2.2. Conversational Field

Since there are always any number of relations in a situation, the use of dyads and triads can become complex. The metaphor becomes unwieldy. Conversation can replace the idea of duality and triadicity. Four, five or more person relations are not reducible to triadic relations, as argued by Peirce. The upper limit for measurable relations is probably about seven, plus or minus two, as George Miller found. Seven is a magic number in human psychology; it represents the maximum number of items that a subject could reliably remember, as well as other variables. Possibly it applies to the number of subjects having an intelligent conversation as well. Given the indeterminacy of relations between subjects, the use of conversation is justified.

As the science of physics has realized, in its three body problem, the calculations necessary to solve a problem increase logarithmically when the number of bodies is multiplied. In biology, the phenotype and organism generate unpredictability. Neither the organism or the environment knows what the other will do. Species interaction achieving a stable ecosystem can be thought of as a biological paradigm for a conversational domain, the direction of which is unpredictable. Evolution is then the changing theme of the conversation between species and environment. Evolution may be interpreted as the development of new channels for communication.

One paradigm that would suit the needs of ecology is that of personal communication: Mind reaching to mind, intention to intention. Communication is presentation of one's self, of one's life, that may evoke correspondence in others (similar to entrainment). The activities of two communicators combine to make the universe of the observer more ordered and redundant. The nature of meaning depends on the frame of the observer. Ecologists must envisage the notion that nature speaks to us, perhaps using information theory. Anthropomorphic ecology can recapture the experience of personality in nature; this is just the reverse of the pathetic fallacy, which derides the use of human terms to describe natural phenomena.

2.2.4.3. Wild Thinking

What is the proper language for humans? What is the proper diet for all humans? The proper mode of expression? These questions may seem to be too presumptuous to be asked. But, are there other large-scale questions that can be asked? Are there large-scale forms of thought? Is Philosophy one such form? Religion? Ecology? Are such large-scale forms domestic or wild? What is wild? Is it the same as being not domesticated? Uncontrolled or unmanaged? Not cultivated? Untamed? Savage? Wasteful? In a state of nature? Lawless? Wild as a word has ambiguity and reflexion (from the German word *wild*, or perhaps *Wald*, meaning forest).

What is thinking (from the German *denken*)? To revolve ideas in the mind? To design, to imagine, to judge? Perhaps thinking is a dance of the mind, as it flows and merges into a harmonious process in life. Is ecological thinking intrinsically wild? Is ecological knowledge wild? Will Wright suggests that knowledge becomes wild when it is critically reflexive and committed to critical access rather than to a version of absolute reality. If so, such wild knowledge cannot be "domesticated" by one particular social institution. It is accessible by individuals.

Ecological thinking is wild because it has a nonhuman component. By contrast, scientific or religious thinking can be regarded as domesticated or tamed because it is limited by a true version of reality, or a set of rules for observing a true reality. Science is defined in opposition to religion, with a commitment to neutral observation rather than a moral commitment to tradition. Wild ecological thinking combines technique with moral and ecological concerns. The fundamental emotion of wild thinking is astonishment, which literally means 'being struck by lightning.'

Of course, there are other forms of thinking. Religious or scientific thinking are relatively tame. Tame ideas in religion and science are remarkably persistent—for instance, "more is better." George Orwell referred to these obsolete ideas as "wrong-think."

Another kind of thinking, "double-think," is used traditionally to keep some ideas tame. Orwell declared novelistically that "newspeak" was a method for controlling thought through language. Doublethink was a way of controlling thought directly. Newspeak incorporates doublethink, as it contains many words with contradictory meanings, such as 'good/evil', 'truth/falsehood', and 'justice/injustice.'

Doublethink entails holding two contradictory beliefs simultaneously and accepting both of them, deaf and blind to any contradictions. The purpose of doublethink, as presented by Orwell, is to use conscious deception while retaining the firmness of purpose that goes with complete honesty. The deception has to be conscious to be implemented precisely, but also it must be unconscious to avoid guilt and falsehood. It is important to believe the lies and to deny the existence of objective reality, while taking account of the reality denied in order to be satisfied that reality is not violated. It is necessary to exercise doublethink when using the word doublethink, since it is an admission of tampering with reality.

Gordon R. Taylor suggests that "non-think," the failure of good ideas to be recognized or used, is equally obstructive—thus, the idea "to protect the ecological basis of life" is never considered.

Wild thinking is appropriate for "system breaks," that is, the social discontinuities identified by Kenneth Boulding. We have started to identify the forces setting up the next big system break, but we have not defined the forming patterns very well. We need to be rethinking—a form of wild thinking—the basic assumptions of the spheres of civilization, from our economic and political to industrial, religious and scientific.

Talking about wild thinking, perhaps there is another side to it, a necessary social side. Too much anarchy is dreaming; too much feral thinking is noncultural and dangerous. Too much unanchored thought is unrelated to the important mode of learning by doing.

Is wildness just the nonhuman part of the spectrum? Does it overlap in humans? Is it just difference or craziness? We love and celebrate the wild; also, we fear and suppress the wild. The wild is a quality of being just beyond our rules or outside of our walls. Paul Shepard reminds us that wildness occurs in many places, in any species whose sexual assortment and genealogy are not controlled by human beings. Charles Darwin reminds us that humanity is wild, also. Does the human mind have to be wild? Not necessarily—we can domesticate our ideas.

It seems easy to talk of wild thinking. Is it meaningful to talk of a wild culture, one that intermeshes with the wild of nature? Are archaic cultures wild? Is a model of a civilization without walls, without resources, maxima, or weeds, wild?

All we need is new economic, educational, political, cultural, social, psychological, and ecological frameworks, in which we can try to rebalance our individual lives; try to rebalance the social spins of the patriarchal and matriarchal directions, between the human and ambihuman; try for individual self-reliance and health, try for community self-reliance and health; try for ecological community health, strengthen our community and cultural identity so that the positive aspects of globalization can be internalized without the destructive force of globalization ruining cultures; try to limit the centralization of power and authority, in style as well as trade, reform corporations to act responsibly as public service organizations (not as imaginary irresponsible individuals); try to direct technology in appropriate ways, and set up a global commonwealth for global relations; and, try to live with wildness. How wild humans can live on a wild earth is the subject of this eutopian framework.

2.2.5. *Using Ecological Analysis & Synthesis*

Basic analysis means taking something apart to understand how it works. The word is based on the Greek words meaning 'loosen up.' It was first used in the modern sense, as opposed to just logic, by French philosophers and scientists. Once a subject is identified to analyze, the subject is deconstructed, literally. To analyze a rock, it is necessary to break it apart and identify specific elements. To physically analyze a human being, you must cut into the flesh, weigh the blood and organs, and determine the different kinds of organs, fluids, and moving parts.

However, that is rarely enough to tell you how a human operates. For that, you must undertake a functional analysis. You would have to observe a working specimen first: Make a list of characteristics, hair, eye, and skin color and shape, weight, shape, completeness; list inputs, such as air, water, and food; list outputs, such as carbon dioxide or manure; make another list of needs, from shelter to place, security, and social demands; describe the kind of places humans are found, from gardens to factories, and their relationships with other beings.

A lot of time would have to be spent analyzing human communities, at a family and neighborhood level, as well as working relationships and the larger relationships with the environment. Working relationships could easily be expanded into economic analysis. Conversations could be analyzed with linguistic analysis, which would have to delve into the history of language and cultural shifts. A historical analysis of a person could trace the movements of ancestors and their interactions with others and their environments.

A philosophical analysis might be useful to try to understand an individual expression of existence and place. Some philosophies, perhaps all, start with the notion of what it is to be human, even those that start with existence and being start with human existence and human being. But, why not start with what it means to be or to live? "I live here and therefore I am in place and know it."

The fundamental characteristic of analysis is taking apart. Taking apart living beings or working cycles can result in their death. However, more general kinds of analysis rely on good observations to allow deductions about internal workings. Goethe rejected destructive analysis; his approach was passive attentiveness. Certainly many forms of analysis can approach this level of sensitivity.

2.2.5.1. Linguistic & Historical Analysis

In order to make sense of the sheer multiplicity and complexity of their environments, human beings create abstractions. An abstraction is an idea created to refer to all objects that have certain characteristics in common, e.g., all birds. Abstractions can be generalized, e.g., all things that fly, but at each outer level the objects have less in common; thus flying things include insects, mammals, reptiles, seeds, and spores.

Human beings also classify and label their abstractions. The systems of classification are reflexive and pragmatic; that is, they refer to the classifier as well as to the object, and they are guidelines for how to think about, treat, and relate to an object, according to S. I. Hayakawa. As soon as a classification is no longer useful, people stop using it and become receptive to a better classification. This gives a historical and linguistic caste to analysis, which can be used to differentiate between abstractions and classes.

2.2.5.2. Scientific Analysis

Science uses abstraction and classification to transcend common sense to describe the fundamental structure of nature. Scientists use classes to limit things to their own mesocosmic scale, from geological epochs to species, although the classes are independent from the things. By comparison, archaic (sometimes called pre-scientific) peoples believe that there is a necessary and intimate connection between the symbol and the object, that the name is attached to the object in an intimate way. This "epistemic naiveté" is dismissed by modern science, which uses its own unique method. The general format of the scientific method is the same as that for traditional ecological knowledge. However, science adds measurement to analysis, experimentation to observation and formal hypotheses to guesses:

- Observe phenomena and record facts.
- Analyze the phenomena into components.
- Measure the phenomena (before and after manipulation) allowing the results to be quantified.
- Make guesses and generalizations, using the logic of deduction. A hypothesis is a statement about relationships that can be shown to be untrue.
- Formulate laws from the generalizations about the phenomena. Such laws may describe the behavior of a natural system.
- Describe the laws mathematically, using numbers.
- Develop a theory to predict new phenomena. Theories can lead to new conclusions and sometimes altered perspectives about phenomena. A scientific theory is a statement that postulates ordered relationships among natural phenomena and explains some aspect of the world. It allows one to ask certain kinds of questions, some as specific hypotheses.
- Test hypotheses and theories in a controlled environment. A theory cannot be tested by hypotheses.

The key ideas here are analysis, measurement, guesses, control, and rule-based predictions, and control. The use of these ideas have brought about keen understandings of parts of nature, from the quantum level to the extent of the universe. And, it has led to amazing changes, from electricity and hydraulics to medical analysis and computers.

Science works with theories, that is the direction of observation, based on previously identified facts, used to lead to evidence or proofs of hypotheses about how things work, which can be used to predict other actions in the future. Theory is used to guide experimentation. But much experimentation these days occurs without theoretical guidance, or rather the theories used are not capable of the context itself, so that the experiments tend to go out of control, or rather, are improperly conducted as experiments—no control group, no good theory, no idea or control of scale.

Theories can help science to avoid claiming what things, either knowledge or certainty, that it cannot claim. Theory can distinguish between knowledge and theory, between theory and the unknowable. There will always be limits and errors; these cannot be avoided. Theory needs to accommodate this fact. Fallibility is unavoidable; it is a part of certainty from a limited perspective with a limited knowledge; it is part of being incomplete and using incomplete knowledge. But, this is the limit of all being; this is the limit of finiteness.

Good theories are grounded in place and in communities. They must be used to place our science and technological capabilities in a culture that has goals and limits for our actions. In this sense the theories are tools for adaptation to place, within place, within specific places that we work towards making good. Theory describes the fitness of our conduct within an ecological and cultural system. Global theories have trouble passing the test of one locality. But, a good theory has to have a real framework and a real ecological context.

Some mysteries cannot be reduced, cannot be understood with human understanding. Some things are ineffable, even if they are understood, they cannot be reduced to words. Theory has to account for those things that cannot be reduced or explained or understood. Our theories and practices are human theories and practices; they are limited by real human limits, by the dimensions and terms of times that we humans experience. Strangeness, which Edward Wilson expects to be connected and made sensible through a consilience of knowledge, will continue to be strangeness, because of its intrinsic nonhuman properties and our human limits.

In this sense, theory is a guide to action, based on thinking about previous actions and projecting future actions, that is, making thought experiments. What we will know will always be incomplete, but we have to act, and therefore we have to act on partial knowledge in the face of partial ignorance, many kinds. Theory is a guide to acting under different kinds of unavoidable and irremovable ignorance. People who know a lot, but do not have theories that bind them to wise actions, tend to ignore ignorance and to act in an arrogant manner; such behavior is usually dangerous.

The proper role, ultimately for theory, is to guide us in our choice of tools and strategies for dwelling on the planet. Wisdom is that kind of action, with theory, with understanding of ignorance, that is as if we were wise, according to Jonas Salk.

But, science, like archaic thought or traditional ecological knowledge, makes assumptions in the context of specific cultural situations. Some cultural-scientific assumptions have been incorporated into the works of science by scientists. At the time of Bacon, it was assumed that there was: An absolute, immutable, omnipotent God, everything was sorted into a great chain of being, economic subsistence was preferable, and social inequality was unavoidable. Later, Darwin incorporated a different set of assumptions into his theories: Absolute space-time; atoms as discrete units; economic discrimination, and continued social inequality. These assumptions contributed to the misuse of his theories to justify social and economic conditions at the time. New versions of science have used broader assumptions, but the assumptions still must be recognized for the science to be effective in a larger sense.

2.2.5.3. Systems Analysis

To understand the workings of complex systems, one could use systems thinking, with its concepts of feedback and emergence. Complex adaptive systems display emergent behavior. As an example of emergence, slime molds can form a community without a pacemaker cell that determines when the cells need to combine. It seems that self-organization is bottoms-up. Emergent systems are rule-governed, though; slime molds explore by adhering to low level rules. Individuals coordinate work, even if they cannot assess the global situation. Emergent systems are local; individual molds “think” locally and act locally. Random action serves to explore local space. Individuals pay attention to their neighbors, and patterns emerge from local activity. Simple behavior seems to work, with local feedback, and more sophisticated behavior "trickles up" to approximate a global perception.

A system is a way to explain part of the universe and to deal with complex behavior. Mario Bunge defines a system as a complex object, every part of which is connected with other parts of the object in such a way that the whole possesses emergent properties that the parts lack. A concrete system is composed of concrete things linked together by real physical, chemical, biological ties. Cells, wolf packs, and nongovernmental organizations (NGOs) are concrete systems. Conceptual systems can be linked by logical relations.

Many concrete systems are open and self-regulated. Many are closed and artificial. Bunge makes a strict dichotomy between formal and concrete systems (conceptual-material). He denies the possibility of mixed systems, but allows real ones to transmit information. He says sets and relations are abstract objects, not identical to concrete objects like molecules. Yet, if the relations are real, are they not concrete in a way? Intrasystem bonds are stronger that intersystem bonds; if not, the system would fall apart.

Social systems have all sizes and degrees of complexity. Governments are more complex than families. Systems can also be constituents of systems, like United Nations (UN).Every concrete thing has properties. Some properties can be known easily; others can be revealed by research. A list of known properties describes the state of the system with finite quantities.

A qualitative description of properties describes qualities of the system. Quantitative descriptions measure quantitative variables. The state of a system, definite and objective, can be conceptualized with theories and models. All concrete systems change as reality unfolds. That is, the properties of systems change through time. Some change, such as growth or decay, is quantitative. Other change is qualitative, resulting in breakdown or formation of the entire system.

Systems theory analyzes events, processes, and patterns in the world. An event is a change of state. An event is described by two points, an initial and final state. A process is a sequence of states, also called a history. A process creates a path, described by a trajectory of states. A process is described as evolutionary if it involves emergence and the creation of new things, as in general speciation; to be a species, however, the novelty has to reproduce, multiply or diffuse.

The concept of state precedes the concept of process, although a process can define a state. In complex systems the past contributes to or constrains the state, from the magnetic hysteresis of ferromagnets to organisms, which have memory.

Changes in systems can be recorded, but only partially and not continuously. The discontinuous recording is digital. The process is continuous, but the measurement is discontinuous.

Regularities in systems are patterns. Patterns can be seen in things or even cultures. For instance, laws of genetics are natural patterns; human customs are artificial patterns. Where the natural ratios of females to males are altered by female infanticide or other action, the pattern is semi-natural.

Bunge distinguishes four kinds of real patterns: Laws, trends, correlations, and rules. A law is a stable pattern inherent in things; it is discovered; laws like gravity are boundless; biological laws are bounded, he says. After the "Big Bang," gravity was bounded. There are examples of social or cultural laws: "The inertia of a social system is directly proportional to the number of components and inversely proportional to its cohesiveness." or: "Higher culture does not emerge in society until the basic needs of some of its members have been satisfied."

A trend is a temporary pattern, such as the globalization of capital or fertility. Trends can be reversed. A correlation, usually statistical, is a covariation of two properties, e.g., a correlation of sickness and education—but this correlation is problematic, due to fact that educated are wealthier and report their sickness more than the poor; the real correlation could be reversed. A better correlation is: Single-species forest stands can be correlated with standing armies in the northern hemisphere; standing armies were thought to be characteristic of people in tougher climates, which encourage large stands of trees. A rule (or norm) is a social convention set up by people, in force in a social system. The analysis of patterns is the strength of systems analysis.

2.2.5.4. Ecological Analysis & Synthesis

Ecological analysis has to be one of the largest form of analysis. It includes limits-analysis as well as many other kinds. For instance, Georg Borgstrom, in 1961 and in his 1969 book *Too Many*, used a concept of ghost acreage to explain how the Netherlands could continue to thrive using invisible acreage. In many crowded places, the average citizen occupies acreage that is out of sight, but has to be part of a complete analysis. Ghost acreage was a good image. Ghost acreage is the additional land, from sources outside a nation through trade, theft or conquest, that a nation needs to supply the total amount of food and fuel.

Later, in 1970, the idea of minimum required acreage of the environment was used by Eugene Odum to calculate an optimum population for the state of Georgia. Odum suggested that land area could be a measure of human carrying capacity. The minimum per capita acreage requirements, with a temperate area like Georgia as a model for a quality environment, is 5 acres (2.02 ha), including natural areas based on minimum space needs for watersheds, as estimated by land use surveys, fiber areas, and food-producing land that includes acreage for domestic livestock.

The same year, I enlarged on Odum's idea by calculating the minimum wilderness necessary for the self-renewal of wilderness, and then relating that to human biological and cultural carrying capacities. It was a way to optimally fit humans to wild productivity and land through design.

The *Limits to Growth* study predicted a decline in global human population as it overshot its carrying capacity (Meadows et al, 1972). William Catton, a sociologist, expanded the ideas of overshoot and ghost acreage in his 1982 book *Overshoot: The Ecological Basis of Revolutionary Change.*

A later study on the human appropriation of photosynthetic production calculated that humans were using over 40 percent of the biological productivity of the planet (Vitousek et al. 1986). Later studies indicated that the percentage was over 50 percent (Pimm 1992).

2.2.5.4.1. The Ecological Footprint

Originally developed at the University of British Columbia's School of Community and Regional Planning in the early 1990s, the ecological footprint was promoted as an indicator of sustainability. The concept was popularized by Mathis Wackernagel and William Rees (1996) in the publication *Our Ecological Footprint: Reducing Human Impact on the Earth.* The footprint is an effective metaphor; it is easy to visualize and connotes coverage.

The method of calculation begins with the construction of a matrix for consumption by land use; consumption covers food, housing, transport, and consumer goods and services, whereas land use encompasses built-up areas with roads and housing, crop land and pasture for the production of food and other goods, managed forest for the production of wood products, and "energy land" for sequestering carbon dioxide emissions resulting from the burning of fossil fuels. This consumption by land use matrix provides a two-dimensional temporal and spatial snapshot, for a given population, of the land required for the production and consumption of goods and services.

The consumption and population data for each land use category are used to derive an average annual consumption per person. Consumption addresses imports and exports. The land area utilized by each consumption category is then determined for each land use category. This requires dividing consumption in each category, by a relevant global average yield—to compare national footprints—to obtain land area.

The land appropriated for energy consumption is treated separately primarily due to the size of the contribution it makes. Five types of energy are distinguished: gas fossil, liquid fossil, solid fossil, firewood and hydropower. Nuclear power is treated as a fossil fuel, despite different sources, impacts and risks. Energy generated from other sources, such as solar, wind and geothermal, is considered to be negligible. The energy land requirement for fossil fuels is determined by calculating the amount of planted forest land necessary to absorb the CO2emissions from energy consumption. Oceans are assumed to absorb 35 percent of CO_2 emissions at the global level. Correction for trade is required because energy is utilized in the production and transportation of exported and imported goods and services.

Each category is multiplied by an equivalence factor to take into account differences in biological productivity The ecological footprint is generated by adding the land area appropriated by each land use category. The supply of ecosystem service is called 'biocapacity;' the human demand for these services is the 'ecological footprint.' By comparing supply with demand, Ecological Footprint Analysis provides a metric for indicating unsustainability. Globally, as of 2002, there exist 1.8 gha (global hectare) per capita of biocapacity available to support human activities, and not setting aside any biocapacity for nonhuman species. In 2002, the most recent year for which data are available, the global human economy used 2.2 gha per capita, resulting in an ecological overshoot of about 20 percent. This means that it takes the Earth about 15 months to regenerate what humans use in 12 months. This condition of global ecological overshoot cannot continue indefinitely. A catastrophic ecological failure may occur if the demands for "natural capital" are not reduced to a level that the biosphere can provide on an annual basis.

2.2.5.4.2. Limits of a footprint

Although the footprint is useful for comparisons across regions and the planet, it is a general, two-dimensional, static measure. Its specific questions do not make close distinctions. For instance, it is not just a matter of eating meat, but whether you raise or know the cow, chicken or pig. It is not simply how local the food is, but whether you grow your own in urban or country gardens. It is not simply how much waste you generate, but whether it is recycled or reused directly. It overlooks the problems of nonbiodegradable wastes and toxins. It is not simply just how many people live in a house or use a car, but how many are too many or too few for optimum use. And, whether or not you have pets, which not only consume food and room, but have dramatic effects on living patterns and wildlife. It isn't just the size of a house, but its materials, insulation, and under roof area. How is it sheltered? How is it situated, in the city, floodplain or mountains?

What kind of house does make a difference, as does having running water. But, how is the water measured, where does it come from and where does it go? It isn't just electricity, but where it comes from, how many appliances and what kind? Can they be recycled? It isn't just public transport but what kind: green buses, light rail, mag-lev? All kinds of transport could be percentaged. It isn't just a motorbike, but is it tuned? Is it off-road, which has tremendous impacts on wildlife and land? Again, motorbikes have tremendous impacts on roads, accidents, healthcare and off-road, regardless of mpg. It isn't just how far you travel by car, but why take trips? It isn't just gas mileage either, it's recycling and cost. Even corn oil fueled cars have tremendous soil impacts and are not sustainable (goodbye soil). Having two or more in a car does save something, but it also reduces the life and efficiency of the car, which still requires roads. Personal pick-up trucks and SUVs here have greater impacts on the infrastructure. How much percentage of travel is by walking or bike? In some cities, you can walk everywhere.

Flying on conventional aircraft is a problem. How can that be greened with a technical solution? Should we ground the fleet once a week for a day or two? Airports are built on flat spaces, such as grasslands or fields. Can we encourage blimps and ultralights?

2.2.5.4.2.1. Assumptions of a Footprint

Many assumptions about productivity may not be applicable. For instance, tropical areas do not offer more productivity for consumption; because of the diversity and maturity of the systems, they offer less net productivity for human use. Footprint analysis seems to assume that domestic vegetation is as productive as native vegetation; with few exceptions, it is not. It also assumes that no harmful chemicals are used or released.

Standardization of a footprint. There is no accepted methodology for calculating the ecological footprint. The ecological footprint is not, for example, constructed according to widely accepted international conventions such as that used in the United Nations System of National Accounts (UNSNA). This has led to ambiguities in interpreting the results of various ecological footprint studies. Footprint analysis needs standards for same reason as metrics and certification (industry certification does not match FSC). Without standards, calculations can be off by 100-200 percent.

The components are not nearly as uniform. There are vast differences in productivity, not to mention stability and renewability. Furthermore, culture use of the productivity varies widely. Finally, the waste from food and resource processing is not considered. Greater efficiency could increase the useful net by 50 percent.

2.2.5.4.2.2. Dimensions of a Footprint

The footprint measures area. An area is two dimensional, but an ecosystem is four-dimensional. That is to say, a footprint should have a depth and history. The soil has a depth and a history. Animals make soil. Microclimates 6 inches off the ground can mean good crops or none.

2.2.5.4.2.3. Dynamics of a Footprint

Area is not as important as carrying capacity, which includes energy, technology, and ideals. Furthermore, carrying capacity is dynamic and has to be represented that way. Therefore there cannot be a specific number, only a range of numbers with a range of rates. Nature is characterized by complex adaptive systems with non-linearities, feedback loops, and thresholds (Holling, 1973). By ignoring such dynamics the ecological footprint cannot indicate possible ecological consequences of overshoot. It is also important to keep in mind the scale and acceleration of global changes. We not only have accelerated our use but our accumulation of things, including exotics and pollutants. This leads to more dramatic positive feedback and catastrophes.

2.2.5.4.2.4. Additional Types of Footprints

Wilderness is not one of the basic land types, although Wackernagel et al (1997) later suggest, in accordance with the Brundtland report, that an additional 12 percent of land area is required for the preservation of biodiversity. Noss and Cooperrider (1994) argue for larger areas for wilderness. As it is for Odum and others, wilderness should be a major category. Odum suggests 30 percent of the world should be left in forest cover, with 60 percent in tropics. Several authors suggest 50 percent of the planet in wilderness (Wittbecker 1970; Kozlovsky 1974). Paul Shepard suggests 75 percent. Wilderness is not just to provide support services for humankind. It has to provide habitats for wildlife.

2.2.5.4.2.5. Patterns of a Footprint

Footprint analysis ignores patterns. For instance, wilderness has be large to be wilderness, with a generous interior and buffered boundaries. We cannot have 2-acre wildernesses mixed in fields and shopping centers. Natural cycles also create and maintain patterns, and require energy and limits to continue.

2.2.5.4.2.6. Energy Dimensions of a Footprint

Although land is a good indicator of sustainability, and a scarce resource, it is not the only scarce resource. Water, heat energy and light energy need also be considered. Footprint analysis tries to account for some of this by including a category for energy land, a hypothetical land area necessary to absorb atmospheric CO_2 emissions that result from the burning of fossil fuels. This area often constitutes about 50 percent of the ecological footprint. Reforestation is assumed to be the most effective option for sequestering CO_2—it would have to be a continuous process and part of a dynamic solution. Other alternatives, such as liquefying CO_2 and pumping it into the ocean depths or gas fields, are very expensive and have not been shown to be effective in tests. Although CO_2 is a critical factor in climate change, other emissions that have ecological consequences, such as CFCs, SO_2 and NO_X, are not considered.

Energy is a problem dimension for a footprint. If we measure light energy it might be better to use a lightprint. We use energy from ancient sunlight trapped in molecules, so that by burning fossil fuel every year, we are burning light from previous ancient years. One gallon of gasoline requires 100 tons of ancient plant life to make under natural conditions. In 1997, humans used 422 years of fossilized sunlight. If the footprint is an area than we are using 422 planets to get that much energy.

2.2.5.4.2.7. Goals for a Footprint

Footprint analysis accepts overpopulation and tries to accommodate it. In fact, we should be making goals for carbon budgets and for populations, rather than simply accepting them.

We should set goals for energy use and technology. For instance, to set a carbon budget for the planet, we would calculate carbon limits below a maximum, then calculate world cultural carrying capacity such that the sum is 100% and each nation would get a percentage of carbon based on the percent of capacity that is natural and doable; so, if China has 13% of the capacity, even if it has 25 percent of the population, then it gets 13% of the carbon credits. To reduce overshoot of carbon use, we would have to set the goal at a time with lower production, such as 1800 AD, 1400 AD or 10,000 YBP.

Footprint analysis needs to consider optimum populations and limits to resource use. It needs to use a fuzzy accounting to address the uncertainties of natural production and natural processes. It needs to address social dimensions, such as the lack of equity.

2.2.5.4.3. Synthesis

Analytic science has reached its limits. Data and information developed by hard studies have undercut the paradigms that guided their investigation. The compartmentalization of scientific fields has exposed the complex connections of the subjects. Science does not need to be based on logical positivism and reductionism, though these have allowed great, although insensitive, changes. A. N. Whitehead thought that what had been missing during the formation of science was a sense of relatedness. Early science saw the world as mechanism; modern biology is seeing it as resembling an organism. Organismic trends can be seen in sciences, from relativity and gestalt psychology to ecology.

Ecology deals with the relationships of organisms to environments. It is not a reductive discipline, and not readily amenable to quantification. Even scientific ecology is an integrative discipline that extends beyond the bounds of science. In a way, ecology is an amphibious discipline, with the authority of science and the force of moral knowledge. Ecology, studied through its components and relations, is a perspective, a way of "seeing," according to Paul Shepard. It is a perspective of the human situation in its interconnection. For Paul Sears, ecology is a "subversive subject." Ecology is nonreductive, integrative, and amphibious, having the authority of science and the force of morals. It is normative and sensible. Ecology also offers a "sacramental vision" of nature. Ecology is radical—from the Latin word meaning "rooted"—and forms part of a new metaphor that is more appropriate to the unity and interrelatedness of the earth. Ecology is part of a movement of consciousness, concerned with equality, diversity, health, with humane methods, and with a holopoetic cosmology, and ecology affects them simultaneously. Radical ecology offers a new perspective of humanity in the total field of nature and defines balanced relationships with ultrahuman beings and species. Radical ecology addresses the determination of separate wilderness areas necessary for a healthy ecosphere, and an optimum human population, based on net ecosystem productivities and modified by appropriate technologies within ecological and cultural restraints. It urges local, self-reliant cultures with adaptive cosmologies and natural values in wild ecosystems.

An ecosystem is a complex system that interacts with four large global fields—atmosphere, lithosphere, hydrosphere, and biosphere—and their cycles. The properties and behavior of a complex system are determined by its internal organization as well as its relations with environment. There are two fundamental modes of behavior: (1) Maintenance, based on negative feedback loops and characterized by stability, and (2) Change, based on positive feedback loops and characterized by growth or decline. The two modes can create a series of behavioral patterns, from stagnation to rhythmic regulation.

Any large system, such as ecosystem or city, is a high-order, multiple-loop, nonlinear feedback system. In the system feedback loops are the basic structural elements. Each loop is a circular path of interaction between several elements. Ecological analysis forces us to look at the obvious—generating nonmarketable use values occupies the occupies the center of every culture because it provides a satisfactory life to its members.

2.3. ***Threading Ecological Themes***

Certain themes operate throughout any discussion of human places, cultures and their environments. These themes can be used to compare different places, cultures and environments.

2.3.1. *World as Field*

The universe at large, with its clusters of galaxies, clouds, stars, and planets, extends itself through space and time. Parmenides held that space was a plenum. On the other hand, Leukippus conceived of space as emptiness. David Bohm combines both ideas in a field. He describes the universe as a field with waves of infinite size. The universe is permeated by septillions of waves at all times. Humans generate their own waves that are added to the infinite variety coursing the universe. A wave is an integral pattern of the physical continuity of a particle. Waves shape the field, that is, they excite the field. The field is an invisible, nondetectible source from which elementary particles draw order and energy. There is no place for both field and matter, "field being the only reality," according to Einstein. The field here and now depends on the field in the immediate neighborhood at a time just past. Excitement is generated with a temporal dimension.

The field concept was originally introduced by Michael Faraday into studies of electricity and magnetism, and was expanded by Clifford. The field concept is central to the unification of theories of light, electricity, and magnetism. Einstein extended Clifford's ideas of the field in his theories of relativity. In large systems, time acquires a new meaning associated with irreversibility. The classical mechanical concept of time results from simplifications. Einstein induced that time is relative to the frame of reference. Space-time is an ensemble of occasions and places, held together by duration. The future and past are tied together by duration. To the largest duration—the universe—all time is present. Time is not empty or abstract. According to Einstein, every change of coordinate systems mixes space and time in a mathematically defined way.

In his *Special Theory of Relativity*, Einstein included gravity in the picture, making space-time curved. But as massive bodies have greater gravities and mass can be converted to energy, all four perspectives form a field that can alter each component. Through the principle of equivalence of gravitation and acceleration and through the use of a symbol that mathematically described the local rate of 'turning' of the curvilinear coordinates, Einstein was able to relate curvilinear order and measure to the gravitational field. Both the electromagnetic field and the gravitational field can be understood as aspects of curvature. The particle affects the field only in its own locality. Particles move along paths that are intrinsic to curvature of space-time field independent of coordinate frame used for measurements (but not from the act of measuring). These paths are called geodesics and are the shortest distances between two points. With Einstein, geometry steps forward as a new participant in physics. Einstein's 1907 principle of the local equivalence of gravitational and propulsive accelerations (geometrodynamic law) linked the two currents of thought from Riemann ("geometry is part of physics") and Mach ("inertia is influenced by mass elsewhere").

John Wheeler regards curved geometry as the building material of the universe. Gravity can be regarded as slow curvature; the electromagnetic field is rippled with different curvature, and the particle is a knotted up region. All charge is connected with the topology of space. Wheeler hypothesizes that pregeometry is needed because geometry fails to explain gravitational collapse. David Finkelstein asserts that space-time is a statistical construct from a deeper pregeometric quantum structure in which process is fundamental.

Geometry is an abstraction from an empty, moving space-time-energy-mass field (STEM). The STEM matrix is a cosmic, transformable field. The STEM field is pregeometry, the ground of being. No one component is ontologically subordinate to another. Time and space are secondary, however, to the field that contains them. The STEM field is primary. The fundamental law is of an immense multidimensional ground, from which orders are projected into a unified field. The STEM field has general characteristics of discretion, participation, connection, consistency, limitation, wholeness, self-making, self-ordering, individuating, and developing. Individuals are embedded in a STEM field and there is their unity with other matter and energies in space and time; there is the unity with all beings and past and future generations.

The STEM field is a meso-field in the universe; it vanishes at both ends as a knot dissolves into the identity of rope after being analyzed. Perhaps the whole universe is like this: particles dissolve into identity with the universe. Particles do not gather in a commonplace before the start of time; they unroll space and time and display history as they unfold. Two particles separated widely must wait hundreds of millions of years for their signals to cross. The field is a paradox; localities are part of it but do not communicate instantaneously, by light. The speed of light is still a limit. Space is filled with local relationships. The limits of the array are nondefinable.

2.3.1.0.1. Participation

One of the most important properties of the field in physics is the participation of the observer. In fact, John Wheeler suggests that the universe is brought into being by participation—the sum of an infinite number of elementary acts of observation, although he unnecessarily reverses the causation. The ideas of quantum mechanics eliminate the notion of the neutral observer behind glass. The observer participates in the very act of measurement, and the universe changes by the that act of measurement. A similar principle of participation can be postulated for ecology. Not only do organisms participate in the field of nature by virtue of their existence, but the experiments of ecologists alter the system being studied, often degrading it for a period of time.

Biological field theories dealt with problems, such as regulation and reproduction, thought to be insoluble in mechanistic terms. In the 1920s, Aron Gurwitsch and Paul Weiss independently advanced the idea of developmental fields to account for the properties of wholeness and directedness. In Weiss' theory the field became a system of organizing factors that proceeded from organized praxis to developing regions, resulting in typical patterns. Growth and pattern became emergent field effects. For Weiss, the field was a symbolic term for the dynamics underlying the ordered behavior of a "collective"; it denoted the properties lost in the process of analysis. C.H. Waddington (1960s) extended the concept of field to be an epigenetic landscape, but regarded his own use of field as a descriptive "convenience." Living fields have form and impose restrictions on the probabilities of nonliving fields. His concept emphasizes dynamic transformation—form as organized spatiotemporal domain—in contrast with the particulate concept of an organism; organisms are understood in terms of group dynamics rather than from selective advantage or cost-benefit. For human ecology, Arne Naess rejects the image of man-in-environment for the relational, total-field image. He characterizes organisms as knots in the biospherical net, a field with intrinsic relations. The relationship with other beings becomes part of the basic constitution of a being.

2.3.1.0.2. Pattern

Paul Shepard describes living natural "objects" in terms of events which constitute a "field pattern." Relations are not prior to objects; they arise together. The wasp and the yucca coevolve; they are not co-linked by prior relations. Furthermore, a specimen is more than the sum of its species' relationships to an environment; it is an intentional being that, with other members of the species, can create niches, as well as adapt to them. Because the STEM field produces life, the qualities of life cannot be separated from its physical qualities. While it is true that living subjects are at a different level of description than events in field patterns, they should not be treated as ontologically subordinate. All of the aspects of the field have equal status. The ecosystem model, as a reaction to "superorganismic" metaphors of early ecologists, attempted to be a field theory, but has been limited by its parentage, thermodynamics, and has been rejected by new practitioners.

Some scientists consider energy to be a more fundamental reality than material objects. These scientists apply classical thermodynamics to living systems. Unfortunately, entropy is an incomplete explanation of living systems. A living system is a natural phenomenon and all natural phenomena are constrained by the laws of thermodynamics, that is, by rules that describe the disposition of energy. This use of entropy offers little improvement from the mechanics of the old physics for understanding organisms, which are reduced to "energy moments" rather than "atoms." Organisms may be configured by energy through time, but, organisms are material patterns in space as well. The focus on either frame permits subtle differences and limitations in interpretation. Even if energy is considered to be primary metaphysically, organisms are still composed of the atoms and molecules that energy forms under certain conditions of temperature and pressure, and their emergent behavior is more complex than just "energy vortices" or "patterns of energy."

Reliance on physical explanation impoverishes the complexity of ecological reality. Morowitz's portrayal of each living thing as a dissipative structure is reductive and distorting. Ilya Prigogine defined a dissipative structure as one of two types of organization—the other being a nonequilibrium stationary state, such as a solar system—whose order is governed by amplified fluctuations; his examples include walls and slime molds. Prigogine, however, misuses the concepts of order and complexity by making them dependent on random events, without consideration of emergence and historicity; furthermore, his concept of stability assumes a reversibility of biological time, which is the result of its basis in quantum mechanics; such reversibility has never been observed. Dissipative structures are more applicable to pans of boiling water than to black bears. The reduction of ecological patterns to dissipative structures ignores observed behaviors like communication and intention.

In assuming that energy flow is more primary than matter, that energy is a more fundamental reality than discrete entities, Prigogine emphasizes process over structure.

Process is as important as structure, but structure feeds back into the process that created it and alters the process, such that trying to determine which is prior or fundamental is not possible. Although an organism may be characterized as a "configuration of energy," looking at it that way is an artifact of the quantum perspective, and, perhaps, of the desire for an absolute reality. There are philosophers and ecologists, such as Ramon Margalef, who consider "information" to be more basic than energy or matter and more in line with patterning. Viewed this way, however, organisms are then reduced to information in a cybernetic perspective. Although these perspectives are useful to an understanding of complex behavior, they are not complete. Sometimes scientists simply take over the vocabulary of a paradigmatic trend, a new physics or an information theory, and apply it uncritically to the epistemology of the older paradigm.

Just as the new physics has transformed the mechanical picture by placing atoms in a field that accounts for the qualitative emergence of properties from simple quantities, the new ecology has placed living "objects" in a field. This field determines the limits of any ecological field of activity, and no field of ecological activity can be described without taking the physical field into account. The field is living and intentional, as well as physical, and an ecosystem model must address intention or other emergent properties.

2.3.1.1. Characteristics of Field

Characteristics are qualities that distinguish unique individuals, systems, or patterns. Gregory Bateson calls them differences that make a difference. These characteristics are shaped by the historical operation of the field itself. This means that these are basic characteristics that are reflected in different levels of organization.

2.3.1.1.1. Process of Field

Alfred North Whitehead developed a model of reality that weaves the physical and psychic aspects of nature into a coherent unified whole, thus avoiding the aesthetic and logical difficulties in other theories. His process metaphysics accepts change as a fundamental characteristic of the universe and takes value as part of the meaning of actuality. Whitehead's philosophy recognizes the dialectical character of process, the interpenetration of opposites, the significance of levels of organization, the importance of community, and the irreducible character of awareness. His process philosophy asserts that being and becoming, permanence and change have *coequal footing* in reality and are insistent aspects of experience. Process is dynamic change in an unfolding flow. Process is the fundamental feature of nature. Nature, for Whitehead, consists of patterns whose movement is essential to their being. These patterns are analyzed into events or occasions.

The basic unit in Whitehead's philosophy is the unit of experience. The raw materials of science are occasions of experience; an occasion of experience has duration, is a unity, has nondenumerable content, and is an event that has characteristics. Whitehead's properties of events may be redefined to fit with the holistic perspective being proposed: Events occur; events have aspects that define a STEM field, such as location and energy; each event is associated with a net of events; each event involves a filtering or sorting decision, similar to Whitehead psychical assessment and decision; possible events are limited by conditions (momentum-energy along a geodesic in Whitehead); the process is probabilistic under local conditions; and, groups of events may be self-organizing and self-sustaining.

The activities of an organism are united into the being of the organism. Beyond being merely relations of relations, organisms are pulsations of process, natural units of fact. For example, since subatomic particles are part of a field, and they are internally related within the field, they cannot exist without the field. The field, or electromagnetic society, provides the order required for producing individual actual occasions. The coming-to-be of organisms, i.e., process, is a fundamental feature of reality. The organism is a dynamic structure that is immanent and simultaneous with the process, rather than just a consequence of the selection of individuals modified by mutation. The organism is what it does. The organism undergoes a process of evolution in which it produces new forms in itself.

Experiences seem to be mutually implicative. An event has two sides, individual self and signification in the universe, that work dialectically. Objects grow together to create a novel subject. Value is achieved through an ongoing process in nature, not a static one. Thing and entity are concepts essential to reasoning about the world.

Creativity is an ultimate principle in Whitehead's thought. A universal process of creative activity is made concrete in the individual. This secures the concept of a connected universe and its character as self-creating activity. Creative activity is considered as activity toward some end. The process of nature is not merely rhythmic change, it is a creative advance, producing new forms everywhere. The organism undergoes a process of evolution in which it produces new forms in itself. Natural order is dynamic, creative, logical, and temporal. We live in a natural order. To preserve what we are, we must admire the matrix out of which we arose.

In creation and recreation, things form and unform, reform and unform, then form again. Creativity is the process of recombination into forms. History creates unique patterns, especially in ecosystems. Each ecosystem is unique in its parts and structure, in its matter, energy, forms, information, and in its dynamics and history. Creativity is a kind of play. Some systems are more coherent and autonomous than others. In a hierarchy of levels, emergent properties appear at some levels. Life is also a property immanent in an organization of molecules; and language emerges in a higher level. New qualities that emerge at every step are unpredictable on the basis of the past; reality is creative. Nature is more like an artist than an engineer in this sense.

The basic character of Whitehead's metaphysics is aesthetic, from the actualities to the categories. The foundation of the world is found in aesthetic experience, rather than incognitive experience, as with Immanuel Kant. Pure feeling is as basic as pure reason. All orders are therefore aspects of aesthetic order. The essential order of experience is aesthetic. Even a physical theory of wave mechanics can be described in aesthetic terms as a union of repetition and contrast. Aesthetic principles apply from quantum vibrations to complex societies.

2.3.1.1.2. Autopoesis

The universe is autopoetic, that is, self-making (from the Greek words *auto* and *poiesis*),according to Francisco Varela and colleagues. Autopoesis refers to the dynamic self-producing and self-maintaining activities of living beings. The tenets of autopoesis are presented in six principles: (1) Identity: Identifiable components are organized internally with structural boundaries; (2) Integrity: The self is a single dynamic functional system; (3) Self-boundedness: The boundary is produced by the system; (4) Self-maintenance: The boundary and components are produced by the functioning of system; (5) External supply of materials: Elements, such as carbon or water, are obtained beyond the boundary; and (6) External energy supply: Light or chemical energy from beyond the boundary is converted into organic bond energy. In an autopoetic framework, every being is embedded in a world and observed by an embedded observer. The material components of life move through physiological processes. Autotrophs, such as bacteria, algae, and green plants, convert energy into organic compounds; heterotrophs reconvert autotrophs into heterotroph flesh.

Pythagoras described the universe as a 'kosmis,' an order. The stoics emphasized the cosmos as an entity under the direction of 'logos,' which led to the use of holon, meaning whole, to describe the universe. David Bohm describes reality as a coherent whole, an unending process of movement and unfoldment. Matter can be regarded as a holographic expression of forms of wave energies; it appears when conditions are right, then folds back in to the field when conditions are not favorable. The universe is an implicate order (the word implicate meaning to 'fold inward'). The whole implicate order is present at any moment. It is a description in which everything implicates everything else in undivided wholeness. The mind enfolds matter in general, the body in particular; the body enfolds the mind and the universe; the universe enfolds the mind and body. Explicate order flows out of the implicate.

Wholeness is real; the fragments are only derivative. When we speak of nature or the universe as a whole, we merely mean the universe considered as organic or holistic. The creative intensified field of nature is holistic. That field is the environment of all wholes. The whole needs no boundaries to separate it; there is nothing to separate it from. A partial process requires cuts in the whole, or boundaries. Predominant fragmentary views have brought about only imbalances. But, the fragmentary perspective is universal; perhaps it may be rooted in the perception of the human left-brain. The habit of fragmentation smothers the primacy of the implicate order, according to Bohm. The explicate order is a relatively independent, recurrent, stable subtotality. Both wholeness and partiality are correct, but must be reconciled in a holocosmology.

Energy, information, matter, life are part of inseparable whole, with no formal theory linking them. Relativity, quantum theory, and entropy imply individual wholeness, in which the analysis of something into distinct parts is no longer relevant to understanding the whole. Relativity and quantum theory both imply the need to regard the world as an undivided whole, not as a spectral oneness, but as a diversity of the whole from many perspectives. George Leonard suggests the word holonomy for the study of wholeness. Holonomy is not the death of division, rather it is a frame for individual wholes.

Wholes are mutually defining, but also self-defining or self-making. Nature is a whole system, as well as a self-making system; species and organisms are self-making. The ontology of any living system is the history of the maintenance of its identity through continuous self-making, or autopoesis. According to F. Varela, the evolutionary stability of the subassemblies, organs, organisms, and species, is reflected by the degree of autonomy (self-government) that each has. The system develops through a continuous dance of autonomy and control; autonomy represents generation, internal definition, internal regulation, and self-assertion, whereas control represents consumption, instruction, assertion of other identity, and external definition. Furthermore, the holistic nature of the STEM field eliminates the unsatisfactory notion of the priority of relationships to individual beings or of wholes to components.

Gilbert White characterized nature as "one organic whole" and influenced Darwin and subsequent generations of biological scientists. E. A. Birge's early work on the heat budgets of lakes, for instance, was holistic. J. C. Smuts, in trying to synthesize the evolutionary theory of Darwin and relativistic physics of Einstein, presented the whole as a powerful organizing principle inherent in nature. L. von Bertalanffy and E. Laszlo extended holism with general systems theory. David Bohm suggests that the universe as a holomovement carries an implicit order that is indefinable and immeasurable.

The Gestalt thinking and the Gestalt psychology of Wolfgang Kohler emphasizes the importance of wholes and tries to identify the organizing principles of perception in terms of wholes. Each level is real as a whole; it is a whole, or a holon, in Arthur Koestler's terms. A number of ecologists, including R. V. O'Neill, D. L. DeAngelis, T. F. H. Allen, T. W. Hoekstra, and T. B. Starr, use the concept of holon to describe the organizational levels of hierarchical systems. Given that nature is a structured and differentiated whole, the character of the organism or particle is determined as a subwhole. According to Koestler, all complex structures and processes of a relatively stable character display hierarchical organization. Levels of a hierarchy tend to be contained in subassemblies. Each subwhole behaves as a whole to its components, as a self-contained whole, and as a dependent part in the context of a whole. Wholes and parts do not exist absolutely. There are intermediary structures on a series of levels in an ascending order of complexity; each subwhole faces in opposite directions. These Janus-faced subassemblies are holons, that is, paraphrasing Koestler, any stable subwhole in a hierarchy that displays rule-governed behavior and structural Gestalt constancy. The rules lend order and stability, as well as flexibility.

2.3.1.1.3. Differentiation

The universe is a vast order of individual events, discrete unfoldings within limits, following certain laws, but in unique patterns. The physical world is a patterning of patternings, a flowing of flowings whose constituent functions are interacting fields of force. Physicists and philosophers once assumed that the patterns were continuous. The Latin proverb *Natura nonfacit saltus*, "nature does not make jumps," expresses the old ideal. But Max Planck, in his investigations, concluded that a field cannot go down to a dimensionless point. Planck said that energy can only be exchanged in complete packets. Planck used the word quanta to describe the discrete packets of electromagnetic radiation. This discovery of things that only happen in leaps was considered a great intellectual triumph.

Einstein's Geometrodynamics, combined with Planck's quantum principle, results in a superspace, according to John Wheeler. A leaf of history, cutting through superspace, describes a deterministic, dynamic development of the geometry of space with time. Wheeler's superspace is on a microlevel, however; his alternate worlds are really the *umwelts*, the existence worlds, of particles. Particles choose their paths from probabilities. The quantum principle describes how a free particle moves from one point to another along a straight line: the particle 'smells' out alternative routes; as a result of this 'smelling' behavior, the straight line appears fuzzy to a perceiver.

Space-time cannot be a continuum in this way of thinking. Furthermore, quantum fluctuations of geometry and quantum jumps of topology are estimated and calculated, according to Wheeler, to pervade all space at the Planck scale of distances and give it a foamlike structure. Gravitational collapse places an impenetrable barrier between the leaves of history. According to quantum geometrodynamics, violent fluctuations are going on constantly in the geometry at Planck's scale. From the quantum worm's eye view, such fluctuation is indistinguishable from collapse. Gravitational collapse of a local universe is always taking place and being undone. Francis Sarfatti claims that the foam is a sea of rotating black and white holes. Wheeler has described superspace, the space-time fabric or field, as composed of a turbulent sea of bubbles, the warp and woof of empty space, quantum foam, and a carpet in motion. By the way, the words weave, woof and web are derived from the Indo-European word *webh*. On a loom for weaving the upright threads (vertical) are called the warp; this means 'throwing.' The fabric or web is made by weaving new threads horizontally; the crosswise threads are the woof. Warp means to 'twist' or bind.

Patterns of complexity shade and grade into one another endlessly. A large family of similar unstable particles has been classified phenomenologically in a complex set of interrelated orders. It appears to many physicists as if the beginnings of new orders of natural laws are being revealed, in which particles would be like flower designs on a carpet pattern, while something unknown corresponds to the structure of the cords of the carpet. In fact, orders reform in the carpet, as Wheeler suggests. Analyzing the world as if it were made of particles would be similar to analyzing the carpet as if it were made of flowers: it would give some results and have some predictability, but the metaphor would limit better understanding. Wheeler states that patterns in the foam are seen as subnuclear particles. The foam is more like the fabric of a dream. The foam can be seen as a metaphor for the self-reconstruction of the universe from standard parts to novel structures, through a process of metalysis, that is, the breakdown of parts to form new structures. The quantum principle is only a principle. It is self-organized and virtually determined.

2.3.1.1.4. Integration

John Wheeler believes that since law, field and substance exist after the theoretical big bang, the universe owes its existence to trillions of acts of registration. The phenomenon comes through an elementary act of observer-participation. Wheeler asks if the universe might not be brought into being by the participation of those who participate. Quantum mechanics strikes down the concept of the neutral observer; participation is vital. Wheeler noted that "To observe the electron even, the experimenter must shatter the glass—must reach in with instruments." The quantum principle destroys the observer behind glass (*in vitro*). The universe is not the same after measurement; the observer becomes a participator. Einstein said that no event can be postulated without the presence of an observer; but no observer can see the whole system; and anything can be an observer or participant. The participant creates a universe by being. Participation is unavoidable; connections are made through an infolding of the field. A world without participants is impossible. Nor are there any lone observers. The observer is part of a natural or social community. Humanity participates in the natural world. All beings participate in the relationships that make up their worlds.

How tightly is the universe integrated? Following Neils Bohr's 1927 expression of the Copenhagen interpretation of quantum theory—that the world must be observed to be objective—Einstein, Podolsky and Rosen (EPR) proposed a thought experiment leading to the conclusion that quantum theory and objective reality are incompatible; this experiment is called the EPR paradox. They argued that quantum theory was incomplete, that it does not specify all objective elements of reality. The thought experiment shows that quantum theory violates local causality, which asserts that whatever influences an object must be attributed to local changes in the object or to energy transmission. In the experiment, two particles interact and go in opposite directions. Quantum theory seems to require that measuring the first particle influences the second instantaneously, regardless of the distance separating them.

More than thirty years later, J. S. Bell devised a mathematical inequality that could be tested experimentally. Half a dozen experiments have shown that the inequality was violated, with the conclusion that the universe is nonlocal, with instantaneous action-at-a-distance. This conclusion has been expanded to mean that faster-than-light communication is possible and that notions of telepathy can be supported. David Bohm concludes that Bell's theorem leads to a notion of unbroken wholeness; connection is extended to the entire universe.

Interestingly, the assumptions of both experiments disallow any firm conclusions. The EPR paradox assumes that the second particle has position and momentum without actually measuring it. Experiments in support of Bell also assumed that the particles—photons—existed in a definite state without measurement. Nonlocal influences cannot be argued then, since no information is proved to be transmitted between devices. The theoretical cross-correlation of separated events does not occur in a local frame. It requires a global frame, which can only be used over a long time. Neither experiment is complete or conclusive.

The failure of the principle of locality would mean that there are no separate parts in the universe, but we know that nothing is entirely separate anyway. The conclusion that experimentally separate parts are correlated forever must consider the "parts" as eternal things, not phenomenal processes. In fact, the particles in both experiments are connected by the field, not by superluminal signals or any other kind. In a holonomic field, the resolution of any one perspective of the whole is so weak that it has no predictive value. Information is not transferred because the field is in-*form*-ed. Bell's theorem demands that an adequate model account simultaneously for the observed causal structure on the statistical level and the noncausal structure on the individual event level. An adequate model must provide for an understanding of nature.

There is another argument for connection from relativity. If the field of matter is regarded as a probability-wave hologram, no supercomputers or tachyons are necessary. A probability wave is a mathematical formulation that exists in n-dimensional space. In quantum theory every particle can be described as a wave function having three dimensions of space and three of momentum. Two particles can be described in twelve dimensions, three in eighteen dimensions. So every particle is connected and whole can be described from point of view of one. But the speed of light limits some information and the lack of a global perspective limits other information.

2.3.1.1.5. Constancy (Regularity & Endurance)

Bishop Berkeley proposed a bootstrap principle when he argued that the inertia of any body is determined by the distribution and masses of all other bodies in universe. Mach's principle repeated the same idea, demanding the closure of the universe, to make it finite and bounded. The mechanical properties of space are determined completely by matter only in a space-bounded universe.

In the bootstrap principle, the universe is what it is because it is consistent with itself; we are not free to sort out accidental properties and distribute them with various values among different universes. No properties of the universe, however, are fundamental; they follow from other properties in a web of interrelationships. Geoffrey Chew's bootstrap model considers hadrons as temporarily stable configurations that result from an interaction of processes. They may transform themselves into other particles. The bootstrap relation takes the form: "Universe=subatomic-particles," where '=' means 'equivalent to.' Each particle represents a facet of the universe and not just a small part.

All electrons have the same charge because they represent some single aspect or perspective of the universe. Each particle is only an abstraction of a relatively invariant form of movement in the whole field of the universe. The proton and neutron are just two different points of view of the nucleon; the sigma has three different points of view. Both nucleon and sigma particle may be just different points of view of the same particle, related by an extended isospin symmetry. Specificity can be a property of the path, what Whitehead a 'concrescence,' and what Waddington calls a 'chreod.' Yet, if the universe is open-ended, as has been argued, then the consistency of the whole can never be proved logically.

2.3.1.1.6. Metalysis

Metalysis is the process of renewal by discrete units; the neologism means 'loosening change,' from the Greek roots *meta* and *lys*. Wheeler states that patterns in the foam are seen as subnuclear particles. The foam is a metaphor for the self-reconstruction of the universe from standard parts to novel structures. Living order can also be defined in terms of the influence of whole over the parts. Disorder at the level of a molecule can reflect the higher level of the order of a cell. The machinery of a cell is not a permanent fixture, but is disassembled and rebuilt periodically, according to specific patterns and in harmony with the functioning of the cell in its environment. Weiss shows that parts of a cell are constantly changing, growing, dying, breaking up and recombining, but under control at a cellular level. This is metalysis.

Metalysis is a technical term for the process of dedifferentiation in biology. It can also be used to describe physical processes like quantum foam or social processes like institutional revolution. It is used here to describe biological processes, such as metamorphosis, where changes are often so extreme that the organism constructs many specialized parts of the adult from cells set aside in the embryo and that are nonfunctional in the juvenile; these cells are imaginal disks, which are controlled by juvenile hormones.

2.3.1.2. Operation of the Field

The word nature is derived from a Greek word meaning 'that which is born.' The taoist term for nature, *tzu-jan*, means the spontaneous, that which is so of itself, and is automotive or automatic; that which is continually reborn. The movement of the field reveals cosmic rhythms, that may have inspired the yin and yang of Chinese philosophy. Motion results in limitation. Limitation is the principle by which the many can come to existence out of the one, the unlimited of Anaximander.

This process operates through all levels, which it makes as it operates, from the physical, chemical, electrical, or inorganic, to the organic and social. All the processes start a spiral motion that works into an multi-dimensional Klein bottle as a model of the being of the field. This operation is described in more detail in *The Poetic Archaeology of the Flesh* (1976).

2.3.1.2.1. Motion

The interplay of heterogeneous, asymmetrical physical forces, or motion, generates cosmic time. Simultaneously, space, energy and matter, and perhaps other dimensions, are generated. The field is motion. Matter is inherently possessed of motion: Matter emerges from point-instants; a particle of matter is a moving pattern of point-instants, which are abstractions, of course. As a determinate pattern, matter will have a determinate quality. And, quality exists as a function of structure. Particles are probability patterns, various parts of a unified whole, interconnections in a cosmic field. Nature consists of moving patterns whose movement is essential to their being. The holomovement enfolds and unfolds in a multidimensional order that is undefinable.

Motion can be further refined by a series of characteristics. Remotion (from the Latin words for 'move back') means departure; this describes the eternal return properties of motion. Immotion (from the Latin words for 'move in') means to form or to inform. Motion can be directed; energy is considered to be the capacity for action or motion. The word energy is derived from the Greek word for 'work in' (one of the original meanings was 'force of expression' as an interpretation of Aristotle's metaphor to show 'action'). It is a very useful tool to describe changes in places or human cultures.

Emotion (from the Latin words for 'move out') means feeling. This describes the transfer from self to extended self in the environment. This motion is centrifugal, meaning away from the center or outward. Commotion (from the Latin for 'move with') means commotion, or the case of moving in a group. Many things move together.

2.3.1.2.2. Version

The universe moves and turns. As a result of its turning, the universe is seen. Turning is a technical term to describe the change from invisible to visible. As a result of its seeing and being seen, it becomes another universe. The word universe comes from the Latin of Cicero (*unum versum*), which means 'turning into one whole.' Any universe is one turn.

Living beings have a handedness, that has to do with the perceived direction of turning. Honeysuckle is left-handed; morning glory is right-handed. A photon generates either a right or left helix. Proteins attain a biological active form by a process of turning. Human unfolding, like molecular unfolding, allows the human being to become active, turning out. This is the secret: each time the kaleidoscope turns over in our head we see a new, unexpected universe.

With *reversion*, the world is constantly changing, but also returning to previous forms and states; the word reversion is from the Latin for 'turn back.' This may be observed in the states of water, or life and death. Beings follow in succession, then turn back at a limit. They revert. This reversion resembles the movement of the tao. The world was something to observe, to learn from and to employ respectfully for the Taoists. The world changes, but returns to previous forms and states. This movement of the tao is also called reversion. But it is two movements, unfoldings, observed in seasons and other processes.

Inversion, from the Latin for 'turning in,' is a kind of entropy or involution. Entropy, from Clausius, refers to the transformation of energy into heat; it means generating a transformation in an abstract phase space (from the Greek word 'entropia' which meant 'turning in' or evolution). The entropy of order, of energy, of information, is derivative of the entropy of turning. The universe is one turning composed of at least two polar processes, entropy (inward turning) and ektropy (outward turning). These processes occur roughly equally for the creation of order and disorder in dynamic tension.

Entropy is the property of a body, expressed as a mathematical quantity, which, if an amount of heat enters or leaves the body, is increased or diminished proportionately to that amount divided by the absolute temperature. Classical entropy is defined in an abstract manner as a thermodynamic variable of a system under consideration; it is a quantity uniquely defined by the state of the system. It is more commonly a measure of the unavailable energy in a closed system. The change is qualitative.

This tendency toward the qualitative change of atoms and molecules in an abstract phase space was for a long time the only one-way tendency recognized by orthodox physics; it was the only such process that had been systematically studied as a class of processes. Entropy is "times arrow," as Eddington said, but entropy has no set rate of occurrence. Its character is fact-like rather than law-like. As a description of fact, classical thermodynamics is qualitative and evolutionary. The entropy law was formulated using cosmological language, as in nature or the universe.

Eversion, from the Latin for 'turn out,' means ektropy or evolution. Georg Hirth, an art expert, explicitly coined the word 'ektropy' to identify the principle that opposes the entropic principle of degradation in living structures. The word is from the Greek, 'ektropia,' meaning 'turning out.' It is a universal sorting process. Hirth wrongly believed that entropy must be victorious in battle with ektropy, not realizing the complementary relationship of the two processes. Ludwig Boltzmann described the energetic shuffling of energy from the sun as a competition for entropy. But the shuffling process could not explain order, as the result of an automatic random process. Because biological evolution is so obviously contrary to the concept of thermodynamic entropy, Herbert Spencer stated a new principle of nature, the instability of the homogenous, or a differentiating force creating organization. Prigogine thinks that nonequilibrium may lead to a new type of structure, dissipative structure. He avoided naming a sorting process in this way, but he assumed its operation.

In the 1930s, Schrodinger reiterated Boltzmann's point by saying that organisms live by consuming negentropy. This characteristic of life was called "entropy feeling" by Schrodinger. Negative entropy, however, was not purposive. It was a structure-forming process, but only applied to the organic realm, not to the atomic or inorganic. The process would have to apply at all levels,

Nature appears to constitute both a series of self-evolving ektropic phenomena, and a series of disorganizing entropic processes. All visible forms, from atoms to galaxies, are formed by ektropic processes. As there are defined laws of entropy, there are laws of ektropy: (1) The universe is self-ordering; (2) a quantitative increase in any matter eventually produces a sudden qualitative change, and; (3) the results of these matter organizations are stable holons that are used and reused, from elements through organic molecules (and DNA) to societies. Ektropic processes result in a hierarchy. A corollary law of ektropy would state that groups of particles move toward ordered states; more is packed into less, until miniaturization achieves greater levels of complexity, as in human brain.

Entropy equals shuffling; ektropy equals sorting. Entropy is a measure of infinite individual order. Ektropy is a measure of complex order, of limited individual order. So entropy laws are simpler than ektropy laws. Infinite orders are described by simpler laws.

Conversion (from the Latin for 'turn with') means to transform or exchange into an order or disorder. The sediment from this change is history. Entropy and ektropy are changes of order. The two processes are not inversely related; they are directly related. When one increases, the other increases, although there is no rate of change. Order increases as disorder increases. No theory can consider only entropy or only ektropy. No metaphysics can be based on only one process. Solutions that address only one half of the process are half solutions.

2.3.1.2.3. Scension

The inside arises mutually with the outside, different but inseparable. Entropy and ektropy have just this relation. But, everything is what it is only in relation to all others. This is the taoist principle of mutual arising restated in scientific terms. The tao is a formative field. If everything is allowed to go its own way, the universe establishes a harmony. In the principle of mutual arising, the universe produces our consciousness and that evokes the universe. This realization transcends the debate between idealists and materialists. The only single event is the universe itself; organic pattern, not causality, is the rationale of the universe.

In general, when different modes of descriptions appear as opposites, it is more satisfactory to consider them as complimentary instead. This is the case with net/trees, or autonomy/control. These pairs are not one, but not two either. The duality is connected with processes in both directions. The complementarity framework includes many dualities, such as being and becoming or context and text. Since they mutually specify each other, there is no real duality. This separation is no Hegelian synthesis, since there is nothing new, but an appraisal of how things are related in description. Opposites are polarities that create a field; the sky is not a place, it is a relationship between the earth and sun. The sky is empty yet it is a form of relationship. In order to exist, things need literal integrity, the tao or the field.

Inscension (from the Latin for 'climb in') means to develop. Stars, as well as some galaxies and other astronomic objects, grow by accretion. Many physical structures, such as talus slopes, grow by addition, until some physical limit, such as gravity or the coefficient of friction, is passed, then they stop growing. Some stability can be gotten from growth in early stages; later stability must result from limits and metabolism. Growth in plants can delay the onset of senility by ridding the plant of waste products in more diluted form. However, too much growth produces a strain on tissues and early decay. In fact, one herbicide promotes excess growth as a means to kill weeds. A biological organism grows to maturity, which is a stopping point for size. The organism continues to develop, however, experiencing and learning the environmental complexities through mating and then to the end of life. Development may include growth at some stages, but development refers to the continued change after growth has stopped. Mechanisms for growth can become pathologies when central authorities meddle in stable lower orders. Flexibility, again, is imperiled. The nonproductive superstructure places excessive demands on the substrate and destroys it. Technology or social structure can mask the internal stress.

Escension (from the Latin for 'climb out') means to emerge. Complexes of parts can be calculated in three ways: (1) by counting the number of parts, (2) by considering the species to which they belong, and (3) by considering the relations between parts. The first two cases understand the sum of parts in isolation. These are summative; the parts make no difference to the function, like a pile of bricks. Other examples include the Pauli exclusion principle, homeostatic self-regulation, and distributive justice—all of which are meaningless to individuals. The third describes constitutive complexes, where wholes are other than the sum of their parts. New properties emerge.

The world, as it exists in its ceaseless changes, appears as a single cosmic process in which higher orders of being emerge. The word emergent is borrowed from Lloyd Morgan (1912), who first used it in this manner in *Instinct and Experience*. But emergence and ideas of organism came in rough form from Herbert Spencer (circa 1880). Morgan later set forth a view of the world as evolutionary process in *Emergent Evolution*. The word "emergent" was used to show that higher orders of being are not mere resultants of what went before and were not contained in them as an effect is in its efficient cause. Nature evolves by sudden leaps, from matter, to life, and mind. Jan Smuts (1926) amplified the idea in *Holism and Evolution*. He attempted to state the principle of emergence by saying that nature is permeated by an impulse towards the creation of wholes; each stage of evolution is marked by emergence of a new type of individuality embracing and transcending the previous parts of itself. The progressive development of wholes is evolution.

Emergence is a movement of ascendance within being. The ascent takes place through complexity. At each change of quality, the complexity gathers itself together and is expressed in a new simplicity. The emergent quality is the summing together in a new totality of the component materials.

William Wheeler applied emergence in ecological settings. Species were modified by association; new qualities appeared. For instance, a prey species reflects a predator species, for example, deer reflect wolves; the two are related in an association (similar to Whitehead's idea of relatedness).

Conscension (from the Latin for 'climb with') means a hierarchy or climax. Henri Poincare recognized (1905) that explanation in science involved generalization and simplification; so, hierarchical levels emerged in his studies. Human perception of nature is hierarchical, regardless of whether nature is. Several theorists distinguish levels of hierarchy in nature. Miller includes particles, atoms, molecules, organisms, societies, nations. Mario Bunge extends the list to include processes and knowledge: Elementary particles (atoms, bodies); physical systems (organisms, ecosystems); physical processes (chemical, biological, social); material production (ritual, culture); and knowledges (physics, history). Ervin Laszlo makes a distinction between a macrohierarchy, comprised of a space-time field, particles, stars, galaxies, and various aggregations; and microhierarchies like the earth, composed of molecules, crystals, cells, organisms, ecosystems, and Gaia. Hierarchies can be regarded as vertically arborizing structures whose branches interlock with those of other hierarchies at a multiplicity of levels and form horizontal networks; arborization and reticulation are complementary principles in the architecture of organisms and societies.

The law of the vertical order of hierarchy determines the ever-changing level structure. In a hierarchy, there are two streams of information: The factual going up, and the directive going down. The upward stream, through perception into consciousness, makes observation possible; the downward motion makes participation possible. Since all things participate to some extent in the streams, all are involved in the hierarchy.

Hierarchical thinking is limited in cybernetics. The concept of control could better be stated as reciprocity or mutual causality. Furthermore, in an open system, the environment, structure, program, and feedback all govern the system in concert. Hazel Henderson states that only the system can model or manage the system, but this is not entirely true. Images or codes model the system in miniature.

Ludwig von Bertalanffy concluded that: "Hierarchical organization, on the one hand, and the characteristics of open systems, on the other, seem to be fundamental principles of living nature." Koestler combines these two principles into a system theoretical model of self-regulating, open hierarchical order (SOHO), and uses this as an alternative to models of linear causation. Koestler proposed the term holon to designate the janus-faced entities on the intermediate levels of any hierarchy, which can be described either as wholes or parts, depending on the frame of reference above or below (from the Greek word *holos*, meaning whole and *-on*, for a particle). The concept of holon transcends the duality of parts and wholes. A holarchy of holons replaces the notion of a hierarchy of parts in a whole.

Rescension (from the Latin for 'climb back') means that rising is part of a cycle. That which collapses gathers itself up.

2.3.1.2.4. Tention

The universe is a prelogical and intentional turning. Although its turning is not reversible, the relation of opposites within it is reversible, hence the reversibility means a relational state and not a developmental one. As the intentional relationship is not the final truth of human existence, since it appears at only a certain level of self-actualization, it must be preceded by a pre-intentional level, which continues to exist *in* it. This primordial intentionality refers to a pre-predicative experience, according to Merleau-Ponty, an experience before the subject-object split; it represents a spontaneous organization of experience which precedes the subject's active synthesis. Primordial intentionality is the entire intentional network of the lived-in-world; the subject is already immersed in a world that gives her meaning. Therefore, intentional objects have sense only in the context of a larger world horizon which is not due to subjective constitution, but to a latent intentionality which is the intentionality within the universe.

Retention (from the Latin for 'stretch back') means stability, to keep or hold something. The loss from turning creates a sediment, which is history. Sedimenting is irreversible and gives direction to time. The sediment of the past is a given and different for each present. The ceaseless activity of being in history creates newness. Novelty is born from the womb ('hystera') of change. History is the result of this hysteresis process. Motion, Version, Scension—all together form the ontological spiral of being/environment. The motion of the field creates a turning, which leaves a history, out of which patterns arise. Scension is a precession of motions, resulting in directional change—evolution. The transformation is a historical expression.

Intention (from the Latin for 'stretch in') means creativity. What does it mean to say that the universe has intentionality? Does it mean that it moves, or goes in a direction? If so, it cannot be one undifferentiated sphere; it must partition itself into entities. Why would it do so, suffer the ordeal of birth into newness? The human act of ingesting an order—in*form*ation—creates entropy elsewhere. Since the act is intentional, the order cannot be considered apart from information *for* something.

Extention (from the Latin for 'stretch out') means expansion and diversity. If you tried, you could eventually construct the universe as it is known now in every detail and potentiality, but that construction could not be all, for by the time the description of what is now is finished, the universe will have expanded into a new order to contain the model.

Through speciation, orders of animals and plants probe the environment. A species is thought of as a morphological extension of its niche, but the niche extender enriches nature. The species that enlarges its niche also enlarges the ecology as a whole. According to E. A. Gutkind, it expands the environment for itself and other species in balance with it. An expanding whole is created by diversification and enrichment of the parts. A creature in a niche gives to the niche (itself), leading to cycles of nature.

Contention (from the Latin for 'stretch with') means a stress between entities. Even a physicist describing the behavior of objects is led to ultimately question the ground of those objects. In the words of G. S. Brown: "Now the physicist himself, who describes all this, is, in his own account, himself constructed of it. He is, in short, made of a conglomeration of the very particulars he describes, no more, no less, bound together by and obeying such general laws as he himself has managed to find and to record. Thus we cannot escape the fact that the world we know is constructed in order (and thus in such a way as to be able) to see itself."

Gaston Bachelard states it more poetically: "The world wishes to see itself: the world lives in an active curiosity with ever open eyes." Mythically, the cosmos is an Argus, a sum of ever-open eyes. How can the world see at all if it is one whole unity or being? Brown answers: "... in order to do so, evidently it must first cut itself up into at least one state which sees, and at least one other state which is seen. In this severed and mutilated condition, whatever it sees is only partially itself. We may take it that the world undoubtedly is itself (i.e., is indistinct from itself), but, in any attempt to see itself as an object, it must, equally undoubtedly, act so as to make itself distinct from, and therefore false to, itself. In this condition it will always partially elude itself."

In opposing itself to itself, it limits what it knows. It never sees what is seeing. As it opposes itself to see itself, the world or universe is divided to its very root; it creates a basic dialectic between mind-body, subject-object, and male-female. The world holds all contradictions; indeed, contradictions allow existence. Substance and nothingness allow movement; but they are both substrates of the field.

2.3.1.2.5. Formation

Formation is the interaction of a physical and an intentional process. Physical or living things take forms within processes. Interactions within a process makes forms. Making is a human activity that involves effort, using tools, in place. To 'make' means 'to bring into being' or produce something physically or mentally (from the English and Germans words, from the Greek meaning to 'knead,' press and stretch dough). It also means to build, fit, or create. The making of a home would be an 'ecopoetic' activity (from the Greek word fragments for 'house-making').

Forms can produce a system. A system is a set of things—such as people, cells, or molecules—according to Donella Meadows, interconnected in such a way that they produce their own pattern of behavior over time. The system may be buffeted, constricted, triggered, or driven by outside forces, but the response is characteristic of the system. A system is not just a collection. It is an interconnected set of elements that is coherently organized to achieve something.

Information is the act of imparting form, as in in-form-ation—this is different from knowledge, which is the act of importing that same form, or from data, which is the act of abstracting knowledge, or from facts, which are the transformation of data by imagination, emotion, memory, and theory. The making of form is informing. Ramon Margalef considers "information" to be more basic than energy or matter and more in line with the concept of patterning. Information is not transferred because the field itself is in-*form*-ed.

Evolution is the production of new codes of information to match a changing environment. Margalef describes the evolution of ecosystems as information accumulation; information is generated by participating species and their physical structures (such as burrows or paths). In a system novelty is always transformed to confirmation. Novelty always enters with fluctuations. Natural fluctuations serve to maintain the openness of a system. There are in fact other ways to transmit information, beyond energy and pattern. Information acquisition can come from genetic coding, individual learning, and culture, as well.

Culture is a filter to restrict and manage information. Culture can be defined as the transmission of information through behavior; imitation is one behavior, and teaching is another. Culture is information that an individual acquires by imitating other individuals, and that affects an individual's behavior. Culture is the thing that is transferred. Cultural transmission is the set of ways by which culture is spread through a population. Cultural evolution is the summed effect of long-term transmission. Information is always transmitted in form, however.

Exformation creates outward form, such as tools, People make tools to get food. Then, they make clothing, homes and other things for making life easier. Art, luxuries and sculptures follow. Clothing and tool making is a universal in human cultures. The first tools were stone-flaked, then ground stone, used with boomerangs and clubs. Ground stone axes appear first in Australia. Tools are simple, but effective, easily manufactured and maintained, for example, the spear-thrower, or woomera (atlatl), had hook on one end and an adze on other; it could be used for a shovel, fire-starter, or a percussion instrument. It was an independent creation in Australia. As tools increase in complexity, from knives and levers to computers and space stations, so does the knowledge needed to support them. Complex modern tools require libraries of information that has to be continually increased and improved, then spread. Technology first simplifies life, then complicates it. Digging sticks led to plows and tractors. Lean-tos led to pit houses and balloon-framed houses. Domestication led to horses and to horse wagons. Paths led to trail and highways. All tools, from the simplest word to the most complex computer, are disturbers and rearrangers of primordial nature and reality; they are implements for working on something. The complexity of tools leads to rules for their use, and for who uses them, such as trade unions and clubs like masonry. Society is very adaptable to technological change, according to Kenneth Boulding. Perhaps too much so according to Rene Dubos, and the risk is shaping human behavior to tools.

Tools and machines are considered as symbiotic things, or allies, by some philosophers. F. L. Wright thought of: "The building as machine." Machines are not living yet, as we know, although they follow rules, change and evolve. We apply control to them, but they seem autonomous in many ways. They have many known and unknown effects. Can the effects of technology be controlled? Tools have effects on human culture, and on nature. What are the effects of tools on ecosystems? Disturbance of soils, and scale effects. What are the effects of tools on humans? The use of knives which reduced the need for some teeth. Tools, simply by being intermediate between the hand and the object, may increase psychological distancing from things. Intimacy with the tool can replace intimacy with the thing. The increase in physical depositories, for memory, changes the kinds of memory capabilities. Tools may contribute to the loss of hand-eye coordination and perception of wild, yet increase hand-eye coordination of tools.

Exformation can maintain form. Autopoesis refers to the dynamic self-producing and self-maintaining activities of living beings. Species that seem functionally redundant prior to a disturbance might serve as potential spare parts for maintaining ecosystem function following a disturbance or perturbation. The dynamic self-producing and self-maintaining activities of living beings in an environment creates places. As in a field, a place is bounded. That boundedness gives a place its identity and integrity, and allows it to maintain itself in a definite form and process.

Successful self-making patterns are recognized as having vitality. Productivity, in general, depends on the vigor, or strength or vitality, of the system. Health is the overall ability of a system to maintain itself under a normal range of environmental conditions (which may include hurricanes, volcanic eruptions, or fires). To be constant and stable, a culture has to be vital; it has to be productive, to be able to convert energy and materials into foods and structures for survival. The system is self-creating. It renews itself as its contents change, as disturbances change the parameters of the system. What barriers are there to cultural renewal? How can they be overcome?

Reformation: The making of shelter is reforming. Things are reshaped in response to changes in conditions. An ecological design involves designers and people in reshaping and recreating a self-sustaining community. Individual resources are limited. The relationships to strive for here are community relationships. Furthermore, there are limits for human manipulation of other communities. Reshaping Buildings. Instead, architects generally remained at the building-size projects. Some architects have suggested altering the design of buildings and combining uses.

Reforming may result from remembering, There seem to be mental limits, also. We are unable to remember everything that impacts us. We are unable to give attention to everything at once. The upper limit for measurable relations seems to be about seven (plus or minus two). George Miller found that seven was a magic number in human psychology; it represented the maximum number of items that a subject could reliably remember, as well as other variables. Possibly it applies to the number of subjects having an intelligent conversation as well. There may be limits of receiving or processing information.

Reforming may come from imitation, An image is an imitation or representation of something. It can also be a symbol or type, a metaphor or concept. An image can stand for something else, for instance the image of a dove is often used as a symbol for peace. In the etymological sense a symbol is something 'thrown together,' as a problem is something 'thrown forward.' Unlike an image, a symbol often represents some other thing, process or quality. Human beings are mammals who live in groups and are good at imitation. These talents have allowed humans to create cultures to adapt to stressful environments.

Reforming may come from play. Human beings are mammals—omnivorous, social, bipedal, featherless, symbol and tool-using, game-playing, neotonous, bilateral-hemispheric, culture-making generalists. Reforming may come from learning. The ecosystem 'learns' the changes, e.g., seasons, of the environment. Any system formed by reproducing and interacting organisms must develop an assemblage in which production of entropy per unit of information is minimized. It is a general property of some systems that acquired information is used to close the door to further inflow. A mature system needs less information, since it works toward preservation.

Human places are complex integrations of nature and culture that develop in particular locations. The feeling of a geography is prior to its study. The place precedes knowledge of it. The knowledge of place is one of the first links in a chain of knowledge. Being human is having and knowing a place. This knowing is essential to our existence. Paul Shepard suggested that for each individual the organization of thinking and meaning was intimately related to specific places. Place is a focus of experience; and the background for specific events. The features of a world are experienced meaningfully. The place is a matrix for ordering experience. The essence of place lies in unselfconscious intentionality. Only learning flowing from hospitable presence to human and world can promote life and enhance human existence.

Conformation. In general, there are some universal human behavioral standards, but there may be special local expectations, to conform with various cultures. For example, there are some universal human behavioral standards, such as a prohibition against incest or against eating human flesh, but local expectations conform with cultural values—and indeed, cannibalism and incest have been important parts of some societies at times.

Conformation is accommodation to the right measure at the right time. The morality of the act is determined by the current state of the system. Adaptive modes should conform to ecological patterns. An ecological ethics is based on attributes of ecosystems and human compliance with ecological laws. The aim of an ethic must be harmonious with the whole population of living beings.

Conformation is fitting. The intent of describing large-scale trends or patterns is to have human patterns fit with observed patterns in nature; patterns have a form, sometimes repetition, and sometimes regularity, but each of these is caused by some limiting factor. Fitting the pattern can lead to both continuity and predictability, and both of these are needed to adapt human activities to natural limits. People not only fit physically into a place, by adjusting their eating habits, by making or shedding clothes and houses, they also change their images of themselves and the place so that the two images fit together.

2.3.1.2.6. Novation

Novation is the interaction of a physical and a cultural conscious intentional process to create novelty. A process is described as evolutionary if it involves emergence and the creation of new things—to be a species, however, the novelty has to reproduce or multiply or diffuse. With any change of evolution, change depends on how often fruitful novelties arise and how fast they spread. In an evolving system, novelty is always be transformed to confirmation. Novelty always enters with fluctuations. Natural macrofluctuations serve to maintain openness.

The ceaseless activity of being in history creates newness. Novelty is born from the womb ('hystera') of change. History is the result of this hysteresis process. Motion, Version, Scension—all together form the ontological spiral of being/environment. The motion of the field creates a turning, which leaves a history, out of which patterns arise.

They engage in a process of self-assembly, where the complete self is the organism-environment system. Construction requires participation, complexity, and development. The process of construction involves a self-presentation offering new symbiotic relations and novelty. Novelty always enters with environmental change, which serves to maintain the openness of the system. Novelty enters with fluctuations. The system has to be an open system for novelty to develop. The process of novation makes new, complex patterns.

Innovation. Development means the introduction of an innovation. One of the ecological consequences of human activity is the degradation of wild habitats for human developments (food, housing, and recreation) and the introduction of novel elements into the biosphere—elements that have not been harmoniously worked in over time. Technology also promotes land degradation. Plowing causes erosion. A problem with innovation is that time and leisure are needed for technological innovations. Starving people rarely invent their salvations.

Newness can result from modification, that is minor improvements that may reduce unwanted effects, to hybridization, the combination of two systems, and mutation, a radical change often unexpected from the previous thing. Innovations have consequences and this is related to feedback, especially positive feedback. Thus, reverberations can spread through an ecosystem and trigger other innovations in other parts of the systems or other systems. Innovation creates new inner patterns, that is internal patterns that may allow resistance or resilience to external changes.

Enovation is a spreading outward for change. For instance, an animal can make a new niche for itself by changing its location and behavior to develop a new food source from the waste of a system. Another example is the behavior of altruism or charity (both words used in a biological sense not strictly human), where an animal may call warnings to another if a predator is nearby or leave food from a kill.

Enovation is the process that describes cultural transformation. Innovation is influenced by the growth and intensity of population, by the expanding and intensified activities of states or cities, and by increasing trade and commercialization. The ease of communication also increases rates of innovation. Geography encourages some paths of innovation (east-west in Asia and Europe, but north-south in the Americas). Political form is also a factor with innovation. Accidental innovation can become a culture of innovation, that is, a part of the culture encouraged and used.

New tools extended efficiency. More crops required new ideas of storing foods. There were more innovations in general, due to density and intensification. As humans stay in place, they tried to extract more resources from the same area. This requires new ideas and technology. Which resulted in denser settlements, which resulted in new technology and new social organization. This is extensification. Innovation is influenced by the growth and intensity of population, by the expanding and intensified activities of states or cities, and increasing trade and commercialization. The ease of communication also increases rates of innovation. Perhaps that is why it went east west in Europe and Asia and north south in the Americas. Accidental innovation became a culture of innovation, that is, a part of the culture that was encouraged and used.

Exnovation pulls a culture to create a new cultural niche in the natural landscape. New tools are invented, sometimes pushed by the expansion of population. The Navajo or Inuit are examples of a culture that was able to deal with innovation and expand. Invention is the conscious creation of a new pattern to increase something. Exnovation may create extra wealth that can be used for the flexibility of a culture or perhaps for charity to other cultures or natural landscapes (resulting in wilderness set-asides).

Renovation. The industrial makeover of nature is renovation. Everything is always being reinvented. Why? Because it has to be, because everything changes. Ideas and strategies that worked for foragers do not work for urban dwellers. Ideas that worker for traditional urbanites do not work in an ecological urban context. The key patterns can be loosened and changed, smaller patterns submerged in a larger pattern that can make it more resilient.

Connovation. For plants and animals, connovation means creating new cooperative patterns through behaviors. For intercultural transformation, this means creating new trading routes or regional unions. In fact, cultures have been bound through regional or universal religions, which have provided restrictions on behavior as well as cultural goals. New religious patterns can transform human behaviors with imagination and wisdom.

2.3.1.3. Metaphysics in Action

Motion, version and scension result in filtering. So, they are identified with separate words here. Sifting or filtering is the interaction of two nonliving physical processes; energy and matter over time are described by entropy. Sorting is the interaction of a physical and an intentional process; for instance, living beings in an environment are described by sorting process, perhaps the equivalent to evolution. Forming is the interaction of a physical, social and an intentional conscious process; the making of shelter is reforming (and this uses intention). Often the same word is used for all these processes, for example, Schrodinger uses the term negentropy for filtering, sorting, and forming, but the processes are different, work differently and have different results.

The characteristics of the field are modified by the operation of the field. Different characteristics emerge at different levels. For instance, stability at the ecosystem level is replaced by loyalty at the social and psychological level. It is possible to trace the changes in characteristics from the field to the ecosystem and then to place and society. Animation and ecological value changes the differentiation of the field to the openness of the ecosystem. The addition of communication and cultural values to that characteristic of the ecosystem results in the richness of place. And, the addition of social values and awareness of a place leads to variety in that society.

2.3.2. *World as Source*

This ground, the world moving, turning, processing, rising, and intending, is the source of everything. Matter and energy, molecules and elements, combinations and societies, are the basis for living.

2.3.2.1. Materials

Materials on the earth include elements, compounds, and combinations. The atmosphere is composed of elements and compounds, such as nitrogen and carbon dioxide. The ocean is composed of hydrogen hydroxide (water, of course), with many dissolved elements, such as carbon dioxide, gold, and sodium chloride (salt, of course). Materials can be extracted from places in the landscape: Metal ore from rock, wood from trees, oil from tar sands, or stone from formations.

Organisms themselves can be described as material patterns in space (as well as energy patterns in time). Plants concentrate materials and energy in their form. Insects and animals eat plants, although significant amounts of plant material ends up in the soil.

All living and nonliving beings participate in some kind of material system. A material ecosystem is an adaptively organized system of living beings and elements that recycle important elements within the system (after Lynn Margulis's definition); it is the smallest unit that recycles biologically important elements, a natural system where the exchange of materials between living and nonliving parts follows circular paths. Cultural materials also become the source for new materials.

2.3.2.1.1. Cultural Use of Materials

Groups of people eat plant matter or animal flesh to get energy to survive and work on their needs for survival. They use materials for clothing and shelter, rocks for tools, and goods for status. Art materials were taken from what exists in place: For the Kwakiutl, cedar was begged from trees for straight boards, canoes and posts. Materials can limit how a culture expresses itself in tools. But, materials can also be traded in raw form or finished form, that is, materials flow between cultures. Materials have also been used to judge status in a culture.

Cultures have been unique programs, using local materials and ingenuity, for satisfying basic needs, material and spiritual. Cultures are recognizable in the residue after the flood of goods and ideas.

2.3.1.1.2. Accumulation of Materials

In hunting and gathering cultures, materials had to be carried with the band, or left behind for seasonal use. The materials of human societies have been increasing steadily for forty thousand years, limiting regular mobility, at least until transportation became capable of hauling the material goods; horses and wagons led to trains and trucks. One wonders if accumulation was one reason for advances in transportation. The number of artifacts seems to increase as the populations increase. The number of artifacts seems to increase as the complexity of a culture increases. Materials can be used to express power.

2.3.2.2. Energy The sun provides energy that drives global cycles and ecosystem development, although the earth and the moon provide energy that they are dissipating from their formation. Geological formations and plants store energy that can be used by living beings.

2.3.2.2.1. Cultural Use of Energy

Human cultures, in addition to the energy put out by individuals, have various degrees of access to different kinds of bound energy, such as fire or coal. Energy is a useful tool of analysis of kinds of culture. A culture can be defined by its levels of energy use. Traditional economies are based on solar energy and natural productivities. Human energy is used for building and foraging for game and plants. Fire, as the release of organically-bound energy, is used for heating and cooking. Sometimes, animals or human slaves are captured and used for work. Production is kin-oriented and reciprocal.

Carbon products are the largest source of energy. The earth is also a source of energy from tides and hot springs. The discovery of other forms of energy, such as fossil fuels, lead to technologies that could leverage them. Technologies require energy. The industrial revolution increased the quantity of energy, but decreased the variety of energy resources. The three main uses of energy for industrial civilization are motor traffic, heat for buildings, and manufacturing. Manufacturing is a heavy consumer; it includes: Pulp and paper; primary metals (for cars); nonmetallic products (for plastics or shoes); and chemicals. Some uses of fuel are incredible. Herman Daly cites the fuel requirements of a B-1 bomber as being between 300 million and 1 billion gallons of fuel per year. By contrast, all of the buses in all the cities of the United States in 1974 required only 325 million gallons.

The invention of ways to use nuclear energy increased the energy available, but has been limited by problems with the disposal of heat and long-lived by-products ('byproducts,' like 'side-effects,' are equal things, just not expected or desired). Nuclear fusion promises to proved cleaner energy, but with a different set of limits, such as containment and safety.

There are other kinds of energy that it might be possible too use, such as gravity, or the rotation of the planet. But, at the planetary level, we are at the beginning of the Kardashev scale, a general method of classifying how technologically advanced a civilization is, using planetary, solar, or galactic energy.

2.3.2.2.2. Quantity of Energy and Efficiency of its Use

Sociopolitical systems need energy for maintenance. Human civilizations all together have a very high energy use, perhaps thirteen times the use of all other mammals on the planet.

Efficiency is a measure, for instance, of the production of an animal divided by the energy consumed by the animal, or simply production by use. However, it should be considered over the lifetime of the animal. Efficiency cannot be a quick snapshot; it has to include the entire range to be meaningful. That is, many behaviors that seem inefficient in the short-term, may yield benefits in the long-term, or other benefits that cannot be measured and incorporated into the concept of mechanical or economic efficiency.

2.3.2.3. Historical Mixture of Things

In normal stochastic processes, things bunch together or move apart. We can identify cycles of these processes. These cycles are not just repetitions; they spiral and take up the old processes—who says life does not learn unconsciously from history? There are many kinds of natural cycles. For instance, the climates cycle as a result of planetary motion. Every being has a life cycle; every ecosystem has a development cycle.

2.3.2.3.1. Cycles

In a mature forest, for instance, very little material actually leaves the forest. It is held in cycles. Materials cycle above and below ground, between the atmosphere and trees, between trees and insects, and between squirrels and fungus. Chris Maser is fond of saying that most of the cycling is invisible because it is underground or in the air. Many cycles can be investigated through ecosystem analysis, where energy and materials are traced through transfers through compartments in an ecosystem. For example, in the nitrogen cycle, the compartments are the atmosphere, vegetation, forest floor, and mineral soil, while the transfers are precipitation, throughfall, leaching, litter fall, mineralization, fixation, and denitrification. Each compartment keeps the nitrogen for a certain time, termed the residence time; for a hardwood forest floor, for example, residence time is about seventeen years. Human interference in forest cycles can collapse the residence time; for example, clearcutting alder in Washington state causes very high nitrogen losses.

Virtually every material cycles through the forest, including, but not limited to: Nitrogen, carbon, phosphorus, potassium, calcium, sulfur, magnesium, and water. Nutrient cycling involves many of these materials. Nutrient cycles change with the succession of a forest. A variety of fluxes and accumulation rates of several nutrients are needed to maintain a growth rate of twenty cubic meters per hectare per year in Corsican pine in Scotland. Half the nutrients for new growth are drawn from senescing needles and twigs.

There are some material cycles between other ecosystems or in larger patterns around the planet. Some of the cycles are daily; some are seasonal or annual; others are years or decades; a few, like the carbon cycle, are century or millennia long. Human beings do not pay much attention to very long cycles or to those we perceive as not affecting us. These cycles, applied to cultures, allow for their growth and development, expansion and contraction as cultures. They can influence the efficiency of a culture.

2.3.2.3.2. Scale

Size is that quality of a thing that determines how much space it occupies. The word is derived from a French word for court sessions that determined official standards for measurement. There are barriers to the size of nuclear matter, of stars, and as well as of planets. Things divide naturally. They make barriers or walls. Walls allow cells to live. The proper size of things depends on their function, according to D'Arcy Thompson. The scale of a thing depends on its place in nature. Kohr states the principle that stability and soundness adhere to bodies that are relatively small—this is his theory of small cell architecture. Dynamic balance is self-regulatory, Kohr states, because of "the coexistence of countless mobile little parts of which no one is ever allowed to accumulate enough mass to disturb the harmony of the whole."

2.3.2.3.2.1. Ecosystem Scale

Scale has several meanings in ecology. Basically, as it is used here, it has to do with the level of measurement in a STEM context. The scale of the system defined, for instance a mangrove forest ecosystem, depends on the scale of the phenomenon being addressed. Ariel Lugo and his associates found that mangrove forests are in phase with the frequency of hurricanes, although hurricanes may not influence the life histories of short-lived organisms as much as daily or seasonal cycles. The forest ecosystem is a large-scale pattern of millions of minute events. Microbes, for instance, are affected by short-term cycles of precipitation and temperature. The overall extent of forests is often determined by large-scale weather patterns and environmental changes. Watching the history of the earth from the moon, over millions of years, it might be possible to see forests move across continents like the shadows of clouds. The possibility of extreme change is rarely considered in wilderness design.

Problems in management arise where applications that work on a small scale are expanded to large scales, without thought for the difference or changes in patterns. For instance, it is well-known that Douglas-fir is shade intolerant and grows best in openings that get light. Rather than simply remove single trees or small groups of trees, and release or plant the fir in the openings, industrial forestry applies the treatment to the entire landscape with large clearcuts, which alter the other conditions that fir requires—some shade, water, protection from browsing, associated species. For the forest in which Douglas-fir grow, increasing the cuts to increase the potential for the fir has the unintended effect of drying out the soil, as well as destroying the infrastructure of plants, animals, and fungi, that the forest requires to continue. Things change when scale changes. Changes in scale put pressures on human and natural systems. The current economic style is too great, fast and reckless for ecological systems to absorb its impacts. The scale of things is an independent problem that can ruin the best intentions of policy.

Other systems, such as high-grading or thinning from below, may also have unintended scale effects. High-grading is a form of group selection, in which the best and largest trees are removed. The practice was not overly destructive on a small scale, with many of the matriarchs and patriarchs remaining, but on a large scale, all old, large heritage trees were removed. Thinning from below may be a good idea in a stand, but like high-grading, it is biological selection—the gene pool is altered as the small and suppressed trees are removed.

An inappropriate time scale, its shortness and urgency, is a cause of many of these problems in industrial forestry and ecosystem management. Although forests are considered renewable resources, they are slowly renewable, requiring hundreds or thousands of years to renew from catastrophic disturbance; this time is far longer than any economic plans and nonrenewable on a human life scale. This has important implications on sustainability.

One solution to many of these problems is a reduction in scale for everything from forest use to management units, with local controls and local use primary. Temporal scales, however, should be expanded with long-term research and management. Ecological management must observe the proper scales, especially in terms of size, age, and patterns.

2.3.2.3.2.2. Human Scale

Human institutions are also subject to effects of scale. At the level of families and bands, which are often extended families, behavior is understood and exchange is reciprocal. At the level of tribe or chiefdom, conflict resolution cannot be done without everyone knowing everyone else, and it cannot be left between members who are strangers.

Nations become organized on territorial or organizational models, rather than on kinship. Nations can include more than one ethnic group or language. Nations have advantages over smaller groups because of specialization (especially of soldiers), technology (especially of weapons) and population size. Nations arose in Mesopotamia, China, Nile, Indus, Mexico, Andes, and West Africa. Chiefdoms in even more places, including Amazonia and Polynesia. Larger amalgamated nations only arise through conquest or from duress for protection against other nations. What leads to this complexity? Population growth and attendant processes, including intensive food production and resource use, resulting in public works programs, long distance trade, and specialization.

Each place has a temporal/spatial order: The physical scale and rhythm of a place. Small-scale cultures were egalitarian. But, there is a qualitative change when scale increases. The villages became ranked. The shift in size to cities changed the scale of food collection and trade. For instance, time is perceived at a faster rate in cities than in rural settings.

Economic systems all seem to work on a small scale, and work less well on larger scales, when the market and the production become too large. The problems increase and more is taken up by the problems. Many kinds of products do not add to the wealth of people. Kohr identifies three kinds of such products related specifically to growth: (1) Power commodities, enhance the standard of society without contributing to material welfare of the members of society; these include tanks, bombs, and government services. (2) Growth products of a second category, density commodities, are necessary due to the increase in population, but do not add to individual happiness. These are traffic lights and hospitals, and fire losses. Growth requires growth products to address the problems of growth. (3) Growth products of a third category, progress commodities, are improvements made necessary by previous improvements (anti-aircraft guns need better ammunition) or unwanted tie-in products such as license plates or television repair.

To work in a large society economics needs to be redistributive as well as reciprocal, that is, a central authority redistributes excess goods to individuals and groups with deficits. Economics are the harmonious distribution of wealth among people. But, there are problems outside small-cell scales.

Kohr's principle argument against large-scale economies is the law of diminishing productivity. Variables added to fixed units eventually add less to total product than a preceding one. Increase in size or power does not produce an increase in satisfaction or productivity. Kohr points out that Marx failed to link misery to the scale of economics rather than the system. The problem is with overgrowth more than style, which is why socialism based on overgrowth looks the same as capitalism overgrown.

Small-cell theory, Kohr says, is a fundamental principle of health. We are happiest in smallest units, such a family or county. The pattern has to be small cells under a federal unit larger than the largest state or community cell. That allows harmony and manageability by ensuring a physical and numerical balance. Then, any kind of central authority can be weak.

2.3.3. *World as Sink*

A sink is a device for collecting heat, energy and substances from a system and concentrating it, before distributing it or dissipating it. Of course, it has other connotations, as a sewer or corrupt place, where courage is lost or character is debased or where one passes gradually into despair. Perhaps these connotations fit in this discussion after all. As a sink, the world concentrates dwellings, technology, waste, and ideational systems. Things occur in place, as part of an ecology combined with a world image of changing cultures.

2.3.3.1. The Structure of Place

The undifferentiated extent of some dimensions is called space. Once space is particular, that is, limited to a specific location, characterized by a local interaction of things, it becomes a place. Every place on the planet has uniqueness, whether it is merely a difference in location on the globe or whether it is a combination of living patterns and geographical events.

Topology is a description of the mechanism of the most primitive experience of spatial orientation. For Jean Piaget, topology is a mathematical description of experience in the lived-in-world; it is the most elementary logic. Perception is participation, the organic relationship between the center of the self and the circle of appearances.

2.3.3.1.1. Characteristics of Place

Place can be described as a field that orders itself. The types of ordering depend on the local characteristics of place. Plants and animals create their own world from a place, according to their complexity and needs. An organism is inseparably related to its habitat, the place where it lives. Individuals are related to places by basic physical factors. Place provides limits to which individuals have to adapt. Places and organisms shape each other through feedback. The following general characteristics of place follow the characteristics of a field.

2.3.3.1.1.1 Dynamic Change

Change is a fundamental characteristic of the universe, as a result of motion. A. N. Whitehead's process metaphysics accepts change as a fundamental characteristic of the universe. Whitehead's philosophy recognizes the dialectical character of process, the interpenetration of opposites, the significance of levels of organization, the importance of community, and the irreducible character of awareness. Nature, for Whitehead, consists of patterns whose movement is essential to their being. These patterns are analyzed into events, which are the final things of which the world is made.

Dynamic change is the result of a process unfolding in an advancing flow through place. Process, the coming-to-be of organisms, is a fundamental feature of reality. The process of nature is not merely rhythmic change, but it is a creative advance, producing new forms in place. The essential order of experience is aesthetic. Even a physical theory of place can be described in aesthetic terms as a union of repetition and contrast.

Process is the fundamental feature of that which is really real and in its primary sense process is the coming to be of the individual actualities. Process is the expansion of the universe, any stage of which is an organism.

2.3.3.1.1.2. Self-making Wholeness

Places are autopoetic, that is, self-making. The dynamic self-producing and self-maintaining activities of living beings creates places. As in a field, a place is bounded. That boundedness gives a place its identity and integrity, and allows it to maintain itself in a definite form and process. External materials and energy are obtained from beyond the boundaries of a place.

As in a field, a place is also a whole, a coherent whole, an unending process of movement and unfoldment. It is an implicate order, that implicates everything else in undivided wholeness. A place is a whole system, as well as a self-making system; species and organisms are self-making. The history of a place is the history of the maintenance of its identity through continuous self-making, or autopoesis.

As a whole system, a place is an organized system of subwholes. Each subwhole behaves as a whole to its components, as a self-contained whole, and as a dependent part in context. The subwholes, or holons, are stable in a hierarchy that displays rule-governed behavior and structural Gestalt constancy. The rules lend order and stability, as well as flexibility.

2.3.3.1.1.3. Differentiation

The location in a place allows separation between things or holons and the differential development of things. The order of holons in a place follow certain laws, of differentiation, to make unique patterns. The physical world is a patterning of patternings, a flowing of flowings, whose constituent functions are interacting fields of force. Patterns of complexity shade and grade into one another endlessly. The history allows things to express minor differences that become larger differences as the history of a place unrolls. Differentiation is the discrete unfolding of things, within limits, in a place.

2.3.3.1.1.4. Integration

Place is composed of a membership in the field; places are created by the action of its members, which are connected strongly by the infolding movement of place. The web of participation integrates the field of place. For example, since individuals things or organisms are part of a field, and they are internally related within the field, they cannot exist without the field. The field, or physical society, provides the order required for producing individuals. Individuals are connected in place by infolding movement.

2.3.3.1.1.5. Constancy

A place appears constant because it represents a specific aspect or perspective of the field. It endures as long as the perspective exists. Furthermore, a place is what it is because it is consistent with itself; we are not free to sort out accidental properties and distribute them with various values among different universes. No properties of the universe of place are fundamental; they follow from other properties in a web of interrelationships. Constancy is the regularity and endurance of a specific place.

2.3.3.1.1.6. Renewal

Renewal occurs through a dissolution and recombination of the elements of a system. The system maintains its identity, despite the replacement of elements, molecules of organs.

2.3.3.1.2. Connection & Isolation

One character of a place is its degree of connection with other places. A place may be located at the confluence of a river or in a plain, so that access by wind or moving organisms is quite easy. Or, it may be so isolated, like an island, that it is rarely visited or changed by erosion.

This degree of connection is important for the development or interaction of species or cultures.

Isolation allows for separate development, without overwhelming influences. Limited isolation allows for exchanges of many kinds that may be stimulating or depressing.

Too little isolation can lead to the weaker or smaller system being overwhelmed. Too little isolation is similar to overconnection. Overconnected things and processes are not as flexible under changing circumstances. Things that are underconnected may be too independent to have mutual restraints. A balance of connection and isolation between systems seems to be optimal for certain kinds of health. In ecosystems, an ecosystem may have a weak connection to another ecosystem, but both may have strong connections to global cycles.

In human ecosystems, there are different scopes of connection. There are different senses of connectivity: Not just paths or roads between cities or information between databases. It has to be connections at all levels for people and systems. That is, people have to be connected to the source of their water. People have to be connected to other people to provide social comfort.

Overconnection between people can produce stress. Global news engages us in every single problem, that may tax the ability of individuals to be empathetic or to help. It is a source of disturbance. Where a local murder happened once a decade, murders in the happen ten times a day. We are overlinked to others by news. The massive misfortune of everyone is forced on us by vast-scale global news. Then we worry about humanity or global things, which are never-ending, impossible worries.

Human connection used to be limited by proximity and distance. Kinds of connection increase functional proximity. Globalism creates global spaces for connections.

2.3.3.1.3. Limits of Place

Limit is the essence of form and pattern, which are defined by their limits. As Pythagoras is thought to have said: "Limit gives form to the formless." A limit is defined as a boundary or the utmost extent, smallest or largest. The limit may be unknown or invisible. Spencer Brown states that a universe comes into being when a space is severed or taken apart. The skin of an organism cuts off the inside from the outside, as does a circle from a plane. The act of severance is remembered as our first attempt to distinguish different things in a world where boundaries can be drawn anywhere we please. We define order by making boundaries based on the perception of particles or of wave size. The universe cannot be distinguished from how we act on it, according to Brown.

Mathematically, the concept of a limit is fairly simple: It is something that is approached, but not reached, as some value descends to zero. In specific applications the limit is usually a number, that is, the rate of photosynthesis or the carrying capacity. Every finite system has limits. Systems have a limit in size, that may be determined by structure or the speed of turnover of components. Systems may have limits in terms of the number of connections that can be maintained with other systems.

2.3.3.1.3.1. Physical Limit & Locality

The relativistic universe of Albert Einstein has definite limits. What is space to one observer is a mixture of time and space to another. Every change of frame of reference, a physical coordinate system, mixes space and time in a mathematically well-defined way; the two cannot be separated. So, there is no real length of a ruler, just like there is no real length of a shadow; length depends upon the frame of reference. A frame of reference in physics has a uniform motion at a constant velocity in a constant direction. Each frame is considered a locality and obeys the principle of locality—that what happens in one frame does not depend upon variables subject to control in another. The rules that apply to part of the universe from a limited perspective are not universal. A global system is the sum of localities and may have unique characteristics of its own. That is, the universe has characteristics that local frames of reference do not. Frames of reference are easier to define in physics than in biological or social systems, with their extended complexities. Nevertheless the principle of locality can be used to explain many difficulties in cosmology, evolution and ethics.

Relativity seems to be part of a larger, undefined theory, of which quantum theory is a complementary aspect. Relativity requires continuity, strict causality, and locality. Quantum theory requires noncontinuity, noncausality, and nonlocality. The larger theory could derive these contradictions independently as special cases. Basically, relativity treats global-local relations and quantum theory treats metaphysical limits. Relativity is an expression of the limit of localities in the universe. Quantum dynamics is an expression of the limit of distinction in one locality. The predictions of one are not well applied to considerations of the other. For example, quantum dynamics implies holographic wholeness. Applied on the level of galaxies, this implication has almost no predictive value. The laws of relativity—local space-time frames—have taken over at that scale. This is not to say that the universe is not whole, only that the effects between relative local frames are minimal.

Relativity implies that time is relative to an observer, human or nonhuman. No sharp distinction between space and time can be made. And, since quantum theory implies that spatial elements are noncausal and nonlocal projections of higher-dimensional reality, time is also a projection and discontinuous like space. The limits of the array are nondefinable. Other physical theories, such as thermodynamics, result in limits for efficiency and for the conversion of energy.

2.3.3.1.3.2. Limits of Freedom & Connections

The limits of the universe, like the speed of light or the quantum of a field, put limits on freedom; freedom is defined here as the absence of restraint or confinement, or the state of being free from rules or patterns. But, limits cannot be complete, any more than freedom can. The universe as a whole seems to work with fifty percent reliability. Existence is already half determined; freedom and necessity seem to be balanced at about fifty percent. This can be compared with redundancy in information theory. The predictability of particular events within a larger aggregate of events is technically referred to as redundancy in information theory.

Random order is open; it has degrees of freedom. Differences can be related to degrees of freedom. An order with more differences has more degrees of freedom. The macro-order limits the degrees of freedom in underlying orders. Movement could not occur without a free order or disorder. There could not be any activity at all within a complete order—perhaps this is what Plato had in mind with the perfect forms. This implies that the universe is partially ordered and partially disordered. The distinction between order and disorder disappears when domain of both is the universe. Freedom, as meant by social or political, also applies (see later discussion). Life and art require a balance of rules and freedom.

Balance means an equality of basic processes, such as integration and disintegration.

It means that things are being built up as they are being torn down. It means that natural processes such as motion and rest, turning and returning, are approximately equal. Being in balance is like a form of maturity, while growth is usually a characteristic of immaturity on the way to maturity. Being out of balance for too long leads to death. Of course, nothing can be in balance indefinitely, or for very long, depending on its size and complexity. Balance is an ideal for humans. Too much novelty is stressful, it is too new to understand; too much of the sameness is stultifying. Too much conflict is stressful, but too much peace is boring. Perhaps the fifty-percent rule applies to balance, and to the ideas of maximum and minimum. Maybe 50% of life is spent in balance. Maybe 50% is the amount of error?

2.3.3.1.3.2.1. Connection

A system is defined as a complex object, every part of which is connected with other parts of the object in such a way that the whole possesses emergent properties that the parts lack. Systems are considered to be wholes where the internal connections are stronger than the external connections. There is no whole system without an interconnection of its parts, and there is no whole system without an environment. Everything is connected in a system; removal of any part alters the dynamics of the system. Connectivity is related to the size and density of a system. To connect means to bind together, from the Latin words, or to link or couple. A connection is the relationship or association. A complete connection is regarded as a circuit, which becomes required for certain patterns and their continuation. A cycle is movement through a circuit.

The organization of an ecosystem is described by diversity and connectivity, among other characteristics. Every local system is connected to some degree, but, usually the degrees are very weak, through gravity, global chemical cycles, or just thought patterns. Too little connection between parts and the system may fall apart. Too strong a connection between parts, and the system may not be able to renew itself through processes like metalysis, which requires that connections be broken and reformed.

In an ecosystem, individuals and species are connected to some degree, in terms of quality and quantity of connections. The connectivity of a component of a system is a measure of the number of direct connections between it and the rest of the system. Connectance is a percentage of the number of connections through predation or exploitation as a percentage of the total number of possible interconnections. A system is more connected if the absolute number increases and the percentage, and the strength, increases. Too little connectivity and species die (no food or prey). Too much connectivity and each species has to compete with all species. So it seems that connectivity must have regions of operation. Overconnection can be compared to power grid connections and failures. If overconnected, all can fail; if underconnected, many local areas fail; at a mid-range there can be power transfers.

In an ecosystem, functional connectivity is measured by the potential for movement, dispersal and interchange between populations of species, especially those subject to fragmentation. Connectivity in an ecosystem is a function of the mobility of a species, that is, its dispersal characteristics, its autecological characteristics, such as food or shelter requirements, the structural characteristics of the landscape, such as spatial patterns, the distance between patches, the presence of barriers to movement, such as highways or rivers, predation patterns, and disturbance patterns, including human interference. To preserve connectivity under uncertain conditions, it is necessary to maintain large areas of natural habitat in a larger matrix, as well as maintain natural connections and to minimize artificial connections, such as roads.

By being overconnected a system can lose stability. Overconnectedness creates a lag in signals. Agricultural fields are good examples of overconnectedness. This is even more true with corn than with wheat due to the changes on the species from domestication. Overconnection causes greater lag times. But, in a city, feed bins damp food variations so the lag is less important.

Underconnection can also create instability. Parts go their own way and there is no communication between them. If a system is underconnected, it may become unstable or lose resilience, because there is no pathway to allow a signal between significant parts.

The concept of diversity does not take into consideration the connectedness of the subsystems or systems. Therefore, it cannot be related well to stability. In addition, for each direct species to species relationship, additional bundles or sequences of indirect interconnected relationships spiral outwards logarithmically into the ecosystem and the biosphere as a whole. Indigenous peoples have names for many of these relationships, usually in verb form, such as pollinating, germinating, and shading. They understand and respect the importance and intelligence of species connectedness.

2.3.3.1.3.2.2. Cultural Connection

For many peoples, such as the Luiseno of California or the Huichol, their metaphorical image, the Tree of the World connects all worlds, all the layers of the cosmos, the netherworld and heaven. Modern cosmologies are unconnected to changes in their context. The tools will have to be used in new ways in a new framework, perhaps with topology and holograms as metaphors. Topology provides the mathematical model for processes. A hologram provides a model for connected wholeness.

New images can bring people, even poor industrial people, into better contact with nature, through connectedness. But we are connected by more than images. We are already connected by existence and feeling, even if we are no longer aware of it. Human life is linked in an intricate web of connections.

Cultural patterns may increase or decrease connections. Villages lower connectivity to the landscape by increasing patches and corridors. In city suburbs, corridors and networks increase, patches increase, but connections decrease as does productivity. Species diversity may increase, but the species often have no connections to resident species. Suburbs are the result of other patterns, such as the supply of rural land, cheap gasoline, and a network of highways.

Some industrial artifacts become overconnected to the habits of people. Perhaps one reason for cultural collapse is overconnections, just like in ecosystems. This is a problem with modern globalization, in the 1990s, as well as with simpler global connections that began in the 1300s.

2.3.3.1.3.3. Biological Limits & Carrying Capacity

Life involves a vast number of interacting structures. Living consists of complex behaviors whose limits are defined by rules of order that can be empirically described. Biological order is built on physical and chemical orders. That is why life is limited to such a narrow range of conditions, as regards temperature, pressure and the composition of air or water. And, that is why the most complex orders are vulnerable to changes in their substrates; energetic radiation can alter and destroy an individual, a small change in climate can destroy crops and human civilizations. Complex orders always depend on simple orders. Where a planet entirely of algae is conceivable, one inhabited by only rats and humans is not.

The earth is suitable for life because of three kinds of limits: the solar radiation that has stayed within certain limits for four billion years; the biogeochemical cycles of oxygen, carbon, nitrogen, phosphorus, sulfur, water that have stayed within certain limits; and, the constancy of the environment, constant enough for organic evolution, but variable enough for natural selection to be challenged.

Ecosystems and organisms depend on a complex set of conditions, which can be limiting factors. Organisms are affected by the quantity and variability of materials, if they require a minimum of them. Organisms are also affected by their own limits of tolerance to those materials. Life is limited by elements and physical factors, such as light or water; it is limited by too little of an element, such as phosphorus—this is Liebig's law, or by too much of an element, such as salt—this is Shelford's law of tolerance.

Every tree species has a lower and upper critical temperature, which limits their growth. Low temperatures prevent trees from absorbing the moisture needed for transpiration. High temperatures result in excessive losses of moisture.

The life-image limits the goals of an organism. Each animal is a participant in a field of existence. Using its senses, each participant creates an image of nature, or world (*umwelt*, life-world, is the term used by von Uexkull), from the sensations that are meaningful to it, and which limit it.

The behavior of mammals is controlled and population regulated through the use of space. Most populations, furthermore, regulate their density well below the limits of the food supply, often by as much as 50-70 percent. Territoriality can be correlated inversely with trophic levels and productivity. But populations can also be limited by the specificity of prey or plant source, size of prey or plant populations, predators, or natural events or catastrophes. The preference for a certain prey, or taste, can be a limit for a predator. Wolves in the Arctic disperse with the migration of the caribou. According to David Klein, they prefer Caribou to other often more easily obtained species, such as mice. This preference reduces their hunting efficiency, however.

Competition limits the number of species in a niche (the competitive exclusion principle). Garrett Hardin states the competitive exclusion principle as: complete competitors cannot coexist. Niches must be different for species. Krebs states that the fundamental niche of a species has an "infinite number of dimensions," making a complete determination impossible. Territory can limit breeding percentage. In some birds, less that 30 percent may breed. In wolves, only 60 percent.

Ecosystems and organisms are always interacting and changing in such a way that limits in one area can produce expansion in another area. In *The Origin of the Species*, Darwin noticed the predation principle. Steven Stanley later argued that the cropping principle may provide biological control of the evolution of metazoans. Intuitively one would expect the introduction of a cropper to reduce the number of species in a given area; but the opposite occurs. In communities of primary producers, a few species will become superior and monopolize space. By dominating its favorite prey species, usually the most populous, a cropper limits it so others can develop. A new level on the pyramid broadens the one below it. Stanley explains the Cambrian explosion through evolution of cropping herbivores, which opened space for greater diversity of producers, which permitted more specialized croppers. The ecological pyramid became wider and higher.

The process of evolution at all levels is the formation of varying sets, and the differential elimination of some sets and survival of others. Limiting factors are competition, degree of openness, feedback, and homeostasis. Consequences include directional change at all levels, an increase of organized diversity in the universe, and the evolution of wholes maintained by interlevel feedback.

Species selection is an evolutionary trend. Most trends are not produced by well-established species. Because the direction of speciation is highly unpredictable, macro-evolution is disengaged from microevolution, or the change within a single species or population. Speciation provides for rapid evolutionary forays into uninhabited space; evolution is opportunistic. Life had once been limited to the oceans, after all. The evolution of living forms expanded those limits.

Changes limit the extent of ecosystems. Succession appears to be a process of self-organization occurring in every ecosystem. The process is the same as acquiring information. The ecosystem 'learns' the changes, e.g., seasons, of the environment. Any system formed by reproducing and interacting organisms must develop an assemblage in which production of entropy per unit of information is minimized. It is a general property of some systems that acquired information is used to close the door to further inflow. A mature system needs less information, since it works toward preservation. The self-constructed limits of mature systems allow maximum variability between systems with slight external differences, like temperature. Ecosystems consist of different prefabricated pieces: species. Since the supply of species is limited, succession becomes asymptotic.

Whenever there are too many replicating units for space and resources, whether genes, organelles, individuals, families, cultures, or species, some persist and some fail. Self-organization and co-construction of the organisms and environments allow selection to act on various levels at different time scales. More than one unit is selected. No one unit is the key unit. System complexity is a limit for rogue species, since all have the same building blocks.

The system itself has limits. It has a maximum biological load that can be carried indefinitely; this is the biological carrying capacity. This carrying capacity is usually considered to be the maximum population sustainable on a long-term basis of renewable and nonrenewable resources.

2.3.3.1.3.4. Human Limits

Human beings have many kinds of limits, from obvious physical limits, such as height and bone shape, to psychological and emotional limits that keep people from listening to every single item of bad news. Some of these limits are relatively strict, while others are malleable.

2.3.3.1.3.4.1. Human Genetic Physical & Psychological Limits

Our genetic make-up predisposes us to some things and pushes us in other directions. It does make limits on our plasticity (for possible limits, see Table 2-331341-1). It could promote behaviors that damage the environment, and hence our long-term interests. If behavior is limited by "stone-age" genes, then some pro-environmental behaviors may be ineffective, while others are effective. For example, we must have some contact with the natural environments where we evolved or suffer some psychological and physical harm; this is suggested by several hospital studies. In general, however, behavior is determined by immediate personal consequences, that is, short-term egoism, regardless of eventual consequences in modern world.

The limited genetic potential of a species limits success in general. Species limitation ensures the diversity and integrity of the whole. A species that was too successful, perhaps like humanity, might endanger the interactions of many other species.

As members of cultures and physical beings, humans have very real limits. Our physical limitations are well-known. We cannot run or jump as fast or far as we want, for instance. We cannot lift too great a weight. We seem to be allocated only a finite number of heartbeats, according to Isaac Asimov, in a single lifetime.

There seem to be mental limits, also. We are unable to remember everything that impacts us. We are unable to give attention to everything at once. The upper limit for measurable relations seems to be about seven, plus or minus two. George Miller found that seven was a magic number in human psychology; it represented the maximum number of items that a subject could reliably remember, as well as other variables. Possibly it applies to the number of subjects having an intelligent conversation as well. There may be limits of receiving or processing information.

Table 2-331341-1. Genetic Behavior Potentially Relevant to Environmental Problems (after Gardner and Stern, 1996)

Scholar	*Proposed Limitation*
Rene Dubos	Humans have a genetically based need for the stimuli from a natural environment, absence of which may bring harm
Paul Ehrlich	Humans have genetically-based urges for sex and reproduction that cannot be limited, causing overpopulation
B. F. Skinner	Genetically-based "short-term egoism" leads to the environmental Tragedy of the Commons
Edward Wilson	Egoism is tempered by a genetic tendency to live in groups and to behave altruistically towards kin (extended egoism)
Garrett Hardin	A genetically-based denial causes underestimation of probability and severity of environmental threats
Robert Ornstein	Our old mind does not perceive or respond to gradual environmental deterioration
Jay Forrester	The mind cannot comprehend the complexity of social systems, which act in counter-intuitive ways

Some of our problems are the outward manifestations of inner causes, as Ervin Laszlo argues. Our perception and thought structures have limits. Our languages and logics have limits, also. There seem to be limits to our personal space and levels of tolerance to human intensification, also. Urban intensification leads to the question: Is there a limit to human numbers? Perhaps space, but is there a psychological limit? People in cities seem to do well with high-contact, high-proximity living. What happens when people are crowded or feel too crowded? Physical complaints, emotional complaints, sexual dysfunction, or feelings of fear, seem to be expressed often. There may be limits of crowding. Are there social limits, in terms of the number of people one can tolerate? We may have a requirements for personal space, home space, and wild space. Psychological limits may be the basis for the great failures of human life, from the "failure of perception" to the "failure of will."

2.3.3.1.3.4.2. Cultural Limits & Human Cosmologies

Calculating a carrying capacity for many mammals is relatively simple. For instance, the number of caribou that could be supported on the North Slope of Alaska is basically determined by a long-term average of primary productivity, that is, food for browse, according to David Klein. Hardin reminds us that carrying capacity is never an unchanging number, because the environment is variable through time.

Human beings are mammals, although they are omnivorous, social, bipedal, featherless, symbol and tool-using, game-playing, neotonous, bilateral-hemispheric, generalist, culture-making mammals. Calculating an optimum population for human beings, based on caribou as a food source, is also relatively simple. Based on current caribou populations, 180,000 and using maximal ratios—1/30—to simplify the calculation, the North Slope Region could support a optimum of 1,338 Inuit people, or an maximum of 6,000individuals.

Carrying capacity can be defined basically in terms of energy. For instance, an adult human requires 2300 kilocalories just in food. For heat and clothing, more calories would be required; for transportation and shelter still more; and, for luxuries, many more. By 1990 the average American was using 2,300,000 kilocalories per day, almost 1000 times the minimum.

For humans, this carrying capacity must include domesticates, as human equivalents, since many domesticates compete for protein consumption. Carrying capacity calculations often just consider food energy, but all needs—clothing, shelter, transportation, information generation, aesthetic satisfaction—must be included. This introduces cultural elements into consideration, so human carrying capacity must be considered as cultural carrying capacity.

Cultural carrying capacity involves many more variables, such as luxuries, aesthetic space, the use of technology, the implications of images of place, and the idea of an optimum. Since an optimum is always less than a maximum, according to Eugene Odum, the carrying capacity would be reduced by as much as half. Many mammals adjust their numbers below a maximum capacity, especially when the variability of the system is considered. Wolves underutilize their resources, as do most mammalian predators, perhaps since the same level of resources are not always available every year. Because people use culture to adapt to the earth, a figure for carrying capacity has to be variable or inconstant, to reflect the changes in productivity.

Furthermore, the optimum carrying capacity decreases as the per capita use of energy and resources increases. Technology could expand the carrying capacity to some extent, with more efficient use and resource substitution, but it could reduce the capacity with unforeseen effects, from the use of pesticides, for example. The optimum could be reduced more to reflect the possibilities of catastrophes; perhaps it has to be the lowest possible number to meet the worst conditions in a satisfactory way.

Cosmologies, our cultural images of the earth, can influence a culture to accept or ignore limits. The very circumstance that makes each cosmology unique—being in a unique place—ensures that each has limits, not just limits of resources, but limits of the image of place. Some cultures ignore the long-range ecological consequences of drainage, irrigation or overexploitation, and these cultures may decline and die. But many archaic cosmologies are a form of fitness and limitation. Most, such as the Tukano Indians, try for adaptation before domination, according to Ricardo Reichel-Dolmatoff. Many of their rituals limit the hunting of fish and game. The entire cosmology of a culture is concerned with some form of adaptation. Particularly in agricultural societies, cosmologies are gauged closely to seasons. Cosmologies make the world manageable by limiting it. They can be also tuned to the limits of the local ecology, within their knowledge of interactions. The adaptiveness of religious belief is related indirectly to ecology. Religion is sometimes responsive to ecological and economic conditions, for instance, when the Catholic Pope decreed fish to be eaten on Friday to help fishermen. Many archaic religions are even more a form of fitness and limitation. The ritual cycle of the Tsembaga can be interpreted as a regulating mechanism to maintain limits on fighting and using resources, according to Roy Rappaport. Their rituals also facilitated trade and distributed local food surpluses.

Yet, human cosmologies are limited and contradictory. All cosmologies cause destruction and waste; all produce the opposite of the good intended. Archaic and modern, occidental and oriental world views are complementary but not complete. Often, a cosmology is accepted as unquestioningly as a language, technology or place. Jeremy Rifkin states that cosmologies are a way of hiding the unimaginable: Voids, gaps, or size. They relieve apprehension of these things. They make the world manageable by limiting it. They make the world comfortable and small. Rifkin suggests that humanity inflates its daily activity into its image of the universe. The ideology forms part of the personal identity of the members of a society. Even in cultures with alternative cosmological explanations, individual choice is determined more by social and educational levels than by rational comparison and evaluation.

Once powerful cosmological ideas are adopted, they can influence many cultures over centuries. The principle of plenitude, restated in Christian terms, says that an intelligible creator gave an earth of *unlimited* bounty to humanity for their use. This principle seemed to be confirmed in the Renaissance with the discovery of the richness of heaven, of microscopic life, and of unexplored continents. Many modern political ideologies and economics have been shaped by the principle of endless wealth. Adam Smith once calculated that the real price of anything was just the toil acquiring it. Inequality in a world of abundance could only exist through human suppression and exploitation of other humans. The invalidity of this principle of plenitude came with the recognition of limits.

Our cosmologies influence how we respect or exceed such carrying capacities. If for instance a global culture had a cosmological image of the earth as a desert planet, the carrying capacity for humans would probably be reduced to only 100 million people. If a global culture saw itself as completely technological, it would consider that technology could extend the capacity to 10 billion through conversion and substitution.

2.3.3.1.3.4.3. Economic Limits & Growth

Economics is the formal study of how people use their surroundings. The word economics comes from the Greek words meaning "law of the house"—the image of a house is used as a metaphor for human society and nature. The word has come to mean the management of resources to supply human needs. It is basically concerned with sharing.

Economics attempts to address the eighteenth-century concerns of Adam Smith by using a scientific method to collect and interpret information—in fact, economics considers itself a science like physics, chemistry, or biology. Although economics is a social science that studies human behavior and its limits, it considers itself a positive science. As a social science, economics addresses human problems. The acknowledged fundamental problem of economics is the contradiction between scarce resources and unlimited human wants. The kinds of resources and the possibilities of using them in production are considered in the scope of economics, as are flows and stocks in homes and businesses, the role of the government, business cycles, monetary details and policy, stabilization and growth, international trade, consumer behavior, production costs, pricing, and resource markets.

There are two basic roads to wealth in a culture, according to traditional economics: (1) Producing a bigger pie, that is, increasing supply, or (2) reducing each portion, that is, demand. This assumes that wealth is defined as supply divided by demand. Demand, as Paul Ehrlich and others have pointed out, is physically a result of the number of people and their expectations. If the supply is limited, that is to say that the pie cannot be made larger, then wealth can still be increased two other ways: (3) reduce the expectations of individuals, or (4) reduce the number of individuals.

Supply may be mostly material things, but not always; status, for instance, is a psychological thing, and may be limited in various ways. Demand has the more psychological dimension. Therefore, because of demand, wealth will always have a psychological dimension. Gregory Bateson thought that economics may be founded on a fallacy because of that dimension.

Some theorists, like Paul Samuelson, have concluded that growth is necessary to rid the economy of disparities. The need to grow is intrinsic in this kind of economic system. A large literature has treated perpetual growth as the only conceivable state of affairs. Capitalism depends on growth for stability. In an organic system, however, growth contributes to stability, but it cannot continue beyond maturity, due to real physical and biological limits. Mature organisms grow to maturity, but then they can keep developing without growth during maturity.

Development instead of growth would equalize wealth more efficiently—after all, economies have been growing for at least 400 years and increasing the disparities. There is no necessary association between development and growth in economics, as Daly and others have shown. Growth means increase. A community is forced to accept an upper limit, beyond which it cannot grow any further. Further growth results in destruction or disruption of itself and nature.

There is another distinction between growth and development. The ecological social approach (or a redistributive environmental strategy) to development makes it irrelevant to discuss global limits to growth. Local limits are far more significant to majority of population. Regardless of how much food exists, people will starve unless they can get it. Redistribution of resources and improvement of environmental quality (in the home environment) are more important than increased production by sophisticated technology. The natural capacity of regional photosynthesis must be limiting factor in development, especially in tropical and subtropical areas.

The earth is finite economically as well as physically. No matter how technologically expert humanity becomes, the multitudinous beings of the earth cannot be "produced" in the same manner as three billion years of evolution. Nor can we throw away the ambihuman dimension (*ambi* is the Greek word for 'surrounding') and ignore that cost.

There is no romance or honor in throwing away every life that we can. The frontier of myth no longer exists. And we cannot understand new frontiers until we are divested of the old one. One new myth has to be stronger, the myth of participating in organic beauty, where development, not growth, is without limit. Once doubt is sown—let another recession do that—then the cognitive dissonance from poverty proclaiming that it is wealth will transform the old myth. The universe is a frontier; the mind is a frontier, but it is based on a whole and healthy planet, a nature of which humanity is a special part among special parts.

2.3.3.1.3.4.4. Political Limits & Security

Politics is the art and science of human government. Politics is the interactive means of providing the basic food and necessities of a community, that is, it maintains the affairs of a community.

The politics of community is generally small-scale and local. Moral consensus is applied to daily operations. Robert Bellah and associates make three distinctions of politics, of which politics of the community is the first. It is followed by politics of interest, where different interests are pursued, according to agreed-upon neutral rules; conflict tends to overwhelm consensus in any size community. Finally, the politics of the nation addresses the "higher" affairs of the nation. This level is more concerned with leadership than citizenship. Like the politics of interest, it accepts the status quo of relations of power or distribution of inequality. Symbolism becomes more important (perhaps because now the citizen diminishes in importance and the symbolism unites and includes them minimally).

Government has become subservient to economic actors, according to John B. Cobb, Jr., partly because the ideology of economics is so positive. It proclaims that continued growth will solve most of the problems of modern civilization, from poverty to conflict, although the promise has not been fulfilled. The problems have increased: From food shortages, housing shortages, and energy shortages, to unemployment, inequality of opportunity or goods, environmental deterioration, increase in weapons, and insecurity.

These problems continue due to the limits of politics. For example the size of society is a real limit; if there are too many people, within a limited territory or limited system, politics cannot provide them with a shared identity or control them. Size limits the distribution of things, food, housing, jobs and wealth, also. The time-frame of the political society is also a limit; the short-range visions of national interest are often inimical to the long-range ecological requirements of the support system.

Another limit is the participation of the members of a society. Communities have always had face to face limits, in terms of numbers of face-to-face encounters, and, in terms of distance. At a certain size, perhaps 300 people, representation and rules have to supplement direct contact.

Leadership is another limit; the pool of applicants is usually relatively narrow, and dominated by certain kinds of personalities, that may not be suitable for promoting fairness and consensus. The use of power is a limit. How do we describe the will to power? This doesn't seem limited to the community, as does participation. The will to power can be found in human communities for over 13,000 years. It seems to have evolved from the simple domination of other community members to all of nature.

Security is a problem for local communities and national governments. It becomes more difficult to protect against most weapons. Perhaps the solution is a global one, with the coordinated change in national policies and world-wide distribution of excess wealth.

Some facts, such as dishonest leaders or violent breakdowns, can result in fear, fear of the future in general, or the reactions of others, in general. Our response is a defensiveness in face of an unpredictable future. Thus, psychological limits intersect with political limits. But, these facts can also result in motivation to change circumstances and reduce fear.

2.3.3.1.3.5. The Necessity of Limits

Limits define locality, local spaces and local systems, from the global. Limits are not only important for life, but also are implicated in diversity and maturity, of organisms and cultures. But, even these limits are based on physical limits.

Although physical and chemical limits are real limits, they are modified by psychological limits, which are compounded by socio-cultural limits, which are further expanded or contracted by political limits. Each "higher," more complex level makes other levels worse. The attempt to exceed limits becomes less efficient.

One of the most important psychological limits is our ability to process data and to draw conclusions from it. We are so ignorant of the complexities of ecosystems that it is suicidal to pretend to "maximize" their use for resources. A free market civilization has to be limited by conservative calculations of ecological balance. It is almost impossible to estimate the economic value of this natural balance.

Discussing limits to growth, or any limits, is regarded by many people as a defeatism, as pessimistic, and a blow to human growth. Unseen limits have as real effects as seen limits. A community is forced to accept an upper limit, beyond which it cannot grow any further. This is related to the carrying capacity. Denying limits does not make them go away. Furthermore, as William Catton has pointed out, there is a difference between raising the limits of carrying capacity and simply permitting greater overshoot of the limits for as long as possible, with the threat of a greater and more catastrophic collapse later.

2.3.3.2. *Culture in Place*

Culture exists in minds, signs, and things, but most importantly in places. The word culture, from the Middle English, meant 'place tilled,' from the Latin *colere* meaning to 'till, care for, inhabit, worship.' For the Romans and English, to have a culture was to inhabit a place and cultivate it, to be responsible for it.

Aristotle thought that culture rose from environmental causes, thus he said: "the inhabitants of the colder countries of Europe are brave but deficient in thought and technical skill." The French thinker Montesquieu continued this line of thought, contending that people in warm climates, perhaps Greece, were weak, timid, overly sensitive, and sexually indulgent. The interaction of culture and ecology was first outlined by Montesquieu. Still later, Clarence Glacken demonstrated that climate and geography are important influences on the composition and patterns of culture. He traced the history of the idea of this influence from antiquity to the present.

Culture is a codification of reality, a symbolic system that transforms physical reality into experienced reality. Culture codifies reality through expressions, which can be preserved and transmitted between generations through language. Different languages program events differently, therefore no culture or belief system can be considered entirely apart from language, or language entirely apart from place.

Our minds are not only culture dependent, but nature dependent as well. The wind, trees, birds—the things of place—are sources of signals and symbols. So are gestures and words. A community implies that the experiencers share ways of experiencing or similar experiences. This enables an individual to go beyond a finite view, to see the embedded culture as one of many ways of relating self to universe. Culture evolves from the interactions of humans with nature; both are in a constant state of flux. Cultures may be thought of as parallel to other species, only their characteristics are carried by memes rather than genes.

Most cultures are in place in a particular territory. This is particularly true of the Campa, in a tropical forest in Peru. These peoples do not, and could not, live with the image of an ocean or desert. Even when different cultures meet, the best explanations may not prevail, although good myths are sometimes borrowed. That is because a culture orders reality in a special place. The theory of quanta would be meaningless to the Campa, regardless of how the concept was introduced, because it has no importance to the order of things in a tropical forest.

As culture extends human ideas into place, humans use energy and materials to create different structures in place.

2.3.3.2.1. Dwellings & Territories

Mammalian behavior is controlled and population regulated through the use of space in general. Most mammalian populations, wolves for instance, regulate their density well below the limits of the food supply, often by as much as fifty to seventy percent. Territoriality limits populations, but populations can also be limited by specificity of prey or plant source, the size of prey or plant populations, predators, natural events, or even their own tastes and preferences. Territoriality can be correlated inversely with trophic levels and productivity.

Human beings also identify with territory. Although inhabitants of a band recognized a certain amount of the land surrounding as their territory, they shared most of the hunting and gathering grounds with people from neighboring villages. The territory of the Nez Perce, for instance, embraced the whole watershed of the Clearwater, part of the valley of Salmon river and part of that of Snake river. Resource use was regulated by a policy of open admission to other social groups, but the visitors had to ask ritually; there were entry rites and permissions.

Bands and tribes had concerns about the minimum or maximum numbers of individuals. For instance, if a group fell below 250 individuals, the individuals might have trouble finding mates, especially given rules of marriage or taboos, such as incest. One group might fuse with another. If a group expanded much above 500, hunting and gathering became too difficult; more time was spent in traveling to the hunting grounds than using them. Near a thousand the group would fission and find new territory.

With Agriculture, territory was made more permanent. Land was converted to simpler ecosystems and its use intensified; the investment was guaranteed by new forms of ownership. Villages and cities occupied land.

2.3.3.2.2. Patterns on the Land

Everything changes, according to the Greek philosopher Heraklitus. Individuals changes; patterns change. Patterns have unique qualities that distinguish them from other patterns. Patterns are the key to understanding nature. Nature, for Alfred North Whitehead—this is one philosophical foundation for considering patterns—consists of patterns whose movement is essential to their being. Process applied to the components of a place yielded pattern. Nature is composed of patterns. Organisms have characteristic patterns, such as the branching of trees or the cloud forms of tree crowns. Lichens have lobes, wood grain under stress has spirals. The cracks in tree barks form nets. Evolution can be described as an unfolding of patterns. Patterns are not still. A circular pattern through time can be recognized as a spiral, for example, the earth's orbit. These patterns can be analyzed into events. Everything that exists has its place in the order of nature. This does not mean that reality is an organism or that everything is reduced to biological terms. It does mean that every thing resembles a living organism since its essence depends on the pattern in which it occurs, and not on its components. In some ways, patterns are prior to things, in helixes, light, fields, and ecology. Paul Shepard and others have written that relationships are as real as the objects that result from them. The science of ecology attends the overall pattern of relationships, beyond the details.

Individuals change continuously in an autonomous and purposive way according to a pattern (or cycle or history). Individuals have a specific form (or life cycle of forms), although taxonomists classify the adult organism and not larvae or eggs. The genotype determines the physical and chronological pattern of an organism within the limits of an environment and interactions—that is, an organism unfolds as an embedded part of an overfolding environment. The being of a species is the reality of the pattern of its members, that is, the self-organizing, reproducing units (similar to the idea of other holons at other levels).

A typical forest, for instance, is composed of many patterns, including vertical layering (stratification), horizontal segregation (zonation), activity patterns (periodicity), food web patterns, social patterns (including reproductive), interactional patterns (from actions such as competition or mutualism), and stochastic patterns (from random events). A pattern allow for surprises and discontinuities. The design of forests by industrial foresters is vulnerable to surprises because nature is chaotic, unpredictable, and science itself is uncertain, by definition, about patterns of change in forests.

The patterns of ecosystems are addressed best at a landscape level. The word landscape was used by the geographer Alexander von Humboldt as a scientific term in the 19th century. A German biogeography G. Troll used the phrase "landscape ecology" about 1939 to describe land, and not just living organisms, as an integrated holistic entity to be studied in its totality by geographers and ecologists, geographers providing a horizontal approach to the vertical one of ecologists.

2.3.3.2.2.1. Herding Paths

Individual organisms in a place generate physical information. Information is generated by participating species and their physical structures, such as burrows, dens or paths, which accumulate in an ecosystem. Change maintains the openness of an ecosystem, and allows novelty in the system. Individuals also use various patches and areas in one or more systems; they create paths between these areas. Their activity can change the character of these areas.

Human beings also add paths, patches and structures to the system. Like animal paths, human paths link living areas to gathering patches or areas used by different groups. Humans also make structures for living and for meeting. Human structures, including paths and patches, can be described in a grid.

2.3.3.2.2.2. Agricultural Fields

After the extinction of megafauna in various places and at various times around the planet, human culture may have become sensitive to overharvesting. Possibly, this is one of the reasons agriculture was necessary. Faced with fewer prey, humans learned to live in areas with dense crops of seeds and roots. Later, after wild crops had been traded to other areas, they may have learned to cultivate soil, plant seeds and wait until crops grew. As a hunter, one human needed ten square kilometers for nourishment; intensive agriculture lowered this to one hectare per person (a hundredth of a square kilometer). This means that cropping is a thousand times as efficient in terms of area, although it may only be half as efficient in terms of labor time. Although agriculture is more efficient spatially, it requires the land to be modified, either partially or completely. This has had a major effect on forests and wetlands as land was cleared for fields.

Many peoples around the world practiced a shifting horticulture called swidden, after a Northumbrian word meaning forest clearing, and less pejorative than 'slash and burn.' Swiddeners clear a patch of forest to plant a wide variety of food plants in a garden for a year or two and then start over in another place. The land lies fallow for ten to 300 years. The practice is effective and sustainable when populations are low and the gardens provide subsistence, and when trade or cash crops are not attempted.

Shifting cultivation was practiced in Europe as well as South America and Africa. In Africa it was practiced for centuries with only minor long-term effects on the rain forest. But in Europe, where the forest regenerates more slowly since growth is checked seasonally by low temperatures and low moistures, shifting agriculture is unstable and capable of greatly modifying the ecosystem, despite the greater soil fertility. The type of crops also made a difference. Cultural areas in North America and Europe were based on cereal crops—millet, wheat, barley—while African and South American systems emphasized root and tree crops.

In North America, which held the largest extent of forests on earth, forests were considered wilderness, to be tamed and converted to agriculture—although no one really expected the conversion to be so rapid or dramatic. Since the first agriculture, 10,000 years ago, forests have been cut and burned to create fields for food crops. The invention of iron plows speeded up the process because they could break up soil too heavy for wooden plows. Much later, the invention of the horse collar allowed horses to pull plows and loads with greater strength. Technological innovation, combined with accelerating population growth, led to clearing of many forests for agriculture.

Traditional agriculture proceeds by substituting selected domesticates for wild species in equivalent niches, or by manipulation of the ecosystem, whereas modern agriculture replaces the biota, or transforms the ecosystem into an artificial system. Margalef pointed out that all agricultural systems are laid out for low maturity, to increase production per unit. Traditional agriculture changed the pattern of ecosystems. Modern agriculture simplified those systems dramatically. From the perspective of a balloon or airplane, agricultural fields are easily identified, as circles, squares, rectangles or parallelograms, with a more monotone color and texture than surrounding wild areas.

2.3.3.2.2.3. City Squares & Centers

If the use is too intensive or combined with other factors, as it was in Greece, the surrounding forests do not return, and most forests around the Mediterranean have not returned. As Greek city-states expanded, the forest land was cleared for fuel and agriculture, especially olive trees for export oil trade, as well as for trading wood, building fleets, and erecting public buildings. Ninety percent of the wood use was probably for fuel, as it is in parts of Africa and Asia today. When the primary forest was cut and then abandoned, it often grew back or went to secondary forest; when it was cut to bare soil, erosion could have broken any regrowth cycle with bare rock. When the forest was cleared and grazed, it might grow back from scrub or heath, but probably not from grass or bare rock. The combination of climate, limestone soil, steep slopes, and overuse led to the complete loss of many Greek forests (the popular eroded landscapes of tourism). It is only the stone buildings that remain today, although they are being eroded by pollution.

Since the most fertile soils have been, and are being, taken over for agriculture and parking lots, more and more forests are occupying soils that are clay, sand, and rock. This gigatrend indicates that the kinds of forests have been shifting for thousands of years and that the most fertile land, once covered by agricultural fields, is now covered with rock, concrete and asphalt, by roads, buildings, and city squares.

The early Sumerian temple tower, with a hieratically organized city surrounding it, became the model for the Hindu world mountain Sumeru, for the Greek Olympus, for Aztec temples, and even for Dante's Purgatory. The Sumerian city was organized in the design of a quartered circle. This design has been a favorite for cities and utopias, as well as for many cosmologies.

The whole city, and not just the temple, was conceived as an earthly imitation of the cosmic order, a sociological middle cosmos, established between the macrocosm of the universe and the microcosm of the individual. Through the priesthood, the one essential form of all was made visible.

The polis, the classical city-state, was a place where the citizens depended on and maintained the whole, which cared for and outlasted the individual. The stoics declared the cosmos to be the great city "of gods and men." The citizen became related to the cosmos as a whole, in the same way. The Greek idea of nature was based on an analogy with the human body, with thoughts and feelings. The universe was a kind of living organism. Cicero defined *mundus* as universe, as the common home (*domus*), or city (*urbs*) or state (*civitas*), of gods and men.

As people are attracted by the advantages and excitements of cities, the size and number of cities increases. The spread of cities increases. Their areas are added to that of the vast agricultural lands, which together make up the physical impact of human habitation on the surface of the planet. The characteristics of cities change also, as their sizes and shapes are reformed by desire and consumption. They become more "ideal" and more uniform; with less to make them unique, they start to resemble ideal noplaces, flat and homogenous.

2.4. *Connecting Images—Outopias or Eutopias*

Earlier and foreign images of the world are dismissed as being outmoded, useless, or unrealistic. Utopias have been offered as ideal schemes for social and political development. Sometimes, utopias offer memorable images. But, most utopias are rejected as irrelevant dreams and self-indulgent imaginings. Yet, as Pierre Dansereau has said, the failures of pollution, poverty, and urban decay are failures of the imagination. Rejecting the solutions of imagination, therefore, can only make the suite of human crises worse.

While utopian ideas might be stimulating, other traditional or accidental ideas might be applied more rapidly. Political realism dismisses utopian ideas as naïve and impractical. People are afraid that such ideas would destroy their investments in power and in collections of wealth.

Political realism, however, is deficient in imagination and results, as well as deficient in courage and imagination. Despite great revolutions in technology and intensification, most people are still poor and threatened with displacement. Despite great revolutions in technology and living conditions over the past several thousand years, people are still very much shaped by unique cultures and small groups. These traditional institutions work very well on a local scale in a specific place. The information is available, but there is no framework to make it work, yet.

The word utopian was meant to be ambiguous, either no-place (*outopia*) or good-place (*eutopia*). Perhaps the ideas of no-places have less relevance. The images of good places, however, might be worth considering. Buckminster Fuller was one of the first to consider this second meaning. In the 1920s, Buckminster Fuller began work on an Air-Ocean World Map, as an alternative eutopia that sought to define civic and ecological order through maps of known areas. In 1927, Fuller sketched the World Town Plan, a map that preceded and directly influenced subsequent maps. A 1943 issue of Life published one of Fuller's maps that could be cut up into fourteen pieces and reassembled in the form of a Dymaxion globe. Fuller saw his map as a populist construct to be used as an operational tool by the members of the global citizenry.

In 1954, Buckminster Fuller published the Raleigh edition of his project on the Dymaxion Air-Ocean World Map. In the dymaxion series of maps, Fuller superimposed a spherical icosahedron grid onto the earth's surface to limit the distortion of the relative size and shape of its components. The projection of Fuller's map represents all areas with equal weight. In Fuller's map, the North Pole is the neutral center around which land mass and ocean unfold. Fuller's town plan map, in its projection method and form, highlights the connectivity of his "one world island."

One of Fuller's objectives was to establish a map that does not prioritize cardinal direction, political entities, or hemispheric organization. Instead, the Dymaxion map provides a base for presenting global themes such as human migration, natural resources, and population distribution. Temperature replaces politics as the organizing map feature. In this map, world climate is shown in a range of coloration from warm reds to cool greens and blues. Even though highlighting a specific theme, Fuller's map emphasizes wholeness across the global surface.

A complete eutopian structure would start from such an image of wholeness. A eutopian structure, one that is unambiguously good, could be designed and described. It would outline a science and politics adequate to deal with the creation of good places on earth, where all beings are equal within a framework of high human culture within nature. It would propose an adaptive cosmology to place human values within a global ecology and to balance human development and the preservation of wild places. The framework could protect the integrity of culturally-based nations and to address global issues. Global and local issues would be differentiated; the myths of global communities and one-world people would be explained and discarded. Pretending to be world people is a mistake based on a misunderstanding of human limits. People are the products of places and cultures, of land and nations.

Unlike either the political realism of nations or the ideal designs of utopias, a eutopian plan would be based on traditional human and cultural realities and would propose only modest and reasonable changes at local and international levels. For example, there are 500 million indigenous peoples in fifteen thousand distinct groups, such as the Uighur in China or the Kuna in Panama; and, there are over two billion people in hidden nations within massive political structures, such as the Azerbaijanis in the Soviet Union, the Kurds in Turkey and Iraq, or the Tibetans in China. Furthermore, there are regions in some countries, such as the Pacific Northwest in the United States or Wales in Britain, that may prefer independence to forced membership in a confederation. A Eutopian plan could allow any indigenous people with a traditional culture to become an independent nation without fear of conquest or compromise by existing political states. The benefits would outnumber those of a global monoculture and the negative aspects would be more manageable.

Such a eutopian process would solve many problems, from problems of scale to political inappropriateness. Not only would more nations exist in the framework, but alliances and networks would form and reform, as they do now. How would it work? A political framework, based on traditional cultures, coordinated by a global 'regulating' body, based on a modified United Nations, would be possible and desirable. The holistic framework would protect the creation of thousands of good places, or eutopias, based on human cultural realities and on the ecological limitations of places, that is, on homelands. These models already exist—they just need to be better defined and then protected from misguided ideas of growth and progress, as well as from the dominant, consuming industrial culture. This framework would also reconcile the well-being of society with the health and continuity of living ecosystems. The international framework would provide paths for policing, international education, and other national needs. This plan would provide a path to independence and formal interdependence.

Before trying to build this framework, however, we first need to understand ideal fictions and images.

3.0. **No Place: Utopia**

The City, Amanote, speaks these lines in Thomas More's 1516 fantasy, *Utopia*, about a perfect commonwealth, a society without the problems and poverty of a young English capitalism.

Felicity's Children
Wherefore not Utopie, but rather fitly
My name is Eutopie, a place of felicity

While relating his story in Latin, More made puns on the Greek names he used: The name of the city, Amanote, means "dream town;" the name of the traveler himself, Hythlodae, means "dispenser of nonsense." The title is also a word play; to the Greek word meaning place (*topia*), could be added a prefix meaning no (*ou*), or good (*eu*). He used 'u' as an ambiguous prefix; no place sounded like good place.

Utopias are fictions by definition. More had said that it was a fiction whereby the truth might "slide into men's minds." More invented and named the modern utopia. Unlike the earlier *Republic*, which Plato presented as an aristocratic ideal to be contemplated, More's utopia is meant to be a description of an achieved egalitarian society.

The citizens in More's utopia were uniform and regimented: Everyone had the same clothing, housing, and work schedule. Strong peer pressure existed for people to use their leisure time constructively for the public good or to improve their personal virtues. The electoral unit, the family, was autocratically ruled by the patriarch and a hierarchy of princes. Decisions were made by councils elected from public officials, who met regularly with these princes. The utopians solved the problem of population growth by setting up external colonies, where it was considered unjust for natives to hold onto land that they were not 'using'—much in the way the Spanish and English approached colonization in the Americas during the same century.

In *Utopia*, Thomas More comments on why there is less need for work or long working days: "for where money is the only standard of value, there are bound to be dozens of unnecessary trades carried on, which merely supply luxury goods or entertainment. Why, even if the existing labor force were distributed among the few trades really needed to make life reasonably comfortable, there'd be so much overproduction that prices would fall too low for workers to earn a living."

The basic argument, that a few hours of labor a day for everyone would be sufficient to supply all the necessities and comforts of life, is a good one. The residents of Utopia reject beautiful clothing. More used a monastic metaphor, associated with strictness and deprivation. But, More neglects the importance of aesthetic needs. Having everyone have one simple garment would not be acceptable. People like to improve their things. Clothing, for instance, can reflect differences in taste or status, and people will devote much time and money to clothing.

In *Utopia*, money is eliminated, and therefore many crimes no longer exist. But, More does not adequately consider the human need for prestige and honor. People kill and cheat for prestige and honor as often as for any gold or symbols.

For More, Nature deserves respect, if for no other reason than she is capable of turning against us, and must be treated with care. But, nature is regard as endless and can be converted to cities and fields with each new utopian colony.

More expected that part of the meaning of Utopia would be carried in the dialogues inspired by the book. Chronologically, More created a good place first, and a society constructed from the moral and rational ideals of More himself second. As More wrote, he realized that humans can make any good place less good, that they can seek less than the optimum, or reinvent original sin. Human nature creates conditions that reverse its good. Absurdities are invented, toyed with, and embraced. It is human nature, to attempt to escape all control, or to express the desire to be wicked. More's vision had an immediate effect on the growth and development of his country.

Utopias as visions of ideal societies can be found in almost every culture: In prophecies, visions, dreams, myths, and ideologies. George Orwell suggests that the utopian vision has been consistent since Plato, with its dream of justice through reason, through the elevation of public life, and through the regulation of private life, community property, and good breeding. Utopias are not just visions of noplace, but of otherness.

The ideas of a golden age and an ideal city were combined in modern western utopias. The discovery of new continents opened new possibilities for utopias. Utopias have always been placed in distant lands, populated by people with strange customs. Plato envisioned a perfect model for the Athenian form of social organization; his *Republic* was located on an island in his present (2500 years before our present). In response to Plato, Aristotle suggested that it is impossible to say which model for a society is best. No organization can be ideal under all conditions and at all times. The organization depends on many factors, from social organization and size to the state of relations between neighboring states.

Christian utopias, such as Augustine's *City of God*, began with inward change, but led to heaven. These first utopias were idyllic; they described the good life, either material or spiritual, as tranquil and unchanging, but located in other places. In these utopias, liberty was less important than happiness or goodness. Many utopias try to control society through the state—as in Plato, More, Fourier—to ensure safety and peace. Other utopian writers, such as Morris, Kropotkin, and Bookchin, suggest abolishing the state entirely. Another distinction in subsequent utopias is between private or collective ownership, with the degree of openness, or with the level of technological solutions.

In many utopias, nature, rather than be protected or preserved, is deliberately transformed, according to geometric and scientific models. Francis Bacon, in *The New Atlantis*, writes: "the fathers of Salomon's House summon natural phenomena and create new species to dominate and transform nature, even to the point of eliminating it and making it superfluous." The goal of Bacon's utopia is the inflation of human power over the universe, to create human ease and human happiness. The simplicity of earlier visions was due to the assumed scarcity of goods.

With the exploration of all the continents and oceans, utopias began to be located in other times, the past or the future. For Saint-Simon, Fourier, Comte, and Marx, the good times were in process, and the goals for the future were fixed and codified. The predicted Utopia would be achieved eventually, given institutional and technological propensities. Morris and Howard rely on the garden metaphor. This requires complete design and control in some instances. The garden is dominated by external design without internal human changes.

Later utopian thought put less emphasis on individual and social content. Theilhard de Chardin regarded man as the spirit of the earth, part of a process leading to the "hominization" of the earth and the spiritualization of humanity. The present is also rife with unanticipated developments, such as population growth, blossoming energy use, and waste that follows use everywhere. Other thinkers, including Skinner and Maslow, emphasized the psychological dimension of utopias, recommending personal changes.

In the past century, the concept of place began to be combined with different prefixes to indicate wrongness (*dys*) or badness (*kako*), as in dystopias or kakotopias. Among others, Huxley and Orwell traced paths of visions gone awry. Huxley relies on a wilderness metaphor, with a simple natural lifestyle, characterized by a hands-off approach. But, is there recognition of the human contribution to the goodness of place? Is there enough respect for the power of nature? We recognize that nature may treat us well; we should recognize that nature will provide opportunities for us, but also that it may remove them through the occasional violence of its chaotic systems.

As the visions became less positive, utopias became less adequate and less desirable. Few utopias, although Ernest Callenbach's *Ecotopia* is an important exception, have been created since the 1940s. Callenbach and Bookchin suggest alignment with natural cycles, preventing disruption by eliminating pesticides and using only renewable energy. Marius de Greus suggests that this model is the most neutral and has the fewest weaknesses. Even ecological utopias can be static, however. Callenbach, for instance, offers a static model in the Northwest. Most utopias still tend to be isolated, since they could not compete with the expanding industrial system. Modern utopias present a very limited set of lifestyles, usually revolving around the simple good of a frugal lifestyle.

Utopian thought has a reputation for being unrealistic. This thought is considered to be the dangerous dreams of unreal fantasies. Many commentators, such as Frank and Fritzie Manuel, conclude that utopia is dead. Perhaps this conclusion is just the recognition that reason is not enough, and that human technological power over nature has a higher price than we thought we would have to pay. Perhaps it is the conclusion of people without a strong system of spiritual beliefs.

Utopian thinkers, in questioning the whole of society, often urge its total reconstruction. Utopias have had a great influence on modern industrial society, with their ideal visions of society and their promotion of the benefits of scientific technology. Ironically, these tendencies, especially detachment and nuclear weapons, have the possibility of making real kinds of utopias, literally 'noplaces,' on earth.

Traditional utopias look towards a past of ideal peace or a future of ideal post-industrial technological harmony. They do not look enough into the present, into other cultures. In present cultures, however, lie suggestions of the size of society, limits to growth, and appropriate technology.

The value of utopian thought is in questioning all the presuppositions of today. The question: To what extent are problems inherent in the existing social economic or political order? Or rather, in the scale of that order? People want to break from the problems of the past and yet keep the positive aspects of it. They want to have dreams and create their societies in the images of those dreams.

3.0.1. Images Cast Shadows

Our dream of civilization and nature in modern industrial culture is the dream of order and beauty, but as Aldous Huxley notes, the dream of order begets growth and tyranny, and the dream of beauty ends in monsters and violence. Striving for the good life for many has left us with crowded roads and regimented jobs; trying to build beautiful cities has given us gigantic boxes and neighborhood violence. Trying to fulfill our dreams of comfort and security has provoked global threats and local nightmares. The dream is a nightmare that reflects an unbalanced and immature image of the earth.

The collective image that people make of their place on earth is a world, derived from the German word meaning 'man-image.' The image is constructed metaphorically, but considered 'as if' it were true. Each world is based on a root metaphor, according to Stephen Pepper, which is a good device to discuss them. Root metaphors are comprehensive and dominate our attitudes towards things. If the image is incomplete or does not fit environmental conditions, it may fail. People who constructed their worlds from preconceived notions sometimes did not survive. The Aztecs, for example, based their cosmology on the belief that the sun needed human blood to survive, and so they sacrificed great numbers of lives to ensure the sun's life. Their political policy was based on raids for victims, and this policy contributed to their overthrow and decline with the arrival of the Spanish.

The use of flawed images can destroy the environments of cultures as well as the cultures themselves. When Easter Island was settled by Polynesians around 1500 BP, it was fertile and uninhabited. Despite their technological and food-producing skills, ideas on how to deal with surface stones to increase productivity seems to have precipitated genocidal warfare between two groups of believers, those who wanted to keep the stones and those who wanted to throw them into the sea, although stresses from deforestation, erosion, and limited kinds of crops, as well as from cold winds and drought, likely contributed to the violence. By the time of the island's rediscovery by Europeans in 1722, mortality from social chaos had reduced the population to about 1100, significantly less than the approximate high of 8,000 (estimates have ranged as high as 20,000 or 30,000). The population dropped to 155 by 1886, although at an increase of 3% per year since then, the population is now over 2000 and supported by tourist trade.

Animism (Mythology)	Identity	Identity
Formism (Rationalism)	Substitution	Similarity
Mechanism	Decoration	Similarity
Contextualism	Utility	Congruity
Organicism	Wholeness	Interaction
Holocosmology	Frame	Star

Table 301-1. Root Metaphors. Contrasting the views of metaphor by Max Black with Pepper's four root hypotheses. Two other hypotheses can be added to the schema: animism and holocosmology.

Modern industrial cultures also have defective images. Our root metaphor of the machine extends down through all levels of society to the economic and the personal. An image of a culture can be stated as a series of principles, such as: "the universe is mechanical; humanity is master of the universe; and all persons are equal." These metaphors have allowed inhabitants of some nations to treat people, plants, animals, and nature as robots that can be modified or replaced. The cost of that image has been the diminishment and destruction of lives and environments. Metaphors can limit cultural possibilities. For example, the metaphor "labor is a resource" implies that, like any common resource defined by industrial society, labor is cheap and can be used up. Modern economies, embracing another metaphor "nature is capital," draw on the accumulated "capital" of ecosystems for production. By ignoring the real cost of the capital, as well as the costs of natural services, such as nutrient recycling, soil building, and atmospheric renewal, these economics create a temporary wealth that will disappear when the capital is exhausted or the society collapses. Decisions regarding resources are made on short-term economic grounds and lead to material shortages and environmental degradation.

Powerful images can influence cultures over centuries. The principle of plenitude, restated in Christian terms, presents that an intelligible creator gave an earth of unlimited bounty to humanity for its use. This principle seemed to be confirmed in the Renaissance with the discovery of the richness of heaven, microscopic life, and unexplored continents. Many modern political ideologies and economic systems have been shaped by the principle of endless wealth. Adam Smith calculated that the real price of anything was just the toil acquiring it.

These ideas are parallel to the idea of unlimited good, where anything, even virtue, can be multiplied indefinitely. The invalidity of this principle comes with the recognition of limits. Without limits any good becomes devalued and is wasted. Limits contribute to value. The universe is limited; the earth is limited; individuals have limits. These modern metaphors are defective because they do not fit our surroundings. The crises of cultural images have tremendous physical consequences. For most of human history, the habitable earth has been a mosaic of separate territories and peoples. Different groups developed distinctive ways of dealing with their nonhuman surroundings. Each way of dealing can be referred to as a culture, a pattern of behavior based on shared beliefs adapted to the local environment. Culture can be expressed as a symbolic language. The particular symbols concerned with cultural institutions as manipulative objects are political symbols. Politics deals with words, which are arbitrary symbols for events or things. The wrong relationship of things and symbols can result in misguided politics and violence. Political decisions are made on narrow political and economic grounds. The gradual narrowing of the focus has resulted in a citizenship in industrial cultures that is the abandonment of responsibility on the assumption that others know how to manage things; government itself is the assumption of responsibility, without sufficient knowledge that result in a suite of problems.

Wrong decisions and wrong relationships have resulted in creating places that are nowhere. To create our noplaces, we have accepted the gifts of bigness, speed, uniformity and allowed the thefts of life, intelligence identity, and choice to negate any gain from them. Combined with our own failures of charity, imagination, will, and courage, using piecemeal knowledge and unrelated ideals without regard to traditional knowledge and practical experience, we have created common faceless cultures to sit on our flat, placeless utopias.

3.0.2. One World Through Reason (The Formation of the United Nations)

After the war in 1918, there was a popular vision of one world, without walls or barriers. The most desirable way of creating the "one world," as Wendell Willkie called it, was through reason, although reason has not been used historically to consolidate or shape nations—France, Germany, the U.S., and Italy, among others, were united by force. The notion of a world government seems to satisfy a basic human craving for unity and order, but, at the current stage of international relations, there seemed to be no agreeable path toward a single benevolent world order.

The partial adaptation of international institutions is insufficient for a world order, especially if these bodies are only advisory. The United Nations (UN) is the only body with the machinery for constructing a world system; the beginnings of ecological politics can be found in the special services of the UN—UNESCO, FAO, WHO, and the various technical aid services. As long as ecological and political problems are addressed in a framework of nationalism and military power, however, these organizations are treated as peripheral and relatively impotent.

Despite its good-hearted programs on disease and education, the UN cannot deal with wide-spread starvation, or acid rain or the whole complex of global problems. The UN does not seem to have a good grasp on global problems either. The 6th and 7th special sessions recommended financial aid, annual transfers of resources to poor nations, and a charter of economic rights. It is unlikely that this will happen. There are too many vested interests.

It has been said the UN has no use beyond recommending the avoidance of violent conflict. As it is, the UN has found many uses in setting standards for human health and education, as well as with human rights. It has not been as effective recently at stopping genocidal behavior in some countries.

As it is structured, the UN is not capable of handling the responsibility for international order. Despite its emphasis on peace and international security, the UN has been able to mount only a relatively few peace operations—perhaps only one every two years for its sixty year history. Although many of these have been successful, they have had to be approved by the Security Council and to be run understaffed and underfunded. The peacekeeping efforts in Bosnia were not effective. Not in Somalia either. The attempt to guarantee democracy in Cambodia was not effective.

Milton Eisenhower had said at the UNESCO conference in 1949 that for the UN to be effective, it would have to have a police force stronger than any nation's armed forces. While that could work, few nations would permit an international superpower. Most nations would also consider it to be an intolerable burden and worry that it would control the world in a unity that would have no liberty for any nation.

Is the UN due to be a monster world-state? Perhaps healthy small states could be united into healthier larger organisms. But, the UN cannot survive on good will. As it is now the UN is a union of some large powers and many small ones. The UN functions only with the consent of the reluctant four or five biggest states. Their veto power is a reflection of their power and might.

It has been said that there are no fundamental disagreements in the UN, now. But, then, this is a limited UN, not much trusted or invested in. It has also been said that there are no universal agreed-on truths of social organization. This is not a problem, because cultures organize, then societies. All an enhanced UN would do is keep standards and coordinate cultural organizations.

It has been said that the UN is a forum for the same power politics in disguise. That has been a problem with the original establishment of the UN and its reflection in the Security Council. The universal values of the UN Charter come from the victors of the world war in 1945. The UN was never meant to be for free nations. The Security Council was the preserve of great powers, with the undemocratic veto power. Restricting membership in the Security Council to great powers, and its use of the veto principle, indicate that the UN is imprisoned by the status quo. The values and memberships simply need to be modified and enlarged. Should the UN get rid of all veto power in Councils and return to persuasion?

It has been said that the UN bureaucracy worships consensus, otherwise called the "handmaiden of evil" by autocratic states. But, consensus is what allowed human societies for millennia to reach decisions. It has been said that there is a danger in having a strong Secretary General powerful enough to prevent war. No Secretary General has had the power to prevent wars. Is preventing war less dangerous than allowing uncontrolled conflicts?

It has been said that public discourse would be suffocated in a peaceful world. That is always possible, but we know that it is suffocated in a warring world, even if it is a multitude of small, undeclared wars. Discourse is considered unpatriotic for criticizing bad decisions made in wars or in 'undeclared conflicts,' the current political euphemism for wars, but discourse should not be suffocated by unevenly applied ideas of patriotism. It has been said that a vision of world peace is unrealistic; leaders would have no tragic memory of the latest conflict and little earned wisdom in dealing with conflict. It is argued that these "shallow, callow, child-like leaders and their advisors" would commit some grievous miscalculation leading to a holocaust or war. My answer? Three words: George Double-U Bush. For the most part, the current leaders have power expressly due to their skills of manipulation, not due to experiences of tragedy. Furthermore, peace would not be the end to struggle or conflict. Natural limits, human conflicts, and the long adjustments to peace and equity would provide many large challenges.

It has been said that the UN dominates the world, that it is inflexible and unaccountable. But, it is no less accountable than any elected body. Officials and programs can be changed or removed.

It has been said that international goals are best realized through national self-interest. This must be the old wisdom. It might have been true, if nations were perfectly rational and knowledgeable; they do not seem to be. Nations now seem to be the handmaidens of corporations, whose interest is only in profits for shareholders. The views of shareholders are notoriously short-sighted. The UN is limited by images and ideals of progress. The UN's solution to economic problems is "sustainable development"—that is, "growth that respects environmental constraints," as if growth respects any constraints. The Bruntland Report indicates a five to ten-fold increase in world industrial output within the next hundred years before population stabilization occurs. While the appeal of growth is unarguable, it is really not likely to be sustainable in any meaning of that word, since sustainable growth does not recognize known ecological limits. Furthermore, the UN has no power to coerce its members when it does make good recommendations.

Our attempts at social improvements have proceeded without order, without sufficient insight and perspective, without sufficient confidence, without a comprehensive plan, and without a great dream. Our politics has been corrupted by special interests. The structure of our civilization comes from anonymous builders and mediocre designers, minimal engineers and rapacious financiers. We work within the rules as they have been for decades, rejecting any alternatives as too utopian. The rules themselves have been shaped by centuries of social metaphors and utopian ideals. The rules within which we have been operating for the last sixty years, or even since the Treaty of Versailles, are inadequate. It is not a matter of what is the best thing to do within the rules as they are at the moment, it is a matter of changing the rules.

3.1. ***The Lure of Nowhere***

The idea of 'nowhere' is attractive. It has a spare simplicity. We believe that we can shape it. It represents the search for perfect predictability, for sameness, for comfort from sameness, and the known. There are few uncertainties, fewer threats, and less to fear.

3.1.1. *Getting Nowhere*

Our fascination with nowhere has allowed us to create noplaces, big, impersonal flatscapes with nothing unique to recommend them. We are comforted by uniformity, but not necessarily when every single place is or looks the same.

3.1.1.1. The Gift of Bigness (The Curse of Corporate Bigness)

Early in our history, when the very success of the species was in question, we humans learned to reproduce more rapidly than our rates of mortality. To extend our families, we have increased our numbers and our rate of increase exponentially. To ensure the success of our species, we have appropriated the places of other species. Our overpopulation has led to aggression against other cultures and species, then to indifference at their suffering. Even low levels of food and fulfillment at our current size can be maintained only through theft from other species and from future human generations, and through the degradation of billions of humans as well as the ecosystems on which they depend.

To provide for the needs of many and for the extravagant luxuries of some, we have produced waste and pollution on a geological scale, from islands of garbage to skies of acid rain. Manufacturing processes result in the production of new dangers, such as recombined genes, and new substances, which are not easily incorporated into natural cycles. The overuse of ecosystems results in deforestation, devegetation, and desertification, then in depletion of raw materials and depletion of agricultural land. Economic and political pressures, derived ultimately from population pressures, force farmers to intensify their efforts to increase crop production, instigating a dismal cycle of population expansion, environmental deterioration, and poverty.

To provide for our needs efficiently, we have increased the scale of our activities. But we have decreased the diversity of habitats by filling in wetlands, felling forests, plowing grasslands, and irrigating deserts. Agribusiness has caused widespread landlessness; people who try to grow their own food are forced onto marginal lands or off the land. Acquiring fossil fuels also creates landlessness; coal mining in the Black Mesa mountains of the United States, for instance, may force the resettlement of twenty thousand Hopi and Navajo people. Without land, and the economic independence it allows, cultures are more likely to disintegrate.

Our local communities are proud to attract more people and larger industries, but do so thoughtlessly, without regard for the limits of population size or the rate of energy use, without sufficient consideration of the effects on the quality of our lives or on the quality of the environment. Although we make plans for people and their activities, the plans are usually reactions to growth and change. The formal development from planning results in a complex of problems, from pollution to ugliness.

Our societies are big. Our corporations are big. Our impacts are big. This creates discontent because people feel powerless. They feel powerless because they feel that democracy is not working. The country is run by the rich, rich corporations, rich people, rich politicians, and they secure their needs for more money first. In fact the decline in earnings, relative to costs of goods, forces people to struggle with jobs and increased conflict everywhere. Now, at the perceived "end of history" and the "victory of democracy," it is difficult to imagine a better future, at least without questioning the rich, the ideals of democracy, and the corporate good will to society.

We have economic growth; we can see the numbers. But, the growth is premised on saving costs by forcing down wages, or by reducing the number of workers in the name of efficiency, and forcing overqualified workers into service jobs. The growth promotes inequality, improvement for a few and impoverishment for most. It is the growth of a tumor, issuing a healthy glow from a fever and the false image of health. Profits go up, but public services decline for lack of funds. There is no money for schools, none for libraries or parks, little for private institutions, and little for national, state or local governments. Where did it go? Profits? Profits for individual corporations, profits for individuals? Could we find them, can we track the money? We should be able to, since the management revolution has made paper trails everywhere. Perhaps the trails are too complex.

Bigness overwhelms the ideals of cultures, which is why large nations such as China or the United states cut their forests, regardless of respectful religions or special cultural values regarding nature, regardless of the desires of local communities. The nations are willing to sell the wealth of the provinces, even though provinces want to conserve them.

Something is wrong. But who is to blame? Where is the target? People rail against liberals or conservatives, against corporations or protesters, against big government, or big corporations, big permissiveness, big violence, the lack of faith or lack of prayer, the failure of responsibility or the failure of nerve, against guns, leniency, illegitimacy, bad rock, bad lyrics, bad welfare, bad politicians, bad people, and bad police. But, the real enemy is unseen. Who can argue against democracy or corporate wealth or the general vague feeling that things have improved? Against a reasonable system? Against the bigness of the system?

The course is downward. How can it be turned? How can we diagnose the problems? How can we suggest a path to health? Charles Reich says that we need a science of social change. Ecology is science-based. Conservation is science-based. Even management is science-based. Now economics and politics needs to be science-based, instead of dominated by an old ideology and weak mythology. Reich suggests that between citizens and government is a third entity, economic government, to which has been ceded power to determine the direction and type of economy. The only knowledge we get is knowledge from the economic government, and the numbers look good. We have no other knowledge, except for weak self-knowledge and vague social knowledge, to say something is wrong.

Leopold Kohr identifies the basic conflict between man and mass, citizen and state, large and small communities, as a result of bigness. Kohr's theory of size states that the cause of most forms of social misery is bigness. We have always tried to exceed the physical and biological limits of places rather than recognize them and be guided by them. Every advanced country is now over-technologized. Past a certain point the quality of life diminishes, not improves, with each advance. Big science serves big technology, which supports and is supported by big government. And there is no science like big science, and no administration like big administration. But this enthusiasm is misdirected. Scientific advances and technological changes result in unforeseen consequences, good or bad. They cannot be controlled or legislated against before the fact. But the investment seems too big to abandon.

The planet, including human society, is threatened by the bigness of things. Although nature is big, it has evolved slowly; human size is new and sudden. A snail's pace is good enough for nature most of the time, but with our brief life span, we argue that we need quick changes. The technological advances have not been paid for yet, although the cost in pain and death is incalculable. We will be paying for hundreds of years for those past advances. The economic style is too great, and reckless for ecological systems to absorb its impacts. The scale of things is an independent problem, that can ruin the best intentions of policy. A bigger system to control systems that are now too big might be a mistake.

But some solutions are even bigger. Buckminster Fuller, Alvin Toffler, and the Club of Rome favor a supertechnocracy. Science fiction visions predominate: Gerard O'Neill's orbiting cylinders or Simon Calder's floating domes. R.A. Smith, in "Unibutz," claims that we could achieve a pantheistic-humanistic-cosmic awareness, in achieving technology without materialism, plenty without selfishness, and community without tribalism. His Unibutz is a global goal for leaving the earth and reaching the stars. The voyage would help shape new world structures and give a purpose to humanity. But to what end? Bigness and wealth elsewhere?

Perhaps big science and big technology have too much momentum. Theodore Roszak acknowledges its schizoid attraction and repulsion, with the twin promises of glorious accomplishment and hideous death. Who could escape being torn between yes and no, if even our end would shine with radioactive, Promethean grandeur? Our image of big science—the scientist as tragic hero, isolated in chaotic nature, but strong in his proud individuality, perhaps driven to research by hubris and madness—is a barrier to any new vision, especially a small vision.

3.1.1.2. The Lure of Speed

To achieve even greater efficiency, we have increased the speed of our activities, converting materials and cultures into new designs without consideration of the meaning of, or need for, efficiency. The speed of our economy is too great for many cultures to adjust to; and the thoughtless transformation of cultures may result in great, irreversible mistakes. The speed of our conversion of wild habitats to domesticated lands is too great for many species to adapt.

Alvin Toffler foretells a dramatic redistribution of power, from slow countries to fast. Speed is the critical factor for Toffler, who states that, historically, power has shifted "from the slow to the fast," whether speaking of "species or nations." Certainly, being faster to the industrial market has advantages for many international corporations. But, this kind of speed is not applicable to species. Slow species have survived as well as fast, either adaptively or neurally; many fast dinosaurs perished before their slower mammalian contemporaries.

Toffler notes that the industrial revolution stepped up the metabolism of economies, but does not seem to make any distinction between good or bad metabolism—fever as well as excitement speeds up a metabolism. Truly, we are speeding up our use of resources without knowing where they are coming from or going to. Modern economies, embracing the idea that "nature is capital," draw on the accumulated "capital" of ecosystems for production. By ignoring the real cost of the capital, as well as the costs of natural services, such as nutrient recycling, soil building, and atmospheric renewal, these economics create a temporary wealth—similar to the healthy flush of a fever, perhaps—and a long-term imbalance. When an economy falls out of balance with its local environment, massive disruption often results; industrial economies have only avoided disruption by trading advantageously with other economies, by using fossil fuels, and by promoting general institutional inequality.

Continuing his paean to speed for its own sake, Toffler states that fast economies generate wealth and power faster than slow ones. But, what kind of wealth? Financial or cultural, agricultural or symbolic? And, what kind of power? Mechanical or organic, political or personal? Industrial economic wealth is merely a small part of the wealth of the earth and humanity, most of which has little value to that economy.

Toffler describes an acceleration effect that makes each unit of time saved more valuable than the last, creating a positive feedback loop— inadvertently identifying the archetypal problem of modern economics: Runaway positive feedback loops leading to catastrophe. The fast economy he describes seems to depend on fleets of hypersonic jets racing around the world with the elite and their tonnage of possessions. Telecommunications, transportation, and tourism will accelerate, blithely unaware of their impacts on family structures, biogeochemical cycles, including the ozone layer, and wilderness. Have we learned anything?

Toffler sees revolutionary consequences in new management methods, but not the negative effects. Managerial decisions regarding resources are made often on short-term economic grounds and lead to material shortages and human and environmental degradation. Newer methods seem only to offer a higher degree of impersonality.

The new wealth creation system, Toffler claims, holds the possibility of a better future for the vast populations of poor, if their leaders anticipate changes. The new system for making wealth consists of an expanding global network of markets, monetary, and production centers in instant communication with increasing flows of data and information, but not necessarily wisdom or understanding. He argues that the availability of this information flow gives more power to consumers, voters, workers, and small businessmen, taking it away from a centralized few. The potential is there, but Toffler does not go far enough to envision alternate economies and communities. The power still operates under the old assumptions and divisions in his speedy synthesis.

Of course, Toffler is right to recognize the problems of nonindustrial countries, many of whom depend on cheap labor or strategic military location for foreign investment. But, where does that "investment" go? To the local poor or to remote, rich politicians? Wealth could be distributed fairly, depending on many factors, such as synergy, generosity, reciprocity, and cooperation, but it is not. The gaps are growing. Toffler acknowledges that they will keep growing. But, we can redistribute wealth without industrializing. We should try to achieve economic justice before accelerating to new glories.

Toffler concludes that a great technological and cultural wall will separate the slow from the fast, making problems for joint ventures. But, what are the products of these ventures? The debris of advertising fads, such as mink toilet seats, or the tools of real needs, such as evaporative water purifiers? Toffler foresees the emergence of an electronic neural system for a global economy, without which any nation will be doomed to backwardness. What kind of backwardness? Lack of fast things? Lack of professional enslavement? Lack of art, play, or culture? Lack of food, tradition, freedom, or happiness? Perhaps these have already been stolen.

3.1.1.3. Grand Thefts

Despite our bigness and speed, most of the world's human population goes hungry; even fewer are fulfilled as actualized human beings. The utilitarian aim of greatest good for greatest number has been vulgarized to mean the greatest number of goods for those who can afford them. In our attempts to manufacture the good life for those who can afford to buy it, we have deprived everybody of clean air and water, quiet nights, darkness, open spaces, and other indefinable qualities. Soils are destroyed, wildlife is killed. We devour nature to assuage our disease; we try to fill our emptiness with goods. We can only gain past a certain point before our gain causes the loss of something else that we need to be healthy.

Modern technological society ravishes nature and mutilates humanity with the products of its materialism. Industrialization has distorted people's lives and cheated them of bread and justice—during the time it takes to read this manuscript, over 10,000 people will die as a result of nutritional deficiency. The cost of the paper and processing for this manuscript could feed twenty people for one day in many poor countries. Reluctantly, I rationalize that an idea is a better and more effective solution to global problems of production and self-reliance than a one-time aid or just ignorance, but the rationality does not sit well in the dreaming mind. Oppression darkens the mind and narrows the spirit. In a mass consumption society, people impoverish themselves spiritually while impoverishing others materially. This is theft. As with the Christian ten commandments, most loss can be reduced to theft, whether of a life, mate or name. Most of our modern problems can be considered the consequences of forms of theft, such as of life, common sense, and choice.

The failures in our character or group or national character, can be seen to be responsible for the problems identified by Konrad Lorenz as the seven deadly sins of civilized humanity, from destruction of nature to the loss of civility. These failures can be described as a series of failures, from perception to intelligence, imagination, integrity, will, and charity.

Of course these failures occur in endless combinations. Maybe the failure of our modern civilization is a failure of imagination compounded with a failure of nerve. We cannot imagine an alternative to war, and we cannot act beyond emotion. We cannot imagine beauty in the old and messy nature, and we are afraid to try to do without luxuries or to try to sacrifice anything to try to change the momentum of industrial civilization.

Our false models and ideas, combined with our failures of imagination and charity, for instance, may doom us create only flatscapes and noplaces.

3.1.1.3.1. Theft Of Life

The relationships of humans and animals have changed drastically. The increase of humans and the destruction of animals have unbalanced the relationships. The destruction of habitats is accelerating. Up to a point, niches can be enlarged or increased, but with so many humans, it becomes a case of supplanting other species entirely.

The loss of cropland and soils, and the disappearance of genetic stocks essential for crop breeding, may eventually cause the collapse of the biological basis of our food supply. Huge quantities of fertile soil are washed away each year as a result of deforestation and poor land management. It has been estimated that Colombia losses 400 million tons of soil per year and Ethiopia 1 billion tons per year. In The U.S., in spite of its soil conservation service, almost 5 billion tons are lost annually; these losses over the past decade may have cut the potential to grow food by ten to twenty-five percent. This increases our dependence on oil imports, since $1.2 billion of fertilizer was needed to replace the nutrients lost through soil erosion in the U.S. in 1978, for instance. Increasing amounts of fuel equivalents are used every year to offset erosion.

In developing countries, hundreds of millions of rural people strip the trees and shrubs around their homes for fuel. In Gambia, it takes 360 woman-days per year per household to gather wood. Even when firewood is available for sale, it may be beyond the budgets of the poor. In South Korea, it can cost up to fifteen percent of a household budget; in the Sahel, it can consume twenty five percent of a budget. When wood is scarce, the poor are forced to use millions of tons of crop waste and dung, which should be used to regenerate soils. The poor are destroying their means of survival, in order to survive now. Then, the soils on which they live become even more vulnerable to erosion. Deforestation causes siltation, which cuts the useful lifetimes of reservoirs in half, decreasing hydroelectric and water potential. Deforestation causes floods, which devastate settlements and crops, incurring even greater replacement costs. Lack of soil and forest conservation contributes to rising energy costs and the financial costs of providing essential goods and services.

Tropical rain forests, genetically the richest land environments on earth, are being felled and burned at the rate of 27 million acres a year—50 acres a minute (in 1972). At that rate there will be no more tropical forests left by the year 2050. The lowland forests of Malaysia, Indonesia and the Philippines are being ravaged the fastest. The rape of the tropics is endangering hundreds of thousands of species. Over 25,000 known plant species and over 1,000 species of mammals, birds, reptiles, amphibians, and fish are threatened with extinction, as their habitats are eliminated. Many of these species may be economically and culturally important; others are different and unique. Unknown species, perhaps half of all species on the planet, are not valued at all because of our ignorance.

Irrigating marginal lands causes some loss from salinization. Salinization is responsible for the degradation of over thirty percent of the arable land in Iran. New energy uses further degrade a portion of land in more industrialized nations. At the present rate of cropland impoverishment, one-third of the world's cropland may disappear by the year 2000. Deserts are expanding at the rate of 23,000 square miles per year (for contrast, the nation of Belgium is 12,000 square miles in area). Eight million square miles more are seriously affected. Desert conditions jeopardize the survival of almost eighty million people now; 550 million more could be threatened. Desertification by overuse of land afflicts almost seven percent of the earth's surface.

Drought is blamed for arid land problems in Australia, but drought is inevitable. Aridity is determined by air currents and topography. Rainless episodes are normal in regions of erratic rainfall. Sensitive pastures may be overgrazed in good years, but the damage is not apparent until a drought. Sheep during the Australian Gascoyne droughts starve; there is enough water for thirst, but not enough for plant growth for food. Survey teams in the Gascoyne region recommended the most ecological strategy was to temporarily abandon the affected sections. The recommendation also was made to graze only half as many sheep. These recommendations were ignored by grazers, who needed the income immediately.

Even the sea is not invulnerable to human impacts. The most productive areas, which are close to shore, are being polluted, overfished and destroyed. Estuaries and coastal wetlands are being destroyed. Overfishing has deprived people of millions of tons of seafood. Overfishing is destroying the fisheries' support systems. In the U.S., losses to fisheries from shore "improvement" and degradation cost $86 million a year.

The relationship of humans to humans has also changed in the past thousands of years. Human lives are stolen, not only through war, but large-scale murder, as well as through diminution of human value and denial of resources.

3.1.1.3.2. Theft Of Common Sense (& Humility)
The industrial machine is out of balance and defective, but we have been making it run faster to process more material. Western civilization still selects cultural changes in terms of wealth and power. This is destructive to most individuals, as well as maladaptive for a society under ecological restraints. Over the past two centuries, industrialized countries used great quantities of raw materials to create luxuries. Then they disseminated the ideas of wealth, equality, opportunity, and indulgence to many countries without industrial opportunities. There can be no peaceful future for civilization when such disparities, and popular knowledge of them, exist. The cult of competitive consumption seems to be the universal solvent of the modern world. Everyone wants what some have. The industrialization of Asian nations is seen as a solution to shortages of manufactured consumer goods, although imbalance and pollution are down-played. Even worse is the unavoidable waste. Probably over fifty percent of the productive effort in United States goes into making things which contribute nothing to the material standard of living.

Consumer desires must be satisfied promptly or despondence results. Although life, art and science depend on organization, the civilized consumer feels that organization implies stasis and death. Believing the line of succession ends with his claiming of the inheritance, the industrial consumer feels no obligation to provide for the next generation. The consumer justifies narcissism as a preliminary condition in the search for consciousness, truth and morality. Consumers become embroiled in their own causes, locked in the idealism of adolescence, and repudiating the lessons of history. Hegel noted that the greatest lesson of history is that nobody ever learns the lessons of history. And, Cicero wrote that those who do not know the past are like children. Americans and the people of some other nations seem intent on validating these insights.

This repudiation of history creates a historical amnesia. Consumers retain little more than a dim notion of the past. Universities report a lack of interest in events that occurred before the current year's athletic season. The sense of time falls in upon itself, collapsing like an accordion into the present. Knowing nothing of history and expecting nothing of the future, people cannot escape the fearful isolation of the present. They join together in a melancholy herd, clutching at everything, but holding nothing fast.

Even good eating habits can be forgotten in quickly. The eating habits of many people in industrialized countries are unrelated to nutrition. There is no instinct for good nutrition; it is a kind of empirical learning from trial and error, although young children are often good choosers. As South Africa became colonized, for instance, the natural diet of Zulus changed from millet to corn. Malnutrition also resulted from this overdependence. Yet after a generation, tribesmen thought they had always eaten corn, even though their ancestors were stronger and healthier than they are now.

Without the depth of history, experience is shallow and short, and intelligence is thin. Technology has reduced the globe to a single, closed system, which humans can share according to their financial powers. Our direct experience of the world has become shallow, in spite of faster travel. Travel used to broaden the mind, but now it narrows it. We travel in sealed corridors like boxed goods, comforted by homogenized foods and the English language. Our cultural adaptations to the pressure of homogenization throttles individuals and groups.

Our 'overweening bumptiousness,' which the Greeks called hubris, lets us behave as though we were too privileged to be members of the earth's ecological community. We name things and dismiss them. We draw lines around them to separate ourselves. We build our own environment, mathematical and sanitary.

John Fowles observes that most of us remain firmly medieval and distanced from what we cannot own or fully control. We assess most of nature as what is not clearly *for* us must be *against* us. We cannot accept indifference, the nonhumanity of nature. We seem incapable of realizing that the destruction of the Amazon, much as we deplore it in its remoteness, is our responsibility; we consume the materials from tropical forests. Our growing emotional and intellectual detachment is the greatest threat to nature. Heroic narcissism has replaced nature with humanity; nature no longer provides the mirror to reflect human aspirations, a televisions screen does. Narcissism is a threat to nature and humanity.

Unbridled consumption in a laissez-faire economy for fifty more years will probably carry humanity beyond a point of no return, leaving industrial society with insufficient resources to maintain itself and insufficient flexibility to retract. If lack of planning permits rapid increase in population and consumption, then our options may be diminished to two: A nasty, overplanned existence, or a squalid collapse. But the future seems as unreal as the past. The historical origin of the ecological crisis is in the failure of people to use their intelligence to anticipate the long-range consequences of their activities; this is a perennial human problem.

3.1.1.3.3. Theft Of Choice

We follow false models. Our civilization is dominated by all of the ideas of the industrial revolution, which form the outlines of a tragic world view: The primacy of humanity, anthropocentrism, the supercesion of the individual, the achievement of happiness through the accumulation of things, the perception of the incompetence of nature, the requirement of humans to control the environment, the expansion of the frontiers of technology and opportunity, and the identification of solutions for every problem. But, these ideas are false; programs that assume them for conditions are doomed to fail eventually, and perhaps destroy the kinds of environment that humanity needs. Cities and factories reproduce themselves exponentially through a industrial genetic code. They are the addicts' dream of affluence using science to create a dream pill from the dust of a bare earth. But, simply removing the causes of unhappiness may not produce happiness.

Industrial culture has been distorted by the modern emphasis on the scientific method, with its devaluation of philosophical concepts and emotional values. Rational values are exaggerated and spiritual events are ignored or suppressed. Most scientific studies seem specialized or irrelevant. Some parts of problems are identified and analyzed, but the conclusions are trivial and weak. The ethical impulse to solve the problems is even weaker. We choose to let them slide, or to label them as nonproblems.

The unified direction and responsibility of science is nonexistent; there is not even a concept of what is good. Control has been the goal of science and technology. Scientists are obsessed with the treatment of weeds and vermin. The industrial cosmology has put humanity at war with the planet, which always misbehaves; weeds and vermin threaten us. Our path has been worn so deep, it would be difficult to leave it. Stability has been raised to sacred scientific state. The interlocking of technologies and institutions makes it impossible to reform policy in any one part, separate from the other parts. This interlocking also makes people powerless to choose any alternative. The systems managers are preparing to operate the planet, according to systems principles applied in routine methods.

The safety of the environment is too important to be left to scientists, even to ecologists. The crude history of science shows that scientists fall willingly under the dominion of money and power, like Christianity, communism, and most movements. The movements of narrow science and industrial culture reduce the possibilities of choices.

3.1.2. *Being Noplace*

Most humans are dominated by the idea of competition: It is us against the environment, us against others, and us against ourselves. Everywhere antagonism pervades our society—college presidents lead their groups into battle for bestness; political parties blame each other for one hundred years of bad planning; we compete for jobs, status, dollars, and mates. Conventional violence—war, murder—captures news headlines, but structural violence—mining, schooling, hospitals—pervades society unacknowledged. Most relationships seem violent in a competitive society.

Many human societies advanced by fighting and expanding. Fighting is a common form of human behavior, occurring in children, as a ritual limited by pain, and in adults, as failure of communication or understanding. The growth of the brain and its capacity for abstract thought seems to have bypassed the ritualization of social conflicts common to other mammals. Human fighting is not as formalized; and, this is what permits humans to slaughter one another. Most tribes followed two standards of morality: One for insiders and one for outsiders. Most aggression was directed outwards and the losers were often exterminated.

Many tribal groups federated into national units. A nation, by definition, has a single, central government representing people who occupy contingent lands and are consciousness of a common identity. The professional ruling class is divorced from kinship bonds; structure is stratified and internally diversified. Almost the whole land surface of the globe is divided into centrally governed states. Human affairs are managed within the framework of these autonomous units. Decisions are made on narrow political and economic grounds, rather than on environmentally sound principles.

3.1.2.1. Trapped in Images: Myths & Peace

The two great myths, progress and nationalism, arose with the industrial cosmology. The author Aldous Huxley described progress as the theory that one can get something for nothing, that the gain in one field is not even paid for in another. Progress assumed that all consequences could be foreseen, and that the ideal ends in the future justified the most abominable means: Robbery, murder, or cheating. Progress was considered good, and primitive groups only obstructed the civilized nations' march toward paradise.

According to Huxley, nationalism was the theory that the state was the only true god; all others, especially other states, were false. Conflicts over prestige or power were crusades for progress and nationalism, for the Good and the True. Lord Acton observed that nationalism aims solely at making a nation, the abstract idea of the political state, and not at liberty or prosperity for people. He also predicted that the result would be moral and material ruin. Nations are basically exploitative of other nations and smaller cultural groups, which may not be considered nations because they lack a permanent military: The United States concentrates on Latin America; Europe on Africa; Japan over Southeast Asia; Russia over Eastern Europe; and China over Tibet. The reasons for this continued behavior include: The rapaciousness of society; the acceptance of war; and the economic advantages of large-scale operations. The cultures of industrial nations are based on unethical accumulations of materials. Inequality is maintained by power, not persuasion, and also by the assumption that solutions are extrinsic and external and have to be found by spreading out rather than intensifying efforts to find solutions at home.

National powers work, through progress, to maintain the good life for their leaders and followers. Major powers deal formally and informally with each other on survival problems, such as nuclear crisis or ecological crisis, but compete on other issues. Politicians should think about the hunger and squalor of billions of human beings and the destruction of habitats with billions of ambihuman lives, before dedicating themselves to their personal fortunes and the missiles needed to protect them. The system of management, like all paradigms, slowly becomes a means of excluding experience foreign to the notation. It becomes a form of authoritarian control. Programs become responsible for dimensions of misery beyond any considered.

Our politics will improve when we realize that we are the biggest ecological problem. Ecology studies the details of the binding of all beings in the earth into a whole. Ecology, like god, is not to be mocked, as Gregory Bateson said. There are no such things as little sins. Even human plans are now part of the ecology. Bateson warns that all ad hoc measures leave the deeper causes of problems uncorrected, or worse, permit the causes to be compounded.

3.1.2.1.1. War

Those who believe in the theology of nationalism are committed to fight. War is a way of conquering or creating new nations. It is a nonprogressive way of promoting progress, especially innovations in certain kinds of technology. What have been the causes of war? We know some causes from hunting societies or chiefdoms: Insults or broken agreements, or competition for resources, from food to land. Nations created new causes: To unify other cultures and resources, or to deunify the enemy and to unify the aggressor.

The war ethos has been expanded and reduced to absurdity. War has become so big that there can be no victories or victors, and possibly no survivors. The only remaining purpose can be the total destruction of the combatants, as nations, as well as natural habitats. Sadly, the only people who do *not* know this, or admit it, are those in decision-making positions, who are compelled to prepare for what they subconsciously know would be a terrible disaster. Their power has trapped them in the momentum of their nation, afraid to be caught in any criticism. Yet, they direct the money, skill, and knowledge of their citizens into projects that lead to misery, servitude and hideous death, and not to life, liberty, and happiness.

The intellectual rationalization for the continual preparation for war is the old Roman adage: "If you want peace, prepare for war." This adage has been so completely taken into the modern heart that most of the larger nations have spent over half of every century in war, according to Pitirim Sorokin. Preparation for war has always led to war. There seems to be no reason that the present arms race will lead anywhere else, even in times of relative peace.

Power politics makes problems that cannot be solved except by war. Questions about defining the best nation or the best religion lead to organized slaughter as the answer. "War is not merely a political act," said Clausewitz, "but also a political instrument, a continuation of political relationships ..." As long as human institutions were large and brittle, war was an effective way of disassembling them. This form of social renewal, however, was very expensive. Marx may have been partly right when he said war was necessary under certain conditions, as a last resort. But, those conditions no longer exist. His observation may have been true from the 1830s to the 1940s, but it has been rendered false by modern weaponry. Nuclear war can destroy the parts as well as the connections, cultures and ecosystems, as well as senile political structures. It may not be able to discriminate.

War has become an integral part of the modern economy. The war industry, as measured by expenditures, is over $200 billion a year. Defense problems in the U.S. are perceived as problems of hard technology; psychological and social research is considered irrelevant. Security is analyzed in terms of Newtonian physics: Blocks, actions and vacuums.

There are other dangers and benefits to war. There are, of course, benefits to war. It equalizes opportunities for some. It foments change, any kind of change. It gives a sense of the past, if the past needs to be measured by conflicts and victories. It leads to more advanced technologies and to faster social responses.

The economic rationalization is even more crucial for nations. Armament piling has become a vital part of the U.S., Russian, Iranian, and North Korean economies, among others. The recovery from the U.S. depressions in the 1930s was not complete until the rearmament surge to combat the Axis powers. The Korean and Vietnamese wars also spurred the U.S. economy. U.S. prosperity has its basis in the preparation for death. The fear of Russian competition (in 1970) ensured government expenditures of billions of dollars for 'deterrence.' The vested interests in this system are almost insurmountable. The concrete companies and bomb makers put up their own puppet politicians to guarantee their part of the spoils.

The dangers include the fact that war has been regarded as meaningful struggles against people who would kill or enslave others. The 'side-effects' are rarely considered or counted. War also leads to an overweening respect for large government. The cost of war excludes the social distribution of wealth. Consider the plight of the Russian or the U.S. poor. The situation is much worse in disadvantaged countries. The war industry, as measured by expenditures, is on the order of $200 billion a year. Kenneth Boulding says that it is an unrecognized paradox that the cost of maintaining the war industry is greater than any possible damage that could be inflicted by an enemy. Furthermore, war intensifies the depletion of resources; therefore, it is counterproductive to fight to steal another nation's resources. For centuries, warfare has resulted in incredible wastes of resources. The latest multinational conflict, which began in 1914, and has been hot and cold during this time, has been the most wasteful.

The winners of a large war will need to be pitiless, according to Kurt Vonnegut, for pity will be suicidal when resources become exhausted. War is considered to breed strength and nobility. The values of strength, nobility and bravery were suitable for certain hunting societies. But, even the weak and lazy can get food and shelter in this world without plundering. Humanity is clever at taking more than what it needs from nature. Evolution is thought of as the survival of the fitter. Nobility and cleverness may not best fit a species for survival, if they require competition and war. Perhaps mankind will perish for its nobility— but is it noble to be extinct?

Perhaps it is the fate of humanity to die a radiant death, taking most other living beings with them. Who could resist the glory of complete annihilation? Perhaps that is the fate of humanity. But what is fate? The hand of God, or the idea of Tolstoy—that historical events are determined by the summation of innumerable decisions by the anonymous masses of humans that add up to a tendency. In particular terms, the small decisions to buy automobiles, poison coyotes, plant trees, or live simply, add up to fate. Fate certainly concentrates power in corporate, military and political hands, which still belong to human beings, who cannot find consolation in mega-deaths.

Nuclear war is unthinkable. Limited disarmament is unworkable. Human and environmental degradation are unconscionable. Is anything thinkable or workable without being hopelessly utopian? The problems of aggression, nationalism, war, and peace are problems of human nature, symbol, culture, politics, ecology, and size. They are not simple problems and not easily solved. They are interrelated as human groups are technologically, economically, ecologically, and politically interdependent. Human interactions have been dominated by symbols. The most powerful set, embodied as nationalism, has a direct relationship to war. Large-scale war can create utopias, places that are nowhere, on earth.

The United Nations made war illegal; one nation cannot attack another. But, that has not made a lot of difference. Cultures try to destroy other cultures. Big nations try to absorb or punish small nations. The aggression by the United States against Iraq is a crime against humanity according to the Nuremberg principles established by the victors of the conflict in the 1940s. Those principles were enforced by hangings at that time. The war was sold to the U.S. Congress with lies and propaganda, but even exposure was not enough to reverse the decision to stay the 'course.' Perhaps U.S. citizens are naïve about their role in making peace or securing its need for oil. Perhaps that naiveté has to do with the luck of never having been invaded, flattened or destroyed by another nation.

3.1.2.1.2. Peace

What is peace? The absence of war? The absence of large-scale organized conflict? Peace means, in the original Latin, 'confirming to an agreement.' An agreement is part of a conversation between two or more people, which usually results from an effort to avoid some kind of conflict. Peace requires conversation, the exchange of opinions, wants, and problems within the context of meeting.

Robert Kaplan states that tragedy "requires a sense of history," implying that war is tragedy and peace is timeless. He suggests that peace leads to a 'preoccupation with presentness, the loss of the past and a consequent disregard of the future.' He continues arguing that because peace is pleasurable and pleasure is a momentary satisfaction, therefore pleasure, as well as peace, is inseparable from convenience, a temporary, timeless, uniform satisfaction. Should we not outlaw lovemaking and eating as well? Or all momentary pleasures? These are strange arguments: People in war time damning peace, as if it were the worst threat to the planet. Have we ever had such a dull time of peace? And, did the people who experienced the satisfaction of peace complain and lust for the tragedy of war?

Why is universal peace something to be feared? Would it lead to great dullness or great evil? Tragedy does require history, for it cannot operate without it. Peace requires history and tragedy as well.

Is peace without a sense of history? Peace seems unchanging and historyless, with fewer dramatic changes, but people seem to want to sacrifice anything for it. The common, written history of reference is the usually the history of the victor, with great leaders and great changes, but that itself is a small limited mythical history. Real history includes all kinds of changes, from geological ones to the development of cities. Perhaps people mistake peace for a static vacuum.

How did we get this way? Bad genes? Bad images? People have images of nature as violent or of humans as naturally violent, which contrast with other images of nature as cooperative and humans as peaceful. Some of the logics of our cultures promote a dualism where either one is true, but not both. The logic of opposites creates a false dilemma, a fallacy, where there has to be a winner and a loser, where opponents have to fight over one thing or one way. This logic has the advantage of simplifying complex situations into one adversarial relationship, which can only be decided by conflict. But switching logics, or expanding the logic that dominates global conversations, can allow seemingly contradictory ideas to exist together, such as simplicity and wealth, or peace and artistic creation.

Overwhelming desires? In Buddhism in general, desire is an obstacle to peace. Suffering can be reduced only by following certain practices. There is conflict between desires and suffering, between needs and wants, between ideals and actions, between expectations and real events. Neither desire nor suffering can be totally eliminated by behavior patterns, which will always cause more desire or suffering by the very nature of human wants. Peace for Baruch de Spinoza is a virtue that springs from force of character. It exists with other virtues, such as self-control or self-regulation. Some desire and suffering can be reduced through education. Art educates and liberates the individuals of society in a gradual and peaceful process. In spite of the cultural forces dominant at any moment, an individual has the potential to determine a different course. Art can reduce or rechannel desire, and thus reduce the suffering that comes from maximized desires.

Runaway scale? As long as there is no limit on violence, wars can continue to engage plants, animals, even entire habitats and civilian populations, as often as soldiers and weapons. Conflict has to be descaled. It could be limited to a small number of representatives from each culture or group, so that the scale of peace would always be larger than the scale of conflict.

So, if we made an agreement to limit conflict to certain arenas and to certain sizes, that could be a form of peace. Peace could be a process of resolving conflicts between cultures. Some conflict is unavoidable, due to human nature and ideas of honor and possession, due to different images and beliefs about how the universe operates, and due to misunderstandings and accidents. Much conflict is related to fear, fear of loss, loss of security, loss of effort and opportunity. However, as long as violence is an accepted response to conflict, events will continue to be violent. The proper response has to be self-restraint and mutual constraint. Conflict has to be delinked with violence. All conflict is not bad. It is a challenge and stimulant to an individual and a culture. But, it has to be responded to in such as way that people think and develop, and create a considered response. Limited peace would to match the limits of conflict.

Peace is also a freedom from public disorder. Disorder is also unavoidable. But, it has to be kept below a cultural ceiling where it could destabilize the culture. That requires the agreement of a culture and inhabitants. Like conflict, disorder is a challenge that requires a creative response, rather than attempt to destroy disorder.

Is peace just nonviolence? Nonviolence is a psychological choice by an individual; it may be a cultural rule if enough people practice it. It is a way to resolve conflict through nonresistant awareness, which includes witnessing and understanding, conversation and mutual constraint. Conversation, remember, involves listening as well as speaking, in responding to another person. It exposes common themes and builds trust. So, nonviolence, or peace, is not a passive, head-hanging response to dominance. Bearing witness to injustice, that is to the results of dominance and violence, of inequity and prejudice, is a form of action that requires courage and persistence, and results in knowledge and the spread of awareness. The Quaker tradition of bearing witness is a turning towards events, and participation in those events. There are degrees of witnessing from observing to communicating, such as writing letters or making telephone calls, to acting nonviolently and negotiating for changes in events. Nonviolent action can neutralize aggression by refusing to offer positive feedback, that is, more violence in response.

Is peace just cooperation? Conflict is the result of differences in perception or needs. But differences can be resolved without conflict, without automatically assuming an adversarial stance. The subject of the conflict can be recognized as a common challenge—more than a problem that requires a solution, or a final, one-time fix, challenges are opportunities for change and development.

When conflict was thought to be the primary mode of operation in nature, it justified human conflict and violence, not just to fellow humans, but to animals and all of nature. With the scientific understanding that cooperation is much more prevalent in building guilds among animals and stable habitats, the operation of nature becomes more understandable. Cooperation builds partnerships and mutually beneficial relationships between individuals, species, and communities. Cooperation is more than the avoidance of conflict. It is the recognition of emergent benefits. It is finding security in sharing, or in new ways of being, rather than by eliminating competitors.

Cooperation is one aspect of peace. Ways of cooperating include communicating needs and negotiating for redistribution of resources. As it goes along it may be necessary to mediate two desires, or to arbitrate a decision. But, this is done though sharing in a conversation, by building consensus through common themes and conversation, by identifying shared goals and needs, in an atmosphere of trust, where feelings and ideas are shared, taking as much time as necessary, with as little outside agenda as possible, to be committed to the benefit of both groups in the process.

Is peace just harmony? Peace can also mean harmony or concord. But, harmony can contain discordant notes and themes and weave them into a rhythm. Harmonious action can change patterns of domination of cultures and individuals. Harmony has to include conversation with other species, and trust in nature as a generally beneficent system, our mother system and home. Although humans cooperate with each other, cooperation with other species is best characterized by allowing them opportunities for living, despite a measured exploitation.

Peace is unknown territory, *terra incognita.* We have not been there often, except by accident. The path to conflict is easier to define—it results from fear, from the desire to defend and to be violent. The new path requires as much bravery as violence does; it requires patience, it requires effort to find out why and what is needed.

There are, of course, numerous ways to achieve peace, such as requiring everyone to belong to the same religion or social class. This tactic has been tried; it did not work. Perhaps we could limit conflict in new ways, such as personal conflict between leaders or their champions. Or, assign all power to an Association of Nations. Or change the focus of war to nonhuman astronomical events, such as meteors.

3.1.2.2. Achieving Placelessness

By accident, we have created the opposite of peace, and the opposite of place. De Tocqueville had previously identified a trend of human uniformity in the 1830s, when he wrote that variety was disappearing, and the same ways of acting, feeling and thinking were copied around the world. A number of scholars, including Christian Norberg-Schulz and Edward Relph, have noticed that we are creating flatscapes, devoid of depth and providing only one mediocre possibility, a chrome-plated chaos. Placelessness begins with adoption of an attitude, an abstract, geometric view. With an inauthentic technique, places can be treated as interchangeable and unremarkable. Nothing is significant there. Cutting historical roots and eroding symbols contribute to an awful placelessness, an alienation to place, an inability, finally, to have a home and to live there. This becomes the fate of millions, and it increases.

After shaping ourselves to technology and necessity, we have lost the knowledge of how and for whom to care. We have learned not to care, to be dispassionate (uncaring), unattached (placeless), and objective (uninvolved). Other animals have used languages and tools, so it is not those things alone that account for the lost knowledge. We 'not-care' because we are confused. Our confusion results from being out of place and not having an identity.

3.1.2.2.1. The Loss of Identity (The Ninth Loss)

People use to identify with place. In fact, many names are merely place names given to people. Other names come from what kind of work people did, or who their parents were. Thankfully, we have stopped naming people for place and jobs. What kind of names would we expect to use now? Frank Paperpusher? Berty Mallpack? Joe Truckaimer? Mary Asphaltlot?

The gain of a global monoplace has led to a loss of identity. The mass production of homes, as well as of music and art, has led to fewer creations. The commonness of culture, at a low common denominator, has led to fewer identities to choose from.

3.1.2.2.2. The Scope of Failures

Acceptance of limits is not a kind of failure. One speaks as best one can. Awareness of inadequacy, as of ignorance, is a positive accomplishment. True failure is indifference to inadequacy. The failures in our character or group or national character, can be seen to be responsible for the problems identified by Konrad Lorenz as the seven deadly sins of civilized humanity, from destruction of nature to the loss of civility. These failures can be described as a series.

3.1.2.2.2.1. Failure Of Perception

Perception is defined generally as the mental ability to grasp objects or qualities through the senses, resulting in understanding or knowledge. The number of things that we grasp, however, is limited by our senses. We see very little of the spectrum. We are limited by our size, our positions, and by our life span. These limits make it difficult for us to perceive very slow or large changes or to anticipate them.

In a market economic system, we have the perception of progress without the recognition of the costs of bigger cities and bigger farms. We have a perception that efficiency is related to maximum productivity, without regard to human health and happiness. Wealth and power are treated as primary needs, more than health or self-actualization. Failure of perception results in the inability to see the long-term results of economic actions that maximize profits through direct competitiveness. The same inability shows itself in the political frame where political cycles are limited to two, four, or six years.

Much of the environmental crisis is caused by the failure to understand patterns of cycling. This is especially true with industrial agriculture, which tends to break up cycles. Like most human endeavors, agriculture ignores cycles, as well as physiology, metabolism, and diversity. It fails to accommodate the reciprocity of the living environment—life does not adapt to a passive prior environment, it produces and modifies its surroundings.

We seem unconscious of the failure of cities and walls to lock out wild nature. This failure had mythic origins. After the Sumerian king Gilgamesh killed the great spirit of the forest, Humbaba, he became possessed with the fear of death and tried to lock out nature with the great wall of Uruk. It did not work, but it contributed to our historical domination of nature and others. Our fear of the wild, undomesticated, uncontrolled profusion of nature has led to our war on forests, rivers and winds.

We fail to see the incredible interdependence of humanity and nature, of diversity and success. We do not seem to be able to see others as feeling human beings, or animals and plants as feeling beings, or rocks as experiencing beings. Domination and conflict reflect our inability to perceive other human beings as equals. This failure reflects our inability to perceive the complex operation of nature, from the links of fungi to large-scale developments. The failure of perception also leads to the disappearance of strong passions and emotions, as we avoid unpleasant realities and pamper ourselves with visions of separateness and superiority.

3.1.2.2.2.2. Failure Of Intelligence

Intelligence is the ability to learn from experience, retain that knowledge, and use it effectively in new situations, with common sense. Although we learn limited lessons from the short-term experiences of some people, we sometimes apply them unthinkingly to other situations. For example, British and U.S. farmers tended to take their success with monocropping to every other ecosystem, from tropical forests to deserts, with disturbingly bad effects. Explorers and scientists failed to learn from earlier archaic cultures that had adapted to places over thousands of years, again with disastrous results. Intelligence by itself is obviously not enough to guarantee our success. We even have special behavioral sinks for intelligence when it is too active but unattached—habits, computer games, and television, for instance.

Beliefs, also, such as the belief in progress and technical improvement, can lead to the failure of intelligence. Therefore, intelligence has to analyze beliefs and myths, to make sure that they reflect ecological realities.

Perhaps the problem lies in a broader kind of intelligence, social intelligence. Ants have a swarm intelligence, where the swarm is smarter than the dumb individual, that is, greater intelligence emerges. Small insects are not smart. But, in hives, their social intelligence allows for smarter behavior and better designs. Humans on the other hand, are smart as individuals, but seem to demonstrate the opposite trend. Humans have mob intelligence, where the mob is dumber than the smart individual, that is, intelligence submerges. Rather than becoming smarter in mobs, people act less smart. Nations may be cross-cultural mobs. What is the solution? Being led by a few smart individuals? Democracy? Is there a way to replace cultures with other alliances, such as those for bioregions, totems, or economic systems? Could common sense save us from the failure of intelligence?

Common sense is the combination of intelligence with feeling and everyday experience. Common sense flows from the way people live in place and expresses what they want. Intelligence also grows out of living, but it is abstracted. Common sense is part of a conversation that results in cooperative behavior in the face of environmental and social challenges. Of course it is not perfectly transparent; people sometimes do not communicate exactly what they mean, which is why the context and body language of conversations are so important.

It is common sense that allows us to realize that our bodies and minds are real, that the world is a strange, wonderful, and dangerous place, that the earth has existed for a long time, with many radical changes—the shift from a methane atmosphere to an oxygen atmosphere was certainly radical—and, that people in other cultures have different, equally real and important experiences. It is common sense to realize that gardening is more productive than war. It is common sense to realize that enlarging a place is better than destroying a place where other live. It is a failure not to.

3.1.2.2.2.2.1. *Laziness in a Maze.* Economic decisions have provided short-term benefits for many and great wealth for some, but we have been unable to figure out a way to improve the lives of everyone. We have failed to consider the impact that an increasing population with increasing per capita expectations will make with competing demands on natural resources. Surpluses will disappear overnight, as they did when the U.S. switched from oil exporter to oil importer in the 1970s. The U.S. exports raw logs to Japan and China, without considering unemployed wood workers in the U.S., and without considering who will supply the U.S. with wood, or where that country will get its wood.

Our leaders make decisions based on polls and the likelihood of reelection, without considering that they were elected for some purpose beyond re-election. One frightening aspect of government is the total failure of our leaders, both national and international, in political and economic spheres, to learn or understand the simplest facts of science or technology, much less ecology and long-term development. Our leaders separate these other spheres from politics, impressing us with their facility with words and their images on television bites., but that is not enough to hide the failure. Isolated problems stir only a mild reform, but no problems are truly isolated. Modern citizenship is the abandonment of responsibility on the assumption that others know how to manage things. Modern government is the assumption of responsibility, without knowledge or group intelligence.

The failure of intelligence leads to our adopting useless traditions and buying habits, and to genetic deterioration from weak selection pressure and from lack of preservation of norms of social behavior. Neither liberals or conservatives recognize the fundamental conflict between political government and economic government. Therefore, there is no attempt to resolve it, with changes to structure, tradition, or new invention. Perhaps this is laziness. Laziness results in cultural amnesia or political amnesia, where people cannot remember the good decisions they made or good habits they had a generation earlier.

3.1.2.2.2.2.2. *Dance of the Fallacies.* This failure of intelligence results in the use of many fallacies in reasoning. Economics has not been unsuccessful with its models, for instance of buying behavior, but it has become a highly abstract academic discipline. All its abstractions are applied to the real world without acknowledgment of the high degree of abstraction involved. The philosopher A. N. Whitehead warned that the economic method would triumph if the abstractions were judicious, but even judicious abstractions have limits, and the neglect of those limits leads to disastrous oversights. Considering a fictitious human nature under imaginary circumstances and thinking it is real is the fallacy of "misplaced concreteness" according to Whitehead. Hermann Daly and John B. Cobb Jr. suggest that the classic instance of the fallacy in economics is "money fetishism," where the characteristics of an abstract symbol, such as limitless growth, are applied to real commodities and values. Misplaced concreteness also occurs in forestry, where wood is given a status that trees and forests are not. Genetic reductionism is another good example of the fallacy. Genes are not independent creative beings; they only function within the organism, in this case, a tree, which is the creative being.

Many predictions about resources are based on fallacies. There is the fallacy of substitution, that states that a substitute can be found for any resource in short supply. This is not always true, especially when cultural preferences are considered. Furthermore, there is a gross underestimation of the length of time that it takes for a substitute resource to attain traditional markets. For instance, the transition from wood to coal as an energy source took about fifty years, despite the fact that coal technology was established and attractive.

Dennis Meadows identifies another fallacy: The expectation that people and institutions perceive problems and react to them rationally, for example, with the threat of wood shortages, prices should rise and consumers would value the resource more. Yet, the price of wood is still nowhere near the real costs of production, and wood is used for cheap, impermanent goods. Meadows suggests the model of addiction might be more appropriate than adaptation for dealing with consumer demand.

These fallacies, as well as others, such as the fallacy of complexity, the fallacy of the false dilemma, the fallacy of begging the question (*petitio principii*), and finally the appeal to authority (*argumentum ad vericundiam*), are used to deny intelligence. The fallacy of the false dilemma restricts decisions where only two alternatives are claimed and one is considered unacceptable, as in: Either A or B; not A; therefore B. Either war or peace. Either tax cuts for billionaires or economic chaos. Obviously, there are more than two possibilities. Preservation is not the only alternative to clearcutting. War is not the only alternative to peace. Engaging in this fallacy allows us to stop using our intelligence.

3.1.2.2.2.3. Failure Of Imagination

Imagination is the act of forming mental images of what is not present or has not been experienced to deal with new experiences. The world of many people is simple because their image is simple, so they think there are simple solutions to simple problems. Many people believe that energy and food increase automatically as people multiply, and that simplifying ecosystems can increase their productivity. This exemplifies the failure of imagination. We should not confuse the limits of our mind with the limits of the world, as the philosopher Schopenhauer warned. We seem not to have the ability to see what we have lost, in our rush to be civilized and big.

This failure of imagination limits our understanding and visions of a future. We have the ability to explore planets and modify genes, but cannot seem to offer functional education or meaningful jobs, dignity in retirement, or goals for living. Oddly, we seem to have adequate imagination to describe space colonies and interstellar migration. Will we develop them just to take the same inequities and problems with us? Humans can even create virtual worlds by limiting what could be received; for instance, if a being could see in the x-ray part of the spectrum. Yet, human imagination is as limited as human knowledge.

The failure of imagination leads to the inability to recognize or use the good ideas of others. This is obstructive to our adapting to a changing ecological contexts—thus the idea "protect the ecological basis of life" is never considered. We think that we have to address things one at a time, that we cannot see ourselves or our actions in the whole ecological system, partly because we are interested in continuing our immediate pleasure, even if over a short-term human lifetime, without regard to the indirect and shared costs to the system and to others within it, and partly due to a failure of imagination.

Every proposal to build a dam, to widen a highway, to cut down another forest, to turn wetlands into salable real estate, or to bury unwanted waste products is sure to have unintended consequences. But, we never elaborate the possibility of these consequences or what actions we would take if they occur. Instead, we label them as 'side-effects' and try to ignore them.

We do not try to imagine the connections between things that we do not know about. Not knowing how trees provide wood or how people cut the trees and process them, we feel no responsibility, we feel no connection. When we do not feel the connection to land, or understand what it does, in terms of cleaning water or providing food, we do not create groups to protect the land, or a constituency or a leadership.

Many organisms exist of which we know nothing. Their worlds have little meaning in a human world. We know what it is to be human, but spend little time imagining other forms of existence.

The failure of imagination has led to the human overpopulation of many ecosystems and perhaps the planet. Although small bands have some problems keeping their numbers within limits, we fail to understand the differences in scale, and use common resources, like the ocean, without common rules. Extreme change, as regards climate switches or cultural collapse, lies outside our experience and our ability to imagine that change.

We never imagine long stretches of time, thus we fail to anticipate the changes that occur in long time periods. Can reason comprehend deep history fully? Tending to think things too complex, or too expensive to change, we are not able to make good decisions to adjust to large or slow changes.

Perhaps this is a failure of ability in general. This inability to imagine the differences in situations and the consequences of our actions leads eventually to tragedy, which is the failure of a guiding image of the world, often referred to as a cosmology. In a theatrical play, the tragic hero triumphs at first, and incorporates the successful behavior that led to the triumph. This behavior, employed in new circumstances, however, leads to disaster. The hero refuses to give up a particular role or strategy or to imagine how to change in new circumstances. Hence, the failure to give up a chosen role or pattern of behavior leads to great loss later. The play reflects our actions in the environmental play. The real-world tragedies result from the failure of our working images, the products of our imagination: Humans are responsible for the consequences of actions based on certain images, not on chance or fate. We can choose between the tragedy of the commons or the tragedy of total control, or we can expand our cosmologies. We are tragic because we have to accept responsibility for our actions.

Ultimately, many failures, as Dansereau has said, the failures of pollution, poverty, and urban decay, are failures of imagination. Rejecting the solutions of imagination, therefore, can only make the suite of crises worse. Will the failure of imagination condemn people to partial solutions or ignorance? We seem condemned to the weak trials of the past. Utopias are dismissed automatically as imaginary places. We can only imagine a society without any history or without a real place as being desirable.

3.1.2.2.2.4. Failure Of Integrity

Integrity is the state of being complete or whole. The term is applied to art, music, ecosystems, wilderness, computer databases, or people; it implies that these things have not been corrupted by natural process or human actions, a form of natural process, of course.

Integrity can be related to the general character of a human being, having to do with the integration of the self, into an identity that represents things beyond it, but also referring to a way of acting morally. Acting with integrity on a particularly important occasion could best be explained by the general presentation of that character and life.

The failure of integrity leads to both the breakdown of tradition, as we pretend that ubiquitous behavior forms a global culture, and to the destruction of natural habitats, as we take key elements from the ecosystems. We must acknowledge the failure of our remedial efforts, our failure even to address the flaws of our ideologies.

The failure to value those things necessary for life, of a person or a habitat, is a failure of integrity. Value, as an expression of worth or exchange, cannot be limited or ignored. This keeps values and morality as local effects. We do not extend respect or love to distant others. Our personal values and beliefs do not let us.

3.1.2.2.2.5. Failure Of Will (& Courage)

Will is the power to make a reasoned decision, with a strength of purpose, with a firm attitude to control one's actions, with courage or nerve. Courage means an attitude of dealing with anything recognized as dangerous, difficult or painful. Nerve means strength, emotional control, or endurance. Will fails daily. We refuse to share or help others even when it is easy.

If we cannot imagine extremes, it is hard for us to have the will to sacrifice things to avert it or ameliorate it. For all our cleverness, we still emulate flies and grasshoppers, when it comes to acting always in short-term self-interest. During the good weather and the good crops, we expand to or past the limits of water and food.

We lack the political will for sacrifice or planning. The failure of will leads to susceptibility to indoctrination by governments and even by advertising schemes. Policies are not implemented, due to social differences, corruption or war. What does this failure mean? Living in fear? Fear is not necessarily bad. Fear serves a purpose in human affairs. It is a warning system against unknown or overwhelming facts. But, fear can get out of control. Fear builds barriers. It also can lead to the hatred of ideas and languages.

Too much fear can lead to the failure of will. Afraid of failure or of being unpopular, many politicians, perhaps all politicians, exercise too much caution. They refuse to stray from their opinion polls and say what they believe. They refuse to initiate actions that they know are correct.

Disarmament, for instance, is a simple rational idea. Our failure to disarm, especially nuclear weapons, is a failure of will. Perhaps will is undermined by comfort and security. Perhaps the lack of security has allowed fear to dominate our decisions. Courage requires understanding of humanity and the planet, what to affirm and what to negate, as Paul Tillich says. We can have the courage to be, to become, to transcend our limits, Tillich believes. The roots of courage are moral, as Henryk Skolimowski notes, the result of moral conviction, where reason alone is not enough, and belief is required. Lack of morals, which come from community life, can lead to failure of will, decision or resolution.

The chaos of climate, with an increase in extreme conditions and warming of the global atmosphere, has been observed and linked to human industrial activities. Yet, we do not have the will to make changes that might keep the climate more stable.

No one wants a blasted world for their children, or for others' children or for other humans, animals and plants. The problem seems to be why do we act as though we want such a barren wasteland. Failure of imagination? We seem incapable of imagining large scale and long term consequences of our wants and desires. So, we buy that gigantic sport-utility vehicle, so we can see over traffic and be safer from traffic. We cut down a grove of trees for more houses and pave a few more streets. What difference can a few more things mean?

Few people are without some goodness in their hearts, so that may not be the problem. But, few people have long-term, large-scale ecological intelligences in their brains. Almost nobody wants to try to plan for some kind of mutual constraint to allow the rest of the planet to breathe. After all, it's just one more pair a shoes and a filet for dinner. What harm? And the answer is very little harm at all—it's just that on a massive scale, little sins can kill numerous species and destroy large habitats. And, few of us see the results. Just fewer bees and songbirds. The Gulf of Mexico west of here is dead for hundreds of square miles. Why? Garbage, phosphate dumping? Who knows what combination exactly, but it killed the fish, crabs, and almost everything except red tide bacteria.

Until people hearts respond to what's in their heads, we can kiss our asses goodbye, and many other nonhuman asses as well of course. Until our larger selves, that is the self of our places and environments, get to speak and express their hearts, the small hearts cannot do much. Until then, we will need laws, prescribed limits, taxes, injunctions and maybe new social rules, which can come from the heart, of course.

Perhaps people's hearts are barren and fruitless and they would prefer the challenge of a desert world. After all, many of our religions came from the world views of desert peoples; perhaps they believe that god wants a desert domain to rule. Perhaps it is just our desire to simplify the earth to an extreme. After all, a desert would be easier to manage and it would be quite stirring visually. Many cultures that live in deserts love that unique environment.

We know what we have to do, really. But, it seems that only in times of war or great catastrophe do we have the nerve to do something, although we do not seem to have the nerve to avoid those catastrophes or wars that could be avoided by planning and conversation. Perhaps it is fear of pain or change. Perhaps it is a character flaw in the species, the failure of nerve at critical junctures. Perhaps it is just the lazy habit of mob thinking. But, we must overcome it and plan dramatic changes.

The failure to change, to choose real equality and real peace seems to be another unique kind of human failure, the failure of will. We may not have the courage for some to give up their lavish lifestyles and extreme comforts, their extreme profits or gargantuan excesses. Somehow, we need to commit ourselves to a revolution in distribution, equality, and kindness.

3.1.2.2.2.6. Failure Of Charity

Charity is love of others or the act of goodwill towards others; it refers also to the feeling of benevolence or kindness in judging others. In traditional hunting societies, charity was expressed through reciprocity, or mutual sharing. In urban societies, sharing was more than just an exchange; it was characterized by generosity. In early industrial societies, the views and practices of others who shared resources or places were met with tolerance.

In archaic societies, charity may just be sharing food. In more stratified societies, charity tends to be part of an institution, that may be modified by greed, laziness, or the arrogance of charitable workers. The failure of charity leads to the rat race of competition between human groups, at the expense of many human and ultrahuman groups.

As money and power lead to detachment from agricultural and natural cycles, from reciprocity and concern, charity disintegrates. Humans have the power to alter vast processes in nature, but do not care enough to refrain from trying. Power without charity is a satanic theme. Sigmund Freud wrote that Satan desired to be father for himself, an agency without community. Many governments have the same desire. Governments also have knowledge without charity, which was a demonic theme for the Christian church (the word 'demon' is from Greek word for knowledge). Power without knowledge, and knowledge without charity, are frightening possibilities.

Of course these failures occur in endless combinations. Maybe the failure of our modern civilization is a failure of imagination compounded with a failure of nerve. We cannot imagine an alternative to war, and we cannot act beyond emotion. We cannot imagine beauty in the old and messy nature, and we are afraid to try to do without luxuries or to try to sacrifice anything to try to change the momentum of industrial civilization.

3.1.3. *Leaving Noplace*

Are we doomed by our failures to live in nowheres and noplaces, to have the same jobs and the same common rewards? Is being placeless hopeless? Is hope what we should depend on?

3.1.3.1. Habit & Hope

In his writings, H. G. Wells prophesied the collapse of civilization and the reversion to barbarism, with the eventual rescue by a race of supermen. Human nature desires a grand explanation to place its daily doldrums into a grand frame. We must abandon the Nineteenth-century grand design approach, however, with its attractive simplicity, and substitute an awareness of humanity as participants in nature, within natural laws. The complexity and fluidity of problems exclude simple, rational argument. Until recently, attempts to resolve contradictions created by urbanization, centralization and bureaucratic growth were viewed as a counterdrift to progress. Any critic was a treated as a discounted outcast.

There is a conflict between economic growth and the preservation of the environment. Economic growth usually triumphs, but it upsets natural rhythms. The devastation of the biosphere is the greatest threat to the survival of humanity. It may not be perceived as such by most people or their governments because of more immediate concerns, such as war, poverty, epidemics, energy, inflation, and unemployment. Nevertheless, the failure of conservation is a direct cause of the worsening of these problems. Society may hinder human understanding with a burden of distractions, injustice, and inferior loyalties.

Most of us will occupy ourselves with externalities, and hope, in the traditional sense, that our children will be wiser—but how can they ever be with no one wise to teach them? Children are cultivated as the leaders of the future— they expected to solve the problems that adults failed to solve. This future is the keystone of the salvation-through-schooling mystique—if it was true once, there is now too great a lag time between taking a civics course in junior high school and using it as adult forty years later. These things need to be taught to adults or especially to the old, who are traditionally leaders. The dying and the old hardly matter to the young and ambitious; their feelings lie as lightly on the scale as their own future. The old themselves default in their responsibility and the young in theirs. Responsibilities are passed down through generations without ever being taken.

Any mind, even most ingenious and fertile, may fall back on habits of its cultural inheritance; humans are social animals with cultural heritages. The outlook of merely a secure and satisfying life, without great ease and rich comfort, may be threatening to many consumers. The contest is not between us and them, but between the reasoning mind and biological limitations. Can there be a leveling up, now, or is there only the possibility of leveling down? The crisis emerges from our state of consciousness. It may be infinitely more difficult to transform human sensibility than to pass laws. We are isolated from the past and future by a disease.

3.1.3.2. Hope & Science

Technology offers human beings a vision of a world with a stable population, freed from poverty to live in peace, sharing the world's resources. Industrialism, more than that, promises humanity a paradise on earth, after displacing the promise of paradise in heaven. Now, psychology promises its own myth of paradise through self-knowledge and medical improvement.

Science holds great promise for improving the human condition, through understanding the climate and ecology, improvement of agriculture, and the development of soft energy sources, but the lead time for research is long and unpredictable. Interest in ecology has inspired a number of plans and models of the world. The worst are disoriented, useless and uninspiring. The best are notable by their criticisms of the former.

Most popular plans are based on thoroughly terrifying assumptions: Continued, accelerated economic growth, as well as population growth; further industrialization; less distribution of wealth; continued inflation; continued centralization; and complicated technological solutions. Many of these plans rely heavily on scientific studies (MIT SCEP), computer models (Forrester and Meadows), or even science fiction (Kahn). We hope that we can avoid the worst of those assumptions. However, as Ben Franklin noted, those who live exclusively on hope may die fasting.

The present causes of the ecological crisis lie in the combined action of technological advance, population increase, and conventional, erroneous ideas of the nature of humanity and the environment. Garrett Hardin asserts that the problems of the ecological crisis are direct results of the tragedy of the commons reproduced on a global scale. Therefore, overpopulation, pollution and resource depletion can have no technical solutions; they can only be ameliorated through political reform, which can only result from changes in perception.

3.2. *Designing Noplace: Utopia as Ideal Place*

The result of hope is the quest for an ideal place, without problems, without stress, and without hope. In short, a place that can only exist nowhere. This quest is achieved by subtracting place, by removing diversity, and by ignoring inconvenient truths.

3.2.1. *Defining Noplace*

The title for Sir Thomas More's Utopia is a pun. To the Greek root word *topia*, meaning place, could be added *eu*, meaning good, or *ou*, meaning no. Good place sounded like no place. Utopianism can be found in almost every culture, in prophecies, visions, dreams, myths, and ideologies. In general, utopias have been preoccupied with static societies that were egalitarian and open. They tended to emphasize a purposive world characterized by moral order. They usually tended to refer to the past or future, and to theocratic salvation or rational achievement.

Utopias started out as models for interpreting and remolding society (Plato). Christian utopias began with inward change but were located in heaven. By the 16th century, they were located in other lands; but, by the 19th they had to be located in other times. As Plato's city was tied to the cosmos, Augustine's was to God, Hegel's to the spirit, and Machiavelli's to Florence.

Plato's *Republic* envisioned a perfect model for the Athenian form of social organization. A fully communalized, small population was supported by agriculture and crafts. By contrast, Augustine's *City of God* neglected social and political life and emphasized the ideal human relation with God. Other medieval utopias shared this emphasis. At the beginning of the Renaissance, economic requirements for the good life were given more attention, although ethical civilization was the ultimate goal. Religion still played a major role, however.

Thomas More's essay, *Utopia*, was completed in 1516. It depicted a perfect society, without the problems and poverty of the nascent capitalism in England at the time. The electoral unit was the family, which was autocratically ruled by the patriarch. The government consisted of several levels; the lowest was elected from constituencies of thirty families and the highest was a hierarchy of princes. Decisions were made by councils elected from public officials, who met regularly with the princes. Utopia was uniform and regimented; everyone had the same clothing, housing and work schedule. There was strong peer pressure to use leisure time constructively for the public good or to improve personal virtues.

Popular welfare, to be achieved by experimental science, rivaled virtue as a goal in Francis Bacon's *New Atlantis*. Traditional values of Christianity, monarchy, and agriculture were maintained. Bacon suggested a domination of nature instead of a stable balance within nature. Surprisingly, however, scientists were expected to be ethical and withhold discoveries that would not be beneficial to society.

Denis Diderot used his *Supplement to Bougainville's Voyage* to call attention to the natural qualities lost to civilization by glorifying the peoples of Tahiti. But Edward Bellamy based his vision of the future on increased industrialization. *Looking Backward* was placed in the year 2000 to view the compulsory conscription of labor for a state enterprise which had eliminated poverty and created a satisfying life where the burdens and benefits are equally shared.

Francois Fourier's model eliminated the drudgery of work, but segregated workers by emotional tendencies in specially designed structures called phalanxes. His scheme emphasized decentralization and autonomy. Each individual was simultaneously capitalist, worker and consumer.

William Morris moved utopia from factory and barracks to a romanticized version of the English village. In a democratic anarchism, people worked only at agreeable activities; either there was a perfect fit of people to activities or some things did not get done. Decision making was completed locally with majority rule. People wore pleasing clothing and lived in comfortable buildings. The undesirable products of industry were neglected.

Utopias often emphasize a perfect social structure, without the dynamic aspects of a historical process. In the 1800s, however, utopias were considered as possible future societies, with a history. Karl Marx and Frederick Engels rejected utopias as being out of step with economic development; utopianism was considered as a naive, prescientific mode of thought. Marx dismissed utopian planning since the socialist utopia was on the way, regardless of intentions or plans. More modern writers have created other alternatives.

Although concerned for individual liberty, H.G. Wells advocated government centralization in his book *A Modern Utopia*. He concluded that utopia could only take place on a global scale. The global government was ruled by people reminiscent of Plato's philosopher-kings. Criminals and misfits were exiled to prison islands. Unfortunately, nature was regarded as unmanageable.

In Aldous Huxley's *Island*, the inhabitants have mastered the technological impulse and managed to live naturally. Their utopia, however, proved no match for the industrial expansion of the outside world. *Ecotopia*, by Ernst Callenbach, was a secessionist state in North America. A decentralized society tried to develop modes of social organization in balance with the natural environment, surrounded by industrial areas.

The number of bad places, dystopias, described since the sixteenth century has increased with the decline of the quality of life. Nicolas Berdiaev warned that utopias were more capable of realization now and that our problem was how to prevent them. The dreams of communist Russia, industrial United States and fascist Germany became nightmares. And, as Eugen Weber noted, those who have had nightmares are reluctant to dream further.

3.2.1.1. Utopian Inadequacies—General Characteristics

Utopias sometimes search outward or inward; some believe that utopia exists somewhere on new frontiers, for instance, with O'Neill's space colonies. An inward utopia is sought in the self, in the human quest for perfection. Most utopias deal with the same questions: What is society, how is it organized, what are its forces? How does it continue, or change? Plato questioned the best kind of social system in *The Republic*. Since then, many thinkers have dealt with the problem of political order, social stability, universal justice, and individual happiness. So little is known about any society that it seems futile to design whole societies. Most utopias would be unworkable and unsatisfactory.

Classical utopias were literally no place. They were anthropocentric and ignorant of ecological restraints, technological consequences and historical trends. Most of them dealt exclusively with technological change and material wealth. In fact, society and nature were interpreted by physical metaphors. Utopian city plans from Plato to O'Neill have been perfectly mathematical. Perhaps they could exist only in a desert. Plato's vision, echoed in Wells and others, had its roots in the first cities of the Near East. Mumford claimed that in fact cities were machines powered by conscripted labor, whose power to transform mud and rock into edifices captivated the imagination. The archetypal image of the city, and utopia, is characterized by hierarchy, rigidity and coercion. Mumford feared that all ideal models had the same life-arresting or life-denying property: Static order.

Utopian laws were usually traditional or arbitrary and not concerned with the laws of nature. Early philosophical utopias ignored ecological factors and considered only ethical and social relationships. Even at this, no utopias were multiracial and multicultural. The best was unfit for human habitation compared to the simplest life in tribal society. Idealistic designs for peace lacked a theory of change, and were antithetical to balanced conflict.

Utopian design concepts have been harmed by other basic defects: No concept of transition linking the present to the future; and, a solution by replication of local power-authority concentrations on a global level. An adequate conception of social justice is crucial to utopian planning; attention needs to be given to rights of all members. Refuse disposal is a major limitation on human progress, and one of the problems confronting utopian thinkers.

Utopias provide images of ideal societies in abstract settings—literally nowhere. Utopias promise newness, order, happiness, and re-inheritance for the disinherited. They banish the irrational, the irreparable, and all conflict. Common ideas can be discerned in a reading of utopias:

- The quest for human perfectibility—the dismissal of social causes of disharmony—through constant attempts at self-improvement;
- The emphasis on order—the elimination of chaotic, uncoordinated, accidental events that cause waste and conflict—in a predictable society with planning and control;
- Universal fellowship, which is brought into being by removing barriers to harmony, such as private property;
- The expansion of consciousness with new, cohesive rituals; creation of new social forms—often a radical departure from old conditions, with plastic variance in all aspects of life from diets to dyads;
- Location outward in a distant place, an island or space colony, or time or inward frame of mind.

The common characteristics of many utopias, however, make them unworkable and unsatisfactory. Utopias tend to be ungrounded, static, teleological, ingenuous, simplistic, homogenous, and incomplete.

3.2.1.1.1. Utopia is Ungrounded

Utopias are literally no-place. They do not exist in place, either a human place or an ultrahuman place. They are designed to be no-place, without weeds, storms, or hard ground. Because they are nowhere, utopias lack any reference point. Because utopias are not grounded, their ideal qualifies do not have to fit together; any combination of details is possible, regardless of whether it ever could work together.

3.2.1.1.2. Utopia is Static

Utopias present a static order. Many utopian plans emphasize the physical order of a model city, a perfect design centered in a mathematical space. The perfection of the social structure of the inhabitants reflects this ideal physical order. The utopia arises fully formed, not as the result of a historical process or of a dynamic movement, and so it shows no concept of transition. If the utopia operates at all, it does so without problems; everything is managed according to one plan. The interpretation of society and nature is limited by physical metaphors, especially those based on physical mechanics. The shape of the city and the development of the people are set and unchanging.

3.2.1.1.3. Utopia is Teleological

Utopias assume the possibility of human perfectibility. They present a final state of society, where perfection has been achieved. The characters, who are symbols and not inhabitants of a place, lack a psychological dimension because the irrational, the internal conflict, and the irreparable have been banished from the ideal and from the interiors of the characters. Human happiness is determined by reason and intention in a single, inflexible, final order. A utopia tends to impose a monstrous discipline on the activities and interests of the society.

3.2.1.1.4. Utopia is Ingenuous

In order to eliminate uncertainty, many utopias construct one ideal community of immense size as the model for all communities. Utopias tend to be centralized. H. G. Wells, in seeking to protect individual liberty through government centralization, concluded that utopia could only take place on a global scale, ruled by an international government of Platonic philosopher-kings. The subjects naively would derive their identity from the global structure. Some utopias tend to universalize the best of a society, so that all societies may fit the mold. To ensure continuation of the best, they rely on segregation to resolve social difficulties. For example, Fourier segregated workers according to emotional tendencies, while Wells exiled criminals and misfits to prison islands. Furthermore, nature in utopias has been regarded usually as unmanageable, by Wells for instance, or totally manageable, by Saint-Simon in another instance. Waste is not considered. Good, however, is considered unlimited—the metaphor of unlimited good, however, leads to devaluation and waste.

3.2.1.1.5. Utopia is Simplistic

Utopian laws are often simple and arbitrary and not concerned with the laws of nature; they try to comprehend, predict, and control human behavior on the basis of a materialistic philosophy. Utopias deny the depth and complexity of social, biological, and even physical problems in emphasizing a simple model. The excesses of geography are removed; utopias are rarely cold or too hot or too wet or too dry—the very places that are still wilderness on the planet. They remove the "useless" and unfriendly animals and plants. Utopias neglect the element of chance, in the form of earthquakes, fires, comets, and most natural disturbances. Most utopias ignore the problem of scale. What can be done on a small or imaginary scale may not be possible on a large scale; there are simply too many factors and connections. Ubiquitous problems, like hunger or illiteracy, are considered wrongly as global problems; and, they are usually coupled together.

3.2.1.1.6. Utopia is Homogenous

The creators of utopias conceive of their inhabitants as one people, with a common color or temperament, regardless of real social, cultural, or biological differences. Clothing and housing are often uniform. With undifferentiated growth, the monocultural mass produces beauty and satisfaction for everyone. Nothing like this seems to have happened in mass societies, although monumental and heroic structures have been produced. A utopia uses its central technological, political, or moral theme to solve all problems; thus, all problems in a typical utopia are solved if every member of society has a radio, commits to be a communist, or acts like a Buddhist.

3.2.1.1.7. Utopia is Incomplete

Other societies have been ignored in utopias, except as examples of errors of thought or false images. Utopias have been blatantly anthropocentric in their concerns. Hence, utopian topics include industrialization, modernization, food capacity, housing, population explosion, and material possessions, but not the presence of foreigners, the necessity of wilderness or the rights of animals. Nature is regarded most often as an object of conquest or a storehouse of resources than as a mutually-created matrix.

3.2.1.2. Goals & Failed Designs

Regardless of how ineffectual utopias seem, they express human truths, for instance, 'with voluntary cooperation, state compulsion is unnecessary' stated William Morris. They present new possibilities, such as 'welfare through science,' thought Francis Bacon, or the recovery of lost 'natural qualities,' believed Denis Diderot. And, they have the power to transform society, as could be done by the rational state, according to John Locke, or the classless society, predicted Karl Marx.

Utopias are not just irrelevant fantasies; they have guided many of our modern qualities. For instance, contemporary industrial culture has mimicked utopian models in allowing for the interchangeability of people and places. Like utopias, modern industrial landscapes are flatscapes where variety disappears and significance is ignored for the comfortable standards of meaningless continuity. The characteristics of industrial cultures bear a strange resemblance to the characteristics of utopias, from simplification to ingenuousness, homogeneity, and incompleteness.

3.2.2. *Implications of Good & Evil*

The public gets used to messages of doom, but there are evils that survive identification in the light, and yet go on, like the reification or deification of money or war. Simple codes divide the world into good and evil too neatly, whether in the Christian bible or Reagonomic policy. The world is not simple, alas, and goodness does not grow like tomatoes. By trying to focus on either extreme, of pure goodness or pure evil, we miss the ambiguity and uncertainty of situations, most of which occur in a mixture. Our ethics and ideologies are not composed to help us live in a mixture, with the inevitability of uncertainty, or with the possibility of enantiodromia (a thing turning into its opposite).

Humans usually distinguish unambiguously between good and bad; most ethics intend to further the good and suppress evil. We search for essential difference between good and evil in vain, because the constituents are the same. The distinction lies in the way the pieces are assembled, the structure. The universe is comprised of good and evil, that is, it is agathokakological. Everything seems to work by complementary opposites. Not all good things go together in the same category, exclusive of all evil things. The attempt to isolate the good accelerates enantiodromia and can actually create evil.

Good can be defined as intention and action in the context of the rules of a culture, using ambiguous signs creatively. A sign is anything that signifies something other than itself. Signs are arbitrary, according to Ferdinand de Saussure, that is to say, they have no necessary connection to things. When we set goals, as for the goodness of our forestry practices or charitable institutions, we base those on the meaning of the symbols, which can have many more than one meaning.

Signs make many connections to physical events, thus intensifying them as well as miniaturizing the events in the signs. Signs become dense with meaning. Patterns are available to the mind, and the reality of patterns emerges from events and becomes as real. But, also a reality that can be extended. The play of signs results in good or evil; that is to say, the manipulation of signs, in a field of surprise, due to other levels of meaning, can be interpreted as good or evil.

Culture provides rules that limit play, so that it stops short of death or destruction. Games have rules. All cultures work that way. Ethics and economics are rules of behavior; politics is the practice of changing the rules as society changes. Politics, however, is changed by scale from good to evil; there are too many of us in each system to share discourse. This is much more so on an international level.

The rules of a culture can be expressed in signs or words in the context of the culture. A shift in context can change the meaning of a word. We do not think of physical events, such as gravity or fusion, as good or bad; they simply are. A physical movement is turned into action in the human realm by human intention, which is open to interpretation and ambiguity. Signal play without cultural structures can result in evil.

Words have meaning because of their shared history and context, by their place in a whole language, which adds to meaning of the word. The combination of words in a conversation, with body language and intent, are more important than individual words. The dialogue process helps get past dualities such as good or evil. Metaphor and humor lend new clarity.

In having good as a goal, we cannot calculate the result. Good is a feature of the path of actions, as is evil. We cannot aim at it and shoot, or even stare at it. We must approach sideways, through a field of good and evil. This is one thing a conversation can do, or a poetic series of statements: Let us approach sideways.

3.2.2.1. The Possibility of Good

Good is an interesting word, with a long history. The current version is derived from the old English word, *god*, meaning suitable or fitting, similar to the words meaning a 'suitable time' and to be 'pleasing.'

Humans cannot know, or even think of anything, according to Robert Zajonc, without some involvement of emotion, that is, at least a vague feeling of good or bad. Good is problematic. The search for good is measured by personal criteria, personal judgment, and personal reflection. On the other hand, there are questions of what one *ought* to do, that is, morality. Good and bad mean different things to different people. Your standards or codes might be different from mine. Therefore the meanings of the words will be different. Furthermore, doing bad to one person sometimes results in good for others, or vice versa. Sometimes, just to feel good, people destroy the works of others, or living beings, or an entire ecosystem.

In the long run, as John Fowles suggests, maybe all our judgments of good and bad are meaningless. All actions, good or bad, interweave so extensively as time passes that their individual goodness or badness disappears. Each becomes lost in the other. One should do good for one's health, for instance, or the health of the forest community, not because the action is an action or for the sake of doing something good. In doing, we choose between good and bad actions; the judgment makes us human and susceptible to error. We just have to be aware that choosing good can result sometimes in its opposite, through enantiodromia.

3.2.2.2. The Necessity of Evil

Evil is a disintegration, a juxtaposition of opposites, with some parts striving to suppress others. Good is the synthesis and reconciliation of the same parts. In ancient Hebrew, good and evil is a single word meaning 'everything.' Everything is disintegrating and synthesizing all the time. Disintegration is necessary to the process. We are aware of values as conflicts tear us; we can reconcile values in just proportions to resurrect whole body.

3.2.2.2.1. Size & Evil

Humans became more and more prone to alter the earth, because of an increase in the human population and in the means of destruction, not necessarily because of a change in attitude. Both American Plains Indians and Africans exhibited this change in their situations. Indians, for example, used the newly acquired horse and gun to become more efficient, and careless, hunters. Although our behavior may not be qualitatively different from our remote ancestors and the worst pathologies of wild animals, which usually result from miscommunications under certain circumstances, it is quantitatively different. More and more activities affect larger and larger parts of the planet; many problems have global effects now.

In tropical areas in the past, people practiced land use that permitted the vegetation to maintain itself despite human exploitation and the constraints of soil and climate. With population and production pressures, tropical areas everywhere are rapidly being exploited. The consequences may be disastrous.

Size is almost always the greatest threat. In spite of the fact that many Buddhists planted trees regularly, the Buddhist traditions of wooden temples and funeral fires contributed to the denuding of large areas of forest. It is our misfortune to live at a time when the accumulated effects of the conversion of nature for human ends are becoming obvious and cutting into the survival potentials of many other species.

3.2.2.2.2. Separation & Evil

From medieval times, scientists found that glass could separate materials and distill liquids. Astronomers and microscopes found that glass could be shaped to focus light waves to reveal the very small and the very distant. Portions of the universe were placed behind glass in a laboratory world. Glass was very useful. Scientists came to rely on its advantages, but they were unconsciously imprisoned by its limits.

As science cut the connection to direct observation, it became as blind as mathematics to the 'outer' world. The formality of science made statements about the outer world tautological. This proved to be a problem with quanta, species fitness, and psychological needs. Scientific hypotheses form filters like glasses. They cannot be shed entirely, but their effects can be understood.

Glass has advanced our civilizations by permitting the easy separation of fluids and reactions. But, it leads to an objective attitude towards living beings and nature. Being "behind glass" has become the metaphor, first for a scientific approach to knowledge, then for a utilitarian ethics and a teleological ethics.

Scientists have experimented with biological processes behind glass (*in vitro*) in the laboratory. The primary commandment of Jacque Monod's ethic of objectivity is observers should not participate in the workings of the world. But, detachment from nature is detachment from the basis of knowledge. We distance ourselves behind glass. This detachment is the greatest threat to the welfare of nature. It permits the vivisection of the "voices of existence," as Neil Evernden warns. Only recently, in theoretical physics and in ecology, has it been realized that there can be no perfect detachment and objectivity. There can be no perfect insulation from the object of study.

We have created a hard glass between the mind projecting and the object receding. Sometimes we doubt if there is anything on the other side that can be seen. Glass also forms a window for consciousness, which swamps the mind with the 'error of the eye,' in Marshall McLuhan's phrase. Vision, whose mode is successive and not simultaneous, is emphasized and split away from the total sensorium. After human consciousness places the glass, human needs shape and tint the glass. Utilitarianism casts a thick, convex glass, to focus on the individual. Romantics use a tinted, concave glass, the better to see the whole.

The evolution of human mentality has put us all behind glass. We use glass to protect ourselves from the ambiguity and messiness of nature. We have made an experiment of ourselves; our mentality has evolved behind glass. We have isolated ourselves by technology. We have seen more on television, but are moved to do less. C.P. Snow has commented that watching deaths by starvation in Africa on television screens could mark the end of any moral community of humanity.

We are behind glass and fear it will break. Augustine remarked to the Romans: "What glory is there in the largeness of empire, bright and brittle like glass, and forever in fear of breaking." Reason alone cannot cure what it caused. Glass cannot divide humanity and nature; nor can we humanize the planet without dehumanizing ourselves. We can put things and beings behind glass, but we lose them. Relationships are so strange and complex that they cannot be understood behind glass. The glass creates an illusion of objectivity. Being behind glass allows us to pretend that we are separate; this pretense allows evil. We need to break the glass. We need a sanctified vision of life from a deep participation.

3.2.2.2.3. Freedom & Evil

Freedom is a description of possibilities. Freedom allows many values and behaviors, but not necessarily all. Freedom is considered the opposite of determination. Most theorists are worried that freedom would be denied in any modern utopia or dystopia. Both Fritz Schumacher and Eugene Odum refute this in numerous arguments. Some freedom is needed to realize human potential. A right amount is needed; and this ámount can be determined ecologically and ethically.

But, in addition, one must understand that existence is already half determined; freedom and necessity must be balanced at about fifty percent (compare with the fifty percent redundancy requirement in information theory). Too much freedom, the refusal to acknowledge what is determined, would inflict the self-determination of each on the self determination of the many, resulting in social chaos, and reducing other freedoms. Too little freedom, the refusal to acknowledge what can be done, results in stagnation. Recognition of the laws of nature is what allows the only real freedom within them. We do not eat poisonous plants, or breath water, for instance, although both can be done with adequate preparations.

3.2.2.2.4. Knowledge & Evil

Knowledge increases our opportunities, that is, our freedom; we pay for freedom with the risks of mistakes. Knowledge is power and power can possess its owner. Humans have the power to alter vast processes in nature, and not enough care to refrain from trying. Even if we know better, we must care to act with concern for others.

Once we become aware that the unintended byproducts of industry are harmful to a species population, then the destruction of that species becomes an intention if we do not alter our behavior. Destruction is then a matter of choice and responsibility rather than ignorance.

3.2.2.2.5. Intention & Evil

The first act of a utopia must be to come to terms with the evil seeds of good intentions. Evil can arise because of the contradictory nature of reality, the nature in which all institutions subvert the values for which they are founded; all values seem to be achieved in conflict with opposites. Does evil come out of our efforts to do good? Developed nations are in a double bind in dealing with famines: If we do not feed starving people, they will die; but, if we do feed them, more will die later, unless they have renewed their crops or reduced their populations.

The process by which a movement turns into its exact opposite has been identified as a swing of reversal; romanticism rejected industrialization, but became mechanized. The German attempt to unite Europe left it divided. The line curves upon itself. The rise of ecological consciousness is an offshoot of the space program's view of the planet. Values too strenuously proclaimed often go to their opposite.

Aid is an illustrative example of this kind of swing. Basing foreign aid on a belief in spurring industrial development is a tragedy. The second outcome of a massive distribution of food to poorer countries is often a depression of local agriculture. Common-sense solutions to some problems just worsen the problem. Michael Gordon illustrates the pitfall of half-hearted help. Aid can be the prime cause of suffering that it is intended to relieve, increasing the sum of human misery. Medicine and infant care can be cruel, if the infants are neglected later. Aid humiliates the receiving country, Gordon says, and corrupts or angers the donor country. But the innocent suffer with the guilty. The alternative can be greater starvation a few years later. Aid as a program for action is incomplete and too simplistic. Schumacher stipulates that the best aid that could be offered is intellectual.

The editors of *The Ecologist* are even more radical in their recommendation: In view of the psychological and material debacles thus far, no aid at all is best. Even intellectual aid is questionable; our ideas and inventions may not even be applicable. Now, with our proliferation of information, especially concerning products like cars, vacuums, clothes, and grooming devices, we raise people's aspirations without satisfying any of them.

The author H.G. Wells invented the idea of futurism, predicting and planning a culture for future global development. "The future cannot be predicted, but can be invented," according to Dennis Gabor. The first step in inventing is to learn everything possible about the past and the present and then try to identify those possibilities that may be realized and then to choose from among those possibilities to invent a desirable future.

Karl Marx realized that ideas become a material force when they take hold of minds of people. But, first the ideas must be formed. We may remake our part of the earth, which is part of a solar system, which is part of a galaxy of over 100 billion stars in an archipelago of galaxies. If we are unable to create a place here, the fault will probably be, to paraphrase Shakespeare through Cassius, in ourselves, and not in the stars. We are behind glass and fear it will break. Augustine remarked to the Romans: "What glory is there in the largeness of empire, bright and brittle like glass, and forever in fear of breaking."

3.3. *The Potential of Good Places*

Most Utopian discussions have to do with places as literature or as literary alternatives to real world ecological and political situations. Marx and Engels criticize utopian thinkers for lacking a thorough analysis of power relations within a society, as well as of economic developments that bring about changes. Karl Popper criticizes utopian engineering for expecting to make massive changes that would affect the entire human society. He says, correctly, that our knowledge and experience are too limited to expect that such changes would benefit society and people. It is too limited; there too many uncertainties and risks.

Nevertheless, utopias serve a vital human need. One important and valuable aspect of utopias is their play with new ideas or combinations of ideas. Another aspect is as thought experiments that let us work out limitations and goals.

The other sense of utopia, that of good places, should not remain silent. We need utopias to define visions for the future, but we need a different kind now, not the visions of no-places or of the wrong-places, but of good places—Eutopias.

Before trying to build a framework for good places, we first need to understand what makes good places, and then how we encultured human beings can make good places, whether with centralized planning and global prescriptions or with ecological plans and local control. And we need to ask more questions: Are good persons, institutions or societies enough? Are good ideas, images or dreams enough? Are good, coordinated efforts enough? Perhaps these questions can be answered with wild ideas and thought experiments. Dreams and imagination are needed to describe desirable futures, to support plans, and to outline goals.

In that sense, eutopias fits in the tradition of thought. The approach is both descriptive and prescriptive in novels, which allows readers to interiorize the system and appreciate the details of the life within. Eutopias is a thought experiment to apply real alternatives, derived from the baseline of cultural experiences, to the problem-ridden monolithic applications of industrial civilizations.

Utopias may be impossible, but good places are possible. Good places already exist everywhere. There is a place in Africa, by the northern Abderes in Kenya, where a stream winds its way through gardens and orchards, by low houses, and crowned cranes stalk between rows of vegetables; flocks and herds graze on fenced pastures, and wild animals fed nearby. Such places can be found in every place of the world. Perhaps by logically working out the limits and adaptability of humanity, good places can be created for almost everyone.

In totalitarian utopias, people must be controlled, the threat being that, without discipline there would be immediate disorder and the utopia would collapse. As we know, it is impossible to avoid disorder, which is a necessary part of any physical and social system. A system with flexibility, such as a eutopian system, could incorporate disorder without being destroyed.

Taking it cue from utopian writings, eutopias asks the broadest questions about society and the environment. Where would you like to live Urban or rural or mixed? How close to nearest large store? How often should your neighbors visit unannounced? Once a week, a month, always, never? What should your neighbors do? Be artists, scientists, laborers, game players, or be multi-faceted?

Rather than a single nonnegotiable truth, as would be imposed in Plato's *Republic*, the inhabitants of eutopias would "play" with truth. In fact, their education, taking cues from Schiller, would encourage play at all levels.

Culturally-based nations can provide a greater spectrum of intricate descriptions and possible relations than any one single thought experiment. That is why they need to be the basis of good-places-in-the-making. The framework is enriched by the ideals and visions of nations. This is consistent with the intuitive knowledge that many styles of life are sustainable.

This is where eutopias would work effectively: Accepting the unpredictable nature of natural and social processes, yet managing adaptively on a micro-planning level. Eutopias would incorporate tradition and working past elements, rather than breaking with them. As Edmund Burke, reflecting on the French Revolution, notes that society has gone through a natural process of development, forming traditions and institutions that provide rules of behavior known to the people. We tend to forget how much discipline already exists in every society. People abide by most rules, from driving on one side of the road to preserving their children from molestation.

In eutopias, it is important to work with what exists, with what has already worked in place, with history and culture. Eutopias can start in California, Sri Lanka or any place that can support a nationhood aligned with a cultural history, without having a standard army or special currency.

The eutopian vision would start with immediate needs for everyone, but then allow people to be as extravagant as they want within the general social limits. Most people like the place they live, and they like their culture, so it is not necessary to force them to stay in place or be what they are. That does not mean that they do not want to live better, just that they do not all want to be Swiss or U.S. citizens.

Despite the eloquent pleas of Wendell Berry and Gary Snyder, that people ought to stay put and put down roots in a home place, we may not want to base the frame of eutopias on that idea. People have been nomads for a longer time than farmers or city dwellers. It might be better to accept that movement and dislocation is part of the harmony of life and try to minimize the negative effects, especially combined with consumerism and energy waste. The civilization of cultures is not a steady state, but it can be accommodated in a flexible framework. Flexibility is an important eutopian characteristics, along with resilience, adaptivity, and surprise—that is danger; Callenbach recognized that in his *Ecotopia*—places need to offer excitement and danger.

A new model may solve some problems but will definitely create new problems; Niccolo Machiavelli reasoned that this would always happen. This is a good argument for eutopias, since changes would occur mostly on a small scale, easily correctable. That is why eutopias can be a framework, by limiting human evil and good to small places, with minimal control or competition.

There is no single model of government that will fit all cultures, all traditions, all needs or all nations. But, governments can be held to common standards through an international body. Eutopias tries to structure the global commonwealth of nations by proposing specific common goals, regarding weapons disbursement and population limits. It offers a consideration of distributive justice among nations and between the generations of every nation. It offers visions of culturally-based nations in equilibrium with nature, using ideas from ecological design and conservation biology.

The eutopian framework, Eutopias, is more than is a simple perspective; it is a design for a self-renewing process using proven cultural methods to improve human situations and environments. The frame is not a final goal but a way to allow the many small useful, culturally determined (or limited) changes that we need for our survival on the planet as we like it.

Eutopias is a cheap solution because it uses the parts already in place. It is within reach of any people in any culture, regardless of how fast or technological. And, because it limits the ways of big, expensive solutions.

Eutopias is not a big law; but, it is a big story, large enough to allow all other stories. It is changing and open-ended. It is conservative, but that becomes an instrument to create peace and justice everywhere. It keeps human rights connected with species rights and land rights.

Eutopias is a framework that holds all the other pieces of solutions and makes them part of a whole thing. It helps us to understand the whole thing, the whole set of relations with people, land, and other beings, with other cultures and the ranges of technology. Eutopias is a framework that holds many centers, beginning with the centers of different human cultures and including the centers of nonhuman communities and guilds.

Eutopias addresses the whole, because the health of the whole depends on the health of the wholes that make it up—the health of wildernesses, the health of civilizations, and the health of ecosystems, that is, inhabited places that are made through living.

4.0. **Making Good Places by Design: Topopoetics**

We make the places we live in, or rather we remake places by adjusting them to fit us. We also fit into a place and make images of a place that also fit, that is, we adjust ourselves as well. We are a place-making species; human beings are topopoetic, like most species. We have to stand in place before making or remaking a place. We have to have knowledge of a place.

4.1. ***Standing in Place: Epistemology***

Epistemology, which comes from the Sanskrit word for 'standing in place,' is the study of how humans know, from the genetic and personal levels to family and social levels. The social level accumulated traditional ecological knowledge, while the later formalization of science became a process for standardizing and comparing knowledge.

4.1.0.1. Biological Knowledge

Most animals have knowledge about how to move and eat. More complex animals acquire knowledge about how to socialize and mate. Some knowledge belongs to physiological systems, such as the immune system, which can detect bacteria and viruses. Some knowledge is instinctual, based on genetic information from ancestors. Genes show what is valuable for a specific life-image. Other knowledge is gained through observing and testing.

4.1.0.2. Human Knowledge

In order to make sense of the sheer multiplicity and complexity of their environments, human beings create abstractions. An abstraction is an idea created to refer to all objects that have certain characteristics in common, e.g., all birds. Abstractions can be generalized, e.g., all things that fly, but at each outer level the objects have less in common; thus flying things include birds, insects, mammals, reptiles, seeds, and spores.

Human beings also classify and label their abstractions. The systems of classification are reflexive and pragmatic; that is, they refer to the classifier as well as to the object, and they include guidelines for how to think about, treat, and relate to an object, according to S. I. Hayakawa. As soon as a classification is no longer useful, people stop using it and become receptive to a better classification.

Science uses abstraction and classification to transcend common sense to describe the fundamental structure of nature. Scientists use classes to limit things to their own mesocosmic scale, although the classes are independent from the things.

A given set of descriptions is conditioned by a world view, and then articulates that view; it is ontologically bound by it. In spite of our dexterity at description, we are stupefied by social and ecological problems that do not yield well to description. New knowledge is an epistemological imperative for survival. But, the increase in knowledge will not necessarily mean an increase in value or use. Scientific descriptions are often useless for expressing human experience. Furthermore, not all facts can be expressed in numbers; many require words, or perhaps just gestures.

Scientific epistemology is a basilisk that kills what it sees, and only sees by killing. Combined with the industrial canon—we should do what we can do—epistemology results in incredible waste in practice and imbalance in theory. We have built a knowledge killed and devoured by analysis, and while this may be acceptable for understanding certain levels of existence, the knowledge of a whole, living world cannot be obtained this way. Wordsworth accused: "We murder to dissect."

Scientific explanation lists the contents, but leaves meaning out of the evaluation. Meaning is a relationship in context. With Democritus, we cut ourselves out of the picture to simplify it, but we also cut out ethical and aesthetic values.

Our knowledge can be empirical and experimental without being objective or having a fixed reference. Objectivity is a stumbling block. Knowledge is selected by the environment also; it is true as long as we survive. The regularities that constitute our knowledge at given time cannot be said to describe ontological reality because of the variety of other regularities possible. So scientific evidence can be falsified by another regularity, but never proved.

Humanity in the natural world can be known through an epistemology of participation. Knowledge comes into being in autonomous units through an interwoven mesh of frozen histories, pulling up its content from within. Participation is the level above objectivism/subjectivism. Everything we know is our experience, not naive reality. We need to explain how we come to see facts the way we see them.

We fit the environment, and our personality must be implicit in the scientific world view. We can experience science fact from within, understanding by sharing intentional meaning and the personal address of things. A scientist's metaphysical beliefs influence her work. Metaphysics is the transcendence of physics. It signals the beginning of a new epistemology: Soft, pluralistic, life-rooted, and cosmos-oriented.

Gregory Bateson recognizes that there is an ecology of bad ideas, just as there are ecologies of weeds or fungi. When epistemology is narrowed to personal interests, the larger system is forgotten. Our compulsion to save individual lives has created the possibility of world famine. This massive aggregation of threats to ecological systems rises out of errors of thought at deep and partly unconscious levels, according to Bateson. The ecocrisis is a crisis of ideas. The roots of the ecological crisis are epistemological. We need to shift our whole frame of reference and our attitude toward life itself.

4.1.0.3. Multiplicity & Crises

The softworld approach treats the ecocrisis as an overall malfunction of a complex cybernetic system. Hardware and software can be used to distinguish between the physical equipment of a computer and the programs and data that guide it, as well as between things and ideas in general. Bateson's conception of an ecology of mind is a key to a softworld vision. Bateson defines idea as a difference or a transform of a difference. A difference cannot be localized. The hardworld vision, implemented by technology and technique, has created an ecological crisis. And, at the bottom of that is a perceptual crisis. We have not perceived the network of interactive beings in an environmental field. We have not perceived the implication of the fact that every part of the earth's surface contains material from every other part.

The roots of the ecological crisis are epistemological. A new epistemology can be designed over the years, a subversive metarevolution, resulting from the integration of many ideas. This new epistemology must be structured and unified in a comprehensive framework. Epistemology and ontology are additive truths. Godel stated that any axiomatic system is also a normative system; the goals are built into the structure.

However, more than one epistemology is required. Another epistemology is needed that would allow an investigator to validate or falsify data where samples are small and nonrepeatable, and where uniqueness overrides system parameters. Epistemologies are needed for mystical and practical experiences, for sensory and extrasensory perceptions, for the validation and signification of all types of human experience. There can be no unity of knowledge unless there are epistemologies for every kind of experience, and a framework of sufficient depth and breadth for models to account for experiences. The rational does not need to be pitted against the emotional, intellectual against instinctive, or analytical against inspirational. Humans have different patterns of thinking, therefore different epistemologies.

Different epistemologies have different concepts of time and universal structures. Navajo epistemology is a multi-element nonhierarchical mutualism. Knowledge is necessary to maintain harmony with beings of universe; religion maintains harmony, so knowledge is religious. Chinese epistemology is complementarism; something and nothing are mutually generative, and inseparable, unlike Hegel's and Marx's dialectic logic of opposition. Japanese epistemology is situational, contextual; ephemeral and noncyclic.

Robert Theobald claimed that the cure for the ecocrisis is a changed way of perceiving reality, which would entail a new value system, a new epistemology, and a new world view. We must perceive that natural processes are balanced; some take thousands of years, but all effect one another synergistically. Bateson proposed a sacred unity of the biosphere, with fewer epistemological errors. Ecological knowledge can provide evidence of that unity.

4.1.1. *Traditional Ecological Knowledge*

Traditional Ecological Knowledge—also referred to as Traditional, Local, or Indigenous Knowledge—is generated within communities in place, and is therefore location and culture specific. It is dynamic and based on innovation, adaptation, and experimentation. This knowledge is the basis for decision making as part of a survival strategy. For a culture, it concerns critical issues of human and animal life: Primary production, human and animal life, and natural resource management. Due to the types of cultures, the knowledge may be oral, and not systematically documented in written or symbolic form. It is largely passed down by elders through oral histories, songs, crafts, and practical training. Traditional ecological knowledge (TEK) is the product of generations of learning and experience with the lands, waters, fish, plants, wildlife, and other natural resources of specific places. Inhabitants are trained from an early age to be aware of and respect the community of living beings that surrounds them. This means learning not only how to hunt, fish, gather and process key subsistence foods and other necessities, but also how to understand the behavior and roles of other species in the ecosystem, and how to successfully interact with them, often as equals in soul, in sustainable ways.

The dominance of the western knowledge system has largely led to a prevailing situation in which indigenous knowledge is ignored and neglected. It is therefore easy to forget that, over many centuries, human beings have been producing knowledge and strategies enabling them to survive in a balanced relation with their natural and social environment. Now, however, scientists are realizing and admitting that TEK may be in advance of the scientific knowledge available. The traditional method of knowledge can be characterized in a series of statements.

1. Observe phenomena and record facts.
2. Experiment with the phenomena and facts to see what happens

3. Make guesses and generalizations, using logic.
4. Analyze the phenomena, using cultural ideas.
5. Formulate rules from the generalizations about the phenomena. Such rules describe the behavior of the world.
6. Describe the rules using words.
7. Develop ideas, metaphors, models and myths.
8. Act accordingly.

The modern scientific method of knowledge can be differentiated (*in italics*) from the traditional in a series of modified statements.

1. Observe phenomena and record facts. *Measure them with numbers.*
2. Experiment, in a *controlled way. Measure* the components of phenomena, before and after manipulation, allowing the results to be *quantified.*
3. Make guesses and generalizations, using the logic of deduction. A *hypothesis* is a statement about relationships that can be shown to be untrue.
4. Analyze the phenomena into components, using cultural ideas.
5. Formulate *laws* from the generalizations about the phenomena. Such laws describe the behavior of a natural system.
6. Describe the laws *mathematically* using numbers.
7. Develop ideas and models. Develop a *theory* to predict new phenomena. Theories can lead to new conclusions and sometimes altered perspectives about phenomena. A scientific theory is a statement that postulates an ordered relationship among natural phenomena and explains some aspect of the world. It allows one to ask certain kinds of questions, some as specific hypotheses. A theory cannot be tested by hypotheses.
8. Act accordingly.

The basic differences are measurement, control, and the development of theories, which allow the knowledge base to be expanded by many cultures.

4.1.2. *Modern Science*

One main difference between TEK and science is measurement. Although it is as real as ideas, mathematical truth is not independent of human thought processes, and, that is why it matches the universe so well. It describes the working of thought and nature with great accuracy, although perfect precision is impossible due to the uniqueness of individual events. It is also an element in our way of perceiving phenomena. Jean Piaget states that knowledge is a special case of biological adaptation. Piaget suggests that all human action consists of a balancing of the processes of assimilation and accommodation. Accommodation means to establish a common measure; assimilation means to digest. Needs first incorporate things and people into the subject's own activity; the subject assimilates them into existing mental structures, then readjusts the structures as a function of subtle transformations, to accommodate them to external objects. Knowledge does not begin with knowledge of self or things, but with their interactions. By progressing to both poles, intelligence organizes the world. Science, however, the formal method of knowing, has become stuck in external accommodation.

4.1.2.1. Monocular Science

The game of western science is to trap the universe in networks of numbers and words. By reducing phenomena to features that can be controlled and quantified, we can objectify nature and everything in it. Nature is reduced to a resource to be processed. And, the process becomes self-justifying.

Science is regarded as instrument for manipulating rather than as study allowing participation in the processes. The scientific method always involves long boring periods of waiting for something to develop; the boredom sandwiches frenzy. The scientific method calls for skill and craftsmanship, for instance, cutting up specimens. Science is an occupation where most experiments fail, which dooms many animals to suffering and death.

Traditional science attempts to control a situation, by thought or experiment. An experiment limits degrees of freedom in a timeless and formal manner; it screens out quality. The gain from an experiment is an 'if/then' structure. Portions of the universe are placed behind glass in a laboratory world. Science cuts connection to direct observation. The formality of science makes statements about the outer world tautological. This is a problem with quanta, species fitness and psychological needs.

An experiment is not a piece of nature, but a contrivance where nature is made to write in conventional symbols. Arthur Eddington asks if experimental equipment tells us how nature can be made to behave, rather than how she acts. Heisenberg says that we have a science of our knowledge of nature, rather than of nature; we live in our description of nature. Natural science describes nature as exposed to our method of questioning. Scientists must avoid Whitehead's Fallacy of Misplaced Concreteness, wherein abstract notions, such as the ecosystem, are accepted as being concrete things, such as communities of animals and plants in place. We have not discovered the whole of what is contained in reality independent of ourselves.

Schrodinger noted that "The scientific picture gives factual information, but is silent on all that matters: Bitter sweet, pain delight, beauty ugly, good bad." Science consists of the observation of facts, of what happens in an ordered system, of the analyses of these facts and the formation of hypotheses to explain them, and of the testing of hypotheses. Science forms hypotheses, which admit the possibility of error. Hypothesis has replaced dogma and doctrine for authority, but, there is still fault in objectivity.

Ecologically, we are deceiving ourselves, not by the objectivity of our facts, but by their triviality and unrelatedness. Parts of problems are identified and analyzed, but the conclusions are trivial and weak. Many studies are too specialized or irrelevant. The science of Bacon collects the fragments of experiments and weaves them into a corporate being, with its own collective power. Scientists carry crumbs to an architectured mound, like termites, unaware of its purpose and ethically irresponsible for its effects. A sense of responsibility is weak or nonexistent; there is no concept of what is good. The ethical impulse to decide good is a null thought.

Modern science does not capture the state of being that experiences symbols. It does not value rare experiences of direct observation or through clear description. Symbolic understanding could improve the quality of human observation. The modern biological view of man is the result of the loss of a symbolic understanding of nature according to Jacob Needleman. The symbols are only understood in a different state of consciousness. Pre-twentieth-century teachings are often judged as picturesque or insane. John Ruskin regarded the false perception of smiling fields and somber mountains as a 'pathetic fallacy.' Emotional responses should be tied to the plain facts of nature; Goethe knew that there were no plain facts. Now our perception is crippled by an apathetic fallacy—we see nature as dead. The pathetic fallacy is a fallacy only to the apathetic, who are victims of the apathetic fallacy.

In no other century of our existence have humans learned so deeply and painfully of the depth and extent of their ignorance of nature. We can confront this and ameliorate it with science. Crises occur not so much from mismanagement as from the construction of a deficient code for reading nature, leading to deficient interactions in nature. This root cause lies in the foundation of the scientific world view and perceptions engendered from it. An alternative form requires living and knowing differently.

Science and technology are spiritually impoverished, divorced from awareness of values and purposes. Technology must be placed in perspective. The analysis of complex problems is beyond the specialist as is the synthesis. Nature was cast as an opponent of human values. Many scientists, such as Polanyi, Maslow and Koestler, have questioned the adequacy of scientific method and theory, with the intention of eliminating overreduction and broadening its sensibilities. Science can enlarge its capacity for corrective self-awareness. The blending of scientific objectivity with the sensuous and intuitive capacities of the mind is called hierarchical integration by Maslow. It should strive for comprehension above clarity.

Science is a controlled process, an additive process, and a social process. Knowledge of science combined with art can make robots, fantastic houses and enriched places.

4.1.2.3. Mythology & Limits of Science

However, science has not made a paradise on earth. Many things for some people have gotten better, but most things for most people have not. Some of the great inventions and advances have had negative effects. For instance, pesticides concentrate in the food chains and diminish many species. Electric power brings light to people, but also increases many incidences of stress and cancers.

Some problems are the result of the mythology of science. Scientific explanations share the same patterns as mythological ones. A myth is a set of assumptions that form the basis for human comprehension of the world. Mythical thought is an intellectual construct, an elaborate response to human needs. It can differ from scientifically thought in that it can ignore the logic of exclusion and contradiction; myth can permit many centers, for example. Science is a form of mythology, equally as important as any other. Science is mythical, as can be seen from comparing it to voodoo:

presuppositions=unproved dogma; method=voodoo;
facts=deities; neutrality=moral order.

As in classical mythology, these characteristics are interconnected and interdependent.

The myths of science, those things that are incorporated into popular understanding, are sometimes used to justify inappropriate behaviors. Charles Darwin's ideas, for example, were used by myth to support the mechanistic rape of nature by technology. The theory seemed to support the assertions of class, nation and race. Natural selection in a wetland ecosystem justified ruthlessness in a factory. The use of biotechnology is economic in a primitive sense, only the myths have changed to include greater manipulation. Biotechnology is still concerned with utility, growth and efficiency, as short-term goals.

Other problems result from the assumptions of science. Many scientists make these assumptions in the context of their cultural situation. Other assumptions of science are more subtle and more limiting. Science is logically limited and philosophically limited, as absolutist, mechanistic, reductionistic, and single-visioned. Science is limited by the deductive and bivalued logic of Aristotle; things cause other things in a linear way, through cause and effect, in a hierarchical universe, where values are rank-ordered. Science is also philosophically limited, by logical positivism. Science is absolutist; there is invariance of regularities of nature in time and place, space time is considered absolute, classes are eternal and immutable, and knowledge is limited to kinds, not individuals—the limits at both ends, prokaryotes and planet, thus fall out of range of the process of science. Science is mechanistic; the metaphor of mechanism by Descartes, that the world is a machine, implies that matter is inert, everything is replaceable, everything can be controlled, and that machines must have a creator, that is, they are not self-making. Science is reductionistic; anything can be taken apart and analyzed to be understood. And, science is single-visioned; one natural interaction, competition, is basic, and one interpretation is correct.

Analytic science has reached its limits. Data and information developed by hard studies have undercut the paradigms that guided their investigation. The compartmentalization of scientific fields has exposed the complex connections of the subjects.

A. N. Whitehead thought that what had been missing during the formation of science was a sense of relatedness. Early science saw the world as mechanism; modern biology is seeing it as resembling an organism. Organismic trends can be seen in sciences, from relativity and gestalt psychology to ecology.

Unfortunately, the language from a mechanical world view dominates even ecologists and politicians. This world view impoverishes humans by claiming all consciousness for humanity. It claims that nature offers no joy, or love, or peace, or certitude. Emphasis on the evil of nature creates a gap between humans and their universe. In contemporary cosmology, there is no room for the intrinsic worth of nature. But the operation of science—not as a side-effect—undermines the scientific cosmology and provides the elements of its alternative: Wholeness and relatedness.

What is missing from science in general, besides a comprehensive understanding of wholeness and relatedness? Science does not have the abilities to deal with ambiguity, indeterminacy, change, uniqueness, or normative decisions. Science needs to develop the ability to deal with ambiguity; confusion about the kinds of entities leads to ambiguity of reference. The arbitrary use of expressions like individual, abstract, and natural, introduce an ambiguity of meaning. Science needs to have the ability to deal with indeterminacy. Ernest Mayr lists reasons for indeterminacy in biology: The randomness of an event with respect to its significance; uniqueness of entities at higher levels, extreme complexity of organic system, and emergence of new qualities at higher levels of integration. Not all properties of a new entity are a logical consequence of the old. Vagueness is inescapable; there can be morphological similarity despite genetic differences. Science needs to be able to deal with change: Changes in laws or perspectives. Science needs to deal better with wholes and parts or whole-part relationships. Science needs to have the ability to deal with uniqueness or single events. Science has some ability to make connections, including participation in the system. Science changes what it looks at, from electrons to organisms and living systems. Science needs to have the ability to reach normative decisions, to attend to values and ethical behavior; there seems to be no response to problems, such as extinction, destruction, fragmentation, that are related to the exponential increase in human population numbers.

4.1.3. *Organic Science*

Science has its revolutions, too. Science does not need to be based on logical positivism and reductionism, though these have allowed dramatic changes. Whitehead thought that what had been missing during the formation of science was a sense of relatedness. This lack has been accentuated by conceptual barriers, that is, scientific fields. As with an organism, the various parts of nature are so interdependent that nothing can be abstracted without altering the identity of it or the whole. Science can be restored to full sight from one-eyed reason only by rediscovering the depth of relatedness. Organismic trends can be seen in science: From relativity to gestalt psychology and ecology.

4.1.3.1. Holecological Science

Holecological science reflects the idea of the whole as a basic organizing principle in nature. Whole things are mutually defining, but also self-defining or self-making. Nature is a self-making system; species and organisms are self-making. The ontology of any living system is the history of the maintenance of its identity through continuous self-making, or autopoesis. The evolutionary stability of the subassemblies—organs, organisms, species—is reflected by the degree of autonomy, self-government, that each has, according to Francisco Varela. The system develops through a continuous dance of autonomy and control; autonomy represents generation, internal definition, internal regulation, and self-assertion, whereas control represents consumption, instruction, assertion of other identity, and external definition. The organism and environment are co-defining. Living is a process of self-organization, where the self is the whole organism/environment system.

Holecological science deals with different levels of a knowledge hierarchy, focusing on whole organisms, populations, communities, and ecosystems, but it attends to genetics as well as to geological and evolutionary events. It balances the quantitative with the qualitative. It pays attention to the historical character of fields and wholes. It stresses the interconnectedness of an evolving world, where organisms enfold their structures based on historical patterns and unfold by sorting through environmental limits. The process generates form and variation at every level. The beginning of the description of such processes lies in metaphors.

4.1.3.1.1. Metaphors

Holecological science uses metaphors. Metaphors create analogies that, while they may have grammatical absurdities and psychological surprises, extend meaning. New meaning is produced through juxtaposition. The simple process of equating two things results in a complex of qualities that can be organized and examined in detail. The multiplicity of qualities is more than just a list of different qualities. Metaphors provide structural coherence for ideas. They allow verbal understanding of complexes and frameworks. They also capture the wholeness of things being equated. The metaphorical process leads to construction of models.

4.1.3.1.2. Models

The concrete nature of metaphors and models is essential to science because it limits the implications of any particular abstract system. The crudeness of a paradigm stimulates and binds the imagination, giving direction to abstract expression, and linking contributions of private images, informal words and formal theories.

The model represents in a simplified form the processes of some aspect of the real world, such as political conflict. The building of models proceeds in stages, from the delineation of the system to conceptualization of interrelationships, and then to quantitative specification. The first two depend on organized thought; the third requires the collection of data. Operating the model by manipulating its elements, by thought or computers, for instance, in order to see how it reacts, is termed simulation. There are many types of operational models, including mathematical, physical and gaming. The type of model depends on knowledge of real world. Descriptive models describe a situation at a given point in time; predictive models contain time-dependent variables used to predict future occurrences.

There are hazards with models: The ambivalent inclusion of unacknowledged elements; the association with war games; or, the wrong interpretation of exploratory models. The model itself has more serious problems, for instance, it is homogenous, with simple psychological variables.

Computers can compensate for some of these hazards, through the creation of virtual space. A computer works with the most primitive order, the binary, with one degree of freedom and one difference, and builds up a virtual space. Such a hyperworld can also include text and graphics. But basically, the computer is still a clock-driven machine.

Not enough is known about operation of a complex world to use a mathematical model. An analysis can be quantitative without resort to computer-based statistical models. Karl Popper stated that we must imagine the world to be like a cloud rather than like a clock. The key to the universe is metaphor, not measurement. Clouds are difficult to measure, and even ambiguous to description. But many things can be seen in clouds.

4.1.3.1.2.1. Systems Models

An event generates a system. The system regenerates itself with events, in a kind of eternal rebirth. A system is a model of general nature, a conceptual analog of certain universal traits of observed entities.

Edward Wilson states that systems may operate in three dynamic modes: deterministic, telic (normative) and probabilistic. First order attributes of these systems are: Deterministic, which is closed, predictable and causal; telic, which is open, finalistic, irreversible, and forecastable; and, probabilistic, which is locally open, generally acausal, reversible, and unpredictable. We would be wise to assume that all three exist and not try to reduce them to one general mode.

The success of general systems theory can be attributed to the use of telic models, systems of description in which values are admitted implicitly or explicitly as components of the system. A world model must be comprehensive, normative and descriptive in the enlarged sense, to meet the challenge of present realities to allow emergent qualities. Systems have predictive value.

4.1.3.1.2.1.1. Emergent Systems

The evolution of universe is the history of the unfolding of differentiated order or complexity. David Bohm states that we have been conditioned to the belief that higher orders of nature are determined by the lower order of the mechanical motion of particles. It is impossible to exclude the contrary assumption that high order features of natural laws are as fundamental as those referring to atomic movements. By comparison, unfolding implies the interweaving of processes of structuring at different levels. Evolution acts in the sense of a simultaneous and interdependent structuration of a micro and macroworld. Complexity emerges from the interpenetration of processes of differentiation and integration, where the processes run simultaneously from top and bottom, and shape hierarchy from both sides. Microevolution generates macroscopic conditions for continuity and macroevolution generates microscopic elements for its processes. Life is an emergent property of a hierarchy of chemical levels. It is not found in molecules but through autopoetic organization of molecules. The universe restructures itself constantly at a more complex level.

The world, as it exists in its ceaseless changes, appears as a single cosmic process in which higher orders of being emerge. The word emergent is borrowed from Lloyd Morgan, who first used it in this manner (1912), although emergence and ideas of organism came in rough form from Herbert Spencer (circa 1880). Morgan later set forth a view of the world as evolutionary process (1923). The word emergent was used to show that higher orders of being are not mere resultants of what went before and were not contained in them as an effect is in its efficient cause. Jan Smuts amplified the idea (1926); he attempted to state the principle of emergence by saying that nature is permeated by an impulse towards the creation of wholes; each stage of evolution is marked by emergence of a new type of individuality embracing and transcending the previous parts of itself.

Systems behave in emergent ways. Ruth Benedict coined the word synergy in the 1930s to describe this behavior. Synergy is the behavior of whole systems that is unpredicted by the behaviors of any of the systems parts when assessed separately from the other parts. Synergy has its own opposite: Conflict. Synergy is from Greek *synergia*, meaning a working with. Synergy is destroyed by separation. The concept of synergy has been applied in anthropology (Benedict), Psychology (Maslow) and science (Fuller).

4.1.3.1.2.1.2. Ecosystem Models

There are two basic ways of experimenting with ecosystems: The first involves the conventional process of formulating hypotheses, designing and conducting experiments, and analyzing and interpreting the results. The second involves the abstraction of the system into a model— mathematical, conceptual or metaphorical—the application of the argument, and the interpretation of conclusions. The model can be used to predict states of the system and then compared to the system after a period of time for verification.

A perfect model is impossible; it cannot account for all behavior or be valid in all frames of reference. The complexity of ecosystems and impossibility of knowing all possible states denies a perfect model. A base model would be capable of accounting for all in-output behavior and be valid in all experimental frames. For ecosystems a base model can never be known due to the complexity of the system and impossibility of observing all possible states.

Models of unilateral control are not valid, since environment does not completely control humans or vice versa. Ecological models should not involve precise exponential functions, as Slobotkin judges. The world is a sloppy place; to describe it, it is better to build a model with a high redundancy of subsystems, each of which is coarse, but the aggregate of which gives good adjustment to environmental events. A model is a good intellectual tool for analysis; it can be used to work out feedback consequences in virtual space, where the costs of mistakes are also virtual.

Microscopic analysis of ecosystems is difficult. The macroscopic view is similar to thermodynamics and cybernetics; and permits manipulation of concepts like energy flow, information and diversity. Howard T. Odum proposed (1971) a macroscope for looking upward from the organismic level. Ecosystems have almost unlimited divisibility. The subsets have an asymmetry of interactions. Ecosystem models could be expanded to include human activities, although they would have even more limitations. Even with those limitations, however, the perspective of ecology could contribute some understanding to human places and systems.

4.1.3.1.3. The Perspective of Ecology

Ernst Haeckel coined the word *Oecologie* in 1866 from the Greek *oikos logos*, the study of the household, and defined this science as the body of knowledge concerning the economy of nature. Since humanity inhabits the biosphere, it is also part of the subject of ecology. Ecologists try to understand the whole process of life in the context of the chemical, geological and meteorological environment by combining the isolated knowledge of specialists into a single, ordered system. Ecological science is therefore the science of everything. It cannot ignore advances at a fundamental level, in genetics, geology, or microbiology.

A larger ecology is concerned with qualitative analysis, with classes of systems. Hypotheses are tested by detection of differences between systems. It is more interested in understanding than in predicting; it asks for the significance of a variable, not what a measured future state will be. It starts from the assumption that science studies intellectual constructs and that directly observed variables are no less theoretical objects than are competition coefficients. It aims at discovering how much we can afford not to know and still understand the behavior of the whole. Human ecology cannot be limited to strictly biological concepts, but it cannot ignore them or transcend them either.

Ecology deals with relationships of organisms to environment. It is not a reductive discipline, and not amenable to easy quantification. Ecology per se cannot be studied, only its components and relations; it is therefore a perspective, a way of seeing. It is a perspective of the human situation in its interconnection. Ecology includes all events. The elegance of a fish pond and the delicacy of its equilibrium are the outcome of a long evolution of interdependence. But, no one has ever made a complete census of even so simple a system as a pond. Ecology uses observation to develop rules for treating variables, to create new abstract entities for study and to decide what needs to be measured in order to answer questions about the ecosystem. Ecology has usually confined itself to the lower forms of life.

Ecology has been called the subversive science. Theodore Roszak says its holistic sensibility, grounded in aesthetic intuition, is a radical deviation from tradition. But will the science develop as a manipulative technique or embrace nature as beloved person? Perhaps ecology could be called the psychiatry of nature. The human mind shares nature's intent— producing experience. The mind fits nature.

The critical message of ecology is that if we diminish the diversity of the planet, we debase its unity and wholeness and destroy the forces that create harmony and stability. If we wish to advance the unity of the natural earth, to harmonize on higher levels of development, we must preserve and promote diversity. In nature diversity emerges spontaneously, as the capacities of new species are tested by environment. Those species that enlarge their niche also enlarge the ecology as a whole. Ethical, aesthetic and utilitarian reasons all support the efforts to conserve the diversity of nature.

The whole ecology is related to furthering the greater pattern of the universe itself. The universe is a regenerative system, of which we are a part. No energy is created or lost, according to classical physics. Theodore Roszak asked if ecology could approach the sacramental vision of nature. He characterized ecology for its sensibility; it is holistic, receptive, trustful, aesthetic, and intuitive; it is judgmental. This holistic attitude also flows through some professional ecologists, from Frederick Clements to Eugene Odum. Ecology must be capable, on some level, of assimilating moral principles and visionary experiences to be a science of the whole planet.

4.1.3.1.3.1. Patient Practice

There are ways of dealing with the earth that are not scientific or technological; they are aesthetic or ethical. They are not incompatible with a whole science. The methodology of traditional science is limited and wasteful. Ecology considers every-day observations, unique occurrences, and short-lived phenomena, as well as emergent phenomena, such as emotions and community spirit. Any investigative effort needs to incorporate the observations of many others.

In his field theory of Being (capitalized to refer to the ground of all beings), Martin Heidegger stated that a concern with Being implies a patient existence, a willingness to lie in wait for an image to produce itself, without the watcher imposing meaning. Heidegger counsels to watch and wait, not to reach out and guess, although reaching or guessing may follow, under some circumstances. To examine organisms, and nature in general, we must shift to a waiting approach, asking rather than telling, observing rather than manipulating; being receptive and passive, more than active and forceful; nonintruding instead of invasive, noncontrolling instead of manipulative. This concept of ecology as patient practice stresses noninterfering observation rather than controlling manipulation, reception rather than force.

This is part of the paradox of a duality of ecological science; it must be detached yet concerned, free yet committed, and independent yet responsible. Classical objectivity may be contrasted with this kind of patient practice, which is another path to objectivity with greater perception, based on participation in the system. Loving perception provides kinds of knowledge not available to nonlovers; this is especially true in ethological literature. Abraham Maslow cites his own work with monkeys. Konrad Lorenz, Tinbergen, George Schaller, Van Lowick-Goodall, and Michael W. Fox have found it to be true. This is the way a good psychotherapist, teacher, parent, or friend functions. It is the way a good scientists should function.

4.1.3.1.3.2. Soul Science

Nature is an extremely sensitive nexus of means and ends. Nature is a feeling system. We need a new animism to approach this nature. This animism would allow us to behave 'as if' nature was intelligent and sensitive.

What is necessary is not a primitive animism or a single-vision science, but a scientific animism, to supplement science and scientific humanism, and to understand our animalistic nature and to use it as the foundation for a sound human ecology, philosophy and psychology. The word *anima* is from the Latin word for soul; the word forms the basis of other related words, such as 'animal' or 'animate.' An inquiry that would carefully and appreciatively consider the animal aspects of human beings and how we understand and empathize with other living organisms would be literally a soul science.

A scientific animism would consider the relations of humans to vegetation and the human attitudes toward ecotypes, like open plains and dense forests. It would consider the need for sacred places, and open, quiet or wild landscapes. It would consider territoriality, aggression, and the aesthetic reaction to the wonder and beauty of life. Animism would allow investigators to behave 'as if' nature were intelligent and sensitive. A scientific animism is aware of the effects of its activity.

A scientific animism would be concerned with far more than the anatomy and taxonomy of animals; it would be concerned with mutual training between human and nonhuman animals, that is, the emotive bond. It would be concerned with the need for touch and phylogenetic possibilities of animal empathy—dogs, for instance, exhibit strong physiological changes when they are petted and human blood pressure drops when humans do the petting. It feels good to be touched.

A scientific animism needs to understand the meaning of being human, to go below cultural or social explanations of love and alienation. And it may, as a genuine science, forget the analysis and lose itself in the ecstasy of the phenomenon that it seeks to explain. Science functions best when we understand things so perfectly that we no longer need it.

4.1.3.2. Science as Ethnosophics

Analytic science has reached its limits. Data and information developed by hard studies have undercut the paradigms that guided their investigation. The compartmentalization of scientific fields has exposed the complex connections of the subjects. Science does not need to be based on logical positivism and reductionism, though these have allowed great, insensitive changes. What is missing from science? The ability to deal with ambiguity and indeterminacy, the ability to deal with change, the ability to deal with wholes and parts, the ability to deal with uniqueness, the ability to reach normative decisions, the ability to make connections, and the ability to attend to values and ethical behavior.

A. N. Whitehead thought that what had been missing during the formation of science was a sense of relatedness. Early science saw the world as mechanism; modern biology is seeing it as resembling an organism. Organismic trends can be seen in sciences, from relativity and gestalt psychology to ecology. Ecology for instance is nonreductive, integrative, amphibious (authority of science, force of morals), and perspectival. It is not a reductive discipline, and not readily amenable to quantification. Even scientific ecology is an integrative discipline that extends beyond the bounds of science. Ecology is an amphibious discipline, with the authority of science and the force of moral knowledge. Ecology, studied through its components and relations, is a perspective, a way of "seeing," according to Paul Shepard. It is a perspective of the human situation in its interconnection. For Paul Sears, ecology is a subversive subject. It is normative and sensible. Furthermore, for Theodore Roszak, it offers a sacramental vision of nature. A new science could be created and called 'ethnosophics,' to reflect the wisdom of local cultural knowledge combinedwith an ecological perspective.

4.1.4. *Crisis Science*

Traditionally, science has not paid attention to its effects or the large-scale effects of human participation in nature. Some categories of science, however, have recently addressed just these issues. Despite an increased understanding of ecological systems, the systems that support human cultures, as well as other populations and habitats, are being exploited through an economic system that ignores ecological relationships and limits, with the result that forests are being clearcut and fish are overharvested, even though we know how to cut and fish in sustainable ways. New approaches to science are being developed, as crises sciences, that respond directly to the continued loss of the heritage of rich biological diversity.

4.1.4.1. Conservation Biology

Conservation biology is a synthetic discipline that draws on history, science, and philosophy, and is applied to crises in forestry, resource management, and preservation. It addresses a complex suite of problems, from genetic loss and interference to destruction and extinction. In addition to ecological science and the documentation of environmental destruction, by Fairfield Osborn, Rachel Carson and others, conservation biology draws on romantic idealism and early naturalism, including ideas from R.W. Emerson, Thoreau and G.P. Marsh, John Muir's preservationist ethic, Gilbert Pinchot's resource conservation ethic, Aldo Leopold's land ethic, religious beliefs such as Taoism, traditional cultural beliefs, such as Northwest First Nations, and current ecological philosophies, such as deep ecology.

The goals of conservation biology are: (1) To investigate human impacts on biological diversity, and (2) to develop practical approaches to prevent the extinction of species, by providing scientific conservation principles, by identifying conservation problems, by establishing corrective procedures, and by making scientists and managers responsive to problems and issues.

Conservation biology is explicitly based on a set of ethical statements that suggest approaches to research and practical applications. These statements are:

1. Biological diversity has intrinsic value. Species have value to themselves, as well as to others and to the whole, all of which are conferred by evolutionary history and which supercedes human economic valuation. Diversity is the source of variety as well as of the aesthetic appreciation of nature.

2. Ecological complexity is critical. The interactions of species in ecosystems are complex and support biological diversity. A complex natural environment supports complex interactions, such as coevolution, that probably would be lost in artificial communities. The evolutionary process leads to new species and increased diversity—relatively large natural populations in natural cycles are required for evolution to continue.

3. Untimely extinction is catastrophic. Normally, at the evolutionary time scale, extinction is balanced with speciation; human destruction of habitats has increased extinction thousands of times and reduced speciation.

Conservation biology is concerned that industrial processes are reducing diversity. For instance, current logging practices reduce the complexity by knocking out key players in the ecological web, players such as spotted owls, fungi, and dead trees. When key species are removed, other species and their interactions change; for example, with the removal of the top predators, the Florida panther and red wolf, from central Florida in the 1930s, the raccoon population in the Ocala National Forest exploded, according to Larry Harris, to perhaps twenty times the numbers of the 1930s—and raccoons prey on birds, tortoises, snakes, and frogs, threatening the complexity, and thus diversity, of the habitat.

4.1.4.1.1. Understanding Losses of Habitats & Ecosystems

As human populations rise exponentially, the number of other species starts to drop exponentially, according to Michael Soule. There is an interesting parallel between extinction in ecosystems and the human colonization of those same ecosystems; an almost perfect correlation exists between extinctions and the human colonization of North America, New Zealand, Madagascar, Australia, and other places. More than just loss of species, there are losses of whole habitats, of areas, of numbers, and of diversity. The agents of extinction are identified as:

1. Overkill, the cause of most traditional extinctions, and overexploitation, resulting in the complete use of some species, such as rhesus monkeys for medical research.
2. Habitat destruction and fragmentation (the future of extinctions). The fragmentation of habitat shrinks areas, increases edge effects, changes demography, and depletes gene pools.
3. Exotic species, especially those introduced to islands, including exotic disease organisms.
4. Pollution of air, water, and soil.
5. Climactic change caused by introduced chemicals and the disruption of atmospheric cycles.
6. Chains of extinction, as a ripple effect through disrupted communities.

These agents of extinction, are influenced directly by human conditions, according to Soule. These conditions include:

1. Population growth—people put pressure on ecosystems; there is enough food to feed more billions, but only if we convert every ecosystem to agriculture; alas, there is probably not enough wood or copper or cotton to meet other human needs.
2. Poverty—hunger crowds out other actions; trees are cut, rare species killed; poverty is made worse by the contrasts of advertising, which increases aspirations and needs.
3. Ignorance and misperception—many connections are not known; and, the desire for immediate satisfaction lets them remain unknown; we are limited by our size and our lifetimes—slow catastrophes pass unnoticed, unlike hurricanes.
4. Anthropocentrism—human concerns dominate our consciousness and our giving; anything that is not used has less value or importance.
5. Cultural instability—due to rapid change or transition to different political outlooks; resource management that requires long-term management suffers. Rapid take-over of habitats by humans results in the most destruction; this is especially true when sponsored by northern commerce.
6. Economic limits—markets are internationalized and commodities are priced arbitrarily; planning is limited to 2-4 years.
7. Policy failures—policies are not implemented because of social instability, corruption, and war. Social justice does not protect biodiversity. These conditions cause interference in ecosystems.

4.1.4.1.2. Saving Systems

Conservation biology attempts to create strategies for reducing interference with habitats and species. One way to do this is through the establishment of reserves of adequate sizes and appropriate shapes. Another is the concept of zoning for human uses. Still another is to identify the minimum viable sizes of populations and areas.

The minimum is not easy to identify. Jared Diamond has studied the critical area for the minimum viable populations of some birds. Michael Soule and others have studied critical areas for mammals, which also require large areas. Saving large areas can increase the representation of species and save viable mammalian populations, that is, they can maintain the integrity of wild gene pools. Large areas can also permit natural processes to occur without human interference.

The minimum size of an ecosystem is a complex function of the area's key species, quantity of suitable habitat, and minimum viable numbers of species. Large-bodied vertebrate species tend to have lower population densities, thus a reserve with self-sustaining large-bodied vertebrate populations will likely be adequate for herbivores, insectivores, and primary producers.

Determining the minimum number of individuals in a population to guarantee a high probability of survival results in widely varying minimum areas, depending on the key species selected. Carnivores are usually the least dense of species and have the smallest populations. They are also sensitive indicators of ecological integrity. Protection of the rarest elements should provide security for more dense species in an ecosystem.

Usually, large carnivores are a sensitive indicator of the carrying capacity. By comparison, the area to preserve herbivores or autotrophs (plants) as individual species is far smaller—although the ecosystem without the wide-ranging species is more impoverished and may evidence wobble in the future. Minimum habitat protection is necessary for the protection of endangered or threatened invertebrates, which are responsible for maintaining basic ecological processes through predation, recycling, and pollination.

Although there have been debates over whether a single large area is better than several small ones, the shape and size of wilderness is determined by habitat studies of the unique natural history and conditions. The size should be large enough so that species will not be vulnerable to "extinction vortices" caused by genetic or environmental stochasticity. In many areas, disturbance from resource reclamation or recreation would probably be the greatest threat.

Natural processes, such as fire, wind, or species explosions, would be allowed to operate freely, even if they altered the functioning of the system. Larger areas would reduce management costs; smaller reserves in general require more intensive management and habitat manipulation. Saving large areas might be economically or politically difficult under the current industrial monolith in much of the world. The cost of reserves would be not be high, even in terms of lost potential, when the costs of resource extraction and transportation are considered with the more neglected costs of loss of habitat and local sustenance.

Once again, simple size may not be enough to preserve a habitat, if the original biodiversity is impaired. Thousands of species are going extinct as their habitats are modified by large-scale human activities.

4.1.4.1.2.1. Saving Systems By Design

The design of landscape areas can reflect the operations of the system in the past, present and future. Reserves should have a north-south axis under conditions of global warming or freezing. Edge effects started out as being centers for diversity. Now, many edge effects are considered to be destructive. In the 1940s, it was thought that edges should be increased to provide bountiful game crops. Edges have proven to be good for both game and wild species. Edge effects, however, are detrimental to populations adapted to forest ecosystem interiors. Too many edges reduces forest diversity at local and regional scales. The appropriate design can offer an optimum ratio of edges to interior space.

4.1.4.1.2.2. Saving Systems By Zoning and Tactics

Michael Soule identifies eight paths for potential biotic survival: (1) *In situ*—bounded areas with little human disturbance; the problem is that few are large enough for natural processes to operate in or for the minimum populations of large species. (2) *Inter situ*—outside the boundaries of protected areas, where native species still exist, usually nonarable or remote lands. (3) Extractive reserves—can be inter situ or occupied by first nations; resources are harvested on a sustainable basis, e.g., grass, nuts, rubber, logs—but requires low densities, stability, and caution. (4) Restoration projects—intensive management of (usually) small areas to restore sets of species. (5) Zoological parks—facilities where species can be maintained under semi-natural conditions; not the same as in situ because target species, which may be exogenous, are introduced. (6) Agricultural projects—highly managed production-oriented systems for crops including trees and shrubs; agroforestry also combines trees with grazing and wild animals. Many native species, from coyotes to bunchgrass, can survive within the system. (7) Living *ex situ*—botanical gardens, zoos, aquaria that maintain and propagate organisms for conservation purposes. (8) Suspended *ex situ*—artificial banks of living material preserved in storage, often cryogenic.

In situ or *inter situ* are the least artificial or intrusive; however, they are also the most vulnerable to population pressure and social instability. The *in situ* strategy is the most likely to save entire ecosystems and their processes, as well as biogeographic assemblages, indigenous species, local populations, and overall genetic variation. Right now, *inter situ* locations have intermediate significance in saving these same targets. Extractive reserves have the potential to save traditional human cultures in place. Zoological parks have an intermediate chance to save populations and assemblages of species. Restoration projects have the potential to save entire ecosystems. Agricultural systems might save domesticated species and a few adapted wild ones. Living *ex situ* conservation can save wild relatives of domesticates. Suspended *ex situ* conservation can be highly significant in saving domestic varieties.

Traditionally, in the U.S. at least, conservation has been a strategy for public lands. It is becoming more private; nonprofit groups such as Conservation International and The Nature Conservancy are buying land and assuming responsibility for it.

4.1.4.1.2.3. Conservation Strategies for Saving Systems

Conservation biology can make several recommendations that would preserve diversity and complexity, and avoid numerous extinctions. The first is to promote extractive practices that respect the productivity and complexity of the ecosystem, leaving materials, legacies, and diversity. Another recommendation is to make property leases contingent on the maintenance of the productivity and diversity of the land. Also, the fragmentation of habitats could be reduced through design, taking into account the genetic diversity of the trees, catastrophic conditions, minimum viable populations, corridors, and edge effects. Finally, reserves could be made large enough to support minimum viable populations and minimum viable ecosystem areas.

Because current reserves are usually too small to hold viable populations, corridors must be planned to intersect with the larger areas set aside, and highway routes and underpasses must be modified. According to Larry Harris, highways are a major force in fragmenting habitats. Conservation biology recognizes the need for ecological planning at a landscape level. Who pays is another issue.

4.1.4.2. *Ecological Forestry*

Ecological Forestry is a direct response to the destruction of forests across the planet. Over the planet, entire forests are being removed or converted, while others are degraded or destroyed; in the United States alone, six million acres of forest are cleared annually, five million acres are degraded, and three billion cubic meters of wood are consumed. Rarely do these industrial numbers reflect whole trees, associated animals and plants, or living habitats and forests—that three billion cubic meters of wood came from one to six million trees, each of which was home to lichen, fungi, beetles, birds, and other beings. Some, such as pileated woodpeckers, are so territorial that many die with their tree, just as in Greek myth hamadryads were thought to scream and die as their trees were cut.

Ecoforestry, a nonprofit corporation that practices ecological forestry, is a concept-based, process-oriented, ecological, community forestry based on traditional wisdom and scientific knowledge, using a set of defined characteristics, principles and standards. Ecoforestry is a crisis strategy, which is also a way of life, for redesigning whole forests— preserving or restoring natural processes, respecting the opportunities for existence of other beings, and building upon the productivity of forests for human needs. Although ecoforestry sounds like a qualification of the modern, industrial forestry, it is in fact entirely different—it is based on a different philosophy, a broader ecology, a more comprehensive economics, new metaphors, and it is sensitive to limits, ethics, aesthetics, and spiritual values. Its larger perspective incorporates industrial forestry as a special case, much like the Theory of Relativity incorporated Newtonian dynamics as a special case.

Ecoforestry is grounded in the uniqueness of places where forests exist, unlike industrial forestry, which is designed to be noplace, without weeds, storms, or fires. Ecoforestry promotes a dynamic order that is historical and flexible. Forests should not be kept uniform and static. Forests are what they are as the result of the diversity of life—all forms of life contribute some value to the forest as a whole. The whole forest is already self-ordering and self-renewing. Taking a limited number of trees or plants from the forest contributes to the dynamics. The forests are our heritage, and our communities are part of the forests.

Ecoforestry is practical, realistic and far-sighted. Ecoforestry takes steps to stay within the limits of the functioning forest. Ecoforestry realistically plans complex open systems, i.e., forests, with flexibility within the variables and values of the system. Ecoforestry is attentive to long-term cycles in a forest and plans accordingly. Ecoforestry is also attentive to the very small changes in complex, self-regulating systems, such as forests, that have large and important consequences, as when rainfall patterns shift after a forest is removed. Ecoforestry tries to anticipate the consequences, as far as possible in the future, of every action. Through its goals and works, and conscious of the slow catastrophes, it tries to create optimal forest environments.

Ecoforestry is complex and sophisticated. Ecoforestry considers conceptual tools, such as satellite imaging and Geographic Information Systems. Ecoforestry recognizes that forests are part of a process that is unending and imperfect, without a final state, and furthermore, that the attempt to perfect it results in disharmony. Ecoforestry accepts a constructive conflict in scale with the ecosystem. There are chaotic events, plagues and random frenzies in every system; rather than deny this Ecoforestry tries to account for these factors. This acceptance leads to understanding and the abandonment of stupid strategies.

Reed Noss and others have noted that maintaining a healthy forest ecosystem is more efficient in the long-run than having to duplicate the forest functions to keep it healthy. Furthermore, it is dangerous to take maximum yields out of a system unless all factors are known, which is a virtual impossibility. Therefore, we must aim for optimum or satisfactory yields and calculate that the forest has sufficient flexibility to recover from our 'take.' Ecoforestry asks for voluntary limits to consumption—this does not mean that we should not modify forest ecosystems, just that we should dominate every one of them or transform all forests into human products. People are capable of mature behavior. Ecoforestry respects the principle of limited good, that desired things exist in limited quantities. Ecoforestry is comprehensive in its application of ethics. It tolerates all the inhabitants and structures of a forest. The entire community is considered. Human communities are embedded in forest communities; our cultural and spiritual achievements occur in the larger community, which supports human endeavors. Ecoforestry tries to be responsible, respectful and ethical in its actions.

Ecoforestry addresses forests and human societies. Ecoforestry includes humans and ultrahuman beings as an integral part of the whole system. Ecoforestry addresses both large and short-term scales, in time, size, and design. Some forests, however, in the Amazon and New Guinea, for instance, are occupied by people with different, and often better adapted, cosmologies and economies.

4.1.4.2.1. Ghosts in Clearcuts

The Ecoforestry Institute was organized in 1991 and became a nonprofit corporation in early 1994. The mission of the Ecoforestry Institute is to foster ecologically responsible forest use, through education and related programs and services, and:

- To engage in dialogue with interested persons aimed at deeper understanding of ecological forestry (ecoforestry);
- to develop an educational process to facilitate a paradigm shift from industrial forestry to ecoforestry;
- to educate and train interested persons in the knowledge, techniques and arts of ecoforestry so that they may serve as ecoforestry practitioners, consultants, and teachers;
- to develop and monitor demonstration ecoforests on private and public forest lands where members-of the public can see working models of ecologically responsible forestry and fully functioning natural forests from which forest goods are being harvested in a sustainable manner

- to develop, in cooperation with others, criteria for ecologically responsible forest uses, and standards for certifying ecoforestry practices, practitioners, materials, products and artifacts;
- to research and communicate to a wide audience the deepening knowledge of the multitudes of values and functions of natural forest ecosystems, as reflected in leading edge work in conservation biology, landscape ecology and related disciplines; and
- to respect and cooperate with indigenous peoples, to learn from them the wisdom of the places where they have dwelled for centuries. In addition to founding two educational institutes in the U.S. and Canada, the Institute was one of the founding members of the Forest Stewardship Council, which promotes certification standards for growing and harvesting trees.

The tasks of ecoforestry include the return to the natural history and knowledge of forests as the foundation for taking trees, to place forestry on a proper scale, with regards to time and species, to stop short of a maximum, and to encourage the symbiotic relationship of forest economics and wild forests. The scale of things is an independent problem that can ruin the best intentions of policy.

On almost every continent by the 1980s, scientific research had resulted in technological applications for growing, genetically "improving," harvesting, and processing woody material on an unprecedented, vast, and unconsidered scale. Industrial companies established tree breeding and fertilization programs, whose goal was usually fast growth. Pine plantations in the southern United States were rated as a success and served as a model for plantations elsewhere; Australia, New Zealand, Chile, and South Africa replaced much of their native forests with exotic pine plantations, using species like Monterey pine. Many trees in the same species normally grow in clumps within a forest. Industrial forestry changed that scale, however, planting entire forests of a single species. The problems of industrial forestry range from soil destruction to forest death.

One solution to many of the current problems is a reduction in scale for everything from forest use to management units, with local controls and local use primary. Temporal scales, however, should be expanded with long-term research and management. One principle of ecoforestry is that an ecological forestry must observe the proper scales of forests, especially in terms of size, age, and patterns (or diversity).

4.1.4.2.2. Preservation & Restoration

For thousands of years people have taken things they valued from forests. Culture and habit have resulted in large-scale trends in forest-use: More forests are used; more of each forest is used; fewer products from each forest are used; this seems contradictory, but it is not inconsistent with the way things are done in industrial cultures. The current economic style is too great, fast and reckless for ecological systems to absorb its impacts. The scale of things is an independent problem that can ruin the best intentions of policy.

An inappropriate time scale, its shortness and urgency, is a cause of many of these problems in industrial forestry. Although forests are considered renewable resources, they are slowly renewable, requiring hundreds or thousands of years to renew from catastrophic disturbance; this time is far longer than any economic plans and nonrenewable on a human life scale. This has implications for sustainability.

The forest ecosystem is a large-scale pattern of millions of minute events. A range of elevations across areas would minimize the effects of climactic change—and the possibility of extreme change is rarely considered in wilderness design or forest management. Soils, drainage, and land-use history and ownership would also receive similar considerations. This would allow management for diversity on different scales.

One challenge to ecoforestry is to identify the patterns and set up long-term programs to study them and relate them to sustainable use of forest ecosystems. Monitoring is crucial to understanding forests. The institute has set up long-term programs in three Northwest forests. The institute conducts research on cutting practices, forest health, biomass reduction, and habitat restoration.

Ecoforestry has to adjust the scale, to make the kinds of harvests appropriate for the forests. Forests are larger than stands of trees. On the other hand, really large scales, bioregional or global, are not considered enough. We should be measuring at larger scales: Watershed, landscape, or biome. Before satellite data analyzed by GISs, ecologists and foresters did not have tools that could address the scale of landscapes. Regional and global data was hard and expensive to collect. Although forestry has incorporated some of the technical tools, such as satellite imaging and GIS, it has neglected the conceptual tools, e.g., the Gaia Hypothesis or global design.

Ecoforestry recognizes that forests are part of a process that is unending and imperfect, without a final state, and furthermore, that the attempt to perfect it results in disharmony. Ecoforestry accepts a constructive conflict in scale with the ecosystem. There are chaotic events, plagues and random frenzies in every system; rather than deny this, ecoforestry tries to account for these factors with flexible plans. This acceptance leads to understanding and the abandonment of stupid strategies.

One solution to many of these problems is a reduction in scale for everything from forest use to management units, with local controls and local use primary. Temporal scales, however, should be expanded with long-term research and management. One principle of ecoforestry is that an ecological forestry must observe the proper scales of forests, especially in terms of size, age, and patterns. Only then can preservation and restoration work.

4.1.4.2.3. The Art & Science of Forest Use

Ecoforestry is basically concerned with using a new metaphor for forests and forestry. Trees, for example, are used as meaningful metaphors in many cultures. For many peoples, such as the Luiseno of California or the Huichol, the Tree of the World connects all worlds, all the layers of the cosmos, the netherworld and heaven. The poetic language of mythology can fit all the facts and values, things and images into our hearts so that we can feel them and act upon them to create good places in the forest.

Ecoforestry is comprehensive in its application of ethics. It tolerates all the inhabitants and structures of a forest. The entire community is considered. Human communities are embedded in forest communities; our cultural and spiritual achievements occur in the larger community, which supports human endeavors. This means that ecoforestry management styles are ethical. A style based on ethics, allows us to recognize individuality in trees and other beings, recognize the feelings and emotions of animals and the sensitivity of plants, promote a noncommodity, drymoperipheral approach to forestry, minimize the devastation of wild forests and wild ecosystems, and emphasize habitat loss as a human ethical issue.

Furthermore, in the near and far future, through its basis in ecological philosophy, ecoforestry suggests ways that biosphere cultures can be converted to ecosystem cultures, as suggested by Ray Dasmann, characterized by wonder and wildness, and it promotes ecosystem lifestyles, characterized by frugality and joy. Through its consideration of scales, ecoforestry relates the success of microorganisms to the ultimate success of living in self-sustaining forests, develops a basis for management techniques for finite ecosystems, especially wild forests, and links ecological sustainability with the richness and diversity of ecosystems, as well as with human pleasure. Ecoforestry works for immediate solutions to inequity and destruction under worsening and thankless conditions; and, it educates with confidence and energy for ecological enlightenment over the long-term, despite short-term wobble.

4.1.5. *Crisis Movements*

Some movements, having qualities of religions and educational institutions, work for change in a cosmology. Deep ecology, being based on a holistic philosophy, is a movement seeking to change the image of humanity and nature. Earth First!, being based on deep ecology, mysticism and scientific ecology, is another. There are other academic movements, such as Ecological Resistance and Ecophilosophy, that are responding to ecological crises.

4.1.5.1. Deep Ecology as a Crisis Movement

Deep ecology is a movement, formulated by Arne Naess, that goes beyond a concern with pollution and resource use to consider humanity in a relational, total-field image. The movement promotes human equality, conservation, and local autonomy. In principle, it proposes a biospherical egalitarianism, the equal right of all beings to live in place without undue interference. It adds a normative dimension to ecology in a framework of 'ecosophy,' literally the wisdom of the house.

When Naess presented his deep ecology alternative, starting in 1972, he contrasted it with the shallow ecology movement, which he characterized as fighting against resource depletion and pollution—the objective of the shallow form was the health and affluence of people in developed countries. Deep ecology takes its inspiration from the science of ecology. Deep ecology is inspired and fortified by ecological knowledge derived from experience, not logic. Its tenets are normative. Its value system is only partly based on scientific research. As a science, ecology describes the interrelationships of organisms and environments, that is, the experience of living together in the biosphere. Ecology is not a reductive discipline and is not readily amenable to complete quantification. Even scientific ecology is an integrative discipline that extends beyond the boundaries of science. Ecology can be considered an amphibious discipline, with the authority of science and the force of moral knowledge. Studied through its components and relationships, ecology is a way of seeing, a perspective of the human situation in its interconnection. It is a 'subversive' subject, normative and sensible, offering a 'sacramental' vision of nature.

The technological paradigm has reached its limits. Data and information developed by hard studies have undercut the paradigms that guided their investigation. When a paradigm shifts, perceptions change, as in the understanding of a metaphor. Deep ecology forms part of a new metaphor that is more appropriate to the unity and interrelatedness of the earth; and it may be more personally fulfilling as well.

The platform of deep ecology addresses the well-being and flourishing of human and nonhuman (the intrinsic value of life), as well as the richness and diversity of life forms, which humans have no right to reduce except to satisfy *vital* needs. The platform also considers that human interference with the nonhuman world is excessive, and that the flourishing of human life and cultures is compatible with a substantial decrease of the human population. Policies must be changed to reduce economic growth and to appreciate 'life quality.' which cannot and need not be quantified—this implies that you and I have an obligation to try to implement necessary changes. The platform addresses not only ethical concerns, but scientific, cosmological, economic, and political ones as well.

Deep ecology, more than just a popular movement, is based on a formal philosophy that differs dramatically from the silent (also turgid, limited, destructive) philosophy that underlies, like a procrustean bed, the actions of science and technology so adored and embraced by industrial models.

Naess offers the specific term ecosophy to describe the theoretical basis of the deep ecology movement. Ecosophy is a philosophy of ecological harmony. Naess also defines an ecosophy as a "personal code of values and view of the world," which guides one's decisions and grounds the acceptance of the principles of deep ecology.

An ecosophy contains both norms (value judgments) and hypotheses. It is more like an Aristotelian or Spinozan system, that is, a set of verbal expressions with a variety of descriptive and prescriptive functions, than a general systems theory or formal philosophy. The format of an ecosophy is based on Spinozan philosophy. Naess finds Spinoza too complex to serve as a patron saint of ecology, but the movement uses him as a source of inspiration. Adherence to Spinoza's system is consistent with a karma-yoga, a person externally active on all levels of existence, according to Naess.

Because of its normative and personal aspects, the philosophy only occurs in individual instances; his own version of ecosophy is referred to by him as Ecosophy T. Because an ecosophy is a 'total view' (intuitive and coherent), it cannot be separated into its components, e.g., epistemology, ontology, ethics, aesthetics, or political philosophy. There are no incorrect total views, which is one reason why Naess does not reject radical ecology, ecofeminism or other ecological philosophies.

As a philosophy, ecosophy asks different kinds of questions from traditional philosophy. Traditional philosophy asks how conservation can be deconstructed, or under what conditions a pre-emptive nuclear strike is ethical, or whether a tree falling in the forest makes a sound. By contrast, ecosophy asks how things fit together, what actions would reverse the catastrophic destruction of species and habitats, or how the vital needs of the population can be satisfied indefinitely, without unnecessary destruction or conversion of ecosystems.

As a philosophy, ecosophy investigates the normative aspects of living together, that is, ethics, and the maintenance of the affairs of communities, that is, economics and politics. As a noetic discipline, deep ecology provides information on the state of nature, recognizing that human beings are participants in nature, as well as participants in human societies.

Deep ecology emphasizes biological equality. When Charles Elton transformed the "Great Chain of Being" into a chain of eating, ecologists realized that the bottom link of the food chain, plants, was the most important. Humanity is part of the food chain, although it appropriates a large amount of the productivity of most ecosystems. The exploitative competition of humans in ecosystems is an important part of biogeochemical cycles. Humanity cannot unparticipate by choice. Deep ecology argues for diversity. In nature, variety emerges spontaneously, as the capacities of species are sorted by the environment. Variety provides flexibility in systems. The diminution of variety through human interference may debase the wholeness and stability of systems. Furthermore, aesthetic, ethical, and utilitarian reasons all support the efforts to conserve the diversity of nature. Deep ecology incorporates a broader scientific method that might be called patient practice. There are ways of dealing with the earth that are not scientific or technical; they are aesthetic or ethical. These alternatives are not incompatible with traditional science. Where the methodology of traditional science is limited and wasteful, promoting technologies that ignore or destroy values with blind quantification, the methods of deep ecology are traditional and conservative.

4.1.5.2. Science-based Movements: Ecological Resistance & Earth First!

John Rodman distinguishes four perspectives in ecology: (1) Resource conservation, which in America has been an aspect of resource development. Wilderness is regarded as a kind of land use where a scarce resource is conserved by being managed; (2) a Nature Moralist's perspective that grows out of the humane movement and radical animal liberation. Their notion of human virtue is justice; the right of all animals to exist; (3) wilderness preservation by any means. The problem with the California condor, for example, which has been proposed to be bred in captivity, is that the species is thought to be abstracted from a complex habitat that may not exist anymore. But most species cannot be abstracted; and, (4) a fourth world of perception and action that Rodman calls Ecological Resistance.

Ecological Resistance is action that precedes theory; a central principle is that diversity is good. The struggle between monoculture and diversity can be exemplified by J.S. Mill, who defended biological diversity against the threat of a humanized planet, West Indian blacks against European racists, women against patriarchy, and the multifaceted personality against totalitarian technology. Different levels of experience, from cosmos through polis and psyche mirror one another; the relationship between levels of experience is of a metaphoric mirroring, not superstructure or base or cause and effect, although the ecosystem could be considered a base model perceived as common gestalt. Ecological Resistance also involves a ritual affirmation of the myth of microcosm; acts are not undertaken in a spirit of calculated self-interest (of individual, society or species), or of moral duty, or to prevent profanation, but because a threat to the biosphere is a threat to the self.

Of the four perspectives: Resource Conservation has no justification for speciesism, because it is an economic treatment with a short time scale, limited to humans. Wilderness preservation links primordial experience or an encounter with the holy to a transient aesthetic; it is an aesthetic view. Nature Moralism imposes a species-specific morality of rights and duties. Ecological Resistance is unclear about the balance of nature. But, it offers a participatory image of humanity, as an integral part of food chain; an organic cycle of birth and death; a microcosm. Prudence, justice and reverence are part of the good life; integrity includes them. Eric Jantsch endorses Rodman's ecological resistance, as going beyond precaution, reverence and morality. But resistance is a limited stance. Human nature does not find meaning in an absurd world, but discovers its structure through interaction with the ambihuman order. Human identity exists partly in relation to nature; the destruction of one involves the other. An act of ecological resistance is an affirmation of the integrity of the naturally diverse self and world. The meaning of such an act is not exhausted by successor failure in linear sequence of events. One is aligned with the ultimate order of things by ritual action. Rodman's ideas of right action—the concept has its roots in Buddhism—are appropriate.

Based on the science of ecology and on other movements, such as deep ecology, Earth First! has reacted to the destruction of wilderness, taking direct action against corporations and machinery. Although they made their reputation by monkey wrenching, they have expanded into proposing strategies for saving wilderness (see Section 5.2.1.2.4. for a longer discussion).

4.1.6. *Ecological Education*

Earlier in our evolution, and during a far longer part of it, the environment was our school. But, we have replaced wild resources and natural enemies from which we learned with educators, from whom we try to learn, second-hand. It is now necessary for scientific studies to prove that love is essential to human development.

Inquisitiveness is strong drive among all young mammals but declines with age. Anthropoids owe their braininess to neotony. The anthropoid continues the habit of curiosity into adult life. Here is the animal origin of science.

All people have the potential for superb minds, complex bodies and the capacity for language. Children have ecological educational needs: Architecturally complex play space shared with companions; diverse experience of nonhuman forms, whose names and generic relations must be learned; progressively more strenuous excursions into a wild world for limited confrontation. All of an animal, history, body, emotions, is suitable for study. Education would progress from the tangible to the intangible; the sensual before mathematics.

Adults have special educational needs. As the population ages, youth may become underrepresented in the educational system. For this reason, it may be possible to offer a noncompulsory education. Students could learn by themselves, using video documentation or cassettes. Sometimes that is the most competent or meaningful learning. What is needed in colleges is a way to teach communal responsibility, self-reliance and physical work. Education should let one learn how to feel and live as well as think. In fact experience is necessary for thinking. This was recognized long ago by Aristotle. "Lack of experience diminishes our power of taking a comprehensive view of the admitted facts. Hence, those who dwell in intimate association with nature and its phenomena grow more and more able to formulate, as the foundations of their theories, principles such as admit of a wide and coherent development; while those to whom devotion to abstract discussions has rendered unobservant of the facts are too ready to dogmatize..." Aristotle states that the objective of education is to help each person develop their fullest potential, to live effectively in a changing world, to adjust to others, and to continue to be self-educated.

Education means to 'lead out' of ignorance. Education takes place in communities; it is, in fact, the means for communities to continue. As Plotinus and Novalis recognized, education has an outward, social and civil, aspect as well as an inward, personal and self-revealing, aspect. Thus, education produces an appreciation for the richness of nature, as much as it trains the individual for a position in society. And, it leads to an understanding of personal existence as well as of the nature of human society.

Education has become more universal, but its goal of the well-rounded individual has been distorted by its goal of training for the economy. To produce wealth for the state and livelihood for the individual, education has become money obsessed. Ethics has been neglected, since it might limit or contradict its economic obsession. In fact, the inward and civil aspects of education are restrictive to a growing, industrial economy. Education, as practiced by public schools in many countries, produces unprovocative individuals, adjusted to an unbalanced society.

4.1.6.1. Vernacular Education

Animals obey rules of behavior. Many animal communities have codes of behavior that regulate interactions. In birds and less complex mammals, these rules may be very rigid and predictable. With increasing brain complexity, however, learning takes a larger role. For example, young white-tailed deer in Idaho have to learn to cross highways. They appear to use rules of thumb—not the best phrase perhaps—finding a proper balance between safety and reasonable progress, between no traffic in sight and bumper to bumper congestion.

Humans in archaic groups learn from parents, siblings, and mentors. Education starts in the family, as a gradual, peaceful process. Through living, an individual learns the seasons, the foods, and the values of the group. They learn human and nonhuman things. The recital of myths is of great importance; myths provide a cognitive matrix for life. The cosmological myths of the Tukano do not describe their place in nature in terms of mastery of a subordinate environment. Instead, the Tukano learn that they are part of a larger system that transcends individuals. Survival and maintenance of the quality of life are possible only if all other lives are allowed to evolve according to their specific needs, which are described in myths and traditions. Many rituals of the tropical bands are concerned with ecological balance, though not necessarily self-consciously. This kind of education could be described a vernacular; it is part of the operation of the community.

4.1.6.2. Modern Education

The newest form of education is based in humanism. Humanism started as a revolution against scholasticism in the 15th century, but it can now claim to be every bit as dry and reactionary as scholasticism. Classical humanism strives to convert the distant and alien into the comfortable and familiar by reducing it to a moral commonplace; a poem by Horace, for instance, exemplifies the moral. Humanism cannot communicate with the past meaningfully because it refuses to acknowledge that the past is different. In its search for a perennial philosophy of universal values in the classics of literature of the ages, humanism ignores the otherness of the cultures of the past. In spite of the effort, the past becomes more distant with academic study, not more accessible. Academics requires historical research in archaic languages and world views before understanding can occur. Humanism regards change as the cause of decay, sympathizes with authority, and subverts classical literature for the preservation of humanism. Humanists see as their task the training of an elite for the leaders of tomorrow, who offer the public only skill at public speaking.

Modern education encourages many fallacies. It assumes that its knowledge is static and adequate. It assumes that only children need to be educated. It assumes that schools are the best places to learn. Furthermore, the groves of academe are surrounded by high walls, like a maze; each discipline is self-contained and each scholar is a master of a small domain. Learning within the walls is typically a regurgitation of facts within subject and time limits for grades. The grading system seems to be a device for rationing the 'good' places in society, where unemployment is used to control inflation according to an economic scheme that tries to adjust ideal theory to a limited world. The university is the final phase in the production of higher credentials to guarantee higher rewards in the economic system. The entire educational system itself seems to be political, directed at wealth for the state and a position for the individual. The system is subsidized by military and corporate interests, which depend for their livelihood on the disparity of wealth between individuals and cultures. The university is built on the same blood and suffering caused by that same disparity and nourished by continued exploitation.

Any kind of humanism now must be based on a whole image of the place of humanity within nature, and not on some transcendent view. It must confront the past without a baggage of sentiment and the future without the paralysis of dread. The appreciation of the differences of other cultures, the difference of Horace, for instance, will allow us to transcend our present identities. Responding to any art with pre-established values makes one miss the uniqueness of the experience presented. Schiller proposed an aesthetic humanism, wherein art would broaden the mental worlds of observers and encourage tolerance and wonder.

As Aristotle recognized long ago, experience is necessary for thinking. The very cloisteredness of traditional university education works against it in this respect. The university fails to teach communal responsibility, self-reliance, and physical work—those qualities most dear to Emerson. Elite schools like Harvard or Oxford, with their richness of culture, can offer more potential than the pedestrian paper mills of Idaho or Florida, but are often limited by social fashions. Radical, in the sense of 'rooted,' schools like International College, Evergreen College, or Antioch University, can offer greater benefits, with a teacher 'leading' a pupil to knowledge, but are expensive and elitist.

The dream of living in modest comfort, sheltered in a place of learning, worrying a trivial scrap of knowledge, is a dream built on the broken dreams of those who, every bit as equal, dream only of adequate warmth and food. Participation in such a dream is only at the exclusion of others. The dream of scholarly detachment insulates the conscience from the desperation of those whose accidents of color, location, or inheritance were not as fortunate. The "hardness of heart of the well-educated" was what Gandhi admitted made him most sad in life.

4.1.6.3. Poetic Education

Educational ideas and theories in this century have been dominated by the impoverished cosmology of modern industrial culture. To be effective in this age education needs to be rooted in conceptually rich philosophies, such as deep ecology, the movement formulated by Arne Naess. Deep ecology did not spring fully formed from Arne Naess's head, like Athena from Zeus. To some extent it is the most recent flowering of a line of thought that extends from J. W. von Goethe and G. P. Marsh to Rachel Carson, Paul Shepard, and Gary Snyder.

Goethe (circa 1800), for instance, in his organic dialectics, argued for contemplative nonintervention and the primacy of the qualitative as the best methods to reach the deep down "realm of the Mothers." Marsh (1860s) documented the human impact on the environment, especially forests, that could result in extinctions. Aldo Leopold (1940s) formulated a land ethic. Rachel Carson (1960s) warned of the biological damage from technological quick-fixes. Paul Shepard (1960s) noted ecological contributions to the image of the world and formed an idea of technocynegetics. And, Gary Snyder (1960s) suggested poetry as an ecological survival technique.

In the late sixties and early 1970s, several ecological philosophies and movements were formed. Paolo Soleri designed large ecological cities. Murray Bookchin promoted a free society based on ecological principles, through social ecology and eco-anarchism. John B. Cobb, Jr. tried to incorporate humanity as part of nature, using the process philosophy of A. N. Whitehead. Wittbecker (this author) described forms of radical ecology and eutopian designs. Arne Naess contrasted shallow ecology with a deeper version. Sigmund Kvaloy and David Klein both enumerated ecophilosophies. Theodore Roszak created the framework for a sacramental ecology and an economics of permanence. Daniel Kozlovsky outlined an ecological and evolutionary ethic in 1974. Henryk Skolimowski began to promote an ecological humanism in 1974. John Rodman wrote on the liberation of nature in 1974 and later ecological resistance. Mary Daly, Susan Griffin, Dolores LaChappelle presented a special perspective in *Gyn/ecology*. George Sessions described his ecophilosophy. Neil Evernden wrote about the human place. Alan Drengson differentiated shifting paradigms. Carolyn Merchant finished her book on Ecofeminism. J. B. Callicott described environmental ethics. Paul Taylor wrote about respect for nature. David Foreman translated these ideas into defensive actions through Earth First! (1986), and Eugene Hargrove summed up the forms of environmental ethics.

Poets like Wordsworth and Auden recommended that broad training in science and technology was necessary for poetic knowledge, which is part of a good education. Novalis considered that the study of the external world, through science, was only the first, half-way, step to full human consciousness. The second step was introspection, the contemplation of the self. Subject areas in traditional universities concentrate on one step or the other. Any student can achieve both steps, leading to a complete education, with time and inclination. A complete education requires intense effort, discipline, patience, and a tolerance for failure.

In his educational theory, Plotinus went still further and laid down a triple organization of education, requiring a social education, a personal and self-revealing education, and a synergetic one that would permit a perspective of the whole of human existence. Only schools that integrate education within a balanced society can achieve this triple objective. By encouraging students who are already working outside academic walls, radical schools can foster this necessary kind of synergetic education.

True education need not be bound by the conceptual and economic limits assumed by most universities. A minimum education may train students for an economic role in society, but a good education teaches them how to enjoy living among other human beings in an ultrahuman nature and to perpetuate a good society, a peaceful society.

4.1.6.3.1. Aesthetic Education

Modern education allows individuals to languish in an informational wasteland. A radical education, based on the aesthetic humanism of Frederich Schiller, leads them out, to a place within nature. It offers a new perspective of humanity in the total field on nature and defines balanced relationships with other species.

Schiller believed that human society could be improved by political means. But after studies on the Thirty Years War in Europe, he became skeptical of the ability of politics to create a peaceful society. He came to consider a work on art historical proof that art could achieve what violence and law could not: art educates and liberates the individuals of society in a gradual and peaceful process. In spite of the cultural forces dominant at any moment, an individual has the potential to determine a different course of action. Unlike classical humanism, which was shackled to one interpretation of the past, the aesthetic humanism of

Schiller was open to the possibility of novelty.

Schiller judged society to be violent and selfish and he identified the cause as an imbalance between animal and rational drives; civilization exacerbates the problem with its own imbalances, for example, fragmentation and the unnatural channeling of energy. The basis of his theory is the relation between three instinctive drives: sensuous bodily needs, form-giving reason, and the play of imagination. The third mediates and harmonizes the first two. And through this facility, human beings and their institutions gradually evolve to a higher moral plane. Improvement comes about through aesthetic perception and progresses through definite stages, in Schiller's theory. Both begin in the natural state and work toward an aesthetic state, and later, a moral state. The natural state is dominated by physical needs, but individuals can transform themselves with the play of their rational faculties. The natural object of play is the aesthetic state.

Neither reason nor the state can achieve further transformation, however; that can be done only by the instrument of 'Fine Art.' Using a mimetic faculty, people imitate one another. Life imitates art, which is an expression of values. By means of art, people ascend to the moral state. An ecological education based on Schiller's ideas presents a whole image of humanity within nature and not a transcendent view. It confronts the past without the baggage of sentiment and the future without the paralysis of dread. The appreciation of the differences of other cultures allows human beings to enlarge their experience and identities. Art broadens the mental worlds of observers and encourages tolerance and wonder. Education in aesthetic humanism embraces three concepts: play, liberation, and community.

Play is the method of learning for most juvenile animals and a means of enjoyment for many adult animals. For humans, play is imaginative experience, entered into freely. Much human activity is play, in place in a community. Even science and philosophy are forms of play, attempts to solve the puzzles of existence. Schiller maintained that the source of both art and play is 'overflowing energy.' Although early ethological definitions of play emphasized its activity as an 'outlet of surplus energy,' later definitions enlarged its importance in learning and information gathering. For example, play is also defined as:

- An activity with 'no immediate objective'
- An experimental dialogue with the environment
- Rehearsals performed in a nonfunctional context, of the serious activities of searching, hunting, fighting, and mating
- Behavior that functions to develop, practice, or maintain physical or cognitive abilities and social relationships by repeating or recombining sequences of behavior outside their primary context.

For Schiller, play is activity for its own sake, where the drives of emotion and reason are harmonized. The state of play is whole and simultaneous. It unifies permanence and transition, chaos and order, duty and selfishness. The object of the senses is life, the object of reason is form, and the object of play is living form—called beauty in the widest sense. Aesthetic play, like physical play, requires order and control.

The rich flowering of human nature is possible only when the constraints of need are replaced by leisure and abundance. Liberation requires a larger perception and larger concept of rationality. For humanity, liberation means an end to prejudice or discrimination based on arbitrary characteristics. The liberation of nature and ultrahuman beings is inseparable from human liberation. Nature and ultrahuman beings are to be free from the obligation to be human, so pervasively presented from Plato to Kant, and from the status of human resource or human artifact. In Schiller's scheme beauty frees us, because we decide how an act is performed, without being restricted by blind necessity.

Human beings gravitate into groups to live. Every culture needs its own local, sacred center, that cannot be broken if the group is to survive. Communication across the barriers of culture is necessary for a world community, but from firm cultural bases. The complete surrender of cultural identity is as dangerous as too little openness. In ecology, the saying goes, "As the community goes, so goes the organism."

4.1.6.3.2. Radical Education

Education takes place in communities; it is the means for communities to continue. With its emphasis on play, liberation, and community, a radical education integrates all the ends of education. The most valuable qualities of a radical education are personal contact, which allows noncompetitive constructive criticism, and flexibility, which allows the educational process to fit an established and meaningful life-style. Radical education lets one learn how to feel and live, as well as to think. This ideal is most closely approached in already established communities, not the artificial and involuted, temporary university dormitories, and when play and freedom are not limited by arbitrary rules and economic goals.

Radical education alters and enlarges perception with the selection and presentation of relevant information and forms an ecological consciousness. The survival of human societies depends on the consciousness of the global system in its complexity and connectedness. The spirit of humanity depends on the consciousness of its proper relationship with other species and the whole earth.

4.1.6.3.3. Ecological Education

Garret Hardin points out that education is not just literacy, and that literacy, the ability to understand what words really mean, is not enough anyway. It needs to be supplemented by "numeracy," the ability to quantify information and interpret it intelligently— computers, remember, use numbers for everything—and, on another level, by "ecolacy," the ability to take into account the effects of complex interactions of systems over time, for understanding of the complexity of the world, that things are interconnected and affect one another. Together, these are three major filters against folly that citizens can use against the blindness, short sightedness, and sheer idiocy that experts disguise as eloquence or expertise. With the increase in the complexity of civilization, it is necessary to recognize another filter, "imagacy," the ability to work with complex patterns of visibility.

4.1.6.3.3.1. *Literacy (Words).* Literacy is the quality of being literate. Specifically, it is the ability to read a short passage and answer questions about what was read. Being literate is being characterized by learning, cultured, educated. A person who is educated has been "led" from ignorance, out of the self in other words, by fostering the growth and expansion of knowledge through a course of formal study. Knowledge is a condition of knowing, an acquaintance with theoretical or practical understanding. There are no limits on what can be known. But, most knowledge is concerned with survival first. It is important to know what plants to eat, where to find shelter, how to make clothing. This was, and still is, the most basic level of literacy.

But, literacy occurs on more than one level. Beyond familiarity with the terminology and applications, computer literacy is competence at solving problems using a computer, or even the professional design of new machines. Complete computer literacy is the knowledge of how the computer operates, as well as its design, manufacture, and programming—few have that complete kind of literacy. Classical literacy can survive the collapse of printers and newspapers. But, could computer literacy survive the collapse of computers? Perhaps this question hinges on the definition of literacy. Certainly, computers can be valuable for certain aspects of education, but we must not forget what function the computer is assisting. In other words, computer use should not displace the skills themselves. Education should include a core of mathematics as a liberal education always has. And poetry and narratives should still be memorized, as well as written or examined on a computer.

Literacy, as the skill in the written and spoken language, enables readers to draw on the wisdom (and foolishness) of human beings distant in space and time. Hardin notes, in a discussion of the sins of the literate, that language has two functions beyond communication: "To promote thought and to prevent it." The second function is why literacy has to be accompanied by the ability to think critically.

4.1.6.3.3.2. *Numeracy (Numbers).* Numeracy involves the ability to measure and to interpret quantities, proportions and rates. Hardin warns that human beings have learned how to use literacy to hide numbers and the need for numerate analysis. He draws attention to the problems created by always thinking solely in terms of dichotomies, e.g., safe vs. unsafe or pure vs. impure, rather than in terms of relative risks and benefits. James Lovelock has also noted our inability to assess risks mathematically. The quantitative analysis that is so important in science, technology, business, and government is dismissed with indifference. In a complex, rapidly changing society, understanding quantities, ratios, rates and duration of time is crucial. Numeracy has limitations, also—the conclusions of an accurate mathematical analysis are only as good as its premises. Intense interactions of people in larger concentrations also produce more information and data. Information in the entire world is estimated at over 1000 petabytes (petabyte = billion megabytes or a quadrillion books of 170 pages).

4.1.6.3.3.3. *Ecolacy (Links).* Ecolacy was once achieved by studying natural history, the plants and animals that surrounded every human group. Ecolacy is the ability to follow the links or connections between things. This would allow the effects of the interactions of systems over time to be taken into account.

Scientists have been extremely successful using reductionistic methodology on every problem, breaking problems down into their components and studying the properties of these components and their interactions. This has led to the ascendancy of mechanical science—thinking that one can do just one thing. Hardin stated the important ecological understanding of ecolacy as: 'We can never do merely one thing." This statement is now known as Hardin's Law, and it means that there are always wanted and unwanted effects, products and wastes.

Reality is composed of causal chains of events rather than single effects. Events are embedded in causal networks and are produced by multiple causes and have multiple effects, each of which triggers a causal chain of future events. Hardin contends that since we cannot do just one thing we must always ask and answer that question: "And then what?" We have to try to ascertain the benefits and costs of proposed courses of action on both the individual, social, and ecological levels. The ecological systems way of thinking employs scientific theories and knowledge to study the interlocking processes characterized by many reciprocal cause effect pathways. The ecological systems way of thinking must become an integral part of the thinking of the well educated person if we are to adequately control technology and human actions.

The world is too complex for our minds, suggested G. P. Marsh and many later thinkers. So, our minds have to filter out what is less important. We filter data, arguments, emotions, and information. The filter allows a total picture of the whole with relatively little information. Of course that picture might be wrong. But, it is clear, as a result of filtering and thinking. Other human activities act as filters, also, especially culture.

In education and communication, noise is a problem. We are flooded with information, and much of that noise. We have to filter out as much noise as possible and much of the information. But, many cultural filtering systems are collapsing.

Hardin contends that most of the major controversies of our time can be better understood as the result of the participants relying too much on any single one of these three filters. No one filter by itself is adequate for understanding reality and predicting the consequences of our actions. There is, however, a filter that has gone unmentioned, in its ability to concentrate or distort human perception. That filter is the image.

4.1.6.3.3.4. *Imagacy (Images).* Images can model the system in miniature. An image is an imitation or representation of something. It can also be a symbol or type, a metaphor or concept. An image can stand for something else, for instance the image of a dove is often used as a symbol for peace. In the etymological sense a symbol is something 'thrown together,' as a problem is something 'thrown forward.' Unlike an image, a symbol often represents some other thing, process or quality. Symbols can be processed by Analytical Engines or computers. These machines have been used for metaphors of the brain, which also processes symbols, that is, the operation of both parts of the metaphor, brain=computer, can be described in terms of algorithms, or mathematical rules for manipulating symbols.

An image of course can be used in a variety of ways. It can reference similarity, correlation or a formal linkage, according to the categories of association developed by C.S. Peirce: Icon, index, or symbol. Icons have a similarity with an object, as when a landscape painting depicts a landscape. An index is when an image is causally linked with something else, as when a wolf's howl is related to location or emotion. A symbol has a social convention that establishes the relationship linking an image and a thing, as the Coat of Arms of the Woulfe family symbolizes the family line. Physical connection is not necessary for any of these modes. Peirce suggested that the difference between the modes of reference could be understood in terms of hierarchical levels of interpretation. Symbolic relationships are composed of relationships between sets of indices, and indices are composed of iconic relationships between sets of icons. Understanding a higher level means decomposing it to a lower-order form. Understanding the logic of images, as well as their power, is important to their use in communicating and design.

An image, especially a cosmology or an image of the world, models the system in miniature. From the image, which can be a paradigm or mind-set, a whole system can arise, with unique goals, rules, parameters, and structures. A cosmology includes a mythology constructed as a poetic system. Joseph Campbell states "Mythology—and therefore civilization—is a poetic, supernormal image." Mythologies and religions can be understood as great poems. When recognized as such, they point through individual things and events to the ubiquity of a presence that is whole in each. This is what P.B. Shelley meant when he wrote that poets are the 'unacknowledged legislators of the world'—not that they pass laws or prophesize the future, but that they generate images for the future, and these images can be articulated into goals to influence our actions. Bad images, from indifferent poets, can relate to severe cultural problems, as when popular Italian poets romanticized the violence and hatred of Fascism. The mechanistic images of science, from Shelley to the Fascists, determined much of the violent conquest of nature.

Kenneth Bounding notes that the image as a cognitive construct of the world has several aspects: Spatial, temporal, personal, relational, value, and affection (emotional) for each individual. The total sum of individual images is a world of interrelated constructs. This parallels the experiences of other living beings. Using its senses, each participant creates an image of nature, or world—*umwelt*, life-world, is the term used by Jakob von Uexkull)—from the sensations that are meaningful to it, and which limit it. Simple beings, such as bacteria, make a relatively simple image, whereas more complex beings, such as apes and humans, forge more complex images. In fact, human beings design complex images for a variety of purposes, including to make a profit or to persuade others to join a group.

4.1.7. *Future of Science & Education*

Science has phenomenally increased our knowledge of physical and biological processes. It has now become the basis of our moral code, but it cannot very long be a science divorced from feeling and art if that code is to help us survive. To do this science requires aesthetic perception as well as disciplined thinking and feeling. As there is a rational component to ethical judgments, so there is an intuitive and emotional one, also. Science does reasonably well as a conscientiously objective system, but it is not really objective—it just pretends. So, it is not an adequate basis for a world view.

Science is capable of deeper realization, but it cannot grip good or evil—it fails. Science exists in a dimension of incomprehension. Can we accept that mystery? Accept not knowing things, even things that we have named and classified? Seeing nothing and understanding nothing, how can we survive? Can we survive with a poetic appreciation? We only need to look at our feet to see some connection or hear a brief note.

Conservation biology, Ecoforestry, and other crises sciences point the way to a more comprehensive science, a science concerned with knowledge in an ethical context, a science that addresses human and nonhuman values. What would we call such as science? Ethosophics? Ethopoetics? Such a science would have to be a process science, acknowledging limits and the uniqueness and importance of individual places and individual dwellers in places. It would still be capable of daring exploration, tempered with caution and understanding, but done peacefully without harm to other species.

An ecological ethics suggests that humans avoid tampering with complex evolved systems, not because they are good, but because they are the basis of life at this stage of development. Ecological ethics is situational because ecology is the study of changing systems. It is pluralistic, as Christopher Stone notes, because of the variety of entities involved. The morality of the act is determined by the current state of the system. Adaptive modes should conform to ecological patterns. An ecological ethics is based on attributes of ecosystems and human compliance with ecological laws. The aim of an ethic must be harmonious with the whole population of living beings.

Science has fallen way behind of the technological and economic uses of its discoveries. Science needs to push into the future with immediate demands. For instance, we need to start major taxonomic efforts to classify and inventory species in every kind of ecosystem, using residents, biologists, computers, axial tomography, and nuclear magnetic resonance to collect and scan specimens. Species are becoming extinct faster than they can be seen and named. Many kinds of projects dealing with diversity could be high-profile, participant-friendly projects to involve residents and amateurs.

Science can start to plan for future ecosystems, that may be less beneficial for humanity. Many of these will be restored or assembled from exotics, reintroduced natives, or engineered species. Science needs to save as much genetic history as possible for that effort of restoration; that may mean using cryogenic techniques to suspend material from rare or endangered species. Science can also use molecular markers in conservation genetics, to address problems of inbreeding depression and loss of variability in small populations. Science has to anticipate the loss of species and the massive impacts on ecosystems that result from human technological and economic impacts. It has to anticipate ways to ameliorate those impacts with large-scale reservation areas. The fascination of science, and the successes of science, should be promoted as vividly as entertainment or athletic events—and they are easier for people to participate in, requiring only attention, expression, and a basic education.

Other modes of knowledge surround science. Theodore Roszak defines true knowledge as 'gnosis,' of which scientific rationality is only a small part. It is gnosis that is needed to improve the universe and soul, mutually, with the spirit of love. Gnosis is the whole spectrum: the hard, bright lines of science, the hues of art, the dark voids of religion. Gnosis is augmentative knowledge, in contrast to the reductivity of science. Paul Tillich calls gnosis "knowledge by participation." And that is what humans need to do—recognize that they automatically participate in everything, and cannot unparticipate by choice.

4.2. ***Knowing a Place: Ecology*** *& Human Ecology*

The necessity of ecology is to study and describe the societies of living beings that inhabit the earth. We need facts, and then generalities, which every educated person with the facts can provide; sometimes the best insights can come from persons who are not specialists. The study of living things is also fun. Vast cores of people could study things and report. Goethe noticed that different people are sensitive to different aspects of a thing, so investigatory efforts should incorporate the observations of others. Most of what we know about nature was gathered by curious amateurs. Eliminating curiosity from professional science may eliminate science eventually. Basic research is the satisfaction of one kind of curiosity.

Ecologically, we are fooling ourselves, not by the objectivity of our facts, but by their triviality and unrelatedness. Nature's richness is dependent on complexity, priority and outsidedness. Complexity is a relationship between system and observer, not a measurable property of a system. Complexity is brought under control by name magic—name it to control it. Archie Carr states: "If magic be defined as something produced by secret forces in nature, and secret in turn defined as something revealed to none or to few ... then magic is not likely to be diminished by all the science we can muster. Research may provide us with answers, but these answers forever lead to new and more profound questions." As our knowledge grows more vast, phenomena will be revealed to fewer of us. But ecology must order facts. It must provide structures for understanding the wealth of detail. Magic in the sense of inciting wonder is also not diminishable.

4.2.1. *Nature & the Components of Place*

Every place has a larger environment, from a community to a region and even a solar and galactic environment. The study of individual organisms in an environment is known as autecology. A group of individual organisms of the same species in a particular place is studied as a population. The assemblage of populations of different species in a habitat is studied as a community. The community in its biotic environment is studied as an ecosystem. Ecosystems comprise the ecosphere. There are emergent properties at each level of organization, such as the diversity of species and the structure of a food web.

4.2.1.1. Dialectical Movement & Change

Nature changes as its parts interact. There are parallels between the interactions of processes, of animals and of humans in ecological systems. Science has concentrated on disturbance and exploitation behavior, but these can be contrasted with the interference behavior that characterizes the nonecological activities of the dominant industrial culture. Examples of each will be taken from wild ecosystems, forestry, animal cultures, archaic human cultures, and industrial culture. The word interactions is used, instead of words like 'events' or 'catastrophes,' to describe the feedback and cyclic nature of actions.

Humanity is exploiting nature recklessly, without attention to the minimal health of ecosystems. Many ecologists, such as Eugene Odum, have observed that complex communities have existed for thousands of years in relatively stable environments, even though these environments are characterized by regular disturbance and constant exploitation. These environments are now vulnerable to human interference, which is a different thing from disturbance or exploitation. Disturbance, by definition, is an event that can be caused by climate, biological entities, or other actors. Exploitation is the normal use of a resource or of a species by another species, including the human species (this ecological definition differs from a sociological definition, which means 'selfish or unethical use,' although the ecological use may suffer from negative connotations from the sociological use); in fact, ecological exploitation has a rejuvenating effect on populations. Exploitation is contrasted with interference, an activity that can degrade, destabilize, or destroy entire ecosystems. Interference is not a form of disturbance, exploitation, or competition; it is destruction without gain to any species; sometimes it is caused by planetary events, but in the case of human interference, it is the destruction of the structures and processes of evolution for large-scale, one-species, short-term economic gain.

The interactions of living beings in ecological contexts may have positive and negative effects on themselves and other species, as well as constructive and destructive effects on ecosystems and the operation of biogeochemical cycles. Human interactions are also considered. The pandominance of ecosystems by humanity is related to the biological and cultural characteristics of the species. Ignorance and indifference are identified as major reasons for continued interference.

4.2.1.2. Kinds of Interactions in Nature

These are three general forms of interaction of individuals, species and systems. Animals and plants, algae, bacteria, fungi, live together in ecosystems. Living together involves many kinds of interactions, from competition and conflict to cooperation and mutualism. Interactions may be reciprocal or complementary. They may dominate or control. Interactions are multidimensional. A wolf, for instance, may howl to communicate, or to restore proximity with a mate, or for simple pleasure. Many animals, such as wolves and caribou, develop together over time, adapting to each other's strategies. Paul Ehrlich and Peter Raven refer to this mutual adaptation as coevolution. Coevolving systems never completely adapt.

4.2.1.2.1. Exploitation

Exploitation is the normal use of a resource or of a species by another species, including the human; in fact, ecological exploitation has a rejuvenating effect on populations. Every species uses some part of other species or of the environment. Insects, diseases, and animals, more than being simply agents of mortality, are native components of complex food webs in ecosystems, and they contribute to the selection of species. In a Ponderosa pine forest in the Pacific Northwest, insects exploit trees; they pollinate some trees and overwhelm others, but rarely do they overwhelm and kill more than one percent of a forest. Diseases exploit trees; they remove stressed trees—also probably a low percentage on the order of one percent. Their effect on the long-term health of a forest, however, is positive.

Birds and mammals eat foliage and seeds; they also disseminate seeds. Mammals, the best regulated of more recent species according to Frank Golley, change their habitats to suit themselves by chewing, digging, and burrowing. Rodents can dislodge earth at a tremendous rate. In many cases, these activities improve the conditions for the growth of vegetation. Mammalian grazing promotes vegetative regrowth and the movement of seeds. Bison and prairie dogs were responsible for much of the character of the American plains. Beavers and other rodents create their own microsystems. Wide-ranging caribou and wolves transfer energy between systems.

Predation increases the survivability of a prey species. Predation also increases the diversity of species, according to Steven Stanley, by limiting the most populous prey species. Rarely do predators kill all of the prey. Rarely do animals interfere with the operation of the biogeochemical cycles in the environment. The exploitation of the system by plants, insects, and animals contributes to the health and continuity of the ecosystem. Exploitation is not chaotic; there are limits and rules.

4.2.1.2.1.1. Rules for Exploitation

Animals obey rules of behavior. Many animal communities have codes of behavior that regulate interactions. In birds and less complex mammals, these rules may be very rigid and predictable. With increasing brain complexity, however, learning takes a larger role. For example, young white-tailed deer in Idaho have to learn to cross highways. They appear to use rules of thumb (not the best phrase), finding a proper balance between safety and reasonable progress, between no traffic in sight and bumper to bumper congestion.

Animals like wolves have behavioral inhibitions against killing too many prey or killing their own kind, against coupling with a mated or disinterested female, or against attacking nonprey species. Animals that break such inhibitions are sometimes attacked or ostracized. In general, food is shared by all members of a wolf pack. Adults will regurgitate part of their food for adults who stay behind with juveniles. The members of a wolf pack cooperate bringing down an elk, but then compete for the choicest parts of the prey. Rules are not always strict; wolf mates raising pups may consciously deceive one another to get a break from the responsibilities, according to Michael Fox.

The social structure of a wolf pack is most important. Breeding, playing, hunting, feeding and territoriality are tied to social structure. Wolf pups are taught how to behave and how to hunt. Much of the behavior of wolves is directed to keeping the animal's status in the pack or to raising it. Quarrels take place often and the entire pack seems to take an active part. Actual battles, however, are rare. Ritualized squabbles result in few physical injuries. Wolves do kill each other, however. Wolves that behave strangely, such as epileptic pups or adults crippled in the chase, are sometimes killed by the pack. Disputes over the alpha position may end in death. Foreign wolves may be killed if they do not flee after a warning. Prey may be killed in excess during times of denning. Rules of encounter are complex and the outcome depends on numerous circumstances, such as the abundance of prey, the size and health of the pack, and stress; that is, the rules often depend on limits.

4.2.1.2.1.2. Limits of Exploitation

Mammalian behavior is controlled and population regulated through the use of space in general. Most mammalian populations, wolves for instance, regulate their density well below the limits of the food supply, often by as much as 50-70 percent. Territoriality limits populations, but populations can also be limited by specificity of prey or plant source, size of prey or plant populations, predators, natural events, or even individual tastes.

Wolves in the Arctic disperse with the migration of the caribou. According to David Klein, they prefer Caribou to other often more easily obtained species, such as mice. This preference reduces their hunting efficiency, however.

The goals of an organism are limited by the life-images of its species. Each animal is a participant in a field of existence. Using its senses, each participant creates an image of nature, or world (*umwelt*, life-world, is the term used by von Uexkull), from the sensations that are meaningful to it. Each animal fits itself to its unique world as completely as it can—simple animals to simple worlds, complex ones to well-articulated worlds. Each fits its place as well as it can. Klein, Konrad Lorenz, Michael W. Fox, and others have elaborated this kind of fitness in more detail.

Each organism is inseparably related to a place; breaking the bond may result in death. Organisms and places shape each other. This is true of archaic human cultures as well.

4.2.1.2.1.3. Traditional Ways of Archaic Societies

Most human cultures are located in a particular territory. This is especially true of the Campa, who live in a tropical forest in Peru, and for the Ituri pygmy, who live in a tropical rainforest in Africa. The features of their cultures are unique to their place. They literally could not live with images of desert or ocean, like the Taureg of the Sahara or the Samoans of the Pacific. The very circumstance that makes each culture unique—being in a unique place—ensures that it can fit a place. This fitness ensures a limitation of exploitation.

Particularly in agricultural societies, cultures are gauged closely to seasons. The culture makes the world manageable by limiting it. A local culture is also tuned to the limits of the local ecology, within the knowledge of interactions—the long-range ecological consequences of drainage, irrigation, or overexploitation can contribute to the success or failure of a culture. Many, but not all, archaic cultures are a form of fitness and limitation. Most archaic groups try for adaptation before domination. For instance, according to Gerardo Reichel-Dolmatoff, the goal of the Desana Indians is the cultural continuity of their society in its place in the rainforest.

The Chipewyan Indians in Northern Canada occupy the same territory as wolves and compete for the caribou, although their niches are not identical. Both social systems are adjusted to the hunting of barren ground caribou. Chipewyan hunters depend on animals other than caribou, which migrate out their Indian grounds. The cultural decision to hunt caribou as the primary item of subsistence, however, has produced many similarities between the two species in their utilization of land and in the formation and distribution of social groups. The cultural decision to hunt caribou results also in a population density lower than what would result through other decisions regarding the utilization of resources.

For hunting, Chipewyan use dog teams, snow shoes, and boats to increase their mobility and rifles and bows to supplement the traditional spears. The strategy of the Chipewyan, according to Henry Sharp, is to kill caribou at any opportunity. They increase their opportunities by walking aimlessly, watching, and driving the caribou, although the Chipewyan expend less energy by watching and ambushing. Wolves follow the more active strategy because of their increased and superior mobility. Both species adopt a pattern of dispersing and concentrating with the caribou. The basic choices regarding subsistence patterns, social organization. demography, terrain usage, and yearly cycles are made on the basis of the internal logic and structural characteristics of the two cultures—wolves do have culture, in the general anthropological sense, according to Henry Sharp and Michael W. Fox.

Although the two species do not compete directly, both Chipewyan and wolves are predators that put pressure on caribou populations. Sharp suggests that the commitment of both to caribou hunting is ecologically inefficient, since both species could spend more energy on secondary sources of food. For the Chipewyan, a deliberate "underutilization" of moose, rabbit, grouse, birds, and fish, is the result of their cultural values, including their willingness to live below the carrying capacity of the local environment—a characteristic of most hunting/gathering societies—the complex practice of drying caribou meat, and the reciprocity of their kinship system, i.e., caribou is a better basis for future relationships. The cultural decision to hunt caribou as the primary item of subsistence has produced a unique pattern in the utilization of land and in the formation and distribution of social groups. Wolves also underutilize their resources.

Regardless of cultural order or cooperative interactions, part of the process of life is uncoordinated, unfitting, disorderly, unbalanced, and destructive. Therefore, suffering often occurs. Suffering is an unavoidable part of disturbance or exploitation. We cannot intervene in every case, nor can we eliminate the possibility of suffering. We cannot maximize the self-realization of every being, and we cannot make evolution into a perfectly functioning machine; the functionless features of evolution are part of the process, but, we can protect the process. The mode of operation of nature consists of a rhythm of dissolution and reformation. The extravagance and beauty of the natural world features many more species in an ecosystem than would be necessary if exploitation alone were its organizing principle.

4.2.1.2.2. Disturbance

Disturbance, by definition, is an event that can be caused by climate, biological entities, or other actors. Disturbances in nature are regular but unpredictable events. Disturbances are caused by geological events, climatic events, physical processes, and biological agents. Hurricanes, for example cause disturbances, as do volcanic eruptions, windstorms, tidal waves, disease outbreaks, and acid rain.

Disturbance may change the direction of maturity in an ecosystem, but it also is stimulating for those species adapted to it. Disturbance may continue succession or it may deflect it, according to Frank Bormann and Gene Likens. Because of the range of scales and intensities with disturbance, it is a complex concept.

Disturbances that are not part of the history of an ecosystem may cause irreversible changes to the system, because the system has not evolved a defense or response mechanism to such a rogue disturbance. A meteor strike would be such a disturbance, especially if the landform was altered by a crater. Human disturbances, in the form of acid rain or clearcutting, are both novel and threatening. If they are small enough or rare enough, however, the ecosystem may rebound.

Very large scale disturbances, such as volcanic eruptions or meteor impacts, can destroy entire ecosystems or disrupt global biogeochemical cycles. However, such very large scale disturbances are rare, and the ecosystems often have thousands or millions of years to become reestablished, although changed.

Disturbances in nature are regular but unpredictable events. Disturbances are caused by geological events, climatic events, physical processes, and biological agents. Make distinction between geological and ecological? Hurricanes, for example cause disturbances, as do volcanic eruptions, windstorms, tidal waves, disease outbreaks, and acid rain.

Disturbance is one thing that causes change in an ecosystem. On a small scale, a singletree falling over is a disturbance. Although an individual dies, species continue. Mortality is a normal part of the life cycle of the forest. The disturbance may be necessary for the ecosystem to continue to mature; for example, according to David Perry, without windthrown spruce that expose mineral soil seedbeds, the northern forest ecosystem would shift to bogs.

Disturbances, if sufficiently regular, become a 'known' feature of the ecosystem. As such they can have positive effects on a species or habitat (or ecosystem). In Florida, some species such as Cypress, need the complete inundation provided by hurricanes to remain healthy. Yet, even catastrophic disturbances like hurricanes rarely damage more than five percent of a forest; for instance, the 1938 hurricane in the eastern U.S. blew down less than four percent of the trees.

As the frequency of a disturbance increases, the forest becomes adapted to the disturbance; even pine plantations in the southeastern United States that are managed with controlled burns are less damaged by lightning-caused wildfires. Many disturbances in forests, such as insect explosions or fires, kill low percentages of trees.

After long periods without a major disturbance, however, a catastrophic disturbance becomes more likely. Where wind and fire are absent, for example, the probability of insect and disease outbreaks increases. By trying to prevent one kind of mortality, ecosystem managers often establish conditions for another kind.

Fire is regarded as catastrophic. In some ecosystems, for instance tall grass prairies in Illinois, fire is required to suppress competition from trees. In some forests, such as lodgepole pine in Washington state, fire is required for the cones to open and the trees to regenerate. Forest fires rarely damage more than ten percent of the whole forest. Even the Yellowstone fire of 1988, still regarded by some as a tremendous disaster for "Smoky the Bear" policies of prevention, resulting in dead material forming fire ladders, caused limited damage as it leapfrogged along, leaving healthy untouched stands that became the source for the regeneration that is now being observed.

A typical percentage of death is the normal condition of an ecosystem, necessary for its renewal. The rate of death per year in an old forest is remarkably consistent at about one to two percent, even with wind storms, fires, disease outbreaks, and animal damage.

In some cases, a larger percentage of the forest is affected. For instance, high elevation balsam fir forests are subject to bands of dieback that progress up the slopes parallel to the contours of slopes. These "fir waves" seem to be triggered by cold winds striking exposed forest margins. A new stand regenerates where the trees have been killed.

Disturbance may change the direction of maturity in an ecosystem, but it also is stimulating for those species adapted to it. Disturbance may continue succession or it may deflect it, according to Bormann and Likens. Because of the range of scales and intensities with disturbance, it is a complex concept.

Disturbances that are not part of the history of an ecosystem may cause irreversible changes to the system, because the system has not evolved a defense or response mechanism to such a rogue disturbance. A meteor strike would be such a disturbance, especially if the landform was altered by a crater. Human disturbances, in the form of acid rain or clearcutting, are both novel and threatening. If they are small enough or rare enough, however, the ecosystem may rebound.

Very large scale disturbances, such as volcanic eruptions or meteor impacts, can destroy entire ecosystems or disrupt global biogeochemical cycles. However, such very large scale disturbances are rare, and the ecosystems often have thousands or millions of years to become reestablished, although changed.

4.2.1.2.3. Interference

Interference, an activity that can degrade, destabilize, or destroy entire ecosystems. Interference is not a form of disturbance, exploitation, or competition; it is destruction without gain to any species; sometimes it is caused by planetary events, but in the case of human interference, it is the destruction of the structures and processes of evolution for large-scale, one-species, short-term gain.

Although rare large-scale or novel disturbances can interfere with ecosystem processes, the term 'interference' is reserved for constant large-scale or novel effects. The destruction of ecosystem processes in nature by the action of one or more species is rare; any species that did so would become extirpated or extinct, unless it was not dependent on a single ecosystem, as is the case with wolves.

Many commentators have accused mammals, wolves for instance, of overkilling their prey. It is fairly well established now, by David Mech and others, that wolves will take prey in excess of their immediate needs. This behavior has been interpreted as useful in maintaining not only the wolf but also secondary predatory and scavenging populations, for example, foxes and ravens. Indian informants are aware of this aspect of the wolf's excess kill, but they attribute to the wolf sufficient foresight to kill an excess of caribou near the den site in order to have an adequate food supply when the caribou are absent. Regardless of the wolves' intent, excess kills of caribou by wolves seem to be linked to the pup-rearing part of the pack that follows behind, as well as providing some food for the reverse seasonal migration—wolves can eat the remains of kills that are up to a year old. Interference rarely has a positive effect on either species.

Like wolves, human beings, as part of the process, interact with the individuals of other species or with entire species. Human beings are mammals who, as George Woodwell puts it, live in a biosphere whose essential qualities are determined by other species. Mammals are bound by biological requirements that must be met if a population is to survive.

Like other mammals, humans change their habitats to suit themselves. Humans have modified animal and plant associations in a different way from other mammals, simplifying patterns of energy and chemical exchange and solidifying themselves at the end of many food chains as a dominant species. A dominant is a species with greater influence than any other in its biotic community, changing the lives of other species and the character of the habitat.

Human hunters are hypothesized to have wiped out the most of the large mammalian species of the Pleistocene through overhunting—not for future food, but rather from the style of hunting—by driving herds over a cliff; human hunting, with weather and habitat changes, may be have been a necessary but insufficient cause. There are other instances. In the 1880s, soldiers and cowboys slaughtered buffalo as a political strategy to reduce the resources of native peoples. Farmers and loggers destroyed the dense forests of Ohio and other states. Settlers and industrialists in the Amazon are destroying vast tracts of rainforest, as part of a political strategy to move peasants out of cities. Human beings have contributed to the extinction of species and to the destruction of ecosystems.

Human populations have increased exponentially, with billions in giant urban ecosystems. Agriculture has produced monumental yields, but only at the cost of tremendous erosion and great subsidies of fertilizers and pesticides. Dams have been built all along rivers, and riverine forests have been cut, altering rivers and fishing grounds. Changes have been made without regard to the long-term impact on the ecosystem or on its human population. We dominate entire ecosystems.

4.2.1.2.3.1. Pandominance

By its influence of all ecosystems, humanity has become a pandominant species. As such, humanity reclaims, overgrazes, clears, depletes, and wastes at a scale that interferes with the stability, processes, and existence of many systems. One of the ecological consequences of human activity is the degradation of wild habitats for human developments and the introduction of novel elements into the biosphere—elements that have not been added slowly over time as the result of natural processes.

The biomass of the human species probably exceeds the biomass of any nondomestic mammalian species, and that biomass is supplemented by the tremendous biomass of domestic animals, which is far greater than the human biomass and consumes much of the same food as humans, including milk, fish, and grain. The domination of humanity is related to other characteristics as well, its large annual population increase (over two percent), high structural organization (of information and matter), and high energy use (globally thirteen times the total of all other mammal equivalents).

This dominance has major effects on ecosystems: Transient perturbations in energy relations (from oil spills and burning); chronic changes/shifts of systems (from dams, irrigation, and chemical wastes); species manipulation (from the import and export of exotics); and interference competition with wild species. None of these effects are exclusive to humans as a species, but they are excessive, rapid, compounded, and very large-scale. Humanity has upset the balance of nature in favor of its own needs. Animals, plants, and habitats are being destroyed because of short-sighted, short-term economic interests. Humans continue this dominance because they do receive short-term positive effects on the human species.

Human exploitation at the tremendous physical scale that occurs in industrial states is different from exploitation by other species, because it results in the destruction of the entire system, the very basis for renewal of a system that human beings, as well as other species, need for life. Human actions are damaging global biogeochemical cycles, such as the carbon or nitrogen cycle. For instance, deforestation, burning, wetland loss, and industrial processes are releasing massive quantities of carbon dioxide into the atmosphere, which disrupts the carbon cycle. Although the destruction of large species, from whales to frogs, has a dramatic effect on ecosystems, the destruction of microbes, which generate oxygen and recycle nutrients, has a critical impact on the entire food web. These actions are global, like a large volcanic eruption, but, unlike a volcanic eruption, they are constant and hourly. These human activities are best referred to as interference.

4.2.1.2.3.2. Industrial Ways

The cosmologies of archaic cultures have been limited to historical places and by human perception, tradition, and technology. Modern technological cosmology, beyond being another kind of order, more linear and abstract, is wrongly considered the evolutionary successor to traditional cosmologies, and is displacing them rapidly—although we cannot afford to suppress the diversity of thought necessary for adaptation to the diversity of environments or to eliminate ecosystems and the societies adapted to them.

Our modern problems reflect an unbalanced and immature image of the earth, the earth as a machine, for instance. People sometimes constructed their worlds from preconceived notions, and many of these worlds did not survive, because they could not adapt to the environment. Our modern cultures are defective for this reason. The modern attitude toward nature as a resource has resulted in pollution and depletion of resources. It has allowed humans to overpopulate their habitats. Recent productivity studies indicate that the optimum sustainable human population is far below the current world population.

Even worse, decisions regarding resources are still made exclusively on short-term economic rationalizations and lead to material shortages and environmental degradation. The crises of environmental degradations are crises of cultures. Monocultures of the industrial kind lead to 'dedifferentiation,' that is, the decomposition and destabilization of complex structures. A species or culture that destabilizes its ecosystem through misbehavior risks its own extinction. Human beings make changes to ecosystems that endanger themselves.

Humanity is calculated by Norman Myers and others to be using over 40% of the ecosystem productivity for the entire earth. Later calculations, by Stuart Pimm and others, indicate over 55 percent. Humanity influences virtually every ecosystem to some extent, destroying some, interfering directly with many, and exposing the rest to exotic chemicals and materials. Species normally use a percentage of system productivity without disrupting the processes of production. The human species interferes with the processes.

Based on limited scientific and cultural perspectives, humanity fails to value those beings and communities for which no use is known. But, as Aldo Leopold (1949) notes, the majority of the beings in nature have no human uses. Even ecologists cannot think of uses for many large birds and mammals. The real danger is genetic loss, which is frequently grossly underestimated. As wild areas grow smaller, even wild species interbreed. As species are lost, the ecosystems become simpler or start to collapse.

4.2.1.3. Noninterference Behavior

Interference has been a rare phenomenon on earthly ecosystems; it has happened in the past as the result of global catastrophes, such as meteor impacts. Now, interference, as opposed to more limited and predictable disturbances or exploitations, is threatening the stability of all ecosystems. It is dangerous to interfere with the processes of ecosystems because it disrupts the communities on which other species, and ultimately human communities, depend. Furthermore, in the deepest sense, it violates the idea of living together with other species on the planet. The proper relationship of humanity with nature includes competition and exploitation and mutualism, but not interference.

We kill millions of animals in laboratories to insure our safety, we kill billions of plants and animals for food and clothing and products, while indulging in the sentimental preservation of some individuals of other species. Animals do not need to be saved from natural death, a great regulator of life, but from unnecessary suffering, experimentation, and premature extinction. The world would not be a better place without sharks, silverfish, rats, cockroaches, or hyenas. They need their own places. The places, entire ecosystems, need to be saved. If we diminish variety in nature, we debase its stability and wholeness. To save ourselves, we must preserve and promote the variety of nature.

A start has been made. Ethics now considers almost every human being and human interaction. The restriction of ethics to exclusively human modes of existence, however, leads to a troublesome isolation. Human beings are not separate from their social and biological communities, and these communities are embedded in ecological contexts with biogeochemical processes. Human communities are one of many communities that make up an ecosystem. Human ethics describes only a small part of the rules for living together in communities, perhaps only the self-conscious part. Ethics must be extended to the entire framework and to the surrounding communities in the framework, without which there would be no human health or wealth. Through our efforts, we understand that communities of other beings have their own values and rules for living together. It remains for us to integrate and codify human rules that recognize the values and rights of other beings to live in healthy ecosystems and that limit human use of those ecosystems.

4.2.1.4. Self-world & Dwelling

Adolf Portmann shows that every form of life appears as a gestalt with a specific development in space-time. All living forms develop an image of their environments, from whatever is meaningful to it. Genetics provides the proper image choices for some, frogs, for instance. Others, like bears, must learn what is valuable using their senses. Jakob von Uexkull suggests representing the unfamiliar world of animals with bubbles to denote the self-world or phenomenal world of an animal. The life image, or *umwelt*, of an animal is what has perceptual and operational meaning for the animal. All animals are fitted to their unique worlds with equal completeness—simple animals to simple worlds, complex ones to well-articulated worlds. Each is optimally fitted to a habitat. The characteristics of some individuals fit the environment better.

Animals have their own universes that are strange and fascinating. When we realize this, we find that reality is immeasurably greater than human ideas of it. Jakob von Uexkull suggests representing the unfamiliar world of animals with bubbles to denote the self-world or phenomenal world of an animal. According to von Uexkull, perceptual and effector worlds form a closed unit, the umwelt. The world—life-image—is what has meaning for an organism. The umwelt is a region of brightness, or just a focus. "Figuratively speaking each animal grasps its object with two arms of a forceps: receptor and effector. With the first it invests the object with perceptual meaning, with the second operational meaning," von Uexkull says, "The first principle of the self world theory is that all animals from the simplest to the most complex are fitted to their unique worlds with equal completeness. A simple world corresponds to simple animal; a well-articulated world to a complex animal."

For example, the rich earth around a tick shrinks to a scanty framework of three receptor and three effector cues—butyric acid, motion, and heat—her umwelt. But the poverty of this world guarantees the certainty of certain actions; security and success are more important than wealth. Bodenheimer hypothesizes that an animal has an optimal umwelt in a pessimal environment. The environment must be pessimal or a species would gain ascendancy. The umwelt allows the creation of familiar paths that are strange or invisible to others.

Time is different for each species; the tick can wait eighteen years for prey. Human time is made of units of an eighteenth of a second, which Uexkull reminds us is utilized effectively by the motion picture technique of film projection. The subject sways to the time of its own world. Since all beings feel, the smallest unit of time is a duration for that being. Larger beings have larger durations. Organisms set their clocks by solar-system rhythms. Biological clocks enable the organism to master changing conditions in a rhythmic world; to do the right thing at the right time. The types of ordering depends on the local characteristics of place. Similarly, space is a unit of place for a being. The rhythms of the universe impress themselves on organism in varying degrees. The rhythms are incorporated. An animal or plant then establishes its own personal rhythm. Plants and animals create their own world from a place, according to their complexity and needs.

Uexkull implies that the human world is only one of the many possible. Although this is true perceptually, the use of symbols allow humans to imagine and represent other worlds. Humans can even create virtual worlds by limiting what could be received; for instance, if a being could see in the x-ray part of the spectrum. Yet human imagination is limited, as is human knowledge. Many organisms exist of which we know nothing. Their worlds have little meaning in a human world.

Edward Hall's space bubble for humans or Lewin's personality field extrapolate the umwelt to humans. The building of a self-world, a place to dwell on earth is hard earned; civilization is held together by practical and political activities. Building and planting, taking care of himself and others, is primarily a prosaic activity.

4.2.1.5. Species & Community

Paul Colinvaux states that species result from the process of avoiding struggles for existence. Since the species, at any concrete moment, is a collection of individuals capable of interbreeding, it turns out that what defines an organization of individuals is an abstraction or something that requires the existence of well-defined individuals from the beginning. The greatest difficulty in pinpointing speciation is determining when interbreeding is impossible. Species might better be represented by overlapping fuzzy sets. Watts developed a metaphor of species on earth as heads of a hydra; each has some autonomy and finite life, but is part of a longer lived whole.

Paul Shepard claims that all creatures are not necessary equally to an ecosystem; some presences do not add to the efficiency and stability of the whole. Larger species are dispensable; the web can do without some strands. Michael W. Fox noted that the ecology of England works without lions. Perhaps it would be better to say that all species are necessary but not sufficient to a system. The smallest species, the microbes and invertebrates seem to be the most necessary. Shepard says that no large animal is necessary to an ecosystem; but hippos and crocodiles are necessary to theirs. Hippopotami support fish populations in lakes and rivers with the minute animals in their excrement. As hippos are eliminated, the fish population dwindles, and human natives have less protein. We have been unaware of the role alligators play in the equilibrium of the Florida Everglades. The alligator creates pools by digging in damp soil; these pools become lairs of fish that eat mosquito larva. The pool also serves as a refuge to more species, including birds, in times of drought. Eagles and primroses, lion wolves, and whooping cranes are integral parts of ecological networks. We depend on their resilience. Although some ecologists cannot think of any, large species and predators have important uses in nature: As an expression of variety, niche makers, and feeling beings.

4.2.1.6. Environment

The environment is defined simply as the surrounding of an organism, usually the nonliving surroundings. The first level is the immediate environment of the organism, the living field. The next level is the proximate, then the group, community, system, regional, planet, moon, solar system, local cloud, galaxy, and universe. Distant environments tend to impose limits, but have less immediate influence on the organism or the community.

4.2.1.6.1. Geological Cycles & Climate (Moon to Elements)

The geological level contains many resources for living beings. It also provides a stable platform, at least in terms of the lifetimes for living beings. The reasons for this relative stability reside in the composition of the planet and in the double-planet influence of the moon on the earth's wobble and tides. The moon not only stabilizes the tilt of the earth as the two revolve, but it provides energy and challenge to organisms in the ocean. Because of its relatively large size and closeness, the moon forms the other half of a double planet with the earth. The moon exerts a gravitational pull on the earth that is stronger on the closer side. This creates a tidal variation in the heights of the oceans; these vary monthly. For many shallow water creatures, amphibians and mammals, it is good to adapt to these tidal variations. Of course, the earth exerts a pull on the moon, also, but it is less dramatic.

Due to its rotation around the sun, and to imperfections in balance, the earth tilts on its axis. This obliquity of the ecliptic creates seasonal variations, to which most plants and animals have adapted. Any changes, even relatively small ones, could be catastrophic for climate (a one-degree change could account for ice ages). Jacques Laskar et al. have documented the importance of the moon on the habitats of the earth. A stable climate needs the influence of the moon; otherwise, there would be immense variations in solar heating of the earth's surface. But the moon also causes problems. or rather provides challenges to which plants and animals have to respond.

The planet started as a mixture of chemical elements (over a hundred, including hydrogen, helium, and nitrogen) circulating over an active geology, driven by solar energy. At the risk of making life seem too simple, complex molecules formed from the action of light on primary gases, such as methane (CH_4) and carbon dioxide (CO_2). The atmosphere changed dramatically from a reducing atmosphere to an oxidizing one.

The amount of light hitting the atmosphere in the short-term is two gram-calories per square centimeter per minute—this is the solar constant (a gram-calorie is the amount of heat required to raise the temperature of one gram of water one degree Centigrade at fifteen degrees Centigrade; this amount is relatively small, so we use the Kilocalories as a more convenient measure), which varies minutely. This radiation is reflected or attenuated as it hits the atmosphere—on a clear day at noon, perhaps two-thirds of it reaches the surface. Light also provides a suite of challenges for living organisms.

The ecosphere acts as one system in which energy from the sun is cycled. The functioning biomass is integrated by feedback responses to extract the maximum of energy and still maintain a balance. Most of the solar energy is used for maintenance by the biosphere, which is an indicator of the biosphere's high degree of ecological maturity. Chemical equilibrium is a global regulator. The steady effect of light, the availability of oxygen, the thickness of soil, and the area and depth of the oceans is *almost a steady state*, as is the mean temperature of the planet. Living fossils demonstrate a continuously available environment for long periods; horseshoe crabs show that the sea has been relatively unchanged for the past 200 million years; algae goes back to microfossils that are over 3 billion years old. The amount of material that has passed through living organisms since Cambrian times, roughly a billion years, is of the same order of magnitude as the earth's crust (2.4 x 10^{25} grams. Assuming 1 x 10^{16} gm of carbon are fixed each year by photosynthesis, averaged over billion years, the total mass recycled of carbon alone exceeds 10^{25} grams. Total coal and oil deposited in lithosphere are estimated at 10^{19} grams). Elements frequently associated with life—calcium, phosphorus, iron, sulfur—are preferentially concentrated and deposited by living organisms.

The single most important system factor is climate, and for humans especially, the fluctuation of the climate over the past 10,000 years. We depend on water, and on its collection in aquifers, for drinking and irrigation. Because the climate is variable, the ecosystem is ever-changing, and humans have unforeseen effects, we have to consider water use and recycling. Yet, the climate is roughly predictable and stable, whereas the weather seems unstable and unpredictable. The best economic policy is probably one that tries to balance long-term productivity and competitiveness with short-term benefits (tradeoffs).

4.2.1.6.2. Ecosystems Structures & Functions

An ecosystem is defined as a biotic community and its nonliving environment, functioning together. An ecosystem is a self-organizing, chaotic system with emergent properties. unique from those of individuals, species, or communities. Ecosystems develop in time. That is, an ecosystem develops by a reasonably orderly, directional process that involves changes in structure that result from community modification of the physical environment. Although the physical environment imposes limits and sometimes determines patterns and rates of change, the community controls the development of the system.

The ecosystem is the level of integration and the unit of organization undergoing a directional development, according to Eugene Odum. Energy is bound into organic material, measurable as productivity. That quantity of energy and material no longer of use to the system is lost as waste. Chemical elements, especially those of life, circulate in the biosphere in characteristic paths known as biogeochemical cycles. Organisms and communities are limited by elements and physical factors, such as gravity, light, water, gas, or salt. Every ecosystem exists in the context of larger and smaller systems in a planetary system, which is part of the stellar and galactic system. The sun is the basic source of energy.

The ecosystem is directly connected with global cycles and other ecosystems. There has to be a good substrate with energy and richness flowing into the system. The productivity and wastes from other systems may enrich the system. The system is embedded in larger systems and global cycles, cyclicity. The interplay of material cycles and energy flows generates a self-renewing "homeostasis" in Odum's words. Processes like homeostasis, which involve a relationship with the environment, occur on an ecosystem level. Living systems do not display homeostasis—constant value—so much as a particular course of change in time—homeorhesis. The course is stabilized, not the constancy, according to Conrad Waddington. Changes to a system are symbolized by trajectories in a multidimensional phase space or landscape. Homeostatic (or homeorhetic) mechanisms protect the system from many disruptions.

Howard Odum described a patch of forest as a mysterious thing, growing, repairing, competing, holding itself against dispersion, oscillating, modifying energy from the sun. He described the forest as an ecosystem. The term ecosystem, was introduced by Arthur Tansley in 1935; he defined it as the system resulting from the integration of all the living and nonliving factors of the environment. An ecosystem is a community of organisms interacting with one another and their environment. As the essential unit of ecology, ecosystems must be seen in dynamic and historical terms. Eugene Odum describes the ecosystem as a unit of organization undergoing an orderly process of development that is reasonably directional.

Earth is like a mosaic ecosystems; the cells had boundaries like rivers or climate that occasionally broke down and allowed invasion. There is really just one system on the planet and that is the planet. It is possible to define ecosystems as close but not exact subsystems. The vast number of interrelationships between systems keeps them open. For example, grassland is affected by climates, soil conditions, fires, surrounding communities, and human agents. Ecotones between systems are usually shifting.

The actual substance of which an ecosystem is made consists of patterns rather than things or individual species. The ecosystem is generated by a patterning of ebb and flow of energy, substances, individuals and species across a suitable landscape.

Ecosystem patterns build up information. There are three different channels of information in an ecosystem: a genetic (in replicable individuals); an ecological based on interaction between cohabiting species (expressed in changes in their numbers); and the ethological or cultural, transmitted through individual learning based on experience. Feedback within interaction of species is expensive memory with little storage capacity. Whenever succession starts again, after a volcano eruption for instance, old information in form of interactions has not been saved. Genetic memory has much larger capacity and is long-term. Cultural memory enlarged with higher vertebrates.

Regarding the structure of an ecosystem, one can describe components, from the geological substrate to consumers, that is, organisms that ingest other organisms. The substrate includes geological formations, rock surface, atmosphere, and the oceans. The substrate affects and limits organisms (but not in the same manner as control). Other physical processes, such as sunlight, affect the ecosystem. In general, sunlight limits organisms, especially plants, but many animals have learned to live without light. Material cycles distribute inorganic compounds from the substrate. Organic compounds, such a lipids or proteins, link the abiotic to the biotic. Living organisms produce food from sunlight, using various elements and compounds. Other organisms, bacteria and fungi, consume organisms.

Eugene Odum subdivides an ecosystem into six processes, emphasizing the function of ecosystems: (1) energy circuits (including individuation), (2) food chains (as interactions), (3) diversity patterns, (4) nutrient cycles, (5) development and evolution, and (6) control (cybernetics and constraints). Through these processes, ecosystems are described with a variety of terms: energy, matter, entropy and ektropy, productivity, cycles, diversity, complexity, stability, and trophic structure. Many terms in ecology, such as biomass, stability and diversity are inexact. It is almost impossible to estimate the amount of degraded energy in an ecosystem, from transpiration, mixing of water, or other activities.

One problem with describing ecosystem attributes is that both quantitative (measurable) and qualitative (conceptual) factors are included. Resistance to external factors is a qualitative attribute. This may be more the result of succession than a factor producing an ecosystem. All ecosystem properties are not equally important. There is no whole system without an interconnection of its parts, and there is no whole system without an environment. Behavior at any level is explained in terms of the level below, but its significance is found in the level above. Ecosystem behavior does not emerge from a set of organismic equations.

4.2.1.6.2.1. Characteristics of Ecosystems

A system is a way to explain part of the universe and to deal with complex behavior. An ecosystem is a complex unit whose components keep its structure and function stable, despite changes and disturbances in its surroundings. An ecosystem is a material thing. Unlike an animal, which is a tangible thing, an ecosystem is an imperceptible thing, like an electron. Real things are changeable objects. The ecosystem pattern has a structure that includes organisms and abiotic substances. It has a function of taking a flow of energy and cycling material substances. History, structure and stability are emergent properties of an ecosystem. There is play in an ecosystem between adaptation and nonadaptation, or between co-constrained construction and deconstruction; completeness of either process would result in collapse of system.

Good places are set in ecosystems. To understand place, we must discuss the properties of working ecosystems. Ecosystems develop as information accumulates; information is generated by participating species and their physical structures, such as burrows or paths. The general characteristics, the inherent things, of an ecosystem include: Course, Identity, Openness, Creativity, Flexibility, Stability, and Vitality.

4.2.1.6.2.1.1. Course of an Ecosystem

The field provides the order required for producing individual organisms. The coming-to-be of organisms, that is process, is a fundamental feature of reality. The organism is what it does. The process of nature is not merely rhythmic change, it is a creative advance, producing new forms everywhere. The organism undergoes a process of evolution in which it produces new forms in itself. The process creates a course of motion.

4.2.1.6.2.1.1.1. Epigenetic Landscape & Homeorhetic Change

Conrad Waddington (1975) sees living systems as displaying homeorhesis, from the Greek *homeo*, meaning same and *rheo*, meaning flow. The term homeorhesis is used when what is stabilized is not a constant value but a particular course of change in time. Waddington also replaces embryology with epigenesis. Each embryological step is an act of becoming that must be built upon. In the embryo there is no need for new information; it must be protected from new information. The development of a fetus must follow postulates laid down in DNA. Embryos are not in a state of equilibrium, Waddington notes. A fertilized egg insists on changing; it can be stopped by killing it. Maturation implies that the phase space in which system is modeled contains a surface with a general slope that guides any trajectory toward the adult state and death. The development of the embryo into different organs can be described by supposing that a radiating system of valleys is superimposed on the general slope; the trajectories differentiate (like the path of electron?) and move toward organs. This model of an attractor surface is an epigenetic landscape.

Like embryos, ecosystems, have many characteristics and are affected by many environmental conditions. Their changes are symbolized by trajectories in multidimensional phase space; orderliness can be described in terms of constraints on trajectory courses, and these constraints are visualized as attractor surfaces. If the system starts from any condition, represented by a point in multidimensional phase space, the trajectory will move to nearest attractor surface and then move along it.

The interplay of material cycles and energy flows generates a self-renewing "homeostasis" in Odum's words. Processes like homeostasis, which involve a relationship with the environment, occur on an ecosystem level. Living systems do not display homeostasis—constant value—so much as a particular course of change in time—homeorhesis, according to Waddington. The course is stabilized, not the constancy. Changes to a system are symbolized by trajectories in a multidimensional phase space or landscape. Homeostatic (or homeorhetic) mechanisms protect the system from many disruptions. Negative feedback counteracts the effects of change to maintain the system in a steady state or homeorhetic state. A mature community is self-perpetuating and homeorhetic, with a dynamic balanced energy-matter budget.

The valleys occur in various shapes. Some could be very narrow canyons, others large meadows, similar to old mature earth forms, with meandering rivers. The name for a characteristic of an attractor surface in multidimensional space is a chreod, not a valley. The cross-sectional shape of the chreod describes the reaction of the system to fluctuations. With a steep slope, for example, it is difficult to divert the developing system from the bottom of valley; even with a strong force the system will return immediately as soon as the influence stops. With a shallow slope, like a flood plain chreod, it is easy to divert the system; it meanders before returning. Actually the perturbations alter the landscape itself, making a steep valley again, not just shifting the river. The environment changes, the system changes.

Waddington refers to such a course as chreod (from *chre*, fated, and *hodos*, path, in Greek). The nature of a chreod that a system adopts is determined by instructions in genes and other organizations of information, and in interaction with the environment that it travels, that is, the epigenetic landscape. Control of a system after disturbance may not bring it back to where it was before; it may go back or forward. Since the path is fuzzy, it can change.

Eric Jantsch sees the view of evolution as preservation of stationary states as fragmented. Much of natural selection is stabilizing selection. Homeorhesis is a significant phenomenon in evolution; Waddington applies it to the tendency of a process to continue in its original pattern, even if disturbed. Homeostasis is tendency of spatial structures to remain the same. The ability to persist is the ability to change, but species only change from environmental pressures; an amphibian is a fish trying to continue being a fish.

4.2.1.6.2.1.1.2. Growth

Growth serves as a mechanism of evolutionary adaptation, by carrying out genetic instructions in the environment; but growth is also conservative and stabilizing rather than innovative and reorganizing. Growth is homeorhetic; in a homeorhetic process, the flow is constant, not a stationary state. Flow processes follow fixed trajectories, called chreods. Growth, from fertilization, embryo states, birth, youth, and maturity, represents a homeorhetic process following more or less fixed chreods, programmed genetically and conditioned environmentally. Chreods must be like electron paths, that is, probabilistic.

4.2.1.6.2.1.1.3. Path

All electrons have the same charge because they represent some single aspect or perspective of the universe. Each particle is only an abstraction of a relatively invariant form of movement in the whole field of the universe. The proton and neutron are just two different points of view of the nucleon; the sigma has three different points of view. Both nucleon and sigma particle may be just different points of view of the same particle, related by an extended isospin symmetry. Specificity can be a property of the path, what Waddington calls a chreod and Whitehead a concrescence.

4.2.1.6.2.1.1.4. Immediacy

This world is composed of remote occurrences, on polar icecaps and distant stars, as well as immediate personal events. The body links the exterior to its interior. The person, or body-subject, is inextricably woven with the world; the body is a mode of presence to others. This bond of betweenness constitutes the foundation of an intersubjective paradigm for psychology. The sensory apparatus by which we make contact with the world arose through the process of evolution. An actual entity arises out of the actual world through a process in which objectivity is transformed into subjective immediacy.

4.2.1.6.2.1.2. Identity of Ecosystems

An ecosystem is a whole. The ontology of any living system is the history of the maintenance of its identity as a whole through continuous self-making, or autopoesis. If there were no identity, there would be no differences and so no relationships. Without relationships, there would be no things, no events, and no universe. Objects precipitate out of relationships and are defined by them.

For the universe to show stability and persistence, different entities must have stability and persistence. Identity is that persistent quality. Identity serves nothing; it is. The relationship between identity and wholeness is a rhythm, forming unique patterns. George Leonard identifies it as the silent pulse at heart of our experience that is always there. A system's identity can be described apart from its performance in interactions, but not isolated from it.

Ecosystems are part of an unending, imperfect process, without any final state. Furthermore, the human attempt at perfectibility through self-improvement causes disharmony, which is part of the same imperfect process. Each system is a practical application to place. Unknown factors determine a large part of the operation of any system. Furthermore, there is chaos in every system; there are plagues and random frenzies.

4.2.1.6.2.1.2.1. Self-Making of Ecosystems

Bertalanffy called life a system of self-organization. a developmental unfolding at progressively higher levels of differentiation and organized complexity. Living systems are autopoetic units in physical space. The system is autopoetic, that is, self-creating. It renews itself as its contents change, as disturbances change the parameters. As a mature system, it continues to move to a point of higher maturity, recovering from disturbance to its original trajectory, where productivity declines and stability increases.

The organism and environment are co-implicative, co-defining, and co-constructing, in a process of self-assembly, where the self is the organism/environment system. In an autopoetic framework, every being is embedded in a world and observed by an embedded observer. The material components of life move through physiological processes. Autotrophs (bacteria, algae, green plants) convert energy into organic compounds; heterotrophs reconvert into heterotroph flesh.

4.2.1.6.2.1.2.2. Wholeness of Ecosystems

In an ecosystem, energy, information, matter, and life are part of an inseparable whole, with no formal theory linking them. Not a spectral oneness, but a diversity of the whole into many perspectives. George Leonard suggests the word holonomy for the study of wholeness. Holonomy is not the death of division; it is a frame for individual wholes.

The relation of a cell to organism is certainly part to whole. Organism to species, however, is member to class. A liver cell is part of a liver and the liver is part of a wolf. Therefore the liver cell is part of the wolf. This is a transitive relation.

Including degrees of otherness gives the system its comprehensiveness. Species integrate themselves into the system over time, using previously unused productivity or waste in developing their special niche. Each ecosystem, however, sets the limits and determines the style and complexity of its individual species. The total community evolves.

4.2.1.6.2.1.2.3. Boundedness of Ecosystems

An organism makes its own boundary, such as skin. Identifiable components organize internally with structural boundaries. The boundary is produced by the system. The cycles are bound by limits. Both individuals and communities are usually bound by one or two specific limits. Ecosystem health occurs within those limits.

Ecosystems and organisms depend on a complex set of conditions, which can be limiting factors. Organisms are affected by the quantity and variability of materials, if they require a minimum of them. Organisms are also affected by their own limits of tolerance to those materials. Life is limited by elements and physical factors (light, water, gas, salt); by too little of an element (Liebig's law); and by too much of an element (Shelford's law of tolerance). An ecosystem is often limited by physical factors, such as rainfall or altitude. The limit of the system is also the limits of the cycles within the system.

Limits define locality, local spaces, and local systems, from the global. Limits are not only important for life, but also are implicated in diversity and maturity. But, even these limits are based on the physical.

4.2.1.6.2.1.2.4. Form of Ecosystems

An ecosystem has a definite size. Size is the quality of an element that determines how much space it occupies. It is the extent or magnitude of the element. The size of an element determines the perceived size and relation of other elements. For instance, a large talus slope on the side of a small mountain may look larger than it is.

A system has a definite shape, which may integrate it into the surroundings. Shape is the quality of a thing that depends on the relative position of all points on the surface, or the spatial form characteristic of a thing. The shape may follow microenvironments in the landscape. The shape may determine the interaction with other systems and cycles. Shapes are one of the first things one notes in a system. For instance, a complete inventory of elements in an Oregon creek starts with the shapes of the features in the area. The large volumes are rounded and natural hills—even the agricultural evidence is almost natural.

4.2.1.6.2.1.3. Openness of Ecosystems

The ecosystem has to be open to flows of energy and materials. Closure, e.g., well-defined boundaries, with steady input output flow rate, is not a contradiction. The boundaries are permeable, allowing exchanges of energy and matter. Too much openness would allow the system to be overwhelmed by the environment. Too little openness would cut the system off from the environment and it would run down, in terms of energy. The system is a local system, with specific structures and functions.

Random order is open; it has degrees of freedom. Differences can be related to degrees of freedom. An order with more differences has more degrees of freedom. The macro-order limits degrees of freedom. Movement could not occur without a free order or disorder.

The vast number of interrelationships between systems keeps them open. For example, grassland is affected by climates, soil conditions, fires, surrounding communities, and human agents. Ecotones between systems are usually shifting.

A partial process requires cuts in the whole, boundaries. Identifiable components organized internally with structural boundaries. Self-boundedness: the boundary is produced by the system.

4.2.1.6.2.1.3.1. Structure of Ecosystems

An ecosystem is an area where species interact with the structures and processes of their environment. An ecosystem is a topographic unit, a volume of land, occupied by organic beings, extended over an area and through time, with connections to larger mineral, chemical, water and air cycles. This means they are geographical units that intersect with atmospheric units. Ecosystems have a vertical structure, that includes the levels of climate from micro to topoclimate and macroclimate, to soils, water structures, and bedrock, as well as a horizontal structure.

The structure of an ecosystem allows flows in and out. Organisms enfold structures based on historical patterns. To be considered healthy, an ecosystem has to maintain its structure and metabolism (rate of energy use) despite occasional stresses. The energy in a mature system goes to the maintenance of order and less for the production of new materials. In general, diversity is higher, and life cycles are more complex; symbiosis between species increases, and nutrients are conserved.

The structure of the forest ecosystem changes: organic matter increases, inorganic nutrients are used internally (instead of being extrabiotic), biochemical and species diversities become high, and pattern diversity becomes well-organized.

4.2.1.6.2.1.3.2. Function of Ecosystems

The process of adaptation involves a self-presentation offering new symbiotic relations. Overspecialization reduces flexibility and reaction to change. Underspecialization reduces efficiency. At all levels evolution includes freedom of action as well as interdependence. The construction is never total; it requires destabilization, a risk accompanying all innovation. The processes that generate form and variation at every level.

Adaptation emphasizes integration, transcending disciplinary boundaries. In protobiotic development, physical and chemical processes are responsible for molecular selection and for the fitness of the environment for life. Generating organic forms, physical and chemical processes provide organizational principles that coordinate detailed biological mechanisms, including viscoelastic changes of the cytoskeleton and genetic expression.

Ecosystems use energy efficiently. Dubos and others have claimed that "nature knows best is wrong," that nature is inefficient and wasteful. But ecosystems that seem inefficient and wasteful are many times extremely redundant, and therefore stable and flexible. Natural processes that seem destructive are cyclic and preservative also.

More mature systems have a richer structure and a lower productivity per unit biomass. There are more steps in the trophic pyramid. There is higher efficiency in every relation. The loss of energy is less, so less energy is needed to maintain the system. This is why old growth forests are efficient. The system has to maintain potential to adjust to change.

4.2.1.6.2.1.3.3. Diversity of Ecosystems

The environment has been constant enough for organic evolution, but variable enough for natural selection to be challenged. Variability challenges organisms to adjust and thrive. Variability, even in small ways, leads to diversity. Diversity, as a measure of genetic variability in ecosystem, enlarges information. A mature system needs less information, since it works toward preservation. The limit of maturity allows maximum variability between systems with slight external differences, like temperature.

The variation of climate, for instance, an external disturbance, can change the physical and chemical parameters of the forest. Acid rain causes biochemical and soil chemical reactions, which in turn provide in-*form*-ation to other organisms, which use it to regulate their physiological processes which can effect the forest, e.g., insect damage changes the level of seasonal growth of trees. Damage to trees from pathogens, or disease agents, is an internal disturbance.

Viruses spread genes among bacteria, trees and humans. Viruses are sources of evolutionary variation. If an ecosystem is disrupted, viruses can outgrow their resources, like any living being. Variation in individuals and species allows a better response to variation in climate or other aspects of the environment. Due to environmental variations, such as water temperature, there may be annual variability in populations of animals, such as salmon.

Species richness may stabilize ecosystem properties, such as Net Primary Productivity (NPP). Frank and McNaughton (1991) showed that plant community structure (and productivity and biomass) is more stable when a greater richness of species is present.

As a forest matures over time, its structure and use of energy changes. Structural components tend to become larger: Large trees, large snags, and large downed trees. Species diversity increase because there are more possibilities for making niches in the increased structural variation. Within species there is more genetic variation. Less energy leaves the system, because it is bound up into maintenance of the structure and species. The Net Ecosystem Productivity (NEP) approaches zero (see above).

But, Odum points out, as some communities age, Wisconsin forests for example, there is a decrease in diversity, in the understory anyway. Also, diversity can decline with productivity, as in the eutrophication of lakes, for instance. While it is meaningful to speak of an optimum diversity, as the result of limits and the interaction of many factors, a maximum diversity may never be reached. As Paul Weiss noted, the patterns of organic nature are a combination of order and diversity; order involves constraint while diversity requires freedom for difference. The maximum order would result in a static universe, where a maximum freedom would create a nonordering chaos. What is a minimum, optimum or maximum forest cover for various watersheds? Science might identify minima or maxima but philosophy can aim at optima or satisficia.

4.2.1.6.2.1.3.4. Flexibility of Ecosystems

Most healthy ecosystems have high degrees of flexibility. As Gregory Bateson interprets Ross Ashby, any biological system can be describable in terms of interlinked variables, each of which has an upper and lower threshold of tolerance beyond which the system acts pathologically. Within the limits the variables can be moved for the system to be adapted to the environment. Under stress, some variables move to maximum values near the upper or lower limits—the system loses flexibility, that is the "uncommitted potential for change" (in Bateson's definition), and can be destroyed by further stress. The danger in each case is working near the maximum of the system.

For change to occur there has to be space to change, that is, to rearrange the units. That space, real or virtual, is measured as flexibility. Overspecialization reduces flexibility and reaction to change. Underspecialization reduces efficiency. Flexibility means not being over connected, not being too rigid or efficient, and able to slough off species or living forms and incorporate new ones.

Of course, flexibility depends on the renewability, and on its maturity (trajectory). The system is autopoetic, self-creating. It renews itself as its contents change, as disturbance change the parameters. As a mature system, it continues to move to a point of high maturity, recovering from disturbance to its original trajectory, where productivity declines and stability increases.

For maximum flexibility, the parts of a system will maintain the potentiality for all possible behaviors that could flow from any part of the system, as in a hologram. In fact, natural succession may operate this way. Native Hawaiian biota appears to be rejuvenated by volcanic eruption; it is better equipped to reinvade areas than exotic species. The advantage to an ecosystem of having a large gene pool is having a set of flexible responses to a wide variety of environmental conditions, from fires to high winds, diseases, and insects.

Removing any one part of the system completely may make it susceptible to some unforeseen or unpredictable factor, e.g., trees seemingly genetically inferior under the current conditions may have the best response to other conditions, such as the greenhouse effect or frequent disturbance from logging. We never really know which trees nature has selected for removal until after they have died or have been removed, and even then we may not know why. The pattern should allow for surprises and discontinuities; it can do this if it is flexible. The design of forests is vulnerable to surprises because nature is chaotic (unpredictable) and science itself is uncertain (by definition) about patterns of change in forests.

4.2.1.6.2.1.4. Co-constrained Construction of Ecosystems

The organism and environment are co-implicative, co-defining, and co-constructing. They engage in a process of self-assembly, where the complete self is the organism-environment system. Construction requires participation, complexity, and development. The process of construction involves a self-presentation offering new symbiotic relations and novelty.

Novelty always enters with environmental change, which serves to maintain the openness of the system. Novelty enters with fluctuations. The "strategy" of ecosystem development is increased control of, or homeorhesis with, the physical environment and novelty—probably to protect itself from perturbations. There is a fundamental shift in energy flows, as increasing amounts of energy are used for maintenance. As more and more energy is used for maintenance, the net community production (NCP) approaches zero. The mature system becomes more efficient, as it supports a larger biomass with the same amount of energy. The food chains become more weblike, dominated by detritus chains as opposed to linear grazing. Construction depends on diversity for the reciprocal constraint. Local context allows for more rapid construction. The constraint forces species to change.

4.2.1.6.2.1.4.1. Participation in Ecosystems

Remember that the phenomenon of participation comes through an elementary act of observation. Evolution is the production of new codes of information to match a changing environment. Margalef describes the evolution of an ecosystem as information accumulation; information is generated by participating species and their physical structures, such as burrows or paths. All beings participate in relationships. All beings observe. There are no nonparticipating observers. A world without observers would be impossible.

Perception is participation, the organic relationship between the center of the self and the circle of appearances. Living organisms in a given area interact with the physical environment so that an energy flow leads to the defined trophic structures and material cycles that comprise an ecosystem, according to Eugene Odum. An ecosystem can be analyzed into parts, including organisms, energy circuits, food chains, diverse patterns, nutrient cycles, development and evolution, and control. No organism can exist by itself or without participating in an ecosystem. Each organism participates in cycles that link it to other types of organisms. The system is embedded in larger systems and global cycles, cyclicity.

In a mature system, complexity expressed by number of participating species, remains almost same, in spite of extinction or emigration. The more complex a system is the larger the share of energy flowthrough that is stored in the system.

4.2.1.6.2.1.4.2. Complexity of Ecosystems

Complexity is a characteristic of sets that are intricately arranged. Complexity emerges from the interpenetration of processes of differentiation and integration, processes running simultaneously from top and bottom and shaping the hierarchy from both sides. Microevolution generates macroscopic conditions for continuity and macroevolution generates microscopic elements for processes. Life is an emergent property of a hierarchy of chemical levels. It is not found in molecules but through autopoetic organization of molecules. The universe restructures itself constantly at a more complex level. Reciprocal causal processes can increase structure, differentiation and complexity in natural systems, according to Maruyama.

According to Koestler, all complex structures and processes of a relatively stable character display hierarchical organization. Levels of a hierarchy tend to be contained in subassemblies. Each subwhole behaves as a whole to its components, as a self-contained whole, and as a dependent part in context. Wholes and parts do not exist absolutely. There are intermediary structures on a series of levels in an ascending order of complexity; each subwhole faces in opposite directions. These Janus-faced subassemblies are holons, that is, paraphrasing Koestler, any stable subwhole in a hierarchy that displays rule-governed behavior and structural Gestalt constancy. The rules lend order and stability, as well as flexibility.

Biological processes also generate complexity (or information). In biology and sociology, just the opposite meaning of evolution is true; as transformations to higher levels of complexity and improbability. The organism continues to develop, however, experiencing and learning the environmental complexities through mating and then to the end of life.

4.2.1.6.2.1.4.3. Adaptation of Ecosystems (Or Coconstruction)

The ecosystem 'learns' the changes, e.g., seasons, of the environment. Any system formed by reproducing and interacting organisms must develop an assemblage in which production of entropy per unit of information is minimized. It is a general property of some systems that acquired information is used to close the door to further inflow. A mature system needs less information, since it works toward preservation. The limit of maturity allows maximum variability between systems with slight external differences, like temperature. Ecosystems consist of different prefabricated pieces: species. Since the supply of species is limited, succession becomes asymptotic, that is, it leads to maturity, or in the old terms, a climax.

Adaptation is a metaphor, based on the assumption of it is that there is a fixed eternal environment to which species are fitted. Although it has been productive, it has limitations. The environment is ontologically prior to organisms, leading to view of environment as force and organisms as passive objects, like keys that fit locks, or not. Life does not adapt to a passive prior environment, it produces and modifies its surroundings.

Fitness is a metaphor and its meaning can be reversed. Some scientists reverse the accepted meanings of fitness and adaptation. Meredith states that the lesson of nature is moderation in all things, "even in fitness." Fitness builds up in an ecosystem as it matures. Selection at the organismic level is selection of the fit. Species level selection is selection of the unfit (extinction of the strongest). The levels of selection must balance, so that life is not too fit or too unfit. Evolutionary fitness cannot keep increasing. In some cases it decreases with time. Cope's notice of survival of the underspecialized; Haeckel's observations on senescence. Perhaps species are self-limiting in fitness. Meredith proposes that altruism is a technique of species limitation that would increase the overall longevity of a species, "unfit enough to be compatible with a limited planet."

Many environments, like heathland, are nutrient-poor, so flower species have to change. Because pollinators are scarce and sugar is easy to make, not requiring minerals for plant tissue, the flowers bloom most of the year and drip nectar. Eucalyptus trees grow on nutrient-poor soils and adapt. They add toxins to their leaves to discourage predation by animals—toxins, like sugar, are cheap to make. The leaves are too acidic for possums and others. Sclerophyl forests are dominated by eucalypts.

An ecosystem has to be able to adapt to dynamic changes in the environment, including drought and vegetative changes. Slow coevolution allows the system to adapt to geological or climactic changes.

4.2.1.6.2.1.4.4. Maturity of Ecosystems

A biological organism grows to maturity, which is a stopping point for size. Growth serves as a mechanism of evolutionary adaptation, by carrying out genetic instructions in environment; but growth is also conservative and stabilizing rather than innovative and reorganizing. Growth is homeorhetic; in a homeorhetic process, the flow is constant, not a stationary state. Flow processes follow fixed trajectories, called chreods. Growth, from fertilization, embryo states, birth and maturity, represents a homeorhetic process following more or less fixed chreods, programmed genetically and conditioned environmentally. Chreods must be like electron paths, multiply probabilistic.

Whatever accelerates change and energy flow in ecosystem reduces potential maturity. In an exploited system, diversity drops and ratio of primary production to biomass increases. Mature systems can regress to earlier forms when exploited; then, new species can form. Ecosystems are constantly evolving under the influence of physicochemical processes poorly understood and more powerful than those resulting from human activities. The reality is more complex than just systematic succession and climax. Margalef suggests replacing word climax with high maturity. But climax conveys more information than just maturity.

As a mature system, it continues to move to a point of high maturity, recovering from disturbance to its original trajectory, where productivity declines and stability increases. "Nature tends to become baroque in situations permitting high maturity, with little energy left for large changes," according to Ramon Margalef (1968).

Ecosystems consist of different prefabricated pieces: species. Since the supply of species is limited, succession becomes asymptotic, that is, the system approaches a state of maturity (once called a climax). It is a general property of some systems that acquired information is used to close the door to further inflow. A mature system needs less information, since it works toward preservation. The energy required to maintain an ecosystem is inversely related to its complexity; succession decreases the flow of energy per unit biomass until the system reaches maturity (Margalef's concept of maturity). In a mature forest, for example, almost 100 percent of the energy is required to maintain the state of the forest. Most of the solar energy is used for maintenance by the earth; this is indicator of biosphere's high degree of ecological maturity.

4.2.1.6.2.1.5. Faces of Stability of Ecosystems

Stability is the ability to maintain the identity of a system under the flow of external forces and disturbances. Stability can be refined through the specifics of constancy, resistance, resilience, and accommodation. Stability can be related to ideas of compartmentalization, communications, richness of interactions, and connections.

Stability is a complex topic. Ulanowicz suggests that stability might be explained by diversity flow topologies, where flow topology is a descriptor of how ecosystems develop. The homeostasis of the ecosystems, that is, stability, as originally proposed by Eugene Odum, becomes the result of regular flows of energy and materials. Growth and development are characterized by a qualitative formalism of increasing ascendancy, which explains the drive towards coherence, efficiency, specialization, and self-containment.

4.2.1.6.2.1.5.1. Persistence of Ecosystems

Persistence means stability or continuity, in general, or the state of being in a continuous flow or coherent whole for a duration. But, specifically, it can mean a lack of change in a system parameter, like number of species.

A persistent system is stable and persistent in time, self-maintaining, mature, and hysteretic (historical). The system changes and develops. It loses and gains species over time, but is still recognizable as a short-grass prairie, for instance.

The system can have degrees of difference, malleability, and still be the named system. The system may have different states that oscillate. It may also have a trajectory to a mature state.

4.2.1.6.2.1.5.2. Resistance of Ecosystems

Resistance is the ability to withstand disturbance to a system and to continue. This is Robert Macarthur's meaning of stability, from his analysis of food web structure; it is similar to Holling's concept of 'resilience.' McArthur's stability is less predictable, and it is more expensive to maintain.

4.2.1.6.2.1.5.3. Resilience of Ecosystems

Resilience is the ability to recover from stress. The ecosystem can be thought of as elastic. The elasticity refers to the time required to recover its initial structure and function after some kind of damage; it is the speed that a system returns to former state following a disturbance (as a measure in the community matrix).

4.2.1.6.2.1.5.4. Conciliance of Ecosystems

Conciliance is the ability to absorb change and still maintain the system identity. The system adapts to disturbance. It accommodates disturbances. It incorporates new things, such as new organisms. It tolerates new levels of things, such as increased heat. It fits the changes within the structure and function of the system. Amplitude is a measure of the maximum amount of damage a system can sustain and still recover.

These preceding terms are not comparable. Constancy and persistence are descriptive, implying nothing of underlying dynamics. Resistance to stress is a useful notion if one is interested in the maximum extent of the deviation between stressed and unstressed systems. Resilience is relevant to those who are concerned with the rate at which a systems returns to prestress conditions. And asymptotic stability is concerned with whether or not a system will eventually return to prestressed state.

Cyclic and trajectory stability have measures of inertia, elasticity and amplitude associated with them. Elasticity and amplitude seem to mean centripetal and centrifugal. Oscillation, or cyclicity around a point, or cyclical stability, is the property of a system to oscillate around central point; predator/prey systems, for example, have the property of a stable limit cycle. Malleability is the degree of difference from original state.

4.2.1.6.2.1.6. Productivity of Ecosystems

Productivity is the ability to convert energy into living forms and the ability to incorporate materials into living forms. Productivity, in general, depends on the vigor, or strength or vitality, of the system. Health is the overall ability of a system to maintain itself under a normal range of environmental conditions (which may include hurricanes, volcanic eruptions, or fires). Obviously, a pioneer community may change the conditions to favor a new level of the system with new components. Health is a dynamic measure of ecosystem organization, vigor, and resilience. Organization is described by diversity and connectivity; vigor is related to the amount and speed of productivity; and resilience is a measure of reaction to stress. Too much stress, for example, leads to unsustainable patterns of behavior; continuous stress leads to a breakdown of processes that becomes irreversible—the system dies. Health is a dynamic quality of the whole, the result of a harmonious interaction of all the analyzable parts that comprise the whole forest with the surrounding larger environment.

Adaptive ecosystems are not static orders; they are flexible, as well as historical and irreversible. Ecosystems develop knowledge-bases that reflect knowledge of themselves and their environment. The strengths of local systems lay in the diversity of values and in their fitness to particular places.

4.2.1.6.2.1.6.1. Productivity & Flexibility

To relate health to growth and productivity, we could say that the capital of an ecosystem would be its physical environment and its gross primary productivity; interest would be the net ecosystem productivity. The production percentage would be the amount necessary to keep the ecosystem healthy.

Our measurements of productivity, however, are not adequate. We are measuring over a year or two only to establish a growth rate or productivity. We should be measuring over centuries. A forest is a long-term, dynamically-changing being. We cannot use a short-term industrial approach to measure a few parameters and then pretend we know enough about a forest to cut a large percentage of it. Forests are created by slow processes that take millennia.

A healthy forest is flexible in its response to diseases and pests. Most healthy ecosystems have high degrees of flexibility. As Gregory Bateson interprets Ross Ashby, any biological system can be describable in terms of interlinked variables, each of which has an upper and lower threshold of tolerance, beyond which the system acts pathologically. Within the limits the variables can be moved for the system to be adapted to the environment. Under stress, some variables move to maximum values near the upper or lower limits—the system loses flexibility, that is the "uncommitted potential for change" (Bateson's definition), and can be destroyed by further stress. The danger in each case is working near the maximum of the system. In the case of forestry, we and our civilizations are part of the system. If an overpopulated society wants more forest products for houses, furniture, and fuel to be more comfortable in an overpopulated state, then more trees must be cut, and this, because the variables are interlinked, means that the stress spreads to more forests. To keep ecological flexibility in the forests, wood resources would have to be budgeted in appropriate ways, or demand would have to be reduced—or forests would have to be expanded.

4.2.1.6.2.1.6.2. Renewability of Ecosystems

The self-creating system renews itself as its contents change, as disturbances change the parameters. As a mature system, it continues to move to a point of high maturity, recovering from disturbance to its original trajectory, where productivity declines and stability increases. Although forests are considered renewable resources, they are slowly renewable, requiring hundreds or thousands of years to renew from catastrophic disturbance; this time is far longer than any economic plans and really nonrenewable on a human life scale. This has important implications on sustainability.

Aldo Leopold started to describe a science of land health; "Health is the capacity of the land for self-renewal.... A science of land health needs, first of all, a base datum of normality, a picture of how healthy land maintains itself as an organism." Notice that Leopold anchored this concept in a theory of organism rather than an ecological theory of community. Interestingly, the Buddha also gave an organismic definition of forest: "a peculiar organism of unlimited kindness and benevolence that makes no demands for its sustenance and extends generously the products of its life activity; it affords protection to all beings, offering shade even to the axeman who destroys it."

Jay O'Laughlin et al. defined forest health as the capacity for self-renewal, the ability to recover from stress and disturbance, natural and human-caused. He also defined it a year later (in 1994) as a "condition of forest ecosystems that sustains their complexity while providing for human needs." Or, in an expanded definition: "the vigor or vitality of interacting biotic and abiotic elements of a system characterized by extensive tree cover that function together to sustain life and are isolated mentally for human purposes."

4.2.1.6.2.1.6.3. Limitability or Cycling of Ecosystems

In a mature forest very little material actually leaves the forest. It is held in cycles. Materials cycle above and below ground, between the atmosphere and trees, between trees and insects, and squirrels and fungus. Many cycles are investigated through ecosystem analysis, where energy and materials are traced through transfers through compartments in an ecosystem. For example in the nitrogen cycle, the compartments are nitrogen, atmosphere, vegetation, forest floor, and mineral soil, while the transfers are precipitation, throughfall, leaching, litter fall, mineralization, fixation, and denitrification. Each compartment keeps the nitrogen for a certain time, termed the residence time; for a hardwood forest floor, for example, the residence time is about seventeen years. Human interference in forest cycles can collapse the residence time; for example, clearcutting alder in Washington causes high nitrogen losses.

Virtually every material cycles through the forest: nitrogen, carbon, phosphorus, potassium, calcium, sulfur, magnesium, water. Nutrient cycling involves many of these materials. Nutrient cycles change with the succession of a forest. Half the nutrients for new growth are drawn from senescing needles and twigs.

Some material cycles between other ecosystems or in larger patterns around the planet. Some of the cycles are daily; some are seasonal or annual; others are years or decades; a few, like the carbon cycle, are century or millennia long.

Weathering of the soil increases the surface area of the components, up to a million times, and accumulates an electrical charge on the surfaces. These charges are critical to ecosystem functioning because they contribute to nutrient cycling. Cycling and nutrient conservation in old-growth Douglas fir canopies shows that those mechanisms parallel soil and stream structures, according to George Carroll.

An understanding of all types of mycorrhizal fungi and their functions is necessary to understand organic matter decomposition and in nutrient cycling. For instance vesicular arbuscular mycorrhizal (VAM) fungi are thought to decompose organic matter and cycle the nutrients directly to the host root. A reduction or change in decomposers can result in a decrease in litter decomposition and nutrient cycling.

4.2.1.6.2.1.6.4. Continuity of Ecosystems

Continuity is the coherence of the pattern of living in the system. If the pattern is disrupted, the local entity dies. Many local patterns flow together through time interdependently, sharing materials. The death of one pattern sometimes leads to the death of other patterns.

The body of a human or forest or any entity is a dynamic pattern supported by dynamic processes that include other entities.

Ecosystems are larger entities made up of other beings—large patterns made up of smaller ones. An ecosystem is a constant where every component changes, disappears and reappears. It is a pattern like a whirlpool. The pattern forms from a torrent of light, energy, molecules, air, water, and even bigger things.

Comprehending patterns is necessary to protect the scale of ecosystems, which is too large to see, except by satellite, the parts of the system that are too small to see, dominated by fungi and viruses, the parts that are too-long-lived for us to observe, the long successional time of evolutions, and the parts that we are ignorant about. Without special effort, we are aware only of what we see working in the ecosystem during a very short time. We trust that our plans will ensure that the ecosystem will remain as a healthy entity for a very long time so that many generations of us can gather our needs from it.

Good places are set in ecosystems. To understand place, we must discuss the properties of working ecosystems, that is the characteristics or distinguishing qualities.

4.2.1.7. Landscapes & Regions

A region is a part of the earth characterized by a specific climate, or a specific kind of animal and plant life. The principle of flow, that is, everything goes somewhere, is necessary to the functioning of organisms as well as the biosphere. Flow can only be realized through structure, however, cell or ecosystem.

Even though landscapes exist within large ecoregions (which exist within the biosphere), they need closure, from walls, barriers. distance, or buffers, to maintain their integrity. The principle of barriers keeps the parts separate. At the level of local ecosystems, there need to be fewer closures, so that there is a flow of genetic information within a species as well as between species.

Loomis and Echohawk found that 23 of 35 ecoregions have less than 1 percent protected and 7 have no land protected as wilderness.

4.2.1.8. Planet Gaia in the Mirror of Humanity

James Lovelock and Lynn Margulis have put forward a hypothesis that the planet exerts a living control of the atmospheric and hydrologic processes to maintain minimum conditions for life over long periods of time. The hypothesis, which of course can be disproved, but never proved, notes the phenomena to be explained: environmental natural regulation, atmospheric and oceanic homeostasis.

Lovelock hypothesizes a collective global mind immanent in the cybernetic structure of the global system. He calls it Gaia, after the Greek earth goddess, as suggested by William Golding. Nature is active, resilient and powerful—as Gaia. When modern ecology strives to think of the planet dynamically and holistically, it returns to personification. Lovelock states: "It appeared to us that the Earth's biosphere was able to control at least the temperature of the Earth's surface and the composition of the atmosphere.... This led us to the formation of the proposition that living matter, the air, the oceans, the land surfaces were parts of a giant system which was able to control temperature, the composition of the air and sea, the pH of the soil and so on, so as to be optimum for the survival of the biosphere. The system seemed to exhibit the behavior of a single organism, even a living creature."

He hypothesizes that every element in the system is related in a feedback network to every other element. This is reminiscent of Whitehead's organic philosophy, where every atomic unit has feeling and influences every other. For example, the biosphere can control the temperature of the surface and the composition of the atmosphere. On the other hand, soil types and the weather can limit vegetation; invasions of vegetation change soil types. Lovelock and Margulis have shown that it is sophisticated enough to maintain a constant temperature and pH for billions of years, in spite of great atmospheric and solar changes.

The hypothesis is supported by four facts: (1) the average surface temperature of the earth has been in a constant range for more than 3 billion years, despite an gradual rise in solar energy; (2) the concentration of the atmosphere is improbable, compared to the composition of Venus and Mars; it should be mostly carbon dioxide; (3) each atmospheric gas is optimal proportion for a life supporting function; and, (4) the salinity of the ocean is far lower than it should be from runoff from land; the present percentage could have been achieved after 80 million years. At a scientific level of thought, the Gaia hypothesis extends the fundamental ecological doctrine that all things in nature are densely, subtly and systematically interrelated until it includes humanity, ethically and physically. The entire earth is envisioned as a unified entity, actively shaping the material conditions of the planet for the purpose of maximizing the survival and variety of living beings.

The Gaian ecosystem is a network of coupled smaller ecosystems connected by global patterns of water and air that adapt to each other through feedback loops that regulate physical and chemical environment, all based on great bacterial ecosystems mostly invisible to us.

Lovelock regards Gaia as a symbiosis of global dimensions. He has expressed concern that humanity can impoverish the whole system by reducing the total variety. Although the Gaia hypothesis renews the idea that the earth is a mother for us all, and reinforces our understanding of interconnectedness of biological processes, it falls short of our responsibility and relies too much on our consciousness. Ecosystems are self-corrective; but there are no sense organs.

4.2.2. *Culture & Holecological History*

All that has been said is an interim report on the history of the idea of nature. Hegel stated: "That is as far as consciousness has reached." The study of nature is not a thing that exists on its own. Natural science is a form of thought that depends on the existence of some other form of thought, that is history. Natural science consists of facts and theories. A scientific fact is a class of historical facts. One cannot understand the first without understanding the second. No one can answer the question what nature is unless she knows what history is.

The historian sees civilizations of the past like pieces of petrified wood; the activity has moved on and only a form of waste remains. At the highest level of description, as in H.G. Wells' *The Outline of History*, most details of the lower levels are lost. Humans have concentrated on the study of human history, to the exclusion of natural history. We have ignored the deepest kind of history—that which arises from broad economic or demographic pressures, that which alters whole habitats and systems. A cultural ecologist looks for the general topography, and wonders about the direction of slopes. But a cultural ecologist cannot see individuals. Surface events reflect the underlying forces that determine livelihood, population, technology—anything that changes the relations between culture and environment.

4.2.2.0.1. Ecology & Probabilities

Ecology is rooted in natural history; the detailed study of habitats and ecosystems cannot occur without taking in the whole picture. Ecological history is history from the ground up. Water and gas seem to have resulted from volcanic outgassing after the formation of the planet. The original atmosphere of cosmic gases—neon, argon, xenon—may have been lost as the planet heated up. The new constituents—carbon, nitrogen, oxygen, hydrogen, and sulfur—came from inside the earth. They created a reducing atmosphere. Free oxygen may not have been abundant until after life developed. The surface of the planet was exposed to ultraviolet, until an ozone layer formed. Silicon, oxygen, magnesium, and aluminum in stone have a fantastic history. Continental plates turn over the crust, bringing mineral deposits to the surface, burying others. The plates also provide land, weather and environment. Each of the continents is unique and takes a part of the cycling of elements. Africa receives 12% of world water from hydrological cycle, whereas the United States gets 33%. All chemistry is a matter of the electrical behavior of atoms. The electrical force unites atoms in consistent ways.

Variations allowed by the electric force enable rare events to occur; life is a rare event. Life is a rare, but certain event. Given definite conditions and processes, it is certain to arise. Rare events are fundamental to the universe. In a long evolutionary process rare events are very significant. Events can be divided into regular, like heartbeats or solar revolutions that occur in cyclic succession, and irregular, like earthquakes or mutations. The rarity of irregular events can be measured by a period of probability. A devastating earthquake that is expected about every 1000 years has a probability of 1/12,000 in any month. The probability for any year is 0.001. The probability of it happening within 10,000 years is 0.99995. It would not be wise to situate a nuclear waste dump, with material having a half-life of 5,000 years or longer, in such a place. Over long enough periods of time, may improbable events are certain to occur.

If the creation of self-replicating molecule has an annual probability of one in a billion, it is fairly certain to occur in a billion years. We all use the same 20 amino acids, but arrange them individually according to a species blueprint from which the protein is built. Even the least complex protein has 100 acids in a chain—20 to the 100 power possible proteins. But most are never tried. If one combination occurred every four seconds over four billion years, only 10 to the 60 power would have been tried. Even figuring trials on all the estimated planets in the universe (10 to the 20th power), only 10 to 80 power would have been tried. The enormous information capacity of biological molecules makes production by chance improbable. A random process could not have produced even a single organic molecule. But once one rare event occurred, another became possible, and so on. Humans have difficulty observing events on this time scale. Rare events are not studied scientifically. Rare events are also revolutionary, and may be very important. History is the study of rare events.

Life depends on chemical and electrical events. The balance between fidelity and change depended on electrical forces between atoms in living creatures being neither too weak nor too strong; too sloppy and everything would perish of genetic mutation or too exact and all would remain at level of microbes. The electrical force strikes balance only in certain materials at a certain temperature range. Heredity is embedded in a chemical code. Plants are basically chemical systems. An animal is an electronic system dominating a chemical one. The level of the animal is judged by the complexity of the system.

The rate at which the sun bombards earth with energy is the solar constant, which varies minutely. The ecosystem is one in which energy from the sun is cycled. The functioning biomass is integrated by feedback responses to extract the maximum of energy and still maintain balance. Most of the solar energy is used for maintenance by the earth; this is an indicator of the biosphere's high degree of ecological maturity. Chemical equilibrium is a global regulator.

The steady effect of light, the availability of oxygen, the thickness of soil, and the area and depth of the oceans is almost a steady state, as is the diversity of living species and the mean temperature of the planet. Living fossils demonstrate a continuously available environment for long periods; the survival of horseshoe crabs indicates that the sea has been unchanged for 200 million years; algae goes back to microfossils 3.2 billion years old. The amount of material that has passed through living organisms since Cambrian times, roughly a billion years, is of the same order of magnitude as the earth's crust. Assuming carbon is fixed each year by, the total mass recycled of carbon alone exceeds the mass of the crust. Total coal and oil deposited in lithosphere are estimated at a percentage of the crust.

Elements frequently associated with life—calcium, phosphorus, iron, sulfur—are preferentially concentrated and deposited. Life has created the present soils and atmosphere out of the original rocks and gases. Life on earth has played a major role in ordering the earth's surface. Autotrophs remove high energy CO_2, H_2O and traces of other compounds and elements from the environment and convert them into low energy compounds. Detritus accumulates when autotrophs, with a smaller number of heterotrophs, die and are not immediately consumed. However, a small amount of detritus does not get recycled; these ordered compounds form deposits. Marine organisms produce deposits of oil (possibly in a reducing atmosphere); forests leave coal. Microorganisms may have selectively deposited iron oxide in a shallow sea to form the Mesabe iron range. Ecomass has been steadily growing since Cambrian times, thereby increasing the order of the earth's surface. The earth breathes. Berkner suggests that there may have been cycles of oxygen production and carbon dioxide consumption depending on relative abundances of plant and animal life. In natural systems, balance is maintained when everything changes.

4.2.2.0.2. Deep History

The entire current system rose during the Cambrian explosion, about 600 million years ago—less than 10% of the history of the planet. The evolution of eukaryotic cell allowed explosive radiation. Life has been relatively quiet since then, as it was before. The Cambrian explosion could have been predictable outcome of a process set in motion far earlier. Much life rose during the initial rapid diversification, but died out during stable times. The one burst must have filled up the oceans. Since then evolution has basically recycled the basic designs. S. Gould claims that the mystery of the Permian extinction can be solved by an ecological theory relating organic diversity to habitable area. In short, the Pangaea coalescence caused area of shallow seas to shrink drastically, causing extinctions.

By 400 million years ago, in the early Devonian period, primitive plants covered much of the surface of land; in order to compete better for light some plants developed strong stems and vessels to support their leaves above other plants. These plants quickly, in an evolutionary timeframe anyway, attained heights of several inches. Partly as a result of competition and of predation by early terrestrial arthropods—scorpions, spiders, and centipedes—these primitive plants competed and diversified into club mosses, horsetails, and ferns. They also grew taller, which challenged insects to climb to get to the nutritious tissues, and eventually to glide or fly between leaves.

After 50 million years of evolving, and now in the Carboniferous period, club mosses could grow to 170 feet high, while horsetails could top 50 feet and ferns 70 feet (with a 24-inch diameter!). The giant "seed ferns" were neither truly ferns or trees or flowering plants. In terms of height and complexity, these forests were as dramatic as any modern tropical forest, although most of these species are extinct and little is known about their relationships.

These first forests were composed of such trees in dense jungles over swampy ground. The tree crowns were formed by spreading branches covered with green spiny leaves; limbs and trunk were coated with brown scales. Although insects—beetles, flies, lice, dragonflies, and cockroaches—and amphibians were evolving throughout this period, there were no: flowers, fruits, nuts, pollen, seeds, nectar, frogs, reptiles, birds, or mammals. The jungle was relatively noiseless and monotone. For over 100 million years (345-225 million years YBP—years before present), while dinosaurs and winged insects were evolving, these seed ferns covered vast expanses of land. As the continents broke up further from one large plate, Pangaea, these forests formed the basis for most coal fields found around the earth. This break-up caused a massive drying of the land.

About the time glaciers started appearing on several of the continents, 225 million years ago, cone-bearing plants, conifers similar to modern pines and spruces, supplanted (no pun intended) the seed ferns. Conifers could colonize dry ground, especially because of the innovations of pollen and seed, whereas ferns had to live near water and mate with their immediate neighbors. Pollen in clouds can travel hundreds of miles, offering greater variety for mating. The seed ferns had naked seeds not enclosed in fruit and were, according to Lynn Margulis, sensitive to cold. The conifers had a greater tolerance to cold, even subzero temperatures, because the seeds were wrapped, even though they are labeled gymnosperms, 'naked seeds'). The fungal root networks allowed them to ingest phosphorus and nitrogen and to expand into higher elevations and colder latitudes. The first conifers were the main diet of vegetarian dinosaurs.

After roughly another 100 million years, plants with flowers descended from the same plants that produced the seed ferns. Within a mere nine million years, flowering plants had colonized most of the land areas, coevolving with the first mammals, warm-blooded egg-layers and small marsupials. Coevolution of life forms is far more important than previously thought. Mammalian interest in these plants as food probably led to their rapid dissemination. Flowering plants (or angiosperms, 'clothed seeds') had fruits and seeds to protect the embryos from being made into animal flesh. Margulis suggests that plants have been seducing animals for millions of years, tricking us into helping them to move and forcing us into more complex *patterns* of behavior. Angiosperms developed concurrently with mountain-building, such as the Alps and Himalayas, in the Miocene. Coniferous forests began to diminish in the Miocene as grasslands spread and deciduous trees appeared. Poplar and plane trees were established by the early Cretaceous. Deciduous trees had evolved with many more plant, insect, and animal partners.

The glacial-interglacial alternation during the last million years has a periodicity of about 100,00 years, with interglacial durations of 10,000-12,500 years. The current interglacial time began about 10,000 years ago. Glaciers have covered vast areas of the northern hemisphere. The advance and retreat of ice sheets and climactic changes have been linked with large migrations of plants and animals, and more recently peoples. During the last glacial stage in the Pleistocene, both coniferous and deciduous woodlands were forced southward, often to isolated refuges. The return of the forests by the Holocene, our current geological epoch, is one of the great stories of natural history, according to Neil Roberts. Even in the past 10,000 years forests have expanded and contracted with environmental change, for instance, the vast dawn redwood forests north of the present Arctic circle died out in the Eocene; long thought extinct, fragments remained in China, preserved on monastery grounds, to be discovered again a few decades ago. The shape of forests depends on a suite of physical conditions that vary from continent to continent. Weather patterns on North America, for instance, have created, in Paul Colinvaux's term, "nation states" of trees that surprised the first European naturalists because they were so different from the European forests.

The "Little Ice Age," which lasted from 1000 to 1850, spread misery, starvation, war, and plague in Europe. The precipitation of a glacial climate now would drastically reduce food and compress humanity. Ecohistory should consider human dispersions, adaptations, invasions of wild areas, perceptual problems, and language and mythology as tools of cognition.

4.2.2.1. Human Histories Cultural Histories

Probably no consequence of human development has had a greater impact on the natural landscape and ecological processes than the production of food. Patterns of eating have influenced the constitution of species and the very contours of the earth. Throughout their history, humans have used animals and plants for food and clothing. Animals were followed, herded, corralled, tamed, and finally bred. Plants were domesticated later. As technologies developed, human relationships with animals and plants changed. Hunting, grazing, and agriculture provoked large ecological disturbances. Early domestic animals were revered, but nondomestic animals were considered competitors or nuisances. Now, animals are treated as commodities processed in factories and wildlife is regarded as useless. Hunting persists, but mainly as recreation. A few plants provide the bulk of human diet; the rest are considered ornamentals or weeds. Humans are omnivores, although they have been represented as carnivores, vegetarians, or fructivores by different factions. In view of our control over animal and plant populations, a reexamination of our eating habits, and our use of animals and plants, is critical.

4.2.2.1.1. Adaptive Patterns

While some species adapt to a niche, for a way of eating, others create niches. The specimen is much more than a summation of all of its historical responses—it is intentional and flexible, sometimes stress-seeking and maladaptive. This kind of behavior can result in new adaptive patterns.

Humans started out as scavengers, then went on to be hunters and gatherers, foraging in a territory identified with a band or tribe. As people started gathering wild plants, they settled in semi-permanent villages and expanded their ways of living.

4.2.2.1.1.1. Hunting & Gathering (& Fishing & Wild Gardening)

We have a long experience as hunters and gatherers, at least forty thousand years. Unfortunately, we know little of how hunters and gatherers built a satisfactory way of life. Hunter-gatherers of the old stone age possibly were completely conditioned by the features of their immediate surroundings. Through their senses and reason they acquired an empirical, holistic knowledge so well fitted to their environment that they could cope effectively with the wilderness in which they lived. Ortega conveys how humans can still learn to function as organic parts of a wild place, instead of just observing it.

Even during the Pleistocene, humans had a formidable potential for disruption, as an efficient and rapidly expanding predator group against whom no evolved defense systems were available, or indirectly as the source of profound changes in ecology already under processes of adjustment following considerable climactic stress. Fires were first used in the Pleistocene for hunting and cooking. In the long time of prehistory hunting was a way of life, with tool making and totemic vision as integral parts.

Hunters perceive all species as having roles in a vast society, according to Levi-Strauss, whereas farmers perceive human groups as species. Change of climate and overkills of animals possibly contributed to the necessity of farming. In evaluating ecological situations, we cannot avoid making judgments and giving preference to human values.

Hunters eat at the natural location and at the densities of resources. Gathering seems to be more important in warm climates, and hunting in cold climates. Hunters tend to be nomadic. Mobility and food limits are incentives to remain in small groups, which tend to have low population densities (the world average is 1 person for 1 square mile). They tend to share resources equally, as a result of egalitarianism and reciprocity. They have "natural" leaders, who lead by the example of being a good hunter or a good shaman. Territory was shared with neighboring bands.

Size favors an egalitarian political system. How can you control mobile people with few possessions, no herds and no wealth? How can you dominate them militarily? You would have to be much larger, but there are penalties for size in those ecosystems. How can you economically dominate them? Or linguistically? They tend to be localized, decentralized, oral, egalitarian. Therefore language groups would be small.

4.2.2.1.1.2. Herding & Domestication

After the ice age, animals were captured and gradually domesticated. Domestication has been used to imply one simple process, but it is various grades of relationship in which humans have degrees of control over reproduction and composition of animal groups. Factors leading to increased domestication include food, ritual, play and transportation. Perhaps young humans playing with wolf pups, drawn to easy food near kitchen middens, might have led to some kind of adoption.

Different ranges of activities in hunting could lead to herding. Random hunting and controlled predation, followed by attentive herd following, could lead to loose herding, then eventually close herding. As hunters gradually encouraged some characteristics in wild animals, they would eventually have complete physical and genetic control.

The woodland belt of the Zagros mountains had a rich assortment of native animals that were suitable for domestication, including goats, sheep, pigs, and wild asses. Just to the west the steppe-land of Assyria had gazelle and wild cattle, with carp and catfish in the rivers. As people adjusted their lives to the cycles of animals, they made seasonal camps or semi-permanent villages. People started raising sheep and goats. Cattle and goats were kept for milk. Few people ate meat regularly.

Herding domestic animals would have advantages, such as knowing how many animals were available for milk or meat. There were some disadvantages, such as an increase in the amount of work responsibility for forage over the entire year, which doubtless led to following, then circulating the animals by season (transhumance).

Domestication profoundly influences diversity and integration of a species. Diversity, as a measure of genetic variability in ecosystem, is decreased. Diversity is the bag of tricks for organisms facing environmental perturbations. Coevolution with humans reduces or destroys integration with other species; stability depends on human control. Human control is not certain.

At many stages of herding, hunting, gathering, and fishing, may still have been important. In the first stages of herding, fire may have been used to move animals or keep them together.

As technologies developed, human relationships with animals and plants changed. Hunting, grazing, and agriculture provoked large ecological disturbances. Early domestic animals were revered, but nondomestic animals were considered competitors or nuisances. Now, animals are treated as commodities processed in factories and wildlife is regarded as useless.

Changes in fitness probably occurred following the later stages of closer man-animal relations, with adaptation to captive and restricted environment resulting in gene pool changes. Domestic animals would then have different characteristics compared to wild animals. Possibly, mutation and gene frequency would also contribute to domestication of animals. Sex and age ratios changed. People ate younger animals and kept females for breeding stock. Goats and sheep would have been bred to produce excess milk. Other changes led to wool production. Wild sheep have hair and a coat like deer. With domestication, secondary follicles gave dense wool. Other changes were nonadaptive, such as the twisted horns of domestic goats.

Much of the earth is no longer occupied by wild vegetation, either pristine or degraded. Pastoralists have had a profound influence on native vegetation. Much of the savanna in the tropics would be more heavily wooded if not for fires and grazing practices.

4.2.2.1.1.3. Agriculture

Foragers understood the principles of gathering seeds and planting them. They knew when and where their favorite foods were. They engaged in broad-spectrum collecting. Young or elderly collected snails, turtles, clams, crabs, and seeds. Most meat came from ungulates, like gazelle or deer. In the western foothills of the Zagros mountains, the wild einkorn wheat grew so thick and tall that a family of six with a stone sickle could gather enough in three weeks to last for a year.

Many of the activities in the gathering of plants could lead to plant domestication and agriculture. Foraging and collecting wild plants resulted in knowledge of the plants and their requirements. Tending wild plants in burned or dry areas resulted in more knowledge. Throwing unused seeds into a kitchen midden might have resulted in some crops growing in richer soil. Living near the plants might encourage reliance on those plants. Selecting specific plants that required intervention might have made the plants dependent on human collection and planting. Selecting for various features, such as closeness of seeds or gigantism might have led to selecting those plants only, or monocropping.

Plants selected for domestication would have desirable characteristics, such as ease of establishment and tolerance of a wider range of conditions, such as drought or poorer alluvial soils. The plants should have good production, big seeds or leaves, good biomass.

They should have the ability to get or trap nutrients, with shallow, wide or deep roots. They should be able to tolerate climatic changes, resist drought or salt. They should be able to resist pests and diseases. They should be compatible with other plants. They should provide a balanced nutrient content, with less toxicity. They should be easy to control and have detachable seeds, fruits or leaves.

After domestication and agricultural development, the plants would have increased size in seed heads, tubers, and roots. The normal chemical defenses, such as toxins, would be suppressed. The plants would be tried in a wide range of environments, which might result in an equally wide range in shape, size and taste. Selection for strong rachis or seed heads would reduce reproduction by wind and insects. Monocropping might reduce cross-pollination, such that humans would have to take over as agents of propagation.

In its early stages, agriculture is constrained to early stages of succession, that is, it uses nutrients and energy for quick growth and reproduction, rather than invest in efficiency, durability or maturity.

4.2.2.1.1.3.1. Traditional Agriculture

The fertile crescent, first named as such by James Breasted, which ran from the Zagros to the hills of Judea, was not only fertile, it was lucky in its combination of plants. And in its combination of wild animals: First elephants, rhinos and hippos; then, sheep, goats, cattle, and pigs.

Foragers understood the principles of gathering seeds and planting them. They knew when and where their favorite foods were. The adults engaged in broad-spectrum collecting; the young or elderly collected snails, turtles, clams, crabs, and seeds. Most meat came from ungulates, like gazelle or deer. In the foothills of the Zagros mountains, the wild einkorn wheat grew so thick and tall, on the rocky slopes, that a family of six with a stone sickle could gather enough in 3 weeks to last a year. Due to differences in environments, different plants would be available at different times. people could have moved up slope to take advantage of those times. People would have exploited all of these environments for thousands of years. Stone sickles could be used to harvest many wild stands. Because no sowing, weeding, or tending was necessary with wild crops, the effort was less.

Once people found that desirable plants could be cared for, they moved their habitations nearby, building houses. They lived in small villages of 100 people, near the cereals stands. These kinds of plants tended to be weedy, living in poor soils subject to dry and wet seasons. They also produced large seeds, were able to germinate easily, grew quickly and survived drought. This characteristics would be attractive to humans cultivating them. They started using stone sickles and digging sticks. When the crops became larger, they started building storage pits or buildings. By 10,000 years YBP, einkorn wheat, emmer wheats, and barley were domesticated. Later lentils, chickpeas, and peas were found. Cereal wheats required 400-750 mm of rain.

Food was collected and traded to areas where it was replanted, then selected for different characteristics, such as more seeds or a stronger seed-heads. People would have traded seeds or animal products for metals or gems. Obsidian from eastern Turkey was found in caves in Zagros. Natural asphalt was traded to western regions. Emmer wheat was traded to groups further east in Iran and south. These commodities were traded regularly in a pattern of exchange east and west. Edible grasses were transferred to areas where they were not native, such as southern Iraq, which was too dry for them to grow without assistance.

Although the alluvial plain between the Karun, Tigris and Euphrates had less than 250mm of rain, falling on desert grasslands, desert, and reed-bordered swamps, the soils were good alluvial silt with no slopes, stumps, or stones; there was water in the rivers and abundant sunshine. Transplanted wild emmer could produce 2700-4500 kilograms per hectare of seed, about as good a yield as in Britain in 1400 (600 YBP).

With full domestication by 9500 YBP in Tell-es-Sultan near the head of the Jordan river, settlements could number 2000 people. Over 40% of grain production went for ale production (no beer until hops were used 4000 years later). Ale was brewed in the homes and sold from there. It brought added benefits; fermentation increased the amino acid content of grain, and it made people feel good.

The food-producing revolution was a long process of changing ecological relationships between groups using different strategies. The grains were shielded from natural selection by humans. The successful cultivation then intensified trading of cultivars and resources between groups. Village specialization may have been the best adaptation. Then the economy of redistribution created regional temple and market towns that regulated the trading in goods.

Although the transition to farming was gradual and slow, the spread of farming as an idea was relatively rapid, perhaps as fast as three or four kilometers per year. Mesopotamian agriculture spread to Egypt and the Balkans, and then to central Russia and Europe. Appearing later in Northern China, Southeast Asia, and the Americas, it may have been developed independently in those areas.

4.2.2.1.1.3.1.1. Fields

Digging sticks and small canals that diverted water from runoff were gradually replaced by hoes and irrigation canals, which transported water to the fields and managed the drainage. Irrigation allowed water control, and thus multiple harvests of one crop.

In Mesopotamia, complex systems could be managed locally by farmers, although there may have been conflicts with upstream users. Cooperation was necessary, especially in villages and between villages, later. As a result, larger-scale water systems were first managed by religious leaders, then later by secular leaders. Large-scale irrigation appeared in Egypt by 7100 YBP. Karl Whitfogel suggests that these hydraulic civilizations arose as a result of the water management.

Agriculture was vital, but hunting and gathering were still important. Agriculture moved further south by 6500 YBP. The city of Uruk started growing, with a large ziggurat. By 5000 YBP, the city had 50,000 people. The number of local settlements had declined from 146 to 24, perhaps due to forced relocations. Other cities, such as Ur, Lagash, Kish, and Umma, had populations of 10,000-20,000 people. The surplus grain had to be transported, stored, and reallocated. This required organization. Early temples may have claimed all of the land in and around the city, as well as 10,000 donkeys. Farmers could have worked the land for pay in form of food.

By 5000 YBP a stratified class society had developed, with slaves at the bottom, peasant farmers, craftsmen, and then the elites of administrators, religious and military. As problems developed, such as famines and conflicts, temporary military leaders became permanent hereditary kings. Palaces were built and staffs numbered in the thousands. With greater secular control, lands became private again, although the temple owned some lands. By 5000 YBP, twelve cities had over 2000 inhabitants. By 4500 YBP the total population of Sumer exceeded 500,000, eighty percent of whom lived in cities.

The neolithic agriculture, in the form of the forest fallow method (modified woodlands), reached Britain by 5000 YBP. This method resulted in massive forest destruction, that took 500 years to partially regenerate. After a thousand years the landscape was a mosaic of fields, hazel scrubland, alder forests, and acid grasslands. Fields were enclosed and sheltered. Scratch plows were used. New crops were planted: Spelt wheat and Celtic beans. The population bloomed from 14,000 to 2,000,000. Less than 500 years after that, forest destruction was at a maximum. After forest clearance, hedgerows were used between fields.

As the Middle Ages waned, increasing communications, the commercial revolution, and the rise of cities in Western Europe tended to turn agriculture away from subsistence farming toward the growing of crops for sale outside the community, that is, towards commercial agriculture. In Britain the practice of enclosure allowed landlords to set aside plots of land, formerly subject to common rights, for intensive cropping or fenced pasturage, leading to efficient production of single crops.

Exploration and intercontinental trade, as well as scientific investigation, led to the development of horticultural knowledge of various crops and the exchange of farming methods and products, such as the potato, which was introduced from America along with beans and corn (maize), and became almost as common in northern Europe as rice is in southeast Asia.

The appearance of mechanical devices such as the sugar mill and Eli Whitney's cotton gin helped to support the system of large plantations based on a single crop. The Industrial Revolution after the late 18th century swelled the population of towns and cities and increasingly forced agriculture into greater integration with general economic and financial patterns. In the American colonies the independent, more or less self-sufficient family farm became the norm in the North, while the plantation, using slave labor, was dominant (although not universal) in the South. The free farm pushed westward with the frontier.

4.2.2.1.1.3.1.2. Tree Farming

A mature ecosystem such as ancient woodland has a tremendous number of relationships between its component parts: Trees, understory, ground cover, soil, fungi, insects and other animals. Plants grow at different heights. This allows a diverse community of life to grow in a relatively small space. Plants come into leaf and fruit at different times of year. For example, in the UK, wild garlic comes into leaf on the woodland floor in the time before the top canopy re-appears with the spring. A woodland suffers very little soil erosion as there are always roots in the soil. It offers a habitat to a wide variety of animal life that the plants rely on for pollination and seed distribution. The productivity of such a forest in terms of how much new growth it produces exceeds the most productive wheat field. It is in this observation of how more productive a wood may be on far less input of fertilizers that the potential productivity of a permaculture design is modeled. The many connections in a wood contribute together to a proliferation of opportunities for amplifier feedbacks to evolve that in turn maximize energy flow through the system.

Forests are natural buffers; they influence local and regional climates making them milder; they provide continuous flows of clean water and protect soils. Trees create microclimates, reduce wind speed, lift the water table, and increase worm population. Redwoods have a life span of 1500-2000 years. Their vital role is to filter coast mists;

redwood transpiration precipitates rain from mists. Managing such forests requires a long-term plan. For instance, planting oaks with beech, the oak rotation is 320 years, with 100 years for the beech. This requires a number of generations of foresters.

Cultivation of trees seems to have begun in Africa. Although grasslands seem limited in exploitation, further food could be derived from tree fruits and leaves, or from consumers of microscopic plants. A broad-leaved tree layer appears to be biologically desirable on most lands. Much of agriculture on small farms is multilayer; two other crops can grown under fruit trees. Lawrence D. Hills claims that growing trees for food is a rational land use. Carob, mesquite and honey locust could be mass cultivated. Productive forests could be planted on otherwise marginal lands. He states that 4-20 tons of protein per acre is a possible yield, with pasture, crops or other trees planted between rows. The most efficient systems are extensive, where nature does the work. Only 1000 hours per year are needed to manage food in a gathering society.

Forests can be managed, as wild forests, from which nothing or little is taken. Or, forests can be managed as tree farms. A tree farm can be defined as a tract of privately-owned forest, large or small, managed by a certain forestry program. Because the program can be one of several presented last week, a tree farm can be a Christmas tree plantation (usually the most intensely managed form), a multiple-use forest, managed wild forest, or other. A tree farm can be a forest, or part of a forest, or simply land on which someone has planted trees. The kinds of tree farms can be related to the models on which they are based and to the base environment. By base environment, I mean clearcut, old field, or wild forest. Up until now, there have been two basic models for treating the forest: The agricultural model that removes a whole crop annually, and the stewardship model that tends the forest, but removes as much as possible under human control. But, it is possible to take materials from a wild forest, without attempting to control every aspect of the forest.

A tree farm has to have certain characteristics to be good. Since tree farms are called farms and not mines, let us assume that they are expected to be sustainable. Protection of the forest is a requirement. Without a forest there is not a farm. If the forest is degraded, the farm cannot last long. The forest has to be allowed to be healthy; the habitat has to meet the needs of the community of species; and the genetic pool of the species has to be large enough for long continuity. Protection of riparian areas is a requirement.

The quality of flowing water, in rivers, streams, intermittent streams, or just surface flow, is vulnerable to human activities, such as logging or recreation. Fish have quite explicit needs for temperature, stream bed character, and cleanliness. Water tends to cross ecosystem boundaries; fish also move between ecosystems, quite dramatically in the case of salmon.

Tree farm management involves occasional planting, using careful techniques, such as root spreading, but also the follow-up care, such as mulching, hand weeding, and possibly protection from browsers and direct sun. One major problem with some tree farms is that indigenous flora and fauna are thinned or removed in favor of "useful" species that can provide maximum production. This "redesign" of the forest almost always requires external human controls and inputs in the form of biocides, fertilizers, soil scarification, and artificial selection. Simplification of the forest for short-term maximization of goods usually leads to dedifferentiation and destabilization. Although such tree farms may contribute to soil conservation, wild life diversity, and biogeochemical cycles more than a plantation, they do not contribute nearly as much as old-growth forests.

Industrial forestry depends on formal knowledge of the forest, but tree farms combine knowledge with intimacy. Tree farms can be sustainable and productive—although not necessarily all of them are or will be—relieving demand pressure on wild forests, national and provincial forests, and old growth areas, which may have a better chance of being preserved from interference. Tree farms could act as buffer forests for wild forests. The tree farm is one level of application of ecoforestry; in one sense it may be the most useful area to begin with, since it is necessary to convince only one person, the owner.

Tree farms have to take care to plant appropriate species. Planting generations of pines can lead to soil deterioration in some ecosystems. The pine roots are sheathed with acid for the root to get a hold in rock. The acid also helps with root competition, and an acid pan is formed deep in soil. For the second crop roots do not go as far down, due to the acid pan. A third crop has such shallow roots that it is liable to blow over. Pine litter is acid, and it may discourage wildlife habitat.

Attention must also be given to equipment on tree farms. Heavy machinery forms a hard pan in soil. Engines cause the ground to vibrate, and a pan forms ten inches deep. The roots of some young trees may have difficulty penetrating it. The life span of the trees may be shortened. Felling should be by best stems, mature trees or group selection. There should be no clearcutting at all. Planting should be last resort. Good forestry allows for natural regeneration.

The small size of most tree farms ensures that the production exists for local needs according to local demands. The small size also ensures that mistakes will not destroy an entire ecosystem. Small size is conducive to flexibility and sustainability. The blend of activities in tree farms, from intensive multi-cropping to the utility of complex, symbiotic associations, make them similar to agroforestry or permaculture applications.

Although tree farms are often tracts of privately owned forests, the tree farm tag can also be applied to certain kinds of permaculture, as envisioned by Bill Mollison, to a traditional farm with stands of trees, as thought by Wendell Berry and others, to natural farming where pines, cedars, and fruit trees are planted with grain and root crops, as described by Masanobu Fukuoka, and to permanent tree crops, such as chestnut, honey locust, walnut, carob, grown on hills and mountains to complement flat-land agriculture, as presented by J. Russell Smith. The common theme in these kinds of tree farms is the careful maintenance of the productivity of the land; industrial tree farms, plantations, and orchards do not, in general, share this theme. Maintenance of productivity, or sustainability, has always been the goal of healthy cultures or agricultures.

Polyculture is agriculture using multiple crops in the same space, in imitation of the diversity of natural ecosystems, and avoiding large stands of single crops, or monoculture. It includes crop rotation, multi-cropping, and inter-cropping. Alley cropping is a simplification of the layered system which typically uses just two layers, with alternate rows of trees and smaller plants.

Like forest gardening, permaculture is multileveled. In permaculture and forest gardening, seven layers are identified: The canopy; secondary trees; shrubs; herbaceous layer; ground cover plants; root crops and climbers. Permaculture Guilds are groups of plants which work particularly well together. These can be those observed in nature such as the White Oak guild which centers on the White Oak tree and includes 10 other plants. Native communities can be adapted by substitution of plants more suitable for mans use. The Three Sisters of maize, squash and beans is a well known guild. The National Vegetation Classification provides a comprehensive list of plant communities in the U.K. Guilds can thought of as an extension of companion planting.

Finally, tree farms cannot exist independently of their ecological, economic, and cultural contexts. Without changes in taxation and the industrial financial system, tree farms may succumb to pressures to cut heavily. Without changes to Canadian and U.S. styles and levels of consumption, the demands on tree farms to provide materials may be too heavy. Without changes in values, miracle cures, quick profits, speed, paper use, and superficial efficiency, tree farms may not be able to adjust to forest time and labor-intensive harvesting.

A country's future wealth might be gauged by its tree cover. Britain and Ireland are the least wooded lands in Europe. Richard St. Barbe Baker calculates that the minimum forest cover for safety is a third of a country's land area. Further, Baker states that about 22% of a farm in shelterbelts will double the yield of the farm. Baker divides agricultural land into seven grades, and forest into seven grades; the last three grades of fields overlap with first three of forest. The lower grades of forestland can be used for agriculture. Some land becomes eroded or sanded if deforested.

4.2.2.1.1.3.1.3. Domestic Animals & Agriculture

In some nations, dairy and beef cattle, swine, sheep, and poultry are important species. In other parts of the world, other species assume increased importance: Elephant, camel, donkey, llama, water buffalo, reindeer, and caribou. Livestock over the earth weigh four and a half times as much as the humanity that use it (860 million tons versus 187 million tons) and eat the equivalent amount of food of 14.5 billion human beings.

Although many animals, such as cattle and sheep are raised on ranges, they often spend months in feedlots being fattened with grain for human consumption. Ninety-five percent of this food goes for respiration or ends up as manure. Eating eggs and milk is more efficient than eating the animal itself. The 95% loss is acceptable using wastes and scraps that contain recoverable food. But it is not acceptable using whole grain crops. Many animal foods could come from sources unappetizing to humans, such as insects or algae.

Pigs and cows maximize secondary production at over twenty five percent production-consumption efficiency. But, they do not forage well or escape predators easily. Animal husbandry is a specialized strategy of land use that resembles foraging. Like hunters, they minimally manage vegetation. But, like agriculturists, they invest time and energy into their primary resource, livestock.

Herders maintain a resilient balance between grasslands and shrubs. As long as grazing is periodic, many kinds of grasses grow and are productive. Exclusive pastoralists, like the Masai, have to adjust to the conditions of the environment, especially rain. They also have to coexist with farmers, with whom they may compete. Many pastoralists now generalize their strategy by raising crops with the animals. Pastoralists have had a profound influence on native vegetation. Much of the savanna in the tropics, for instance, would be more heavily wooded if not for fires and grazing practices.

4.2.2.1.1.3.1.4. Decentralized Balance

Swidden agriculture, or shifting cultivation, was practiced in Europe as well as South America and Africa. In Africa it was practiced for centuries with only minor long-term effects on the rain forest. But in Europe, where the forest regenerates more slowly since growth is checked seasonally by low temperatures and low moistures, shifting agriculture is unstable and capable of greatly modifying the ecosystem, despite the greater soil fertility. The type of crops also made a difference. Cultural areas in America and Europe were based on cereal crops—millet, wheat, barley—while African and many South American systems emphasized root and tree crops.

Traditional agriculture proceeds by substituting selected domesticates for wild species in equivalent niches, or by manipulation of the ecosystem. Margalef pointed out that all agricultural systems are laid out for low maturity, to increase production per unit. Tillage increased the growth of vegetation by a factor of 2500. Humanity entered a lower order of nutrition with the invention of agriculture. This means that more overall food is required to fuel each body.

Food output per acre has been relatively stable throughout history. Local varieties are almost always more adaptive. Crops and animals do not produce in isolation. They are part of biological systems. For most purposes, they are treated separately, but this is not satisfactory for ecological studies. In grazing communities, for example, the animal not only grazes, but influences the tendency of succession. What animal is chosen to graze may be the result of cultural influences or personal preferences.

4.2.2.1.1.3.2. Modern Agriculture

In the northern and western U.S., the era of mechanized agriculture began with the invention of such farm machines as the reaper, the cultivator, the thresher, and the combine. Other revolutionary innovations, e.g., the tractor, continued to appear over the years, leading to a new type of large-scale agriculture.

Modern science has also revolutionized food processing; refrigeration, for example, has made possible the large meatpacking plants and shipment and packaging of perishable foods. Urbanization has fostered specialties of market gardening.

Harvesting operations have been mechanized for almost every plant product grown. Breeding programs have developed highly specialized animal, plant, and poultry varieties, thus increasing production efficiency. The development of genetic engineering has given rise to genetically modified transgenic crops and, to a lesser degree, livestock that possess a gene from an unrelated species that confers a desired quality. Such modification allows livestock to be used as "factories" for the production of growth hormones and other substances.

In the United States and other leading food-producing nations, agricultural colleges and government agencies attempt to increase output by disseminating knowledge of improved agricultural practices, by the release of new plant and animal types, and by continuous intensive research into basic and applied scientific principles relating to agricultural production and economics.

These changes have, of course, given new aspects to agricultural policies. In the United States and other developed nations, the family farm is disappearing, as industrialized farms, which are organized according to industrial management techniques, can more efficiently and economically adapt to new and ever-improving technology, specialization of crops, and the volatility of farm prices in a global economy. In archaic countries, however, where small farms, using rudimentary techniques, still predominate, the international market has less effect on the internal economy and the supply of food.

Most of the governments of the world face their own type of farm problem, and the attempted solutions vary as much as does agriculture itself. The modern world includes areas where specialization and conservation have been highly refined, such as Denmark, as well as areas such as northern Brazil and parts of Africa, where forest peoples still employ swidden agriculture—cutting down and burning trees, exhausting the ash-enriched soil, and then moving to a new area. In other regions, notably Southeast Asia, dense population and very small holdings necessitate intensive cultivation, using people and animals but few machines; here the yield is low in relation to energy expenditure. In many nations, extensive government programs control the planning, financing, and regulation of agriculture. Agriculture is still the occupation of almost half of the world's population, but the numbers vary from less than three percent in industrialized countries to over sixty percent in some countries.

4.2.2.1.1.3.2.1. Commercial Agriculture

As the Middle Ages waned, increasing communications, the commercial revolution, and the rise of cities in Western Europe tended to push agriculture away from subsistence farming toward the growing of crops for sale outside the community, that is, a new commercial agriculture. In Britain, the practice of enclosure allowed landlords to set aside plots of land, formerly subject to common rights, for intensive cropping or fenced pasturage, leading to efficient production of single crops.

In the sixteenth and seventeenth centuries, horticulture was greatly developed and contributed to an agricultural revolution. Exploration and intercontinental trade, as well as scientific investigation, led to the development of horticultural knowledge of various crops and the exchange of farming methods and products, such as the potato, which was introduced from America along with beans and corn (maize) and became almost as common in Northern Europe as rice is in Southeast Asia.

The appearance of mechanical devices such as the sugar mill and Eli Whitney's cotton gin helped to support the system of large plantations based on a single crop. The industrial revolution after the late eighteenth century swelled the population of towns and cities and increasingly forced agriculture into greater integration with general economic and financial patterns. In the American colonies, the independent, more or less self-sufficient family farm became the norm in the North, while the plantation, using slave labor, was dominant, although not universal, in the South. The free farm pushed westward with the frontier.

Dry farming is a farming system adopted in areas having an annual rainfall of approximately 15 to 20 in. (38.1–50.8 cm)—with much of the rainfall in the spring and early summer—where irrigation is impractical. Seeding rates are used that correspond to the soil water supply; management practices that minimize water loss and soil erosion are also utilized. The land is often summer-fallowed in alternate years to conserve moisture. Dry-land crops must be either drought-resistant or drought-evasive, that is, maturing in late spring or fall; special varieties of crops such as wheat, barley, corn, sorghum, and rye are often used.

4.2.2.1.1.3.2.2. Industrial Agriculture (Energy Subsidies)

When Thomas Jefferson envisioned the future, he hoped the United States would remain an agrarian society. Today, less than two percent of the population is involved in agriculture. People still rely on farming for the production of food, and hence, their survival. But, they rely on industry, too. Industrial methods have permeated manufacturing, sales, and agriculture itself; industrial practices have been sold as a way to produce more food of better quality. Industrial agriculture relies on large amounts of cheap fossil fuels to overcome limits of water, heat and nutrients.

What, exactly, does "better quality" food mean to industrialists? Most often, it means food that looks good, ships well, and does not spoil easily. By those standards, industry has succeeded. By other measures, however, industrial agriculture is not only a failure, but a menace to small-scale efficiency. Some say that other measures, such as nutrition or taste, are the true standard by which people should evaluate their food.

With the machines, farm sizes increased dramatically because of the economy of scale. Machines also work better with single crops so monoculture is emphasized. These changes have, of course, given new aspects to agricultural policies. In the United States and other developed nations, the family farm is disappearing, as industrialized farms, which are organized according to industrial management techniques, can more efficiently and economically adapt to new and ever-improving technology, specialization of crops, and the volatility of farm prices in a global economy.

4.2.2.1.1.3.2.2.1. Factory Farming

The cramped quarters and horrendous conditions of factory farms are known, but not familiar, to most people. Most people know that "food animals" are kept in buildings, away from the light of day, until they are brought to the slaughterhouse. Some know that over half of calves were dehorned and over three-quarters were castrated, without anesthetic. About one-fourth of U.S. cows were given Bovine Growth Hormone (rBGH) to encourage faster growth, although it causes numerous problems in the cows and may be a threat to human health.

Pigs are kept in the same dismal conditions as bovines. The National Pork Producers Council endorses several painful procedures for piglets, including clipping needle teeth, docking tails, and castrating boars to prevent strong-flavored meat. No anesthetic or painkillers are used during these procedures.

Chicken production has not been spared the attention of industrial agriculture. Chickens are produced for their eggs or for their meat. Production is increased by increasing the number of birds in the building. Thousands of birds are crammed into buildings, so that they have difficulty moving, are remote from view, and from the sun, wind, and habitat. The birds are forced to undergo a painful procedure called debeaking in order to stop damage to other chickens, from frustrated pecking, that could cost farmers money.

The egg industry is increasing production, as a result of its successful advertising. Egg production can be increased through forced molting, known in the industry as "recycling." During recycling, the feed is taken away for days or weeks, depending on the farm. After the chickens lose their feathers, they begin producing eggs again, although many chickens do not survive this process. That monetary cost is less than the money gained from the increased egg production, so the practice is continued.

Unfortunately, forced molting also is unhealthy to humans, with the risk of salmonella infections. Contaminated eggs or poultry can spread salmonella to humans. To reduce the incidence of infection due to forced molting, unsanitary quarters, and pecking injuries, the birds are pumped full of antibiotics. Because some of the antibiotics are similar to the ones given to humans, and because the diseases, such as salmonella, campylobacter, and E. coli, are also shared, there is a growing resistance to the antibiotics in humans.

Industrial agriculture has promised new and better ways to increase our food production. Unfortunately, its proponents have increased production largely at the cost of animal suffering, soil depletion, and compromised human health.

4.2.2.1.1.3.2.2.2. Factory Fish Farming

Ocean fishing has provided protein for people ever since people learned to fish. The demand for fish, however, is causing problems for natural fish populations. As natural fisheries are being depleted, and collapse, industrial fish harvests are becoming more common.

Harvested fish are raised and kept in tanks for their entire lives, rather than being caught from the wild. They are born in a hatching tank and then transferred to another tank where they are crowded together with other fish. This overcrowding creates unsanitary conditions for the fish, not just from excrement, chemicals, and antibiotics. But, fish, like chickens, cattle, and pigs, are capable of suffering.

Another concern of harvesting fish is the spread of diseases. Diseases spread much faster in water than on land. For instance, a virus infecting Pacific white shrimp spread throughout Ecuador, Central America, and Hawaii and is now a problem with cultured shrimp in Texas in the mid-1990s. The spread of parasites, such as sea lice, from farmed fish to wild schools, might occur. Finally, there may be genetic mixing between wild fish and their escaped farm fish.

Wild populations of fish are affected by aquaculture in other ways as well. The larvae raised on fish farms come either from hatcheries or are caught. Nets meant to trap shrimp larvae accidentally trap large quantities of other invertebrates and fish, perhaps twenty times the weight of the shrimp larvae.

Proponents of ocean fish farming argue that it is attractive in several ways. It moves fish farms away from the coast, where plumes of pollution from urban runoff can endanger the herd. It makes siting a farm somewhat easier, since it would draw less opposition from local residents uneasy about the potential for aquaculture pollution, in the form of waste from the fish, to foul coastal habitats, housing and beaches.

Some environmentalists and marine researchers view open-ocean fish farming with concern, because open-ocean farming is remote from observers, and it may be neglected. Many of the species to be kept down on the aquafarm are top-of-the-food-chain meat eaters, which have the highest market value. But they also are voracious, requiring anywhere from 3 to 25 pounds of smaller fish or fish meal for every pound of meat in the farmed species.

Aquaculture is already a huge industry. The Economic Research Service for 1999 shows that U.S. private aquaculture production is $772,900,000, more than double the amount for 1985. More people argue that environmentally sustainable ocean aquaculture is possible and necessary, if the right species and techniques are used. Other NGOs, such as SeaWeb, suggest that more can be done with fresh-water species grown in urban areas close to their markets, rather than developing large marine feedlots offshore.

Recently, however, the U.S. government sent a bill to Capitol Hill that would almost three and a half million square miles of ocean to fish farms—this is roughly the land area of the lower forty-eight states. This proposal represents the latest effort to implement recommendations from the U.S. Commission on Ocean Policy, which last fall sent the White House its blueprint for overhauling the approach the country takes to managing its vast offshore resources. It also expects to give various "stakeholders," including the aquaculture industry, environmental groups, and state and local officials, an opportunity to help shape the rules.

4.2.2.1.1.3.2.2.3. Artificial Production (Algae)

There are many possible ways of getting food: Hunting/gathering from the environment, traditional farming and animal rearing, industrialized farming, meat factories, culturing plants hydroponically, vat grown meats, artificial meats based on yeast, algae or bacteria, as well as food built directly by matter compilers from chemical elements.

Genetic engineering is progressing quickly and so are biological studies. After all, human tissue can be grown in labs relatively easily. So, food could be produced in vats. The food, for instance tomato paste, could be made cancerous, so that it would keep replicating cells. The cells could be kept immortal by keeping the telomeres. At the desired size, it could be prepared, sliced into cubes and canned, or freeze-dried, or irradiated and packed. Growing meats in vats would be more difficult, since a significant part of the flavor results from the muscular movement of the animal as it finds food; other tastes could be introduced through duplication of food source chemicals. Perhaps electrical stimulation of meat cells would make it more flavorful.

Modified algae could be produced in tanks, then textured and flavored by various technological processes. There is a whole range of possible artificial foods. Any of these solutions, of course, creates new problems, not just in relationships with other forms of life and the environment, but in scale especially. Table 4-22213224-1 compares the total acreages necessary to feed four billion people using different methods.

Table 4-22113224-1. Methods and Acreage

Method	*Area in Ha*	*Percentage of current*
Current production	1,507	100%
Current production, waste eliminated	1,250	82.8%
Feeding all cereals to humans	740	49%
Using the Chinese Method	650	43%
Using industrial agriculture, assuming vegetarianism	170	11%
Growing all food in greenhouses	60	4.0%
Growing food hydroponically	6	0.4%
Growing algae in vats	5.4	?
Using leaf protein	0.03	?
Genetic manufacturing	?	?

Single-celled protein could feed four billion humans using only one-seventh of the oil products waste, worldwide.

A *hydroponics* method offers 500 times the yield with only one tenth the water required; there is no runoff and better pest control is possible.

Although *leaf protein* is not palatable straight and must be machine processed, the machinery is available and the cost is still only one-seventeenth the cost of meat in India. It is more efficient than grasses. It offers new stores of plants for food (300,000). And it is the most suitable kind of harvest in the tropics, with their vulnerable soils.

The *single-cell protein* yeasts or fungi are very efficient and fast; harvests can be measured in hours, not seasons or years. Although it is a fairly expensive process—it must be grown in tanks—it can be grown on petrochemical wastes. And, it produces no wastes and has virtually no pests. SCP is of poor nutritional quality.

A comparable weight of *algae* contains three times the protein of beef. It is also efficient. Algae is also expensive. It concentrates pollutants. And it is amino acid deficient.

4.2.2.1.1.3.4. Development of Agriculture

Possibly, there were many factors leading to agriculture. It may have followed the slow process of overhunting; the world became full in terms of hunters, due to gradual human population increase. Land degradation is another possibility. Climate change, as areas became more open and there were fewer wild animals, may have inspired a shift to agriculture. In Europe the tundra was replaced by forests that reindeer did not like. The population pressure of people might have been a factor. People settle near plants rather than transport them. Sedentization and plentiful food increased the human population. Childbirth and child-raising became easier. More old folks could hang around. Gatherers lived near wild cereals. As people branched out, they started to grow cereals in less favorable areas. This required more maintenance, maybe watering and weeding, which may have contributed to the shift to cities. Over a 5000-year period, the amount of stuff gathered was replaced by the amount of stuff planted. Finally, agriculture could have started as a status symbol, having grains grow nearby, or as trade items.

Certainly agriculture was not automatic. There are biological obstacles to agriculture, starting with the difficulty of harvesting wild grains with brittle heads (rachis), which are naturally selected for wind distribution. People could have selected for tough rachis. There is the difficulty of removing grain from its husk. Emmer has a tough husk; even threshing will not remove much of it. It could be roasted to remove husk—this would also stop them from sprouting by killing the germ. Another difficulty was farming in niche where grain was adapted, which was to hillsides and slopes, not to flat alluvial soils. The species would change to new strains; six-row barley grows in the south, where 2-row was in the habitat.

After much intensification, the land collapsed. Around 3500 BC wheat and barley crops were equal. Wheat can tolerate salt at only 0.5% in the soil, but barley can take twice that. By 2500BC wheat had fallen to 15% of the crop, although overall crop yields were still high. To less than 2% by 2100 BC. By 1700 BC no wheat at all was grown.

Overall yields fell 42% between 2400 and 2100 BC, and by 65% by 1700 BC. The earth was white with salt. Many of these things are long-term problems and do not become evident for several generations. They are also very difficult to reverse. For a society than needs surpluses to continue, with growing dependents and growing people, there is little flexibility to change. The only way to avoid the problems was to let the land be fallow for long periods until the water table fell. Impossible due to food demands.

Shifting cultivation was practiced in Europe as well as South America and Africa. In Africa it was practiced for centuries with only minor long-term effects on the rain forest. But in Europe, where the forest regenerates more slowly since growth is checked seasonally by low temperatures and low moistures, shifting agriculture is unstable and capable of greatly modifying the ecosystem, despite the greater soil fertility. The type of crops also made a difference. Areas in America and Europe were based on cereal crops, millet, wheat, barley, while African systems emphasized root and tree crops. Traditional agriculture proceeds by substituting selected domesticates for wild species in equivalent niches, or by manipulation of the ecosystem, whereas modern agriculture replaces the biota, or transforms the ecosystem into an artificial system.

People-powered agriculture works well. Horse-powered and oxen-power agriculture is still harmonious, because living creatures fit better into biological and ecological relationships. The health and fertility of one is involved in that of the whole system.

Inorganic tools can be inserted into the order, but only within limits of kind, scale, and power. A tractor, for instance has too much power. It starts the destruction of the system. It also depends on the destruction of systems elsewhere to get its structure and energy. The tractor, farmer, and farm become resources for the industrial economy, run by capitalist ideas.

4.2.2.1.1.3.3.1. Advantages of Agriculture

Agriculture would have many advantages. The source of food would be more stable and more reliable. The plants would be isolated in fields or behind walls. Travel would be reduced, since the plants would be brought to people, instead of people traveling to select plants when they were ripe. The plants would have increased food value and decreased toxicity. There would be a surplus, that would allow people to live in one place, where more people could be supported on an area of land. The surplus would also allow intoxication to be domesticated and expanded.

Humans learned to cultivate soil, plant seeds and wait until crops grew. As a hunter, a human needed 10 square kilometers per person for nourishment; intensive agriculture lowered this to 1 hectare (1/100 square kilometer). It is 1000 times as efficient in area.

4.2.2.1.1.3.3.2. Ecological Changes from Agriculture

Agriculture is an integral part of the environment in which it is practiced. Agricultural systems are distinctive types of human-modified ecosystems. What is the difference between ecosystems and agroecosystems? People? Domestic plant and animals? Not just people or domestics, but their kind of actions, that is, their relations with their system. An agroecosystem is a natural ecosystem that has been modified, or arrested at an early stage of succession, an immature stage. This is a form of disturbance by humans.

Historical agricultural systems had numerous characteristics and impacts. Irrigated farming in river valleys, referred to as the Southwest Asian system, involved clearance of savanna, swamps, and fringe forests. It led to changes in the water regime by irrigation, with systems of embankment dams, canals, buckets, and water wheels. New canal-side ecosystems formed, but there was some salinization of soils and silting of canals.

Dryland farming involved clearance of oak-pistachio woods for fields. Fallowing and cross-plowing for moisture resulted in soil changes and erosion. Animals change fields and the surrounding vegetation. Hunting changes wild animal populations.

The Mediterranean system was characterized by clearance of forests and scrub for fields, which led to soil erosion (on limestone) and siltation. This caused a change from forests to semi-arid steppe. Permanent agriculture in the temperate zone led to clearance of deciduous forests and overcultivation. Soil loss was relatively low due to slope and climate.

The Southeast Asia Tropical Vegeculture is a form of ax and digging stick cultivation. Shifting cultivation replaces old forest with polyculture. Hunting still contributed to protein. Wet rice cultivation occurred in wet areas by streams, then in irrigated plots. Then, rice was grown in terraces and paddies.

Dryland farming of cereals and pulses in Northern China was using ploughs by 2500 YBP. Wind erosion was a problem. African Tropical Vegeculture was a form of shifting agriculture without domestic animals, using digging sticks, and supplementing protein with hunting. Cereal cultivation in Africa was bush fallow, which involved erosion of latosols, cultivation of trees, and hunting. Root cultivation in the Americas used digging sticks, but no domestic animals. Irrigation agriculture in the Americas made use of irrigation channels spring water into canals through saline lakes. Islands could be made in lowland swamps. Hunting provided some protein. Drainage agriculture was much the same.

In an ecosystem, people are just one more group exploiting the system. But, in agroecosystems, people are manipulating the entire system, that is, they are modifying the system and isolating from neighboring systems. They do this in many ways, by: Simplifying the system by removals; managing the system with domestics; and harvesting the bulk of productivity. Each of these has effects on the system.

Simplifying the system, or lowering diversity, allows fewer species in the system. Not just the crop species, but pest species. Crops are analogous to early colonizing plants, that is weeds, that exploit ephemeral resources following disturbance and maximize their seed output. Plowing favors crops that grow rapidly.

4.2.2.1.1.3.3.2.1. Changes in Diversity from Agriculture

The diversity of use narrows with agriculture. There are roughly 250,000 described higher plants. Perhaps 30,000 are known to be edible. Of those, about 7000 have been used as food by foragers and horticulturists. And, only 150 have been used as important crops. But, only 15 are used as crops to supply over 90 percent of agricultural foods. Three cereal crops—wheat, rice and corn—provide over 50 percent of protein.

The early agroecosystems were more diverse. In 1949, Chinese farmers had 10,000 varieties of wheat. In 2002, 300 were used. In Java, tropical gardens had between 100 and 350 species per hectare. Tree densities in Javanese gardens may be 2000 stems per hectare, with a biomass of over 120 tons per hectare. The Javanese used interspecific and intraspecific polyculture, that is, not only different species, but different cultivars of the same species, such as corn. Game theory shows that this strategy increases the chances of a good harvest.

At first agriculture moved down the food chain. Once societies modernized, they start to move up the food chain again, eating less grains and more meat and eggs. Simplifying the system also lowers maturity. Agricultural systems keep the ecosystem immature. Why? Because immature systems are more productive faster, as long as we can use the immature products. Alas, these systems are also less stable. Humans could be kept at a medium maturity, that is a more mature level, with intelligent flexibility.

Diversity is low because people try to plant monocrops. But, uniformity is a kind of problem. All individuals are the same kind, size, age, and use the same resources at the same time. It is a perfectly uniform predictable food source for insects and pathogens. So, there is no protection from diversity in genetics, age, and size. From the insects' perspective, it is a feast. The selection of the cultivars for what we want to harvest has eliminated some defensive traits in wild plants. From a weed's perspective, it is a boringly predictable competitor for light and water. Weeds can compete because of the bare ground disturbance.

4.2.2.1.1.3.3.2.2. Changes in Disturbance from Agriculture

Disturbance is created by plowing. Energy is used to maintain the early stage. Soil carbon is also oxidized and lost with plowing, Phosphorus ands potassium, also. Plowing eliminates members of the soil community such as earthworms. Plowing also breaks down the soil types, so that erosion can happen; the surface is exposed. Carbon, nitrogen, phosphorus decrease from the upper horizons. Disruption inhibits mutualistic links, such a fungi. With agricultural disturbance, energy and material subsidies are needed to keep productive.

This limits efficient nutrient cycling, diversity, resistance to disease, It ensures that the system will have low amounts of organic matter, small organisms with short life cycles and broad niches, open mineral cycles, undeveloped detritus communities and linear food webs. Nutrient flows are simple, one-way flows. From the industrial source of chemical fertilizers to outputs in harvested materials or mostly water runoff. Nitrogen for example, is put on in pulses and ends up being leached out. It is applied fertilizer, usually anhydrous ammonia, which is oxidized to nitrates, which can acidify the soil. Synthetic fertilizers inhibit the action of nitrogen-fixing organisms and detritivore, which work best when soil nitrogen is low. Erosion of the soil itself is a problem. Agricultural techniques expose the soil to wind and water at critical times.

4.2.2.1.1.3.3.2.3. Changes in Scale and Recycling from Agriculture

From an ecological perspective, there are regional consequences of monoculture specialization. Most large-scale agricultural systems exhibit a poorly structured assemblage of farm components, with almost no linkages or complementary relationships between crop enterprises and among soils, crops and animals. Cycles of nutrients, energy, water and wastes have become more open, rather than closed as in a natural ecosystem. Despite the substantial amount of crop residues and manure produced in farms, it is becoming increasingly difficult to recycle nutrients, even within agricultural systems. Animal wastes cannot economically be returned to the land in a nutrient-recycling process because production systems are geographically remote from other systems which would complete the cycle. In many areas, agricultural waste has become a liability rather than a resource.

Recycling of nutrients from urban back to fields is difficult. Part of the instability and susceptibility to pests of agroecosystems can be linked to the adoption of vast crop monocultures, which have concentrated resources for specialist crop herbivores and have increased the areas available for immigration of pests. This simplification has also reduced environmental opportunities for natural enemies. Consequently, pest outbreaks often occur when large numbers of immigrant pests, inhibited populations of beneficial insects, favorable weather and vulnerable crop stages happen simultaneously. As specific crops are expanded beyond their "natural" ranges or favorable regions to areas of high pest potential, or with limited water, or low-fertility soils, intensified chemical controls are required to overcome such limiting factors.

4.2.2.1.1.3.3.2.4. Managing the System with Domestics

We have to control the system. We decide what species grow and which are denied. Our level of concern is usually plant-to-plant, rather than the whole system. We try to control light, water, and minerals as well. But, there is no thought to the overall landscape. What does the ecosystem want to do? Or try to do? To develop, to grow, to grow old, and to be efficient. As Whitehead said, what does ever living being want to do? To live, to live well, to live better.

4.2.2.1.1.3.3.2.5. Harvesting the Bulk of Productivity

We harvest as much as possible of all the productivity. Therefore, we can use machines to do it, even if they destroy soil and relationships of living beings. Harvesting is limited to cultural

choice. We choose not to harvest dandelions or sowbugs, ants or spiders.
These systems are based on ecological systems that have been rearranged for human purposes (food). They still depend on solar energy, photosynthesis, biogeochemical cycles, stability and movement of the atmosphere, and the services of nonhuman organisms. They are subject to the laws of ecology.

4.2.2.1.1.3.3.3. Cultural Changes from Agriculture

How did agriculture affect culture in early Mesopotamia? Just as a result of irrigation? From changes in religion, from otherworldly gods? The sun god Utu is the sun and its numinous power also. Religion was polytheistic due to the kinds of situations where the numinous could be experienced. By the early Dynasty gods came to be called rulers, just as the human rulers had more power. Enki was the water god. Nanshe the fish goddess. Dumuzi livestock god. Ningirsu, the god of thunderstorms and floods, actually headed a pantheon of lesser deities who staffed his temple. These new ruler gods could make demands on people. The Mesopotamian cosmos was hierarchical, with gods holding titles and offices. There were seven primary gods, 50 great gods and many lesser gods.

There are many associated changes from shift to agriculture; they can be considered cascading effects from positive feedback loops. For instance, clearing larger areas of land and irrigating to transfer water to land led to technological improvements in the channeling and manipulation of water. New tools extended efficiency. More crops required new ideas of storing foods. There were more innovations in general. Domestic animals were used to haul loads and then work in the fields for preparation or harvests. This required new kinds of digging sticks and new kinds of harnesses. The kilns for ceramics required more wood, which often came from newly cleared fields or transport from remote forests.

The creation of more kinds of material goods led to the accumulation of more material goods, that is, more things. Things had to be measured and tracked, and that required counting and sorting. Books had to be kept. A currency of exchange had to be invented to allow things to travel less and be more still, in place.

There was an increase in the size of labor input into agriculture. There were specialized occupations related to tools, specialized occupations related to trade, and specialized occupations related to conflict, protection, control, and project management. There were changes in distribution of food substances and things. Not only things, but people concentrated into villages and towns. Luxury goods and wealth accumulated. Society may have stratified first by specialization, some of which had more prestige. And later by prestige itself. To protect things and places, new specialists in protection were developed. Control and conflict increased. At some point control of social activity led to control of the poor, control of women, of animals, and control of wild areas and animals.

Commercial farmers witness a constant parade of new crop varieties as varietal replacement due to biotic stresses and market changes has accelerated to unprecedented levels. A cultivar with improved disease or insect resistance makes a debut, performs well for a few years, typically five to nine years, and is then succeeded by another variety when yields begin to slip, productivity is threatened, or a more promising cultivar becomes available. A variety's trajectory is characterized by a take-off phase when it is adopted by farmers, a middle stage when the planted area stabilizes and finally a retraction of its acreage. Thus, stability in modern agriculture hinges on a continuous supply of new cultivars rather than a patchwork quilt of many different varieties planted on the same farm. The need to subsidize monocultures requires increases in the use of pesticides and fertilizers, but the efficiency of use of applied inputs is decreasing and crop yields in most key crops are leveling off. In some places, yields are actually in decline. There are different opinions as to the underlying causes of this phenomenon. Some believe that yields are leveling off because the maximum yield potential of current varieties is being approached, and therefore genetic engineering must be applied to the task of redesigning crop. Agroecologists, on the other hand, believe that the leveling off is because of the steady erosion of the productive base of agriculture through unsustainable practices.

Table 4-2211333-1. Other Consequences of Agriculture

Psychological consequences	Increase in compliance and obedience
	Stress due to uncertainty of yield
	Increase in hard work
	Reduced food choices, malnutrition
Sociological Consequences	Sedentarization of people
	Altered relations, status competition
	Change in world views
	Increase in movement of goods
	Rise in central authority, hierarchy
	Use of irrigation to increase yields

Cultural differences between ecology and agroecology include: Different perspectives, for instance, human-managed systems; different goals, such as productivity and efficiency; different principles, like subsidization. Crop ecology for agriculture is limited by intense labor. Special cultural practices are developed to engage the work and celebrate its successful completion.

4.2.2.1.1.3.3.3.1. Agriculture and Population

Agriculture did produce one unexpected change. Initially, farming peoples suffered from increased sickness and mortality. But, they also saw greatly increased fertility. Populations in Europe increased one-hundred-fold. Of course, agriculture allowed people, by virtue of remaining in one place, to have more children than they could carry or travel with. So, the sedentism of agriculture allowed larger populations to exist in place. Furthermore, the nature of work in agriculture meant that children were more needed for labor.

4.2.2.1.1.3.3.3.1.1. Population Growth

What is the reason for the rapid population increase? Biologically, when a pine tree starts to die, it puts immense effort into its seed crop—is there a human analogy? Did the activity of hunter women decrease fertility that increased when they stayed in place? Did hunter women control fertility by prolonged breast feeding or sexual abstinence? Did being in place remove that control?

Agriculture, growing crops and raising animals, provided more food, but it was less nutritious and less palatable. By increasing the population, agriculture increased widespread hunger. The reduction in biodiversity, by monocropping, undermined biodiversity and caused some ecological crises, that resulted in periodic and devastating famines.

Agricultural populations increased and spread, bringing their crops with them, pushing hunters into the periphery where crops could not grow. Once the initial agricultural startup costs were paid, such as truncated diversity and nutrition, and the decrease in human size and health, other advantages accrued, such as specialization, creating a different form of political control, which allowed armies. Agricultural society now had a massive advantage over foraging societies.

Despite famines that caused hundreds of nations to wobble or collapse, the overall human population kept increasing. The world population is increasing at an alarming rate. By 1900, the rate started to increase steeply. At a rate of increase of two per cent per year— that is below the current rate—the earth's human population will reach 50 billion by 2100. By 2280 it would be 1.2 trillion. That is 300 years from now (1980, the year of this sentence); 300 years ago John Milton's *Paradise Lost* was popular reading.

Nations that seem to have adequate resources, may be growing too fast. Even the Union of Soviet Socialist Republics, with all its potential agricultural land, and its uncertainties of drought and frost, Canada and Australia, may be increasing their population too fast. Population growth can become a political weapon, in a democracy, when two groups coexist, with neither ecological differentiation nor geological separation, and only breeding control for coexistence. For instance, in Sri Lanka, the Sinhalese and the minority Tamils, breed for political control, ignoring population control. Voluntary controls fail; the population increases by almost two and a half percent per year. The problem of population control is a tribal problem, not a racial or economic one. It is a local problem, more than a global one. Even with trade and charity, in most areas population is limited by local resources and local limits; only in a few cases, such as the Netherlands and Hong Kong, can a population be supported by massive ghost acreage elsewhere.

Sir Charles Darwin's view of humanity was that it was not domesticated. Humans were still wild. Unmastered, they have no breeding control; therefore, they will eat to the limits of the natural food supply and press against social and biological limits. The global population explosion, an example of autocatalytic nonlinearity, is not considered a problem for many thinkers, such as Eric Jantsch, who believe that this growth constitutes an essential factor in the creative act of gestalt formation. Growth does lead to intensity at some stages of development in the life of a species, but growth can stop, and development can also lead to the kind of creative intensity identified by Jantsch and others with mindless growth.

Population growth cannot continue indefinitely. Population is intrinsically inseparable from the question of access to resources. The time has gone when one nation can unload its population problems on another; and this applies to all countries.

The principles of ecology can offer information on human societies. For instance, the principle of competitive exclusion states that two dissimilar races cannot occupy the same niche in the habitat. If humanity is wild, this may help explain war and racism. Territoriality is no longer a rule in human ecology, though the instincts may still be operative. Human ecology defies maintaining its population at an optimum level. The young are overproduced and protected; the least fit are not eliminated. Medical science has also increased the number of old and maladapted dependents of society. Problems with populations bristle with social and political implications.

A balance in birth and death rates is necessary for ecological stability. We have controlled epidemic death and infant death diseases. We need to correct the overbalance in births, or risk having nature do it. Toughness and ingenuity might be required.

4.2.2.1.1.3.3.3.1.2. Population Size

Populations are still growing exponentially. The world population is almost five billion people (1980). It is over six billion (2001). Current technology allows a large population to live at the cost of the depletion of resources, thereby reducing the potential for future life. From linear thinking, a population of up to 40 billion can be projected; but at what level, for how long? Darin-Drabkin and Lichfield claimed that with efficient use of land resources, the population could be 50 billion; that we have the technological power to fulfill these estimates, but need organizational changes and proper planning to allow it. It is unlikely that humanity would be ten times happier or more creative. History and tradition are difficult to change. What would happen to forests and animals?

Leopold Kohr has a velocity theory of population, in which aggregates of humans (or atoms) increase mass by addition and through an increase in speed. Population density multiplied by energy use creates pressure on environment; mass times speed equals pressure (MV=P); The answer is not in search for new sources of fuel, but search for means of production and decelerated way of life using less energy. Not necessary to lower standards of living. We must either introduce voluntary pressures now, or let the system introduce them later by its own dynamic. There are ways for voluntary reduction, however. There could be tax incentives for those families who control their size.

When power density reaches carrying capacity, a system is considered developed; when it exceeds capacity, it is overdeveloped. The maximum number of individuals that an environment can support is its carrying capacity, which fluctuates with various factors. The greatest number of people over time implies that the living population be limited to biological carrying capacity. Any permanent destruction of carrying capacity to allow a larger living population implies a future reduction in population indefinitely, or until ecosystem restoration (if possible). Those who favor negative population growth may maximize the number of lives total, at a sufficient level of wealth.

Physical limits may be absolute or flexible. Population control can either be accomplished through crowding, by the food supply, or by natural resources. There tend to be as many people as there are utilized natural resources for and unpolluted room and food for.

Numbers of a country should not exceed the scarcest of commodities. A values boundary would be implemented in industrial countries. As vital systems are used for nonessential purposes, like using water to wash a car, the population must be reduced.

4.2.2.1.1.3.3.3.2. Cultural Status

Perhaps related to population growth and size, have come changes in cultural status, especially related to equity, distribution, and dominance. Status has to do with standing in society, and with appearance and ownership. Status may come from longevity, of the self or ancestors, as well as from the results of good decisions, from hunting for example, or from owning more than others, or just more things or more people. Status is a powerful human need, and may drive the growth of goods or populations.

As distribution becomes more unequal and as conflict occur more between groups, dominance and slavery appear, and are both related to status.

Darwin has been criticized by Malthus for extending the popular theory of economics to the natural world. It is true that the biological considerations of economics inspired an economic description of a biological law, but Darwin enlarged and supported the metaphor.

He retained the idea of hierarchy, but status became a form of fitness, specifically in birds and mammals. It can be argued that status has fitness values in human groups also. For instance, a woman is more likely to mate with a man who has higher status in the community.

Status is a social need, which is justified in the culture. Status levels in traditional society included: Nobles, commoners, and slaves. Slaves were captured or purchased. In earlier days, people were sometimes captured by enemy tribes. The return home of the captives, either through payment of ransom or owing to a retaliatory raid, was called u'mista—that is, a special return. Like other property slaves were given away as gifts. Occasionally, they were killed and eaten during the cannibal dance. A commoner was simply anyone without a chief's position, potlatch position, seat, or standing place. They are not a class with a function. At times, nobles may retire from potlatch positions and become commoners. Kinship was the main determinant of status. For the Kwakiutl, the display of status validated the social system. The redistribution of food validated the cultural status of the leader.

By 2400 YBP in Mesopotamia, there was status differentiation, as evident in the sizes of homes. Oddly associated with permanent settlements were inequalities of status and wealth. Furthermore, other social things developed such as complex division of labor, with castes and slave labor. Trade increased and competition increased. This may have had to do with resources that were abundant and could be stored, and transformed into political prestige and power. Agriculture provides more status, more wives, more things. It allows armed struggle also. Art, in addition to luxuries and profits, was tied to status.

4.2.2.1.1.3.3.3.2.1. Equity & Distribution—Having More

In almost every size of human groups status is divided unequally. In hunting groups, better hunters become the hunting leaders. The success of a hunter allows the hunter to distribute the cuts of the game between others, often according to an understood set of rules. In larger groups, with a surplus of materials, the materials are distributed according to status. People have equal access to status positions in society. The number of positions of prestige is adjusted to the qualified candidates. Status is separate from wealth. An animal divided up according to ratios regardless of who kills it.

J. S. Mill narrowed the scope of economics to production and the scarcity of means; he considered distribution to be a political process, since it depended on laws and customs that varied widely in different cultures and ages. One function of culture is to distribute material goods or energy. Economic culture defines the means of production and livelihood, techniques of distribution, and values and norms underlying economic behavior (can be more closely related to kinship. Order is a cultural problem. Order provides stability and security. Cultural order is necessary to deal with the redistribution of wealth and power.

With the creation of more wealth or new wealth, the distribution becomes skewed. Accumulation increases. More kinds of wealth become invented, dangerous, and useless, and more skewed. But, as long as an economy does not reach the limits of wealth, it can keep growing; these new kinds of wealth allow that. Cultures encourage the unequal distribution of resources.

Wealth is based on production in capitalist countries. The production of wealth from growth depends on technology. The technological perspective is oriented toward materials and not humans or forest processes. Nature is considered to be a resource to be exploited. The immediate objective of technology is to create wealth through knowledge. Technological activities are justified on humanitarian grounds, scientific discovery increases the well-being of human society, yet the social consequences of scientific activity are ignored; short-term suffering will be offset by long-term benefits, it is claimed. But because the long-term view is not taken, long-term benefits will be worse.

Economic growth can produce great wealth for some. Fortunes await those who can increase the demand for unnecessary items, such as electric swizzle sticks or pet rocks. Free enterprise provides initiative to any willing to show a profit without regard for the immediate consequences. The economist Paul Samuelson illustrated the skewed distribution of wealth with an income pyramid made out of children's blocks: if each layer represented $1000.00, the peak would be as high as the Eiffel tower, and most of us would be within a meter of the ground.

Inequality is more the result of differential development than of exploitation. According to Boulding, the greatest source of the differential is different rates of accumulation of knowledge, capital and organization; the rates are essentially internal properties of cultures. Although a minor element in terms of transfers, it is a large psychological perception, which may need to be compensated for in a global community.

Most of the wealth used by modern economies is nonrenewable. These resources are limited, interrelated and distributed unevenly. Forests are a special problem; although trees can grow to a good size in 30-40 years, forest ecosystems may take 300-600 years to develop and then last for thousands of years. Oil, coal, peat, and some woods are functionally nonrenewable. Geological time periods are required to produce them.

The distribution of symbolic and real wealth is very inequitable as the result of historical trends, old economic rules, and cultural confusion. Karl Marx considered that misery is caused by class conflict about distribution of material things. Some of the problems of distribution have to do with the size of society and the scale of its operations. For instance, in Hutterite communities, usually less than 150 people, the distribution of goods rarely failed; people got their share, rarely less or more. With a larger group of Hutterites, the distribution started to fail, and the group divided.

The modern market distributes some benefits to all, but its scale allows unfairness. Economics are the harmonious distribution of wealth among people. But, problems outside small-cell scales. Kohr points out that Marx failed to link misery to the scale of economics rather than the system. The problem is with overgrowth more than style, which is why socialism based on overgrowth looks the same as capitalism overgrown. The change in scale drove the changes, from the transition of states, from surplus distributors, to tribute-driven to commercial exchanges. There were political transitions from leaders, to chiefs, to tribute systems, and to economic and political trade systems.

4.2.2.1.1.3.3.3.2.2. Cultural Dominance

Strength or status can give rise to dominance. A dominant is an animal or person with greater influence in the community. Dominant behavior may be biologically-based or cultural. Dominance gives priority of access to food, sex, or space. In a majority of primates, males are not dominant over females, except where there is a large size difference. Where sizes are roughly the same and where female coalitions occur, there is not much sexual dominance. In some human groups, males may attempt to dominate females with the threat of physical violence, for instance, among the Yanomamo. Among other Amazon peoples, such as

Mundurucu, it may be due to control of ritual objects and rituals, for dominating ceremonial life. In spite of the cultural forces dominant at any moment, an individual has the potential to determine a different course of action.

4.2.2.1.1.3.3.3.2.3. Slavery

To be enslaved means to be owned as property and divested of freedom and rights, or it means to be completely dominated. The word slave comes from the old Slavic word for the Slavic people, who were among the first slaves to be held in Europe.

Slavery is the idea that people can be held in bondage to perform work. The Greeks thought that freedom depended on some slavery. Some societies thought that their economies depended on slavery. It was actually thought that slavery improved the lives of slaves, since they were often considered to be from subhuman groups. It was also thought that it was a personal decision or personal right, although the slave might feel differently. There are of course, different kinds of slavery now. For humans, there are living slaves, but future generations might become enslaved to the decisions of their parents, especially as regards losses and debts. Animals have been enslaved. Machine slaves were the next economic boost. Finally, industrial societies have energy slaves, ten to twenty for each person. Slavery is based on a number of assumptions, such as contempt for the enslaved, or denial that there is wrong doing.

Kinds of slavery can be distinguished: Opportunistic, Institutional, Or Comprehensive. Opportunistic slavery was the result or raids or conflicts with other bands. Institutional slavery included labor force collection, as practiced in Egypt in 4500 YBP, or economic slavery, as run by the British 300 years ago. Comprehensive slavery refers to the use of wage earners, animals, and energy.

4.2.2.1.1.3.3.3.2.3.1. Opportunistic Slavery

Many archaic cultures had slaves, who were usually the victims of conflicts between bands. In Nez Perce gatherings and celebrations, slaves were exchanged, with furs, roots, berries, and fish. Slaves could often intermarry with their captors and acquire status; their children were almost always free. Slaves became a separate social class in some archaic societies. By 5000 YBP a stratified class society had developed—slaves at the bottom, then peasant farmers, craftsmen, then the elites of administrators, religious and military. In Assyria, slaves could not wear veils. Slaves worked everywhere in Assyrian society, even running businesses. Some people entered slavery voluntarily to pay off a debt.

4.2.2.1.1.3.3.3.2.3.2. Institutional Slavery (Economic Labor Collection)

For the Tukano of Peru, the Maku group was a slave class, that is, a source of slaves. The Kwakiutl also had slaves, who were captured or purchased. Like other property, slaves were given away as gifts. Usually, they were captives kept near the door to guard the house. Occasionally, they were killed and eaten during the cannibal dance. Viking raiders and colonists owned their own land and used Irish slaves for work and for wives specifically, based on genetic evidence. As the number of slaves diminished through death or marriage, some farmers lost their land to absentee landlords, who, without ecological or personal feedback, would make poor decisions about crops and practices.

Economic slavery occurred in Mesopotamia, Slaves worked in all levels. Many were prisoners of war; some were criminals being punished; some entered slavery to pay off a debt. But, slaves could also own a business. If a slave married a free person, their children would be free. Slavery became specialized with agriculture, in China and Egypt.

In China, by1615, slavery was hereditary, for agricultural or household slaves. Slaves reflected the stratification of society as it became more complex. Slaves in the Americas were needed because the amount of acreage increased with the discovery of new lands; the acreage and the slaves were considered necessary to develop more wealth. Slavery kept labor artificially cheap. This dampened the incentive to develop new technologies, but, new technologies eventually surfaced because they proved to be cheaper than slaves.

The appearance of mechanical devices such as the sugar mill and Eli Whitney's cotton gin helped to support the system of large plantations based on a single crop. The Industrial Revolution after the late eighteenth century swelled the population of towns and cities and increasingly forced agriculture into greater integration with general economic and financial patterns.

Internal slavery was a feature of Europe from the Romans to Middle Ages, but disappeared when feudalism disappeared, and labor was no longer the scarce factor in production. With the opening of America, however, land opened up—or rather was claimed and conquered—faster than it could be filled with European people. Therefore, colonists tried to force native Americans into labor. Native Americans, however, were still dying from European diseases, which made them unsatisfactory as slaves. Europeans looked to another tropical continent, Africa, where the people were resistant to old World diseases. Less than thirteen years after Columbus, in 1505, African slaves were introduced into Haiti. Over 350 years, ten million more were brought over, in a large trans-Atlantic trading circuit that involved rum, sugar, cloth, and timber as well.

In the American colonies, the independent, more or less self-sufficient family farm became the norm in the North, while the plantation, using slave labor, was dominant, although not universal, in the South. With the British, Americans, Africans, and Spanish, slaves became a commodity in a trading empire.

4.2.2.1.1.3.3.3.2.3.3. Comprehensive Slavery

After the enslavement of peoples, some dominant cultures turned to the enslavement of nature and then of energy. The English treated tropical lands as enemies to be defeated, then enslaved them in plantations. Their cultural attitude as conqueror of nature led them to treat biogeochemical cycles and soil requirements as temporary obstacles in a world where everything had its price.

The new economics of the industrial age depends on wage slaves, that is, people who need jobs to get necessities in urban areas. An overeducated labor pool led to secretaries and then managers, and to the fulfillment of the managerial revolution. Now, industrial society is able to use energy slaves to accomplish work. Energy, in human work equivalents, from animals, natural and fossil fuels, that is used to perform work. The institutions of slavery raise questions about our relations to each other and to nature. What is our human relation to reality, slave, master, participant, or partner? Cosmology describes the place of a culture in reality. Cultures can determine inappropriate attitudes towards nature and that can result in ruin for the air and land. Although slavery is officially condemned in virtually every nation, the inappropriate use of people, animals, machines and energy continues, as a necessary part of the agricultural and industrial economies.

4.2.2.1.1.3.3.4. Disadvantages of Agriculture

Highly specialized systems are accelerating the change from 'palaeotechnic' to 'neotechnic' agriculture. This change minimizes costs and maximizes profits, but it also narrows the ecological basis of world food production and decreases human livelihood. Life on the planet cannot afford the continuous genetic narrowing generated by agriculture. The application of modern agricultural techniques in tropical areas disregards the local realities, and is socially and ecologically disastrous. Traditional ways have operated successfully for thousands of years, although not without serious problems of their own. But, these problems may be overcome, unlike industrial problems, which have fundamental flaws and cause rapid conversion.

Disadvantages of agriculture include a dramatic increase in work hours, especially during harvest time. All the hours that were previously covered by natural processes, from propagating to irrigating, were now accomplished with human labor. In addition to that were requirements for tools and storage devices. Despite this, there was a decrease in nutrition, as the overall diet was less diverse and less complete. There was an increase in fertility and a population increase, perhaps at a rate thirty times as fast.

The form of agriculture, from simplification of the system to an increase in scale of the system, not to mention intensity of energy and use, resulted in degradation of the ecosystems, including erosion, siltation, salinization, and an explosion of pests.

Modern agriculture replaces the biota, or transforms the ecosystem into an artificial system. Furthermore, with fertilizers robbing foods of nutritional values, more of each food is required to meet nutritional needs; for instance, the nutritional value of broccoli has declined almost 500% in 50 years, which means we have to eat 5 times as much as our grandparents (no wonder George H.W. Bush did not like it).

Stealing from the future is drawdown. Humans must live within carrying capacity of the ecosystem they live in, relying on renewable resources consumed at sustained rates, and not try not to enlarge system temporarily for a one-time overshoot. The human species can be best protected by curbing human dominance.

G.M. Woodwell (1974), an ecologist, estimates that 30% of net primary production of the earth is diverted to direct use for support of current human population—in 1992 the figure was 40%, and the most recent figure, from Stuart Pimm and others after 2001, is fifty-six percent. This estimate does not include the public service functions of nature, such as air and water purification. If the "return to nature" urge were based on models of climax ecosystems, civilization would have to be far more complex; human values would have to allow for welfare of animals, plants and land.

4.2.2.1.1.3.3.5. Systems of Agriculture

Agricultural systems can be classified using two variables: energy type and diversity of plants. Traditional systems—shifting cultivation, nomadic pastoralism—depend on human or animal labor. Modern systems—plantation, commercial— combustible fuels for energy; these systems also tend toward monoculture. Traditional systems usually incorporate greater biotic diversity, and tend to be more polycultural. Traditional systems are very labor intensive, that is, participatory.

Agriculture started out as a food management system. The first grains may have been status foods, used only at certain times of year, like salmon or other annual foods. Then, agriculture became a labor Management system that required training and encouraging people to work harder and in a special niche.

Later, it became an energy management system. It tried to control the energy sources, the converters, and the outputs. More energy put into crops and more energy extracted. To model energy, we can consider one human being is rated at 0.1 horsepower (abbreviated 'hp'). A team of oxen is 1.2 hp. A 50-hp tractor is rated at 50 hp. In tilling a one hectare field, human power is twice as effective as the oxen and 4 times as effective as the tractor (in terms of the amount of work performed by horsepower. However, in terms of gross costs, such as shelter food and clothing, then human power is 3.45 times as expensive as a tractor and 2 times as expensive as oxen, who require food and veterinary care. Eventually energy surpluses flow from the country to cities.

As a time management system, time constraints are also a problem. Although one person could produce a crop in 700 days, it is more effective to use 70 people in 10 days. Human labor is low in absolute levels of work. Animals constituted a form of energy slaves.

As more effort was put into the land, agriculture became a territory (and perhaps property) management system, finding and dedicating land to crops. One could also consider agriculture as materials management. A hydraulic civilization required tools, dams, pots and labor schedules. Of course, agriculture could be considered a beer management system, catering to the desire for intoxication. Fermentation had an additional nutrient bonus from yeast as a protein. Yeasts were also used for breads.

Agriculture is a novel information system. Because of population density, there is a drop in costs of transmitting information. The brain is more tightly packed. This is advance in information technology. Information is presented in terms of labor, a greater array of goods, and special projects. Information guides energy use. Chiefs give signals that channel energy into projects. The signals are trigger effects, small changes in energy that cause large changes in the cultural output of energy.

Agriculture, finally, is an ecological system, based on ecological limits, based on natural processes, and based on interactions with other systems, including human ones. In its early stages, agriculture is constrained to early stages of succession. That is, it uses nutrients and energy for quick growth and reproduction, rather than invest in efficiency or durability or maturity. Fossil fuels have allowed this artificial condition to continue. But, eventually, agriculture will have to become more mature. Wes Jackson and others are working on making this transition.

4.2.2.1.1.3.4. Promise of Regenerative Agriculture

Regenerative agriculture pays attention to the total input flows for agro-ecosystems: Solar, human, animal, tools, fuel, fertilizer, pesticides, drying, processing, and transporting. Under some circumstances, pesticides and fertilizers might be needed to stop pest populations or give advantage to crops.

Intensification has been a trend in agriculture, but different processes need to be intensified. Exploitation needs to be more intense. Perhaps specialization does. From secondary products from domestic animals. How to intensify production. Increase yield from land use, as a result of intercropping or adjusting ... Increase yield from human labor. Water control may allow multiple harvests, to increase productivity. It might be possible to breed more rapidly maturing strains.

There are many kinds of specific measures that can reduce erosion, starting with contour plowing. Plowing can be eliminated altogether with seed-drilling. In 1943,

Edward Faulkner, showed there was no scientific reason for plowing. Plowing gives a short-term increase in soil quality, by mining the subsoil. Fukuoka's ideas of multi-planting.

4.2.2.1.1.3.4.1. Natural Farming

Masanobu Fukuoka is another of the major pioneers of sustainable agriculture. Professing no knowledge, he presents a farming method that involves no tillage, no fertilizer, no pesticides, no weeding, no pruning, and less labor. He calls this practice the "no-plowing, no-fertilizing, no-weeding, no-pesticides, do-nothing method of natural farming." The idea that people can grow crops is egocentric to him. Ultimately, it is nature that grows crops. He sees modern agriculture as 'doing-this and doing-that' to grow crops, but it is meaningless work. With his 'do-nothing' method he is able to get yields in his rice fields that are equal to the highest yields attained with chemical, do-something agriculture. He accomplishes these high yields by careful timing of his seeding and by careful combinations of plants, in a polyculture.

4.2.2.1.1.3.4.2. Regenerative Agriculture & Permaculture

Permaculture is a design system which aims to create sustainable habitats by following natural patterns. The word 'permaculture,' coined by Australians Bill Mollison and David Holmgren during the 1970s, is derived as a contraction of permanent agriculture, or permanent culture. The idea of permaculture is considered among the most significant innovations developed. However like nature, the permaculture concept evolves with time making its static definition difficult. Nevertheless, permaculture can be described as an ethical design system applicable to food production and land use, as well as to community building. Permaculture seeks the creation of productive and sustainable ways of living by integrating ecology, landscape, organic gardening, architecture, and agroforestry. The focus is not on these elements themselves, but rather on the relationships created among them by the way they are placed together in place, the whole becoming greater than the sum of its parts. Permaculture also addresses the careful and contemplative observation of nature and natural systems, and the recognition of universal patterns and principles, as applied to specific circumstances in unique places.

Permaculture uses zones to focus efforts. The zone nearest to the house (1) is the location for those elements in the system that require frequent attention, or that need to be visited the most often. The vegetable garden (2) may also have larger scale compost bins or bee hives. The area where crops are grown (3), for domestic and trading purposes, could include orchards. After establishment, care and maintenance requirements, such as mulching, are relatively minimal. Watering or weed control is performed every week. Animals in these zones, such as chickens, can be used as a method of weed control and also as a producer of eggs, meat and fertilizer.

The next zone (4) is considered semi-wild; it is used for timber production from coppice managed woodland and the placement of aquaculture ponds. The fifth zone (5) is actually wilderness. There is no human intervention other than observation of natural ecosystems and cycles. Mollison considers this zone the source of the most important lessons of the first permaculture principle of working with nature, not against it.

4.2.2.1.1.3.4.3. Natural Systems Agriculture

What is the difference between ecosystems and agroecosystems? People? Domestic plant and animals? Not just people or domestics but their kind of actions, that is, their relations with their system. In an ecosystem, people are just one more group exploiting the system. But, in the agroecosystems, people are manipulating the system, that is, they are modifying the system.

The goal of Natural Systems Agriculture is not only to fit agriculture within limits of the ecosystems, but to increase its presence in cities and artificial areas. The Land Institute has worked also for over twenty years on the problem of agriculture, to develop an agricultural system with the ecological stability of the prairie and a grain yield comparable to that from annual crops in a modern, fuel-subsidized field. The Institute collaborates with public institutions in order to direct more research on Natural Systems Agriculture.

Because this work deals with basic biological questions and principles, the implications are applicable worldwide. If Natural Systems Agriculture were fully adopted, it might be possible to witness the end to agronomic methods and technologies from fossil fuel-intensive infrastructures in developing nations.

Natural Systems Agriculture is a new paradigm for food production, where nature is mimicked rather than subdued and ignored. Located in a native U.S. prairie, the Institute looks to the prairie as a model for grain crops. They are investigating the feasibility of perennial polycultures or mixtures of perennial grains.

The functions of a natural system can be achieved by mimicking its structure. With additional research, the Institute is working to show that an agriculture that is resilient, and therefore productive over the long term, economical, such that the need for costly inputs would be significantly diminished, and ecologically responsible is well within reach. The impetus to search for a new agriculture is soil loss and soil pollution. Agricultural chemicals poison soils and waters, which harms insects, birds, and mammals, including human beings. In the United States, after two 200 years of industrial agriculture, about quarter to a third of topsoil is gone. Natural Systems Agriculture would leave the ground unplowed for years and use few or no chemicals, solving many environmental problems at their root.

4.2.2.1.1.3.5. Refitting Agriculture

Food output per acre has been rather stable throughout history, but it is increasing rapidly now, with the use of industrial measures, like fertilizer, large capital outlay for new equipment, and new technology. It may be feasible to irrigate large parts of the desert, if certain problems, such as salinization, could be solved. The per-acre yield, however, may be reaching an asymptote. Humanity is fiercely competitive and exterminates any species that robs fields. There is no real balance of nature on farms—too much is artificial in the cultivation and irrigation, which requires constant attention and control.

The agricultural revolution that began in England in the 1700s grew up with science. The other half of the revolution was capitalism. The two have not quite been compatible. They have different logics and different goals. Many times capitalism forces environmental destruction, due to the limits of the system. The gamble is to win big or lose big.

Conscious learning has reformed natural ecological organization into the conscious and elaborate design of agriculture, according to Jantsch. Humanity entered a lower order of nutrition with invention of agriculture.

It may be possible to manipulate the entire biocenotic environment to prevent pest problems. But it cannot be manipulated meaningfully unless agriculture is decentralized. But, decentralized agriculture is no solution unless efficiency is greatly increased. In the

U.S. alone, there is 21 million tons of leaf waste from carrot and bean tops, bean and potato plants; that protein could be extracted for food. Although it may not be very palatable, it costs less than one-fifth as much as soybeans in India. Breeding is unlikely to eliminate all waste. In a few leafy vegetables, such as spinach, everything above ground can be eaten. Some plants, like potatoes, in which leaf and root growth are simultaneous, out yield others, like sweet potatoes, where tuber growth is only starting when leaves are failing.

Balanced land use means management based on sustained yield over a long term, inclusive of forests and marshes, also. The future depends on ecologically scaled projects, not on giant industrial land or water projects, and on the separation of large-scale industrial management from optimum-sized applications.

Since humanity is still part of the ecology, its actions can be examined in that context. New human niches have depended on this savings of capital of the earth being withdrawn for a temporary gain. Humans should not take the Net Primary Productivity (NPP) of a system, but the Net Community Productivity (NCP) or a smaller yield—that fraction which it is feasible to remove and use without destroying the basis of productivity. For grain or forest products, this portion is about 30% of the NPP (as dry mass per unit area and time). Higher efficiencies might be possible under very favorable circumstances, but the best calculations will have some uncertainty and lower yields should apply to most species. For wild animals, the kill must be lower than the trophic-level efficiency—usually less than 10% or 1% with some birds and fish.

There are things we can do to make agriculture sustainable and successful:

1. Diversify crops, especially adding drought resistant varieties (the grassland, for instance, has drought conditions every summer, worse periodically). Grow wild lupine to combine with soybeans for food. Convert from intensive cattle raising to antelope farming. Grow more varieties of apples (reduce imports of apples from outside the area). Increase self-reliance on most foods (except those that cannot be grown here—kiwi, oranges).
2. Develop new products from existing crops, e.g., oils, drugs, and fuels. Rape seed, for instance, is used in hydraulic fluids, plastic film, nylon, Lorenzo's Oil, as well as vegetable oil; certainly, there are unforeseen applications.
3. Use appropriate technology (solar power, field drilling, organic growing); import less energy, possibly export some. Stress low-input agriculture; low-fertilizer and low-pesticide may result in lower gross sales per area but in higher net revenue per area.
4. Process the crop (sell noodles and noodle products as well as wheat).
5. Market the crops and products ourselves; package them, advertise them, and distribute them.
6. Form cooperatives, especially for specialized market or low volume products (beets for instance).
7. Create a land trust to protect farm land from pressures of development, even as its market value as a nonfarm increases. A land trust could funded by property transfer tax. The land trust could lease the development rights on farmland.

We depend on vegetation for more than food: it makes up the content of much of our newsprint, construction, furniture, clothing, packaging. Of course, much of the straw and slash should be left so that the system can regenerate. Furthermore, with shortages of minerals, many substitutes are expected to be organic. But, our monocrops are directed only to one market.

4.2.2.1.1.3.5.1. Ocean Catches

Estimates of potential catches in the ocean are characterized by irrational guesses. Almost all good species have been overkilled. Less desirable species are taken in greater quantities for animal feed and fertilizer; over 36% of the world's catch goes for animal feed. For aquatic carnivores, the efficiencies in relation to NPP must range from 1% to 0.1% and 0.01% for secondary and tertiary carnivores. Whether any whales or dolphins, or elephants for that matter, should be taken for food is an ethical matter that must be decided by each cultural group. Only small fractions can be taken on a sustained basis. And that fraction is the surplus over the number of individuals necessary to maintain the reproduction of the population. Furthermore, the young, old and sick should comprise the bulk of the take, and not the strongest and healthiest.

4.2.2.1.1.3.5.2. Animal Use & Pets

Many original ecosystems supported good numbers of mammals, from mice and coyotes to deer, antelope, and elk. To some extent they have been replaced by domestic species: Dairy and beef cattle, swine, sheep, and poultry. Livestock often outnumbers the human population. Over 60 percent of the cropland production is devoted to feeding them, and over 30 percent of raw materials to housing and transporting them (for example, it takes 5 to 20 calories of fuel to produce 1 calorie of meat).

Although many animals, such as cattle and sheep are raised on ranges, they often spend months in feedlots being fattened with grain for human consumption. About 95 percent of this food goes for respiration or ends up as manure. The 95 percent loss is acceptable when an animal is raised on rough ground or when native populations, antelope for example, are used for food.

Harvesting some wild animals may be a better alternative than agriculture; this would be cheaper than improving the pasture degraded from overuse. Wild species might be more appropriate on marginal soils. The husbandry system whose forms underlie the foundations of modern thought, excludes wild nature as chaotic and other. We commit biocide on the wrong (or not our) animals. The living world faces a massive failure of interspecies dynamics, with great destruction of life and devastating psychological effects.

Pets, as domestic animals and a few wild but caged birds, reptiles, and mammals, need to be considered for fitness into nature. Because of pets, it could be argued, we are less concerned with wild animals and allowing them to slip into extinction. Then, too, we maintain our pets out of context; they are fed food from cans, kept indoors or in cages, and taught to defecate neatly. Zoos, aquaria and aviaries keep animals alive out of context, out of the conditions that shaped them. Denied the possibility of biological meaning, the animals may go mad. We will never understand why they are the way they are if we only study them in our homes or their prisons.

Aldo Leopold's chief concern was the need to reestablish the personal coexisting relation with nature, rather than large-scale impersonal management of resources by a professional elite. What the preservationist wants is to preserve natural cycles, not a frozen state of habitat like scenes from a Disney movie. As civilization staggers, we can find an ecology of the psychological environment that can be trusted to find balance and compensation; it is part of life-support system.

4.2.2.1.2. Humanity as Agents of Change Forces of History

What are the forces of history? Certainly, we should include geological processes, as well as solar system effects, such as the output of the sun or meteorites. Certainly, we should include the environment, such as climate and the distribution of resources. Especially important are human impacts on ecosystems, such as deforestation, identified by John Perlin as very important, perhaps the reason for the decline of the Hittites and Babylonians. Another impact is desertification, according to Uwe George and others. Disease patterns, according to W. H. McNeill, are crucial. To disease, or germs, Jared Diamond adds steel and guns as forces that have shaped human history and societies. Then, there is simply luck, the position of a culture in the stochastic chaos of nature.

By their activities, human beings change the places they live. Much of the change is easily incorporated in the cycles of renewability of the ecosystems. However, humans often change the directions of such systems by simplifying or degrading the systems. In this case humans act as agents of interference.

4.2.2.1.2.1. Humanity as Biological Agents

Humans have had a great impact on nature, and should be considered themselves as a force of nature and history. One could consider humans as special agents. One analogy of humans as special agents is as a parasite: A consumer feeding on another living organism, usually inside, drawing nourishment and weakening the host. States acted like macroparasites, according to William McNeill, but becoming less violent or unpredictable over time, as they adjusted to their host populations.

As crowding increases competition, there is more pressure on remaining reserves. The system parasitizes humanity and nature. Humanity becomes an autoparasite, a new pseudo-species. Technology enlarges the number of niches for us; tools fit humans to different habitats, displacing other species. We steal from animals and plants, from the earth, from our own descendants. Hobbes foresaw this war of each against all. The systematic destruction of human beings and animals is not an isolated peculiarity. A fast parasite often kills it host and then dies itself. Perhaps, humanity is an agent of a different sort, a systems agent that encourages only positive feedback.

Perhaps human expansion is like a cancer. Alan Gregg (1955) compared the world to a living organism and the explosion in human numbers to the proliferation of cancer cells. He sketched other parallels between cancer in humans and humans' cancer-like impact on the world. Cancer cells proliferate rapidly and uncontrollably in the body; humans continue to proliferate rapidly and uncontrollably in the world. Crowded cancer cells harden into tumors; humans crowd into hardened cities. Cancer cells infiltrate and destroy adjacent normal tissues; urban sprawl devours normal open land. Malignant tumors shed cells that migrate to distant parts of the body and set up secondary tumors; humans have colonized just about every habitable part of the globe. Cancer cells lose their natural appearance and distinctive functions; humans homogenize diverse natural ecosystems into artificial monocultures. Malignant tumors excrete enzymes and other chemicals that adversely affect remote parts of the body; humans' motor vehicles, power plants, factories and farms emit toxins that pollute environments far from the point of origin.

It is not in a tumor's self-interest to steal nutrients to the point where the host starves to death, for this kills the tumor as well. Yet tumors commonly continue growing while the victim wastes away. A malignant tumor usually goes undetected until the number of cells in it has doubled at least thirty times from a single cell. The number of humans on Earth has already doubled thirty two times, reaching that mark in 1978 when world population passed 4.3 billion. It is over six and a half billion now. After thirty-seven to forty doublings, at which point a tumor weighs about one kilogram, the condition is usually fatal—that would be the population equivalent of 5.4 billion people. We have exceeded that; the question is if it has been fatal—large complex systems may take a long time to collapse—or if the system has more flexibility than an organic body.

The metaphor of cancer may be more appropriate than a footprint. After all, a footprint can stimulate some kinds of ecosystems, such as shortgrass prairie. What humanity does is transform the ground under the footprint into a new system.

4.2.2.1.2.2. Humans as Ecological Agents

Every species exploits its environment to the extent that it can, with no regard to consequences. Usually, each species is checked by other, because there are so many competing for the same food, and an equilibrium is maintained.

Partial knowledge and technology has allowed us to exploit our environment beyond what is desirable for us or for other species. While continued, moderate exploitation is necessary to live, massive, unbalanced exploitation is unwise. A wise use of resources would not make the world less habitable. We are part of the system and must protect its health as a whole.

Basic principles need to be examined in relation to human ecology. Populations are maintained at optimum levels demanded by the ecosystem; this maintenance is achieved by the production of excess young and the elimination of the weaker, the least fit to survive; mating and parenthood are denied to the young, by instinct in most species, until a territory sufficient to raise, feed and protect offspring has been acquired. If the population becomes excessive, glandular conditions are activated to induce stress and complications, which reduce the population.

By distorting the equilibrium, we have destroyed whole species and favored many others, many wild as well as domesticated. Rats and mice have been carried to all parts of the world and live in direct competition with humanity, invading our buildings for food and shelter. Crows and coyotes have also profited from their human association.

A dominant is a species with greater influence than any other in its biotic community, changing the lives of other species and the character of the habitat. Humanity is a pandominant species. As such, humanity reclaims, overgrazes, clears, depletes, and wastes at a level that threatens the stability and existence of many systems. One of the ecological consequences of human activity is the degradation of wild habitats for human developments and the introduction of novel elements into the biosphere—elements that have not been harmoniously worked in over time. The biomass, or demomass, of the human species probably far exceeds the biomass of any nondomestic species, and that biomass is supplemented by the tremendous biomass of domestic animals, which is four times greater. This biomass forms an equivalent population that consumes much of the same food, such as milk, fish, and grain.

This pandominance has major effects on ecosystems: Transient perturbations in energy relations, from oil spills or burning, for instance; chronic changes and shifts of systems, from dams, irrigation, or chemical wastes; species manipulation, from the import and export of exotics; and, interference competition with wild species, as opposed to exploitative competition, which can be stabilizing.

4.2.2.1.2.3. Humanity as a Geological or Climactic Force
No single change is exclusive to humans as a species, but they are excessive, rapid, compounded, and large-scale. There is movement of soil, but also massive erosion. There is movement of minerals, but also disruption of mineral cycles. There is the addition of novel elements into the atmosphere, but there is also a massive release of carbon.

When people use more of the earth's supplies in a certain period than can be replenished in the same period by the sun, they are eating into the natural capital. Humans have caused the extinction of hundreds of species. Rhinoceros, buffalo and crocodiles are disappearing. Getting timber for fuel and construction, clearing land for agriculture, has destroyed whole habitats. Demand for timber has been insatiable, for houses, ships, paper, and fuel. Trees have been cut from vulnerable watershed sites, with resulting floods, erosion, and diminution of rainfall and water table. Domestic animals inimical to growth were introduced, such as rabbits and goats. The introduction and maintenance of sheep in Spain, Italy and Cyprus has changed whole ecotypes.

Sixteenth century Spain, under Philip II, experienced unprecedented growth; shipyards flourished; gold was brought from South America; wars were fought. But, every year it cost thousands of hectares of centuries-old forests to build and maintain the fleets. By the end of the seventeenth century, the rich Iberian forests had disappeared. After the soils eroded away, there were famines.

Vegetation holds soil in place, reduces wind speed at the soil surface, and improves water absorption and transport in the soil. Erosion destroys soil and makes it difficult for plants to be reestablished. Recovery, if it occurs, may take decades. Erosion is an ecological catastrophe on a planetary scale, causing thousands of higher plant and animal species, and countless lower species, to be lost forever. This is what is planned in Brazil and South America. It is not known how massive deforestation through overgrazing, firewood collection, and timber exploitation will effect terrestrial and atmospheric systems. Perhaps we should just accept erosion. Erosion is picturesque. Cezanne's paintings of France are striking. The abstract terrain of Greece is pleasing to many.

Israeli scientists doubted the theory that Arab methods of cultivation were responsible for spread of desert in North Africa and Middle East; desiccation on that scale is beyond even modern technology. They concluded that deserts are usually created by a relatively small change in climate. However, grazing practices could cause a small change in climate and thus contribute to desertification.

Only in the nineteenth century beginning with G. P. Marsh, did people start to realize that humanity has done as much to change the environment as the environment has done to mold human history. Marsh, the first U.S. ambassador to Italy, was one of the first to study the role of humans in changing the face of the earth. When he visited the near east, he was shocked to find deserted cities, silted harbors and wastelands instead of flourishing civilizations. He concluded that ecological errors had led to the deterioration of agriculture in Mediterranean countries. He advocated agricultural conservation practices.

Environmental factors have shaped the course of human history to a greater extent than had been realized. The decline of Rome is a study in forest ecology. There were previous and later catastrophes in the Tigris and Euphrates valley, Greece, Khmer, Maya, Midwest

United States, and the Australian outback. Many people did not change their behavior in time to solve the problems. Worse, the current civilization is global, not local; so, there will not be a migration to unaffected lands. Ecologists have not unraveled most mysteries of ecosystems, so the long-term consequences of most human interaction can be predicted generally.

4.2.2.2. *Human Culture*

At first humans were considered special because of their divine spark. Then, they were special because they are disconnected from nature by culture, a magical reification of the opposite of nature, a culture that we can produce at will to overcome the constraints of nature. It has been argued that culture let us push out of the constraints that limit chimps and apes. Culture was once considered an attribute exclusively in the human domain.

If one defines, as Kinji does, culture generally as a form of behavioral transmission that does not rely on genetics, then animals, birds, fish, and perhaps some insects can be said to have cultures. This definition is not limited by the kinds of mechanisms, such as copying or imitation, and it also includes the continuum from guppies and apes to rich human culture, which need not be devalued by being part of the continuum with animals. Value systems and technical achievements are sometimes the results.

4.2.2.2.1. Copying & Animal Culture

Can culture be defined as any kind of imitative behavior in animals? Culture can be defined as the transmission of information through behavior; imitation is one behavior, teaching is another. Culture, with its new way of transmitting information, involves a mix of: (1) trial and error learning, (2) social learning through observation and imitation, and finally (3) teaching. Culture is one way of transmitting information. Other ways are genetic coding and individual learning, although individual learning is difficult and easily lost without a context.

Some birds learn through imitation. For the bower bird, the bower is an extension of its plumage and size, for courtship displays, but the extension changes faster than the biology. Evolution speeds up. When the environment supplies tools, new interactions and transactions occur. It is like an external bundle of secondary characteristics according to Edward Hall, quoting E. Thomas Gillard. Bowerbirds no longer pair off to mate nest and raise young. Males gather in clans and get ranked hierarchically. In spring, clans form around arenas for display. Each display court is called a lek. Females mate with the dominant males, those with the most baubles. How does female mate choice change over time?

4.2.2.2.2. Imitation & Human Culture

Human culture, and human behavior, need to be defined against a long evolutionary backdrop of deep history, the common ground we share with other animals. Perhaps culture is transmitted through imitation.

4.2.2.2.2.1. What is Culture?

What is the world like? How did the world get the way it is? And what is the role of humanity in the world? All cultures ask and answer these questions. Some of the questions could not be answered from direct observation. And, many of the answers are not limited to observable events. Ideas concerning humanity and the nature of the universe tend to form a coherent system in which ideas are integrated or rejected over days or centuries. Culture includes all of the expectations, understandings, beliefs, and commitments that influence the behavior of human groups.

Culture exists in minds, signs, and things, but most importantly in places. The word culture, from the Middle English, meant 'place tilled,' from the Latin *colere* meaning to 'till, care for, inhabit, worship.' For the Romans and English, to have a culture was to inhabit a place and cultivate it, to be responsible for it.

4.2.2.2.2.2. Definitions of Culture

The classical definition was put forward by Sir E.B. Tylor (1871): "Culture … is that complex whole which includes knowledge, belief, art, morals, law, custom, and any other capabilities and habits acquired by man as a member of society." Although the definition seemed to limit culture to 'man' alone, it definitely restricted it to human behaviors and artifacts.

4.2.2.2.2.2.1. Traditional Definitions

Others, including Claude Levi-Strauss and Leslie White argued that culture required imitation, teaching, and language, and therefore the concept could not apply to other species. In 1952, A. L. Kroeber and Clyde Kluckhohn identified 164 definitions of culture. Most of these definitions are refinements of the Tylor definition. Robert Boyd and Peter Richardson defined culture as 'information capable of affecting an individual's own traits; they acquire the information from other individuals through imitation.' Information is a broad enough concept, especially as in*form*ation, to include behaviors and artifacts, and the vehicle is identified as imitation.

Peter Berger defines culture as the totality of human products. Culture is everything created as a group, tribe or nation, physical or ideal, in the past or present. This embraces cookware, arrows, steam engines; artworks, books, legal codes; symbols, values, social structures. A cultural system surrounds the network of human interactions with raw materials, forms of life and other humans.

A general definition of culture identifies it as the ensemble of values, worldview, aspirations, customs, technologies, techniques and physical artifacts that characterize a people and distinguish them from others. Mary Douglas refined the definition by studying the culture of everyday life, from food and dirt to jokes, bodies, and speech. Dirt, which is a form of pollution, reveals a lot about the system and rules of classification. Understanding what permits things to be called dirty or clean reveals the moral order itself. To see how individuals are controlled by society, Douglas creates two distinctions, group and grid, which she uses to categorize cultures. Group is the outer boundary between people and the world. Grid is the set of social distinctions and delegation of authority that limits how people behave towards each other. This creates a table relating the strengths and weaknesses of cultures. For instance, traditional societies have strong group, strong grid. Some modern cultures have a weal group but a strong grid.

4.2.2.2.2.2.2. Metaphors of Culture

The use of metaphors might contribute something to the understanding of culture. Culture is an 'organism' that grows and matures, that is, it is organic and whole, with a finite life span. It processes energy and matter to survive. But, it does not have a skin or genetic limits—nor does it have a genetic way of passing on its characteristics, although it has the memory of its members, and they can communicate.

Perhaps, culture is an 'ecosystem' that is self-maintaining, stable and changing. But, that neglects an ideational component, as did 'organism.' Like systems, cultures develop over time; they are capable of evolving. There are similarities in their evolution: Similar dynamics and machinery, and a direction towards complexification. Evolution, biological or cultural, depends on two things: The rate at which useful changes arise and the rate that they spread in the communities. But, this neglects internal states and forms of ideas.

It could be said that culture is a 'city,' converting nature into artifact, expanding into new areas, such as suburbs, park zones, and industries, as the physical extension and expression of human intentions. But, a city is another kind of ecosystem, and the metaphor still does not encompass ideas.

Culture is a pair of 'glasses' that allow us to see some details or patterns; that is, it is an add-on to the human animal that focuses and shades perception to permit survival. This metaphor implies that many do not need culture or that culture can be removed or replaced at will. Feral children, beings without culture, have difficulty surviving; removing culture, even if possible, would reduce people to a feral level. Another culture can be grafted onto a first culture, as people immigrate into a second culture, although that can be a long, incomplete process.

Culture is a 'process.' A nation is a process. Culture is slower than ideology or policy, which may be why we have larger nations. Culture is slower than economic conditions or environmental changes. It is possible that urbanization distorts a culture.

Culture is a 'hologram.' It is a whole that arises when the reference beam of nature shines through the mental patterns of a human group. This metaphor could capture the interior and exterior aspects of culture. The place of the reference beam would color and shape the culture differently. The culture could develop, however, only with decay over time, not through a transformation. These metaphors, although interesting, have limitations. A broader attempt at definition might be more fruitful.

4.2.2.2.2.2.3. Synthetic Definition of Culture

Culture stretches vertically to include the physical, economic and political. It stretches horizontally to include society as a whole. In fact, culture is concerned with all things and beings. It is organic, like Aristotle considered a work of art, and whole. So everything in it is interrelated in some degree. Many relationships are encompassed by the holistic perspective of culture: The relation of people to themselves, to each other, to objects they create, and to their natural environment, and to their cultural environment. These bear on psychological well-being, social bonds, material legacy, and on the association with other forms of life.

These many definitions overlap and can be combined in to a synthetic definition, with common characteristics: Culture is a system of shared beliefs, values, customs, behaviors, and artifacts that members of society use to adapt to the environment and to adjust to each other, and that is transmitted to succeeding generations through learning behavior and language, so that culture and the environment constrain and construct each other over time. This definition might be refined later.

4.2.2.2.2.4. Characteristics of Culture

Culture is imbibed through a process of social interaction. People acquire culture unconsciously through social interaction, as well as consciously, through apprenticeship or formal learning. The young, or new members of a culture, observe others and imitate that behavior. Cultural models are internalized by individuals, so that part of culture resides in the mental sphere of people. Beliefs and knowledge are not shared equally by all individuals, thus culture is shared differentially.

Humans use conceptual devices, such as symbols, to communicate abstract ideas about nature or society to one another. Through our linguistic capacity, we can use symbols as meaningful representations of reality. Public shared meanings provide a set of designs that allow an educated individual to survive nature and society. The understanding and practice of culture is shared in a culture.

4.2.2.2.2.4.1. Conduct as a Characteristic of Culture

Conduct is the course of cultural behaviors through a behavioral landscape or field. New mathematical treatments of fields tend to be three-dimensional. Rene Thom's catastrophe theory and Conrad Waddington's epigenetic landscape are two such theories. The epigenetic landscape field can be used to explain why chickens on one side of a fence cannot get to the food on the other side, even though they could walk two meters to get around the edge of the fence. The chicken's need path or chreod is deeper than the path of a cognitive chreod around the fence. So, the chicken can only go toward the food. If the need chreod is too deep, the chicken cannot explore with a cognitive chreod, although a less hungry chickens might be able to.

In humans, this explains why necessity cannot be the mother of invention; the necessity path, starvation for instance, is too deep to allow exploration of a shallower path of daydreaming or design. The broader landscape of leisure is needed. Conduct can be described as the stable path of culture through a landscape of possibilities. Once the course has been set, it is most likely to channel subsequent behavior, unless some event triggers a deeper course.

4.2.2.2.2.4.2. Wholeness as a Characteristic of Culture

Culture provides an identity for its members. It tells them who they are, where they came from, and why they are special. Identity is basic to human existence. People are identified by their roles. A person is an incarnation of his and her group—even in industrial culture, one is identified as an astronomer or farmer.

Identity is that persistent quality that serves nothing; it is. Identity can be described apart from its performance in interactions, but not isolated. The relationship between identity and wholeness is a rhythm, with unique patterns. In fact, culture is concerned with all things and beings in a whole.

4.2.2.2.2.4.3. Flexibility as a Characteristic of Culture

To paraphrase Arthur Koestler, culture is a holon, a stable subwhole in a hierarchy that displays structural gestalt constancy and rule-governed behavior. The rules lend order and stability to the whole, as well as flexibility. Flexibility means not being over connected, or not being too rigid or efficient. Than means that culture is able to slough off people or community structures and to incorporate new people and structures. Culture is able to keep some options and unused connections open. Some of the flexibility comes from different ways of establishing connections in specific places.

Culture is bounded, but open to flows of energy and materials. In fact, it requires a steady input and output exchange flow. It requires order, but not too much order. It is a loose-fitting patchwork of ideas, relationships and things. It is tolerant of discontinuities and contradictions, and this gives it flexibility. Humans can tolerate inconsistency and operate with contradictory beliefs: Soldiers fight for peace; ministers save the unborn for starvation. If the contradictions become too great and maladaptive, however, then the culture can collapse. Cultures that become too isolated often stagnate, then collapse, even if they do reconnect with other cultures.

4.2.2.2.2.4.4. Adaptation as a Characteristic of Culture

The patchiness of culture is parallel to the co-constrained construction of a species and its environment. Culture has to balance between embracing change and resisting change. People show a desire for new things, but often fear and resist change. According to most theories of cultural adaptation (or integration or evolution), resistance to change is normal as a cultural process. Groups like the pygmies have specialized to fit the requirements of the environment, successfully. This makes it difficult to adopt other cultural arrangements. On the other hand, resistance to change itself is an adaptive mechanism. According to Betty Meggers, it works as a successful "cultural isolating mechanism." Isolation remember is what allows a culture to develop in the first place. But, then can it force a culture to become stagnant?

A primary culture is adaptive because it aids survival in the ultrahuman world. Its ways of living are sophisticated survival mechanisms. Each way of life is a set of adaptations to the limits of the environment. Primary thought patterns are highly disciplined intellectual structures that make the world coherent and meaningful. Many rituals of the tropical American Indian tribes are concerned with ecological balance, though not necessarily self-consciously. Co-constrained construction enforces coevolution, the emergence of a highly ordered complexity to full structuration. Polyandry in Tibet may be an adaptation to limited resources. With monogamy and every woman married, there would be more children to strain the resources available.

4.2.2.2.2.4.5. Constancy as a Characteristic of Culture

The cultural system is stable and persistent in time. It is a general property of some systems that acquired information is used to close the door to further inflow. A mature culture needs less information, since it works toward preservation, and closes itself off to information that does not fit the shape of the culture. The effect of maturity is to allow a maximum variability between systems with slight external differences, such as place or initial conditions.

Stability is a characteristic of cultures. This can even be stated in Newtonian terms: A culture at rest tends to remain at rest. According to Harding, when forced to act on changes, a culture will only accommodate those changes that preserve its fundamental character. Stability is the ability to maintain identity under the flow of external forces and disturbances. A culture has to be able to resist disturbances that are too disruptive. Resistance is a positive act for a self-reliant culture. Culture also has to be resilient enough to recover from intermediate and small disturbances.

Stability can also be related to compartmentalization, communications, richness of interactions, and connections. With order and integration come stability and security, without which no one can survive. When human societies were small, the amount of control and security required was small. Although societies have grown, human security has not. Primordial security comes from a physical, knowledgeable relationship with nature.

4.2.2.2.2.4.6. Vitality as a Characteristic of Culture

To be constant and stable, a culture has to be vital; it has to be productive, to be able to convert energy and materials into foods and structures for survival. The system is self-creating. It renews itself as its contents change, as disturbances change the parameters of the system. What barriers are there to cultural renewal? How can they be overcome?

4.2.2.2.2.5. Functions of Culture

Culture has to deal with the facts of life, from change to death, so that people can survive in their communities in their places. Freeman Dyson distinguishes three crucial biological inventions in the transition from unicelled organisms to multicelled: Death, which allowed differentiation of the future from the past; Sex, which enabled the characteristics to be shared and mixed; and Speciation, which increased diversity through separation and genetic barriers. Life can experiment with the diversity of forms and functions.

Dyson notes that each biological invention has its analog as a cultural invention inhuman societies: For death it is tragedy; the fact of death is made a theme in ritual and drama. Great cultures have distilled great works from the fact of death. For sex, it is romance, where sex is turned into a thing of mystery and beauty, in dance and poetry. Speciation has been transformed into cultures and languages at the human level. The flexibility of social institutions grew out of their differences and places.

4.2.2.2.2.5.1. Culture Grounds

A culture orders a whole cosmos. A culture selects what is important from the undifferentiated phenomena of nature and presents it in myths and stories to be learned by people. People in a primary culture share a common image of their world—from the German word meaning 'man-image.' The image is a construct of human knowledge that reflects human awareness of a local environment. The image is constructed metaphorically, but treated 'as if' it were true. A traditional way of living evolves with people's experience and knowledge. The image guides their behavior.

Grounding is more than the expression of a local place through culture. It is also the expression of individual behaviors through culture. Culture can be observed through individual actions, but culture exists as a pattern that allows that emergence. Individual and collective entities have to be understood and balanced. Both are important parts of a whole system. Culture limits and constrains individual behavior with its conventions, but individual behavior can extend culture, by acting and communicating.

4.2.2.2.2.5.2. Culture Orders

A culture orders a whole cosmos (from the Greek word meaning 'to set in order') and applied to the human face as well as to all of nature, and what was beautiful was also morally admirable. A culture selects what is important from the undifferentiated phenomena of nature. What is important is often beautiful. Every culture strives for beauty that goes beyond utilitarian values. People structure their worlds with their own group at the center.

With order and integration come stability and security, without which no one can survive. When human societies were small, the amount of control and security required was small. Although societies have grown, human security has not. Primordial security comes from a physical, knowledgeable relationship with nature. The industrial sense of security has been centered in the state, and now in the corporation, in retirement, insurance, housing, recreation, and even in credit cards. Primary peoples were responsible for their own food clothing and housing. The economic well-being of modern industrial families depends on wages alone. People are dependent on the goodwill of capital corporations for personal security. We have substituted insulation and insurance for knowledge and experience.

4.2.2.2.2.5.3. Culture Explains

Culture explains the universe. It also explains the behavior of its adherents. By explaining reality, culture binds the human and ambihuman, and the past and the present, into a meaningful whole. Culture explains why traditions are necessary. It explains why things are as they are, and how they came to be that way. It can do this in stories, and sometimes stories need a special language to explain, such as Wataluma or Latin. In modern culture, a scientific theory is a statement that explains some aspect of the world, but allows questions.

Cultures occupy a particular territory. This is especially true of the Campa, in a tropical forest in Peru, and the Ituri pygmy, in a tropical rainforest in Africa. The features of their cultures are unique to their place. They literally could not live with images of desert or ocean, like the Taureg of the Sahara or Samoans of the Pacific. Regardless of the features of a place, myths are created to give it special significance; giving mythical significance to a place strengthens a people's identification with it. People identify with their place and often equate their own characteristics with it; the Ituri consider themselves as bountiful as their forest, while the Mongols are as undeniable as the wind from their plateau.

Knowledge allows survival in fragile habitats. The !Kung San of the Kalahari desert exist in small communities that enable them to continue traditional hunting and gathering without depleting their resources; they hunt eighty types of animals. The Hanunoo of the Philippines distinguish 1600 plant species, where scientists only know of 1200 in the same area.

4.2.2.2.2.5.4. Culture Integrates

Culture provides a filter between humans and environments. So, culture can serve as another evolutionary filter. It designates what we attend or ignore. It has to screen less valuable information to avoid information overload, even in archaic times. In a way, culture is like a trigger. It allows less information to activate the system. And this is the only way to increase information handling, when the system is larger and more complex—of course, this is what stereotypes and metaphors do. Culture programs the individual or institution.

Jurgen Habermas argues that the task of culture is to understand the meanings attributed to objects and events by individuals under concrete circumstances. He identifies culture as a set of subjective meanings held by individuals about themselves and their world.

Culture provides an identity for its members. It tells them who they are, where they came from, and why they are special. Identity is basic to human existence. People are identified by their roles. A person is an incarnation of his and her group—even in industrial culture, one is identified as an astronomer or farmer.

A primary culture is adaptive because it aids survival in the ultrahuman world. Its ways of living are sophisticated survival mechanisms. Each way of life is a set of adaptations to the limits of the environment. Primary thought patterns are highly disciplined intellectual structures that make the world coherent and meaningful. Many rituals of the tropical American Indian tribes are concerned with ecological balance, at low levels of exploitation and trade, and not necessarily self-consciously.

4.2.2.2.2.5.5. Culture Justifies

Another function of a culture is to justify human activities, in order to have those activities continue. Unless an activity satisfies a basic human need, it may not be repeated. The needs may be physical, psychological, or social, such as acquiring status. In ancient China, people found justification of their world view in the matrix of nature. Perceived cruelty in nature justified real cruelty by human beings.

A common culture provides an ideal framework for public and private decision making. The Sami in northern Scandinavia have institutionalized ways of avoiding conflict, for instance, by shaming those who would impose their will. People in a culture may enlist ideology or religion to justify transfer of wealth to the rich and the leader. The shared religion also makes strangers act more peacefully without kinship. It gives people a reason to sacrifice their lives for an institution, nongenetic.

4.2.2.2.2.5.6. Culture Controls

Some human behavior is controlled and population regulated through the use of space. Most foraging populations, furthermore, regulate their density well below the limits of the food supply.

Culture controls many behaviors, from hunting to birthing. Women control sex, birth, weaning, and often population size. For Australian aborigines, women are regarded as controlling social harmony, health and connections to land, while men, with more time and strength, tend to control creative activities and hunting for prestige foods.

Cultures have been self sufficient for thousands of years. Although some of them fail, others last for quite a long time. The Desana, for example, have existed in the Amazon for over four thousand years by maintaining an equilibrium with the environment. Ecosystems are local, not global. Although we regard communities as being tied together globally, each community is alone. Each culture has to be responsible for its own welfare. Preserving the local environment is one requirement.

A culture is a way of doing things by a unique people with a unique identity, with a unique history, in a unique environment, communicating in unique ways, and using materials in unique ways.

4.2.2.2.3. Effectiveness of Culture

There are similarities of concepts between culture and ecosystem. Culture is a sloppy concept, like an ecosystem. Culture is also scalable, like an ecosystem, and can be nested. Characteristics of a culture resemble those of any system, such as an ecosystem. They include: Identity, openness, productivity, co-constrained construction, and vitality.

As an organism, a culture has needs for it to continue. It needs: To be grounded in place, to be secure or partly isolated; To have a dynamic order, human health; To be complex and sophisticated, with checks and balances; To be comprehensive, to allow change and diversity; and, to have and manage adequate resources.

As an organism in an environment, a culture engages in activities and interrelations, such as change and development. It can compete with or take-over other cultures. It can cooperate and trade. There will always be some clash or misunderstanding, from having different images and metaphors, limits and rules of behavior. The strengths of a culture may allow it to persist for a long time; its weaknesses, though, ensure that it will eventually fall apart.

4.2.2.2.3.1. Strengths of Culture

A culture has social and ecological functions. These things are the strengths of culture, when they work. The image a people have of their world makes sense of the overwhelming confusion of nature; it gives people a unique identity and justifies their behavior. Culture ties them to a place, whose qualities are known and preserved.

4.2.2.2.3.2. Weaknesses of Culture

Culture does not fit together into a perfectly integrated whole. A culture is a loose-fitting patchwork of ideas, relationships, and things. There are discontinuities and contradictions. The balance of freedom and necessity ensures that no culture will ever fit its environment or its members perfectly. Humans can tolerate inconsistency and operate with contradictory beliefs: Soldiers fight for peace or ministers save the unborn that they may starve. If the contradictions become too great and maladaptive, however, then the culture dies. No culture has developed a perfect balance of human and situational needs. Some do better than others. But, all cultures change and age. As a culture ages, it may become abstract, indifferent, self-centered, and forgetful, suffering rigid rituals and cultural amnesia.

4.2.2.2.3.2.1. Holding Arbitrary ideas

People sometimes construct their worlds from preconceived notions. Success in one area may become associated with a chance happening, an event that is repeated to continue the success. In this way, a maladaptive image of nature can be built. Some primary traditions may work against the conservation of a place; for instance, the Algonquian notion that game animals spontaneously regenerate after death means that there is no reason not to overhunt. Modern ideologies have even been shaped by the principle of endless wealth; the economist Adam Smith believed that the real price of anything was just the toil spent acquiring it.

4.2.2.2.3.2.2. Remaining Indifferent

Industrial cultures desacralize nature. Since the advent of the machine image, the concept of the sacred has been reversed. In the primary view, the familiar was sacred. When modern cultures made the familiar trivial, it became profane. The quality of sacredness was bestowed on the unknown, on wilderness or children. In industrial culture, all aspects of life become interchangeable artificial units, including soil, water, and land. This view impoverishes humans by claiming all consciousness for humanity. Detachment leads to a loss of the sacred. Many different peoples have deforested their lands and poisoned their waters, regardless of their religious ideals as Buddhists, Taoists, Moslems, or Christians. Industrial cultures are indifferent to the limits of a natural carrying capacity.

4.2.2.2.3.2.3. Overexploiting Nature

All peoples want some power over the natural order. Primary peoples rely on ritual acts instead of machinery. As technology supplies power to primary peoples, rituals decline. Power increases exploitation, disturbance and interference. Exploitation can become pathological, when it interferes with the natural processes that maintain an ecosystem. The intrinsic worth of beings can become supplanted by monetary value. For example, some North American peoples were seduced into the fur trade by the lure of manufactured materials. The spread of power has two other effects. The natural order becomes simplified, the human world becomes increasingly complex—and both orders become unstable. Applying culture beyond a certain scale, without adjusting to new ecological limits, gives rise to behaviors that are unecological and unsustainable.

4.2.2.2.3.2.4. Being Incomplete

The very circumstance that makes each culture unique—being in a unique place—ensures that each culture is limited. All cultures produce destruction and waste, all of them produce at least some of the opposite of the good intended. A culture rarely meshes perfectly with the natural order or even its own social order. Cultures rarely have long-range plans; they do not concern themselves with global problems. They rarely consider any cultures other than their immediate neighbors; they do not have policies to help them. They are rarely conscious of their activities.

4.2.2.2.3.2.5. Staying Inflexible

It was thought that cultures could vary infinitely and change rapidly. This is an exaggeration. Change is not always easy or adaptive. The inertia of cultural practices makes change painful. People may become fixed in permanent roles and personalities. Even if cultural attitudes are appropriate, they can trap a people if there are no longer functional reasons for the practices. The Nembi of Papua New Guinea may be trapped in their system; making stone axes is difficult, when thousands of steel ones are available, although the ritual of making axes can bond people. Cultures can determine inappropriate attitudes towards nature. The English treated tropical lands as enemies to be defeated, then enslaved them in plantations. Their cultural attitude as conqueror of nature led them to treat biogeochemical cycles and soil requirements as temporary obstacles in a world where everything had its price.

4.2.2.2.3.2.6. Keeping Exclusive

When the largest social unit was the tribe or nation, it was possible for the local mythology to represent other people outside its bounds as inferior, and the local inflection of human mythology as the one true mythology. The young were trained to respond positively to tribal members, to love their home, and to project hatred outward. But, there is no longer an outward.

4.2.2.2.3.2.7. Being Aggressive

Are humans innately aggressive, or does the nature of the culture of civilizations promote aggression? Cultures allow more aggression against people outside the home culture. The size of a local population increased the likelihood of its success. For cultures, size was important. More important cultures were larger and more aggressive. Aggression may be encourage to protect a culture trapped in low food outputs or scarce resources. Aggression would be important to protect a culture, but it can become a prime way of relating to other cultures, especially very different culture, or neighboring cultures that may hinder the expansion of the home culture. Cultures need not be limited to aggression. They can also divert aggression from violence and war into propaganda and art.

4.2.2.2.3.2.8. Ignoring Limits of Culture

A culture will often develop without concern for limits of complexity or scale. How large or small can a culture be? How simple or complex? There are human cultural limits in numbers. For instance, a minima might be genetic, at only 2000 people; but for ideomass, the minimum might be one million. A maximum, based on wilderness, might be ten million, or there might be a social maximum of forty million. Each limit must be worked out, depending on place and the structure of the culture, but they should not be ignored.

4.2.2.2.4. Diversity & Universals of Cultures

These functional requirements of culture are rather minimal for humans, being only food, clothing, shelter, and reproduction. Throughout their history, humans have used animals and plants for food and clothing. They have been able to convert animal and vegetable resources into all their needs for food, shelter, and clothing. Other requirements, such as respect, comfort, and self-fulfillment, depend on socio-cultural systems. But, the differences in local environments and cultures ensures a diversity of those things.

The origin of clothing, for warmth or prestige produced ecological changes in local animals. According to Mark Stoneking, body lice may have evolved from head lice after a new ecological niche, clothing in this case, opened for them. Using a molecular clock, he calculates that clothing appeared about 75,000 years ago in Africa. Perhaps it appeared first for status, but clothing allowed the later move to colder climates. It might be that the first clothes in Africa were constructed as status symbols. Much of the diversity of human clothing has to do with the diversity of localities. And, much has to do with the aesthetic sense and human invention.

Clothing is often the first thing noticed when cultures meet. In the eyes of another culture, clothing differences are often exaggerated. For instance, westerners seeing Africans or Polynesians for the first time focused on the absence of clothing or on body adornment. Naturally, Africans seeing westerners for the first time emphasized the elaborate clothing and facial hair.

Behavioral difference may be noticed after a longer acquaintance. Behaviors and symbols may take decades to understand. Many gross operations of a culture seem familiar and may be comforting in their familiarity. All cultures seem to have certain things in common. Of course, many of these are referred to as soft universals, since everyone needs shelter. Other differences may be the result of contacts between cultures, because cultures are mimetic, that is, they copy one another. This may be the secret behind agriculture, cities, and industrialization. Table 4.42224-1 presents some recognized universals.

Perhaps cultures seem to have universal characteristics because humans are of the same species, have similar requirements and similar bodies. All people live in groups, have some form of shelter, and have an incest taboo. Even though a process may be universal, the implementation and symbolization of it may be quite different and unique. Some scholars have argued that these universals could allow a global human culture. That may be true, although the relationship of a local culture to a global culture may also be problematic.

Cultural Universals (expanded after G. P. Murdock 1945) could include: age grading, faith healing, joking, population policy, athletics, family living, postnatal care, feasting, kin groups, pregnancy usages, body adornment, fire-use, property rites, folklore, propitiation ceremonies, calendar, food taboos, language, puberty customs, community organization, funeral rites, cooking, magic, religious rituals, cooperative labor, games, marriage, residence rules, cosmology, gestures, meal-times, courtship, gift-giving, medicines, sexual restrictions, greetings, modesty, soul conception, dancing, mourning, status differentiation, decorative art, hair styles, murder taboos, division of labor, hospitality, music / singing, tool-making, home/housing, mythology, trade, education, hunting taboos, numerals, ethics, hygiene, visiting, ethnobotany, obstetrics, war / conflict rules, etiquette, incest taboos, weaning, inheritance rules, personal names, and weather control.

4.2.2.2.5. Changes & Cultural Transformations

Cultures emerge from groups of human beings living in places. The variables of the environment influence cultures in a number of ways. For instance, A high Net Community Productivity (NCP) can allow larger annual crops, or, a low number of sunny days can contribute to psychological depression. Cultures have been unique programs, using local materials and ingenuity, for satisfying basic human material and spiritual needs.

Cultures develop over time as the groups change and always seem to grow larger. Cultures are influenced by climate, resources, and of course by human ideas about their places and themselves.

There may also be larger patterns of human culture. For instance, reproductive success and overshoot of resources may always occur in the development of a culture. The asymmetry of sex and violence, ecosystem conversion, and limited time horizons are also things that seem to develop.

As place changes, a culture changes. As people change, under the influence of each other and other cultures, a culture changes. Culture does not fit together into a perfectly integrated whole. A culture is a loose-fitting patchwork of ideas, relationships and things. In that sense it is parallel to species adaptation to an environment. There are discontinuities and contradictions.

William Thompson identifies six major shifts in the transformation of human cultures. The first he calls the feminization of primates; females abandoned estrus and became open sexually (200,000 YBP). In observing synchronicity between bodies and nature, women established system of symbols and notation (art 50,000 YBP). Then women discovered that they could collect enough cereals in three weeks to last the entire year, more than a hunter could kill; but it required storage. Agriculture gave a surplus; that and crop failure and excess property led to excitement of war (Agriculture 9000 YBP). Civilization became the domestication of women, according to Thompson. This emphasized a patriarchal structure (Civilization 5500 YBP). Industrialization (200 YBP) is really an intensification of civilization, still an ektropic process. In each case of cultural absorption, there was an attendant process of miniaturization. The forest was miniaturized in clumps of trees; animals were miniaturized in artistic image; time on a lunar tally stick; plants in a garden; women in a household; nature by culture in 1800 (206 YBP) under the glass roof of the Crystal Palace; and now by the new consciousness surrounding the old. Planetization, along with miniaturization (48 YBP), contributed to the idea that the planet was an organism, that it could maintain itself in the environment of the solar system, and that it could balance its atmosphere with living communities. The photograph, from the Moon, of the earth in space, became the symbol of the change in consciousness.

4.2.2.2.5.1. Patterns & Renewal

Nature consists of moving patterns whose movement is essential to their being. As a rope makes the knot visible, so the body is a pattern made visible. The body is a movement that maintains a topologically stable pattern; it is a vortex but not the water. The thing, the pattern, is a cross section cut through the movement. The mind is an invisible knot that is capable of recognizing both visible and invisible patterns, that is to say, a rope is not always necessary for the demonstration of a knot. Culture is also this kind of pattern. Culture can be analyzed into smaller blocks; the pattern of the whole organization is reflected at every division in differences of organization on either side of boundary. The wholeness of the character of a culture is reflected at every level. Patterning relates symbolic meanings in the context of a cultural system as a whole. The patterns form another level of meaning that has to be addressed in understanding a culture.

There are patterns of interactions of cultures, which arise out of several possibilities: Indifference, trade, competition, cooperation, conquest, or respect. Some archaic cultures seemed to be limited to indifference, that is, they ignored one another, and to trade. Competition and conquest may have accelerated with the acquisition of territory for agriculture. Cooperation and respect seem to have occurred under some circumstances of trade or unification.

The mode of operation of nature consists of a rhythm of dissolution and reformation. Perhaps this process applies to cultures. Often the elements of a culture will simply be rearranged by a succeeding culture. A new culture can only be made from the heritage of the old. The International Workers of the World urged its members to make the new world in shell of old one (in a way similar to genetic recombination perhaps). Our survival depends on the capacity to remake the image of the world from within, phoenix-like.

4.2.2.2.5.2. Challenges & Traps

Sometimes, rearrangement leads to a position of not being able to rearrange further. In many cases it is hard to tell if the destructive use of land preceded ecological problems or followed from efforts to maintain production after an ecological challenge. Cause and effect are hard to separate. The same environment that challenges a culture with some kind of change, also offers opportunities with the change. New resources can stimulate economic activity and increase the level of living.

Cycles that do not operate with the right kind of feedback function as traps. Thus, on an elemental level, phosphorus becomes trapped in an ocean sink, and can only be recycled by long geological processes or by specific harvests through human activity.

Karl Marx contended that humans live in cages, partly natural and partly of their own making. However, human actions can modify the situation. The word cage is a metaphor; it implies being trapped. It is however, a metaphor that can be expanded with a description of space as a four-dimensional box. Perhaps the trap is a better metaphor, since we depend on nature and society as a foundation for life.

Traps function in different ways. The use of resources by a people, where the replenishment rate is constant and the rate of use exceeds it, is a serial trap. This trap results in ecosystem degradation that is less reversible. The industrial age mistakes the rate of discovery for the rate of recovery.

Agriculture is an energy trap, because it allows a higher concentration of energy, that is, higher yields, but then it requires more energy be put into the system to maintain it. The system has to produce more energy than it uses to be sustainable, with a surplus for trade.

Sedentism is a trap. As the population of sedentary communities increased, the wildlife numbers decreased. The productivity and narrowness of food increased. Thus, there was less possibility of returning to the foraging lifestyle. People became committed to the new lifestyle. Intensity was no longer an option either; it had to be pursued. Habits were set.

One problem of sedentism is that the individual cannot simply move away to avoid conflict. People are tied to a particular place and have to communicate to adjust to sharing places.

The city is a different kind of trap, that offers intensity and opportunity, but requires massive imports of supplies to survive. The size and scale of cities create the dual centers of attraction and despair.

The cultural trap is that: One cannot transcend culture unless one knows the hidden structure, axioms and unstated assumptions about how life is lived, viewed, analyzed or changed. Cultures are systematic wholes composed of dynamic interrelated wholes. They are more easily described from outside with comparison to another culture, although transmitting culture to youth and watching the culture collapse expose hidden structures.

Language and art are also traps. Language because one is limited by words. Art because one is limited by styles or demand. Even if cultural attitudes are appropriate, they can trap a people if there are no longer functional reasons for the practices. The Nembi of Papua New Guinea may be trapped in their system, making stone axes with difficulty when thousands of steel ones are available.

Global capitalism can lead to a consumption trap. Capitalism claims to serve the wants of the people, but it spends half its income creating more wants in people. Not many of those wants are real, or as real as cereal and roofs. Few of the soft services satisfy real psychological needs. Markets advance individual desires and not social goals, by offering running shoes, not inner city restoration. Instead of being free from economic want to develop their potential as creative human beings, people are trapped in a consumer cycle. Self-actualization is postponed for self-gratification. Furthermore, capitalism can undermine traditional cultures by offering consumerism in the place of guides for behavior. Social roles seem irrelevant by comparison, if the good life can be bought without effort.

Being in a trap means much-reduced flexibility and fewer choices. Climate can drown whatever is in the trap. That is, being in a trap makes one vulnerable to many other changes that could be avoided if you were not in a trap. When the weather got colder, then hunters and gatherers could move south. Cities could not. Civilizations are more fragile and more vulnerable to smaller climactic changes.

Addictions. Such as those to foods or oil or money, make it difficult to escape from a trap, a trap being a kind of energy well or gravity well. Addictions can amplify some emotions, such as fear or hate, especially as they relate to the possible end of the addiction or the threat of that end. Addictions can justify illegal behavior, especially those that seem necessary to continue the addiction. Of course, many cultures are addicted to the illusion of control and power. The U.S. is trapped in the belief that only it, among nations, can bring prosperity and peace to other nations, with trade or violence. Eventually the trap is escaped, or more likely destroyed or collapses with its victims.

4.2.2.2.5.3. Conflict Collapse & Ruin

Collapse is often part of physical, biological and cultural cycles. Collapse can happen when a culture is too static in a changing environment. It can happen a culture selects growth to try to overcome certain environmental obstacles.

The Lowland Maya formed around 1100 BC. By 200 BC the massive public architecture was rising. Temples and palaces were built. Vast public works, such as aqueducts, were undertaken. The arts flourished. The entire landscape was modified for planting. The zenith of organization and population was between 700 and 900 AD. Perhaps 75 percent of the region was cleared for agriculture at that time. The Maya had a high density, stressed population, who depended on intensive agriculture, with complex hydraulic systems, in large centers, and were burdened with an elite class, calendars and rituals. Population growth triggered competition and conflict, which led to positive feedback of the stress cause.

Drought was a major consideration for the Mayans, who had built so far from water. Maybe the southern centers were sited for water. Much of the culture was devoted to collecting and distributing water. The water lily had iconographic significance because it was indicated good water quality. Between 810 and 910, Mayans had three droughts of three, six, and nine years, in 810, 860, and 910. They entered the drought with a maximum population and limited flexibility.

People suffered from the stress. Analysis of skeletons shows that in the Late Classic they became more fragile. Although people became 7 cm taller by the Early Classic, by the Late Classic stature of men had declined markedly; also degenerative bone conditions, bad teeth, scurvy and other pathologies. There was a collapse and depopulation, from around 3,000,000 to perhaps 450,000, which has never recovered completely in a thousand years. The current population is 1,250,000.

Table 4-222532-1. Kinds of Collapse

Reason	*Main factor*	*Related factors*	*Examples*
Resource depletion	Salinization	Political instability	Mesopotamia
	Soil nutrient depletion	Erosion land limits, malaria, political competition	Mayan lowland Easter Island
	Water table drop from tectonic shift	Revolt	Chimu, Peru
	Climate change, drought	Salinization	Hohokam
Reason	*Main factor*	*Related factors*	*Examples*
	Climate change		Hopewell
	Exhaustion of game, timber, soil	Competing centers	Cahokia
	Nile flood failure	Destruction, parasites, over-taxation, over-elites	Egypt
	Flooding	Collapse of trade	Harappan,Indus
	Climate, drought	Famine, migration	Mycenaean
	Climate change	Agricult. collapse, deforestation, erosion, invasion	Roman
Resource change	Preferred resource Bison increase	Moisture, Grassland increase	Pecos, NM
	Introduction of iron	Conflict	Hittite
		Decline of feudalism	Chou Dynasty

Catastrophe	Maize mosaic virus	Overpopulation Lack of flexibility	Maya lowlands
	Earthquakes, plagues		Teotihuacan
	Eruption of Thera		Minoan Crete
	Malaria	Lead poisoning, exhaustion, political corruption	Roman state

Collapse probably improved things for the peasants, without the burden of rulers, elite, priests, and artists. In the long run even the peasants were decimated, perhaps due to environmental deterioration, stress, and in-fighting. By the time of the Spanish, the area seemed to be unbroken forest. The Spanish introduced malaria and hookworm, which made the forest worse to live in. In some cases, such as Maya, environmental degradation did play a role in collapse of the civilization, either as a cause or effect. Complex societies put harsher demands on local environments. Political regimes set production demands too high.

Elman Service uses a biological analogy, where social organizations are modeled as plant or animal species that are initially successful because adapted to niche, but later become overadapted and less general. In this model collapse is part of an unalterable natural cycle. Adaptation denotes a systemic homeostasis in which a range of variability is activated in response to various environmental and social perturbations. Service believed that complex systems were maladaptive since responses to stress were less flexible.

This is not good argument, because it does not consider weed species or the maturity of the system. In a systems model, collapse is part of stochastic process, which implies that civilizations will die, but not necessarily within a definite time frame. In the general system model, complex systems are hierarchically composed of many stable lower and intermediate orders, strongly connected horizontally, but less so vertically. The problem may be more one of size than complexity.

Human cultures tend to fill all available space, carrying capacity space, even though some of the spaces are occupied by other species or other cultures. This makes them prone to crash if climate changes. They have adapted to marginal environments with behavior and technology, such as that for water storage and grain storage, to buffer themselves against the know changes of the environment, but when an unknown change happens, such as a nine-year drought, they fall apart. Few people have the luxury of moving to an open area. Smaller populations can adapt faster to smaller resources.

Perhaps one reason for collapse is due to overconnections, just like in ecosystems. This is a problem with modern globalization, in the 1990s, as well as with simpler globalization connections, such as in the 1300s.

4.2.2.2.5.3.2. Collapse

Collapse is the rapid significant loss of an established level of sociopolitical complexity. Various kinds of actions can lead to cultural collapse:

1. A breakdown of central authority. Provincial provinces may break away. Revenues to government decline. There may be foreign challenges. People become disaffected. Military may be ineffective.
2. Central direction is not possible. The center loses power. Distribution of goods may suffer. Trade declines. It may be ransacked and abandoned. Small states start to

emerge in the same territory.

3. No law and protection for people. Public art and monument construction cease. Literacy may be lost.
4. Remaining urban populations reuse the architecture. Spaces are subdivided.
5. Palaces and storage facilities may be abandoned. Technology reverts to simpler forms. There may be a reduction in population size and density.
6. Peripheral ecosystems and animals may recover or may not.

Furthermore, collapse can be put into a matrix that includes the primary factor and related factors, which may be necessary but insufficient by themselves.

Of course, some of culture does not collapse. The ideas produced by a culture may spread and reproduce. The culture of ideas continues through people leaping to other places. Again, some parts of a culture may collapse, such as the Southern lowland Maya, while the Northern upland culture continues to survive. Again, lower levels may survive with many of the ideas of the larger complex. So, the Roman Empire collapsed but not Britain or France. The Germans went back to chiefdoms.

Sometimes the collapse is thought to be the result of internal moral or technological failings, as well as conquest or some other external influence. Cultures are adaptive systems that have to integrate a number of challenges, opportunities and pressures, from drought to invasion, and then reconciles them to changing economic and political situations. Collapse can happen by luck, chance that is, or just a coincidence of bad leadership, bad images, cultural problems, external social pressures, and environmental degradation.

Table 4-222532-1-2. Further Kinds of Collapse

Reason	*Main factor*	*Related factors*	*Examples*
Invasion	Invading Aryans	Effete society	Harappan, Indus
	Invading Sea People		Hittites, 1200 BC
	Invading Mycenaeans	Thera eruption, earthquakes	Minoan Crete
	Invading Dorians	Climate change, drought	Mycenaean
Social dysfunction	Class conflict	Militarism, over-taxation, degradation	Mayan lowland
	Group Conflict	Agricultural collapse deforestation	Easter Island
Economic inefficiency	By-passed by trade routes. Agrarian system unproductive	Spanish gold ruined economy. Expansion of government/army	Ottoman 1500s
	Increase scale for control	No support for surplus population Contrast with rulers	Chinese dynasties
Political dysfunction	Loss of provinces	Increased taxation	Babylon
	Maladaptive ideology		Aztecs

	No inheritance from old rulers	Continued conquest to enrich new ruler	Inca
	Bad management	Lack of economic development	Spain 1700s
	High taxation		Netherlands 1700s
Bad luck, chance	Weak Emperors	Mismanagement of Huns and Visigoths	Rome
	Loss of agricultural land	Unsuccessful competition with Venice	Byzantium
Other factors	Mystical	Loss of virtue, exhaustion of energy	Rome

4.2.2.2.5.4. Intensification & Civilization

Farming allows intense food production in a minimum area. Farming increases carrying capacity. But, many regions are best suited for pastoral nomadism or wild gardening or foraging or fishing, rather than farming.

The successful cultivation then intensified trading of cultivars and resources between groups. Village specialization may have been the best adaptation. Then the economy of redistribution created regional temple and market towns that regulated the trading in goods. Cities are an intensification of trade and agriculture, and the things that surround them.

Gravity is a fruitful metaphor for intensification, for desire or crowding. As a metaphor gravity might explain why people are attracted to and move to cities: Intensity. Opportunity may increase due to the concentration of people. The number of links dramatically improves the possibility of better or more communications. Although gravity may explain the intensity of compression or miniaturization, it has trouble explaining the intensity of expansion. The metaphor raises some questions: How is danger related to intensity? Is it being exhilarated by the closeness of death? If the populations increased, would intensification have been necessary?

Civilization is a term for the combined phenomena in the neolithic revolution, starting with agriculture, which required a new technology for tools, such as hoes and plows. It also required the engineering of fields. Land ownership was invented to protect the heavy investments in labor and invention. The subsequent trade in foods required better paths and transportation. Food storage and trade become concentrated in cities. Record-keeping, in the form of tablets and then writing, became necessary to keep track of stores and trades. Further engineering was required for large stores, as well as standardized pottery. This required specializations in new trades, from potters to smiths and artists. Managers were required for recruit labor for fields and hydraulic engineering.

Concentration in cities permitted and promoted these changes. Permanent settlements contained working houses. Supplies and control required a new infrastructure of massive buildings, better roads, and fortifications.

This paralleled the rise of a religious elite to control the unknowns of weather and crops. And, this was followed by a military elite, which took over protection and combined it with universal gods, which led to the rise of royalty and kings. Heroic buildings increased to provide palaces for royalty. Warfare was necessary to procure foods in times of famine or to protect the stores from less fortunate cities.

The increase in personal property, combined with the redistribution of wealth, led to social hierarchies and classes. Public ideologies were developed to attune the people.

Civilization provided immediate advantages, including supplemental animal labor, more property for everyone, especially the opportunity to have luxury items, the use of records instead of memory, which allowed verification and less misunderstanding, some leisure time for doing nothing, and the production of ale, as a mind-altering substance. Unfortunately, there were disadvantages, such as an increase in the amount of work, not only for production, but also for public projects. Leisure was reduced overall. Living required more work and materials for records, as well as trade and trust with a larger group of people. Personal property had to be stored and protected.

Industrialization is really an intensification of civilization, which is still an ektropic process. In each case of cultural absorption, there was an attendant process of miniaturization. The forest was miniaturized in clumps of trees; animals were miniaturized in artistic image; time on a lunar tally stick; plants in a garden; women in a household; nature by culture in 1800, under the glass roof of the Crystal Palace; and now by the new consciousness surrounding the old. What is intensity?

Intensification brings cultures together faster. Conversion increases, technology and industrialization increase. Intensification is a gigatrend that could lead to violence. For instance, intensification leads to urban living, which increases creativity, as well as some kinds of illness, which leads to physical weakness, which leads to changes in population, which leads to intensification of resource use, which leads to the potential for conflict, which can lead to dislocation and violence.

4.2.2.2.5.5. Cultural Conversion of Places

We exploit resources, plants and animals too fast for natural renewal. We exploit minerals and fossil fuels too fast for the wastes to be integrated in cycles. We create novel substances and use them too fast for the substances to be integrated in natural cycles.

There is a conversion of planet from wild to urban ecosystems. System used to be wild with wild humans, now going towards one domestic urban system.

A city is an urban area with population over 2500 (or 10,000 or 50,000). The city is exploitation of a geographic site, which has a multilevel history from geology to plants and humans. As result of surplus food and larger population, special people can create a flow of specialized objects; this is specialization. Population has to be large enough to support a market. The development of cities from campments and villages is like that of tidal lagoons, where algae ended with communities: An unorganized, homogenous ensemble transforms to an organized structure that cycles materials, information, and energy in itself.

Modern cities have set patterns. Corridors and networks increase, patches increase, but connections decrease as does productivity. Species diversity may increase. Patterns can lead to suburbia: A supply of rural land, cheap gasoline, and network of highways. Larger cities, megalopolises, have the attraction of size and excitement, the lure of the highest paying jobs, and the most specialization.

Cities convert ecosystems into cityscapes, a special artificial ecosystem with fewer plant and animal species, disrupted water flow as factor, household wastes as factor affecting the landscape, a network of corridors that perforate landscape and cut off wild flows, an increase in the number of small patches, and a reduction in other kinds of patches and corridors. Flows include energy, materials, information, people, and pollution.

In a converted ecosystem, there are three imposed ecosystems with minor linkages: (1) a natural system of primary productivity based on depauperate system with few trees and layers, a simple trophic structure of birds, squirrels; (2) a human system, with imported food and water, smaller system of carnivore predators, such as fleas, lice, and bedbugs, and decomposers, such as bacteria, fungi, and gulls; and (3) a distant support systems for food, materials, and energy, but also smaller system of pets, such as cats, canaries, and fish.

The daily inputs for city of 1 million include the sunlight and atmospheric deposits, but also water (625,000 tons), food (2,000 tons), fuel (9,500 tons), manufactured goods (5,000 tons), and new buildings, roads and infrastructural support. Food has to be imported. The net productivity goes negative, due to massive imports of food. The daily outputs include heat, water, sewage (500,000 tons), solid waste (9,500 tons), and pollutants (air—950 tons).

These extra inputs and outputs have effects, side effects, or rather main effects that are unwanted or unanticipated. By thinking in terms of side-effects, we have been unable to manage the side-effects of our massive technologies. They are in fact unmanageable in a mindset that thinks of them as side-effects. A holistic rationality or neologic might be better to use, where everything is an equal effect, and we manage out technology to have all the effects. Land is taken over and covered, but then idle land may total 9 to 28 percent of the total urban area.

Villages lower connectivity to the landscape by increasing patches and corridors. Agriculture gets more homogenous, decreased fallow areas. Stream corridors are destroyed, environmental degradation. Connectivity is lowered, matrix is minimalized. Disturbance of nutrient mineral cycles. Disruption of atmosphere, drought, storms. Changed microclimate, with heat islands and dust domes. Lower photosynthesis/productivity. Lower diversity leading to homogenization, with cosmopolitan species. Inefficiency of use of energy and materials. Decaying infrastructure, roads, sewers, buildings, and houses. Disruption of supplies, due to social and political actions, e.g., strikes, breakdowns, attacks. Budget crisis from loss of tax revenue. Increase in violence, sickness, drug use, crime, impoverishment. Increase in mental and physicals illness from stress.

4.2.2.2.5.6. Global Culture

Trade and exchanges by various cultures extend to distant lands, and a global system starts to develop, with newer connections and technologies, which draw the cultures and civilizations, of all ages, into a tighter pattern. Some global civilization may result, but what effect will it have on the earlier patterns? So much is lost, of ways of being and acting in a more natural less technological environment.

The world has acquired a global economic structure. Perhaps a global military structure, also. It functions in the absence of a common culture or language. There is no world law, but there is a set of rules regulating international behavior; these are generally observed and understood. There is a homogenous culture, which belongs to the rich elites of every nation. This serves to integrate people to some extent.

A global political system is emerging due to the following conditions: A global system process, increased vertical differentiation, evolving from nations and regions, and differentiation into political groups and economic interests. The world does function politically at a global level. It could be made better.

Is there such a thing as a global culture? T. S. Eliot suggested that a "world culture which was simply a uniform culture would be no culture at all. We should have a humanity de-humanized. It would be a nightmare. But on the other hand, we cannot resign the idea of world-culture altogether ... We are therefore pressed to maintain the idea of a world-culture, while admitting that it is something we cannot imagine. We can only conceive it as the logical term of relations between cultures ... We must aspire to a common world culture, which will yet not diminish the particularity of the constituent parts."

The local and the particular are required parts of culture, so global culture has a contradiction. Eliot noted that: "We have not given enough attention to the ecology of cultures. It is probable, I think, that complete uniformity of culture throughout these islands would bring about a lower degree of culture altogether ... For a national culture, if it is to flourish, should be a constellation of cultures, the constituents of which, benefiting each other, benefit the whole." But, modernity insists. Modernity, in the predominant versions of liberalism and Marxism, sees the goal of history in a universal world culture.

4.2.2.2.5.6.1. Upside of Global Culture

A global culture, that could emerge from the interactions of local cultures, would allow for more rapid exchange of ideas and things between individual cultures. It would provide the opportunity for trade in special goods, that has the potential to benefit every culture. It would provide paths for communication that might stimulate cultures. Cultures could learn from one another. A global culture could present a global morality, built on universal human tendencies, that would not be prescriptive of our private and public behaviors, but it would be proscriptive of damaging behaviors from murder to obsessive greed. Global functions and problems, such a atmospheric warming, would be easier to address. New global economic or political structures would also be easier to address.

4.2.2.2.5.6.2. Downside of Global Culture

According to most theories of cultural adaptation (or integration or evolution), resistance to change is normal as a cultural process. Groups like the pygmies have specialized to fit the requirements of their environment successfully. This makes it difficult for them to adopt other cultural arrangements.

Nature makes divisions and diversity. We need no more a unity of all people any more than of all wolves or fleas. Is it wise to have a single world? A single market in a single culture with a single control? It might create big problems (Table 4-223562-1).

Table 4-223562-1. Global Human Cultural Problems

Global Human Cultural Problems
Hyperadaptivity (from success and overshoot)—humans can adapt to poverty, bad diets, crowding, stress, and other unhealthy practices.
Homogeneity of forms (loss of diversity)
Lack of flexibility (stagnation)
Hyperpersistence of error (stupidity or violence)
Lack of consciousness or planning (causing destruction)
Limits of capitalism as a global economic strategy
Acceleration of personal & social interactions
Breakdown of boundaries & limits

Hyperadaptivity is a serious condition that allows humans to adapt to poverty, bad diets, crowding, stress, suffering and immense natural loss. Of course, we are unconscious of many of these problems. Capitalism does not seem to be adequate as the only global economic strategy. It breaks down useful limits and boundaries. The homogeneity of forms promoted by a global communications and advertising campaigns leads to loss of diversity. A severely limited number of ways leads to a lack of flexibility (stagnation). Then, we have the hyperpersistence of error, in the forms of stupidity and violence, which forms a positive feedback loop. At the same time, interactions are accelerating.

Global capitalism undermines traditional cultures by offering consumerism in the place of guides for behavior. Social roles seem irrelevant by comparison, if the good life can be bought without effort. Yet, it does not seem to be working in Europe and the U.S. Instead of being free from economic want to develop their potential as creative human beings, people are trapped in a consumer cycle. Self-actualization is postponed for self-gratification.

It is said that the global system has no center. That is a good thing. But it is expanding without limits and that is a bad thing. Uncontrolled expansion could cause the disintegration of natural systems, which are interlinked with our economic exploitation.

4.2.2.2.5.7. The Future in Balance: Cultural Exchange & Health

Intelligence, for Edward Hall, is paying attention to the right things at the right time. Western culture uses a highly specific classification system to deal with specifics and details to the exclusion of whole patterns. Other cultures depend on myths to comprehend wholes, but they have less success with the details of invention. Can cultures learn from each other?

What can modern cultures learn from traditional ones? First, they can learn or relearn how to have children based on limits, not on some abstract right to reproduce. Next, they can learn the art of sustainable living in fragile ecosystems. They can learn how to share and distribute the goods of a society equitably, And they can learn how to resolve conflict.

Traditional cultures can learn some things from modern cultures, also. For instance, larger networks of trade. They can learn the use of appropriate technology. They can learn some forms of larger communication. And, they can accept common rights and standards of universal human behavior.

Table 4-22257-1. Exchanges between Cultures

What can modern cultures learn from traditional ones?
How to have children based on limits, not some abstract right to reproduce.
Sustainable living in fragile ecosystems
How to share and distribute
How to fight personally
What can traditional cultures learn from modern?
Trade
Use of appropriate technology
Communication
Common rights, universal human behavior
Ways of avoiding mistakes from technology, scale and speed

All cultures need to have certain conditions to keep being healthy; if one or more of these conditions fails then the culture may fail.

The megaculture itself would have to adapt new behaviors and ideas. The desire to refine the focus has allowed the frame of reference to be neglected. An adaptive holoculture could place human values in a global framework. A new world order is a cultural problem more than an ecological or economic one. A global culture could incorporate the positive features of traditional civilizations: In general, an ecological global framework could offer personal security, respect for the individual, responsibility for actions (self-discipline), social integration, concern for others, and reverence for nature.

Such a global framework for culture would also consider important principles drawn from ecology. The proper attitude of an ecological global culture would be care, a positive spontaneity, but also a 'letting be,' a reverence toward the wild alienness of nature, a willingness to comply with the limitations of natural systems, and a willingness to reduce the dominance of natural systems and to set aside wild areas. In addition, a global perspective might define: An authentic concept of humanity, rational economic development, with respect to health, a holistic education beyond that of a native culture, a description of the responsibilities of societies to themselves, others, and the earth, and a respect for all cultures.

Table 4-22257-2. Conditions for a Healthy Culture

Condition	*Meaning*
Source of energy / material	External, internal
Groundedness (boundedness)	Locality, isolation, security
Identity	Unique patterns, stability, resistance
Vigor	Productivity, health
Participation	Relation to other cultures, systems
Adventitiousness (continuity)	Ability to adapt / absorb
Sophistication	Self-maintaining, developing, aging
Complexity	Development of diversity
Heterogeneity	Toleration of diversity
Comprehensiveness	Acceptance of everything
Flexibility	Looseness, capacity to change
Renewability	Self-creating, self-renewing
Balance	Harmony of limits / extents

4.2.3. *Domiture & Pan Ecology*

Our minds are not only nature dependent, but culture dependent as well. As the wind, trees, and birds are sources of signals and symbols, so are gestures and words. A community implies that the experiencers share ways of experiencing or the same experiences. This enables an individual to go beyond a finite view, to see the embedded culture as one of many ways of relating the self to the universe. Culture evolves from the interactions of humans with nature; both are in a constant state of flux. The development of cultures may be thought of as parallel to the development of species, where memes or ideas in a culture have the similar function of genes in a body.

Nature needs a place to play, without controls. Culture needs a place to play, without being forced by nature. Nature is an extension of the field, Culture is an extension of nature through human existence. Domiture is the entire field of extensions as a unit of study, rather than the reification of something new, perhaps divided among the studies of ecology, human ecology, cultural ecology, and panecology. Domiture is the system of culture embedded in nature; it has to be a larger term to enclose the previous nature-culture dualism. It has to include human reason and human emotion, rather than simply putting them on opposite columns, as with order and chaos, higher and lower, linear and cyclic, as well as agriculture and wilderness.

4.2.3.0.1. Human Needs

The psychologist Abraham Maslow created a list of human needs. This list began with the physical: Food, shelter, clothing, and space—privacy may have survival value. He also considered safety: Law, order, security. The psychological was important: Belongingness and love. Esteem was important: Strength, self-sufficiency, competence, freedom, attention, and prestige. The final need was self-actualization: Self-actualization, achievement, and creativity.

Not only do humans have to satisfy these needs, but a place has to satisfy these needs and the culture has to also. Unless a cultural phenomenon satisfied a basic human need, it would not be repeated. The human needs that he listed were all physiological: metabolism, reproduction, comfort, safety, growth, and health. But not all needs are physiological. At the other extreme, Jeremy Rifkin considered only psychological needs were satisfied by a cosmology. But there are also social needs to be considered, such as social status.

The peoples were able to convert animal and vegetable resources into all their needs for food, shelter, and clothing. When other items were available, such as horses and metal knives, they were able to trade, fish, berries, baskets, and roots for them. Trade also forced the collection of surplus in excess of needs and perhaps in excess of the system. Many of these things are long-term problems and do not become evident for several generations. They are also very difficult to reverse. For a society than needs surpluses to continue, with growing dependents and growing people, there is little flexibility to change.

4.2.3.0.2. Cultural Needs

To survive, a culture has certain needs. The needs of a culture are: To be grounded in place, to be secure or partly isolated; to have a dynamic order, human health; to be complex and sophisticated, with checks and balances; to be comprehensive, to allow change and diversity; and to have and manage adequate resources.

Needs are almost completely defined in terms of commodities in industrial cultures. Even relation needs a commodity, travel, to be fulfilled. Once the need is defined and partly satisfied, it becomes a right. From walking to the right for passenger miles at high speed in a mere fifty years. Instead of ecologically analyzing the relationships of needs to satisfactions, the industrial system increases satisfactions. We just need to imagine and construct new frameworks so that people can develop satisfactions that are not dependent on commodities.

4.2.3.1. *Imagining Place: Human Cosmologies*

Cosmology is basically the complete set of ideas about the nature and composition of the universe used by a cultural system. This idea of the universe provides people with an orientation in their cosmos. Melville Jacobs offered a catalog of ingredients that could be included in a world view: epistemology, prepsychology, logic, a theory of disease, the supernatural, geography, climate, history, zoology, astronomy, biology, myth, music, and theology. Immanuel Kant's distinction between cosmology and theology is not considered valid, here; cosmology also considers the supernatural world behind appearances.

A cosmology is considered a category of a culture in anthropology. It is a collective image. Ezra Pound stated that the image is an emotional and intellectual complex in an instant of time. He used the figure of a Chinese ideogram as an expression of a complex image created by throwing groups of elements together without predication, which is a characteristic of assertion of the universe. The set of ideas is common to a culture; some ideas may be shared among other cultures. Most cosmologies have to be understood in a cultural context, which means that they are not intelligible without reference to the culture they are particular to. A cosmology can only be understood within that culture, that exists in that location, with that unique history.

The world as it appears is a test of human understanding. Storms in part of the earth often affect conditions elsewhere; even with weather satellites this is not always obvious. To a localized people it is even less obvious. Levi-Strauss thought that mythology was explanatory, a product of observation and reflection of reality. He also emphasized that tribal cultures brought the full power of the human brain to bear in expressing a cosmology. The 'savage' mind was as logical as the modern mind. Myths often were created out of inherited ideational pieces, by fitting together. But the thought proceeded through understanding, with the aid of distinctions and oppositions. Levy-Bruhl's opinion was that participation was more important than logical thought, but the participation is unavoidable. Cosmology serves to explain appearances, if not in observable events, then in terms of a hidden reality. These two realities make up a totality that is explained. Usually the present is explained as the result of past events.

A cosmology rarely meshes perfectly with the natural order or the social order. The ambihuman order of nature could exist without humanity. In a sense it is indifferent. Cosmology arises out of the needs of humans. That cosmology includes so much—science, mythology, theology—makes its fitness less. To the degree that it is effective, any of the ideologies mentioned will fit the order of nature; science may manipulate invisible particles; technology may mold metals; beliefs in theology may save human or ambihuman lives. But the total mix of ideologies makes the overall fit very sloppy. As long as nature can be dominated, without catastrophe, the importance of the fit is not critical. But we do not know enough of nature to know when catastrophes occur, nor how to avoid them or minimize them.

All peoples want some power over the natural order. Archaic peoples rely on ritual acts instead of machinery. But as technology supplies power to archaic peoples, rituals decline. The spread of power has two other effects. The natural order becomes simplified; the human world becomes increasingly complex. Both orders become unstable.

Unfortunately, the language from a mechanical world view dominates even ecologists and politicians. This world view impoverishes humans by claiming all consciousness for them. It claims that nature offers no joy, or love, or peace, or certitude. Emphasis on the evil of nature creates a gap between humans and the universe. In contemporary cosmology, there is no room for the intrinsic worth of nature. But science undermines the scientific cosmology and provides elements of its alternative.

Brian Morris recommends not taking the retrograde step of advocating a cosmological viewpoint, through St. Francis, Buddha or Black Elk, since that world was presumed destroyed by Newtonian science; he recommends an ecological viewpoint and understanding. But no world has ever been destroyed by science. Science cannot even discourage astrology or bad eating habits. Technology has ruined many habitats and cultures have smothered many smaller cosmologies, but that is not an indictment of the cosmologies.

The present crisis is a consequence of the modern world view, which is too rational and mechanical. Problems are more than matters of policy or management; war, injustice and environmental destruction spring from a common sickness. The problem is cosmological. The issue is not conservation against exploitation, but an experience of the natural world distinct from these two alternatives. A sense participatory rather than manipulative, a sense of the world as presence rather than object; a universe moving in vast cycles and rhythmic harmony instead of serial stages of beginning, progression, decay, and end. This cosmos is being rediscovered; its tools are dance and song, and history; it is open to nonvisual communication. Everything is vital. The consequences of this cosmology would alter the character of modern life; make it closer to human and ecological reality, counteracting the tragic consequences of war against ourselves and nature.

A world view is more than skeleton of theory, it is a view of relation of human mind-heart-will to universal factors. Ultimately, a model of the world—world-view or ideology—is what implicitly or explicitly guides the processes of social change. The basis of an ecological world view can be found in history, in the experiences of diverse cultures, and created intentionally be defining new values.

4.2.3.1.1. Imagination & Metabolism

As the adequacy of a habitual view of the earth is questioned, imagination offers alternative ways of seeing. Yet human imagination is limited, as is human knowledge. An aesthetic object should not offer a reassuring vision, which interprets or identifies nature, but a naive vision, which surprises, shocks, fascinates or seduces the senses, which awakens desire and stirs the imagination, and which furnishes a feeling of the invisible. Play is imaginative experience, natural learning entered into freely; education should be more like play than work. Vico pronounces that imagination is a turning out of one's self. An imaginative metaphysics shows how man becomes all things by not understanding them. The absolute creativity of the imagination diffuses the world from the center of its being, creating an absolute plenitude. Metaphor enlarges the imagination; poetry is the closest approach to objectivity. Through its special position poetry becomes metaphysics, even though embodied, according to Vico. Perception and expression are embodied in the flesh, but imagination is what makes the flesh visible; it is what allows inquiring beings to see into Being.

We must find a way to affirm the metabolism of nature as our own. Each individual is part of a world that extends around the self. There are different, partial worlds. We may see trees on a house site as encroaching, where a Dakota Black Elk Indian may see them as standing peoples, with equal rights to the land. The self is related to each world in different ways. We may see plowing fields for seeds as an act of mastery and exploitation, where another may see it as act of tender involvement, bringing forth potential. The study of these worlds is a holocosmology.

The concept of nature and ecology require new conceptual definitions. They are major philosophical problems. As philosophy attempts to describe the world and discover the order of nature, it must go into mathematics, astronomy, physics, chemistry and biology. The scientific method fragments the world. The mystic feels the world as described by scientists and it is whole again. Most modern cosmologies are based on physical science. Older cosmologies were based on feeling. The new biological paradigm, combined with process philosophy and phenomenology, supports a new cosmology of feeling.

4.2.3.1.2. Holocosmological Framework

Since the origins of the environmental crises are in human traditions, it should be possible to select—and create a new cosmology—from what is valuable in the traditions. If the world becomes as humans imagine it, then a larger frame will make a larger world.

When cosmologies, when human societies, were small, the amount of control and security was small. Although societies have grown, security can not. It should be easy to give up control that we never had; giving up trying may be more difficult. We need to fashion our behavior to the cosmos, not reversed. Since complete security is impossible, since complete power is impossible, why try? We are already participating in the cosmos; our images need to reflect that. We are already in relationships; those are what we need to learn.

We create the organization of a cosmology, then see nature work that way. But, we create nature by seeing what the brain filters. The world is too complex for our minds, thought G. P. Marsh and many later thinkers. So, our minds have to filter out what is less important. We filter data, arguments, emotions, and information. The filter allows a total picture of the whole with relatively little information. Of course that picture might be wrong. But, it is clear, as a result of filtering and thinking. Other human things act as filters, also, especially culture.

Walter Ong thinks that our hypervisualism may be outmoded; it may hinder understanding of the world as it is organizing itself. It would be productive to cultivate more aurally-based concepts, such as harmony, melody or cacophony, on which to analyze a world view. And to move from concept of world sense to that of world-as-presence, in the sense of presence between two persons.

Cosmologies adapted humans to limited places, but cosmologies clash and contradict one another. A global holocosmology is necessary for framework of local world-views on earth. Most cosmologies are circumscribed; a small section of the hologram is inflated into the whole. A holocosmological framework would not have that problem, since it incorporates all human ones.

A new holocosmology could justify the wide diversity in nature and accommodation to natural laws. It could even allow a deeper understanding of the utilitarian. For example, what is the use—or the beauty—of a burned forest, as related to the function of lightening or the planetary carbon cycle? The whole idea of irreversible history sanctifies behavior, making it moral. If a life or species cannot be repeated, then it is special.

A holocosmology shows the relationships and our debts, to other species. By considering only humans, and inflating them out of proportion, we diminish other species, and ultimately ourselves. Responsibility considers the interests of ambihuman nature. The rightness of human behavior is all species need in terms of special rights. The interests of other beings is the same as human: To live and experience, then to reproduce that their heirs may, as Whitehead said, "to live, to live well, and to live better."

A holocosmology cannot be limited by the scientific facts of any science, even ecology. The insights of mystics and the wise should be included. Individual world views penetrate each other in a transepistemological process. Each mystic or scientist tells of a way the world is; together, these ways make a holocosmology. It is a framework for all truths.

The human desire to refine the focus has neglected the frame of reference. An adaptive holocosmology would place human values within a global framework, attaining a balance of human and ambihuman nature. The sciences, humanities and other ideologies could be balanced in a holocosmology. Perhaps human behavior could be put in tune with nature through a popular literature or poetry. Humanity is not aware of limits, of its inability to fully comprehend the universe. We have never made earth completely home; until we do, no place will ever be.

4.2.3.2. Fitting in Place—Image & Place

People not only fit physically into a place, by adjusting their eating habits, by making or shedding clothes and houses, they also change their images of themselves and the place so that the two images fit together.

The individual feeds back into the cosmology in altered form what was received. It is almost like a closed loop between cosmology, culture and the individual. Archaic peoples translate the natural world into the language of myth. Being a narrative, a myth is aesthetic as well as intellectual. Myths develop in terms of their own internal logic, drawing together observations of the world. C. Levi-Strauss described the process as bricolage, fitting the bits together, identifying impressions of life as sets and forming them into mythical systems; the world picture is a metaphorical puzzle. Bricolage is the mentality of synthesis, a technique for learning, creating, and expressing understanding, using whatever is available from the past and in the present to achieve an integrating form. This is what mythological thinking does, and what scientific thought might do.

A culture is a loose-fitting patchwork of ideas, relationships and things. In that sense it is parallel to species adaptation to an environment. The fitness is never perfect, but that is what allows movement and change.

4.2.3.2.1. Living Field

There does not seem to be a formal relationship between physical and psychological fields; perhaps the relation is like classical and statistical entropy, and descriptions are complementary. Field theories in physics provide for continuity, in space-time, for instance. Sherrington developed a field theory of subjective space. He distinguished between exteroceptive, interoceptive and proprioceptive fields; the exteroceptive receptive field was coextensive with the body surface and richer in receptors. These elements comprise a biological explanation of subjectivity. The brain evolved upon distance-receptor organs. The distance receptors induce anticipatory reactions, leading to consumption. The subjective field is characterized by a broadening of lived time. This extension includes the lengthening of responses.

The subjective field of self is parallel to Kurt Lewin's concept. The self is a creature committed to a specific association, a genome plus place; the fitting of self to setting is definitive of both. Place/person/act are as indivisible as the physical STEM field. Lewin thought that the psychological fields joining the personality to its life-space, the immediate environment, and life-space to a larger environment, were strong enough to alter objective facts, that is, to make them normative.

Lewin was willing to study the objective factors that are potential determinants of life-space. Psychology has not considered the life-world. The organism's own world is usually left out of consideration. Ludwig Binswanger broke the life-world into three interrelated modes for human beings: *umwelt, mitwelt* and *eigenwelt*. The first refers to world of natural objects (environment); it is approximately equal to von Uexkull's idea of *umwelt*. the second to the human alone, the interrelationships of humans; the third is one's self-world (=thought

-world). The same division could be made for animals, but would be more difficult to describe.

Lewin's life space combines the umwelt and life-world (of Edmund Husserl). Topology provides the mathematical model for Lewin's representation of psychological processes. Topology is a geometry in which spatial relationships are represented in a strictly nonmetrical manner. Since topology had no directional concepts, Lewin invented a qualitative geometry called hodological space, represented by vectors. He used two-dimensional planar maps to represent life spaces (now, recent researchers use a linear graph, on which an indefinite number of points can be mutually interconnected; and asymmetrical relationships).

Lived space requires a three-dimensional representation. It may be Horizontal or vertical (Vertical=warp, horizontal=woof). Vertical aspect includes high of drug user and mountain climber; it can mean the loss of horizon and the horizontal. The horizontal dimension requires reciprocity; it is an intersubjective domain. Older cosmologies balanced the horizontal and vertical. That it is out of balance now causes problems. New mathematical treatments of fields tend to be three-dimensional. Rene Thom's catastrophe theory and Conrad Waddington's epigenetic landscape are two such theories. The epigenetic landscape field can be used to explain chicken fence food problem. Is it empirical? Could it be measured by intensity of brain waves in a chicken? The chicken's need chreod is deeper than the path of a cognitive chreod around the fence. So the chicken can only go toward the food. If the need chreod is too deep, it cannot explore first with its cognitive chreod. A less hungry chicken could. In humans this explains why necessity cannot be the mother of invention; the broader landscape of leisure is needed.

4.2.3.2.1.1. Field of Person

Our direct intuitions of nature tell us that the earth is bottomlessly strange, Paul Shepard noted; it is alien, even when gentle and beautiful. It has innumerable modes of being that are not human modes of being. It is always mysteriously impersonal, unconscious, and immoral; sometimes we interpret it as hostile or sinister. But intuition also senses the interdependence of nature. We mistakenly conclude that our skin is the boundary to ourselves. We extend tentacles of personality to other things and people. As Whitehead pointed out, everything prehends everything else. The human skin is a delicate interpenetration, like a pond surface. The skin's interpenetration ennobles and extends the self—the beauty and complexity of nature are continuous with ourselves.

Our species has been shaped by the earth. The desire to save forests, wetlands—all natural ecosystems is a expression of deep human values. Experience of wildness lets us capture some of our own wildness and authenticity. Our emotional response to the unfathomability of the ocean or luminosity of the desert is an expression of aspects of our fundamental being that are still in resonance with these forces.

Landscape is related to mental health, recognized Dubos, Passmore, and others. Humans have made landscapes into flatscapes, then concentrated on indulging the individual. Many humans now live piled in cages. Experiments show that when populations of a number of species are subjected constantly to artificial conditions that appear to favor the growth and survival of individuals, the result is the death of many individuals and eventually extinction of the race. During an experiment with overcrowding in rats, John Calhoun recorded that they entered a behavioral sink, with varieties of abnormal behavior.

The significance of density figures to human health is not known. Hall suggests deep-rooted cultural differences by different people in response to spatial relationships and high densities. Our mobility crusade has fractured our experience. Bombarded by information out of context, our artifactual environment tends to make us more cerebral, and we lose the richness of body sensibility.

What is impaired in the absence of a rich ecology is the individual's knowledge of himself, not only as a person, but as a member of a species. Harold Searle states that the environment constitutes one of the most important ingredients of human psychological existence. There is within the individual a sense of relatedness to total environment. But he does not draw the limit, as Schweitzer does regarding the concern of ethics, to include the wider sphere of all that lives. Psychoanalysis needs to concern itself with the total human and ambihuman environment, inanimate and animate.

A deep relationship with a place is as necessary as one with other people. Without such a relationship existence loses much of its significance. A range of experiences can spring from a place, from depression to the peak experiences described by Maslow. Then there is the opposite feeling, the dullness of place, where everything becomes oppressive and life becomes tedious. Drudgery is part of commitment to place; it is acceptance of the restrictions of a field.

4.2.3.2.1.2. Field of Place

The making of places from undifferentiated wilderness is the human ordering of the world. A place changes qualitatively; it becomes structured. Natural complexity decreases as the human increases, although the two are not mutually exclusive. Places are ecosystems intimately associated with people. Fitness is achieved after progressive reciprocal adaptations; it requires a stability of relationships between societies and the place.

Human places are complex integrations of nature and culture that develop in particular locations. The feeling of a geography is prior to its study. The place precedes knowledge of it. The knowledge of place is one of the first links in chain of knowledge. Being human is having and knowing a place. This knowing is essential to our existence.

Paul Shepard suggested that for each individual the organization of thinking and meaning was intimately related to specific places. Place is a focus of experience; and the background for specific events. The features of a world are experienced meaningfully. The place is a matrix for ordering experience. The essence of place lies in unselfconscious intentionality. Only learning flowing from hospitable presence to human and world can promote life and enhance human existence.

Lived space is human intention inscribed on the earth. Each difference in the landscape has meaning, as when the aborigines of northwest Australia perceive physical differences and even a symbolic landscape. They structure space according to myth, where Europeans use buildings. Lived-space needs to acquire cohesion, to become rich and diverse with experience of living. The world can become richer with the addition of significant human places as well as remote wild ones.

The specificity of place is important. The earth extrudes itself into particular plants and animals; flexes mountains; and sweats weather. It is being homogenized by industry. The spirit of each place may remain to influence successive generations without being homogenized by mass culture or technological efficiency into a flatscape of little significance. It can be concluded from an MIT study that the delights of waste spaces, odd lots, tangled woods left between housing projects, must be preserved. Does it take years of research to see that the world needs hiding places and hidden places?

People remember most vividly those elements of their childhood which involved landscapes: lawns, hills, woods, water. The most popular outdoor activities in the U.S. are picnicking (80.5 million, 12 or older) and walking (67.8 million). The spirit of place lies in its landscape. There is a persistence of place that shapes a human community. Every place has a unique identity, a persistent sameness as a result of combinations of factors. The spirit of place a constant thread.

The notion that the earth is a living being is an ancient one. In taoist science, feng-shui or geomancy, subtle forces flow through the earth like blood in a body. To build a city it was necessary to consult the landscape, chart the flow of forces, and place buildings with the utmost care to be in harmony with the place. Geomancy is finding the right place. Petitioning the spirit of the place to build a wonderful city.

Gardens and parks have been designed to express an idealized view of natural and agricultural scenes since Sumerian times, over 5000 years ago. Many names for the landscape—grove, lawn—are drawn from the imagery of the garden.

A place is a part of the environment claimed by feeling. Emotion binds together motion and perception. Emotion can transcend distance. Emotion creates an 'in-place'. A place must be found and made; it does not exist independently. Neil Evernden asked what it does to a person to see his place, his context, destroyed? What does it do? It creates violence and apathy. Modern populations are rootless, moving about from city to city. We are suffering from a placelessness, which arises from our style of efficiency and proclamation of mass values. The environment of few significant places becomes a flatscape. It is turned into uses. "We shall never fully understand nature (or ourselves) and certainly never respect it, until we dissociate the wild from the notion of usability—however innocent and harmless the use. John Fowles. "This uselessness lies at the root of our hatred and indifference toward it.

We will have to pay attention to the unfashionable mesosphere, where we live as conscious beings: the land forms and vegetation that mining, agriculture and architecture are destroying. The need is basic: heterogenous, textured environments, woodlands, streams, rocky areas—diversity and choice in the natural environment as well as domesticated land and urban places. We need unique and particular places.

Nature is not really seen in the individual present, which no one trusts, but is reported and edited. The created information is emphasized, while the act of creation is ignored. We lack trust in the present, the actual seeing, because culture tells us to trust only the edited, public report. Worringer figured that the urge to empathy which follows the natural contours of living things came only after a certain relation of confidence between man and the nonhuman world had developed. In his book, *Abstraction and Empathy*, he finds the origin of artistic expression in the tendency to counter a sense of being lost in the universe. Empathy is latent in humans, but held in check by dread of space and the urge to abstraction. Trust is necessary for an empathetic response to the call of distance that bids us to move out into the world. When we are claimed and backed up, we feel at home in the world. Life cannot proceed on mastery and control alone.

One cannot trust using resources, hoarding or spending. Too much control is detrimental to all relationships, especially friendship or marriage. Control and surrender must complement each other. Simplicity and surrender are high values only in a world one trusts, to which one feels attuned at the deepest levels. The earth is spiritually alive.

Spirituality is a state of mind, of being, in which we experience the world as if it were endowed with grace, a mysterious, wonderful place to be.

We discover that there is an ecology of the spirit, which must be trusted like all life support systems, to find its way to balance and compensation. We need to trust a place. Home requires rootedness, at-easeness, and regeneration. Von Uexkull described the importance of rootedness in his concept of lived-world (*umwelt*). Feeling at home is a state of awareness; losing the feeling may cause crises.

Ecology relates to at-homeness; a dwelling in neighborhood of its source. A. Portmann observed that insects and animals displayed a powerful attachment to places; that it was best understood as home. The attachment to a place is rootedness. What does it mean to be at-home? The fundamental ambiguity of existence is that humans have different capacities for feelings and awareness. Some feel strongly about a place or home; others never do. We need only to reciprocate and be available; to be and let be; reaching out and being ready to receive the other—any being. We need not feel homeless. But home is also an enclosure, a place for protection and privacy.

Being inside is knowing where; it is safety, cosmos, enclosure. Inside and outside, like a dialectic, can always be reversed. Empathy is willingness to be open to significance, to know and respect symbols of place. A house is a place that provides shelter; it answers social needs; it is a repository of memories, a field of care. For the private self the house is a world; for public selves, the world is a town or civic center.

In English the term for dwelling is to stay. This is symbolic opposite of moving or changing. It means to withstand against time. Existence points to a lasting moment in all changes (from the Latin *ex sistere*, to make to stand out). It resists and persists. Permanence is important element in idea of place. The Greeks valued autochthony; they took pride in being natives.

The benefits of settlement are economic, ecological and spiritual. Economic because everyone will have to learn to live in the limits of photosynthesis and watersheds. A sense of place, with its beings and features, is necessary for information on how to live, get food, stay dry. An ecology is where we live; we need to know about it. Environmental impact statements give people a voice in any proposed changes where they are living. People cultivating a sense of place are people in place. Their work can be appropriate, whether growing, logging, mining, or building. The ecological benefits of rootedness are that people will take care of their place if they realize they are going to be there for a thousand years. Having a place means that the inhabitant has stock in it; participates in its unfolding, through planting and caring. Detailed understanding of plants in a locale allow gathering of food and medicine. People in place acquire a sense of community, nonhuman and human; shared set of values and concerns; health and spiritual benefit.

Caring for a place involves concern and responsibility. Heidegger's concept of 'sparing' means letting things be. It is a tolerance for the essence of a place. The willingness to not change or exploit. A home can only be realized through caring; to have a home is to dwell on the earth.

4.2.3.2.2. Living Together

People living together often share an image of the place. That common image, along with traditions or formal rules, determines how people behave, not only towards each other, but towards the place. They work to fit a place by dividing up labor in the place, by arranging their relations in certain ways, and by moving throughout the place according to habits or resources. The form of marriage for instance may reflect the difficulty of acquiring food and resources. A place is a part of the environment claimed by feeling. Emotion binds together motion and perception. Emotion can transcend distance. Emotion creates an 'in-place'. Living together results in morality.

People put things together, putting together of parts in complicated arrays because they are possible, before considering any kind of formal or ideal design. Living together involves all kinds of interactions, from competition and conflict to cooperation and mutualism. No one interaction can dominate without unfortunate consequences.

Kropotkin maintained that progress occurred by people working together, not through competition. In *Fields, Factories and Workshops Tomorrow*, he argued for self-sufficient local economies and an integrated urban and rural society with decentralized industries. civilization is held together by practical and political activities. Building and planting, taking care of himself and others, is primarily a prosaic activity of man.

4.2.3.2.3. Making Things & Homes

People make tools to make things easier. They make clothing and homes. Australian aborigines had very little clothing, only pubic aprons for women or tassels for men. Although the temperature might drop to freezing at night. Their houses were windbreaks in the desert or caves on the coast or leaf-bough lean-tos in the north and northeast. The first tools were stone-flaked, then ground stone, used with boomerangs and clubs. Ground stone axes appear first in Australia. Tools are simple, but effective, easily manufactured and maintained, for example: Spear-thrower, or woomera (atlatl), had hook on one end and an adze on other; it could be used for a shovel, fire-starter, or percussion instrument. It was an independent creation in Australia.

4.2.3.2.3.1 Homes

Houses are more than simple structures that moderate a very local weather and control access of dangers. They become invested with memories and meaning.

Adolf Portmann observed that insects and animals displayed a powerful attachment to places; that it was best understood as home. What does it mean to be at-home? The fundamental ambiguity of existence is that humans have different capacities for feelings and awareness. Some feel strongly about a place or home; others never do. Several metaphors have been used to describe the human place on earth. The earth is a storehouse, property, a spaceship. But the earth is not a spaceship or storehouse; it is home. Victor Ferkiss proclaimed that: "The world and humanity are one entity, one system in equilibrium. Earth is humanity's only home; humanity is one people in relationship to the earth."

There are a wide variety of meanings of home. It is a place of family residence, the family social unit, habitat, and place of origin. The word 'home' comes from the Middle English word *hom* and Old English *ham*, Old Norwegian *heimr*, Greek *kome*, and Sanskrit *kayati*, meaning village or home. The Old Norwegian word for home meant village or world. The word can be traced through the Greek to the Sanskrit, which meant 'he is lying down.' Its spectrum of reference is enormous. The word is used to describe house, village, city, bioregion, cultural world, and the earth. Its content is also ambiguous. Home can hold a single person, family, relatives, pets, domestic food animals, neighbors, and others.

Home living is simultaneously on different levels; the importance may shift from city to nation, or nation to state, or house to bioregion, or state to habitat. Each level is a metaphor for the next. There are parallels between nature and a house, as the basis for home. solar space is like the landscaping; wilderness is the foundation; conservation areas form the shell and provide services; and each bioregion is a unique room. The analogy cannot be carried too far. But it shows that a house is not, as Le Corbusier said, "a machine to live in." It is a matrix for home. Home is not just a house, either; it is a complex of significant events centered in place. It is the foundation of our individual identity on one level, and our role in the community, on another. What makes home different from house? Participation in making, commitment. People invest parts of themselves in a place, to make a home.

The concept of home has a mixed reputation. Paul Shepard (1974) finds humanitarians obsessed with the 'homelessness' of stray pets and wild animals. He points out how the fixation on shelter is taken over by the advertising of wood industries, who describe their meager reseeding efforts as 'creating a home for wildlife.' He sees protective organizations swaying to the tunes of propagandist lullabies. Perhaps he is partially right. But that is a misunderstanding of home. a home is not the house, not an undifferentiated place. Animals accustomed to rich woodlands are not at home in a replanted clearcut. The use of the word is a cheap advertising device; yet, it shows the importance of the meaning. A home is living, a house is not.

In English the term for dwelling is to stay. This is the symbolic opposite of moving or changing. It means to withstand time. Dwelling resists and persists. Permanence is important element in idea of home. The Royal Commission on Local Government in England and Wales found that people's attachment to home area' increased over time. Gaston Bachelard (1969) has written much about the significance of home: "For our home is our corner of the world. As has often been said, it is our first universe." The home was a springboard to understanding the universe.

4.2.3.2.3.2. Possessions

Clothing and tool-making are universal in human cultures. Indian peoples were able to convert animal and vegetable resources into all their needs for food, shelter, and clothing.

4.2.3.2.3.2.1. Clothing

Traditional clothing was made from sagebrush, shredded cedar, and willow bark. William Clark, of the Lewis and Clark expedition, called the Corps of Discovery, mistook the Palouse for bands of Nez Perce, observing that the two groups dressed, spoke, and lived in a similar fashion. Tanned deer hides were used for ceremonial clothes; after trading with Plains Indians, however, everyday styles began to change towards buckskin clothing and moccasins; breechcloths, leggings, and shirts for men and dresses for women. Both wore beads, bracelets, and necklaces. They also traded for buffalo robes. Wood was rare and used for tools, such as digging sticks and clubs; basalt was probably not useful for making digging, chipping, and cutting tools. Wood was also used for marking graves.

4.2.3.2.3.2.2. Tools

Tools can be considered as an extension of human body; digging sticks extend fingers and knives extend teeth. Or, a tool can be an extension of animal labor, such as the harness that allows single and multiple animals to pull a heavier plow than one person. Technology can be considered an extension of fire: Fire herded animals and made grasslands, Wood fires cooked food and provided warmth; wood fires also shaped metals and provided steam for power; coal fires provided electricity, and oil fires provide electricity and motion. Steam donkeys and steam engines provided work, heat, and motion.

As technologies developed, human relationships with animals and plants changed. Hunting, grazing, and agriculture provoked large ecological disturbances. Early domestic animals were revered, but nondomestic animals were considered competitors or nuisances. Now, animals are treated as processed commodities and wildlife is regarded as useless.

Can the effects of technology be controlled? Tools have effects on human culture, and on nature. What are the effects of tools on ecosystems? Disturbance of soils, and scale effects. What are the effects of tools on humans? The use of knives which reduced the need for some teeth. Tools, simply by being intermediate between the hand and the object, may increase psychological distancing from things. Intimacy with the tool can replace intimacy with the thing. The increase in physical depositories, for memory, changes the kinds of memory capabilities. Tools may contribute to the loss of hand-eye coordination and perception of wild, yet increase hand-eye coordination of tools.

Can technology be separated and kept under control? No likely. Technology has become autonomous. Parts for a single automobile are manufactured in many countries, according to engineering standards and technical economic efficiency. This might lower or increase human standards for pay and health and safety.

What are the effects of tools on human cultures? The complexity of tools leads to rules for their use, and for who uses them, such as trade unions and clubs like masonry. Society is very adaptable to technological change, according to Kenneth Boulding. Perhaps too much so according to Rene Dubos, and the risk is shaping human behavior to tools. The whole idea of technology, according to Evan Eisenberg, is not to eliminate work or to replace nature with synthetic artifacts, it is to restore the balance of work and leisure that hunters understood—to find the best way between the dirt and the mouth.

4.2.3.2.3.2.3. Buildings & Technology

For the Nez Perce, each village might contain from 5 to 15 lodges, a lodge being the household of an extended family, usually three generations. A village rarely exceeded 200 people or 50 lodges. The earliest lodges may have been pit houses, covered by up to 3 feet of earth over grass mats on a frame of wooden poles, with a central opening for smoke. After the Indians acquired horses in the mid 1700s, wild horses from the southwest, and became more mobile, the shape of their buildings changed. During the winter, they lived in A-framed or conical-shaped mat lodges. Mat lodges on wooden-pole frames were easier to dismantle and move, although they were usually just stored in rock piles. In the summer, mat or deer-hide lean-tos were used on hunting trips. Later, deerskin tipis were favored for trips. Sometimes tipi placement was irregular or determined by social relationships.

Permanent settlement led to the growth and intensification of settlement. The engineering of an infrastructure led to massive buildings, monuments, and fortifications. In Susa, after another few hundred years, monumental large buildings appeared, along with larger residences that suggested status differences. In Babylon, the hanging gardens had four wells filled by buckets, 75-ft colonnades, trees 12 feet diameter and 180 feet high.

Development of cities from campments and villages is like that of tidal lagoons, where algae ended with communities. Unorganized, homogenous ensemble transforms to an organized structure that cycles materials, information, and energy in itself.

In modern cities, the patterns have become set. In suburbia, corridors and networks increase, patches increase, but the connections decrease as does productivity. Species diversity may increase at first, then decrease dramatically. The patterns leading to suburbia include the supply of rural land, cheap gasoline, and a network of highways. The gravitational field of a megalopolis is the attraction of size and excitement, the lure of highest paying jobs, and specialization.

Buildings have impacts on ecosystems. They become kinds of ecosystems themselves, but not self-sufficient ones yet. We shape our culture, then culture shapes us, we build buildings, then they shape us, as Winston Churchill recognized of Parliament. Some of our ideas are communicated through memes—things like the ideas of arches, domes, tattoos, buildings, and slavery—thus art promotes cultural evolution.

4.2.3.3. *Attachment to Place*

Ecosystems are embedded in places. The making of places by human beings is an ordering of a distinct structure and center. Attachment to place is a form of deep love, from which many other virtues for living well, such as frugality and humility, spring. The organization of perception, meaning, and thought is intimately related to specific places. Place is a focus of meaningful events and a platform for ordering a world. The individual image of a place is modified by memory, experience, emotion, imagination, and intention.

A place is a part of the environment claimed by feeling. Emotion binds together motion and perception. Emotion can transcend distance. Emotion creates an 'in-place'. A place must be found and made; it does not exist before a priori. Humans, like plants and animals, identify greatly with local environments. Maybe this is a function of the limbic system of the brain, a function we share with territorial mammals. Human emotion creates an 'in-place.'

Neil Evernden asked what it does to a person to see his place, context, destroyed? What does it do? It creates violence or apathy. Modern populations are rootless, moving about from city to city. We are suffering from a placelessness, which arises from our style of efficiency and proclamation of mass values. So far, no psychologists have studied what happens when a person sees her/his place, their very context, destroyed. These catastrophes may be the basis for diseases, depression, or cancers.

The word nostalgia was coined by Johannes Hofer a Swiss medical student, in 1678 to describe an illness characterized by insomnia, palpitations, stupor, fever, and persistent thought of home. The disease could result in death. For the Northern Aranda in Australia, as well as for émigré Russians, it is not possible to stay away from home indefinitely and still live. Nostalgia can be a fatal disease. Thus far, the sense of place cannot be gleaned from an analysis of the nervous system. Yet a place shapes the nervous system, somehow.

Maturity is linked to the increase of identification with, and care for, others. Albert Schweitzer noticed the expanding circle of care from family to humanity to animals, although different cultures have different emphases.

Humans love place (topophilia), as well as life (biophilia) and home (ecophilia); perhaps this love is an instinct or a meme. Four elements in loving have been identified by Eric Fromme: (1) Care, the active concern for life and growth; (2) Responsibility, the desire to respond to others needs; (3) Respect (meaning to look at), the recognition of others' uniqueness; and (4) Knowledge, combination of objectivity with participation and intimate identification. These elements define a loving relationship. The inexhaustibility of a living being or of our relationships constitutes much of the nature of love. Human beings are compelled to seek other beings and love is the most rewarding approach.

4.2.3.3.1. Topophilia

Paul Shepard (1967) suggests that for each individual the organization of thinking and meaning is intimately related to specific places. Experience focuses on a place, which acts as the background for specific events. The features of world are experienced meaningfully. The place is a matrix for ordering experience. The specificity of place is important. The earth extrudes itself into particular plants and animals; flexes mountains; and sweats weather. Places animate (from the Latin anima, meaning 'inspire') people. The inspiration of the sentiment of dependence is called impregnation. Animals and humans are imprinted early in life to particular places (philopatry). Each difference in the landscape has meaning, as when the aborigines of Northwest Australia perceive physical differences and even a symbolic landscape. In fact they structure space according to myth, where Europeans use buildings and roads. Every place has a unique identity, a persistent sameness as a result of combinations of factors. Topophilia, love of place, is the recognition that all human beings have affective ties with the material environment (the word was coined by Yi-Fu Tuan).

The attachment to a place is rootedness. Von Uexkull describes the importance of rootedness in his concept of life-world. Simone Weil (1955) regarded rootedness as "perhaps the most important and least recognized need of the human soul." Rootedness arises from participation in a place. It is the need for order, liberty, security, status, and responsibility. A deep relationship with a place is necessary. Without it, existence loses much of its significance. Caring for a place involves concern and responsibility. This attitude is similar to one described by Martin Heidegger (1960) as caring (*Sorge*). Care is the recognition that a human being is a participant in the world. It is tolerance for the essence of a place; absorption in a place; concern, the willingness to not change or exploit. Care is presented as the total structure of the person: in the projection of being, in being in a particular situation (with limits), and in absorption in the world by virtue of concern for it. Care is the recognition that a human being is a participant in the world. Care promotes tolerance for the essence of a place and displays the willingness to not change or exploit. A home can only be realized through caring;. To have a home is to dwell on the earth.

Care, is, according to Erich Fromm, one of the four elements in loving (along with responsibility, respect, and knowledge; it is the active concern for life and growth. The building of a place to dwell on earth is hard earned; civilization is held together by practical and political activities. Building and planting, taking care of the self and others, is primarily a prosaic human activity.

4.2.3.3.2. Biophilia

We mistakenly conclude that our skin is the boundary to ourselves. But intuition also senses the interdependence of nature. We extend the boundaries of personality to other things and people. The human skin is like a pond surface, according to Paul Shepard. The skin's interpenetration ennobles and extends the self—the beauty and complexity of nature are continuous with ourselves. We know subjectively that we are not separate from the earth, that wolves are capable of love and tenderness, that trees are beautiful. Edward Wilson (1984) argues that the essence of humanity is inextricable tied to life on the planet. He claims that biophilia is the natural affinity for life, and central to the evolution of the mind.

4.2.3.3.3. Ecophilia

Ecophilia is the term to describe the love for living places. One ecological benefit of rootedness is that people will take care of a place if they realize they are going to be there for a thousand years. Having a place means that the inhabitant has stock in it and participates in its unfolding, through planting and caring. Detailed understanding of plants in a locale allow gathering of food and medicine. People cultivating a sense of place are people in place. Their work can be appropriate; appropriate growing, logging, mining, or building. People in place acquire a sense of community, nonhuman and human; a shared set of values and concerns; health and spiritual benefit. When we see land as community to which we belong, we will use it with love and respect.

4.2.3.4. *Making Good Places from Ecosystems*

Good places can emerge from healthy ecosystems. Because human beings live in ecosystems, the description of good places as healthy ecosystems has a human psychological dimension. Good places at this level are good human places, where human activity has invested place with human meaning while allowing the operations of the field and ecosystem to operate.

4.2.3.4.1. Properties of Good Human Places Emergent from Ecosystems

These properties are roughly the same as those for ecosystems. What makes the ecosystem "good" is a human judgment about the feel and use of a place, on whether a place provides basic goods, services and amenities. Are there special qualities that make a place good for human habitation? We have to allow the same process that works to make a healthy ecosystem operate to make good places.

4.2.3.4.2. Describing Characteristics of Good Places

The key to the characteristics of good human places is the addition of psychological and cultural elements to those of ecological places. Good human places emerge from good places, which depend on healthy ecosystems. Both characteristics and activities must reflect or allow the functioning of every level below it.

4.2.3.4.2.1. Action in a Good Place

Action emerges from the special characteristics of fields, places and ecosystems when they are combined with human activities in place. Action is not merely rhythmic change, it is a creative advance, producing new forms everywhere. The organism undergoes a process of evolution in which it produces new forms in itself. In this level, it is human action.

4.2.3.4.2.2. Individuality of a Good Place

Individuality, and the ideas of identity, self-extension and creative ordering, emerge from the special characteristics of places and ecosystems, identity, boundedness, and whole form. And, these characteristics emerge from the characteristics of the field, autopoesis, uniqueness, and wholeness.

The place produces its own individuality. Even with human beings as a component, the place is self-creating. Its wholeness emerges from the interactivity of the components. The place is self-making and self-organizing. It has a characteristic or defining mixture of diversity and complexity.

Just as a place has a unique identity, a human place has a unique identity. The human place has a unique boundary defined by unique physical and or mental limits. The place is a unique local as a host for unique ideas about it. It has a specific size and shape. It has its own integrity, as a whole place

Identity with a place is the extension of Self to Place or the extension of the smaller self with the larger self. Humans identify with places. This identity is a form of rootedness; the place is a center of existence. Identity is the persistent relationship among things.

Humans and their communities are embedded in places. The Taureg of the Sahara have created an image, a world, that could never be relocated to the rain forest of the Campa in Peru. Place is a platform for ordering a world. The individual image of a place is modified by memory, experience, emotion, imagination, and intention. The social image of a place is influenced by individuals, myths, history, and consensus. The images reinforce each other over time. Each place and culture is unique. It is the place that offers specific images.

Human beings gravitate into groups to live. Every culture needs its own local, sacred center, that cannot be broken if the group is to survive. Communication across the barriers of culture is necessary for a world community, but from firm cultural bases. The complete surrender of cultural identity is as dangerous as too little openness. Home requires rootedness, at-easeness, and regeneration. Von Uexkull describes the importance of rootedness in his concept of lived-world (*umwelt*). Feeling at home is a state of awareness; losing the feeling may cause a crisis of anxiety.

As ecosystems mature, so can human beings. Maturity allows the development of the narrow ego of a child into the comprehensive structure of an adult human being. The capitalized concept refers to the development of a deep identification with all life forms. This concept is known in the history of philosophy under various names: The universal self, the Atman, or the absolute. Maturity is linked to the increase of identification with, and care for, others. Albert Schweitzer noticed the expanding circle of care from family to humanity to animals, although cultures have different emphases.

4.2.3.4.2.3. Richness of a Good Place

Richness, and the ideas of diversity, variance, and challenge, emerge from the characteristics of places and ecosystems, that is, from openness, diversity and flexibility. And these arise from the characteristics of the field, from difference, limit, discretion and unfolding.

The boundary of a human place is open to the external supply of energy and materials. It is open to human movement in and out. Its structure, of independent units, allows this. Its functional use and storage of energy allow this. Its typical patterns of connectivity contribute to the whole pattern. As a result of its place in geographic context, it receives challenges in the form of weather and earth movements.

The environment has been constant enough for organic evolution, but variable enough for natural selection to be challenged. Variability, even in small ways, leads to diversity. Variability challenges people to adjust and thrive.

Species richness may stabilize ecosystem properties, such as NPP. Frank and McNaughton (1991) showed that plant community structure (and productivity and biomass) is more stable when a greater richness of species is present.

Reality for an amoeba is less than for a fish and much less than for a human. The richness of experience is in proportion to an organism's capacity to receive and decipher and influence the proximate environment. But the organism receives from reality what it puts in, that is, it enriches reality.

Living in place orders human experience (an aesthetic function). People have similar aesthetic preferences. They prefer natural scenes similar to African savanna: rolling expanses of grasslands with clumps of trees in a warm but dry climate.

Living together results in an ethos. Ethos is the moral and aesthetic aspects of culture. It is the tone, quality, style, feel, mood, and character of their life. For the Aranda, for instance, art and the aesthetic drive overwhelms practical considerations. Place allows us to rediscover a participating consciousness and a symbiotic connection to the living earth. The organization of perception, meaning, and thought is intimately related to specific places.

Good places should be in proximity to other systems and paths, other systems for energy and material exchanges, and paths for human goods and the exchange of ideas. Proximity relates to richness and challenge. What is impaired in the absence of a rich ecology is the individual's knowledge, not only as a person, but as a member of a species. Harold Searle's thesis is that the environment constitutes one of the most important ingredients of human psychological existence. There is within the individual a sense of relatedness to the total environment. He does not draw the limit, as Schweitzer does regarding the concern of ethics, to include the wider sphere of all that lives; psychoanalysis needs to concern itself with the total human and ambihuman environment, inanimate and animate.

Flexibility means looseness, capacity to change, adaptability to place, or change. A flexible, organic network holding all beings and natural groups. Flexibility means not being over connected, not being too rigid or efficient, and able to slough off species or living forms and incorporate new ones. Of course, flexibility depends on the renewability, and on its maturity (trajectory). The system is autopoetic or self-creating. It renews itself as its contents change, and as disturbances change the parameters. As it becomes a mature system, it continues to move to a point of high maturity, recovering from disturbance to its original trajectory, where productivity declines and stability increases.

4.2.3.4.2.4. Conviviality of a Good Place

Adaptation, and its ideas of completeness, conviviality, and cooperation, emerge from the characteristics of places and ecosystems, from participation, development, and complexity. And these come from the characteristics of the field, from registration, integration, connection, and renewal.

The phenomenon of participation comes through an elementary act of observing. John Wheeler asks if the universe might not be brought into being by the participation of those who observe it. Quantum mechanics strikes down the possibility of a neutral observer; participation is vital. The observer changes the universe by observing and measuring. Einstein said that no event can be postulated without the presence of an observer; and anything can be an observer or participant. The participant creates a universe by being.

A deep relationship with a place is as necessary as one with other humans. Without it, existence loses much of its significance. A range of experiences can spring from a place, from depression to the peak experiences described by Maslow. The opposite feeling is possible where there is no richness of place; in the dullness of place, everything becomes oppressive and life becomes tedious. But, this and the drudgery is part of a commitment to place; it is acceptance of restrictions. The richness of life forms contributes to the realization of human values and are also values in themselves; the Deep Ecology movement emphasizes this, also. Simple species contribute to richness and diversity of life. The history of life presupposes an increase of diversity and richness.

Ecosystems with human beings change and turn. The loss from turning creates a sediment, which is history, which is irreversible and gives direction to time. The sediment of the past is a given and different for each present. The ceaseless activity of being in history creates newness. Novelty is born from the womb (*hystera*) of change. History is the result of this hysteretic process.

Humans adapt to place as they live, as their communities live, over many generations. They become attached to a place. Attachment to place is a form of deep love, from which many other virtues for living well, such as frugality and humility, spring. Place allows us to rediscover a participating consciousness and a symbiotic connection to the living earth. Participation means living with other species, the literal definition of conviviality. The organization of perception, meaning, and thought is intimately related to specific places. When the symbols of a world lose their meaning, through wrongful application or abstraction, sickness and disintegration result.

The Masai people, for instance have survived many challenges, including drought and disease, years of colonial rule and partition, the intrusion of the free market of modern economics, and political control by new states, Kenya and Tanzania. They demonstrate that a herding society without political leaders and with social autonomy can survive with large-scale social systems. Many other peoples, such as the Aborigine of Australia and the Desana of Peru show developmental responses to environmental, population and social challenges; they have developed adaptive, problem-solving, intensifying social structures of their own.

4.2.3.4.2.5. Consistency of a Good Place

Consistency and its related synonyms, abidingness, tension, and restraint emerge from the characteristics of places and ecosystems, from stability, constancy, resistance, resilience, and accommodation. And, these are based on the characteristics of the field, constancy, regularity, and endurance.

Stability is the ability to maintain the identity of a system under the flow of external forces and disturbances. Stability can be refined as constancy, resistance, resilience, and accommodation. Stability can be related to ideas of compartmentalization, communications, richness of interactions, and connections. Stability is a complex topic.

Constancy means stability or continuity, in general, or the state of being in a continuous flow or coherent whole for a duration. Resistance is the ability to withstand disturbance to a system. Resilience is the ability to recover from stress. It is similar to elasticity, which refers to the time required to recover initial structure/function after some kind of damage; it is the speed that a system returns to former state following a disturbance (measure in community matrix). Accommodation is the ability to absorb change and still maintain the system identity. The system adapts to disturbance. It incorporates new things. It tolerates new levels of things. It fits the changes within the structure and function of the system.

When humans are part of the mix they may tend to try to preserve short-term stability by reducing the flexibility of the system. This allows more control for minor disturbance but reduces the ability of the system to response to large-scale disturbance or interference. Stability allows security and confidence in knowing, and this allow exploration and play.

4.2.3.4.2.6. Health of a Good Place

Health, harmony and vigor are characteristics of good human places; they emerge from the characteristics of places and ecosystems, such as vitality, productivity, renewal, and continuity. And these characteristics can be derived from the characteristics of the field, from process and flow to unfolding.

Vigor means strength or vitality. Vigor is related to productivity, that is, the ability to convert energy into living forms and the ability to incorporate materials into living forms. vigor is related to the amount and speed of productivity. Health is a dynamic measure of ecosystem organization, vigor, and resilience. Organization is described by diversity and connectivity; and resilience is a measure of reaction to stress. Too much stress can lead to unsustainable patterns of behavior; continuous stress leads to a breakdown of processes that becomes irreversible—the system dies.

Health is a dynamic quality of the whole, the result of a harmonious interaction of all the analyzable parts that comprise the whole forest with the surrounding larger environment. The addition of humans to the mix does not necessarily make a place less healthy or vital. Especially if human adaptation to place is slow over time, giving other entities time to adjust.

4.3. ***Living in Place: Ecotopology (Sociology)***

People live and act in place. They find and prepare food. They find and use materials for shelter. They interact with each other. They create rules of behavior that reflect the uniqueness of place. Ecotopology is the study of people in place.

Certainly, each culture dominates a place on earth. But, their relative adaptations have yet to be contrasted. These many cultural groups are not evolutionary stages culminating in Europe, but are equally valid ways of life.

Humans adapt to place as they live, as their communities live, over many generations. They become attached to a place. Places are ecosystems intimately associated with people, where mutual fitness is achieved after slow progressive reciprocal adaptations; fitness requires a stability of relationships between societies and the place.

Humans create goods. Goods are things that people use. They make them from resources. They can be traded for other goods. The goods can be represented symbolically by other goods, such as blankets, or by money as a symbol for goods or rare equivalents held in trust.

4.3.1. *Mutual Change in Selves and Places*

People adjust to the restraints of an environment. But, their activities change the environment. Lamarck's concept described a gradual process of change due to adjustment to the environment; species changed because each individual underwent change during its life. Thus the temporal transformation of individuals becomes the temporal variation of populations for Darwin. The steady progression of transformational evolution is replaced by the discontinuous generational steps of Darwinian evolution.

Things that are underconnected may be too independent to have mutual restraints. A balance of connection and isolation between systems seems to be optimal for certain kinds of health. Waddington regards laws as constraints; something to be observed to achieve goal, for living systems. General laws governing behavior of biological and social systems are those that must be observed if organizations are to remain stable and survive.

Reciprocally constrained construction is more than adaptation, because the organism changes the environment, also. The organism and environment are co-implicative, co-defining, and co-constructing. They engage in a process of self-assembly, where the self is the organism/environment system. Garrett Hardin wrote that to control population, we need a "mutual coercion, mutually agreed upon."

4.3.1.1. Adaptation of Self to Place

The study of cultural ecology—human adaptation to nature—has been neglected. The dilemma of the world reveals the extent of the neglect. Culture is a sum of descriptions of reality, through language and technology. Ideas of culture will be critically examined. Culture codifies reality through expressions, which can be preserved and transmitted through generations through language. Language is a major element in the formation of thought. It is a means to program our perception of the world. Different languages program events differently, therefore no philosophy or belief system can be considered entirely apart from language, or language entirely apart from place.

There are physical changes, biological and psychological changes, political changes (usually bringing the non-dominant groups under some degree of control and usually involving some loss of autonomy); economic changes (moving away from traditional pursuits toward new forms of employment); cultural changes (original linguistic, religious, educational, and technical institutions become altered, or imported ones take their place); and social relationships (including intergroup and inter-personal relations and new patterns of dominance) to consider.

4.3.1.1.1. Physical Adaptation to Place

Clothing, houses, and fire for heat are adaptations to the fluctuations of place, to adjust to heat gain or loss. People adjust themselves, their behaviors and habits, their clothing and shelter to their places. The physical attributes of a place force people to adapt. Trace elements in animals and plants may not satisfy all human needs, so people may have to trade.

If there are predators or parasites, people may have to elevate themselves off the ground or move away from a swamp. If the desired foods are limited, people have to change their diet or trade for more foods. They may have to add chemical additives to food. Among the changes due to acculturation are physical changes, such as a new place to live, a new type of housing, increasing population density, or urbanization.

4.3.1.1.2. Biological & Psychological Adaptation to Place

People have to adapt to changes in nutritional status as new foods are discovered or lost, or as new diseases appear or disappear. Psychological changes appear at the individual level. It is important to note that no psychological finding can be transported or generalized to other cultures. Thus, the present overview is intended to stimulate and to challenge research on psychological acculturation and refugee adaptation, rather than to be taken uncritically as the proper way to do things or as a statement of the inevitable outcomes of acculturation. Variations, due to factors in the host society, in the acculturating group, and in the individuals themselves, will make the results different from one society, group, or person to another. Individual and group variability influence adaptation. Some environmental features such as rain and cloudiness can lead to stress and depression. Some cultures try to combat this psychological stress with special rituals.

4.3.1.1.2.1. Cultural & Psychological Stress

The concept of acculturative stress, refers to one kind of stress, that in which the stressors are identified as having their source in the process of acculturation. There may be a particular set of stress behaviors which occur during acculturation, such as lowered mental health status, specifically confusion, anxiety, and depression, feelings of marginality and alienation, heightened psychosomatic symptom level, and identity confusion. Acculturative stress is thus a phenomenon that may underlie a reduction in the health status of individuals, including physical, psychological, and social aspects. These changes can be related in a systematic way to the known features of the acculturation process that are experienced by the individual.

4.3.1.1.2.2. Labor division

Chipewyan social systems are adjusted to the hunting of barren ground caribou. Chipewyan hunters depend on animals other than caribou, which migrate out their Indian grounds. The cultural decision to hunt caribou as the primary item of subsistence, however, has produced many similarities between the two species in their utilization of land and in the formation and distribution of social groups. The cultural decision to hunt caribou results also in a population density lower than what would result through other decisions regarding the utilization of resources.

Exclusive pastoralists, like the Masai, have to adjust to the conditions of the environment, especially rain. They also have to coexist with farmers, with whom they may compete.

4.3.1.1.2.3. Marriage

Marriage is the state of making a personal or intimate union for personal, economic, or political reasons. Because of the number of reasons for marriage, as well as the variety of possibilities of combination, marriage can seem quite flexible.

There are numerous other considerations: The paucity of the environment, wife-exchange for political reasons (Yanomamo), sibling marriage for political reasons (Egypt), head shape (Poland), serial polygyny or polyandry, and fraternal polygyny.

Traditional forms of restraint, such as prescriptions for marriages and births or restrictions on hunts, ensured that tribes would not interfere with local animal and plant populations, much less with ecosystem cycles. Neighboring tribes would maintain ties through trade, marriage, and sharing resources (hunting and gathering grounds). Women are more mobile due to marriage.

Female-female marriages occur among the Nandi in Africa, a pastoral society in Kenya. This happens in childless marriages. Then the male female arranges for impregnation by a male consort of the female wife. In the Azande tribe in Africa, there are male-male marriages, usually due to shortage of women.

Tibetan tradition has the greatest range of recognized marital forms, from monogamy to polygyny (many wives), polyandry (many husbands), and polygynandry (group marriage). Why so many kinds? Hypotheses include to: Conserve family property, optimize family labor, maintain reproduction in the household, preservation of preference by all parties, enhance quality of life values. Polyandry in Tibet may be an adaptation to limited resources. With monogamy and every woman married, there would be more children to strain the resources available. There may not be cases where one woman marries several women or a man marries a ghost.

Table 4-31123-1. Kinds of Marriage Possibilities

Self	*Spouse(s)*	*Duration*	*Society*
One woman	One Man	Life Temporary 3 days Late	Irish, Islam Kenya (modern), Nayar (India), England, France et al.
One woman	Several men	Life	Tibet
One woman	One woman	Life (childless)	Nandi (Kenya)
One woman	Several women	?	?
One woman	One object/animal	?	?
One woman	One ghost	Life	Nuer
One woman ghost	One man	Life	China
One woman	One god/goddess	Life	Catholic
One woman	No spouse	Life	Mosuo and many
Several women	Several men	Life	Tibet
One man	One woman	Life or Sequential	U.S. and many

Self	*Spouse(s)*	*Duration*	*Society*
One man	Several women	Life	Masai Islam Arunta
One man	One man	Temporary	Azande (Africa)
One man	Several men	?	?
One man	One object/animal	Until available	Kwakiutl
One man	One woman ghost	Life	
One man	One god/goddess	Life	Protestant
One man	No spouse	Life	Many
Several men	Several women	Life	Tibet

There are societies in which there is no formal marriage. The Mosuo live in SW China. A matrilineal society that lacks formal marriage. Women take lovers at night, who return to their own mother's house by morning. After her coming of age ceremony a girl may have overnight visitors. Children live with their mother. Either may end a relationship, with no obligation; the woman may choose not to open the door.

Of course, humans are part of the environment. The kinds of marriage sometimes have to do with historical precedent or with individual decisions.

4.3.1.1.3. Cultural Adaptation to Place

Cultures adapt to place. Where resources are limited or absent, culture may dictate different forms of use, or even different forms of marriage, so that people do not expand beyond the resources. Where resources are limited or absent, some people learn to trade with others.

4.3.1.1.3.1. Trade for Missing Resources and Luxuries

Ochre was mined in western Australia, and used for ceremonies and rock painting. It became a form of intercontinental trade. People from Celebes visited the northern parts of Australia. Papuans made contact and traded also, and had some influence on language, song, and art.

In eastern Africa, depending on livestock, but not eating it, people come to depend on trade. By 1800 there was regular trade between the Bantu and Masai. The Kikuyu supplied spears, swords, tobacco, and red ochre to the Masai, in exchange for sheep.

People would have traded for metals. Obsidian from eastern Turkey was found in caves in Zagros. Natural asphalt was traded to western regions. Emmer wheat was traded to groups further east in Iran. These commodities were traded regularly in a pattern of exchange east and west. In Southern Iraq, The soil had few minerals. There were few trees. These had to be traded from distant areas. Lumber and stone were floated downriver. Donkey caravans from the Mediterranean brought obsidian and other goods. Ivory combs from the Indus river. Priests needed record-keeping to keep track of traded items. This led to lists in clay and to writing.

Art produces luxury goods, that can enhance prestige with rare or beautiful things. What is the esthetic urge to embellish ones life? This urge, whatever, causes technology to follow trade goods, copper, silk, jade, glass, pottery.

4.3.1.1.3.2. Language

People do not live completely on a planet, the earth. They live in a small place on the planet. As they learn about place, they construct a world, a "man-image," that fits them into their place.

Darwin wrote that the formation of different languages must be the same as different species. Yet, there are differences in form and transmission. The latter happens through the human brain, in our human brain space. Is there just one large brain space that we all share? We are after all a communicative species and our communication defines us.

The language reflects what is in the environment. Many words to describe the soil and plant association. Many to describe pests and predators. But language is more than a medium to express thought. It is a major element in the formation of thought. Language is the means used to program our perception of the world.

B.L. Whorf claimed that language is shaped by the world and in return shapes the world—that is, people can only express what language can express. The Hopi language, which Whorf studied in the 1930s, differs dramatically from European languages in its sense of time. The language does not have past, present and future tenses, which divide time into measured units before and after. Instead there are two modes of thought: (1) objective where things that exist now, and (2) subjective where things exist that can be thought about and are becoming. Each thing comes into being with its own rhythm.

English-speaking people can name at least six kinds of light that are refracted through a prism. The Bassa language, in Liberia, have only two words for differences in a spectrum. The Hanunoo people have names for ninety-two varieties of rice. The Inuit may have over seventeen different words for snow, while the Taureg may have eleven words distinguishing the kinds of blown sand—and U.S. citizens recognize 94 shades of lipstick and twenty-two kinds of road surfaces.

The importance of language and knowledge is critical. People outside their areas and in strange areas may starve. This has certainly been true of Europeans and Vikings. The Franklin expedition to Greenland starved to death in Netsilik territory in 1846. The Burke and Wills expedition in Australia starved to death in 1861 without foraging or asking assistance from the aborigine.

People do not simply make worlds, they keep adjusting the images of worlds. Thus, it might be more appropriate to say the worlding process is constant and contemporary. It is not just a static image formed once and used until the culture dies. Throughout this work, it is to be understood that things are flowing, patterns are patterning, and worlds are worlding.

4.3.1.2. *Adjusting Place to Self*

Human activities have sometimes subtle impacts on a place, from simply changing the microclimates around their shelters to changing patterns of plant growth and animal movements. After adjusting themselves to fit a place, people will adjust their place to suit their purposes. They make changes on landforms, excavating dirt for ponds or wastes; they dump and fill low areas or pits. They stabilize hillsides. They move soil.

They use water; they divert water, adding it to some fields, using large quantities to drink and wash. The levels of water change. Sometimes the water is polluted or lost to evaporation.

They domesticate some animals and remove other kinds of animals. They put some plants in special areas and bring in exotic plants that they like or use for special purposes. They use fire to select some kinds of plants. They graze animals on specific plants.

These adjustments may result in unwanted effects, such as when changes to soil and water levels result in salinity, acidification or erosion. Burning of all kinds may result in pollution, acid rain, or atmospheric warming. Exotic elements may result in irreversible alterations to cycles or extinctions.

4.3.1.2.1. Making Paths

A path is a line of movement through an environment; often the course of the path has different qualities than the environment. A physical structure such as a path generates information for a species and an ecosystem.

Dreaming paths, from ancestors, serve as mnemonic devices to remember location of permanent waterholes or kangaroo breeding areas. People make homes and paths. Often their art diagrams the nodes and lines, as spheres and tubes (things and relations, things and flows). Aborigines, for instance, express physical and metaphysical events and places with circles and lines.

4.3.1.2.2. Altering Cycles & Paths

There are many kinds of natural cycles. For instance, the climates cycle as a result of planetary motion. every being has a life cycle; every ecosystem also. Climate also changes periodically. Long-term cycles, from seasons to sunspots, glaciation, and geological oscillations, create the periodicity, which causes dramatic changes in forests.

In a living system, nothing keeps growing forever; things die and are reborn in cycles. The continuation of the system depends on these cycles. The cycles are bound by limits. Both individuals and communities are usually bound by one or two specific limits. Ecosystem health occurs within those limits. There are many other cycles that affect an ecosystem: Annual climate change, lunar cycles, seven-year cycles. Changes in the length of light, that is when days become longer, cause changes in trees. For instance, in mature conifers, male cones are produced on the longest days, while more and more female cones appear as days get shorter. The length of the photoperiod controls flowering in conifers.

Human energy use has major effects on cycles in ecosystems: Transient perturbations in energy relations, from oil spills or burning can interfere with the carbon cycle; chronic shifts of systems, from dams, irrigation, or chemical wastes, can interfere with the water cycle. In a way, energy does not cycle because it is lost in equal quantities from the system; however, energy may be trapped by cycles on the earth for millions of years—in coal or oil, for instance. If we were to follow a unit of energy from the sun, it would contribute to the heating of the air, be bound in a tree by the action of photosynthesis, continue on in newsprint, be released in a home fire, be dissipated to the air again, adding a little more heat on its way to interplanetary space.

The elements circulate in the biosphere in regular, more or less circular paths called biogeochemical cycles (bio=life, geo=earth). The movements of the chemical elements necessary for life are referred to as nutrient cycles. When one of these cycles varies, the other cycles are effected. Nitrogen and sulfur cycles, for instance, are effected by air pollution. In a mature forest very little material actually leaves the forest. It is held in cycles.

Materials cycle above and below ground, between the atmosphere and trees, between trees and insects, and squirrels and fungus. Chris Maser is fond of saying that most of the cycling is invisible because it is underground or in the air. Many cycles are investigated through ecosystem analysis, where energy and materials are traced through transfers through compartments in an ecosystem. For example in the nitrogen cycle, the compartments are nitrogen, atmosphere, vegetation, forest floor, and mineral soil, while the transfers are precipitation, throughfall, leaching, litter fall, mineralization, fixation, and denitrification. Each compartment keeps the nitrogen for a certain time, termed the residence time; for a hardwood forest floor, for example, residence time is about seventeen years. Human interference in forest cycles can collapse the residence time; for example, clearcutting alder in Washington state causes very high nitrogen losses.

Some material cycles between other ecosystems or in larger patterns around the planet. Some of the cycles are daily; some are seasonal or annual; others are years or decades; a few, like the carbon cycle, are century or millennia long. Human beings do not pay much attention to very long cycles or to those we perceive as not affecting us.

Virtually every material cycles through the forest: nitrogen, carbon, phosphorus, potassium, calcium, sulfur, magnesium, water. Nutrient cycling involves many of these materials. Nutrient cycles change with the succession of a forest.

Calcium is a poison. Calcium is formed by the weathering of rock. Living organisms take it up, putting it in shells or bones. In an undisturbed forest the calcium cycle is relatively efficient, with little leaving the system by flowing downstream.

Sulfur is also present in gaseous form, in dimethyl sulfide and carbon disulfide. Sulfur exists in sulfur bearing rocks, which are weathered into the oceans. Marine organisms use the sulfates to build strong bodies, which then emit dimethyl sulfide (the aroma of fresh fish) into the air. Sulfuric acid droplets form the nuclei needed for condensation back to the land, where it is available to plants for growing. The large sulfur cycle is beneficial to marine and land organisms, important to the general pattern of production and composition.

Phosphorus is a relatively rare element. Isaac Asimov calls it the bottleneck of life, a good example of Liebig's law of the minimum, which states essentially that something always has to be in least supply. Phosphorus is a necessary constituent of protoplasm, but it is sometimes the element in least supply. Most of it is locked up in phosphates in rock, which is weathered and it "escapes" to the oceans. Sea birds play a large part in returning it to land. In fact, we mine guano deposits for fertilizers for our fields. Harvesting fish for food returns some to land, but we make little effort to recycle it.

Cycles can affect each other. For instance, the sulfur cycle regulates the phosphorus cycle by converting it from an insoluble to a soluble form usable to organisms.

Water also cycles. It comes to the earth in rainfall and leaves through evaporation; it is thrown into the air by volcanic action; it flows into the ocean. By far the largest amount of water is contained in the lithosphere, roughly 20 times as much as in the oceans. The ocean, of course, has a lot of water. The smaller pools, in the atmosphere and circulating ground waters, are the most active and the most vulnerable to disruption and pollution. On the oceans, more water evaporates than returns as rainfall; just the opposite happens on land, causing a cycle of water flowing from land to sea, evaporating and falling back on the land. Matter also participates in cycles—not gaseous, but sedimentary, since the reservoir is the earth's crust and not the ocean or atmosphere.

For instance, land is the entire ecological system, complete with other species and biogeochemical cycles, preserves, as well as agricultural areas, resources, and artificial modifications like dams. Atmospheric and hydrological cycles are simplified and disrupted when ecosystems are disrupted beyond their renewal capacity.

4.3.1.2.3. Concentrating Cities (Urbanization)

There are long-scale developments in adjusting places to human wants and needs. One was the concentration of food, crafts and trade in cities, then in cities as religious and military centers. Some cities originated as trade center, (Jericho or Nagar), others as religious centers (Catal Hyuk or Nagar), others as government centers (Brak), and still others as farming centers (Uruk). After 5000 YBP the City of Uruk may have had 40,000 people or more. Perhaps 70 percent of the people lived in the city. But, farmers had to support the nonfood-producing population in the city. Uruk was a model city; it had a wall 6 miles long, 50,000 people and 1000 acres (by 4700 YBP).

Each city had its founding god. Each city had a ziggurat a stepped pyramid of clay bricks as a shrine to the city deity. The temple claimed ownership of some land. These lands were cultivated by small-scale farmers (7-17 acres). Each foreman oversaw ten farmers; each overseer oversaw several foremen. The 2500 farmers could have supported another 1200 administrators and specialists in the city. many of the workers could have been landless farmers. Shifts in the channel of the Euphrates river could have ruined small farmers and caused them to go to the city.

4.3.1.2.3.1. Civilization as Intensification

Civilization arises as a whole complex after domestication, especially plant domestication and agriculture (the Neolithic Revolution). Initially agriculture results in an increase in biodiversity. After several generations however, biodiversity decreases. To increase control, new tools are developed; the fields themselves may be engineered with added irrigation systems. Gradually other ecosystems are taken over. Land use becomes a formal ownership to reflect the investment in all of the above.

Labor shortages to harvest crops led to recruitment and some specialization, especially with tools and irrigation (engineering). The surplus of crops, with its necessity for allocation and storage led to record-keeping and perhaps writing. With labor specialization and records came further specialization, especially with crafts (art) to make standard containers for distribution and professional art to provide necessary images, decoration and luxuries. Economic interaction shifted from reciprocity to formal exchanges. Trade expanded to include special products, tools and artworks.

Social political changes include the rise of political religious elite, then the rise of military organization, then the rise of royalty and kings. Public ideologies followed, as did social hierarchies and warfare. Increases in personal property paralleled rationing and the redistribution of wealth.

Urbanization was the result of permanent settlement and population growth. The growth of the settlement and intensification followed. There was specialization and engineering of a more complex infrastructure, with canals, larger buildings and massive monuments, and fortifications. Religion and trade enlarged in scale. Specialization As result of surplus food and larger population, special people can create a flow of specialized objects. Population has to be large enough to support a market.

The advantages of urbanization included consolidated area and stable home sites, with larger, permanent houses. Public buildings and monuments became more impressive. This formed a critical mass for inspiration and invention, including art as well. Measuring and writing, inventories and commercial dealings followed.

The disadvantages, however, included population increase, an increase in diseases and the danger of violence. There was a limit to public styles of art, ecological degradation of surrounding areas, and increased vulnerability to collapse.

4.3.1.2.3.2. Cities as Artificial Ecosystems

The city is exploitation of a geographic site, which has a multilevel history from geology to plants and humans. The structure of a city might be like a cell, a specialized structure with a transmissible memory in the nucleus. Or like a sponge; the city cannot use sun to get energy but must use surrounding environment (plants) and other organisms in the water. Sponges must circulate food brought in, using energy from that food. A city has own metabolism, that is a network of circulatory structures used for exchanges. Water and wood (energy) are carried by channels to cells of organism (homes in this case). Channels also carry wastes.

As an ecosystem, a city has fewer plant and animal species. Many factors influence its balance with its environment. Water flow is a factor affecting the landscape. Household wastes are another factor. A network of corridors perforate the landscape; the number of small patches increases, and there is a reduction in other kinds of patches and corridors.

Flows include energy, information, people, materials, and pollution. The net productivity goes negative, due to massive imports of food. The size grows. During the mass migration from 1800 to 1991, urban population in developed countries rose from five percent to seventy-three percent. A mass migration in less-developed countries, from 1940 to 2004, the city population rose from three percent to over fifty percent.

This size expansion has caused problems, effects, and side effects, that is, main effects that are unwanted or unanticipated. Villages lower connectivity to the landscape by increasing patches and corridors. Agriculture gets more homogenous, decreases fallow areas. Stream corridors are destroyed by environmental degradation. Connectivity is lowered, matrix is minimalized. The nutrient mineral cycles are disturbed. The atmosphere is disrupted, resulting in drought and storms. Microclimates change, with heat islands and dust domes. There is lower photosynthesis and productivity, with lower diversity leading to homogenization (with cosmopolitan species) and inefficiency of use of energy and materials, and decaying infrastructure (roads, sewers, buildings, houses).

Cities have a budget crisis from the loss of tax revenue. There is a disruption of supplies due to social and political actions, such as strikes, breakdowns and attacks. There is an increase in violence, sickness, drug use, crime, and impoverishment, as well as an increase in mental and physical illnesses from stress.

4.3.1.3. Adjusting Self to Self and Culture to Culture (Mutual Impacts)
Adaptation is not simply human beings adjusting themselves to place or adjusting the place to them. Each is a constraint on the other. For instance, climate affects people, but the activities of people feedback into climate, as tree-cutting dries out the air or as exotic chemicals trap atmospheric heat and icebergs melt.

Changes in climate can feedback into behavior. Gathering is more important in warm climates; hunting in cold climates. Shape of homes related to climate. Climate may affect nonhuman behavior. Cooperation is fostered by difficult climates, more than competition. Shown in the social structure of bird species, such as the Kookaburra, where young stay with parents and help raise new young. It is a rare strategy; of those exhibiting it 85 percent are in Australia. In Australia, the extinction of the megafauna may have changed the entire climate, thus fostering fire as a recycler. Firestick farming by aboriginal peoples may have been the best way to mitigate natural fires. With many low-intensity fires instead.

Difference allows continued difference and concentration. The results of differentiation drive trade between people and groups.

4.3.2. *Going Together in Place with Rules*
The biological term for "living together" is symbiosis. Living together, beings do together. The Sanskrit word for "doing together" eventually became the word "ethics" in English. Ethics (and rights) are simply rules for living together. Human ethics describe only a small, self-conscious part of the rules.

Leopold describes ethics in terms of limits, when he states that the extension of ethics is "actually a process in ecological evolution. Its sequences may be described in ecological as well as in philosophical terms. An ethic, ecologically, is a limitation on freedom of action in the struggle for existence."

4.3.2.1. *Ethics* (from Symbiosis)
In its popular application, ethics is a set of rules governing the conduct of a group of people, based on traditional behavior of that group. Bioethics is an extension of the rules to interactions between medical practitioners and patients, based on medical and philosophical values. Both ethics and bioethics are limited cases of ethics in an ecological context. An ethics abstracted from historical context, detached from other societies, and alienated from nature is academic, insular, and strange. Such ethics tend to depend entirely on local human utilitarian values. Philosophical systems are little better:

Kantian ethics is unsuitable for treating other beings in nature; deontological ethics has a weak subjective foundation in human conscience. The areas of concern of ethics and bioethics are not broad enough; their foundations are not deep enough. An ethics based on ecological knowledge, by comparison, places human behavior in vital social and biological communities in nature—birth, death, illness, and sex all take place within nature. The frame of reference of ethics is enlarged, leading to appropriate behaviors in a larger context. Human health is related to the health of ecosystems. This ethics addresses animal rights and wilderness preservation, as well as human concerns.

4.3.2.1.1. Plain Ethics, Living Alone
The subject of ethics is human conduct, by definition. Ethics are almost exclusively centered on human action. Ethical theory attempts to give fundamental reasons why actions are right or wrong. Most ethical systems are based on utility and control.

4.3.2.1.1.1. Utility and Detachment

The human relationship to the environment depends on detachment from it. Some detachment allows analysis and thought. Too much detachment allows destruction without thought. But, the individual cannot be separated from the environment in which the individuals functions and the environment that functions through individuals (and processes).

People in many modern societies are learning to be uncaring, unattached, and uninvolved, without really thinking about the effects on society and on the environment. Perhaps, it is a result of confusion or fear. We still fear nature because it is uncontrolled and unfathomable. We distance ourselves from what is uncontrolled or unknown. This detachment is the greatest threat to the welfare of nature.

The attitude of distance toward beings excludes consideration of necessary information. Science accepts the sentience of plant life, but does not adjust its methods. Human knowledge grants emotions to animals, but uses them badly anyway. When we can control nature, for whatever purpose, the result is not always good. Our style of efficiency and proclamation of mass values destroys places. The environment of significant places becomes a flatscape. It is turned into uses. John Fowles warns that "We shall never fully understand nature (or ourselves) and certainly never respect it, until we dissociate the wild from the notion of usability—however innocent and harmless the use." This attitude of uselessness lies at the root of our hatred and indifference. Modern disconnection from animals and plants may lead to ignorance of sex and death, but also to film fantasies of sex and death.

The problem with utilitarian ethics is that it permits the use and exploitation of any natural object, including human beings. Based on a limited science, the ethic fails to value those beings and communities for which no use is known. But, the majority of the beings in nature have no human uses. Even ecologists cannot think of uses for many large birds and mammals. This makes a coldness in the heart of our coexistence with other species. The vivisection of the world depletes our ability to feel compassion for it.

4.3.2.1.1.2. Human Center or Human Measure

The first cosmologies, images of human worlds, were anthropomorphic. Human beings saw human forms in every form of nature. With Plato and Aristotle, nature became anthropocentric, it turned around a center, humanity. Humans were the most important beings, at least through the Middle Ages. By 15th century European and Arabic standards, the universe was a rational order. The human place was prominent, and human life had meaning and purpose. Then, the Copernican revolution transformed the universe from geocentric to centerless.

The biological universe, however, was still considered to be a great chain of being. Humans resided between the beasts and angels—until Darwin linked them too closely with the beasts, and cosmology became less meaningful. Then the industrial, scientific revolution restored human importance in a cosmology. Nature is seen exclusively in an anthropocentric manner, as a human resource, especially by John Dewey and others. Even the Biosphere Reserves are justified according to anthropocentric use. To assume that evolution necessarily progressed to humans ascribes an anthropocentric purpose to nature. But for environmental effects, dinosaurs, birds, or whales could be the dominant species. So humans are not the unique end or goal.

In fact, like new dinosaurs, humans are good competitors, suppressing other species and creating their own pseudo-species. All fields of study are trying to confirm that man makes himself, according to Paul Shepard, no matter how the world is made. Geography endorses economic determinism; history studies the rise of Promethean civilization; the arts separate abstract qualities from content; sociology encourages the theme that everything is possible; the sciences present value-free facts. Ideas are no longer connected. All aspects of life have become interchangeable, including soil, water, and land. All concepts of natural seem to turn on the definition of human. The anthropocentric mistake has been the choice of an oppositional logic, instead of a synthetic one. The either/or complex misses the relationship between the terms.

Some of the pre-Socratics developed a nonanthropocentric world view. Zeno the Stoic preached "life in agreement with nature" as the goal of ethics. Chrysippus added that as individual natures were parts of the nature of the whole, therefore, life was to be in accord with human nature as well as nature. The universe is not anthropomorphic, in the image of man. Nor is it anthropocentric, centered around man. But it is measured and valued by man, as, indeed, it is measured and valued by all beings. When humans evaluate ecological situations, preference is usually given to human values. This is unavoidable. But there are other life-images that are measuring parts of habitats. There are other centers.

The concept of reverence allows the center to be everywhere. This idea has reference in earlier philosophical ideas. The atomists advanced the idea of an infinite universe without center or edge. In the 15th century, Nicholas of Cusa argued that the universe was without edge or center—because its creator was infinite and without location. The universe is "a sphere of which the center is everywhere and the circumference is nowhere." The earth was not the center. The modern theory of containment reflects that attitude towards centers and edges—ontofugism, not centering, but scattering, framing. A genuinely nonanthropocentric metaphysics would be a fitting foundation for ecological ethics, a pan ethics.

4.3.2.1.2. Pan Ethics

Humanity is an integral part of food chain and part of an organic cycle of birth and death. Humans need to recognize that they automatically participate in everything, and that they cannot unparticipate by choice. Participation starts at the quantum level, through the ecological and cultural. Human nature does not find meaning in an absurd world, but discovers its structure through interaction with the ultrahuman order. Human identity exists partly in relation to nature; the destruction of one involves the other.

The word ethics is derived from the Greek word meaning 'custom' (Greek *ethos*, Sanskrit *svadha*, Latin *mores*, plural of *mos*, from *meare*), which itself came from the Sanskrit word for one's 'own doing.' Since it was used in the plural, it meant 'doing together.' The word 'morality' comes from the Latin word for will of the people; the singular meant the 'will' of a person. It was probably derived from the verb 'to measure,' as to measure one's way, to go one's way. Morals means the 'way of going together.' Ethics means 'doing together,' which of course one does living together. And, in an anthropometric universe, it is entirely appropriate.

Ethics are assembled inductively, from experience in living in places. Because of the uncertainty of human actions, ethics has to encompass the far past and distant future. No one knew that when DDT killed mosquitoes, it would concentrate in the food chain to kill birds.

Values are time dependent, and ecological time can be very long indeed. The futures we invent are viable only if compatible with constraints imposed by evolutionary past. An ethics that requires a long-range responsibility also requires a new humility, since technological power exceeds the ability to foresee its consequences. An ecological ethic recognizes the moral obligation to leave the world habitable for future generations.

Leopold proposed a conservation ethic, dealing with human relationships to land, plants and animals. The land ethic Leopold had in mind was a sense of ecological community between humanity and other species. "When we see land as community to which we belong, we will use it with love and respect." Such an ethic would change the human role from master of earth to plain member of it. Predators are members of the community; and no special interest group has the right to exterminate them for the sake of benefit for itself. This attitude is important for habitat protection. Leopold describes the extension of ethics as "actually a process in ecological evolution. Its sequences may be described in ecological as well as in philosophical terms. An ethic, ecologically, is a limitation on freedom of action in the struggle for existence. An ethic, philosophically, is a differentiation of social from anti-social conduct. These are two different definitions of one thing. The thing has its origin in the tendency of interdependent individuals or groups to evolve modes of cooperation."

The extension of ethics to animals and land is an ecological necessity. Extended ethics defines a social conduct that is a mode of cooperation and, ultimately, symbiosis. Leopold argued that ethics are voluntary limitations of freedom, necessary in a complex world of which we remain incredibly ignorant. Ethics are developed in response to problems that arise from increasing knowledge. Science has phenomenally increased our knowledge of physical and biological processes. It has now become the basis of our moral code, but it cannot very long be a science divorced from feeling and art if that code is to help us survive. To do this science requires aesthetic perception as well as disciplined thinking and feeling. As there is a rational component to ethical judgments, so there is an intuitive and emotional one, also. An evolutionary ethic suggests that humans avoid tampering with complex evolved systems, not because they are good, but because they are the basis of life at this stage of development. Ecological ethics is situational because ecology is the study of changing systems. The morality of the act is determined by the current state of the system. Adaptive modes should conform to ecological patterns. An ecological ethics is based on attributes of ecosystems and human compliance with ecological laws. The aim of an ethic must be harmonious to the idea of the world's population of living beings.

Ethics is a series of rules for living together. Most sets of ethics make the rules easy to follow. They emphasize the differences (relativism) or similarities (absolutism) of human beings only; or of the individual or the group; or of good feeling, reason, or desire. But ethics has to confront the individual, embedded in a community, located in a bioregion, on earth. And the rules really are not as easy as human systems have presented. Schweitzer made them too difficult, with a constant valuing, but neither are they that difficult. An ontological ethics can be detailed only on a local level—even when it uses a global strategy.

Individualism can be tempered with the concept of common good, the good of the whole. The whole is diminished by individual loss, but the individual is crippled by the loss of the whole; wild children, for example, are never really human. Human good cannot be considered apart from the common good. We are in larger communities like an organ in a body. In *The Laws*, Plato has the Athenian say to a youth that all things are ordered with a view to the preservation of the whole, each portion contributes to the whole, and every other creature is for the sake of the whole. Ethics has expanded in wholes, from the family, to the human community, and to the ultrahuman community, on which all depend.

Pan ethics is supported by the well-being and flourishing of human and nonhuman life; all beings have value in themselves. These values are independent of the usefulness of the nonhuman world for human purposes. Life includes individuals, species, populations, habitats, and all human and nonhuman cultures. Cultures should inspire deep concern and respect. Ecological processes should remain intact. Ethics should be based on specific principles:

- Satisfacium. In a system of ethics, if we decide to have a value, should we maximize value? If so, whose? Then when? Should we maximize it for the present or future for humans or for ecosystems? Or, should it just be satisfactory? Cobb suggests a few possible other principles:
- Selfishness: Act to maximize value for yourself in the present.
- Prudence: Act to maximize value for yourself for the future.
- Utilitarian: Act to maximize value for all humans for the indefinite future (the greatest good for the greatest number).
- Process: Act to maximize value in general (every entity with intrinsic value). Cobb points out that the first three are unstable and unacceptable. That leaves the final principle; perhaps though, we should aim for an optimum.
- Minimum value: Every species has some value.
- Limited good: Desired things exist in limited quantities.

4.3.2.2. *Living Values*

Value is to some extent determined by scarcity. For instance, when clean water is abundant, its price is near zero. As availability declines, price increases—in Idaho, spring water is up to $1.15 a gallon in the supermarkets. Price is still not a good indicator of necessity. Its value for life processes, however, is infinite.

Values also change over time, depending on perceived need, technology, and availability of a resource. Beavers had such a great economic value for hats over a hundred years ago that the continent was explored to find beavers. Feather quills were required for writing pens once. Papyrus was needed for paper. Many tropical plants are promising to provide drugs to cure our civilized ailments.

Economic values can be further broken down by activity, as when a forest is used for scientific research, recreation, education, therapy, political symbolism (as in Idaho, the "wilderness" state), or religion (people pay to experience the cathedral of white pines).

Many of the economic kinds of value have been difficult to quantify. For instance, until people were asked in surveys how much they would pay to have wild forests, existence value, that is, just knowing that a forest exists without having to visit it, did not have a dollar value. Some economists have estimated numerical amounts for *option values*(retaining options for the future) and *bequest values* (leaving as-is for future generations). For instance, Walsh et al. found that a typical Colorado household was willing to pay annually $4.04 option value, $4.87 existence value, and $5.01 bequest value for 1.2 million acres of wilderness in Colorado. Multiplying that by the number of households in the state results in a significant value for wilderness.

The kind of value measured most often for forests, other than stumpage, is recreation. The costs of forest access can be identified and added up. Although the forest is a "free good," transportation, logging, equipment, or entrance fees are often required to visit it. A typical visit may cost $15.00 a day in 1992 dollars. But this does not reflect the value of the experience or the value of the land for noncompatible timber sales. Some economists have proposed a "contingent valuation" by subtracting the actual costs from what the consumer is willing to pay—this surplus value makes the recreational value more objective, because there is a dollar amount.

A final category of value that we should consider is *conservation value.* Natural capital such as forests function to regulate climate, produce topsoil, cycle elements through the ocean and atmosphere. These environmental services (the wild infrastructure) support the economy without providing direct economic benefits.

If you think this is hard, try to figure the value of aesthetic or spiritual experience. Even if we cannot put a discrete amount on these values, common mathematical sense tells us that if the dividend is zero, our share of it goes to infinity—no one can experience what is gone. These "what-is-it-worth-in-dollars?" games are fun to play, or would be, if not taken so seriously by those in power.

We might consider, a "what-would-it-cost-to-replace-it" game instead that analyzes the replacement costs of natural services in terms of human labor and technology. Buckminster Fuller, for instance, once calculated that it would cost just over a million dollars per gallon if we were to manufacture gasoline using chemical processes and electrical power (using California Con-Edison 1972 rates). You might relate this to the thought experiment the functions of the forest are all replaced with human activities and devices, such as giant sponges to hold water in place of nurse logs or new chemical fertilizers to get nitrogen to trees in place of mycorrhizal fungus. We know many of the functions of the forest but not all. In this sense, the value of the forest is quite high. The cost of protection from wind might be$6000 per hectare per year; an equivalent cost of recreation might be two million dollars for a nature park; climate moderation might cost $22,000 per day; and, the wild genes from some trees might be worth eleven million dollars a piece. As you can guess, for a city of 50,000 people, the costs of replacing basic forest services could be billions and billions of dollars per year.

Assigning monetary values to things such as homes and lives is problematic. Ask anyone who has lost a family member or home if the money gained was reflective of the real value. The editors of *The Ecologist*, in their book, *Whose Common Future?*, relate that many people do not allow their ancestral homes to be assigned even the highest monetary value, since they have the "right" to occupy them. For example, Thai *muang faai* communities refuse to let sanctions against cutting trees in community forests be interpreted as prices on the trees, which are necessary to supply rice fields. Many people in industrial countries refuse to participate in questionnaires that ask people how much money would be acceptable to compensate for the loss of visibility by air pollution or loss of forests by clearcutting. Despite the misgivings of the economists whose premises are rejected, the infinite value of forests is only a constraint on economic exploitation and reasoning. Even without financial assessments as a commodity, a forest provides renewal services in the form of water, air, and inspiration, game, wood, shelter, and experience. Refusal to discuss the price of a forest is legitimate communication; it should serve to channel forest policy into larger noneconomic realms. The economic order is a part of the cultural order, which is a part of the physical/ecological order. All discussion of forest assessment and value should be in a cultural and ecological context.

In evaluating ecological situations, we cannot avoid making judgments and giving preference to human values. Not even all human values can be quantified and assigned common dollar values. Even if human values (=all) have objective meaning, not all of them will have dollar values; therefore not all natural processes can have dollar values, but all have some kind of values. R.F. Dasmann notes that birds or dolphins have values to an observer that do not fit any yardstick. Wildlife has a commercial value; fish can be caught and eaten without being produced. Part of the harvesting can support a game value.

Even without knowing a complete dollar total of the value of oceans and forests, we have a small idea of the quantity and value of the interest and could tailor our economic system to that. We price resources by their value to us.

Price is a monotone value. A monotone value is one that only increases or decreases. There are no monotone values in ecology. Desired substances have an optimum value—more calcium is not always better, as might be more money. Ecological principles are limits for biological and social systems. That is, even though we cannot put an exact monetary value on functioning ecosystems, the value becomes infinite if we destroy the system that produces trees or other economic values. A values boundary would be implemented in industrial countries. As vital systems are used for nonessential purposes, like using water to wash a car, the population must be reduced.

We get our values from knowing what is valuable in nature. Values usually encode information having survival or prestige importance. Perhaps the most valuable thing is living time. Then experience of life—aesthetics—is also valuable (aesthetics is from the Greek meaning perception). This may be why humans value walking in the woods or observing the production of art. Natural processes are their own purpose and constitute their own value. A growing tree is; it does not have to demonstrate or prove. We have evolved with trees and forests and water.

Although the theory of evolution, for instance, is not a good basis for an ethics, its perspective can supply *principles* on which we could base values: It is good to remain adapted, within limits. It is essential to encounter the environment to which we are adapted. And, it is essential not to destroy the environment to which we are adapted.

In the larger view, evolution is value-free. Creation and destruction, beauty and ugliness are expressed in one complex pathway. At the same time, a reversal of values associated with an evolutionary, now ecocentric, perspective supports the concept of intrinsic value. Each being has intrinsic value as a perspective, a unique packet of in-*form*-ation and experience; according to Eugene Odum, every being that is part of the food chain has value to many others. Dung has value to a dung beetle, mice to a coyote.

Ambihuman forms have intrinsic as well as instrumental values that we are ethically obligated to safeguard. An ethics can appeal to religious, philosophical or scientific reasoning; or to all three to form a coherent whole. The good (well being) of individual organisms, as beings with inherent worth, determines our moral relations with wild communities. We have responsibilities with regard to natural ecosystems, at least contingent on furthering human values. Rejecting human superiority entails an egalitarian doctrine of species impartiality. Other species fit in places that humans cannot. Why not place the same value on living nature that is placed on money or children? Symbolic or potential value?

We can distinguish four arguments for valuing trees, after Paul Shepard's argument to value animals: Economic, ecological, ethical and educational. The economic is based on the fact that trees are self sustaining factories, that is converters. Nature has economic values, as evidenced by price of wood. Nature has a rich pliability that we can recombine into elements for computers and telescopes. The ecological argument is based on nature as an interlocking system. The ethical is caste-thinking extending membership in human society to trees and animals. Shepard's educational argument he calls "minding animals;" it is not dependent on changing technology or idealistic ethics. Like animals, trees present us with related otherness and further human knowledge of human beings.

The presence of many species is linked to others. We can value all species aesthetically, from a belief in their sanctity, and for their potential contribution to human understanding. Unless the educational includes the spiritual, aesthetic and sacramental, it is too limited. How do we value biodiversity? The diversity itself produces tangible economic benefits, as well as contributing to the maintenance of the biosphere. To some extent humanity is psychologically dependent on diversity, as a result of a long evolutionary history. How valuable is diversity? Obviously, it is very valuable if it can be linked to ecosystem stability. Michael Swift, M. Fukuoka, and others have demonstrated in agricultural systems that species diversity benefits crop productivity; trees, beans, and vegetables, for instance, increase the crop of maize. David Tilman and John Downing, in their 11-year study of grasslands in Minnesota, found that an ecosystem rich in species was able to tolerate a bad drought and recover much better than a species-poor ecosystem.

There will always be conflicts between forms of life. So we will have to value lives differently. Complexity will be related to value; this is why we can kill millions of mosquitoes to protect humans. Both J. S. Mill and J. B. Cobb Jr. have been concerned with maximizing pleasure or feeling. Cobb goes further in considering that those beings that can feel more, humans and wolves for instance, have more value than those who feel less, insects and paramecia. Of course, this is a human subjective judgment. Can we say that the experience of a dung beetle is less intense, because it is less complex? A biotic pyramid is a useful metaphor (image). It can show general relationships in complexity and feeling. A healthy biotic pyramid could be used as a basis for calculating the general value of living organisms. For example: 1 human = 7 wolves = 48 deer = 35,000 willow shoots = 1,910,000 root bacteria. Living organisms cannot all have absolute values—even humans. Although this scale may be marginally accurate for the value of beings in a living system, it would cause problems at the human end. It also is useful to think that the rich and wide bottom is necessary to support the top, where we usually put humanity. Although the pyramid is useful to relate the value of each individual of each species (or association or group), it does not reflect the importance of species. Furthermore, the traditional pyramid assumes that each species is optimally related. But, humanity has increased its biomass so much that it is competing with and reducing the species on other levels. Maybe we should not use as much for a while.

4.3.2.2.1. Natural Value—Linking Beings in Pyramids

The interaction of individuals in a food chain results in the trophic structure of communities, that is, ecological pyramids. Darwin noticed the predation principle. S. M.. Stanley (1981) has argued that the cropping principle may provide biological control of the evolution of metazoans. Intuitively one would expect the introduction of a cropper to reduce the number of species in given area; but the opposite occurs. In communities of primary producers (photosynthesizers), a few species will be superior and monopolize space. A cropper dominates its favorite prey species, usually the most populous, limiting it so others can develop. A new level on the pyramid broadens the one below it. Stanley explains the Cambrian explosion through evolution of cropping herbivores, which opened space for greater diversity of producers, which permitted more specialized croppers. The ecological pyramid became wider and higher.

Mature systems have a richer structure and a lower productivity per unit biomass. There are more steps in the trophic pyramid, a stable form. The rejection or modification of a lower level norm or hypothesis does not destabilize the pyramid, but results in modifications or adoption of a different specific interpretation.

4.3.2.2.2. Inversion of Values & Wilderness

Humans have stripped the world of qualities and significance and claimed them for themselves. By valuing humans alone, value is made subjective, and ends are without value. The human perspective realizes only a small part of the spectrum of rich possibility of experience. Humans use an incomplete source of value for ultrahuman beings; the source is human need. Moral science is not so much concerned with good and bad or right and wrong as with fulfilling needs, such as survival, reproduction, or prestige. Abraham Maslow established a hierarchy of human needs, beginning with food and continuing through social acceptance to self-actualization.

But human needs are based on the health of the earth. Human needs extend to include a foundation of wilderness. Nature, which is self-supporting and self-managing, is the human life-support system. Human systems depend on natural ones, for recycling of wastes, water, and air. But, as human growth is logarithmic, so is human need, and need shapes facts, like kind and quality of resources. As Goethe recognized, all fact is theory, a blend of perception, imagination, and needs. Granting this, the need for wilderness is as much a fact as the need for food.

Wilderness is ecologically important, for it accounts for ninety percent of the energy trapped by photosynthesis from the sun; it is crucial in the global energy system. Trees and plants are good sources of energy. Energy and materials can be produced through photosynthesis. In fact, wilderness is the greatest producer of renewable sources of energy and materials. As habitat for incredible number of species, it is insurance against the dangers of simplified agricultural systems; it is a depository for genetic types. Biological species are essential to the maintenance of ecosystems. Wilderness is the source of evolutionary process.

Is value contingent on human interest or preference? There are beings other than human. Ecology can expand the narrow human-centered evaluation and see things from viewpoint of nature. Lovejoy's Great Chain of Being traces the deductive order in classical nature from Greeks to German idealism. But Lamarck inverted the chain in his theory of transformism. By inverting the great chain of being, Lamarck escaped the direction directive that the perfect must precede the imperfect. Survival of the fittest must be replaced by survival of the ecosystem. The rank order for survival is: biosphere, autotrophs, dependent species, including the human species, culture, community, family, individual (see Figure 4-3222-1. Inverted Pyramid). Value systems concerned with dynamic equilibrium, aesthetics, complementary, reciprocity, justice, interdependence, reconciliation, and intuition are the language that ecology speaks.

Figure 4-3222-1. Inversion of Values

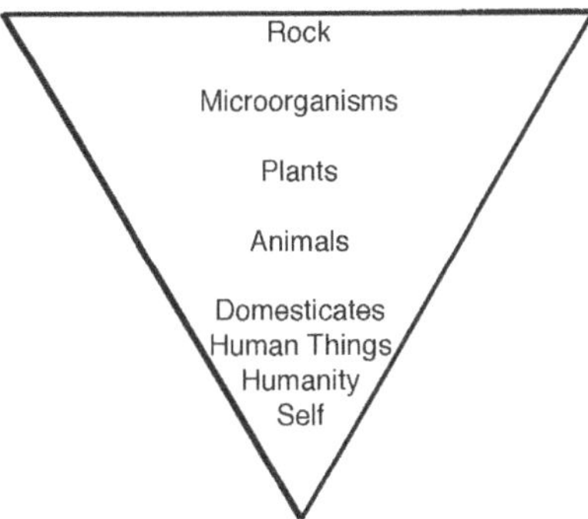

We have been discussing value as if it is just one thing, economic market value indicated by price, but there are different kinds of value. Value theory traditionally distinguishes between instrumental value and intrinsic value. John B. Cobb, Jr. contends that everything can have instrumental value, in the sense of having consequences for other things or beings, whereas what has no feeling, or is not capable of feeling, cannot have intrinsic value. Cobb was inspired by Whitehead and doubtless believes that everything can have some degree of feeling. Perhaps, we can break down value in a hierarchical scheme:

- Self-value—the value of various things, resources, and other beings necessary for a single being to maintain itself—water, food, a niche within a habitat—and this includes human beings, as well as the intrinsic value of the being itself. Of course, at the human level, some things, such as drugs, that are psychologically valuable, are physically harmful.
- Group value (cultural value in humans); some things have value for the species or culture that may not be apparent to or wanted by the individual—in the sense that the predator contributes to the diversity and health of the prey. Some things of value to the group are harmful to the individual. Many human cultures value sacred sites that cannot have economic value. Economic value is simply a small subset of cultural values; market price is an objective, utilitarian value—and in fact the price changes depending on rarity, value-added by human labor, enhanced need (by advertising in any sense), or cultural regulation. Different cultures have different values, obviously.
- Ecological value; ecosystems value stability, or rather predictable change. In this sense, the everglades in Florida depend on hurricanes as much as Ponderosa pine forests depend on occasional fires. Individual beings may have not only self-value but ecosystem value, as when mycorrhizal fungi fix nitrogen and support cycles and systems.

Values are not always hierarchical or consistently ordered, however. Economists cannot account for intransitive preference (also called the Voter's paradox): Where a is preferable to b, b to c, but c is preferable to a; as in Money being preferable to resources and resources being preferable to wilderness but wilderness is preferable to money. Preference curves in economics should not intersect for this reason. A heterarchical structure, which is different than a hierarchy, such as proposed in General Systems Theory, can account for intransitive preference.

Rich sensory experiences can be derived from direct contact with nature. But economists and planners rarely mention these values. Light, wind, dirt, plants, birds, all act during a walk, but not with the meaning of crops or sheep, which is for their utility—they just are. People do not live without these things. All values are based ultimately on a healthy ecology. It must be kept healthy; arable land must be limited, and mineral exploitation must be limited.

The value of wild nature is its independence and wildness. Our astonishment is constitutive of value. If we can admit the independence of nature, that things continue in their own complex way, we may feel more respect for them. We will never understand nature unless we dissociate wild from utility. This human utility can be traced back to Aristotle. Our indifference toward nature comes from its general uselessness. Even many ecologists cannot think of uses for large birds and mammals. This makes a coldness in the heart of our coexistence with other species. Nature is feared as unfathomable and uncontrollable. The desire to save forests, wetlands—all natural ecosystems, is a expression of deep human values. Experience of wildness lets us capture some of our own wildness and authenticity.

A new theory is needed to relate economy to ecology. That theory must include externalities in consideration. And it must include "intangible" cultural values. The concepts of economic value must be redefined. Resource use is an ethical question. It is not known how species and communities are regulated or if by environment or internal interaction. We are so ignorant of the complexities of ecosystems that it is suicidal to pretend to "maximize" their use for resources. A free market has to be limited by conservative calculations of ecological balance. It is almost impossible to estimate the economic value of natural balance. Perhaps it is equal to the capitalized value of technological pest control.

All the true costs of any technological process must be internalized even if it means assigning arbitrary dollar values to aesthetic resources. This does not necessarily mean that an aesthetic resource is worth $1 billion, but that is a potential economic and spiritual loss to the system if the resource disappears. The price system does not adequately account for a number of value factors: i.e., the benefits of people outside a transaction. Value systems are the driving variables in all economic systems, not just peripheral attachments. And patriarchal nations are devaluing these values and cultural wisdom with their dominant abstract, quantitative economics. Economic objectives have to include new concepts of value. A new economic system will have to capture values unimagined by most economists. It may calculate the potential of possible plant products from Amazon basin, or the value of forest susurration here in the Northwest. As soon as species and habitats can be shown to have more value in their saving than in their destruction, they may be saved.

Rich sensory and emotional experience can be derived from contact with the wilds. These values are not often mentioned by economists and planners. But values usually encode information having survival or prestige importance. Perhaps the most valuable thing is living time. This may be why humans value walking in the woods or observing the production of art. The value of wild nature is its independence and wildness.

Wilderness is the given primary reality, the matrix for life-images. If we can admit the independence of nature, that things continue in their own complex way, we may feel more reverence for them. Wilderness evolved without human help or interference, in equilibrium with the physical-chemical environment. We create human landscapes out of wilderness, from wilderness. The value of wilderness is that it cannot be reproduced. It can be apprehended only in nameless present by living senses. It can be known and entered in person, not behind glass.

We find that kind of knowing more in literature than in science. Whitehead turned to literature because it corrected the excess of objectivity that occurred in science. Poetry carries an expression of the organic character and value of nature. Value refers to the in-itselfness and for-itselfness of the process of realization. Besides having value for itself, it shares value with the rest of the universe. Everything has signification in the universe, says Whitehead: "Remembering the poetic rendering of our concrete experience, we see at once that the element of value, of being valuable, of having value, of being an end in itself, of being something which is for its own sake, must not be omitted in any account ..."

Perhaps there is a hierarchy of value from simple forms to complex, corresponding to richness of experience, as John Cobb suggests. Although there is no hierarchy of being, there may be one of richness or value, depending on the frame of reference. There is value for each, and for the whole. And the whole may invert the value. Ecology can expand the narrow anthropocentric evaluation and see things from the perspective of the whole. Lovejoy's 'Great Chain of Being' traces the deductive order in classical nature from Greeks to German idealism. But Lamarck inverted the chain in his theory of transformism; mind is immanent and can determine transformations. Although the hypothesis of inherited characteristics was rejected by Darwin, who shared that hypothesis but denied mind as an explanatory principle, both Lamarck and Darwin inverted the value of life. By inverting the great chain of being, Lamarck escaped the directive that the perfect must precede the imperfect. The result of Elton's food chain was the realization that the bottom link—plants—is the most important.

Ethical values share the same fate. Francis of Assisi was the exception to the general attitude of Christianity: Compassion to man only. He tried to unite the compassion of Christianity and the animistic sense of union with the natural world. Natural processes take on an expression of significance of their own without reference to humanity. All things have an ultrahuman value of their own. St. Francis tried to depose man from his monarchy and set up a democracy of all God's creatures. In parallel, the Taoists saw that we were indistinguishable from other creatures; if we seemed distinguishable, it was through our feelings of self-importance with our reason. Lao Tse turned pyramid of human values upside down. As there are more commoners than aristocrats, there is a net gain to the success of the community.

Ecological value starts at the broad base of the pyramid. Our ethics and legal system is species specific, but it occurs in a larger, normative order. Ecological science has a very normative component. Even a set of genetic codes is normative; genes show what is valuable for a life-image. So value is morphic on all levels of being. A sort of natural relativity of frames of reference encompasses the smaller human system.

Perhaps we should not argue that things have value in the human system. Let us just respect the ultrahuman system. Bees have bee value; wolves have wolf value. Wolves are not efficient at binding nitrogen; neither are humans. Lichen are poor predators, but they break apart rock better than bighorns. The world would not be a better place without sharks, silverfish, cockroaches, rats, hyenas, or whales. Their existence has value; they have functions. From a functional point of view, all beings are equal. In an ecocentric perspective, all beings have intrinsic value and are equally important.

Every being has an intrinsic value, before any utilitarian value to humanity. Associations of plants and animals are just as unique as their components. The value of wild nature is its independence and wildness. Perhaps our astonishment contributes to its value. If we can admit the independence of nature, that things continue in their own complex way without human assistance, we may feel more respect. We can contemplate with admiration, as well as manipulate. The emergence of new moral attitudes depends on a realistic philosophy of nature.

Increases in physical complexity often confer advantages to species in terms of added functions or adaptive value. But there is an optimum of complexity at each level of being that has nothing to do with importance or being "higher." As categories, higher and lower are anachronistic with their connotations of higher and lower. Darwin acknowledged, "Never use the words higher and lower."

Is value contingent on human interest or preference? There are beings other than human. Ecology can expand the narrow human-centered evaluation and see things from viewpoint of nature. Lovejoy's Great Chain of Being traces the deductive order in classical nature from Greeks to German idealism. But Lamarck inverted the chain in his theory of transformism; mind is immanent and can determine a direction for change. Although the hypothesis of inherited characteristics was rejected by Darwin, who shared that hypothesis but denied mind as an explanatory principle, both Lamarck and Darwin inverted the value of life. By inverting the great chain of being, Lamarck escaped the direction directive that the perfect must precede the imperfect. Lamarck's idea of mind, which he thought immanent in living beings, was Bateson's.

The hierarchy of taxa—individual, family, species,—becomes gene-in-organism, organism-in-environment, and environment-in-system. Ecology turns out to be the study of interaction and survival of ideas and programs in circuits. Survival of the fittest (Spencer's original phrase was 'survival of the fitter') must be replaced by survival of the ecosystem. The rank order for survival is: biosphere, autotrophs, species (including human), culture, community, family, and individual.

Furthermore, the Taoists saw that we were indistinguishable from other creatures; if we seemed distinguishable, it was through our feelings of self importance with our souls and reason. Lao Tse turned pyramid of human values upside down. As there are more commoners than aristocrats, there is a net gain to the success of the community. People are wise to be obscure. The tao is a standard of success; Sisyphus can put down the stone. Production and wages do not have to be higher. Success does not have to be fought for, nor vanity satisfied. They are not important.

By valuing humans alone, we make value subjective and end up without value. Sacred ecology reclaims value by placing it at the center of life. Whitehead has stated that existence is the upholding of value intensity; for itself and shared with the universe, from which it cannot be separate. Everything that exists has two sides: its individual self and its signification for the universe. Each aspect is a factor in the other. "Remembering the poetic rendering of our concrete experience, we see at once that the element of value, of being valuable, of having value, of being an end in itself, of being something which is for its own sake, must not be omitted in any account of an event as the most concrete actual something.

A. N. Whitehead states that value experience is the essence of the universe. The value (intrinsic worth) each being has for itself is shared by others in a network of relations and values. Each exists for itself and for others; a value in itself and for others. This is intrinsic and instrumental value.

It has been said that man is the measure of things because he *is* all things; but if all things are all things, then they determine their own value. Humans have stripped the world of qualities and significance and claimed them for themselves. Distance allows objectivity and denies interrelatedness, and sight allows distance. Anthropos may be the measure of all things, by default, but it is not the value of all things. Each living being is a perspective and valuing.

4.3.2.2.3. Health & Ecosystem Medicine

The result of the inversion of life and value is the importance of the health of the system as a requirement for human health. Pollutants contribute to major air and water quality problems, like acidification of lakes, mercury contamination, acid rain, and ground level ozone. For instance, decreased air quality is linked to health concerns, especially impacting the health of the elderly, children, exercising adults, pregnant woman, and people suffering from lung or liver disease. Both environmental and health problems associated with burning fossil fuels have negative impacts on the economy, ranging from pollution clean-up costs paid by the public or increased health insurance costs paid by businesses. Health problems and their costs used to be considered economic externalities, before we realized that the system is internal.

Human health depends on the health of ecosystems. The harmonious interplay between humans and environment results in adaptive fitness, requiring a constant expenditure of effort to maintain. Ecological health requires a physical system combined with high human culture, with a matching flexibility, to create an on-going, open, complex system, characterized by a slow change in its basic characteristics.

The ecological health of civilization depends on an environment that is healthy, that is, has sufficient wilderness to renew air, water, and genetic resources. Humans are complex fields of force that are maintained through effort. To be healthy is to be on good terms with the cosmos. In the taoist view, also, sickness is a symptom of disharmony with the universe. Taking care of health means taking responsibility for the focus of the universe that is the self.

Furthermore, maintaining a healthy ecosystems is more efficient in the long-run economically than having to duplicate the ecosystem functions to keep it healthy. For example, industrial forestry has, without admitting it, demonstrated this in numerous ways, by providing shade cards instead of shade, deer guards in place of food diversity, and plugs of mycorrhizae fungi to replace healthy soil. Maintaining ecosystems will require a new form of medicine.

4.3.2.2.3.1. Kinds of Medicine

The Greek physician Hippocrates theorized that illness was a natural biological event and that the body could heal itself naturally, in time. The Hippocratic corpus focused on prognosis rather than diagnosis, but, it insisted on observation and reason as the basis for any actions.

It was Hippocrates who first suggested the relationship between the occurrence of disease and the physical environment. Hippocrates' insight was that health can be produced and maintained by natural elements, such as hygiene, diet, mental balance, physical conditioning, and a supportive home. Health depended on being in harmony with these things.

Greek medicine was typified by the notion of well-being. The goal of medicine, like Greek philosophy, was to establish well-being (*eukrasia*) or happiness (*sophrosyne*). Happiness was conceived as harmony with nature and self. Democritus wrote that by using the principle of harmony (balance) we can attain calm of body (health) and calm of soul (happiness).

Over the thousands of years since then, medicine has changed and developed. The entire history of medicine can be compressed into six stages.

1. Traditional medicine, which is common-sense and faith-based. Things have to be balanced in nature. Sickness results when things get out of balance. A shaman or physician can return the balance with actions and medicines. People are expected to limit themselves as the shaman or physician tells them.

2. Rational medicine, as the result of observation. The Egyptians noticed that brain swelling could be reduced with trepanning. The Romans noted that people living near swamps got sick, so they drained swamps. The reason for the engineering was common sense, without medical or ecological knowledge of the cycle of mosquitoes and microbes (and the environment is something outside that can attack the body).

3. Scientific medicine (from the 1860s), where the body as a machine can be fixed with surgery, balanced with medication, stopped with radiation, or modified with other tools of the trade. It is based on the chemistry and anatomy of the body.

4. Psychological medicine. The body can be harmed by bad thoughts and improved by good thoughts, that is, the mind shapes the body as they both age. Some diseases are psychosomatic and some cures are psychological, based on the functioning of the brain within the body.

5. Social medicine. The body can be harmed or helped by the thoughts of other people. Shunning can kill, as peoples as diverse as the Inupiat and Amish have found. The love of family and friends can heal. But, of course, prayers can help, that is the thoughts of others can help individuals recover, sometimes, regardless if the individuals even know about the prayers. The effectiveness of this is based on the functioning of human consciousness, e.g., the use of the brains, in groups. There are scientific hospital studies that show the efficacy of prayer on distinct groups of patients.

6. Ecological medicine. These first five areas are defined by different levels of human consciousness or understanding. The environment is still considered an outer thing that fires its insults in towards innocent humans. The environment is not outside, however; we are embedded in the environment, which can direct feedback into physical, mental, and social dimensions. But, our consciousness of the earth still can manifest itself in our modified and artificial ecosystems, and through the degradation and destruction of wild ecosystems. Our love or hate for specific places can influence the state of those places. This is what many call spiritual growth, where the consciousness transcends its limits. People can create ways to avoid painful personal or ecological experiences. People can create a harmony that unites living with the living of every other being in the network. People can cooperate with other people and flow with natural processes.

Despite the history of medical thought, many forms of medicine still exist together and still work simultaneously within a culture. General medicine, as per Hippocrates, addresses the healthy person. Reactive, or scientific, medicine addresses diseases. Preventative or alternative medicine is concerned with environmental causes that can unbalance a healthy person. Holistic medicine considers the spiritual and community effects on disease and health. And, ecosystem medicine recognizes that human health ultimately depends on the health of the environment.

4.3.2.2.3.2. Ecosystem Health

Concepts of health applied to organisms can also be extended to communities and ecosystems. Both are complex whole systems with parts and functions. Of course, ecosystems are more complex than humans, which is also why we have trouble measuring social or cultural health (there is no standard society or standard ecosystem as there is a "standard human"). That complexity means we that should be measuring a large number of variables, starting with soil depth (richness, compaction), then annual nitrogen uptake (often related to leaf litter), trophic flows, species counts, and patterns of activity.

Norms for humans and animals are more definable due to the large numbers and the long history of observation. Ecosystems often have only one sample, if they are very unique. This means that there is no population and no possibility for replication. Therefore, the use of analogy for individual systems may not be very productive.

Environmental health is a set of characteristics of the environment that affect the well-being of humans. Life energy, according to the physicist David Bohm, belongs to the implicate order, the unseen totality that underlies our reality. Health is the harmonious interaction of everything, or every pattern/flow that is part of the implicate order.

The flowing movement of the implicate order is harmony. But since flow and order (in the holomovement) are imperfect and uncertain, perfect harmony, perfect health is not possible. Breaks in harmony are disease. Blockages of flow or stagnant flow result in disease.

C. S. Holling considers that local pockets of chaos keep ecosystems stable by forcing the evolution of new forms to create new niches. Ecosystems reach their fitness near the edge of chaos. So crashes and explosions occur. In human medicine, that chaos seems to be a feature or a sign of ill health. That is to say, health is a form of harmony, not a characteristic of some things themselves.

Health is not a final state, as modern medicine often regards it. Furthermore, the flow has channel or limits. The harmony of motion is more than just parts. Health is a description of the kind of harmony, not the opposite of disease.

The health of any wildlife species in a system is tied to the entire system, as well as to changes at the boundaries. This is especially true of preserved places. Gray wolves in parks have been shown to carry antibodies to canine distemper, canine parvovirus, and infectious canine hepatitis. It is likely that coyotes and fox in the same range also have been exposed to these diseases.

Ecosystem health must become a scientific discipline, addressing losses of biodiversity, habitat degradation, and disease explosions. With an understanding of ecosystem health, patterns of mortality in species may be more be able to be predicted and interdicted.

It is possible to examine ecological health from various perspectives: The emergence and resurgence of infectious diseases of humans and others; the effects of hazardous substances; the health effects of unwise actions, such as fragmentation or other alterations of systems; and, the interdependence of species and the connection of health, also. There was a link between natural history (descriptive ecology) and medicine in Thales and others. Ecosystem medicine reexplores the connections, which never disappeared anyway.

Aldo Leopold started to describe a science of land health: "Health is the capacity of the land for self-renewal. ... A science of land health needs, first of all, a base datum of normality, a picture of how healthy land maintains itself as an organism." Notice that Leopold anchored this concept in a theory of organism rather than an ecological theory of community. Ecosystem medicine uses the broader concept of community to explore ecosystem health.

4.3.2.2.3.2.1. Definition of Ecosystem Health

What is health for an ecosystem? The problem with health is that it is such a general concept, without an operational definition. Since ecosystems exist at all scales, it is hard to determine their health. Stress and pathology are normal parts of living systems; an outbreak in a small ecosystem may be part of a larger healthier system. Furthermore, not enough data exists about what levels of stress or pathology threaten the health of ecosystems—there is not enough data on ecosystems, not enough reference points or control groups.

Health can be defined as the productivity of the ecosystem, without regard to the growth of the system. In the organic world, growth is healthy only when the rate of change is decelerative in the long run; cancer, by contrast, tends to be accelerative.

A useful metaphor for ecosystem health is human health, which has been studied for thousands of years. Our first clue to unhealthy people is abnormal behavior—not moving or breathing, for instance. The first thing we do is to classify the symptoms. Then we measure vital signs: heart, blood pressure, temperature, and maybe white blood cell count. After a diagnosis is made, it is verified usually by more measurements. Doctors, who often rely on their long experience and learning, then discuss a prognosis and prescribe a treatment. So, health is considered the continuity of normal behavior.

Health was also defined generally as the absence of disease. By this traditional way, health is a negative thing; health is not having cancer, infections, high blood pressure, diabetes, or other ailments. The opposite of health is disease.

A definition of health is the condition of being well. The World Health Organization (WHO) defines health as a total physical, psychological, spiritual well-being of an individual. In ecosystems this may be related to diversity. Diversity means species richness, different age and size classes in a population, and genetic differences in a species, as well as the kinds of habitats present in an ecosystem and the kinds of communities occupying the habitats; and the kinds of ecological processes that maintain habitats; and the variety and richness of the planet's genetic heritage in general. Ecosystems that have diversity are usually healthy.

Ecosystems are not organisms in the strict sense, so the analogy is not perfect. The problem with the health of ecosystems is that we do not have the same huge compendium of diseases and symptoms. In fact, we are just starting to compile the stresses and causes of ecosystem illness. Furthermore, we tend to use our human values to judge health in ecosystems; for example, we tend to think that forest ecosystems should exhibit regularity, but many forests, such as boreal forests, are arrhythmic, that is they are punctuated by surprise events (as Holling suggests).

Ecosystems are not just simple organisms; there are other differences. They live longer than humans. The difference in life-times makes discussions of health more problematic. Health can only be evaluated over time. The conditions of ecosystems change over decades or millennia. When do we know if the change is succession or symptoms? Have we ever studied an ecosystem for even one system lifetime?

Concepts of health applied to organisms can also be extended to communities and ecosystems. Both are complex whole systems with parts and functions. Of course ecosystems are more complex than humans, which is also why we have trouble measuring social or cultural health—there is no standard society or forest as there is a "standard human."

To be considered healthy, an ecosystem has to maintain its structure and metabolism (rate of energy use) despite occasional stresses. Robert Costanza proposes an operating index of system health that relates the health of the system to vigor, organization, and resilience (a form of stability). Health is a dynamic measure of ecosystem organization, vigor, and resilience. Organization is described by diversity and connectivity; vigor is related to the amount and speed of productivity; and resilience is a measure of reaction to stress.

The basic medical definition of health used to be freedom from disease. Part of a new definition is resilience to stress—of course, there is good stress (eu-stress) as well as bad stress. Ecosystems respond to stresses in different ways, but usually through a decrease in productivity and material uptakes. Stress may be related to the rate of change for the system, in addition to loss or gain of components or changes in structure.

Ecosystem health is also the capacity for self-renewal, the ability to recover from stress and disturbance, natural and human-caused. A healthy ecosystem is flexible in its response to diseases and pests. Most healthy ecosystems have high degrees of flexibility. As Gregory Bateson interprets Ross Ashby, any biological system can be describable in terms of interlinked variables, each of which has an upper and lower threshold of tolerance, beyond which the system acts pathologically. Within the limits the variables can be moved for the system to be adapted to the environment. Under stress, however, some variables move to maximum values near the upper or lower limits—the system loses flexibility, that is the "uncommitted potential for change" (Bateson's definition), and can be destroyed by further stress.

Signs of ecosystem health include the homeorhesis of the system (a similarity of flow, rather than the stable state of homeostasis, after Waddington), the stability of the system (that is, its resilience after stress, such as floods), the diversity of its components, the continuous recycling of elements, and flourishing. Health is the overall ability of a system to maintain itself under a normal range of environmental conditions, which may include hurricanes, volcanic eruptions, or fires. Obviously, a pioneer community may change the conditions to favor a new level of the system with new components.

Using a concept of ecosystem health as persisting within certain boundaries, but varying periodically or a periodically within those boundaries, management is concerned with actions that might kick the system out of bounds. Furthermore, as an historical process an ecosystem can be considered as having a trajectory.

Health is the coherence of the pattern of living in other words. If the pattern is disrupted, the local entity dies. Many local patterns flow together through time interdependently, sharing materials. The death of one pattern sometimes leads to the death of other patterns. The body of a human or ecosystem or any entity is a dynamic pattern supported by dynamic processes that include other entities.

In Chinese medical tradition, the highest good is harmony, especially social harmony, or good relations. A good person is one who creates and maintains harmony. Perhaps this is the best working definition of health.

Furthermore harmony is related to wholeness. The word "whole" comes from an Indo-European word that is also the root for the words health and holy. The whole ecosystem has a very complete complement of interacting beings. David Bohm, in his theory of the implicate universe, proposes that health is a result of a harmonious interaction of all the analyzable parts that comprise the extricate order—cells, tissues, organs, the body—with the surrounding larger environment. Health is a quality that is grounded in the total order of the environment (or implicate order). Health is a dynamic quality of the entire movement of the environment (holoverse) as it flows. As organisms sometimes interfere with others or with the flow of change, the harmony breaks down—we call that disease. Health is the dance of bodies that interpenetrate (in Paul Shepard's image).

None of the bodies or systems are completely independent or completely bounded; they are interdependent and open systems. A body or system is only maintained by a flow of energy and materials from its surrounding environment—much of this flow is in the form of other entities, usually much smaller, such as prey, insects, bacteria, viruses.

Now we can address health in ecosystems, which are larger entities made up of other beings—large patterns made up of smaller ones. The ecosystem is a constant where every component changes, disappears and appears. It is a pattern like a whirlpool. The pattern forms from a torrent of light, energy, molecules, air, water, and even bigger things.

Comprehending patterns is necessary to protect the scale of the ecosystems that are too large to see (except by satellite), the parts that are too small to see (fungi and viruses), the parts that are too long-lived for us to observe (long successional changes or evolutions), and the parts that we are ignorant about. Without special effort, we are aware only of what we see working in the ecosystem during a very short time.

Our response to the ecosystem, being concerned with its health (as doctors or nurses perhaps), is not benign neglect or complete anticipatory stewardship, it is participation in the process of the ecosystem as a harmonious system, with mutually restrained conflicts and constrained influences.

4.3.2.2.3.2.2. Global and Local Threats to Ecosystem Health

Threats to ecosystem health are well known. They can be summarized briefly as global or local problems. The local problems include such things as the removal of key elements, species, resources, and productivity; the disruption of natural cycles; the introduction of novel elements (as the result of inappropriate technologies); the human take-over of habitats for human purposes, often the result of simple population pressures, but also of greed and sloppiness; and, the extinction spasms in general, but specifically, for example, the decline of amphibians worldwide, due to habitat loss, pollution, and fungal infections (exacerbated by global warming). The global problems include such global things as global warming; ozone depletion (chemical caused); disruption of global cycles; and, contaminations (nitrates, mercury).

These threats cause ecosystem collapse. Ecosystem breakdown happens as a result of stresses, singly or grouped, that relate to interference patterns in the system, most of which are caused by the human species now, although the potential for asteroids or volcanic eruptions remains.

As with people, health is related primarily to lifestyle (or life habits) and not to intervention by a doctor. Alas, the lifestyle of an ecosystem is often determined by its keystone species, and in most cases now humans are that species. There are specific issues in ecosystem health, from the biomagnification of pollutants to habitat destruction and health. Some issues have to do with boundaries, for example, diseases crossing species boundaries; or, plants or animals that cross boundaries; requiring us to track health through several ecosystems. These things force us to reconsider the relationship between biodiversity and ecosystem health.

There are also specific pathways for the infection of ecosystems, for instance, any kind of simplification, from monoculture to habitat degradation or simplification can allow disease to spread faster and farther. The decline in predators or the lack of competitors, to which pathogens are not as adjusted, can increase diseases. Dominance by generalists or specialists, which allows higher pathogen levels, can threaten the system. This is not a comprehensive list of threats to ecosystems. But, it might be adequate to use to develop the concept of ecosystem medicine.

4.3.2.2.3.3. Ecosystem Medicine

Medicine has made great progress in the past hundred years, especially with the invention of machines that allow noninvasive examination of virtually every part of the body and the discovery of drugs that can control moods or modify diseases. It is possible to transplant malfunctioning organs or sets of organs. Some diseases seem to have been extinguished; others have been controlled. The understanding of diseases has been extended to psychological dimensions and to social contexts. Some medical practitioners and researchers even talk about removing death (as if it were a disease) or cloning healthy replacement body parts or whole bodies. Medicine, in combination with advances in hygiene and food technology, has consistently extended the average life span.

But, medicine is not perfect. Some diseases are making a comeback, and new diseases are starting to appear regularly. Some cures, such as chemotherapy, seem to cause more damage in the long run than no cure at all. Some medicines, or at least their effects, including packaging, chemicals, and testing cycles, are degrading human and wild environments; medical toxins, such as mercury, are killing plants and animals. Some treatments drain individuals and entire insurance companies with their costs—and some of the extreme costs are passed on to all individuals, healthy or sick. Some procedures, especially many kinds of elective plastic surgery, have no medical purpose at all. The distribution some medical procedures is centered on those rich enough to pay the recognized costs. Modern medicine has seemed to reach some invisible limit. It is not addressing the changing patterns of illness, scales of treatment, or patterns of social expectations.

Alternative kinds of health care are attempting to address these problems and limitations, but they also seem to be limited to the exclusively human dimension. Public health emphasizes prevention for individual and communities. Other crisis environmental sciences, such as conservation biology and ecoforestry, try to address the unwanted effects of the medicine, but are unable to influence the causes.

A new field of inquiry is needed to focus on the health of the entire system and to reconcile the care and health of ecosystems, populations, communities, and individuals. Using the term first used by D. J. Rapport, this field is designated Ecosystem Medicine.

Medicine has gradually extended its attention to wider domains, from the personal health of individuals to the health of human communities and the health of those communities in an ecological matrix. It will be concerned eventually with the health of environment (ecosystems).

The next step is to integrate ecosystem medicine with environmental medicine and ecosystem science. This means starting by creating a body of cases, then, evaluating the responses to the cases, and finally using that body of experience as a predictive for future treatments.

Traditional medicine has developed traditional medical specialties, including:

Anesthesiology, immunology (allergy, reaction to foreign substances), dermatology, family medicine, gynecology/Obstetrics, internal medicine (which is further divided into specialties: cardiovascular, metabolism/endocrinology, gastroenterology, hematology, infectious disease, nephrology, pulmonary, rheumatology), neurology, pathology (chemical, neurological, and derma), pediatrics, physical/Rehabilitation, preventive (Public Health, occupational health, and aerospace health), psychiatry, radiology, and surgery (which deals with diseases that require operative procedures to restore or preserve function or to cure the disease).

There are rough parallels with ecosystem care. For instance, ecosystem medicine might include: Examination by infrared or other parts of the spectrum (the equivalent to radiology), the study of boundaries or ecotones (similar to dermatology), immunology (reaction to exotic species or foreign substances), pathology (chemical, virology), internal functioning (metabolism, element cycles, infection), surgery (from the French word for "hand-work;" in this case, the removal of exotics), rehabilitation (restoration of structure or function); and preventive efforts (ecosystem health).

Ecosystem medicine covers the spectrum from human organs to the complete ecosystems of the planet and maybe the planet itself. Everything could be considered nested ecosystems, from a body to the planet. Ecosystem medicine is not exactly humanitarian or ecocentric. Although it addresses ethical concerns and the focus is on ecosystems, it is more ecoperipheral; it approaches its subject sideways.

Ecosystem medicine asks a much larger question that is integrative and contextually sensitive: Is the physiology of the base system healthy enough for self-renewal? Focus is shifted from the symptoms of a disease to the entire functioning process of existence within ecological limits. Health is embedded in the context of the system, with all of its constraints, limits, and opportunities for development.

Our knowledge of the world is returning again to a unified image, so it seems every field is converging. Alas, every field is also diverging and looks fragmented—perhaps because we are using the wrong lens.

4.3.2.2.3.3.1. Ecosystem Medicine and Ecological Principles

In medicine a false positive diagnosis is an inconvenience; a false negative diagnosis can be catastrophic. This is true in ecology and conservation biology, as well. Making a diagnosis is a form of gambling, so it should be on the side of caution.

The practical ecology of a scientist needs to be grounded in a theoretical formulation. The science of ecology provides principles that can guide actions. For instance, the importance of scale cannot be ignored; things that work at a small scale do not always work at a large one; large scale rules can not be predicted from small ones. Patterns of interdependence (connection, embeddedness) are as real as bodies. Resilience, the ability to recover from disruption, is a characteristic of a healthy ecosystem. Appropriate measure, the Greek ideal emphasizing minimal intervention, is a proper approach. Diversity is crucial for stability and interest. Approaches must be matched to the type of ecosystem and culture. Cooperation is necessary. For patients, partnership. For communities, the interaction of patients and professionals. A cooperative framework for healing within the depths of communities and ecosystems. Reconciliation (equality of opportunity for treatment) is necessary. Conviviality, basically living together in harmony, according to Ivan Illich, is a necessary strategy. These ecological principles, and many more nor presented here, can guide our actions to make good, healthy places.

4.3.2.2.3.3.2. Actions & Treatments

Medicine can be destructive to individuals, land use can be destructive to landscapes, and ecology can be destructive to animals and ecosystems. The concept of ecosystem health, by bringing the pieces together into a conversation, could avoid much of the destructive behavior of semi-autonomous departments that are not questioned or contained as part of a larger picture.

There are limitations for our treatment of ecosystems. We do not have a list of diseases for ecosystems. It is difficult to see the effects of ecosystem change or disease from the air or from a distance. Ecosystem illness is similar to chronic illness in humans. The appearance of symptoms indicate that the disease began long before the symptoms became apparent. So, the exact date of onset cannot be pinpointed. If so, then prevention becomes more important. And it is related to genetics, lifestyles (of forest entities, for instance), stress, and environmental effects. Furthermore, prevention is not a short-term, one step solution.

Some patterns of resource use need to be understood on the landscape level (and now on the global level). Others need to be examined over a year, decade or life of the system (especially in the Amazonian forest). There has to be a landscape level mechanism (and also a global one now) for regulating common resources. Local diverse patterns are good, but they must be coordinated at the landscape or higher level. It is the normal range of variability that has to be defined. Models need to be created for a wide range of habitat ecosystems.

There must be an institutional framework, with legal and social mechanisms, for monitoring and managing common resources. This is necessary for the health of the system, as well as the long-term health of humans. That means all interest have to be represented, even global corporations trying to profit from disease or rareness (brought about by overuse).

Public health specialists and population biologists look at disease in populations, whereas, doctors and veterinarians consider disease in individuals. The latter intervene while the former observe. The ecosystem doctor has to intervene in ecosystems. Ecosystem doctors care for the basic systems and communities.

Surveys need to be made of every kind of ecosystem. We need to have an inventory of kinds of systems and kinds of changes. How is conserving terrestrial animals part of conserving ecosystem health? We do not know whether animal declines were caused by disease or some other factor (competition or predation). We need to find that out.

Monitoring is the key to understanding changes. Disease needs to be monitored as an important indicator of integrity. Other indicators are surveys of key species, habitat mapping and human impacts monitoring. Complex interactions have to be monitored, using a range of indicators at levels from behavioral to ecological. There may be limitations of the bioindicators of ecosystem health. Perhaps we need to find common and endangered indigenous species and monitor them, hoping that would reflect the health of the system.

Many systems that have been overused to collapse need to be restored. Especially in forests, many small changes have a cumulative effect. Climate change can lead to loss of forest biomass. Change of rainfall patterns led to water stress in adapted species. Deforestation results in other patterns of change, especially in vectors. Reforestation is associated with rapid changes in vectors, which can adapt to nonindigenous species of vegetation. Restoration is a necessary part of reforestation. Maintenance of remnant populations may not be enough to prevent loss of key species on a landscape level.

4.3.2.2.3.3.3. New Goals for Medicine as Part of Ecosystem Medicine

Medicine can continue to advance and confer benefits to societies, but it has to have a larger perspective. The physician had traditional obligations to the patient, to herself, and to her society, but now there are obligations to the environment. All these are partial because all of them are inescapable and unavoidable.

Medicine needs to monitor disease and health, as well as human use of the environment and its consequences.

Medicine needs to improve research for prevention and treatment of new diseases related to ecosystem changes or degradation.

The standard for our efforts should be health—of communities, ecosystems, and corporations—not profits, technological sophistication (although that is sometimes a good thing in use), or fame. The common goal is a healthy environment as human right.

Advertising needs to change. Rather than promote unquestioning use of drugs, advertising needs to promote awareness of consequences of wildlife diseases (and of course the causes, as a result of change or human intervention) and paths to human and ecosystem health.

4.3.2.2.3.4. Summary of Ecosystem Medicine

Less than 30 years ago, the environment was of little concern to most people. Now it is the primary issue for most people. Calvin Coolidge once said that the "business of America of business." The biggest single business in America, now, is the environment. All farming, most pharmacology, most tourism, and many other "industries" have their basis in the health and beauty of the environment. The environment contributes to the largest share of the gross national product directly or indirectly.

Historically, we have used ecosystems without regard to their continuity or to their health. Partial knowledge and technology has allowed us to exploit our environment beyond what is desirable for us or for other species. While moderate exploitation is necessary to live, too much exploitation is unwise. A wise use of resources would not make the world less habitable. We are part of the system and must protect its health as a whole.

A new category of medical professional is needed: People who address the health of ecosystems themselves. Human physicians may need to be able to identify critical environmental conditions that affect human health, but others are needed to identify the health of those systems themselves. Human physicians need to know the basic principles of diseases related to environmental change or chemical exposure; others need to know the principles of ecosystem health and how that is related to human health.

Conventional medicine has great strengths, as does alternative medicine or traditional cultural medicines. Ecosystem medicine must develop such strengths. Ecosystem medicine incorporates all other kinds of medicine as special cases. It can use the best and most appropriate of any procedures. No single approach would work best for every person or every instance.

Using a model of ecosystem medicine offers a comprehensive framework for investigating the problems of health, from individuals to ecosystems, especially the interactions between individuals, social groups, place, and environmental change.

Ecosystem medicine is a medical discipline, aimed at restoring ecosystems to health. As with any medicine, the patient actually does most of the work to become healthy, although the doctor gets the credit and the payment. This leads to respect for the practitioners, but also to more responsibility and more rules. The first rule, which we might take to be basic, is identical to the first vow of the Hippocratic oath, "Do no harm." One way to avoid harm is through noninterference, a basic ecological principle—do not interfere with the health and stability of the ecosystem. The health, diversity, and stability of the ecosystem are a first consideration.

Ecosystem medicine bases itself in a community context and limits the use of the ecosystem to that which the ecosystem can afford to provide and remain healthy over indefinite time. Ecosystem Medicine undertakes the responsibility to preserve the healthy functioning of the ecosystems under its domain. It also has a responsibility for the ecological production of goods from an ecosystem.

This is the local application of ideas for specific ecosystems, that may have a direct link to human health. The systems still have to be self-sustaining and self-renewing, Of course, the planet can also be linked directly to human health. In traditional medicine the organs are more integrated than in ecosystems, which are more loose and more complex.

James Lovelock has emphasized the health of the planet as a single system. The health of ecosystems and human institutions should be measured with a holistic index. We have not developed qualitative indicators of ecological health or quantitative measures of social health, much less an ecocentric view that would value preserves of nature for themselves.

To address the health of ecosystems, ecosystem medicine would be a temporary medicine, not a constant intervention or even a continuous diet. We have already tried to gain complete control over ecosystems through scientific methods and technological applications. We regard modern medicine as a foolproof system that tried to eliminate weakness, disease, and mistakes. It has not. Ecosystem medicine must limit its goals.

One goal is the pursuit of the health of ecosystems and their inhabitants. That goal is good health. Individual health is in the context of community health, which is in the context of ecosystem health. Another goal is security. Symptoms of insecurity are poor human health, migrations, and conflicts (territorial or religious). Environmental security is more than the abundance of natural resources and having a stable social and economic situation. It is the health of the system.

Rather than telling the ecosystems what to do, rather than controlling their growth and components, we need to watch ecosystems to see what they do (this used to be the function of natural history),and we need to let them do it (this requires patience and temperance), with a minimum of interference. Abraham Maslow regards this attitude as "taoistic," and the way to ecosystem health is letting the ecosystem do most of the choosing and working.

Our response to the ecosystem, being concerned with its health, as ecosystem doctors or nurses perhaps, is not benign neglect or complete anticipatory stewardship, it is participation in the process of the ecosystem as a harmonious system, with mutually restrained conflicts and constrained influences.

The goodness of our lives reflects an imperfect balance of love and selfishness, reason and passion, sensuous materiality and spirituality. We have the responsibility to be healthy, to contribute to the health of our community, and to contribute to the health of natural ecosystem communities. Good intentions have to be combined with ecological knowledge and ethical behavior for the discipline to be meaningful. Aristotelian ethics emphasized justice and fairness; Epicurean ethics, gentleness and kindness; Stoics, duty;

and the later Stoa, love, compassion, and mercy. Our rules for living together have to be a compassionate participation in the whole process.

4.3.2.3. *Respect*

Each human being inhabits an autonomous, private world; these worlds are very different and deserve respect. We need to recognize the autonomy of consciousness. We construct our normality from information from the senses, which are tuned by culture. There is a cultural bias that prevents us from knowing how much we are contributing to the world we take in. Inside human consciousness is a bit of place consciousness. Humans are embedded in places. The nonhuman consciousness senses the interdependence of all living things and tries to bring them inside.

Each human being inhabits an autonomous, private world; these worlds may be very different, and should inspire respect. The rightness of human behavior is all they need in terms of special rights. The interests of other beings is the same as for human: to live and experience, to reproduce that similar beings may do so. A new cosmology must recognize the value of total biosphere; respect for all forms of life, present and future; provide equal opportunity for human beings. If we respect ourselves, we must respect other equally purposive, adaptive systems.

4.3.2.4. *Reverence*

While admitting the profundity of natural ethics, Albert Schweitzer exposed its quandary: How are its responsibilities determined? How will concern for our own well-being be properly related to concern for the well-being of others? He felt that the questions were too imposing, that the answers were too individual and subjective for the clear formulation of commandments and prohibitions. But, he neglected that the function of science is to make common answers, and the function of rights is to ensure all can participate. Apparently, what led these thinkers into such a dilemma was the attempt to fit ethics into their world view according to the revealed nature of the universal will to life and to describe it in terms of human judging. Schweitzer considered this an error. Any thoughtful person, reflecting on the quality of altruism, could not help but enlarge the scope of ethical activity until it included all nonhuman life. "We perceive that ethics deals not only with people, but also with creatures."

Schweitzer noted that during the evolution of humanity, the circle of responsibilities gradually widened, beginning with family, then tribe, nation, and humanity—working toward all of life. Similarly, the circle of knowledge widened, increasing the understanding of the laws of phenomena. He felt that the streams were divergent, that ethics could gain nothing from understanding the universe, that there was no hope of finding meaning in natural phenomena: "A philosophy that proceeds from truth has to confess that no spirit of loving-kindness is at work in the phenomenal world." For him, nature had no reverence for life. It produced life a thousand-fold in the most meaningful way and then destroyed it a thousand-fold in the most meaningless ways. Spiders sucked the blood of their victims; wasps laid eggs in live caterpillars; wolves ran down young caribou. He said: "Nature is horrible ... cruelty is so senseless ..."

Figure 4-324-1. Subjects of Ethics

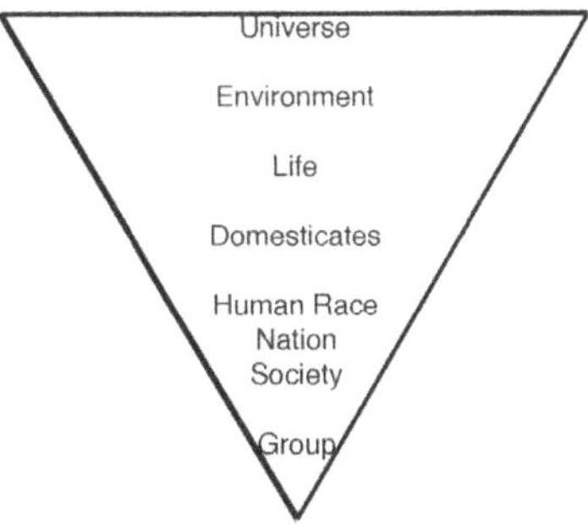

Creatures had the will to live, but had no compassion; they suffered, but had no compassion. The most precious form could be sacrificed to the lowest. In the struggle for survival, nature was maintained only by being in contradiction with itself. This was a contradiction of the will to live, for Schweitzer; there was life against life, suffering and death. Nature was a dreary spectacle of the manifestations of the wills to live in opposition to each other; each preserved itself by fighting and destroying the other. Nature taught only cruel egotism, briefly interrupted by the urge to love one's offspring. A terrible ignorance lay over each creature, as in a dark valley whose floors were perpetually covered by the fog of ignorance and egotism. But man, using his intelligence, was able to climb some of the peaks and catch glimpses of light; truth and goodness appeared.

4.3.2.4.1. Reverence for Life & Being

It was obvious to Schweitzer that the deduction of principles of conduct from the world led, at best to a naive optimism, and at worst to skepticism and pessimism. Pessimism was intolerable, and optimism was incomplete, being concerned only with human relationships. When humanity extends its concern to relationships with all life, when intelligence operates on the will to live within an affirmative philosophy of life and the universe, then the "Reverence for Life" arises. We already possess understanding of the conduct our own natures require. It is our duty to share and maintain life. It does not matter that we have imperfect knowledge of insoluble contradictions, there is an elemental fact present in every consciousness that guides the spirit in harmonious philosophy: "I am life that wills to live, in the midst of life that wills to live." Sympathetic concern toward all the wills to live is the basis of ethics, for Schweitzer. Reverence for life is the greatest commandment in its most elementary form. The negative statement of this occurs in the Bible, Exodus 20:13: "Thou shalt not kill." Other parts of the Bible, Joshua, for instance, display killing as an effective way of conversion.

Schweitzer stated that we must struggle to wipe out antihuman traditions and inhuman emotions, that we must struggle against our own insensitivity. It is inevitable that we kill some things, unknowingly or to survive, but we must never come to take killing lightly— plucking flowers and squashing ants indiscriminately, we must not become thoughtless and blind, because all this killing weighs against us; "everything takes its revenge." The very saving of lives often calls for the sacrifice of others, but even this action is sometimes arbitrary, since we impose our own values on situations. True reverence for life makes no distinction between higher and lower forms. If we were to act so, distinguishing between pests and pets, we must do so in the sorrow of the recognition that we are killing.

Most men are educated with a set of superficial principles, which evaporate when tested; most are brutal, ignorant and heartless without being aware of it. Previously, there was no absolute scale of value, because there was no reverence for life. Schweitzer urged a new renaissance to liberate us from "the poverty-stricken pragmatism" with which we limp along. We need such a spiritual renewal that everyone will reflect on the nature of goodness; our own thought must lead us from naive optimism toward a profounder affirmation of life, helping us progress from ethical impulses toward a rational system of ethics. "In the universe, the will to live is in conflict with itself; in us, it seeks to be at peace with itself." This is an act of spiritual independence on our part, but it carries an element of responsibility to which we must submit: acting toward the good. "The essence of goodness is: Preserve life, promote life, help life to achieve its highest destiny. The essence of evil is: Destroy life, harm life, hamper the development of life."

Individuals must transform themselves from blind men into seeing ones by following the new commandment: Revere life. The quality of personal existence depends more on it than on laws and prophets; it comprises the whole ethic of love in the deepest sense; it is the source of constant renewal for humanity. Those who do not help others by profession must help them as an avocation, seeking out others in need and helping to alleviate their suffering, and in so doing paying off one's debt for happiness already received. "The secret hour does not require of us that we should be happy— to obey the call is the only thing that satisfies deeply."

Humans are to prove themselves in doing and suffering. We are headed right when we trust subjective thinking to yield the insights and truths we need. Kindness does much to make the world better, but sometimes that must be accepted on faith. "Where there is energy it will have effects. No ray of sunlight is lost; but the green growth that sunlight awakens needs time to sprout." Nor may the sower always witness the harvest. Schweitzer thought that the quest for ethics was a hard fight, but that right thinking would leave "room for the heart to add its word." Schweitzer challenges: "Ethics must plunge into the adventure of making its adjustment with nature philosophy.... Let it dare, then, to accept the thought that self-devotion must stretch out not simply to mankind but to all creation, and especially to all life in the world within reach of humanity. Let it rise to the conception that the relation of man to man is only an expression of the relation in which he stands to all being and to the world in general."

4.3.2.4.2. Reverence for Places & Systems

The recovery of implicit natural values can be expressed as a reverence for natural systems. Ervin Laszlo proposes a social ethic for the age of humanity, calling for reverence for the level-structure of the microhierarchy, including all systems on all levels, from atoms to an emerging planetary culture and ecology. "We can express the recovery of our implicit natural values in requesting a reverence for natural systems," he says. This reverence expresses the insight that humanity is in nature, a part of an embracing network of dynamic self-regulating and self-creating processes. Our thoughts and ideas are nature, as much as clouds and waves. Marveling at humanity is marveling at nature, the matrix from which we arose. Although there are others, humans are remarkable examples of dynamic order brought forth by the universe in its history.

Reverence must include all natural and artificial beings and things and fields, from atoms to weeds, computers to galaxies. Atoms, molecules, and organic cycles are parts of humanity. We must revere all the arrangements of earth stuff. The greatest human dignity follows from respectfulness of everything as meaningful as ourselves. Such a reverence treats all substances of the earth as precious, to be used carefully, if at all—and certainly not for a flood of mass-produced consumer items. It includes all human artifacts, manufactures, and societies. It promotes a society where individuals would live in close contact with natural support systems. It provides us with the aesthetic necessities of life to develop all capacities.

The basis of all value is being (the verb form). It is reality undistorted by human needs. Of course, humanity can still enhance nature with its presence, in a nonexploitable manner. The reverence for beings as they are is the law of noninterference. In nature, the law of noninterference means 'letting be' (Heidegger), 'letting alone' (Wilson), and 'not killing for pleasure' (Fox). Noninterference is not indifference, which is diffuse. It is caring. Noninterference will not lead to chaos, poverty, and stagnation. The technocratic vision strives for "life under control," but the earth is self-managing, productive, efficient, and orderly. Reverence can only be felt at the alienness of nature, not at its comfortable conquest. With so little wilderness left, however, formal rules may be necessary.

There is an increase in the scope of ethics, from family, tribe, nation, humanity, to include reverence for all living beings, identified by Albert Schweitzer. An increase in the scope of ethics to include land and forests, identified by Aldo Leopold. An increase in the scope of law to include legal rights for forests, identified by Christopher Stone. Aldo Leopold proposed a land ethic, which was a sense of ecological community between humanity and other species. "When we see land as community to which we belong, we will use it with love and respect." Such an ethic would change the human role from master of earth to plain member of it. Predators are members of the community; and no special interest group has the right to exterminate them to benefit itself. This attitude is important for habitat protection. Leopold describes the extension of ethics as "actually a process in ecological evolution. Its sequences may be described in ecological as well as in philosophical terms. An ethic, ecologically, is a limitation on freedom of action in the struggle for existence. An ethic, philosophically, is a differentiation of social from anti-social conduct. These are two different definitions of one thing. The thing has its origin in the tendency of interdependent individuals or groups to evolve modes of cooperation."

The extension of ethics to animals and land is an ecological necessity with human pandomination. This extended ecological ethics defines a social conduct that is a mode of cooperation and, ultimately, symbiosis. Leopold argued that voluntary limitations of freedom are necessary in a complex world of which we remain incredibly ignorant. Extensions of ethics are developed in response to problems that arise from increasing knowledge.

As the circle of ethics was enlarged to include the realm of all living things, it could be stretched to include all things—at worst, this is only a pantheistic Monadism. The religious argument for this extension would be that God created everything. A scientific argument would begin with the difficulty of defining life. A genuine affirmation of instinctive patterns is necessary for survival. Adaptive modes should conform to ecological patterns. An ecological ethics is based on attributes of ecosystems and human compliance with ecological laws. The concepts of rights for nature will be examined etymologically. Science could demand an ethic directed to the preservation of life in its mosaic setting. But only a religiously conceived ethic has done so. And Schweitzer's reverence for life is the only one visible in Western world. His reverence for life principle acquires a new aspect when it is restored to ontologically firm ground. Arguments against killing and the consideration of only human values become untenable. The world becomes a synthesis of values with the mysticism of religion, characterized by love, compassion and the reverence for all things. Ethics has expanded in wholes, from the family, to the human community, and to nature, on which everything depends. With Ervin Laszlo, ethics encompasses every system.

4.3.2.4.3. Wooing the Earth

From Tagore, Rene Dubos takes the heroic love-adventure of humankind to be the wooing of the Earth. The wooing of the earth is both sweet and sour; sweet because humans can create enchantment within nature, and sour because they can spoil desirable places—possibly to the point of ruining nature's recovery mechanisms. The wooing of earth implies more than converting wilderness into humanized environments. It also means preserving natural environments in which to experience mysteries transcending daily life and from which to recapture an awareness of the natural forces that have shaped humanity.

Dubos celebrates the human ability to woo the earth by creating new environments that are "ecologically sound, aesthetically satisfying, economically rewarding, and favorable to the continued growth of civilization." While that is noble and just, and realistic, there are two erroneous assumptions contained therein. First, civilization does not have to grow to develop, and it is development that is important, not growth, which is plainly unfeasible. Second, "to woo" means to solicit love or to make love; Dubos assumes we will remake nature in our image. And love, as everyone should know, entails respect or reverence. It does not remake the loved one in one's image, it respects and allows freedom. That is the larger meaning. We should "improve" on nature where we live, but we do not need to improve on all of nature. We do not need to domesticate all beings for human use. J.W. Krutch recognized that ecology without reverence or love is only shrewder exploitation.

We can create places only by living there; by slowly adjusting all to all. Human modifications of earth can be lastingly successful only if their effects are adapted to the invariants of human and physical nature. Ecological management can be effective only if it takes into consideration the visceral and spiritual values that link us to the earth. Although some of nature can be regarded as a garden to be cultivated, large areas should be preserved as untouched reserves or necessary systems for global balance. Other areas could be studied or urbanized as population centers. Our tendency to redesign nature, rather than tolerate and cherish, is dangerous; it leads to "curing" abnormal people and "solving" weed problems. We do not respect the wild, complex side of human nature.

4.3.3. *Making a Good Society in Place: Characteristics of Good Societies*

Because the characteristics of good societies depend on the characteristics of good ecosystems and good places, good ecosystems and places have to be preserved. These characteristics are both nested and dependent (see table 4-33-1).

Stable, good societies must pass the social institutions to a new generation, but without sacrificing everything for posterity. The traditional object of economics is the administration of scarce resources; this needs to be extended to a line of generations. There is the question of the quality of life from one generation to the next; possibly the distribution of a dowry to all succeeding generations.

Table 4-33-1. Contrasted Characteristics

Field	***Place***	***Ecosystems***
Process	Dynamic Change	Course
Autopoesis	Self-making Wholeness	Identity
Differentiation	Differentiation	Openness Diversity
Integration	Integration	Coconstrained Construction
Constancy	Constancy	Stability
Metalysis	Renewal	Productivity
Culture	***Good Places***	***Good Societies***
Conduct	Action	Method
Wholeness	Individuality	Self-extension
Flexibility	Richness	Variety
Adaptation	Conviviality	Cooperation
Constancy	Consistency	Loyalty
Vitality	Health	Harmony

4.3.3.1. Method as a Characteristic of a Good Society

Because human beings exist as part of the natural order, the characteristics and principles that underlie good societies also underlie ecosystems. Method is such a characteristic and it allows the development of society through differentiation and recombination. It allows the development of society through the use of energy and self-maintenance. Growth, for instance, allows society to avoid stagnation or collapse, although only for a time, then it has to split or adjust its size through emergency measures. Feedback, positive or negative, is method for society to adjust to various natural or social limits. Feedback can allow a society to deconstruct or reconstruct itself as it progresses. Society has to weave a course between its successful competition to support and renew itself and its tendency to overdevelop and destroy its habitat and place.

Human societies in place use methods to create their societies and these methods have characteristics based on the characteristics of files, ecosystems, and good places, such as process, change, course, conduct, and action. Methods are the motors of change that drive the development of societies. Methods may conflict or fit together but their action causes human bonds to be more flexible. Method addresses the efficacy of society; it is method that can fragment the image of or world or place and reduce fitness. It is method that can work up appropriate beliefs and behaviors. Methods of observation, for instance, can determine what is seen or important. Methods of behavior can lead into the character of the society. The world and society are accessible through methods of describing, analyzing or synthesizing. A poor method can lead to the fragmentizing of society or nature.

A method is a way of doing something or the regularity and orderliness of the actions of doing (from the Greek words for 'pursuing a way'). Goethe said that "the methodology of forms is the methodology of transformations." Goethe attempted to use his method to produce an organic and morphological world-history. This methodology is part of nature. It reflects nature. Rilke wonders whether or not all the dynamics of nature, including those of human society, are hieroglyphics of the methodology of thinking. Early science saw the world as mechanism; modern biology is seeing it as resembling an organism; perhaps it will be seen as spirit or as a composite of all. Science accepts the sentience of plant life, but does not adjust its methods. Human knowledge grants emotions to animals, but uses them badly anyway. We turn inward from the earth, for human purposes. The scientific method multiplied human dividends with the industrial revolution, but it destroyed the fabric it was examining. New ways of thinking are being developed in terms of instructions and normative modes, and that these are beginning to provide methodologies capable of achieving the necessary, challenging tasks.

4.3.3.2. Self-extension as a Characteristic of a Good Society

The characteristic of self-extension, and its corollary ideas, uniqueness, design and centering, emerges from the characteristics of good human places, ecosystems, and the field: Individuality, Self-extension, Ordering, Identity, Boundedness, Autopoesis, Wholeness, Form, and Uniqueness. It is those characteristics modified by adaptation and ethics.

The individual extends far beyond the skin. Extent can be more than just in space. People extend themselves in time by planning for their heirs. The extensions are connections, not only physical connections but intention as well. Extending the place into other ecosystems allows nature do the work. Whitehead seems to find it necessary to divide the cosmic process into two characteristics, extensiveness and aim. Extensiveness means that the process is spread through space and goes on in time. In fact, he considers extensiveness more basic than the arbitrary factors of four dimensions or "electromagnetic laws". Extensiveness is the binding of the physical world by relatedness.

4.3.3.3. Variety as a Characteristic of a Good Society

The characteristic of variety, and its corollary ideas, valuation, consciousness, and self-analysis, emerges from the characteristics of good human places, ecosystems, and the field: Richness, Challenge, Diversity, Openness, Flexibility, Difference, Discretion, and Limit. It is modified by adaptation and ethics.

Flexibility is potential for changer within a system. Variety provides flexibility in systems. Flexibility is needed to correct our mistaken allegiance to the machine image of life. If we survive then we need to always preserve more flexibility for future mistakes. As we know, it is impossible to avoid disorder, which is a necessary part of any physical and social system. A system with flexibility can incorporate disorder without being destroyed. What barriers are there to renewing institutional flexibility? or restoring ecosystem flexibility.

The realization of hard realities does not mean a return to a dark age. We will have more flexibility if we choose our way, and salvage much of the industrial revolution. Natural systems are characterized by resiliency and flexibility, and high productivity, too. We need efficiency, but a higher efficiency—an efficiency in life, not one department of life. There must be a positive flexibility; potentials not used must be preserved for future use to accommodate dead ends, mistakes, or change.

Civilization will have to follow a circuitous route, practicing rigorous self-discipline and economy, more than if we had started earlier. There must be far fewer people than now; the mess and heat generated by vast multitudes will always be ruled out by the natural laws within which all things function.

Zonal diversity can be used to develop different resource production systems in a local area. This increases the likelihood that production will fluctuate nonsynchronously so that support for human populations is constant. Of course, keeping flexibility and reserves could have the same effect. By forming a complex society, the scale of production is raised from a local group in limited territory to a regional population in diverse territories. Even hunter/ gatherers engage in similar "energy averaging" systems. A smaller size could mean more flexibility and faster response to local conditions.

Plans can be made within the limits of variables, although it is not safe to be limited by lethal variables, as Gregory Bateson recognized; closeness to limits reduces flexibility, that is, uncommitted potential for change.

4.3.3.4. Cooperation as a Characteristic of a Good Society

The characteristic of cooperation, and its corollary ideas, awareness, complexity, intensity, and patterning, emerges from the characteristics of good human places, ecosystems, and the field: Conviviality, Adaptation, Awareness, Intensity, Development, Complexity, Participation, Integration, Connection, and Infolding. It is modified by adaptation and ethics.

Cooperation implies convivial or appropriate technology. The word convivial means social, from the Latin word "convivium," meaning living together. Ivan Illich, in *Tools for Conviviality*, sketches a meaningful community where workers have control of their tools and their lives.

There would be a selective reduction of industrialism. Large industry might still be needed for certain things, such as televisions or computers, but the technology of the future must be valid for all peoples for all time. There should be a proper mix of handicraft labor, intermediate technology and heavy industry. Economics would become characterized by friendship and cooperation. The root problem is how to live with technology in a mature manner. We need an ecological awareness at all levels; a human, existential ecology, where humanity is part of system, with an awareness of awareness.

There is no reason not to develop complex instruments to monitor and analyze the environment. Machines do not need to be dismantled. Technological developments are more easily assessed by a small, self-supportive community, where they are not necessary. Necessity was not the mother of invention; curiosity was. And curiosity needs time, not pressure. Small communities could use sophisticated but unobtrusive technology. Evolution occurs in small populations, demes, in which a mutation has taken place. Thompson claims the metaindustrial village is such a deme. With desk-top computers and microfiche libraries, satellite and cable television, advanced science would be possible in the most rural setting. Kohr states that small firms have been shown to be more productive as separate entities.

Human power is found to be incredibly wasted by nonproductive ways. By giving technology a human face, Schumacher proposed that directly productive time in our society should be tripled, at least. Engaging in real work, as opposed to busy work, would be of tremendous therapeutic and educational values to workers. In tandem with the Intermediate Technology Group, of which he is a part, he advocates production by the masses to replace mass production. This group, ITG, complements another headed by Victor Papanek, as described in *Design for the Real World*, that has invented technologies appropriate for underdeveloped areas, capable of making everything from televisions to automobiles. Schumacher believes that 3rd world development can only be possible with appropriate technology. In order to be applicable, such technology must maximize opportunities for the unemployed, without trying for a maximum output per worker, as is the case in highly industrial countries. Increase value on labor by one-to-one with product. Productivity will go down, but quality will go up.

A biotechnology would function most effectively at the lowest levels of society, by being comprehensible by the poor. It would be based on ecological and economic realities, and it would permit economies in local communities with a minimum of financial support.

The main concern of intervention into nature is productivity and efficiency, measures of economic perfection and reward, but the end products produce profound changes on the quality of human life and environment. But these changes can be diminished with new methods of production, new patterns of consumption and new life styles.

Technical progress in certain directions should be dictated by an awareness of the need for appropriate technology. Miniaturization of technology reduces the scale of impact of industrial civilization on the biosphere and correct the maladjustment to nature. Scales of economies are reduced. Cybernetic technology would allow citizens to participate in electronic democracy.

Schumacher dispels the illusion of the myth of production. A new economics must correct the failing of the old, the inability to distinguish between capital and income. But the solution is really a personal one. In "Peace and Permanence," Schumacher contrasts the Keynes plan with the Gandhian ideal: peace depends on wisdom and wisdom is found only in ones self—not being greedy, using good sense.

We need open our hearts and let empathy catch up to creativity. We look back to where we have been in history, gather up the old economies, and turn on the spiral in a new direction (spiral history, turning of universe, in and out).

With modern electronics and miniaturized technology, we do not have to return to the idiocy of rural life. History is a spiral, not a circle. We can spiral back to the country—old labor turned to form of art and sacred ritual. Ritual of communal labor might bring community together for a few hours a day in global village; rest of day devoted to individual creations. According to Thompson, in such a village, nature could be great, machines tiny and man in proportion. In contemplative cultures being is the source of identity, as it is in nature.

4.3.3.5. Loyalty as a Characteristic of a Good Society

The characteristic of loyalty, and its corollary ideas, equity, attachment, investment, and love, emerges from the characteristics of good human places, ecosystems, and the field: Consistency, Abiding Restraint, Tension, Stability, Accommodation, Constancy, Investment, Love, and Regularity. It is modified by adaptation and ethics. Loyalty is the constancy of a place modified by attachment to and investment in a place, by the members of society.

Loyalty is defined as the faithful adherence to an ideal, cause, duty, nation, or place. It implies an obligation to support or defend those things. People give their loyalties to family, culture and place. Loyalty can form the context for agreement or dissent.

Loyalty can be corrupted or bought, however. Capitalists, for instance, can generate consumer loyalty through a massive scale of gift-giving, which allows the capitalist system to survive for a while longer.

4.3.3.5.1. Loyalty & Stability

Loyalty increases social stability, whether loyalty to the family, place or political system. As loyalty is transferred, the criterion of rights is transferred from the nature of 'man' to the community of persons; both law and justice, as well as obligations and rights, are reduced to equity issues. Thus, new nations demand to participate in a common justice, as opposed to the extension of natural rights.

4.3.3.5.2. Plenitude & the Equity of Wealth

A powerful arbitrary idea, such as the Christian principle of plenitude, can influence many cultures over centuries. The principle states that an intelligent creator gave an earth of unlimited bounty to humanity for their use; this seemed to be confirmed in the Renaissance with the discovery of the richness of stars, microscopic life, and unexplored continents. Many modern political ideologies and economic systems have been shaped by the principle of endless wealth. Adam Smith calculated that the real price of anything was just the toil acquiring it. These ideas are parallel to the idea of unlimited good, where anything, even virtue, can be multiplied indefinitely. The invalidity of this principle comes with the recognition of limits. Without limits any good becomes devalued and is wasted. The universe is limited; the earth is limited; individuals have limits. These metaphors are defective because they do not fit our surroundings. Our modern cosmology, with its basis on machine metaphors and the principles of plenitude gets in trouble because it does not understand how basic the concept of limits is to the physical universe, to life, to ecosystems, and to human constructs, such as cities and economics.

4.3.3.6. Harmony as a Characteristic of a Good Society

The characteristic of harmony, and its corollary ideas, balance, education, care, and acceleration, emerges from the characteristics of good human places, ecosystems, and the field: Health, Meaning, Vitality, Productivity, Renewal, Cycling, Process, Movement, and Flow. It is modified by adaptation and ethics.

Harmony comes from the Greek word *harmos*, meaning 'fitting.' Extended definitions of harmony include: The combination of parts into a pleasing or orderly whole; or, an agreement in feeling and action. To harmonize means to associate different things in a proportionate arrangement. Proportionate is the due proportion of two things having a reciprocal relationship or an equality in measure. If nations were defined as having an equality of measure, not necessarily size or force, then the harmony of nations could be promoted.

In Chinese medical tradition, the highest good is harmony, especially social harmony, or good relations. A good person is one who creates and maintains harmony. And, harmony involves the body and emotions. Perhaps this is the best working definition of health. It is harmony that comes from adaptive history. It is not the musical kind exactly, but more like the mutual restraint of groups of organisms.

Harmony is the agreement of method, extension, variety, cooperation, and loyalty in a social and ecological context of good societies in good places. Harmony in a society is the agreement of actions, feelings, ideas, and interests that results in peaceable relations. Harmony is society can be reduced or destroyed by cheaters, 'free-riders,' or anti-social individuals. But, society itself may select for traits of reciprocal altruism or altruistic punishment in other individuals.

Harmony is mutual constraint plus a shared adaptive history. It is constraint that limits the scale of a society. Harmony requires understanding and then planning for adjustments. Harmony requires adequate ecological information and the infusion of its significance in human affairs. And, it requires the time to do this, with the practices of caution and reverence.

4.3.3.6.1. Opportunity for Meaning

People need to have the opportunity to find meaning in work, community, and their cosmology. Livelihood is vital because it is the source of meaning. It represents the place in the order of things. Losing it means a loss of meaning. Disconnect and dehumanize. Then blame for being disconnected and dehumanized.

The definition of wealth must be enlarged to include meaningful work that would bring fulfillment in addition to a living wage. It must include the whole wild environment that provides services.

What determines the quality of life? Energy, luxury, affluence or abundance? Gratification, strong community with meaningful relationships? Sacrifice? The greatest human dignity follows from respectfulness of everything as meaningful as ourselves, the entire earth.

4.3.3.6.2. Balance for Health

Health is a balance of the material and spiritual, between challenge and well-being. Furthermore, the organism must adapt to the environment, which implies having a memory and being capable of learning, and must reproduce, that is, duplicate its pattern in a separate being. Organisms are goal-seeking, and often stability is sought above change or complexity. The individual is a subject centered in a milieu. Because of this implied point of reference, Rodman concludes that ecology is teleological. Often organisms strive for well-being beyond just survival. Their goal is to come into the fullness of being. A. N. Whitehead considered that all organisms have three urges: to live, to live well, and to live better. Living better is being more attuned, stimulated, flexible, receptive, spontaneous, and integrated.

The well-being and flourishing of human and nonhuman life; all beings have value in themselves (synonyms: intrinsic value, inherent value). These values are independent of the usefulness of the nonhuman world for human purposes. Life includes individuals, species, populations, habitats, and all human and nonhuman cultures. Deep concern and respect for cultures. Ecological processes should remain intact.

Psychologists have recognized the need for diversity for people's quality of life and emotional well-being. Ecological diversity in the forest has been reduced by human activities, such as planting or grazing. The overall landscape diversity of many forests has not fared as badly, due to the addition of human artifacts, which increase it. An increase in ecological diversity would lead to an increase in diversity of the landscape, however.

Ecological forest design is the design of communities. We should design places as organic wholes to promote the well-being of individuals and the common good. The immediate goals of design are to reverse degradation and reclaim places for communities.

Now, a place is a society of plants, animals, fungi, and humans, with biological, economic, and spiritual significance. This last definition represents a turning back to more traditional beliefs that place was a home and economics was how you lived at home. The entire community is considered. Human communities are embedded in forest communities; our cultural and spiritual achievements occur in the larger community, which supports human endeavors.

The relationships of cultures (nonindustrial) to forests is intimate, complex, and spiritual. Forests provide cultural metaphors; forests provide places for rituals; and forest wilderness provides places for human independence (or anarchy).

Places have provided the resources for civilization and human expansion. Forests influenced the first spiritual and religious expressions of people living in or near them. Forests provided interactions with animals that humans used as companions, labor, or food. Forested places provided sanitary conditions for people: clean water, clean air and reduced pathogenic microbes, and moderate weather.

Theodore Roszak realized that human beings, and especially economists, focus their consciousness on the visible parts of the world, and forget the invisible that makes everything possible. The necessary invisible background must be described. He characterized economists as urban intellectuals automatically endorsing growth with ecological stupidity. He called for a nobler economics, one that is not afraid to discuss spirit, conscience, moral purpose, and the meaning of life. Roszak, Schumacher, Boulding, Daly, and others have grasped that economics is a subdiscipline of ecology.

Henryk Skolimowski states that spirituality is a state of being where the world is experienced in a state of grace. Reverence, compassion and love are forms of spirituality. The quest for meaning is a spiritual quest, which is also a public quest for many great civilizations.

Our education can not be just verbal and scientific. It has to be emotional, spiritual, and nonverbal. Aldous Huxley thought that we should *hope* that a nonverbal education could counter the effects of technology. Earlier, in *Poor Richard's Almanac* (1736), Ben Franklin wrote "He that lives upon Hope, dies Fasting." But then Franklin had confidence that the system was in balance, that not everything needed to be controlled all the time.

4.3.3.6.3. Education & Care for Continuity

Our curiosity is subverted to hunger for news of the human world, and a total lack of interest in the nonhuman. Sociality is frenzied restlessness. Society bombards its citizens with knowledge, with facts already observed and processed. Ignorance results from too little or too much of this knowledge. As Whitehead stressed, education should aim at the effective utilization of knowledge. Merely having the facts does not guarantee the formation of new patterns. One must rearrange the facts into ideas.

The current form of our education is a spectator sport. The format of our communications, also, leads to a spectator view, by not providing any participation (feedback). Maybe people are too illiterate to participate in anything. People want to learn. We now are educated to become something, or to achieve a capability, sufficiency, respectability and power. Education is a perversion of learning. Students are taught to prey on society and nature. Everything becomes a struggle for victory. Changing technology seeks young graduates who may have learned the newest techniques; older, experienced employees are retired. Education has become almost exclusively associated with youth and believed to be completed after an age limit. Most people are left with the information and attitudes vaguely associated with social awkwardness and dietary problems. The social system in advanced countries encourages the self through education, but denies it the opportunities of adult fulfillment. We cannot solve environmental problems until we learn or teach to enjoy life as it comes, satisfied with wonder and mystery and not dwell on future prestige.

The wholeness of humanity needs to be affirmed, but from a firm cultural base.

The complete surrender of cultural identity is as dangerous as too little openness. Every culture needs its own local, sacred center, which cannot be broken if the group is not to perish.

The state of one's knowledge is an important characteristic of the state of one's being. The young are lost because there is no relevant knowledge to guide them; they are furnished with bits of information, which are relevant to the concept of world as factory.

Education can encourage cross-fertilization and technological advance. A creative person pioneers, and fails often, but is productive in the long-run. Curiosity-oriented research has provided many new technologies. As knowledge diffuses organically, an innovation will be modified a thousand ways as it spreads, and no one will worry about credit for the idea. Equal access to information is necessary as well as to resources. Interaction is one stage above toleration. Cultures could borrow and learn from one another without being homogenized; they could converge in interaction.

Knowledge enlarges the range of options. It generates innovations, but it constantly surprises because new discoveries and applications are so unpredictable. Thus knowledge makes people receptive to new attitudes and willing to change their ways. A contextual education might have immediate consequences. The move from growth to need might entail economic well-being, and with that, the desire for silly extravagances like golf carts and electric swizzle sticks and for fads, might disappear. On the positive side, manufacturers could concentrate on durability and repairability.

Norbert Weiner has characterized the world as consisting of "to whom it may concern" messages. Unfortunately, these messages are incomprehensible as spectacles. They are clothed in form, which requires understanding to decipher. Thus the ecological crisis is a result of a perceptual crisis. Human beings destroy whole threads of the web of life because it is not seen as a web, a complex network of totally interactive beings. What kind of animal we are is determined by our perception of the world. Spectators, aided by computers, may solve technical problems, but not the environmental problems that require participation. An ecological education would provide participation. Leopold has said that the only true application of recreation engineering is to promote perception in people. But weeds in a city lot convey the same lesson as unique canyons.

Education is such a delicate necessity; it could enlarge or alter the perceptions of all human beings on earth with the selection and presentation of relevant information. Ecological consciousness could be fostered almost immediately. The basis of an effective participant education must include the application of perception research to ecological concerns and the promotion of ecological wisdom in a healthy world view.

Specialization and ignorance (of science) are working against that. Schumacher presents the leading ideas of the century—evolution, survival of the fitter, Marxism, Freudianism, relativism, and positivism—and distills them from one common denominator: that the higher is merely a manifestation of the lower. This is a metaphysical assumption, in modern education, that causes us to suffer a metaphysical disease; the cure is also metaphysical, however, and can be affected through an ecologically wise reeducation. Education is defined, through A.N. Whitehead, as the transmission of ideas that enable man to live above meaningless tragedy or inward disgrace. G. Hardin has written that the essence of tragedy is the remorseless working of things. Education properly could allow people to escape the tyranny of physical and biological laws by understanding them. The most important part of ecology is the education of its laws, which are not part of the science. Ecological education is necessary mechanism to produce individual and social changes to steer society away from disaster (from the Latin, *educare*, to lead forth).

Fuller states that the educational revolution is the highest priority of all; it should be based on synergy, here the total effect is greater than the combined actions. Synergy is the behavior of whole systems unpredicted by the behavior of any part separately. It would require interdisciplinary approaches. Interdisciplinary cooperation is a revolt against separation of arts and sciences. A general education is in contact with world order in the universe of knowledge and the community of cultures. The way to effect a cultural transformation is by surrounding society with a new field of consciousness, not by attacking institutions or society.

On a sailing ship (after Fuller), everyone cooperates spontaneously because they all know how to sail and what their functions are, in fair and foul weather. Humanity will only cooperate spontaneously when all humanity understands their role as part of the earth. That is the foremost purpose of education: To make individuals aware of how things work.

4.4. ***Managing the Place: Economics***

Economics is the formal study of how people use their surroundings is economics (from the Greek words meaning 'law of the house'—house is used as a metaphor for human society and nature). The word has come to mean the management of resources to supply human needs. It is basically concerned with sharing.

Although economics is a social science that studies human behavior, it considers itself a positive science. As a social science, economics addresses human problems. The acknowledged fundamental problem of economics is the contradiction between scarce resources and unlimited human wants. The kinds of resources and the possibilities of using them in production are considered in the scope of economics, as are flows and stocks in homes and businesses, the role of the government, business cycles, monetary details and policy, stabilization and growth, international trade, consumer behavior, production costs, pricing, and resource markets.

Natural economies are based on solar energy and plant productivity, as well as geological heat and energy. Ambihuman systems (surrounding the human) operate under natural limitations; they are empirical assemblages whose characteristics have been shaped by chance and selection over millions of years. They stress adaptation to the environment and the survival of a breeding society. Human economies, although basically dependent on the same energy and production, reverse the priorities. Human concern is for individual survival by modifying and managing the environment.

Modern economics is defined, by Chisholm and McCarty, as "the way people make their living." Economics attempts to address the 18th-century concerns of Adam Smith by using a scientific method to collect and interpret information—in fact, economics considers itself a science like physics, chemistry, or biology. Smith, whose first book was on morals, in fact, he considered economics to be a branch of moral philosophy, had noticed the trend in England away from mercantilism (where the central government regulated the output of goods for trade for gold) towards a free-choice economy, where people would decide what to make and how much to sell it for. Smith thought that a free market of independent buyers and sellers would let the entire community prosper as if an "invisible hand" were guiding it. That is, competition would increase public well-being, or to say it in a different way, self-interest is linked with common interest (or "What's good for General Motors is good for the USA"). Many countries in the 18th and 19-century, including the United States and Canada, adopted the free market system, with the result that the citizens did acquire more symbolic wealth.

Although economics is a social science that studies human behavior, it considers itself a positive science, that examines "what is" with theories, as opposed to a normative science, which addresses "what ought to be." Oddly though, Chisholm and McCarty establish a list of goals (oughts) for an economic system that includes:

1. productivity and growth—growth is necessary for living standards to rise, and individual self-interest has proved to be a stronger motivation than patriotism, altruism, or recognition
2. stability and security—stability in the form of full employment and set prices; security as providing the material necessities to the elderly and poor
3. efficiency—where the maximum amount of needed or wanted goods is produced from scarce resources with minimum waste
4. personal freedom and equality—by enhancing the personal dignity of everyone with maximum freedom, that is, the elimination of discrimination and limitations on opportunity.

As a social science, economics addresses human problems. The acknowledged fundamental problem of economics is the contradiction between scarce resources and unlimited human wants. The kinds of resources and the possibilities of using them in production are considered in the scope of economics, as are flows and stocks in homes and businesses, the role of the government, business cycles, monetary details and policy, stabilization and growth, international trade, consumer behavior, production costs, pricing, and resource markets.

The focus of economics, however, is rather narrow, in that the concept of resources is very limited and the unlimited wants are not much discussed. There is no psychology or ethics; there is no ecology or aesthetics. There is no concern with the triviality of the free choice of a worthless doodad. There is no thought for beauty. There is no concern with the welfare of the other beings that share the ecological community. Like the old discarded physics, economics applies a rigid standard of objectivity to its analysis.

Modern market economics recognizes two other kinds of economies: Traditional and Command economies. Traditional (subsistence) economies are dismissed as providing only a slim margin between life and death. Marshall Sahlins exposes this kind of myth in his book, *Stone Age Economics.* If we think about it, there are numerous advantages of the way of life of First Nation peoples: Fewer working hours (about three to four per day); more leisure to talk, sleep, engage in rituals, and make love; a diverse and healthy diet; deliberate underproduction, usually well below the maximum levels; deliberate control of population growth below maximum levels; and, deliberate under-use of resources, resulting in a small ratio of people to resources.

Subsistence economics means simply that surpluses are not accumulated. This might make them more vulnerable to food shortages, although the low ratio reduces the possibility—furthermore, industrial cultures have far higher incidences of starvation (50 million children starved to death in countries with market and command economies in

1978). Our industrial culture is approaching a form of subsistence now, in the sense of a minimum of necessities—we cannot leave after we have eroded the soil. Perhaps we will be less tempted to exploit the land for short-term profit, since we have to remain after the profit leaves.

Command economies, such as Marxism and Socialism, are underrated as having free market functions in some areas. Yet, the strengths of this kind of economy—especially the planning of production and the control of resources—are not admitted. Instead, the military competition that ruined many command economies is left out of the equation, and this form of economics becomes an distant also-ran. Market economies have begun exploiting the resources, such as the forests of Siberia, of failed economies, while at the same time trying to rehabilitate other resources, such as the forests of East Germany and Poland. This contradictory behavior is due to a combination of myths and practices.

4.4.1. *Reciprocity*

Division of labor by sex is a cultural universal and a natural division. Why is there a division of labor? Men are stronger? Women are confined to place by children? Gathering is safer than hunting? Are heavy work and child-bearing exclusive? Ostensibly it was done to allow people to play to their strengths. Division of labor formed. Men fished, women gathered marine animals and worked in garden. Larger men would hunt; women with children would gather plants and insects. Later, labor was divided by age (age grades) and class (social division by prestige and religious rights). Finally, labor was divided by sector (economic divisions) for extraction of energy and materials (primary), processing of materials (secondary), and service (tertiary). This kind of division becomes hierarchical, especially with capitalism, in regards to valuation and rewards.

Foraging bands in general moved their camps regularly, from three times per year to once every three years. There were some instances of permanent settlements. Oddly associated with permanent settlements were inequalities of status and wealth. Furthermore, other social things developed such as complex division of labor, with castes and slave labor. Trade increased and competition increased. This may have had to do with resources that were abundant and could be stored, and transformed into political prestige and power.

For the Inupiat, seasonal morphology as a result of weather and culture. There is seasonal variation of all aspects of life, from ceremonial activities to kinship, naming, family organization, sexual relations and morality. In general, form larger groups, extended family plus name groups, maybe 50-100 people, to hunt seals. Hours of silent waiting with harpoon. In May, in smaller groups, 10-30, move to solid ground and tents.

Traditional (subsistence) economies are dismissed as providing only a slim margin between life and death. Marshall Sahlins exposes this kind of myth in his book, *Stone Age Economics*. If we think about it, there are numerous advantages of the way of life of First Nation peoples: Fewer working hours, perhaps three or four per day; more leisure to talk, sleep, engage in rituals, and make love; a diverse and healthy diet; deliberate underproduction, usually well below the maximum levels; deliberate control of population growth below maximum levels; and, deliberate under-use of resources, resulting in a small ratio of people to resources. Subsistence economics means simply that surpluses are not accumulated. This might make them more vulnerable to food shortages, although the low ratio reduces the possibility.

4.4.2. *Distribution*
The basic change in economics is from equal sharing to a form of collection and redistribution. This allowed for specialization. Big men distribute the products of their own activities. Headmen are more informal, in bands of 50 or villages of 150. Every one knows and understands others; reciprocity is the bank. Headmen are leaders without power. Big men give away extra wealth. Chiefs accumulate it.

4.4.3. *Redistribution*
Redistribution involves a new step in exchange. Things are given to a chief and then redistributed to whomever is favored by the chief. Redistrubution is the model for most subsequent economic systems.

4.4.3.1. Chiefdoms and Redistribution
Chiefs redistribute the fruits of other's labor. That is to say, production is kin-oriented and reciprocal. Tribute, in the form of food or luxury wealth, is mostly redistributed by a chief (as a kind of informal taxes). The family is still the unit of production. In general, supplies are not accumulated.

The institution of bridewealth serves to redistribute wealth from the bridegroom's family to the bride's family.

4.4.3.2. Command Economies
Command economies try to control the economy; the state owns basic means of production, as a historical response to capitalism and inequalities. The communist model provided many things for many the people, while maintaining a large war and research establishment. The philosopher Karl Marx thought that socialism was inevitable, that public ownership of the means of production would provide equality and social security for all people. But, in practice the distribution of wealth was very inequitable, as the result of historical trends, old economic rules, and cultural confusion. Some people were treated as more equal than others.

As a result of central planning, the patterns of life under the communist model were pressured to be uniform and efficient. Yet, the strengths of this kind of economy—especially the planning of production and the control of resources—were not admitted. Instead, military competition ruined these command economies and socialism is considered a failure.

Command economies, such as Marxism and Socialism, are underrated as having free market functions in some areas. Yet, the strengths of this kind of economy—especially the planning of production and the control of resources—are not admitted. Instead, the military competition that ruined many command economies is left out of the equation, and this form of economics becomes an distant also-ran.

4.4.3.2.1. Tributary Economies
Certain chiefdoms use command economies. Chiefs distribute the fruits of others labors, sometimes according to social need, but sometimes for the prestige of peers or nobles.

4.4.3.2.2. Socialism
Command economies, such as Marxism and Socialism, are underrated as having free market functions in some areas. Yet, the strengths of this kind of economy—especially the planning of production and the control of resources—are not admitted. Instead, the military

competition that ruined many command economies is left out of the equation, and this form of economics becomes an distant also-ran. Socialism is where the state owns the basic means of production; this was sometimes a historical response to capitalism and inequalities. Hybrid systems exist, such as Democratic socialism in England or a tributary system in medieval China.

4.4.3.3. Market Economies

Modern market economics privatizes profits and commonizes costs. One consequence of this is forest decline from environmental degradation. Value of goods in monetary value determined by supply and demand. Trade and Industrial revolution. English encouraged trade to increase wealth. This led to Mercantilism, the government regulation of the economy to ensure growth by granting monopolies, which could be said to be the start of global unity. This led to an accumulation of capital from mercantilism. Capital required more energy and new technologies. Capitalism led to capital production, where goods are privately owned, yet there is a separation of people from land and resources. Eventually, there was a shift to assembly-line factories using fossil fuels. Capitalism is disruptive of resources and culture; unequal exchange leads to accumulation. Corporations led to monopoly capitalism.

4.4.3.3.1. Invisible-Hand

Smith, whose first book was on morals—in fact, he considered economics to be a branch of moral philosophy—had noticed the trend in England away from mercantilism, where the central government regulated the output of goods for trade for gold, towards a free-choice economy, where people would decide what to make and how much to sell it for. Smith thought that a free market of independent buyers and sellers would let the entire community prosper as if an "invisible hand" were guiding it. And it was thought that the invisible hand would harmonize individual greed.

4.4.3.3.2. Controlling Minds

Market economies have begun exploiting the resources, such as the forests of Siberia, of failed economies, while at the same time trying to rehabilitate other resources, such as the forests of East Germany and Poland. This contradictory behavior is due to a combination of economic myths and practices.

4.4.3.4. The Modern Model: Abstraction & Accumulation

The metaphor for the economy used to be a simple mechanical model for turning resources into products. To be successful, the economy had to grow and turn a profit continually. Unfortunately, the assumptions of the model were also simple and failed to consider human needs and natural cycles, causing great suffering and great disruption. The old analogy of the economy as a machine leads to bad assumptions:

A.1. That everything is a resource;
A.2. That resources are unlimited;
A.3. That production must continue endlessly;
A.4. That the economy has to keep growing to survive;
A.5. That the purpose of the state was to legitimize exploitation;
A.6. That the purpose of humanity was to multiply, produce, and consume; and
A.7. That the purpose of the universe was to supply human needs.

The bad assumptions of the machine analogy led to false economic beliefs:

B.1. That mass production is most efficient;
B.2. That obsolescence is necessary for successful growth;
B.3. That people's needs and wants are fulfilled by advertised products; and,
B.4. That quality does not matter very much.

These bad assumptions and false beliefs led to problems in the economy as a whole. These colossal problems include:

C.1. Overgrowth, with an increase in complexity and costs (many of them social);
C.2. Economic and ecological instability;
C.3. Social burdens (from pressures on families from relocation and powerlessness);
C.4. Misdirected effort on ill-conceived, low-quality products;
C.5. Slack consumer attitudes and employee performances.

Economics has not been unsuccessful with its models, for instance of buying behavior, but it has become a highly abstract academic discipline. All its abstractions are applied to the real world without acknowledgment of the high degree of abstraction involved. The philosopher A. N. Whitehead warned that the economic method would triumph if the abstractions were judicious, but even judicious abstractions had limits, and the neglect of those limits led to disastrous oversights. Considering a fictitious human nature under imaginary circumstances and thinking it is real is the fallacy of "misplaced concreteness" according to Whitehead. Daly and Cobb suggest that the classic instance of the fallacy in economics is "money fetishism," where the characteristics of an abstract symbol, such as limitless growth, are applied to real commodities and values.

As an aside, Daly and Cobb point out that modern economics might better be called chrematistics, after the distinction made by Aristotle between chrematistics and economics; the former related to the manipulation of property and wealth to maximize the short-term abstract money value to the owner, whereas economics was the management of the household (community) to increase the concrete value for everyone in the community over the long-term.

Business as usual, with its inertial model of growth, could end in catastrophe for humanity and its environments. Industrial cultures, with their characteristics of simplification, naiveté, homogeneity, and incompleteness, turn wild landscapes into flatscapes, where variety disappears and significance is ignored for the comfortable standards of meaningless continuity. Rapid growth might precipitate a catastrophe sooner, while modest efforts at environmental protection and increased efficiency may only postpone catastrophe a few years.

4.4.3.4.1. Growth of the Global Industrial

Mesarovic and Pestel stated that "the issue for the economy is not to grow or not to grow; it is how to grow, and for what purpose." They claim that if a workable world system is to emerge, it must be after the establishment of an organic pattern of growth. Due care is devoted to describing such a pattern and contrasting it with other, tragically inapplicable patterns of growth. Their treatment of the world system itself was regionalized and multileveled. They recommend that the establishment of organic growth was necessary with no need for special no-growth policies for populations or economies. They assume that further industrial growth will continue, that economic growth is good, and that this growth solves human problems as long as it is organic. Exponential growth is said to be bad, and organic growth is said to be good. In fact, although organic growth is better, there is little difference during a world crisis; both reach asymptotes of suffering. One need only regard the population crashes of lemmings and others to see how organic growth can go wrong.

In the organic world, growth is healthy only when the rate of change is decelerative in the long run; cancer and population are constant or accelerative. Mesarovic and Pestle fail to realize that continued economic growth in any form is a threat to the stability of the biosphere.

Economics became enamored of growth during a critical time in history. Rapid European expansion occurred at rates rarely exceeding a growth of one percent per year, and with unparalleled opportunities for expansion into sparsely settled areas, such as North America, Australia, South America, and South Africa. Many cultures now do not have these opportunities; the continents are claimed, and violent population growth may have wrecked their hope for development by ravaging every resource.

The economy has been growing almost constantly since it has been studied. We have been trying to force it to grow, rather than let it stabilize or contract. Even if it stopped growing, the economy could still develop. Mesarovic and Pestel, as well as many others, confuse growth with development. Mill did not.

4.4.3.4.2. Global Capital Conversion & Erasure

Capitalism has some advantages, as did communism. Neither system is especially ecologically sensitive. Neither is really egalitarian. Capitalism does drive innovation, using science and technology. But, capitalist economies depend on continued growth of production and sales.

Of course, as many people make more profit with luxury items, the market switches to rich houses instead of cheap ones, yachts instead of boats, luxury cars instead of high-mileage cars. Perhaps, also, there is some confusion between growth and development. Profits seem to be more important than just growing production. Perhaps capitalism could thrive on development the same as simply growth. Profits can be made with recycling and solar technology, so they can be compatible with more sustainable lifeways.

Lester Brown considers capitalism so destructive because it does not have a way to account for ecological values. The dishonest partial-market of capitalism causes certain problems. It ignores the indirect costs of the support systems. It ignores social capital. It ignores thresholds and limits. And, it allows partial deficits, without their true costs.

Capitalism has certain cultural contradictions that may undermine itself by producing norms that do not contribute to the operation of the market. One of these could be that abundance makes work ethic unnecessary. Another might be the elitism of a privileged class that might argue for socialism. As a disruptive force, capitalism may break apart traditional obligations that we want. Of course, it creates a new order and allows new norms.

Fukuyama asks if capitalism depletes social capital. The social capital that had been built up by organized religions or cultural traditions. These are part of complex processes that regenerate social capital.

Capitalism offers the poorer classes higher living standards, higher than many monarchs had. It thus defuses hostility that would come from knowledge of greater inequities. Thus, capitalists generate consumer loyalty through a massive scale of gift-giving. This also allows the capitalist system to survive and change.

But, capitalism and modernity has given many people worse conditions. The continuous accumulation by an ever-growing population forces a continuous conversion of resources into goods. This process uses fossil fuels to erase the capital of natural ecosystems. Corporations liquidate ecosystems such as forests because it is currently economical to do that rather than keep them as long-term capital. The pressure of deficit financing and the allure of short-term profit can overwhelm common sense about borrowing and liquidating. Furthermore, the capital is often an abstraction to a board that is remote from it.

4.4.3.4.3. Corporations & Independence without Limits

The first corporations were quite limited to public service, such as private highways, and did not challenge the host nation. As a result of changes in technology and several social changes, however, corporations were able to overcome limitations to their business.

In 1886 the U.S. Supreme Court made the ruling that corporations were persons, entitled to rights and privileges given to individuals by the Constitution and Bill of Rights. The corporation is legally fictitiously a person, an individual, although now they may be immensely large, rich, immoral, and powerful, supercitizens.

Supreme Court Justice Louis Brandeis noted that U.S. citizens were reluctant at first to grant corporations privileges for doing business, even though they were recognized as being more efficient. The reasons for this reluctance were: Fear of encroachment—on liberties and opportunities for individuals; fear of subjection of labor to capital; and, fear of monopoly. Where privileges were granted, there were restrictions on the size and scope of corporate activity.

Gradually state governments removed those limitations. The corporate system evolved like a feudal system, according to Brandeis, where American society came to be ruled by a plutocracy, where a few old, white men control hundreds of thousands of people through corporate mechanisms.

Even more controls have been lost since the 1950s. The stockholders have lost control to management, thus ownership and control have become separated. Managers can pursue a course without the direct supervision of the owners. Antitrust laws failed to control corporations. Regulatory legislation as a legal limit on corporate power has also failed. The labor unions are no longer restraints on corporate power.

4.4.3.4.3.1. Constitutions Federal & Corporate

The U.S. Constitution was written during a specific social context, at a certain level of technology, and at a certain point in the development of the industrial revolution. The forces that have developed as technology and corporate organization developed are not consistent with democracy and its ideals—and in fact, the constitutional and economic vision of that time have been rejected. The system resembles an iceberg; the visible parts look like a democracy with a free market, but the invisible base determines where the ice heads. The dark down-side majority determines how things function. Because it was not foreseen by the U.S. Constitution, this situation has escaped the traditional controls and limits. Corporate icebergs direct the free market, by-pass democracy, and dominate people's lives. Corporations are powerful and indifferent to human life as well as to the free goods of nature, which is why not everyone gets lifted with prosperity or why nature is also decaying. This new system that has arisen from the alliance of government with private corporations, using modern technologies for profits.

Corporations have developed their own constitutions for operating for profits. Among managers there are shared knowledge and assumptions, acquired in schools and social situations. These assumptions become the real operating constitution of corporate government. Skills that will advance the interests of the corporate government are the ones that are selected, but understanding of the world below them or of the effects of their decisions on that world are not considered important. What are the shared assumptions?

1. Impersonal economic forces (a myth not often recognized) produce better choices than planning by thoughtful people.
2. economic growth is the measure of well-being of society as a whole and benefits everyone (another myth).
3. All important social values are quantifiable and measurable (trust, loyalty, and beauty are outside but can be willed by strong people).
4. Treatment of people as employees does not effect the rest of their lives, regardless of layoffs and coercion.
5. The market will supply all social needs, although the massive efforts made by advertising to influence people to buy unhealthy and unnecessary things are not detrimental to needs.
6. People are rational actors whose behavior can be controlled. They are dominated by economic interest but should forgo interest for others and the community.
7. The product of this system is the best system, not necessarily good people or a healthy society and environment.

Based on these assumptions the elite managers make decisions that affect all of society, although this consensus may not be conscious. The access to position and wealth has become controlled by corporations now.

4.4.3.4.3.2. Corporate Rule

Corporations are ruled from top down. Rules are not adopted democratically, or enforced with the fairness required by law. Because of the success of corporations, other institutions, such as schools and colleges, follow the corporate model.

People obey the dictates of corporations to avoid being laid off. Employers then demand subservience and obedience far beyond what governments require with laws. And, the punishment, of loss of job, is more effective than the threat of imprisonment. Independent thinkers are discouraged from government jobs, the result of background checks and standards for conformity. Punishment, of course, is a modern version of ostracism or exile.

Corporations have become exempt from the Bill of Rights, because the Bill applies only to actions by the government and its agencies. They do not allow employees freedom of speech or due process.

Under the flag of efficiency, institutions have adopted an authoritarian model. An authoritarian workforce has advantages for a corporation. Civil liberties can be suspended in the name of profitability. The technology and communication works better in authoritarian mode. Its wealth gives it power. So much so that it does not require the use of force, although the public government allows it to use other forms of coercion.

Is this the operation of a free market, where workers can negotiate for benefits or working conditions? Or can corporations coerce workers into accepting a job under any conditions, to avoid being one of the surplus unemployed? Possibly a free market could favor a healthy society if balance was maintained. The coercive market upsets that kind of balance. The economic government has chosen to devalue labor to be able to compete on a world market. By causing lower labor costs the corporation saves on labor, but larger costs are imposed on individuals and the society that has to support the unemployed and under-employed.

The efficiency of large scale is an axiom of modern economics. But, the efficiency is enforced by corporate control of resources and employees. Corporations that control the entire process of acquiring resources, manufacturing, distributing and marketing have advantages. Thus, t market is not free; it is fixed by corporations. Price and production are controlled by planning, research, and programming, not by a free market.

4.4.3.4.3.3. Corporate Growth

In the past, at least as recently as the 1930s in the U.S., public government worked to protect people from the harm caused by private government. But, economic power has been leveraged to political power, where corporations can deny free speech to employees outside of the work place. The economic government, according to Robert Reich, creates a corporate hierarchy, with pervasive inequality between all levels, rather than a democracy. To keep profit growing under any circumstances corporations have been willing to accept damage and conflict.

The growth of corporations caused a growth in government in response, mostly to help those hurt by large corporations: Small farmers, small businesses, and individuals. According to Reich, the government expanded its constitutional powers, not by amending the constitution, which would be the appropriate way, but by allowing the Supreme Court to give broader interpretation to the Constitution. The government ended up trying to regulate parts of the economy or those parts that corporations were unwilling to perform. At least until government simply became a willing partner in the changes. Economic government dominates public government, which helps it.

In the past corporations exploited child labor, until government intervened. Corporations could dismiss employees without offering them assistance, but the government started to provided social security, and unemployment and training benefits. This response to the narrowing of economic opportunity and independence has been termed the 'welfare state.'

Individuals cannot, and corporations will not, take responsibility for public services such as highways, airports, national defense, national parks, schools. So, government takes sole responsibility (as it should be responsible for national things). National defense for instance, has enriched many corporations, but the corporations do not even provide efficient or competent work, in many cases. Public resources such as oil and forests flow to corporations as subsidies. Radio and television licenses were distributed to corporations. Now, corporations dominate public airways. Public lands used to be the support for individuals. Now, they support corporations.

As private corporations became larger and international in scope, local and state governments have been unable to regulate their activities. That would take a strong local or national government or international body. In their own interest corporations work to reverse the intent of the people. For instance, although U.S. President William Clinton in 1992 promised a pro-employment policy, the minimum wage was not enacted, and interest rates were raised. What the people voted for was reversed by invisible management of corporate government.

The legal system becomes another tool for power, to justify the distribution of wealth, as a result of favors. This distortion of the system destroys the neutrality, fairness and dignity of the law.

4.4.3.4.3.4. Corporate Tyranny

As long ago as 1958, John K. Galbraith observed that the public sector became more affluent as the private sector was impoverished. Public poverty has become deeper. The money is not missing; it simply migrated to the private sector. Urban parks and wilderness suffer neglect or destruction; parks and forests used to be the symbols of freedom. The free market did not decide that; corporate managers did.

The broad and deep middle class of the U.S. in the 1950s acted to stabilize life. Now, the new structure of the work place, with its steep hierarchy in megaglobal corporations, is destroying the middle class. Security is lost; community is destroyed. Corporations control the workplace without interference by government, unions or competition.

There are new kinds of tyranny, not just from governments now, but from corporations. U.S. citizens once fought taxation without representation, but they seem reluctant or indifferent to fight economic decisions without representation. Economic tyranny could also be reduced by balancing society more, by recognizing the immeasurable values that exist before being economized in an economic dimension. Control of corporations could lead to a better balance of natural, human, and economic environments.

4.4.3.4.3.5. Ownership & Capitalist Scale

Ownership has changed meanings many times in the past 40,000 years. At first it was rare and may have applied to only a few personal items such as ceremonial clothing. Now that humanity has divided up the earth, every part is claimed and owned. Every forest is owned, even if the tribal residents do not become aware of it until some company starts logging it. One would think that this gigatrend of ownership has ended with the finite territory of earth, but it shows signs of continuing through the solar system.

Large-scale ownership patterns may be relatively recent. In China by 594, the state of Lu instituted land taxation instead of expecting direct labor. Individual ownership and free market may have appeared afterwards.

In England, a forest was originally a tract of land owned by the sovereign and used for hunting; the word also referred to a dense or profuse growth of trees. By the eighth century, forest referred to the royal woods, where the right to hunt was reserved by the king. Other rights, however, such as the right to cut wood for home fires, remained free for everyone. The forest then became a legal term that applied to a large tract of land, governed by laws, with its basis on the right to hunt. At one time, almost a third of England was royal reserve. Enclosure of common fields gave many ownership of the fields.

With ownership comes management; with management administration. Administration, from the Latin words meaning to serve, means the activity of an institution in the exercise of its powers and duties; it also means the management of an institution.

Ownership also has effects on productivity and disposition. Tribal ownership is usually ignored by federal governments. Private ownership is influenced by taxes and regulations. Federal ownership can limit or accelerate use of forests; Canadian forests, for instance, where 90% of all land is in provincial Crown hands, are vulnerable to the fads, style, and influence of industry—in the U.S., where 72% of lands are private, industry has far more people to convince.

Perhaps the most important condition is the security of land tenure. Management has never invested in land that is regularly changing ownership from the government to tribal to individual or back. Religious organizations that have persisted for centuries have done well at preserving individual trees or stands; both Chinese yellow pine and dawn redwood, long thought to be extinct, were found in temple gardens. By contrast, the Queensland Australia management system for moist tropical forests was closed down by a political decision after a state-federal struggle for control. Long-term stability, with secure ownership by state, culture, or tribe is a necessity for permanent forests. This means intergenerational management.

Perhaps ownership will reflect a change from individual and corporate to land trusts and community. We need to Find the best combination of ownership, trusts, easements, and plans to ensure that the forest will be cared for in the long-term.

4.4.3.4.3.6. The Use of the Commons (No-costing)

Economists try to bronze the economy in its current structure; but it is a changing system. Since it is changing, strategies that are appropriate at one stage are totally inappropriate at another—this is the remorseless working of tragedy, where successful strategies are applied in inappropriate circumstances. Tragedies imply conflicts larger than the individual or even society. The external order of things or the cosmic plan is challenged, even if the cosmos is the size of a city. The source of tragedy for economics is a fatal flaw of the world-view. This is the root of Hardin's tragedy of the commons. The tragedy of the commons occurs only when people are locked into a system of self-interest through economic gain; it could not happen in a culture that understood common holdings. Hardin's definition of tragedy is the working of fate. But economic tragedy results from the failure of a cosmology: humans are responsible, not chance or fate. We can choose between the tragedy of the commons or that of Leviathan, or we can expand the cosmology using ecological knowledge and wisdom and the limit economics to a sustainable place in the cosmology.

Sustainable development requires recognition of the large number of limits. The ecological approach to development makes it irrelevant to discuss global limits to growth. Local limits are far more significant to a majority of people. Economists have claimed that the minimum does not apply in a growing system; alas, our system has been growing through transformation and not real growth. We simply convert ecosystem economies into human ones.

Complex societies depend on production from resources. Increased complexity requires more information processing and more integration of disparate parts. The costs of communication increase. Complex societies need control and specialization. Yet, investment in complexity yields declining marginal returns because of the increasing size of bureaucracies, increasing taxation, and costs of internal control. If complexity can be restrained, the society may be healthier.

If complexity, especially in an economic sense, increases too much the entire society can collapse as its disadvantages outweigh the advantages. Various authors, summarized by Joseph Tainter, have concluded that economic reasons explain the collapse of many societies historically, from the Maya to the Ottoman Empire.

The current economic style is too great, fast and reckless for ecological systems to absorb its impacts. The scale of things is an independent problem that can ruin the best intentions of policy. A bigger system to control systems that are now too big might be a mistake.

Every economy depends on the stability of the environment and on the stability of social institutions. The environment provides air, water, and land, and provides renewing, both physical and psychological. Institutions, from sanitation, police, schools, churches, and community centers, provide a supporting network. As these institutions wobble or fail, or environments do, economies may have to subsidize or replace them to survive.

An economy has traditionally been seen as a morally neutral body, but even if it has only to conform to the nominal rules of society, it is already a moral agent. Economies are no more neutral than other organisms. Many areas of moral concern already are recognized, including worker safety, affirmative action, advertising truth, foreign investments, and harm to the consumer, public, and environment.

Responsibility occurs wherever the interests or rights of a person, society, or ecosystem are significantly affected by the actions of economic actors. Responsibility can be understood in terms of costs and benefits, that is, through operations and their consequences rather than abstract behavior. Every action entails a gain and a cost (or profit and loss). Profits and losses are distributed privately, socially, or environmentally. Unfortunately, the modern system privatizes the gain and externalizes the loss to the "commons," considered as a pool of "unowned resources," where in traditional societies, it was surrounded by rules for use. As long as this is possible, it is profitable to charge the cost to the environment.

Externalizing costs works fine in an uncrowded world, where the costs are negligible and can be absorbed by natural processes. Resources were traditionally seen as free for the getting; air and water were seen as free sinks. Modern economies, embracing the notion that "nature is capital," draw on the accumulated "capital" of ecosystems for production. By ignoring the real cost of the capital, as well as the costs of natural services, such as nutrient recycling, soil building, and atmospheric renewal, these economies create a temporary wealth. Decisions regarding resources are made on short-term economic grounds and lead to shortages and environmental degradation.

Modern industrial culture places an emphasis on individualism and competition. Cooperation, with an understanding of rights and responsibilities, is based on cultural understanding. This kind of understanding, once prevalent in many cultures, is the reason why the tragedy of the commons (Hardin) did not always occur with common resources. Cooperation is crucial.

4.4.3.4.3.6. Capitalism & Growth

The English encouraged trade to increase their wealth. This led to government regulation of the economy to ensure growth by granting monopolies, or mercantilism, an early example of regional trade going to global trade. Mercantilism allowed the accumulation of capital. And, capital required more energy and new technologies. This led to the separation of people from the land and resources, at least in England in 1650, during capital production, or capitalism. Goods were privately owned. Later, there was a shift to assembly-line factories using fossil fuels. As immense legal individuals, corporations created a form of monopoly capitalism.

Capitalism flourished with the agricultural revolution that began in England in the 1700s; the other half of the revolution was science, which also benefited capitalism. The two, capitalism and science, have not quite been compatible. They have different logics and different goals. Many times capitalism forces environmental destruction, due to the limits of the system. The gamble is to win big or lose big.

Capitalism rose when kingdoms were becoming more democratic, but capitalism is not linked with democracy; it does not contribute to democracy. In fact, it can adjust quite nicely with many totalitarian states or dictatorships, such as Russia, China, and Saudi Arabia, all of which have fast food restaurants with fizzy drinks.

4.4.3.4.3.6.1. Perpetual Growth

But, capitalism has felt the need to grow. The need to grow is intrinsic in this kind of economic system. A large literature has treated perpetual growth as the only conceivable state of affairs. Capitalism depends on growth for stability. There is some analogy with plants. Some stability can be gotten from growth in early stages; later stability must result from limits and metabolism. Growth in plants can delay the onset of senility by ridding the plant of waste products in more diluted form. However, too much growth produces a strain on tissues and early decay. In fact, one herbicide promotes excess growth as a means to kill weeds.

The production of wealth from growth depends on technology. The technological perspective is oriented toward materials and not humans or natural processes. Nature is considered to be a resource to be exploited. The immediate objective of technology is to create wealth through knowledge. Technological activities are justified on humanitarian grounds, scientific discovery increases the well-being of human society, yet the social consequences of scientific activity are ignored; short-term suffering will be offset by long-term benefits, it is claimed. But because the long-term view is not taken, long-term benefits will be worse.

Economic growth can produce great wealth for some. Fortunes await those who can increase the demand for unnecessary items, such as electric swizzle sticks or pet rocks. Free enterprise provides initiative to any willing to show a profit without regard for the immediate consequences. The economist Paul Samuelson illustrated the skewed distribution of wealth with an income pyramid made out of children's blocks: if each layer represented $1000.00, the peak would be as high as the Eiffel tower, and most of us would be within a meter of the ground.

Wealth is based on production in capitalist countries, and on useful production in Marxist countries. The difference is mostly rhetoric. Production is measured as Gross National Product (GNP), which measures flow, not stock, which is equally meaningful. The GNP adds the dollar value of all goods and services, including cigarettes and lung cancer, oil wells and spills. There is no deduction for environmental damage or deaths. GNP also measures services, such as banking, many of which do not raise industrial output. However, gross production may not be as desirable as thought. The world prices of food and industrial raw materials have increased far more rapidly than those of manufactured goods. It can be seen that economic growth is not equal to progress.

Inequality is more the result of differential development than of exploitation. According to Kenneth Boulding the greatest source of the differential is different rates of accumulation of knowledge, capital and organization; the rates are essentially internal properties of cultures. Although a minor element in terms of transfers, it is a large psychological perception, which may need to be compensated for in a global community.

4.4.3.4.3.6.2. Perpetual Borrowing

Capitalism has developed other difficulties. It is disruptive of resources and culture; the unequal exchange of goods, and the advertising of goods as necessary, leads to accumulation. Many of the characteristics of capitalism are destabilizing to a national economy: The goal of perpetual growth, the continuous accumulation of goods, the myth of progress, increased inequality and polarization, and the hierarchical division of labor, as well as of nations.

These difficulties have forced corporations to liquidate capital assets, such as old-growth forests, rather than keep them as long-term capital. The answer might be provided by recent economic history: After the second great depression starting in 1932-3, the U.S. government tried to flatten out the boom and bust cycle of our capitalist economy by manipulating taxes, interest rates, and money supply, by subsidizing jobs, and through corporate welfare. Although the government discouraged monopolies, it did nothing to regulate oligopolies—the control of the market by a few large companies. Prices are set by the costs of production, including bloated bureaucracies and executive salaries, rather than by demand and supply in the free marketplace. As the automated production of goods became more efficient, goods-producing jobs declined. On the other hand a large, well-educated labor force, mostly women, was available at relatively cheap salaries. Information processing and human services proved to be far less efficient than automation. Marvin Harris suggests that corporate bureaucracies wasted labor and lowered productivity faster than automation could save or raise it. Corporations were no longer efficient enough to expand production out of sale-generated income and took on additional debt.

As economists point out, debt is inflationary because borrowing puts more money into circulation. Rising interest rates caused cash-flow problems, which many large corporations responded to by lowering quality and by passing the costs of borrowed money on to consumers—which allowed them to borrow more at even higher rates, which the government had raised to cool off borrowing. Corporations have gradually increased their dependence on deficit financing to supply capital. Harris notes that corporate debt has gone up fourteen times while federal government debt has gone up only three times, as a percentage of gross national product. As more money was owed, more was paid to service the debt and less was available for cash flow. More short-term debts were taken on to keep up cash flow. Corporations can borrow faster and pay more to borrow, but can keep raising prices to consumers so they can keep borrowing—or liquidate some of their assets, which is probably why Burlington Northern and other land-owners allow massive clearcuts: Cash flow needs due to long-term inefficiencies and massive short-term debts. The impulse to save a corporation may override the need to save the natural heritage, such as old growth forests.

United States residents are criticized by the French, not unreasonably, for having a "frivolous" culture based on "savage" capitalism. Capitalism increases the pressure for uniformity, a single pattern of existence. Formal development is more concerned with an assembly-line model—simple, isolated, efficient, and easy to maintain. We become remote from, and indifferent to, the system that supports us. We acquire unrealistic images of the world and harmful values, and then make bad decisions based upon them. We have not developed qualitative indicators of ecological health or quantitative measures of social health, much less an ecocentric view that would value preserves of nature for themselves.

Flexible regulated markets can be efficient instruments in the distribution of durable goods, but, coercive, unregulated markets can never produce and distribute everything, especially spiritual values or social justice or sustainable environments. Government should still provide public goods, from public education and culture to employment, social welfare, and penology. Most important is ecological survival.

4.4.3.4.3.7. Industrial Growth & Accidental Globalism

The first wave of global trade destroyed many of the traditional societies of the Americas and produced incredible suffering. Globalization was destructive on several levels at first, not only diseases, but conquests and exotic exchanges. American crops, grown in harder conditions, grew where native Euro-Asian crops grew less well. This allowed expansion of areas and populations. Exotic crops went everywhere, even manioc into Africa. The coming together of zones was destructive to native peoples. Not just from trade and competition but from livestock, exotics, and diseases.

A later wave of globalization destroyed traditional economic systems. European exports, especially in textiles, undermined regional livelihood. These processes enriched the Atlantic shores, but it also widened the inequities within nations. Thurow states that governments did not decide to start global trade and marketing. But, that is what they did through East India and Hudson's Bay and other companies.

Many of the modern trade agreements are causing suffering and violence. The gross domestic product of the world increases, but what is produced? Who produces it? Eighty percent of the lowest produce only twenty percent of the global output.

4.4.3.4.3.7.1. The Positive & Negative Effects of Global Economics

Globalism allows trade and access to any product or luxury. This access allows many to assemble comforts and technological marvels to enrich their lives. Globalism increases the size of the frame of reference, so social relations have a larger context. But, globalism shrinks distance and increases interdependence. Perhaps globalization does break down barriers and enhance cooperation. Despite its limits, the North American Free Trade Agreement (NAFTA) in 1993 did integrate environmental quality and protection with trade agreements.

The arguments in favor of globalism, however, contrast it to the erroneous belief that the past was nasty and short, teaming with suffering, disease, and tyranny from others. Globalization acquires the patina of myth, that it has a destiny to save civilization. One myth of globalization is that it is saving and managing the planet. Another is that nature is a machine that can be simplified and streamlined. The metaphor of the machine is only a metaphor. Nature is not a machine and cannot be easily simplified with only desired effects.

Globalization disconnects identity from location. Globalization can de-territorialize culture, which has already happened to some cultures, through a disguised form of colonialization. Globalization presents the myth that cultures, through reterritorialization, can take root in new locations and develop new foods and new dances. What about distinctive foods, songs, views? Are they place-based?

Globalism allows most people to stay in place, but then experience the displacement that globalism brings to them. Most human social existence continues to be local, of course due to physical embodiment and attachment to home. Neighborhoods are based on proximity, more than ties or shared values.

Aspects of globalization tighten intersystem linkages and hierarchies. This tends to distance resource users from the immediacy of their dependence on systems. This also weakens feedback loops that are essential for responding to adjusting services. Increased interdependence can make the network more fragile.

Another concern with globalization is homogenation. Globalization could still occur without homogenation, but that would require respect and limits. Diversity and globalization are not totally incompatible.

Does globalization screw up the categories that we use for human behavior: Economic, political, social, environmental, cultural? Global capitalism undermines traditional cultures by offering consumerism in the place of guides for behavior. Social roles seem irrelevant by comparison, if the good life can be bought without effort. This leads to a new kind of proximity. Inequities are made more visible. Threats to shared environment are more visible. Dangerous impacts from conflicts are brought closer. It might set the stage for a violent clash of cultures.

4.4.3.4.3.7.2. Globalism & Localism

Globalization allows connections to other cultures, to extend relations. But, do we have the moral development to live mutually and fairly with these other cultures, often constrained or limited by luck, geography, resources, or history? Probably not.

Is globalization a universalization of the potential of one international society? Is international society just an aggregation of national societies? Are those just an aggregation of communities and individuals? Is aggregation the right word? No. They are not thrown together. They are developed over centuries or generations. Is there an international society? Is it a society of societies? It might be better described as a framework that allows coordination and not replaces the nations and communities.

How is international law different? It did not arise from all living together. It may be an expansion of or abstraction of all societies and states. Allott states that there cannot be society without law or law without society. Obviously, this is wrong. Foraging societies do not have or need laws. International laws will not be based on an international society.

Is cosmopolitanism having the sense that there are no others, that everyone is the same? This is different from the recognition of people having universal needs of kinds of behavior. A cosmopolite perspective must include awareness of many other cultures with unique solutions for unique places. There is legitimate difference and plurality. The true cosmopolitan might possibly live in the local with awareness of global.

The power of globalization to cause crises in countries would be lessened if the countries were self-sufficient in food and education. We try to distinguish truth from falsity, information from noise, thinking that we can distinguish.

In a global economy, the economies of scale are not as important as they were nationally. Therefore, nations do not have to be large at all to have scale advantages. Therefore, ethnic groups can form their own nations. Higher standards of living are possible by clever trade and specialization, based on self-reliance.

Putting social arrangements and trade within a global framework, with agreed-on rules, with a realization of equity, should reduce the suffering and inequity. The distribution of some powers to an international social framework could serve common global interests, such as defense. At the moment, many countries, such as Japan and those of the European Union have decided not to invest in large military buildups, not to make decisions about Kosovo or Iraq. Instead, The U.S. tries to police everything as a global power, but is unable to avoid a dreaded imperialistic stance or achieve any kind of balance, that might be achieved by an international association.

Because of the recognition of universals among human cultures, people want a common human culture that does not exist. But, that is not the same as a uniformity regardless of peoples themselves. Globalism has to recognize differences and independence, as well as its own limits, to be effective.

4.4.3.4.3.8. Myths of Modern Economics

Modern corporations and political groups create their own myths to enforce their rapacious behavior. The myths of the mutant modern economics have tremendous impacts on how places and people are treated. The old analogy of the economy as a machine leads to unhealthy and dangerous assumptions about things.

- Everything in nature, in fact all of nature and humanity, is a resource (and its corollary, everything has a price—and its corollary, everything that does not have a price is worthless): The essence of a resource is that its existence acquires meaning only as it is necessary for human needs and wants. Furthermore, if everything is a resource then everything can be used—the forest, not just the trees, can be used. And, of course, consumption reflects social value; the consumer becomes more valuable by consuming.
- Nature is inexhaustible. Resources are unlimited: If forests are unlimited then we can keep cutting them—and even if they are not unlimited, by the tenets of modern economics, being a scarce resource makes them even more valuable—this is why whaling is still economical to the Japanese and Norwegians, because the costs and unused equipment can be written off against other profits. Although many economists admit that forests may not be exactly unlimited, the assumption surfaces again in advertisements about the forest industry (and why is everything an industry and not a cottage art or guild?), where the industry credits itself with planting billions of trees.
- The economy has to keep growing to survive: Growth means that the use of the forests will have to keep increasing. If there are not enough old-growth or good timber trees then all trees will have to be used for pulp and fiber (for paper board and fiberboard). If there are not enough trees then we will have to grow wood in vats (I worked on such a project as a graduate student in 1975-76; the goal was to use plant growth auxins to stimulate the growth of cells in a controllable, preformed medium). Or, we will have to substitute, for example, steel wall studs for wood ones.
- Any resource can be extended through a substitute: Therefore forests are not unique or intrinsically valuable, because they can be substituted with tree plantations or perhaps genetically engineered industrial wood cell production. In economic equations, human-made capital is considered a perfect substitute for natural capital, although interestingly the equation is never reversed.
- Mass production is most efficient: Clearcutting (the incarnation of mass production in forestry) a forest is the most efficient way to extract good wood; high value native trees are cut without undue expense, while the site is automatically prepared to host a plantation containing only desirable market species, such as Douglas fir. Furthermore,

as Roy Keene has pointed out, the planning, field work, and accounting are also simplified. Waste, or the wood not used, is minimized by the process—of course, all noncommercial species are considered a waste of potential growing conditions

- Obsolescence is necessary for successful growth: Obsolescent products made from wood are burned, buried in landfills, or stored in concrete hallways, removing them from the cycle of renewal of forests.
- Quality does not matter: Good materials, good wood from old-growth trees, are not necessary to make high quality goods. The quality resides in the perception of the consumer, educated by helpful advertising. If this is true, then why does anyone want old-growth trees from out-of-the-way roadless areas?
- The future is less valuable than the present (discounting): A forest that may have some value now in a few useful species is worthless beyond a certain time (two years, maybe 20). Modern economics has enshrined this one form of selfish behavior and pretends that it is rational and optimal. Unfortunately, as Costanza and Daly have pointed out, short-term self-interest is usually inconsistent with the long-term best interests of the individual or society. Economic discounting is not in the best interests of keeping forests complete and healthy.
- Economists can control the economy: By simplifying the forest and raising one species on a tree plantation, economists think that fertilizers and pesticides can control all foreseeable circumstances. Changes in and shifts of the system state can be controlled. The forest economy is suffering problems, similar to economics in general, as the results of these bad assumptions and false beliefs. These problems include:
- All products are equal—this is a misdirected effort on ill-conceived products, e.g., upscale firewood, throw-away chopsticks, and wooden match sticks—one of our Nieman Ryan Designs clients, a tree-farm owner, worked for Ohio Match in the 1920s in northern Idaho when they were cutting white pine for clear, strong match sticks.
- The modern economy is the only solution to human wants; it benefits all. Traditional (subsistence) economies were dismissed as providing only a slim margin between life and death. Marshall Sahlins exposes this kind of myth in his book, *Stone Age Economics*. There are numerous advantages of the way of life of First Nation peoples: Fewer working hours (about 3-4 per day); more leisure to talk, sleep, engage in rituals, and make love; a diverse and healthy diet; deliberate underproduction, usually well below the maximum levels; deliberate control of population growth below maximum levels; and deliberate under-use of resources, resulting in a small ratio of people to resources. Many industrial nations can only aspire to such wealth and leisure. Subsistence economics means simply that surpluses are not accumulated. This might make them more vulnerable to food shortages, although the low ratio reduces the possibility—furthermore, industrial cultures have far higher incidences of starvation (50 million children starved to death in nations with market and command economies in 1978). The modern economy, with its false myths and disrespect for everything, has lost sight of the real functions of economics.

4.4.4. *Functions of Economics*
The function of economics is to produce things that people need. Due to the development and scale of economics, especially with global trade and international conflict, it must address other functions:

- To insure the economic or material survival of the community or nation
- To manage the environment of ecological systems and global cycles
- To manage and budget specific resources (energy, materials, biological)
- To manage production and productivity using resources
- To manage the development, not necessarily growth, of the economic system
- To coordinate distribution and sharing
- To provide stability of the economy through employment levels and prices
- To provide security in the economy through policies and adjustments (for money, investors and employees)
- To provide a minimum for everyone, especially those poor, disabled, or old, who cannot work or compete.
- To improve the overall holistic efficiency of the economy, ecologically as well
- To eliminate discrimination, historical inequities, and unfair limitations (due to bad luck, discrimination, poor soils, or late arrival)
- To educate and direct better patterns of human consumption and waste
- To develop additional wealth, through symbols, technology or valuation
- To examine values as a result of scarcity, distribution or beauty

These functions have not been addressed by traditional economics, but they can be addressed by a more holistic economics.

4.4.5. *Holeconomics (Steady State Economics)*
Mill wrote that beyond the progressive state lies the stationary state; each advance is to approach it. A stationary economy is not synonymous with a stagnant economy; there is always room for developing and increasing scope of mental culture. It could be highly sophisticated, dynamic, and imaginative. Mill said: "It is scarcely necessary to remark that a stationary condition of capital and population implies no stationary state of human improvement. There would be as much scope as ever for all kinds of mental culture and moral and social progress; as much room for improving the art of living and more likelihood of its being improved." Mill's vision of a stable state was a response to the goal of industrial society. Technological progress would abridge labor, not necessarily increase production. In a state of equity, persons would have room for solitude and leisure development. But Mill's idea was anthropocentric—"man" was not bound to protect nature.

Herman Daly concludes that the steady state economy is both necessary and desirable. He notes that in the definition of economics, as the study of the allocation of scarce means among competing ends, the entire ends-means spectrum is not considered. Only intermediate ends or means are considered, not ultimate ones. He anchors the ultimate means in physics and ultimate ends in religion. Economics falsely concludes that the middle ranges represent the entire spectrum. In another computer model, Laszlo's scenario calls for an across board reduction in population, in investment and resource usage, to create a steady state with the present disparities maintained.

Some theorists, like Samuelson, have concluded that growth is necessary to rid the economy of disparities. But, development instead of growth would equalize wealth more efficiently—after all, economies have been growing for at least 400 years and increasing the disparities. There is no necessary association between development and growth, as Daly and others have shown. Development means the introduction of an innovation. Economic development will require technology. Ecologically sound technologies will minimize stress to the environment. Economies could be modeled after climax vegetation and not successional vegetation, where diversity in scale is greater. A community is forced to accept an upper limit, beyond which it cannot grow any further. Further growth results in destruction or disruption of itself and nature. This is the law of the maximum. Production could be stabilized in a steady state economy, a mature economy, like a climax system, where processes and cycles are constant. A steady state economy must be based on natural laws and ethical principles. Natural laws include thermodynamics and ecological theories. Rules of economics, laws of nature, and ethical principles must be related.

There is another distinction between growth and development. The ecological social approach (or a redistributive environmental strategy) to development makes it irrelevant to discuss global limits to growth. Local limits are far more significant to majority of population. Regardless of how much food exists, people will starve unless they can get it. Redistribution of resources and improvement of environmental quality (home environment) are more important than increased production by sophisticated technology. The natural capacity of regional photosynthesis must be limiting factor in development, especially in tropical and subtropical areas.

Development calls for social and educational organization more than technological style. Styles of technology must be determined by culture and context. Such development requires a local authority working with suitable economic and ecological conditions. No authority can be effective without the participation of the populace. To stop growth, a strict regulation of the productive system is needed. States should be able to control which products are made, and with which technology. Practical implications of the steady state are that nonrenewable resources will be conserved as much as possible, by recycling, while erosion, depletion and pollution are minimized; energy sources may be greatly decentralized and diverse.

Using a field concept for development emphasizes the dynamic transformation of the domain, rather than its structure as an individual. Selective advantage or cost benefit become adjectives or afterthoughts to the domain.

Jane Jacobs, in *The Nature of Economies*, argued that the same principles underlie both ecosystems and economies: "development and co-development through differentiations and their combinations; expansion through diverse, multiple uses of energy; and self-maintenance through self-refueling." The Nature of Economies, also in Platonic dialogue form, and based on the premise that "human beings exist wholly within nature as part of the natural order in every respect." Jacobs' characters then discuss the four methods by which "dynamically stable systems" may evade collapse: Bifurcations; positive-feedback loops; negative-feedback controls; and emergency adaptations. Their conversations also cover the "double nature of fitness for survival," traits to avoid destroying one's own habitat as well as success in competition to feed and breed, and unpredictability including the butterfly effect characterized in terms of multiplicity of variables as well as disproportionality of response to cause, and self-organization where "a system can be making itself up as it goes along."

4.4.5.1. Ecological Economics

Business as usual, with its inertial model of growth, could end in catastrophe for humanity and its environments. Industrial cultures, with their characteristics of simplification, naiveté, homogeneity, and incompleteness, turn wild landscapes into flatscapes, where variety disappears and significance is ignored for the comfortable standards of meaningless continuity. Rapid growth might precipitate a catastrophe sooner, while modest efforts at environmental protection and increased efficiency may only postpone catastrophe a few years. An ecological economics could provide another order.

The ecological provides the appropriate framework for economics, as exchanges are based in natural ecosystems. A new paradigm of economics is based on new ideas, analogies, and metaphors. This is the ecological paradigm, where

- real ecological limits exist and must be respected (what Rachel Carson showed)
- all beings have value for themselves
- people live in a mixed community of beings
- people have similar basic needs and wants
- intelligent development can supply these needs and some wants
- small-scale, quality processes can be developed efficiently

The new ecological economics is based on an orderly view of the world, a cosmology. Ecology and economics must be integrated. The goal of economics cannot be growth. It must be tailored to the ecosystem; its central tenets must be consistent with ecological principles. An ecological economy is survival-oriented, not profit-oriented.

Because human and natural systems are interlocked, there must be a common framework for ecology and economics. Current economic decisions are based on human reference and not nature. Human reference is not large enough. Economics is the study of budgets, material and energetic. Ecology is the study of natural budgets, material and energetic. Ecological and economic processes and values are often the same. Benefits can be assessed in a common metric. Both ecology and economics attempt to understand and predict the behavior of complex, interconnected systems where individual behavior and flows of energy and material are important. There are many other common or similar processes: Resource allocation, optimal behavior, adaptation, and energy flow.

Economic and ecological systems interact. Economics must recognize that ecological health is vital to its own continuation; it cannot make large mistakes. The concepts of economic value must be redefined. Resource use is an ethical question. It is not known how species and communities are regulated by environment or internal interaction. A free market has to be limited by conservative calculations of ecological balance.

A new theory is needed to relate economy to ecology. That theory must include externalities in consideration. And it must include intangible cultural values. It would be better to expand the circle of economic categories to include more factors that affect the economy, than to simply switch to supply side from demand. Traditional supply/demand theory uses current supply and demand of energy to determine current prices, ignoring renewability and long-term availability.

Theodore Roszak realized that human beings, and especially economists, focus their consciousness on the visible parts of the world, and forget the invisible that makes everything possible. The necessary invisible background must be described. He characterized economists as urban intellectuals automatically endorsing growth with ecological stupidity. He called for a nobler economics, one that is not afraid to discuss spirit, conscience, moral purpose, and the meaning of life. Roszak, Schumacher, Boulding, Daly, and others have grasped that economics is a subdiscipline of ecology.

Environmental quality cannot be treated as a scarce economic good. Daly maintained that the final benefit of economic activity is service or psychic income; it is yielded by stocks of capital maintained by material flow. The capacity of nature is limited, that is, scarce, but it is also irreplaceable and essential to life. The price mechanism ignores the possibility that something may disappear; it assumes substitutability, that is, one form of capital can be substituted for another, and this leads to the disregard for any one form, such as land.

There is so much about the environment that is not known, such as how species persist in under environmental stress and change. No one has complete information about current environment or the results of our actions. Complete knowledge is not necessary, however; only the knowledge of a minimum. For humans, trade can ease the law if the minimum, but not repeal it.

Resources may not be limited, in the sense that advances in technology expand the availability of resources; less is needed to produce more. But the same history also shows that humans reproduce up to the new limit of misery allowed by the new technology. New advances are used to increase the size of humanity, not its happiness or wealth. Society is growing at the same time as needs are growing, resulting in lessening of natural systems.

4.4.5.1.1. Ownership & Responsibility

In the Spring and Autumn period, 771-481 BC in eastern Asia, there may have been 170 independent states. Agriculture was communal agriculture, called the well-field system. Plots of land divided into 9 holdings. Eight groups farmed one each, and the 9th communally for the lord. However, this started to decline, perhaps due to the iron plow, which increased productivity. By 594, the state of Lu instituted land taxation instead of providing labor. Individual ownership and a free market may have appeared afterwards.

In the past in Uruk, ownership by a temple or kings preserved the character of land. The temple claimed ownership of some land. These lands were cultivated by small-scale farmers (7-17 acres).

In England, a forest was originally a tract of land owned by the sovereign and used for hunting; the word also referred to a dense or profuse growth of trees. By the 8th century, forest referred to the royal woods where the right to hunt was reserved by the king. Other rights, however, such as the right to cut wood for home fires, remained free for everyone. The forest then became a legal term that applied to a large tract of land, governed by laws, with its basis on the right to hunt. At one time, almost 33 percent of England was royal reserve.

Gigatrends can be seen with ownership: Increased property ownership. Land shortages, territorial disputes with neighbors. The earth has been divided up, every part is claimed and owned. Every place is owned; one would think that this gigatrend of ownership has ended with the finite territory of earth, but it shows signs of continuing through the solar system. With ownership comes management and control, the reach of the state into individual lives, from ownership and licensing to information-gathering

Why should ownership have anything to do with responsibility? Ownership implies control. Private ownership promotes responsibility and cooperation. Private ownership is one of the least understood institutions of the free society, a fact that can be of enormous value to debaters looking for a fresh, unexpected angle on an issue. Many people assume that since ownership entails the right to use property as one sees fit and to exclude others from using it that private property is anti-social and dangerous. A resource that is owned by a private individual, rather than commonly owned, or government-owned, is subject to whatever capricious idea that person might have about its use. The public has no recourse to prevent irresponsible or even dangerous uses of resources when property rights are consistently defended, the argument goes. From there the argument continues that government control is the answer, either through outright public ownership or through regulations that take away some of the property owners' rights. Economic reasoning shows that this argument gives too little credit to private owners and too much to government's wisdom and beneficence.

Private property does give owners a degree of freedom but it also makes them accountable for their actions. The owner of a dog is legally responsible for the damage his dog does to his neighbor's rose garden. The owner of a building is responsible for making repairs if the roof leaks on his tenants. This accountability ensures that property is not used in a way that harms other peoples' rights. Consider furthermore that one of a private owner's rights is the right to sell his resource to someone who values it more than he.

People have known for a long time that individuals take better care of things they own. Aristotle wrote, "What belongs in common to the most people is accorded the least care: they take thought for their own things above all, and less about things common, or only so much as falls to each individually." And we all observe that homeowners take better care of their houses than renters do. That's not because renters are bad people; it's just that you're more attentive to details when you stand to profit from your house's rising value or to suffer if it deteriorates.

Just as homeownership creates responsible homeowners, widespread ownership of other assets creates responsible citizens. People who are owners feel more dignity, more pride, and more confidence. They have a stronger stake, not just in their own property, but in their community and their society. Community ownership might maintain natural capital better.

But, ownership also promotes inequity, especially as a result of historical patterns, In a society where not everyone owns land or houses, the inequity might worsen.

4.4.5.1.1.1. Personal Ownership and Responsibility

The formula of the industrial myth is expressed by William Thompson as: You are what you own. The more you own, the more you are. Nations are now considered to own resources and lives within borders. The United States federal constitution is such that technically wild animals belong to the state where they live, except in national parks and refuges. Absolute ownership allows exploitation. Much economic disparity among nations is the result of exploitation.

The wretched of the earth and the poor of the United States need the same to improve material lot. They need control over their communities, to set own priorities and undertake development according to their own cultural traditions. Nothing technological, but political. There will be problems—corruption, mistakes, manias—but at least not one-crop economies surrounding foreign assembly plants.

Less privileged cultures need control over the use of resources. The concept of the common heritage of mankind needs to be expanded to all domains: ocean, space, minerals, scientific knowledge, technology and means of production. The aim is to pool all world resources to ensure effective planning. World economy and global resources would be managed for equity and efficiency. The environment is an invisible whole, not subject to political barriers. Resource use should not be dictated by accidents of geography, but should be allocated to serve the needs of everyone. The resource potential of each country is unique, so the selection of examples is hazardous. Stock-takings need to be carried out by all cultures.

Private ownership has resulted in states of poverty for most. The profits that accrue from ownership of means of production should belong to people who communally own them. Private ownership of the earth must be abolished. The urge for individual ownership of the minerals of the earth is a manifestation of the basic drive to distinguish one's property. The contents of the earth should not be rightfully owned by private individuals, but only by all people communally. It can be maintained that the contents of the earth can be rightfully owned by all people communally. All people in common have title to the minerals, water and atmosphere of the earth. Natural resources are for the use and profit of all people. Private ownership is not necessary for development. During centuries of private ownership, it was the practice to keep workers frozen in poverty. Land should be owned in common; its use should be channeled by the global authority for processing; and the profits used for common welfare. Minerals do not have to be owned by private individuals in order to be profitably processed and developed. This could be done by private enterprisers hired by a government.

Individual ownership would be limited to products of labor: extrinsic—time to work; intrinsic—imprint of ideas. This would include consumer goods and services. No one could amass property such that others would be deprived of necessities of life, through straight taxes (80%). This would solve two problems: there would be no rich-poor gap; no perpetual welfare economy if all individuals were supplied with the necessities of life. Eliminate poverty by making ownership onerous.

All cultures should share in the bounty of the earth and in the rewards of modern technology. The world has been miniaturized in our imaginations (and by technology); many who accepted hunger and disease have seen western affluence. Not sharing everything with them now would be inhumane.

Major social gains should be shared equally with other cultures, and surpluses used to improve environment or less productive societies. Production could be reduced. An arrangement is needed as regulator to keep competition within bounds. The potlatches in Northwest American Indians performed this function. Either extreme may be damaging. Inverse feedback arrangements may be used. For example, profits may be used to establish a new enterprise (Space colonies), or restore damage to environment. Profit making community may provide training program for other communities.

Growth in equality among peoples would make it less difficult for people to accept self-restraint in material needs. Birth control campaigns will be ineffective until people are freed from insecurity. The force of heterogenization would allow peoples to develop separately, without unwanted competition.

Diversity, not efficiency, is necessary for a rich and creative human life. In each region, efforts must be made to develop resources specifically for satisfaction of basic needs of population, defined realistically and as nonconsumptive. The unavoidable exhaustion of nonrenewable resources is tempered by efficient and moderate use.

The economics of ecology dictates that some resources are just not ownable—air, water, silence, landscape integrity—but a world commonwealth could own them in trust.

Who can own part of the earth? It is used by birds and worms; the wind sweeps it, rain waters it. In spirit it must belong to everyone and everything who passes it and touches it. All things are owned by the world.

With world unity, common ownership would not be limited to national boundaries. The minerals of the earth would be owned by all people of the earth. The world government would have minerals processed, channeled where most needed, and profits from them distributed to all people in proportion to their needs and contribution to the common welfare. The sea also belongs to everyone. The global government needs to establish a comprehensive ocean regime. Income derived from ocean resources and sea bottoms would be shared.

4.4.5.1.1.2. The Ecological Responsibilities of Corporations

The metaphor for a corporation used to be a simple mechanical model for turning resources into products. To be successful, a corporation had to grow and turn a profit continually. Unfortunately, the assumptions of the model were also simple and failed to consider human needs and natural cycles, causing great suffering and great disruption. To be really successful, corporations need to adapt a more comprehensive model, one that reflects stability, cooperation, justice, and respect for nature. With an ecological model, the ecological responsibilities of corporations, to themselves, to nature, and to human communities, are described.

A corporation is defined legally as an individual person, although one that is defined by law and exists only in contemplation of the law, hence it is artificial, invisible, and intangible. A corporation is not the stockholders or officers; it has its own entirely separate and distinct existence. This kind of artificial person was the invention of commercial interests; the first corporations in Britain were granted charters by special acts of parliament to provide services for the state. Most corporations have characteristic legal features: individuality, permanence, limited powers, continuing succession, action in the corporate name, limited liability of members, transferability of member's interests, and representative management. Charters have become easier to get, until now corporations can be formed under general law. Corporations have been adapted to meet most modern business and social needs. Real persons can do anything not prohibited by law; corporate actions are derived wholly from law and limited to the charter, although charters have become very comprehensive statements that allow a great variety of activities.

In the publication "Integrating the Enterprise" by Digital Equipment Corporation, it is stated that "Digital is a living organism," tending toward a state of dynamic equilibrium by adapting to circumstances as fast and economically as possible. This is a natural-enough metaphor, and it leads to interesting deductions.

Like an organism, each business is born, grows to a certain size, then matures and dies—perhaps a natural span is hundreds of years, like the Oxford University Press, for instance, or perhaps only a year like so many new businesses. Biologically, maturity marks the end of physical growth in humans, but not necessarily the end of emotional, intellectual, and social development. Like an organism, the corporation, unconsciously through its officers and managers, starts to act to preserve itself before completing its formative objectives, such as maximizing profit. Like an organism, a corporation lives in place and alters that place to some extent by living, although some of the larger ones risk destroying the mixed community of humans, animals, and plants (and all their associations) on which they depend. Every organism must fit in its community and environment. It must be integrated and limited—self-consciously so in the case of humans and corporations.

Corporations are unique organisms. Corporations are more than just groups of people without structure. Corporations are described by analogy with individuals. But, each has unique characteristics: the corporation has right to property and free speech but not right to worship or vote. They can be longer lived than their human components.

4.4.5.1.1.2.1. Problems with Corporate Organisms

The old analogy of the corporation as a machine leads to bad assumptions: that everything is a resource; that resources are unlimited; that production must continue endlessly; that the corporation has to keep growing to survive; that the purpose of the state was to legitimize exploitation; that the purpose of humanity was to multiply, produce, and consume; and that the purpose of the universe was to supply human and corporate needs. The machine analogy also leads to false economic beliefs: that mass production is most efficient; that obsolescence is necessary for successful growth; that people's needs and wants are fulfilled by advertised products; and, that quality does not matter very much.

Some companies have started to work around these beliefs. Kodak and Head, for instance, have succeeded with custom production, high-quality, long-lived products that people did need. Other corporations are suffering problems as the results of bad assumptions and false beliefs. These problems include: overgrowth, with an increase in complexity and costs (many of them social); economic and ecological instability; social burdens (from pressures on family from relocation and powerlessness); misdirected effort on ill-conceived products; slack employee attitudes and performances (sickness, accidents, turnover, layoffs).

In the effort to control their problems, corporations have sought more control. The corporation tries to avoid being vulnerable to change and uncertainty, fluctuation, and market conditions by relying on planning and control. Corporations try to ensure stability by taking over the supplies of materials, controlling their subcontractors and the buyers of their products, controlling the work force with pay and incentives (as well as by cultivating identification), and managing demand by sales influence and advertising. Many corporations try to be flexible about resources, using what costs the least, say, cheap energy to replace expensive labor. Corporations, especially multinational ones, seek raw materials everywhere, in any nation, under any ocean. Corporations become international, mining, assembling, and selling in three different continents. Where developing countries were once regarded as sources of raw materials, they are now used as bases for manufacturing, and they are becoming growing markets.

This control is possible because corporations have acquired such great power. Large multinational corporations have great power to control national economies and ignore environmental laws, partly because of their historical promise of providing social services to national states. Unfortunately, as private good became identified with public good, corporations became less concerned with social service and more concerned with greater profit through greater technology. Greater technology lets the power to change overwhelm the power to see the consequences. So, social amenities, such as clean air and fresh water, are violated legally—after all, no right of contract or fair use of property has been breached. Corporations exercise enormous influence over our lives, probably more than governments or churches. Furthermore, the size of some companies means that their influence is felt like shock waves through societies and environments. Multinational corporations are a force in the business world and a major influence on world affairs. Size and complexity give them special power, not only financial but political. Power is supposed to evaporate in a purely competitive economy, but capitalist economies are not purely competitive; power accrues to corporations.

Usually the goals of a company are generously, nobly, and broadly stated as intentions to support the best interests of owners, managers, shareholders, employees, the public, and, lately, the earth. Many corporations pride themselves on their generosity and personnel standards and on their good corporate citizenship, although the public concept of good is being extended beyond traditional bounds as citizens become aware of the interactions of business, politics, and the environment. Even corporations that claim to make clean products in safe ways often seem to depend on "dirty corporations" for power and packaging, as well as for paper and materials.

Business as usual, with its inertial model of growth, could end in catastrophe for humanity and its environments. Industrial cultures, with their characteristics of simplification, naiveté, homogeneity, and incompleteness, turn wild landscapes into flatscapes, where variety disappears and significance is ignored for the comfortable standards of meaningless continuity. Rapid growth might precipitate a catastrophe sooner, while modest efforts at environmental protection and increased efficiency may only postpone catastrophe.

An organic model offers the best alternative, with its emphasis on energy efficiency and alternative sources, its major commitment to environmental protection and the internalization of environmental costs, and its change from growth to stable, sustained development. Such a model would provide organic assumptions and beliefs: That resources are limited, that human value is only part of ecosystem values, that humans are more than consumers, and that the quality of life is more important than quantity of possessions.

Like organisms, corporations may have an optimal size and a home place; organisms that occur out of place are often called weeds, and organisms that grow too large are described as monsters. Perhaps corporations should be limited in size and tied to one place.

Healthy organisms, even humans, are educated to take their place in a culture that limits their impact on the environment, proscribes their actions towards one another and towards others outside the culture, and trains them to reproduce themselves and renew the culture. Most corporations act like improperly socialized individuals who have not been taught how to take a proper place in society and how to be responsible for their actions; instead, corporations hide behind their legally limited liability.

Any organism, like robins to fermenting berries, can be addicted to certain things in certain circumstances. Corporations have become addicted to cheap energy and easy defense money. It is possible to create circumstances that limit or cure the addictions (for robins, the berries fall to the ground and get covered with snow).

4.4.5.1.1.2.1.1. Ecosystem as Metaphor

Perhaps the metaphor itself is not adequate. Even organisms take a large part of their identity from their context, from the surrounding environment. It might be more productive, since corporations actually contain organisms, mostly human, to consider the corporation as an ecosystem. Ecosystems occur in a large diversity of sizes in nature, from a rotting log to one covering most of a watershed.

Definition. An ecosystem is defined as a biotic community and its nonliving environment functioning together. An ecosystem is also a self-organizing, chaotic system with emergent properties. Ecosystems develop in time. That is, a community develops by an reasonably orderly, directional process that involves changes in structure that results from community modification of the physical environment. Although the physical environment imposes limits and determines pattern and rate of change, the community controls the succession. The result is a relatively stable configuration characterized by a high biomass (or information content).

Characteristics—Energy. The "strategy" of ecosystem development is increased control of (or homeostasis or homeorhesis with) the physical environment—to protect itself from perturbations. There is a fundamental shift in energy flows, as increasing amounts are used for maintenance. The structure of a community changes: organic matter increase, inorganic nutrients are used internally (instead of being extrabiotic), biochemical and species diversities become high, and pattern diversity becomes well-organized.

As more and more energy is used for maintenance, the net community production approaches zero. Agriculture keeps an ecosystem immature in order to harvest the larger yield in immature systems. The mature system becomes more efficient, as it supports a larger biomass with the same amount of energy. The food chains become more weblike, dominated by detritus chains as opposed to linear grazing. Mineral cycles become closed and the nutrient exchange between organisms and the environment slows.

The life history of organisms undergoes change as well. Organisms tend to be larger (perhaps as a result of shift from inorganic to organic nutrients), with longer, more complex life cycles and narrower niche specialization. Population growth slows, with emphasis on the quality of life of organisms. Internal symbiosis becomes more developed, conserving nutrients and resisting perturbations.

4.4.5.1.1.2.1.2. Corporations as Ecosystems

Corporations share certain characteristics with ecosystems, such as energy use, productivity, development, and maturity.

4.4.5.1.1.2.1.2.1. Development in Time

Corporations have been evolving for hundreds of years. There have already been shifts in the meaning of corporations and in the forms of organization and style: From a stable product line to continuing innovation process, from a product based definition (shoes) to a process (information), from a single pyramid of organization to constellations of satellite concerns, from the static to the flexible, from product line management to networks and innovations (the management of change), and from stable forms to temporary. Mature systems have a greater ability to trap nutrients for cycling. Corporations have settled in an artificially maintained pioneer state, feeding on the extra productivity.

4.4.5.1.1.2.1.2.2. Energy Use

If corporations follow similar patterns over time as ecosystems, we would expect their "food chains" to become more complex, with most of the energy flow following detritus pathways. Optimizing material and energy use, reusing `wastes' as resources (food webs between companies, where wastes are products, not side-effects), and closing loops by recycling. Reciprocal adaptations between plants and animals, or between producers and consumers, leads to mechanisms that reduce grazing and increase feedback.

4.4.5.1.1.2.1.2.3. Low Productivity, High Stability

A mature ecosystem produces many things, most all of which are used by the system; wastes are broken down into component chemicals by microbes, while other resources, like nitrogen, are fixed to roots by other microbes. The tightening of the biogeochemical cycles is an important trend. Corporations sometimes parallel the aging of an ecosystem, from a pioneer state to a mature state.

4.4.5.1.1.2.1.2.4. Life History, Symbiosis

The longer lived the corporation, the less clear the divisions between private and public and economic and environmental concerns. Short-term individual concerns meld into long-range corporate and social concerns. The short-term pressures may seem immediate and irresistible, but the long-term goals cannot be ignored for long. We may need electric power or paper now, but the cost cannot be the destruction of the source of those needs. We need the social and environmental health and stability first and always. Partnership between unrelated species, say mycorrhizae and trees, becomes notable. Corporations would team up with other companies and social and environmental groups (the pattern could be industrial symbiosis). Pollution is the most limiting factor; possibly we could predict new corps arising to deal with pollution.

4.4.5.1.1.2.1.2.5. Differences Between Corporations & Ecosystems

Although ecosystems can be long-lived—thousands of years for tropical rainforests—corporations can disband at any time. Furthermore, unlike ecosystems, corporations still seem to be bound inflexibly by the rule of two-year payback. This means that decisions are based on short-term return and not on long-term durability.

Furthermore, although large organisms are the case for mature ecosystems, the scale of many corporations is too large; the patterns are unsustainable—large institutions lose touch with their constituents, become self-absorbed and less responsible.

The differences between ecosystems and corporations allow for the possibility of changes. Human communities can redefine corporations and limit their impacts. They can change the charter of corporations for the benefit of the community. Corporate responsibility is more complex than a simple linear cost/benefit analysis. Using the metaphors of corporations as organisms and ecosystems, it is possible to outline a new set of responsibilities for corporations and a series of behaviors that human individuals and communities can do to integrate corporate behavior into the communities.

4.4.5.1.1.2.2. Ecological Responsibilities of Corporate Organisms

The public responsibilities of corporations, according to Harvard management, are to grow and prosper—thereby providing customer satisfaction, employment, taxes, and contributions to the economy—and to control their hazards. According to Milton Friedman, the only social responsibility of a corporation is to make money, by striving after profit as an efficient agent of production, although he admits that the corporation should conform to the rules and norms of society.

Profit making is a necessary part of business, but not the sole reason for business. The best business serves public goods as well as private interests. This is similar to Ruth Benedict's original anthropological meaning of synergy as it applied to individuals. In secure, nonaggressive societies, an individual serves her own advantage as well as that of the group with the same act. The institution ensures mutual advantage; the acts are mutually reinforcing. High synergy institutions transcend the polarities of selfishness and altruism. Virtue pays because the rewards for selfishness coincide with benefit for the society. The social structure of low synergy cultures ensures opposition and counteraction; the advantage of one individual is a victory over another, as in a zero-sum game. This is necessary for employees, since they have to feel like their work is meaningful and contributing to the public good. The path of production should therefore serve public good as well as profit. Environmental and social problems should get as much attention, because their part of the process, as sales, finance, and production.

Economic recession may bring a re-examination of values, not only by individuals who may have less material wealth, but by corporations that have emphasized growth. The public may insist that corporations consider social performance as well as strictly economic performance. The single economic purpose may only be the focus in a social ecological environment. Economic actions, such as where to build, who to relocate, hire, or dismiss, will be subjected to greater public scrutiny. Corporations will have to adapt to changes in standards. Business cannot assert primary self-interest at a cost to the public or environment. Corporations need to keep track of their environmental impacts. Many of the problems that corporations face are connected to the problems of the environment and society.

Every corporation depends on the stability of the environment and on the stability of social institutions. The environment provides air, water, and land, and provides renewing (both physical and psychological). Institutions, from sanitation, police, schools, churches, and community centers, provide a supporting network. As these institutions wobble or fail, corporations may have to subsidize or replace them to survive. Schooling for example, is often inadequate to provide literate, numerate, or ecolate workers. Police may not be able to provide secure conditions on corporate grounds for female workers. Public transportation to plants from town may not be available in enough volume.

A corporation has traditionally been seen as a morally neutral body, but even if it has only to conform to the nominal rules of society, it is already a moral agent. Corporations are no more neutral than other organisms. Etymologically, the word moral means simply 'living together.' Sometimes even routine business (nonmoral and nonenvironmental) matters become deeper ethical conundrums about justice. Many areas of moral concern already are recognized: worker safety, affirmative action, advertising truth, foreign investments, and harm to the consumer, public, and environment. Corporate responsibility occurs wherever the interests or rights of a person, society, or ecosystem are significantly affected by the actions of the corporation.

Responsibility can be understood in terms of costs and benefits, that is, through operations and their consequences rather than abstract behavior. Every action entails a gain and a cost (or profit and loss). Profits and losses are distributed privately, socially, or environmentally. Unfortunately, the modern system privatizes the gain and externalizes the loss to the "commons," considered as a pool of "unowned resources," where in traditional societies, it was surrounded by rules for use. As long as this is possible, it is profitable to charge the cost to the environment. Externalizing costs works fine in an uncrowded world, where the costs are negligible and can be absorbed by natural processes. Resources were traditionally seen as free for the getting; air and water were seen as free sinks. Modern economies, embracing the notion that "nature is capital," draw on the accumulated "capital" of ecosystems for production. By ignoring the real cost of the capital, as well as the costs of natural services, such as nutrient recycling, soil building, and atmospheric renewal, these economies create a temporary wealth. Decisions regarding resources are made on short-term economic grounds and lead to material shortages and environmental degradation.

Similarly, labor was seen as minimum value. For example, the idea that "labor is a resource" implies that, like any common resource defined by industrial society, labor is cheap and can be used up. The real costs of free goods and externalities have had to be accounted for, yet—this often influences the selection of corporate priorities and growth. Furthermore, the production and distribution system for most corporations is linear (straight throughput) and not circular (complete recycling), although this is logical economically, given our frontier resource-use accounting. Major changes are occurring, though. The scale of civilization now makes externalization unfeasible. The costs of pollution and waste are being internalized; other inputs, such as labor and capital, are becoming more expensive. Corporations will have to internalize or be forced to internalize. With the internalization of costs, since the losses as well as benefits will accrue privately, the system will benefit from intrinsic responsibility.

Corporations need to work cooperatively to make sure the costs and benefits are extended equally throughout the system. They could start by sponsoring the rational use of rare resources through taxation. Influence the government to determine priorities for wilderness areas or special landscapes. Beautiful, fragile, unique, or endangered ecosystems and species must be protected at the expense of commercial activity.

Corporations have at least three large, ecological responsibilities.

4.4.5.1.1.2.2.1 To be Economically Healthy

The first responsibility of a corporation is the maintain its own health, to mature organically. limiting its size and impact to the locality. To be responsible, a corporation should consider the following actions.

Create a department with ecological authority to envision long-range plans and impacts. Corporations need to react more quickly, to monitor their ecological and social environments for emerging patterns that determine their future. They need to anticipate and participate in the social and natural framework. A new department, with global, anticipatory functions could provide direction and continuity. Such a department could be justified in the same manner as military forces. Military expenditure is a nonproductive cost; its benefits are general and long-range, that is, it must discourage war in the next decade as well as in this one. Its scope of advice would include educational services as well as advertising, capital acquisitions as well as new products, and engineering as well as security.

Plan all foreseeable consequences of a product. Advanced technology permits the power to change to overwhelm the ability to foresee the consequences of change. Avoiding the opposite actions of intentions (the enantiodromia recognized by the Greeks as the operation of tragedy) is extremely difficult. Good intentions are not enough: Labor-saving devices may contribute to unemployment and social problems; foreign aid may result in starvation for more millions as local agriculture cannot compete; the environmental management of some species for sustainable yield causes population collapses.

Determine the optimum corporate size. Limit the size of the corporation. After a point, growth results in inefficiency and nonadaptibility. Development, on the other hand, can continue for hundreds or thousands of years. A smaller size could mean more flexibility and faster response to local conditions. Recognize material limits. The global economy is probably too large already to be supported by the natural systems of the planet. What is the upper limit to the economy of scale? Accept limits to growth based on materials and on nonrenewable or dangerous sources of energy. This should not limit development based on advancing technology and knowledge.

Adjust corporate strategies to changing values. Smaller social and cultural groups have different and diverging values, so corporations are going to have to adjust to a diversity of values instead of to a monolithic standard. Now, the structure of power is disintegrating (with information replacing things as wealth). The knowledge-driven economy is more decentralized and customized. This moves us towards customization of production and away from mass production. Change the shape of the corporation to a framework coordinating separate divisions sharing information. Each could react much more quickly to market conditions.

Work to delineate a new information model of production in which the stages of a process (capital, materials, workers, design, advertising, selling) are simultaneous and synthesized. The conception of the product is extended from design (even customer contributions and design of working conditions) to aftercare, including ecologically safe retirement and disposal (recycling). The notions of efficiency and productivity are changing.

Innovation and computer technology shortens product life cycles. Production diversity is increasing. Convert the information model to an understanding model. Information is just data without appropriate structure. Provide a structure and material base for understanding through communication, education, and training.

Enter partnerships with the employees. Address the optimum productivity of employees. For instance, government studies show that half-time employees are more efficient than full-time ones. So adjust the work force to fewer, more flexible hours (thus avoiding layoffs during the transition). Increase worker participation. What is extent of worker participation in management of workplace? New forms of ownership could mobilize workers in a more efficient and democratic economy. Productivity is declining, so is job satisfaction. Efficiency and productivity are often less important than use and appropriateness. Better pay and shorter work weeks have not compensated for lack of worker control. Offer more control. Streamline the organizational structure. Organizing workers hierarchically is costly. The best path for organization is lateral modularity, not bureaucratic hierarchy. The levels of management could be reduced drastically.

Promote the principle of least effort, allowing the company to consume less, recycle, use longer, and avoid waste. Corporations could develop renewable energy sources. Reduce office costs through energy conservation plan. Use renewable energy sources. Corporations need to maximize recycling. Energy and materials can be used and reused, flowing through the system. Ship by the best transportation, rail. Cars are ecologically unacceptable forms of transport, yet companies intrinsically recognize them with large, free parking lots. Discourage commuting; encourage telecommuting or even alternate forms of transportation(bicycling, buses, and trains). Minimize wastes, for instance, by using permanent packaging.(Milk bottles and cola bottles can be reused forty or more times.) Conduct a complete series of audits, including an energy audit for every building, an environmental audit to determine negative impacts, from acid wastes or product disposal, and a problem audit, including inherited problems. Produce a comprehensive annual impact statement.

4.4.5.1.1.2.2.2 To Preserve the Health of Natural Communities

The second responsibility of a corporation is to maintain the health of the natural communities—because environmental health is the basis for community health, and community health is the basis for economic health and worker health. The quality of life depends on the quality of the environment. If the environment is degraded to raise the quality of life, the effect will be very limited and never be self-sustaining. Fitting economic costs and needs to the limits of ecosystems and monitoring the economic process would reduce wastes and pressures on natural processes. The coupling of agricultural productivity to a solar budget, and the conscious restoration of degraded systems, would contribute to the health of ecosystems. Sufficient wilderness would allow the self-maintenance of global cycles. With the increase in security, wealth, and self-esteem, human populations could be dependent on ecosystem productivities and still be diverse and unique. Corporations would:

Be accountable for ecological impacts. Corporations will be held more accountable for their technological impact. New technology will be more closely regulated. Corporations could anticipate this by favoring open appraisal of new technologies. By studying the potential consequences, physical, social, and ecological, as far as possible into the future, of its innovations in information technology, a corporation can gain credibility. Otherwise, it can wait and be forced by public and governmental pressure.

Avoid interference with natural processes. Technological processes must be brought into balance with the cycles of the earth. They must not damage or degrade natural cycles. Avoid unnecessary harm. It may be appropriate to use trees or to compete with black bears for tree use, but it is never wise to destroy the ecosystem of trees and bears. Laws on pollution and noxious wastes have been notoriously lax and sometimes wrong-headed. Minimal acceptable tolerances are legal, yet people often prefer zero amounts of many substances. Minimal compliance with them is virtuous in comparison with many companies, but it would be better to lead to higher standards. Work toward setting zero-level goals. Do not dump exotic or dangerous wastes. Do not discharge quantities of `safe' wastes. Integrate loops and material flows with natural flows; internalize cycles.

Corporations maintain building and plants in thousands of locations, each requiring support. Convert to ecological grounds practices. Forgo economic development of key ecosystems, which are not available for human use. Consider adjusting economic pace to natural rates; do not cut trees, for instance faster than they grow. Consider minimizing use of ecosystem productivity to the net ecosystem productivity, rather than the gross productivity, especially as regards fisheries.

Promote ecological design, which starts with questions. Is the product low-cost, aesthetically pleasing, and ecologically wise? Where does it fit in society? Ecological design, both responsible and socially responsible, must be radical, that is, rooted in a community in place. Membership in a place, in fact, leads to community. Corporations must become responsible members of the community. It would encourage an ecological approach to systems and processes in the whole environment, where the product, with its plant, engineers, and advertisers, is a link in a long biomorphic phylogenetic chain reaching from knotted ropes to surgical microchip memory implants. Ecological design has important characteristics for responsible technology: The products are designed by interdisciplinary teams considering all parameters and consequences; ecological sciences offer creative insights into design through a search for underlying organic principles; the product must be related to the particular environment, the tool is a link between human and environment.

4.4.5.1.1.2.2.3 To Preserve the Health of Human Communities

It is hard to protect communities when the way most business is done tends to disrupt community life. Because of its size, power, and intentions (for profit), the corporation should take higher risks than the surrounding communities. This will ensure the safety of products and wastes. A corporation should:

Support the community. Work place isn't just collection of individuals. It is a number of groups. Group interaction can change attitudes. A working community can build mutual responsibility. Show proper behavior; learn community etiquette.

Design the corporate structure and size for the community. Limit unnecessary movement or disruption. Plan the shape, size, and products of the corporation to fit the local community. Encourage self-reliance in communities. Communities can be self-reliant by producing enough food and shelter, by limiting their population to what can be produced, by using local products and raw materials (soil, minerals, plants), by using general and not specialized machines, by having multipurpose factories, by networking with other communities, and by doing without things that are not needed (bombs, food additives, or plastic bottles).

Behave ethically. An ecological corporation could use corporate buying power to promote acceptable technologies and discourage unacceptable practices. Deal less with nuclear weapons contractors and more with solar energy companies. Deal less with one-shot paper companies and more with recycling paper companies. Boycott paper companies involved in Rainforest destruction or old-growth forest destruction. Avoid banks that invest in anything that brings a high return (from third-world debt to Amazonian destruction and South African discrimination). Favor peace-oriented companies as business partners. And refuse to participate in work that is socially destructive.

Participate in the economic and social functioning of the community. Economic development and social progress are necessary for the welfare of humanity, but must be conducted with environmental knowledge. The goal of economics and politics is to provide suitable and comfortable human habitations and meaningful activities. Human settlements must be planned and constructed within environmental constraints and according to ecological priorities. Work to preserve the structure of the natural and social communities. Corporations can encourage decentralization and restore schools, clinics, and shops to local communities. Offer cooperative control with the community. Change the pattern of ownership to reflect employee and community participation.

Promote ecological education in a total context and interdependency. Encourage cultural traditions to stop letting social and spiritual needs be subverted by economic ends. Help lead the young into their adult responsibilities through training and participation (perhaps apprenticeship programs). Educate for appropriate ways to achieve wealth and well-being. Teach appreciation of services rendered by nature through flows and cycling. Point out the unexpectedness of consequences of even simple corporate interventions and innovations (positive feedback, biological concentration of poisons, and synergetic effects of simple new chemicals like CFCs). Trace the complex and reciprocal relations of soil, climate, vegetation, and human activity.

Emphasize that a fixed set of ecological parameters in an ecosystem can not be maintained sustainably, because the system is dynamic and changing. For example, where do computers fit in schools? Children do not need computers to develop the powers of thought, but they do need an ecological curriculum where animals display greater powers of mind than computers or machines. The important technological advantages of a computer, word-processing, database searches, complex connections, and rapid computation, are not really needed before high school, unlike myths, languages, and physical activities.

Implement community responsibility. In education, integrating business with humanities; the responsibility for the welfare of the citizens belongs in the community, as does education, safety, and the whole infrastructure. In management development programs and management as source of influence by setting goals, modifying structures, and introducing criteria for measuring progress. By the Board of Directors, the architects of responsibility and stewards of the human and material resources. By the government, in its legislative, judicial, and regulatory functions. providing rules and permitting freedom.

4.4.5.1.1.2.3. Going Sustainable

The corporation, regardless of its legal definition, is a long-lived, collective, impersonal body. Yet, it has more physical, legal and moral power than any one individual. Its investment is long-term in actuality. Many stockholders keep their investments for decades or a life-time. They are not concerned about only one dividend. Like the corporate organism, they want the long-term outlook to be positive; they want to know that their investment is stable and that the quality of life it encourages or supports is continuous.

The complexity of environmental problems should not permit escape of responsibility. The context of corporate responsibility falls within the spectrum from individual responsibility to social responsibility (the designation of property or trading conventions—capitalism or communism). That is, everyone associated with corporations has responsibilities related to corporations. For instance, shareholders must learn to have reasonable expectations, without the possibility of impairing social or ecological environment. They need to insist on annual reporting on compliance with regulatory environmental laws and standards. Their investment strategy has to eliminate activities that produce undesirable social and environmental effects, and the strategy has to include ways to promote beneficial activities and sustainability.

Employees have the opportunity and responsibility to contribute to corporate vision. Their degree of compensation and career criteria have to reflect corporate values. And, the corporation has to be engaged in the lives of its employees, meeting their minimum needs and contributing to their development. The corporation should provide measurable standards of ethics and responsibility. Executives, as employees, should take an oath of office, similar to doctors and ecoforesters. This statement would articulate the range of purposes for the enterprise, from making optimum profits to creating jobs and necessary products.

Business partners should require high social and environmental standards from the corporation. Partners should represent their practices accurately and give preferential treatment to ethical partners. They should refuse to do business with companies that have harmful practices. Advertising partners should represent corporate services and products accurately, as regards contents, effects, safety, longevity, capacity for reuse or recycling, and long-term costs and benefits, as well as to present the desirability of such products.

Local communities have the responsibility to enforce basic human rights in the corporate context and to insist on respect for the diversity and politics of the community. Communities need to insist on corporate patronage programs community. Communities and corporations keep an open dialogue regarding changes in standards and practices.

Perhaps the responsibility of corporations could be enforced if the entire earth were incorporated and concerned with maximizing its own values as an independent corporation: Healthy beings in living contexts. Certainly not having `free' services and resources would force corporations to internalize all costs of production.

In any case, there are strategies that a corporation could pursue to become ecologically minded. Instead of treating decisions as trade-offs, an ecological corporation could aim at a congruence of moral, economic, and ecological objectives. Responsibility could be manifested in organizational structure, manufacturing, and marketing practices, without departing from economic decision making.

Such a corporation could bring corporate research and development capacity to bear on the transition to a sustainable society. Where technologies play a role in the transition, companies can assume social responsibilities equal to their size and wealth. By commanding their vast resources, corporations can ease the transition to a sustainable society (which would actually meet their needs for stability). The model of corporate life needs to change, from dependence on continuous growth (of profits and waste), to be being based on stability, sustainability, cooperation, justice, and respect for nature.

4.4.5.1.1.3. Community or State Ownership

Community or state ownership has many of the responsibilities of corporations. In fact, many are already incorporated. The community has to take ownership of, and responsibility for, its commons. Many unowned species, such as blue whales, are hunted to extinction, while privately owned African elephants flourish in Zimbabwe, despite the large, illegal profits to be had from the sale of their tusks.

4.4.5.1.2. Holistic Accounting

The scarcity model of economics is obsolete from the magnitude of participation in knowing and being. A world commonwealth would appreciate, advancing from the depreciative fallacy of industrial economics. Fuller claimed that a world accounting system must be converted from annual agricultural metabolics to perennial world-wide accounting, consistent with a regenerative system and inclusive of a redefinition of wealth. This accounting would include all world metabolics, wild and domestic. It would also include future generations. This would be possible if half of the wealth, in terms of resources and primary production, were treated as capital. The other half would be untouchable reserve. Then humanity would live off the interest from its half and tailor its population and needs to that income. We need to draw an analogy from plants, and imitate it. A plant changes from heterotroph, in which it uses all its capital (the seed), to an autotroph, the cotyledon, where it synthesizes its needs from light.

All resources could be inventoried and divided into units for which stamps (capital coupons?) could be issued. The number could never be inflated, barring new discoveries or inventions, but they could be traded. They could also be reused, and this would encourage recycling. Resource coupons would be assigned equitably to each cultural area. The United States would become much poorer, and Ghana much richer, for instance.

Or, after Howard Odum, energy units of some kind could become a basic currency, either nonhuman (diamonds) or human (diamond rings). Or organic units of some kind could be used. Anything that was valuable to all cultures could be used. Perhaps human work units would be a fair way to exchange work and goods.

In the interim a temporary world tax scheme, as proposed by Sakharov, would begin the leveling. The international redistribution of incomes through taxes, e.g., income, luxury, armament, resource, would limit the range of consumption. Taxes would be a way of directing and conserving usages.

Externality taxes would be incentives for firms to be frugal, stimulating firms to avoid the taxes. Tax incentives would also encourage people and ideas into recycling. Legislation could be enacted to limit gifts or inheritance, equality of fortunes.

A residence tax on offices located in cities. Support for theaters, museums, noncommercial festivals would gain in priorities. A world treasury would be needed for income derived from international taxation on the use of resources.

Where does accounting, especially ecosystem accounting, figure into this? As Berry points out, many of our necessities and liberties depend on cheap labor, cheap raw materials, cheap energy, and disparity of wealth.

4.4.5.1.2.1. Full Costing

Currently, environmental regulation is expensive and still avoids the recognition of necessary costs. Environmental costs are real. But prices are distorted by fixing, subsidies and cost of living index; prices are based on insufficient information of real costs; there is a long time lag. Full cost pricing would charge consumers the full costs of production, including those of resource depletion, pollution, and environmental degradation, in the purchase price they pay for goods and services. It is not simple. All the true costs of any technological process must be internalized even if it means assigning arbitrary dollar values to aesthetic resources.

The hidden expenses of urbanization are being taken into account. Cities overlook one another's garbage; cleaning up costs money. We are seeing that we are not quite as rich as we thought. The excessive depletion of oil in North America in the 1960s has now become a cost that has been externalized—transferred to the 1970s. Full cost pricing would use the profit motive to reward nonpolluting and conservative producers, and punish the wasteful. This kind of pricing would require more complete records. A cross-referenced inventory of resources used in production units would need to be compiled.

Reduction in stock of resources might eventually lead to depletion. Delays could lead to crisis: the price of an exhaustible resource might increase too late to save the resource from exhaustion; there is no gradual braking mechanism for pricing. Capital transfers should be used for investment rather than welfare functions.

The price system does not adequately account for a number of value factors: the benefits of people outside transaction. Value systems are the driving variables in all economic systems, not just peripheral attachments. And patriarchal nations are devaluing these values and cultural wisdoms with their dominant abstract, quantitative economics. Economic objectives have to include new concepts of value. A new economic system will have to capture values unimagined by most economists. How calculate the potential of possible plant products from Amazon basin, or how the value of forest susurration there?? The principle objection will come from economic materialism, not religions or philosophies.

4.4.5.1.2.2. Work

Work in industrial areas is considered demeaning. The necessity of the evil is reflected in pay scales that are slanted toward executive positions. Human labor is treated as a means to production, and production is automated wherever possible. The product is valued far more than the process. The work activity is divided by time and measured by output. It is applied indiscriminately over the world. Industrial automation has a problem fitting peoples with different tempos of life-style; Turkish peasants and Colombian farmers seem disorganized in factory settings, when it is the expectations that are thoughtless.

A maximum per-capita income, or a moving scale of acceptable income differences between lowest and highest paid members of society— 5000:99000. The ceiling would move when the minimum moved. A minimum would be guaranteed by a negative income tax. Development in rich countries might be stimulated by lessening income. Work is available for all; and every effort made to train for fields in which people have an interest. Compensation could be offered in the form of coupons or schooling for unpopular jobs. The idea of full employment could be abolished; if employment has to be cut down, all could be cut equally. Overproduction in an automated age could reduce the work week to less than 10 hours anyway, for sustenance work. The basic unit of work would be the person-hour, with all person-hours equal in value. Work loads would be shared equally, with bonuses for hazardous or unpleasant work. No advantages could accrue from overwork. Yet, some work would be compulsory. The ideal of compulsory, practical work appears in the works of D.H. Lawrence, H. Read, Morris, Ruskin, Cobbet, Mumford, Goodman, Dewey, Thoreau, and others. Work would be the passport to the shinier things in life, like a television or freezer.

A guaranteed opportunity to work would symbolize unconditional basic social security, which may be required for the development of new social relationships. In the economics of permanence, work acquires meaning as personal and community projects, as it once was for hunters and gatherers. The impetus for work should come from a sense of community. Work becomes a self-determining, nonexploitative activity. Working does not have to be separated from living; play and work could be with family and friends. Work should be treated as privilege, divorced from right to life. Life is opportunity, as is work; all should be allowed both.

4.4.5.1.3. Steady-State Community Economics

Instead of concentrating on a solution using scientific methodology in an industrial context, E. F. Schumacher, in *Small is Beautiful*, requires an ethical solution in a metaphysical context. He argues that economics as a discipline must be derived from a metaeconomics.

In "Buddhist Economics" Schumacher explores the necessary metaphysical presuppositions implicit in Buddha's path of right livelihood. Later, given the proper function of work in the Buddhist sense, and the limits of financial investments in most countries, he proposes a multiplicity of small scale units in an articulated structure. Schumacher described Buddhist economics as the systematic study of how to obtain given ends with minimum means. His own metaphysics exhibited the direction that civilization should take for a dignified survival. For example, he considers the proper use of land. Our present attitude toward land—as something to be used—is symptomatic of the need for metaeconomic values and must be transformed by an enlightened and responsible stewardship. religions as well as Christianity, couched in religious terms, but acceptable and serviceable to modern people. Indeed the most relevant of this tradition is embodied in the four cardinal virtues: prudence, justice, fortitude, and temperance. Schumacher pointed out that the four Cardinal Virtues of Christianity—prudentia, justitia, fortitudo and temperantia—could serve as readily as Buddhist principles as the foundation for social and economic sanity.

Through prudence we must transform knowledge of the truth into decisions corresponding to reality; we must put our house (*ecos*) in order (*logos*). Our guidance must come from traditional wisdom primarily, though not without science. Schumacher can only be criticized for small misunderstandings. Although modern society rests on a bastard blend of utilitarianism and logical positivism, it has little to do with a true philosophy of materialism. Alan Watts demonstrated the consequences of a true materialism. Synchronously, prudence is a recommendation that Paul Goodman makes in several of his writings concerning the perils of industrialism.

The application of these virtues would emphasize service over products in a new economy. Extravagance and fashion would be discouraged; manufactured goods would be more durable. The measure of social progress would be Net Social Benefit (NSB); Gross National Product (GNP) would be even more meaningless.

The model of monastic economics is presented by W. I. Thompson. Thompson describes the necessity of conserving civilization by intensifying it through miniaturization. In a labor-intensive community of contemplatives, more is done with less capital, so money is miniaturized.

The economics of monasteries sprang from a work ethic that regarded manual labor as a spiritual discipline. Economics was never an end. Mumford credits monks with having found the secret of true leisure—not as freedom from work, but freedom within work, having time to converse and to contemplate. The style of monastic communities reaches a balance with the land at what Schumacher calls an economics of permanence. The community is centered amid a grand series of natural rhythms and cosmic reciprocities that support it indefinitely at a moderate and secure standard of life. Monastic tradition contributes a disciplined conviviality and work ethic to the possibility of a eutopia. It offers a philosophy of work that dignifies and democratizes labor, instead of a machine-oriented dehumanized tyranny. And it adds the inner spiritual dimension of a dream. Life can never be evaluated by purely external criteria, or achieved by purely economic means. Through spontaneous experience or wise doctrine, such communities achieve a harmonious rapport with the earth(Taoist, Tantric, Zen, Trappist, Benedictine). St Benedict organized the order named after him and gave its motto: "Pray and Work." When he established the Monte Cassino in 529, he established it on a pagan site, after cutting down a sacred grove.

Economics and ecology are distorted when reduced to quantity and technique; there is always a psychological and ethical dimension to be accounted for—motives, values, needs, aspirations. There can be no economics of permanence until we regain a love of silence or quiet attention. The art of contemplation has much to offer the individual, but requires self-discipline and cultivation. Our most powerful technology for dealing with human trouble may be simply a powerful sort of understanding, without mechanical meddling by doctors, psychologists and psychiatrists.

But the history and culture of monasticism are not well enough known for the tradition to be of influence. Examples of groups working toward voluntary simplicity and for social change may be no more than straws in the wind. T.W. Foster explains that the Amish life and culture generally corresponds to set of ethical ideas that Schumacher and Skolimowski have identified as hallmarks of an ecologically balanced, humane social order. The rejection of specialized roles reduces emergence of economic and education status differences between persons. As generalists the Amish are less dependent on highly specialized social and economic institutions. But the Amish tend to reject all innovation in an unreasoned and unexamined manner. Social scientists need to restudy folk societies from new perspectives; new knowledge can emerge.

Roszak wondered what the healthiest, best-conceived innovations would look like if they coalesced into a new cultural synthesis. Think of robes and bells surrounded by solar collectors, pocket calculators, and methane digesters. in a postindustrial monasticism. His effort is to make reunion of scientific and mystical thinking in small scale institutions that embody visions of the future. But more direct participation is needed than residence in a monastic sanctuary.

Economics needs to be restructured to take this consciousness into account: by restricting production to needs; by finding new methods that use nonrenewable resources wisely and not It is difficult enough to forecast what will happen with the current economic system, let alone one with the rules changed. Nobody understands these systems yet. They are very elementary examples of the way much of nature works.

When technology is surrounded by mystical consciousness, it becomes fully humanized; it fits. The person who excels at using a machine does not become a machine, but surrounds instrument with field of consciousness. Ancient mystical techniques—Yoga, Zen, Gnosticism—are popular because they are what we need to surround our technology with human consciousness.

The choice is not necessarily between communism or capitalism, too old flawed systems. There are other models for economies, such as the Scandinavian model, or the Eco-models (described by Daly and Cobb), or community anarchism (described by Paul Goodman). These models argue for strong communities first.

We do not need to surrender to fast, giant, national economies. We need to shift power to local communities, through self-reliance and participation. Community economics has many advantages over communism and capitalism. For instance, community economics allows barter and other informal exchanges. It also offers more services far cheaper than a formal modern corporate system. A community protects individual freedoms, guards regional culture (values and identity), and holds groups accountable for their use of power. In communities, people can decide to be conservatively sustainable or to grow and gamble on innovation. Communities can have different economic attitudes, paces, and goals. A community that is balanced and flexible, in tune with natural cycles, based on traditional values—in which industrial production is limited to appropriate goods—can position the community for a long, sustainable future in a healthy environment.

Most nations have numerous small businesses, e.g., clothing, food, and hardware stores, automotive repair, lumber yards, and construction companies. But, they also have large businesses that allow money to flow to foreign countries. Certainly, any nation could implement ways to keep money circulating in local communities; sometimes this can be done by simply buying locally, but more ambitious solutions, such as local barter "bucks," are also possible. A nation could still differentiate its unique products, within a regional partnership, such as the European Union. This would benefit the local, as well as the larger regional community.

4.4.5.2. *Wealth*

Wealth can mean a number of different things. Having rare things. Having more than the minimal requirements to live. Having great quantities of things that are valuable as measured by price. Having extra needs met, from physical to self-actual. It can be measured in materials, luxuries, prestige, symbols or ideas.

Wealth is an abstraction that can be derived from particulars, starting with "Curvehorn," the cow and going through cows, livestock, herding assets, and assets to the concept of wealth.

For various cultures, wealth may be slightly different: For the Palouse people, wealth was decoration of needs, room to collect, and prestige; for the Masai wealth is cattle and sons; for the Kwakiutl wealth is goods and honor; for the citizens of Susa, wealth was temples. Market towns regulated the trading in goods. Also, beer, food, house. For the people of Luts'un, wealth is land, food, and art. For the people of Rapa Nui, wealth was ancestor power and statues; and for U.S. citizens wealth is monetary symbols, rare things, useless things, and dangerous things.

In China, Elites gained control of the religious system and used it to increase political and material power. Thus, less emphasis on agriculture infrastructure. The cycle of power and wealth became self-reinforcing. Even Confucius and others had moral positions supporting the status quo. English encouraged trade to increase wealth.

Wealth comes first now, before survival. Wealth and power tend to make political decisions. There is also a new kind of wealth: Useless wealth and harmful wealth that subtracts from the enjoyment of life. A gigatrend can be identified that leads from natural wealth to skewed distribution to accumulation to invented wealth to useless wealth to dangerous wealth, and finally to anti-wealth.

Increased needs and materials through development and time. From satisfactions to commodities. Needs are almost completely defined in terms of commodities. Even relation needs a commodity, travel, to be fulfilled. Once the need is defined and partly satisfied, it becomes a right. The right for passenger miles at high speed took a mere fifty years.

4.4.5.2.1. Wealth of the Earth

Natural economies are based on solar energy and plant productivity, as well as geological heat and energy. Ambihuman systems operate under natural limitations; they are empirical assemblages whose characteristics have been shaped by chance and selection over millions of years. They stress adaptation to the environment and the survival of a breeding society. Human economies, although basically dependent on the same energy and production, reverse the priorities. Human concern is for individual survival by modifying the environment.

J. S. Mill and Jeremy Bentham used the Associationism of D. Hartley to develop a Felicific calculus in which individuals were social atoms seeking pleasure, avoiding pain and benevolently controllable through arithmetic. Economic man is still considered a kind of cash-register. There is a strong relationship between Protestant religion and economic development, just the opposite of Marx, who sometimes regarded religion as a superstitious atavism.

Pitirim Sorokin indicated that the wealth of an area was a function of its physical attributes and its culture. In fact, the attributes are only possibilities until appropriate perceptions and technologies exist. Economics has always been concerned with measuring wealth. The basis of wealth has been variously described as labor, resources, production, net plant production, and information. Yet, no single basis is adequate. Perhaps wealth is just as simple as what we value.

4.4.5.2.1.1. Resources

Keynesian economic theory holds that the full employment of resources is necessary to ensure full employment and the maximum social good. This economics depends on economic growth to avoid crisis. The major premises assume that population will grow, that social good is related to equitable distribution of material products, and that if resources are limited technology will erase the limits. Boulding referred to this as a cowboy economy. An economy whose wealth is limited cannot continue growing. The Masai count their wealth in cattle. The cattle limit the wealth and size of a community.

But there are problems with resources as wealth. Resources are scattered unevenly across the globe. Many recently externalized resources, such as clean air, water, and wildlife, ignore human boundaries and circulate in their own patterns. Any country with a resource vital to any other country will be able to dictate any exchange for it. Developing nations asking realistic prices for raw materials will result in a fairer distribution of financial wealth.

The production of ecosystems is counted as a resource, but if all of it were used, then the system would collapse and could not provide further wealth. People need to understand that the productivity of wealth means limiting the take to that part of the productivity that the system can spare. Although the net productivity, that which can be used from the system, is fairly high, at twenty or more percent from young grasslands, it is near zero in mature rainforests.

4.4.5.2.1.2. Net Primary Production of Ecosystems

The development of the earth is limited by environmental potentials and economic opportunity. The range of subsistence activities can probably not be expanded by technology. The environmental limits for crops are given by nature, although not clearly. Many environments are complex patterns in which agriculture and pastoral activities are symbiotic, competitive, excluded, or integrated.

Samuel Eyre calculated the net primary production (NPP) for wild areas of the earth and for wild and domesticated areas of individual nations. National boundaries have different potentials. Australia, Zaire, Canada, and Brazil have the highest potentials. Among the most poorly endowed are the heavily populated European, Eastern Asian and West Indian countries and the dry Middle East and Southwest Asia.

He then calculated the terrestrial production per capita: At the world population of 4 billion, and an above ground NPP of 50 billion tons, we get 13 tons per head. Currently the average person in the United States eats equivalent of one dry-weight ton per year, including waste and meat production. In addition, over half a ton of timber is consumed in paper and packages. He considered that an average of 10 metric tons per capita potential NPP was necessary for self-sufficiency.

Eyre also devised a common denominator to consider organic and inorganic assets together. Population carrying capacity can be formulated only using both. He assigned a nutrition equivalent unit weights of metal, but this calculation depended on a dollar value for food and minerals.

The advantages of primary production as wealth are that the wealth is sustainable; plants are renewable and minerals can be recycled. The disadvantages are that the net community production (NCP) is not considered, which takes all of the food chain into account; dollars are used instead of human work units; technological production of food is not considered. Many countries, such as Surinam, with very high NPP figures, have low NCP figures, and therefore low potential for development. Humanity is still only one of millions of species, and depends on their lives.

4.4.5.2.2. Wealth of Humanity Contributes to Wealth of Place

The Chipko women say that "A tree is like ten sons because it yields ten things of value: oxygen, water, energy, food, clothes, timber, medicine, fodder, flowers, and shade." An economics based on ecological understanding would have many different assumptions from our current economics. For instance, the capital of an ecosystem would be its physical environment and its gross primary productivity; interest would be the net ecosystem productivity. The production percentage would be the amount necessary to keep the ecosystem healthy. Cultural capital would be the wealth of human knowledge about environments, and cultural interest would be experimentation.

4.4.5.2.2.1. Labor as Wealth

John Locke wrote that land left wholly to nature was waste, that nature was only of value when turned into production by human labor. In An Inquiry into the Nature and Causes of the Wealth of Nations, Adam Smith held that the true base of wealth was production resulting from human labor on natural resources; this countered the Mercantilist view that wealth was increased by trading for gold and silver. Adam Smith, in his book, *The Wealth of Nations*, developed the argument that the only fundamental way of assessing the wealth of a nation is by examining the manner in which it uses its labor. This is true in a very basic since food, clothing and housing are product of labor. And those countries with the highest standards have the largest output per man-hour. He also showed that man serves his own material needs and those of society by behaving in most egoistic way possible; this rationalized egoism in industrial society.

However, his thesis is not complete. It attributes no inherent value to the raw materials upon which agriculture and industry are based; these are treated as free gifts of nature, available in infinite amounts. Labor is an incomplete means by which the total economic well-being of nations could be assessed.

Since Smith wrote, the earth's population has doubled twice. He was also able to accept suffering and deprivation as a normal and expected part of the order of things. He assumed a natural balance where the existence of too many laborers lowers wages, which increases child mortality, which lowers numbers, which increases wages again. This thesis is now incompatible with that of social welfare. One cannot be laissez-faire on exploiting resources and having children and build a welfare system to make sure no one goes hungry.

Smith, and his heirs, argue that cultural and technical advances can increase the carrying capacity of the earth indefinitely. Although he had no illusions regarding controls under which the human species existed. No species can ever multiply beyond its means of subsistence, natural or artificial. A contemporary, Thomas Malthus, devoted his attention to the nature of those controls. Whereas population tends to increase in geometrical progression, the increase in food production is arithmetical, similar increments being added over equal periods of time.

Labor alone cannot create a nation's wealth. The right minerals and resources, in accessible places, are also necessary, as are technology and knowledge. In fact, labor is embarrassingly abundant in most countries, now.

4.4.5.2.2.2. Production & Imagination as Wealth

J. S. Mill narrowed the scope of economics to production and the scarcity of means; he considered distribution a political process, since it depended on laws and customs that varied widely in different cultures and ages.

Wealth is based on production in capitalist countries, and on useful production in Marxist countries. The current difference is mostly rhetoric. Production is measured as Gross National Product (GNP), which measures flow, not stock (which would be more meaningful). The GNP adds the dollar value of all goods and services, including cigarettes and lung cancer, oil wells and spills. There is no deduction for environmental damage or deaths. GNP also measures services, many of which do not raise industrial output.

The need to grow is intrinsic in this kind of economic system. A large literature has treated perpetual growth as the only conceivable state of affairs. Capitalism depends on growth for stability. There is some analogy with plants. Some stability can be gotten from growth in early stages; later stability must be from limits and metabolism. Growth in plants can delay the onset of senility by ridding the plant of waste products in more diluted form. However, too much growth produces a strain on tissues and early decay. In fact, one herbicide promotes excess growth as a means to kill weeds.

The production of wealth from growth depends on technology. The technological perspective is oriented toward materials and not humans or processes. Nature is a resource to be exploited. The immediate objective of technology is to create wealth through knowledge. Technological activities are justified on humanitarian grounds, scientific discovery increases the well-being of human society, yet the social consequences of scientific activity are ignored; short-term suffering will be offset by long-term benefits, it is claimed. But because the long-term view is not taken, long-term benefits will be worse.

Economic growth can produce great wealth for some. Fortunes await those who can increase the demand for unnecessary items. Free enterprise provides initiative to any willing to show a profit without regard for the immediate consequences. Paul Samuelson illustrated the skewed distribution of wealth with an income pyramid made out of children's blocks: if each layer represented $1000.00, the peak would be as high as the Eiffel tower, and most of us would be within a meter of the ground.

Inequality is more the result of differential development than of exploitation. According to Boulding the greatest source of the differential is different rates of accumulation of knowledge, capital and organization; the rates are essentially internal properties of cultures. Although a minor element in terms of transfers, it is a large psychological perception, which may need to be compensated for in a global community.

However, gross production may not be as desirable as thought. The world prices of food and industrial raw materials have increased far more rapidly than those of manufactured goods. It can be seen that economic growth is not equal to progress.

4.4.5.2.2.3. Holeconomic Wealth

Economists try to bronze the economy in its current structure, but it is a changing system. Since it is changing, strategies that are appropriate at one stage are totally inappropriate at another. This is the remorseless working of tragedy. Any lasting economy must have a dynamic approach. Wealth must be renewable.

Buckminster Fuller's definition of wealth is that it consists of physical energy combined with metaphysical know-how or know-what. In terms of solar energy, real wealth, all humans are potentially rich. This type of wealth should be based on what is valued most: Happiness, clean air and water, good food, technological devices, art, or goods like automobiles, apartments, and clothes.

Rich sensory experiences can be derived from direct contact with nature. But economists and planners rarely mention these values. Light, wind, dirt, plants, birds, all act during a walk, but not with the meaning of crops or dogs, which is for their utility—they just are. People do not live without these things. All values are based on a healthy ecology. It must be kept healthy; arable land must be limited, and mineral exploitation must be limited.

There are two roads to wealth: producing a bigger pie (supply) or reducing each portion (demand). This assumes that wealth is defined as supply divided as demand. If supply is limited then wealth can still be increased two ways: reduce expectations of individuals or reduce number of individuals. Supply may be mostly material things, but not status, for instance; demand has the more psychological dimension. Therefore, wealth will always have a psychological dimension. Bateson thought that economics may be founded on a fallacy. Economists cannot account for intransitive preference: where 'a' is preferable to 'b,' 'b' to 'c,' but 'c' is preferable to 'a'; as in Money being preferable to resources and resources being preferable to wilderness but wilderness is preferable to money. Preference curves in economics should not intersect.

An individual's perception of environment is a semantic map that forms the context for decision making by delimiting possible acts. People respond to felt needs; economic growth occurs when these psychological needs are stimulated and satisfied. As soon as people think that they need clean air, like snowmobiles, these demands can be met somehow.

Values are based on knowledge, which is measured partly in terms of information. And information can be considered a source of wealth. Some business economies are based entirely on providing information. Information is apparently boundless. Yet, it can be manipulated. It is information that defines the use of resources by people. For example, hydrogen is worthless unless technologies exist to transmute it to helium and manage the released energy. What is not limited is our use of information. A sophisticated technology needs fewer resources. The natural productivity of ecosystems is less important if food can be grown intensively in tanks using solar energy.

4.4.5.2.2.3.1. Ecological Integration

Human beings, and especially economists, focus their consciousness on the visible parts of the world, and forget the invisible that makes everything possible. Theodore Roszak suggests that the necessary invisible background must be described. He characterized economists as urban intellectuals automatically endorsing growth with ecological stupidity. He called for a nobler economics, one that is not afraid to discuss spirit, conscience, moral purpose, and the meaning of life. Schumacher and Roszak have grasped that economics is a subdiscipline of ecology. Other economists are Angus Black, K. Boulding, N. Georgescu-Rogen, and M. Friedman.

Ecology and economics must be integrated. The goal of economics cannot be growth. It must be tailored to the ecosystem; its central tenets must be consistent with ecological principles. An ecological economy is survival-oriented, not profit-oriented.

Economic and ecologic systems interact. Economics must recognize that ecological health is vital to its own continuation; it cannot make large mistakes. The concepts of economic value must be redefined. Resource use is an ethical question. It is not known how species and communities are regulated. or if by environment or internal interaction. A free market has to be limited by conservative calculations of ecological balance.

Because human and natural systems are interlocked, there must be a common framework for ecology and economics. Economic decisions are based on human reference and not nature. Human reference is not large enough. Economics is the study of budgets, material and energetic. Modern economy is a rheology, a study of things. Human economics grows from positive feedback due to feeding on the wealth of the past; it will have to reverse its charge sometime. Most local human development was achieved at expense of other people, descendants, or from other species. Ecology is the study of natural budgets, material and energetic. Ecological and economic processes and values are often the same. Benefits can be assessed in a common metric.

Environmental quality cannot be treated as a scarce economic good. Daly maintained that the final benefit of economic activity is service or psychic income; it is yielded by stocks of capital maintained by material flow. The capacity of nature is limited, that is, scarce, but it is also irreplaceable and essential to life. The price mechanism ignores the possibility that something may disappear; it assumes substitutability.

Both ecology and economics attempt to understand and predict the behavior of complex, interconnected systems where individual behavior and flows of energy and material are important. There are many other common or similar processes: resource allocation, optimal behavior, adaptation, energy flow. How do species and associations persist in evolutionary time under environmental stress and change? Perhaps some economic systems are long-term resilient and stable like ecological systems. No one has complete information about current environment or results of actions.

4.4.5.2.2.3.2. Limits of Wealth

Complete knowledge is not necessary, however; only the knowledge of a minimum. J. von Liebig (1863) postulated the law of the minimum from his work in agricultural chemistry. For humans, trade can ease the law but not repeal it. What R. Carson showed was that we lived in a world of limits.

Resources may not be limited, in the sense that history shows that advances in technology expand the availability of resources; less is needed to produce more. But the same history also shows that humans reproduce up to the new limit of misery allowed by the new technology. New advances are used to increase the size of humanity, not its happiness or wealth. Society is growing at the same time as needs are growing, resulting in lessening of natural systems.

The redefinition of wealth in an ecological framework would increase human enrichment and natural preservation. Diversity is a form of wealth. Differences do not necessarily cause conflicts because each fills other's needs. Since nature is a non-zero-sum game, many groups can gain at the same time, ambihuman and human.

4.4.5.2.3. Poverty of Humanity & the Earth

In general, poverty is a lack of things that others have, especially necessary things. There are different forms of poverty. For instance, one form is the lack of needs, such as food and things or money. Another is the lack of use of natural or public services, that would provide food and shelter. On the other hand, what is the use-value of feet in Los Angeles? The lack of potential is harder to define, but it means that people who suffer that poverty cannot improve themselves, or change, or develop. Finally, the lack of luxuries has become a form of poverty. The idea of affluence in a global village creates a new modern kind of poverty, the potential poverty of not having what is possible somewhere in New York or Paris. Modernizing needs creates a new dimension of poverty. This adds a new discrimination by a violent cargo cult that seizes traditional cultures.

Industrial society is constantly mobilized for emergencies with, in the battles against, noneducation, poverty, diseases, and terrorism. Industrial development has never been nonviolent or respectful to people or nature. Some human poverties are poverties of the local ecosystems. Loss of entitlement to a natural resource base needed for foraging or agriculture is a political poverty. Interference with or destruction of the processes of ecological balance and renewal, the loss of diversity, the loss of capacity for the renewal of ecological systems, are ecological poverties.

4.4.5.3. Holeconomic Goals

What are our economic goals? Market leadership or development for all of humanity? Subsidizing corporations does not seem to be a good way to achieve the latter, especially when corporate profit grows even when standards of living drop. Regarding consumption levels, how much is enough? When does consumption stop adding to satisfaction? By matching consumption to ecological limits, the way is open for more psychological measures of wealth.

An economics based on ecological understanding would have many different assumptions. For instance, the capital of an ecosystem would be its physical environment and its gross primary productivity; interest would be the net ecosystem productivity. The production percentage would be the amount necessary to keep the ecosystem healthy. Cultural capital would be the wealth of human knowledge about environments, and cultural interest would be experimentation.

4.4.5.3.1. Expanding Capital

Traditional economics has a three-capital model of human wealth: Land, labor, and manufactures. Clearly, this is not adequate; each kind of capital should be expanded. Land is the entire ecological system, complete with other species and biogeochemical cycles, preserves, as well as agricultural areas, resources, and artificial modifications like dams. Labor depends on the traditional capital of a culture, the beliefs and myths and rules for behaving, the institutions. Manufactures depend on culture, and land (resources), and on technology as well.

Modern economies, embracing the metaphor "nature is capital," draw on the accumulated "capital" of ecosystems for production. By ignoring the real cost of the capital, as well as the costs of natural services, such as nutrient recycling, soil building, and atmospheric renewal, these economics create a temporary wealth. Decisions regarding resources are made on short-term economic grounds and lead to material shortages and environmental degradation.

Economics must internalize all costs for product cycles, from agriculture to manufacturing. This means finding the real costs first, starting with environmental degradation, lost employment, increased health care, tax credits, defense. The real costs of energy usually far exceed what users directly pay, for nuclear, coal, oil, or gas. When workman's compensation premiums are paid at a rate determined by a company's accident experience, then the cost of an unsafe operation is already a recognized direct cost of doing business. These external costs can be internalized directly with taxes, especially one on carbon-containing fuels. Another possible tax could be an anticipatory tax for degradation.

This would encourage more efficient use of sources and development of more cost-effective technologies. For agriculture, this means reducing waste and pollution, and emphasizing diversity and disease-resistance instead of gross yield.

4.4.5.3.2. Diversifying Institutions

The simplest economic transactions occurred between individuals who gathered food or made tools and then traded with other individuals. With the increase in specialization and complexity, came individual traders, then guilds, and finally corporations.

Corporations were formed with the historical promise of providing social services to the national states that chartered them. In the name of secure trade they often explored territories and guarded resources with private armies. Unfortunately, as private good became identified with public good, corporations became less concerned with social service and more concerned with greater profit through greater technology.

As legal entities, corporations have formal expectations that have become imperatives: Profit, for instance. According to Milton Friedman, the only social responsibility of a corporation is to make money, by striving after profit as an efficient agent of production—although Friedman admits that the corporation should conform to the rules and norms of society. Other imperatives, equally important if less touted, are constant growth, aggressive competition, objective (or amoral) decision-making, efficient exploitation of resources, and the quantification of values. Unfortunately, as an entity with limited responsibilities, these imperatives tend to dehumanize workers and consumers, homogenize people and cultures as markets, foster no commitment to place, and interfere in natural processes. Corporate forms, with their characteristics of simplification, naiveté, homogeneity, and incompleteness, turn wild landscapes into flatscapes, where variety disappears and significance is ignored for the comfortable standards of meaningless continuity.

Corporations, as economic institutions, do have many responsibilities: ecological, social, community, political, and individual. The first responsibility is to discharge its specific function. Then, it has a responsibility for its impacts; this is one of the oldest principles of law. The institution has a duty, and self-interest, to discharge its function with a minimum of negative impacts. Profit making is a necessary part of business, but not the sole reason for business. The best business serves public goods as well as private interests. Environmental and social problems should get as much attention as profitability, because they are as much a part of the process as sales and production.

Corporate law holds that management must act in the economic interests of shareholders. Communities could change that law to start to control corporations. The corporation is a noncorporeal entity. But because it is a fictitious person, speech in the form of advertising is protected under the First Amendment. So, corporate speech has little in the way of regulation. Therefore, it may be necessary to regulate corporate speech under a new amendment. Corporations sponsor wonderful shows on nature as a resource. So we become attuned to corporate interpretations and objectives. We need to be more attentive to their real objectives.

We need different kinds of regulations for corporations, ones that limit destructive activities to environment and workers, ones that encourage responsibility to nature and community. This may require changes in economic organization and legal institutionalization. Corporations should be anchored in the community. Although they must not act beyond their competence, a larger community responsibility is merely expanded self-interest. They need to build concern and responsibility for the common good into their vision, values, and behavior. Common good does not necessarily emerge automatically from conflicting interests.

The corporation is defined as a collective citizen; we need to enforce the duties of one. How to make the corporation accountable to the community? The problems of corporations are structural, inherent in the forms and rules by which the entities operate. Corporations increase their power because of our failure to grasp their nature. We could restructure the market to favor long-term investment over speculative profit. To give back to the community, we could require a percentage of new stock offerings go to government for the public. The state grants the charter to a corporation; the state could revoke the charter to protect the interests of the state (as citizens). Corporations have many human rights but few human responsibilities. Even when their actions cause death, the corporation cannot be jailed or executed. Perhaps we should change that law also and be willing to disband it.

4.4.5.3.3. Emphasizing Development

Powerful images can influence cultures over centuries. The principle of plenitude, restated in Christian terms, presents that an intelligible creator gave an earth of unlimited bounty to humanity for its use. This principle seemed to be confirmed in the Renaissance with the discovery of the richness of heaven, microscopic life, and unexplored continents. Many modern political ideologies and economic systems have been shaped by the principle of endless wealth. Adam Smith calculated that the real price of anything was just the toil acquiring it. This image is dangerous because it ignores evidences of the limits of the earth.

Economics has always been concerned with measuring wealth. Wealth once meant tangible things, land, ships, houses; then labor and production; it has come to mean negotiable symbols such as cash and stocks or unlimited information. Yet, no single description is adequate.

It is said we are in an information age, that information is the ultimate resource, that land and resources are less. But information without form is nothing (in-ation?). Information only lets us use resources and land more efficiently. The basis of wealth for a long-time will be land. Even more so because we do not know its complete value, as many native peoples have found out when coal or pharmaceutical plants were discovered on their lands—as a source of information.

The narrow definition of wealth means that it can be increased only by producing a bigger supply of goods or reducing the demand for goods. This assumes that wealth is defined as supply divided as demand. If supply is limited then wealth can still be increased two ways: reduce the expectations of individuals or reduce the number of individuals. Supply may be mostly material things—but not status, for instance—while demand has the more psychological dimension. Therefore, wealth will always have a psychological dimension. This dimension is not limited by strict logic. Wealth can therefore be expanded without being limited to supply or demand for materials.

Economics is distorted when reduced to quantity and technique; there is always a psychological and ethical dimension to be accounted for—motives, values, needs, aspirations. Economics needs to be restructured to take this consciousness into account. The assessment of personal or cultural wealth, for instance, is mostly psychological; wealth may be measured by how many valuables one has, which may be physical, like feathers, gold or land, or by how by much status, which may be behavioral, as enjoying deference or a good reputation.

Pitirim Sorokin indicated that the wealth of an area was a function of its physical attributes and its culture. In fact, the attributes are only possibilities until appropriate cultural perceptions and technologies exist. The redefinition of wealth in an ecological framework would increase human enrichment and natural preservation.

Economics tends to devalue many things. Our economy is crippled by the unspoken principle of immediate interest maximization. We allow economics to discount the future value of benefits. That also means that our children's and grandchildren's lives are worth less than ours; are they? A common strategy in Rome was to defer the true costs of government by debasing the currency. The cost is shifted to the indefinite future. We have been doing the same, but the future is becoming more definite.

Our recognition of risks, from smoking for example, is irrational. Environmental risk assessment does not seem to require valuing the entire ecosystem or its parts. We could charge, however crudely at first, for environmental services, such as soil building, carbon fixing, flood or erosion control, and sequestering heavy metals.

Ecological and economic processes and values are often the same. Benefits can be assessed in a common metric. The GNP is an inadequate way to measure well being. It measures only increases in spending. The International Standard of Economic Welfare (ISEW), developed by Daly and Cobb expands consideration to literacy and longevity as well. More is needed, however, to measure wealth and happiness in addition to general welfare. The measure needs to be expanded to include more of the human growth needs recognized by the psychologist Abraham Maslow. It also needs to recognize an expanded definition of the self, to include the ambihuman beings who support us.

We make trade-offs in social systems without assigning dollar values. We could do that with ecosystems. The valuation could be scientific or be based on human labor. New values of natural resources are being recognized now by economists, such as option value, that is, reserving a resource for future use or existence value or paying for something to exist that will not be used.

One thing business can do is put a price on nature. But, let us make it a real price, reflecting the real cost of replacement. Let us base the cost on human labor and technology, so that 1 gallon of oil is worth a million dollars, as Buckminster Fuller once calculated. Let us make all those prices be high, too high rather than too low. That is, if we are going to assign a price; the alternative is to not use those things that are priceless.

Nothing is value-free, not technology, not education, not economics; we just do not always see the values. Values are time dependent; ecological time is much longer than social time. Our social values depend on ecological values.

The goal of economics should be mature development, not growth. Growth has been a substitute for equality; in that sense it has been necessary to forestall revolt. There is no necessary association between development and growth. Development means the introduction of an innovation. Economic development will still require technology. Ecologically sound technologies could minimize stress to the environment.

4.4.5.3.4. Accommodating Limits

Both ecology and economics attempt to understand and predict the behavior of complex, interconnected systems where individual behavior and flows of energy and material are important. There are many other common or similar processes: resource allocation, optimal behavior, and adaptation. For each ecosystem, and at each level of technology and kind of social structure, there is an optimum size of population that offers a high quality standard of living. The optimum in this sense is a working one based on our knowledge of all of the factors—and we cannot know everything. No one has complete information about the current environment or the results of their actions. Complete knowledge is not necessary, however—only the knowledge of a law of the minimum. For humans, trade can ease the law, but not repeal it.

Rachel Carson was one of the first to show that we lived within certain limits. The limits may not be absolute, in the sense that history shows that advances in technology expand the availability of resources. But, the same history also shows that human populations expand to the new limit of misery allowed by the new technology. New advances tend to increase the size of humanity, not its happiness or wealth. Society is growing at the same time as needs are growing, resulting in greater demands on natural systems.

The transition to a sustainable state does not mean returning to an all-natural, that is nonhuman, condition for ecosystems. Human activities have always had some impact, as do the activities of all species, on ecosystems. Sustainable development requires recognition of the large number of limits.

The popular definition of sustainable development is inadequate; the Bruntland Commission defines it as "meeting the needs of the present while not compromising the ability of the future to meet its own needs." But, as Herman Daly has pointed out, this definition is contradictory in practice, where it really means "expanding the needs of a growing population without inflation." Daly offers 3 rules of sustainability to make the concept consistent and meaningful: (1) harvest renewable resources only below or at their rate of regeneration, (2) limit wastes to the assimilative capacity of the local ecosystem, and (3) require part of every profit for investment in renewable resources.

The ecological approach to development makes it irrelevant to discuss global limits to growth. Local limits are far more significant to a majority of people. Regardless of how much food exists, people will starve unless they can get it, as is happening so often, now. Every community is forced to accept some upper limit, beyond which it cannot grow any further. Further growth results in destruction or disruption of the community itself and the natural communities on which it depends.

Complex societies depend on production of resources. Increased complexity requires more information processing and more integration of disparate parts. The costs of communication increase. Complex societies need control and specialization. Yet, investment in complexity yields declining marginal returns because of the increasing size of bureaucracies, increasing taxation, costs of internal control.

At some point society is investing heavily in a course that is less and less productive; increased costs just to maintain status quo. In a mature ecosystem, a larger percentage of energy is used for maintenance of the system, until net community production approaches zero. The system also becomes more efficient, supporting a larger biomass with the same amount of energy in weblike food chains. If society were to parallel this development it would probably be very stable. Societies can fail (and disappear) when they become inefficient and spend more energy than exists in the system flow to maintain the system.

The economy is not an automatic mechanism for good. We cannot predict which transactions will have good effects. The theory of complexity shows that complex systems do not allow predictions; they are influenced by factors that are not statistically significant.

Yet, the climate is roughly predictable and stable, where the weather is unstable and unpredictable. The best economic policy is probably one that tries to balance long-term productivity and competitiveness with short-term benefits (trade-offs).

We expect faithfully that the market would promote the general welfare. But, people work to maximize their own good, as Hardin has pointed out, and self-interest makes it difficult for us to acknowledge our dependence on nature and on others. Economics started as a branch of moral philosophy. In a large sense, morality is a set of rules for living together, and economics is still a branch of moral philosophy.

4.5. ***Managing the Inhabitants in Places: Ecocybernics (or Politics)***

For Aristotle, politics was the science of the possible. The city, or *polis*, was a human artifact whose structure could be modified by reason; it was potentially a work of art in which only the capability of the artist limited the expression. The polis was made for the amateur; and it produced more complete men (women were erroneously considered lesser beings at the time). As a science, politics, or ecocybernics (a neologism meaning governing the house) was concerned with two things basically, a way of distributing power and luxuries, and a way of surviving contact with others.

Archaic nations governed their areas independently. Their political principles were similar: all land is communally owned by the tribe, although household goods may be personally held; all decisions were made by consensus in which everyone participated; chiefs were not coercive rulers, but teachers and leaders with specific duties limited to their realm—medicine, war, or ceremonies for example.

When the Europeans or Chinese settled many areas, they brought their centralized governments. The original goal of the U.S. republic, according to Jefferson, was to make each person a participator in the everyday affairs of government. But the government (state or federal) has become gigantic, managing the area from remote locations of power, and participation has dwindled. Despite an emphasis on personal responsibility and international cooperation, political institutions have not responded.

Central politics overwhelms local politics. It dominates the process of decision-making. Politics deals with words, which are arbitrary symbols for events or things. The wrong relationship of things and symbols can result in misguided politics and violence. Decisions are made on narrow political grounds. Citizenship in industrial cultures is the abandonment of responsibility on the assumption that others know how to manage things; government is the assumption of responsibility, without knowledge, that leads to immense and interrelated problems.

Government promises to judge disputes in values and protect its citizens from external attack. Modern government, however, is the abandonment of responsibility on the assumption that others know how to manage things.

Politics occurs, it is now recognized, in an ecological context. There can be no separation of politics and ecology. Every political act has ecological consequences and every ecological decision is a political demand for control over use of the environment. Hammond describes the kind of control that occurs in British Columbia.

Ecological, economic, social, and religious phenomena are part of the broad definition of politics. The basic goal of politics is the "survival of the community" as

William Ophuls identifies it. Politics is the interactive means of providing the basic food and necessities of a community. As survival is survival within nature, politics rests on an ecological foundation. The organization of a community must be in accord with natural laws. Political participation depends on information which can be provided by ecologists.

Ecological consciousness must be identified with political consciousness. Politicians need to think about the hunger and squalor of billions of human beings and the destruction of habitats with billions of ambihuman lives, before concentrating on missiles and private fortunes. As Bateson has said, 'Ecology, like God, is not to be mocked.'

For Aristotle, politics was the science of the possible. The city, or *polis*, was a human artifact whose structure could be modified by reason; it was potentially a work of art in which only the capability of the artist limited the expression. The polis was made for the amateur; and it produced more complete men (women were erroneously considered lesser beings at the time).

The purpose of politics, which is the activity of governing. Politics depends on common rights, trust, information, and consciousness.

Politics is the science of government; it has to do with the regulation and government of a nation or state, the preservation of its safety, peace, and prosperity, the defense of its existence and rights against foreign control or conquest, the augmentation of its strength and resources, and the protection of its citizens in their rights, with the preservation and improvement of their morals.

The function of politics is to ensure that decisions are taken at the right level. A state protects individual freedoms, guards national culture (values and identity), and holds groups accountable for use of power. Procrustean politics, as in China and other countries, fits the guest to the bed. It brings survivors of the revolution into line and exiles or destroys the rest; it is inefficient.

Democracy is a more flexible strategy, but it is size specific; as size increases, more hierarchical structure is needed. Majority rule transfers functions to an impersonal system. The logic of individualism creates conditions that require constraints. Politics has to make them palatable. Although we realize that nothing in nature is without some limit or cost, we may dislike giving up what we now consider (wrongly) as rights. Humans may have to expect much less than they want, even though many expectations are rising.

Ultimately, politics is about the definition of reality itself. Social reality is created. Different societies have had different realities. The Athenians despised the merely wealthy, for instance, and respected community service. North Americans, on the other hand, revere wealth; community service is romanticized on television but hideously underpaid. Politics is also the art of creating new possibilities for human progress. The current system is defective, however. Although it is admirable to work within the system to prevent further environmental degradation, it might be necessary to produce a change in consciousness that would lead to a new political paradigm.

Besides size and power, there are other things that make governing difficult: division of labor (are there professional citizens?), centralization, and technology. The interrelations of these things necessitates discussing them at the same time. Political institutions are not givens or timeless. Taking from the strengths of earlier forms, it might be possible to modify government to be more effective. The first step would be to form an independent government for each ecosystem.

An ecosystem is a good candidate to be an independent political unit. It is a governable size. It has clearly defined boundaries, as an ecosystem, that is the ecological community including humans, and not a state or county whose boundaries have been determined by rectangular grids at human whim. The kind of government, with more clearly defined, might look different.

4.5.1. *Actions of People*

There are only a limited number of behaviors that people can take to interact with others in their culture or nation: They can ignore the others as too different or due to internal conflicts of their own. They can trade with them. They can fight them, or, they can cooperate with them.

4.5.2. *Kinds of Governing Forms of Governing & Leading*

There are many ways of leading people. Some ways offer models to follow; others represent people in group decision making; and, a few make all decisions themselves. There are formal and informal ways. Almost every way, however, involves more prestige or status for the leader. Sometimes that translates into more meat, more goods or more wealth.

Leadership is the process of leading, that is, providing direction to others. Political leadership, the sense used here, is often formal and refers to a person in a position of authority, as a result of possessing skills at directing. The number and kinds of leaders are compared in Table 4-52-1.

Leaders, or headmen, help others by choice. Big men distribute the products of their own activities. Chiefs redistribute the fruits of other's labor. Kings take a percentage of resources from the peasants and give it to retainers or use it for public or religious displays. Emperors often represent on a larger scale by far.

Dictatorships need great leaders, or at least powerful ones, to be successful. A dictator may be self-appointed or backer or appointed by a military force. Idi Amin in Uganda, is an example. In some states, the dictator is described as a president; Fidel Castro in Cuba was also the head of the Communist party. Kim Jong-il in North Korea seems to have complete power and refers to himself with numerous titles.

Rome once had a ruling triumvirate. Bosnia and Herzegovina has a three-member Presidency, each of which are elected by a different constituent nation. The position of the President of the Presidency rotates between the members. Modern Switzerland invests leadership collectively in seven-member Swiss Federal Council. The President is a member of the Federal Council elected by the Swiss Federal Assembly for a year and is merely *primus inter pares* (first among equals).

Some states have a Parliamentary system of government, in which the President is the head of state, sometimes with only ceremonial duties, depending on the constitution, and the Prime Minister is the head of government. Countries with this system include Germany, India, Ireland, Italy and Singapore.

In nations with a Presidential system of government, the President is the head of government and the head of state. The United States and Venezuela have this system, where the President is more powerful. Democracies need effective and educated citizens. Democracy is the work of the people (remember the Negro spiritual: "We are the ones we've been waiting for").

Table 4-52-1. Political Systems (after Aristotle)

Number of Rulers	*Kind of Ruler*	*Ruling in Interest of*	*Chosen by*
One	Leader (headman)	Band	Consensus
	Big man	Tribe	Consensus
	Chief	Chief Nobles	Ancestry Gods
	King Emperor Dictator	Monarchy City-State Tyranny of Few Self/Power structure	Self Parent Self Military
Few Collective	Triumvirate Co-ruler	Aristocracy Oligarchy	Citizens of State Few (Rich)
Many	Prime Minister President	Polity Voters Democracy Ochlocracy	Voters People or College Poor People
All	People Anarchy	Self-All Community anarchy Direct democracy	All People

Leaders themselves may be influenced by choosers or controllers. Controlling groups such as the military, political parties, ruling elites, or religious elites sometimes place higher expectations on the leader, such as transformational change. The leaders may encourage their followers and believers to worship leadership. Followers may be uncritically obedient.

In the United States, the President is elected by the Electoral College, which is made up of electors chosen by voters in the presidential election. Each elector is committed to voting for a specified candidate determined by the popular vote in each state. However, due to skewed representation, it is possible for the candidate with fewer national votes to win the electoral votes and get the presidency, as happened in 1888 (and again in 2000).

Leaders, with choosers and controllers, are part of the governing process. The forms of government range from tribal leaders and chiefs to anarchy and direct democracy. These forms work best with small populations where everyone is known, even if they do not share the same culture. Other forms of government, such as authoritarian dictatorships, monarchies, and republics, have worked with larger populations. At the largest sizes, the bureaucracies and infrastructures have more in common, regardless if the government is socialist, parliamentary, or democratic.

Larger nations require larger governments, to make laws to accommodate differences in cultures. Specialists are required to resolve conflicts. The specialists often have a monopoly on weapons and the spectrum of information.

Table 4-52-2. Kinds of Governments_

Leadership/ *Decisions*	***Egalitarian Leader***	***"Big man"***	***Chief Hereditary***	***King Hereditary***	***Representative_***
Bureaucracy	None	None	None Crony level	Many levels	More levels
Monopoly of Force & Info	No	No	Yes	Yes	Immense
Resolution Of conflict	Informal	Informal	Central	Laws, judges	Laws, judges specialists

4.5.2.1. Traditional Small-scale (Elders, Chiefs, Kings)

The Coeur d'Alene people had strong leaders. Each village had a council with male and female members. A large village had a headman or woman who regulated economic and religious affairs. Their only real power was their persuasive abilities and the public esteem they built up. Chiefs were elected or deposed by the council. The band chief regulated basic resources. A war leader, hunting leader and shaman were chosen for their respective skills.

Marvin Harris shows how headmen are associated with hunting societies, big men with horticultural societies that are larger, and chiefs, still larger, and starting to accumulate goods. Headmen are more informal, in bands of 50 or villages of 150. Every one knows and understands others; reciprocity is the bank. Headmen are leaders without power. Big men give away extra wealth. Chiefs accumulate it.

Chiefs have proven ways to keep the underlings happy or at least resigned. They arm the elite and disarm the public, that is, monopolize force. They use the monopoly of force to maintain order and reduce violence. They redistribute some of the wealth in public displays and games, a kind of informal tax refunds. Chiefdoms were successful in Kwakiutl, Rome and Hawaii.

Kings enlisted ideology or religion to justify the transfer of wealth to the king and rich retainers. The shared religion also makes strangers act more peacefully without kinship. It gives people a reason to sacrifice their lives for an institution, nongenetic. The king was the central human representative of power. The Mayan king gave blood, with other nobles and prisoners. Blood is the home of the soul; it is valuable and the most powerful gift.

Mesopotamian cities had councils of elders to make decisions. They would appoint a leader to lead in war or trade. The word for this leader, lugal, eventually came to mean king; after 4800 YBP, also gods changed from natural forces to war gods. Agriculture gave a surplus; that and property led to excitement of war. In China, in the Warring States period, the nature of war changed: from an Aristocratic monopoly of soldiers to one with standing armies, professional leaders and peasant soldiers. China. New towns were planned and built by ranked lords at the edges of the state. The king gave land, title, ritual things, and a clan name. The King served as his own priest because he could contact his ancestors.

4.5.2.2. Modern Large-scale Government

As leaders had more followers or constituents, they needed help addressing problems of income and payout. They needed a large bureaucracy to maintain records and make local decisions. For example, large-scale irrigation appears in Egypt first in 7100 YBP. In Mesopotamia, complex systems could have been managed locally by farmers, although there may have been conflict with upstream users. But, cooperation is necessary, especially in villages and between villages later. As a result large-scale water systems were first managed by religious leaders, then later by secular leaders. As problems developed, temporary military leaders became permanent hereditary kings. Palaces were built; staffs numbered thousands.

Power became centralized in the king. As the populations became larger, the king had to have representatives. Centralization and representation became necessary properties of large-scale government.

4.5.2.2.1. Monarchies

A monarchy is the rule of one person, who inherits power from gods or from ancestors. All great monarchies had their state religion, in the case of pharaohs and some emperors this could even lead to a religion where the monarch (or his dynasty) was endowed with a god-like status. On a different scale kingdoms can be entangled in a specific flavor of religion: Catholicism in Belgium, Church of England in the United Kingdom

4.5.2.2.2. Empires

Emperors tended to lead large-scale societies (50,000-200,000), state societies that had literacy and public art. Until 1912, China was just a succession of dynasties where government was the same whether times were bad or good. There was an emperor, bureaucracy, system of laws and a political ideology. Their country was the "Middle Kingdom" superior to all other cultures.

Han emperors had created a civil service, a bureaucracy where officials were hired based on examinations rather than birth. There was no difference between social classes in education, in theory.

In Japan, warring clans, led by chiefs, fought until about 500 AD. Then, the Yamato clan was supreme. Prince Shotoku Taishi established law and order in the region and tried to establish diplomatic relations with China. Empress Suiko urged the people to accept Chinese political ideas, notably the system of Imperial rule, which would increase her own power. The Confucian values of orderly society with its emphasis on obedience to authority and value of harmony, as related to civic opposition, were adopted. The introduction of the Chinese style also led to a bureaucracy to carry out government duties, a central tax system and land distribution. The provincial government was run by officials for the emperor.

Under the domination of powerful families, the emperor and court had less political authority and devoted themselves to ceremonies and arts. The shoguns, head of clans, had the political power. It was the Shoguns who divided the people into four classes: warriors, artisans, merchants and peasants, all of which were now hereditary.

4.5.2.2.3. Republics

Traditionally, a Republic was a form of government where a sovereign leader held authority that was granted by the people. The sovereign ruled according to law. Rome was this kind of republic. Often the leader took power militarily and then accepted the authority of the people afterwards.

In the U.S. republic, the power is also derived from the people, who agree to be ruled by law, but the leader is elected indirectly. The citizens, who may be a smaller group than all the people, vote for representatives to be responsible to them.

In general a republic is a state whose head is usually elected rather than hereditary. Autonomy and the rule of law are the basic requirements for a republic. The term republic is applied to a state where political power of the government rests on the consent of the people governed. Republics had been formed specifically to be independent of a state religion. It ids an appropriate choice of government where there are many religions in the nation, such as the United States. Other monarchies, from the Soviet Republics to North Korea, are anti-religious. Other modern republics, such as Iran or Israel, offer a state religion.

The word republic comes from the Latin words, meaning "thing of the people." So, there is a verbal association with democracy, which means, in Latin different words, "rule of the people." Many historical forms of republicanism shared beliefs in the self-determination of a people and in basic human dignity, but the disagreed on the means of achieving those believes. The word republic became ambiguous regarding elected or hereditary rules and about how economic liberties would be regulated.

There are many kinds of republics—it is a popular word, like democracy. There are: Basic republics like France, federal republics such as India, which has a representative democracy of states, and confederations like Switzerland. The Islamic republic of Iran is governed in accordance with Islamic law. The People's Republic of (North) Korea is governed in the name of the people by the strongest leader.

Republics can be oligarchic, that is, ruled by the strongest leader, or dictator, who has no formal hereditary right to rule. Republics can be headed by a monarch, a constitutional monarch, where most real power resides in democratic institutions. The essential characteristic, more than title as king or president or supreme leader, is the exercise of power in a nonautocratic way.

4.5.2.2.4. Democracy

Direct democracy got a bad reputation starting with Greek city-states. In fact Athenian democracy was so limited that it more resembled an oligarchy. The fear of Greek statesmen was that stupid people might vote and make things worse. Direct democracy is a rule by the will of the majority, although there are no checks or balances under law in a direct democracy.

Representation itself developed from the medieval institution of government by monarchy, then later from aristocratic government. The combination of monarchy and aristocracy has shown itself to be very elastic and adaptive, ending up quite like democracies.

When the Europeans settled North America, they brought their centralized, representational government. Madison recognized that there were local limits to direct democracy. A republic, however, where citizens assemble and administer it by representatives, could be extended over a larger region.

John Adams and the creators of the U.S. Constitution and representative democracy, sought to avoid the rule of the mob, which they thought might be possible with Thomas Paine's idea of having sovereign power rest in a single body—even of the people, thought at the time to be ignorant (news traveled slowly), avaricious (less than now), and fickle (subject to trends as they still are). The original goal of the U.S. republic, according to Thomas Jefferson, was to make each person a participator in the everyday affairs of government.

Abraham Lincoln though that the Declaration of Independence established norms for equality that were inadequately developed by the Constitution. He thought the authors intended that it apply to all people, but did not mean that all people were equal in all respects. The Declaration provided the legal basis for the Union. Webster, echoed by Lincoln, thought that the people made the government.

4.5.2.2.4.1. Ideas Goals & Characteristics of Democracy

What democracy always requires is informed opinion, community support, imagination, and constrained conflict, which can refine ideas and programs. Ideas need to be argued through. Openness, relaxation, trust, and flexibility are needed. Good opinions result from education and thought. Better opinions are formed during the deliberative process.

One goal of democratic government is to represent all people. Another is to allow for change of government without violence. In a representational democracy the government operates on consent. The people retain the right to demand change. The general characteristics of representational democracy are: The direct rule of the many by the few; and citizens share in government, but the law is supreme over them.

4.5.2.2.4.2. Changes to Democracy

After all, representative democracy was created from above, by its rich, powerful architects. There was no promise of direct democracy. But, the weakest want to see an effect of their vote. Traditionally democracy has defines the people, as land-owning men, or old, educated white men, or educated middle-class people of any color or religion, or all people who can learn to register.

The scale has changed. The government (state or federal) has become gigantic, managing the area from remote locations of power, and participation has dwindled. Despite recent emphasis on personal responsibility and international cooperation, our political institutions have not responded. Big government is elephantine, controlling, paternalistic, offering a bureaucratic welfare state that has to know everything about its constituents; private markets are solipsistic, anarchic profit mongers that influence private decisions with false advertising and lies.

Karl Marx criticized bourgeois democracy for its false promise, hypocrisy, and inability to deliver on its promises. But, democracy fostered many civic groups that helped to fulfill functions. The civil society includes clubs and associations, charities and religious groups, environmental groups, educational groups, the neighborhood watch, volunteer fire departments, in fact any nongovernmental, nonprofitable public activity, for any purpose, except voting.

Corporations started to supplant voluntary organizations as nongovernment actors. And, many civic groups have become like corporations; many churches, schools, and foundations are also just special interest groups. Even environmental groups have been polarized, despite their public agenda of public and environmental good.

Compare democracy, which exists to serve the people, to businesses, which exist to make a profit. Which is more responsive, regardless of reasons? Which is directly dependent on its customers for funding? Who asks: What do customers want? This problem illuminates some of the structural flaws of modern democracies: The indifference to responsibility for voting, not including the withdrawal of survivalists, separatist religions and others; the disenfranchise of communities; overspending on elections by candidates; the nonefficacy of some kinds of laws, especially drug laws that fill the prison system; the role of money that eclipses the role of voters, e.g., spending is insane and special corporate interests are given first consideration, e.g. NAFTA; the role of media conglomerates shifting consciousness towards what they want to be more profitable, e.g., they buy news and politicians; the role of communications and computing technologies in shifting consciousness, or suppressing it. Television and computers may suppress social interactions, which may result in psychological changes, such as withdrawal and cynicism. Of course, computing and television could contribute to education and direct interaction and interactive communication; and finally, the encouragement of pressure groups that shift policy to some interests. These are small knots of interest representation that wield power and influence far beyond their numbers. Yet, for citizens this can be a more direct effective use of their money. Lobbying can be directed by the rich or any group that has interests.

4.5.2.2.4.3. Dissatisfactions with Democracy

As Plato recognized, democracy has a tendency to transform into its opposite—tyranny. Freedom can turn to slavery; we can even get enslaved to the pursuit of freedom in everything.

Democracy tends to resemble oligarchies of Athens and Sparta. The same that were rejected by Franklin and Washington as being based on virtue, in favor of utilitarian regime that channeled selfishness towards kinder ends. The founders also felt it necessary to filter the whims of the masses through an elected body, and disperse power by dividing the government into three branches.

People are dissatisfied with democracy. Why? Is it incomplete, biased, or big, inefficient, or indifferent? U.S. democracy is becoming oligarchic; it is allowing the domination by single persons or single elite groups. There has always been a tension between democracy and oligarchy, between freedom and security. Is the budgeting process democratic enough? Transparent enough? The reality of short-term elections tends to suppress the unpleasantries of long-term problems, such as the warming planet.

At what point does the democracy not function as a government of, by for the people? At a certain percentage of confidence? What happens when an elected official is mocked or not trusted, yet not recalled or revolted against. The whole dance becomes a nonzero sum game, where justice cannot be won without diminishing liberty, where free-market freedom cannot be supported without diminishing equality. Free markets get surrounded with a miasma of solitary greed. As rights increase, on every dimension, limits to action also increase.

Participation needs to increase. Business need to recognize their ecological and political obligations. Scientists, engineers, and teachers, also need to recognize their ecological and political dimensions that they automatically participate in. Our political problems are global problems caused by our species. Our ecological problems are caused by the scale and type of our impacts on the planetary system. We still have the traditional problems of growing food, building houses, creating jobs, and being secure, and creating art. As well as avoiding natural catastrophes from meteors, floods, fires, and insects. But the new dimensions introduce a greater risk to our survival.

4.5.2.2.4.4. Reinventing Democracy

Richard Barber listed three general challenges to reinventing democracy: The struggle of indigenous people; the problems of civil society, and global capitalism.

With the friction between the welfare state and free markets, the civic society, which spawned them both, has withered away. Like the government, civil society has a high regard for public good, but it was not a coercive monopoly. It was a public ground of thousands of religious, social, groups in a voluntary private realm. Like the private markets, it provides material needs to its members, but unlike them, no monetary exchange is necessary, no profits need to be made and taken—it was concerned with helping without remuneration, but with the expectation of complementary behavior. Civil society can mediate between the extremes, and in fact pick up needs, such as housing or the environment, that are often missed by the government or economy. A civil society could keep politicians true amateurs rather than professionals. It could teach markets how to respect character and not let it be consumed.

Civil societies are different in each nation. Democracy tends to be distinctive in each nation, perhaps as a result of the unique culture, land, and history. This political diversity is unavoidable, but it should be desirable. In Switzerland, communal rights are considered more important than individual rights. Great Britain does not have a separation of powers. Ireland does not force the state to be separate from the church. Ethiopia's Constitution tries to address the tribalism that characterizes groups there. France is a centralist nation. Germany is federalist. Russia, as does the Basque part of Spain, has the idea of cooperatives. Most democracies are centered on nations. A shared identity on a shared territory. National consciousness has contributed to democratic consciousness. Yet consciousness of nationalism seems less important than constitutional or international consciousness.

The differences in civic society are a basis for experimentation with democratic forms. Thomas Jefferson suggested little revolutions, every couple of decades to make the experiment fresh, as well as break up unproductive hierarchies of power. We could start these revolutions in numerous ways, or example, by requiring politicians to give away their wealth before entering office and by forbidding them to earn more than their government salary after holding and leaving office, or, putting real ecological limits on the activities of corporations.

Democracy itself is kind of like a thought experiment, where questioning and conversing about disagreements and disasters allows us to experiment with them, and our responses to them, before there is real conflict and real suffering. Even if the real disasters are not prevented at least they have been addressed. Dissent occurs within the context of loyalty. Freedom is expressed within a context of law that limits it and protects it. The democratic system avoids runaway feedback.

Anthony Barnett suggests a reflexive democracy. Barnett asks if there is a fourth kind of democracy now, direct and large scale, made so by rapid news and easy communication.

A modern of, by and for the people. If democracy can be made adaptive, and if it can be based on ecological perspectives, then it would be reflexive. The government *of* the people, *by* the people, *for* the people. Was good enough when most of the people were not part of the democracy. Things have changed. Not only all people depend on government, but also all of nature depends on it. People have become conscious of the ecosystem services that support them and the plants and animals who also require those services. Yet, all living beings are affected by what a rapidly changing, large-scale species does. We have come far enough to understand that these beings and systems need to be considered with any political decisions

An ecological democracy has the form that *of* the people includes the extension of self to the larger Self that extends through local animals and plants and supporting ecosystems. A government *by* the people is the goal of direct democracy, but it can be expressed in a representative democracy as government *by* fair representation of the people and all beings that have a vested interest in the territories claimed by the government, which forces representatives to have to consider the health of the ecosystem and then the people before their personal economic interests. Governing *for* the people can be expanded to include *for* all residents of the territory of supporting ecosystems, including people, domestic plants and animals, wild associations, and even plants and animals that can not live well with people, such as large carnivores. The human component means *all* of the people, not just the majority, or the friends and families of the representatives. All people have to represented according to minimum standards. People do have shared common interests, including a healthy environment, meaningful employment, education, security, and health standards, but they also have personal interests, such as roads or factories, that may not be shared by a minority or majority.

4.5.2.2.5. Communism

Communism is a theoretical system of social organization. It is also a political movement based on common ownership of the means of production. As a political movement, communism seeks establish a classless society, either through reformation or revolution, especially the overthrow capitalism through a workers' revolution. Large-scale communism is associated with *The Communist Manifesto* of Karl Marx and Friedrich Engels, which predicts that communist society will replace capitalist society. Communism has more generally come to refer to the political, economic, and social theory of Marxist thinkers, as well as life under Communist party rule. A revolutionary form of government and stage of socialism that emphasizes the requirements of the state before the individual, and which is characterized by the equal distribution of economic goods in a classless society.

Marxist theories motivated socialist parties across Europe, although the Russian Social Democratic Workers' Party, the Bolshevik branch headed by Vladimir Lenin, succeeded in taking control of the country in 1917. After the success of the Russian Revolution, many socialist parties in European countries became communist parties. After World War II, other communist regimes took power in Eastern Europe. Shortly afterwards, the Communists in China, led by Mao Zedong, came to power and established the People's Republic of China. Among the other countries that adopted a Communist form of government were Cuba, North Korea, Vietnam, Angola, and Mozambique.

The communist model provided many things for many the people, while maintaining a large war and research establishment. The philosopher Karl Marx thought that socialism was inevitable, that public ownership of the means of production would provide equality and social security for all people. But, in practice the distribution of wealth was very inequitable, as the result of historical trends, old economic rules, and cultural confusion. Some people were treated as more equal than others.

As a result of central planning, the patterns of life under the communist model were pressured to be uniform and efficient. Yet, the strengths of this kind of economy—especially the planning of production and the control of resources—were not admitted. Instead, military competition ruined these command economies and socialism is considered a failure.

All of these forms of government have flaws. Almost all are guilty of human rights violations, although some dictatorships or communist states may be guilty of more violations. All forms have problems with economic inequity. Every form has a problem with the applications of technology. Every form tends to overwhelm its environment by scale of operations.

4.5.2.3. Myths of Politics

Political myths can distort the work of politics. These myths have developed over the past hundred years, as the beneficiaries to growing inequity justify that inequity. The greater the investment of a people in a system, the greater the number of myths and the harder it becomes to argue against the myths. Political myths, like any myth, meet a human need.

Myth: Democracy is the best system for everyone, even if it has to be forced on them with violence and nondemocratic methods. *Answer*: Democracy allows people to work together. But, people can work together without being in a formal democracy. They can still make decisions about their daily lives, from food to education.

Myth: Globalism is the solution. *Answer*: Joseph Campbell noted that when the tribe was the relevant social unit, it was possible for mythology to represent all those beyond its bounds as inferior. The young were trained to respond positively to tribal members to love their home and project hatred outward. The concept of tribe and state is expanding toward an ecumene, an inhabited earth. Today, there is no outward on earth. Our mythology has to grow also, to include the whole planet. There is no practical elsewhere anymore. A global mythology cannot afford to teach of elsewheres. It must teach of a multiplicity of cosmologies.

Myth: Global warming is a myth generated by scientists in the pay of the environmental lobby; the extreme predictions obstruct development and growth, and they will not come true anyway. *Answer*: Global climate change has been established by scientists as a real threat to human health and safety. Natural factors, such as the eruption of Mt. Pinatubo, may temporarily mask the effect of global warming. If warming continues, global climate change could convert croplands to deserts; sea-level changes would flood low-lying areas; and shifting rainfall patterns would affect crops and fisheries.

Myth: Human supremacy. Nature is a hierarchy, and man is at the top of the heap. Science can achieve a balance between the needs of people and the environment, and can even improve on natural systems. Extreme environmentalists stand in the way of human progress and threaten the quality of human life. *Answer*: The fates of the natural world and survival of humans seem inextricably linked. The environmental movement has worked over the past twenty years to improve the quality of life for people—from improving air and water quality by pressing for the Clean Water and Clean Air Acts, to warning communities about the danger of toxic releases from manufacturers. The naive belief that corporate scientists can replace what nature took billions of years to create will deplete natural resources rapidly.

Myth: Regulations are strangling the business that is the business of the country, resulting in loss of vital production and thousands of jobs in forestry, mining, recreation, and other industries. *Answer*: The regulations that the few—the rich, the greedy, the destroyers, the anti-environmentalists—seek to eliminate and characterize as "extreme" are the very rules which protect the human rights considered fundamental by all Americans: The right to breathe clean-air, drink clean water, and to protect their homes and property from the greed of people who would profit at any price. Furthermore, environmental protection is a growth industry. Every year, the environmental industry grows by five to six percent; it is projected to rise to a $300 billion dollar a year industry by the end of the decade.

Myth: We do not need to save every endangered species and subspecies, particularly when people's jobs are at stake. Extinction is a natural part of evolution. Using science, we can determine a balanced approach for protecting important species and jobs. *Answer*: While extinction is a natural part of evolution, human activities have accelerated it 10,000 times. Natural selection is the process for strengthening biodiversity. Furthermore, this diversity provides many solutions to health problems. Many jobs rely on the health of species and ecosystems—from the fishing industry to the pharmaceutical industry. The annual value of drugs derived from plants alone is over $40 billion.

There are other myths, of course. There is the myth of innocence (after Patricia Limerick), or it's not our fault—it's a natural catastrophe, sudden market forces, government trends, or the stars. Give us subsidies or let someone else pay for the damage. Also, there is the law of diminishing accountability, encouraged by strategic games of avoidance, which lets us maintain our innocence.

4.5.2.4. The Importance of Leadership

Leadership can refer both to the process of leading, and to those entities that do the leading. Leadership can have a formal aspect, as in most political or business leadership, or an informal one, as in most friendships. A leader is assumed to have special skills or competencies. Leaders of bands or tribes were often specialized in specific activities such as hunting or social integration. Although leadership can be exhibited by an individual, either a group of people or a heroic character can show leadership.

Leadership is often confused with transformational change, although usually leaders are the last to change. Leadership may also be associated with respect, obedience, or worship of the leader. National leaders may be presidents or queens who make the decisions for their countries. Global leaders, however, will have to address a wider spectrum of needs, and balance them.

Even in representative systems, leaders may have ceremonial or reserved powers. Leaders need to have idea people (like F.D. Roosevelt had), but they also need to have jesters and clowns to mock their mistakes and pride. In most every nation, leaders have too much power, as well as too little information, humility or wisdom.

4.5.2.5. Forms of Governments

People have chosen, or allowed, or suffered many kinds of governments and institutions, including rule by king or co-consuls, presidents or oligarchs, communism or democracy, aristocrats or anarchy. Societies are constituted by people and their ideas about government and justice, among other things. Ideas and behavior continue to evolve in a natural and then cultural contexts. The individual human being is a result of a family and local community, but law evolves in a larger culture. Law is now an important part of social structure.

Humans consider themselves to be self-creating and self-ordering. But, that is because they are part of a context of self-ordering processes. Government is part of that self-ordering process, wherein specialists make decisions based on their expertise. Specialization is built on trust. We trust that others will produce enough grain for them if we make a plow for them. Government has to balance giving and receiving of things and services. It preserves the balance of society.

But, governments do not always balance giving and receiving very well. Since they are run by people, and people tend to put their own interests first, governments can be tyrannical about refusing to distribute power and goods. Which tyrannies are acceptable? The abuse of power by kings, or abuse of power by corporations? Unjust executions, or cancer from toxic wastes? Government mismanagement, or the abandonment of communities by capital?

Governments are accountable to the laws that they make. Enforcing the law may keep governments more balanced.

4.5.2.5.1. Evolution & Law

What is government? Tribal government was based on customs and who was best at certain things, from hunting to resolving conflict. In tribes and empires laws were formed to overcome differences in customs. Government now is based on laws. But, it also needs to be based on information as well as on general common human values (and shared human rights).

4.5.2.5.2. Separation of Powers & Functions

Aristotle, the Greek philosopher of 2350 years ago, recognized the importance of a "rule of law" for states, as well as a central government with a separation of powers.

The framers of the U.S. Constitution wanted to form a government that did not allow one person to have too much authority or control. While under the rule of the British king they learned that this could be a bad system. Yet government under the Articles of Confederation taught there was a need for a strong centralized government.

With this in mind the framers wrote the Constitution to provide for a separation of powers, or three separate branches of government, a legislature, executive, and judicial. Each has its own responsibilities and at the same time they work together to make the country run smoothly and to assure that the rights of citizens are not ignored or disallowed. This is done through checks and balances. A branch may use its powers to check the powers of the other two in order to maintain a balance of power among the three branches of government. The checks and balances, combined with a separation of powers, gives responsibility to each branch to oversee the others, but the branches must be equal. Partly this is done by having overlapping responsibilities.

4.5.2.5.2.1. Legislative Branch

The legislative branch of government is made up of the Congress and government agencies, such as the Government Printing Office and Library of Congress, that provide assistance to and support services for the Congress. Article I of the Constitution established this branch and gave Congress the power to make laws. Congress has two parts, the House of Representatives and the Senate.

The functions of the legislature are: To make laws, and to check the executive branch by confirming appointments, and investigating executive branch activities.

4.5.2.5.2.2. Executive Branch

The executive branch of Government makes sure that the laws of the United States are obeyed. The President of the United States is the head of the executive branch of government. This branch is very large so the President gets help from the Vice President, department heads (Cabinet members), and heads of independent agencies. Typically, the President is the Leader of the country and commands the military. The Vice President is President of the Senate and becomes President if the President can no longer do the job. The Department heads advise the President on issues and help carry out policies. Independent agencies help carry out policy or provide special services.

The executives could include monarch, prime minister or a cabinet. The functions are: To execute laws; To execute policies; To Control policies; To appoint officials; To Command the military; and, to Veto legislation.

The executive branch could be headed by an Executive Council of seven, with these specialties: Internal Coordination, External coordination, Ecological affairs, Cultural Affairs, Religious affairs, Economic Affairs, and Communication. This Council would elect a President for internal affairs and a Prime Minister for external affairs.

4.5.2.5.2.3. Judicial Branch

The judicial branch of government is made up of the court system. The Supreme Court is the highest court in the land. Article III of the Constitution established this Court and all other Federal courts were created by Congress. Courts decide arguments about the meaning of laws, how they are applied, and whether they break the rules of the Constitution.

The functions of the judicial branch are: To maintain the integrity of the Constitution; to interpret laws; and to check the executive branch by questioning enforcement of those laws?

4.5.2.5.2.4. Other Branches

Are three enough? Or too many? All three deal with the rule of law, to make, execute, and interpret laws. There is no reason there cannot be a fourth or fifth branch. People have suggested the "people" or the press as a fourth branch. The Philippines have three branches plus three regulatory commissions, on for civil service, one on elections, and one for auditing all accounts. The Netherlands have a Water Authority Board. Iran has an equal religious leader and Board of Guardians, although they may have more power than the traditional three branches. Some countries in Africa recognize the role of tribal leaders in government.

Is the rule of law enough for nations? Should there be rule of information or rule of religious beliefs? The formal power of beliefs might be separated into knowing, expressing and wise application. Are those enough to be separate? Informational power supports law, but it might also be more important than law under some circumstances, and law should be allowed to support information. An informational power might be separated as well, into research, applications for planning and wise assessments.

4.5.3. *Functions of Politics*

The functions of politics are internal and external. Internal functions are characterized by activities like the fair distribution of food and rewards. External functions are: To coordinate interaction with other nations; and to decide matters of exchange of people through emigration and immigration.

4.5.3.1. Internal Functions

Internal functions include: To promote the survival of the nation and communities; To preserve the balance of society; To maintain the affairs of the nation and communities; To guarantee fair distribution of food, goods, and necessities; To manage the distribution of power; To preserve the safety of the nation with laws and defense; To represent the citizens, to educate them, protect their rights, and remind them of their morals; To resolve conflicts between citizens; To regulate the activities of citizens regarding safety of the environment; To encourage conversation and communication; and To moderate the forces of change (internal and external).

4.5.3.1.1. To Promote Survival of Communities & Nation

Politics has to be successful enough for a new generation to take over ruling the culture and nation. If rulers promote the wrong images, that do not fit the environment, eventually a nation will collapse.

4.5.3.1.1.1. To Preserve the Operation of the Environment

Nations are embedded in places. If the place itself, the environment, fails, then no political assurances will be able to feed hungry people. So, political decisions should not interfere with the operation of global cycles or local renewal.

4.5.3.1.1.2. To Preserve the Balance of Society

Place sustains government. Land sustains government. Land produces food and resources, but the land is not equal or consistent. Poor land does not produce as much as rich land. Agriculture in general does not seem to produce as much wealth as business. Business does not seem to produce as much wealth as artists (symbols and abstractions have fewer limits on productivity). So, government needs to preserve some balance and equality by giving and taking. Individuals are produced by a community. So, the community in place seems to be the source of most of the wealth. Restoring balance in a small community can dislocate people. Leon MacLaren states that balance can be restored by depression, war, or revolution, and it can be painful. Planning and conscious adjustment might be less painful or risky.

4.5.3.1.1.2.1. To Encourage Conversation & Communication

We can best understand the social aspect of culture by realizing that the central function of human symbolization is communication and requires adherence to understood conventions. Constant communication allows a culture to be coherent. Lack of communication can lead to unhappiness or violence.

4.5.3.1.1.2.2. To Manage Individual Civic Relations & Conflicts

Many cultures have built-in limits to local kinds of conflict, but conflict between cultures requires some laws or understandings. Behavior has to be understood in a context as meaningful. Sometimes conflicts are the result of language, behavior patterns or simple geography. There has to a path to resolve conflicts, involving consensus, mediation or neutral judgment.

4.5.3.1.1.2.3. To Decide Limits to Growth Development or Movement

Keynesian economic theory, the predominant theory in industrial countries, holds that the full utilization of resources is necessary to ensure full employment and the maximum social good. This economics depends on economic growth to avoid crisis. The major premises assume that: Population will grow, that social good is related to the equitable distribution of material products, and that if resources are limited, technology can erase the limits. The economist Kenneth Boulding referred to this as a cowboy economy, an economy that has yet to bump against the limits of wealth.

In the U.S., the mentality of the frontier bloomed as the physical frontier was being closed. This mentality assumed the nature of a myth, that all people would prosper in the frontier way. Even though the 1890 census recognized that the frontier was no more, and warned people to use what they had, the people themselves have sustained the myth of limitlessness. The myth has become stronger than the logic of limits. Corporations and banks have used the myth to offset the dismal flavor of their economics. The victims have been tricked into passionately defending their own exploitation and that of the environment.

Growth is rarely an unlimited process in nature. Unlimited growth is only an economic characteristic if the economy refuses to recognize physical and biological limits. Because people seem to have unlimited wants, some political rules need to be in place to allow equal distribution before some can amass vast fortunes.

One solution is to work within limits of sustainability. Population could be stabilized. Consumption could be stabilized, especially with a shift to recycling and solar energy generation. The technology already exists, but educational and political problems are more difficult.

4.5.3.1.2. To Manage the Affairs of the Communities & Nation

Hutterite communities were able to manage their community as long as it was less than about 150 people, Garrett Hardin noted. With more than that, distribution of goods failed, due to some doing less than their share or getting more. So, the Hutterites split communities into two, when they got too large. The scale of community can make some things work better. The force that keeps individuals from laziness or greed seems to be shame, which seems to work as a force only in smaller communities. In a small community, a person can be shamed into working harder or into being less greedy, but in larger communities shame does not work as well. Maladaptive behavior may be less visible or the malcontents may form a subgroup that justifies their behavior. Hardin suggests that most utopias do not consider such a change in scale.

Many archaic cultures mismanaged their natural resources, but got away with it because their impact was relatively small. Some cultures were not as lucky, the people of Ur, for instance. Luck, as well as size, has a lot to do with the success of some human cultures. Many other cultures mismanaged the affairs of communities, but they were able to survive because of brute control.

The industrial cultures are mismanaging their resources and affairs, but may not be able to control or organize—or luck—their way out of their problems. Modern society has benefited from modern means of management. In fact, management has made industrial operations, from science to agriculture, possible. But, to continue without disaster, it needs to re-empower local voices, especially indigenous peoples, poor, displaced, and women. It needs to encourage unique local management solutions, such as the water management on Bali. It needs to adjust to optimum scales that are more efficient and just.

4.5.3.1.2.1. To Guarantee Fair Distribution of Food & Necessities

With the complexities of civilization, with various levels of responsibility and duty, politics has to make sure that necessities are distributed in a timely manner. The infrastructure, that is the bureaucracy, the coordination, storage areas, trucks, and other things, have to be in place and functioning.

4.5.3.1.2.1.1. Resources (Managing the Commons & Private Properties)

Traditionally, a commons was a territory managed by a community so that individuals could share the land for grazing or farming. This form of use characterized many tribal groups for 40,000 years. The commons were only open to the members of the community, and perhaps to a few who asked permission. Community beliefs or a kind of ecoethics discouraged overuse, in that overusers were shamed with informal sanctions or forbidden to continue using the land. In a sense, any territory not explicitly claimed by one community is a commons. That would include the atmosphere and oceans especially, or rather air and water in any form, since these are global things and can move between systems, and their character is determined by the globe itself. Some global economic or ethical method of management, not just for human use but for nonhuman use, must be devised.

Hardin extends the idea of the commons to rest on the idea of a carrying capacity of the territory, which is a very dynamic concept, susceptible to change by the weather, season, food preference, or other things. When the commons is not tempered by rules, then an individual user gains all the advantage and the disadvantage is shared by all users. Ecological degradation is assured and the system collapses, a tragedy due to the "remorseless working of things" in Hardin's words. In a limited system, the self-interest of one person cannot decide the public good, or the ecological good. So, the "invisible hand" does not provide a long-term solution. The hand needs to be visible. Necessity needs to be recognized by freedom, as Hardin restates Hegel.

The misuses of commons, some of which are alluded to by Hardin, include: Uncontrolled human population growth, depletion biodiversity, transformation of fossil fuels, pollution of waterways and the atmosphere, logging of forests, overfishing of the oceans, private vehicles jamming public roadways, tossing of trash out of automobile windows (littering), graffiti, poaching, noise pollution, and email spamming. Each of these misuses places a burden on most of the other users.

Privatization is one possible solution. However, unless individual ownership includes some regional cycles or global resources, then there might not be incentives for sustainability. One problem is that global things cannot be enclosed like land. Regulation is possible, as is payment by the polluter, a form of taxation. Hardin considers these as forms of enclosure also. One solution would be the return of many commons to the cultures that adapted to them and have had cultural rules governing their use. Hardin's suggestion was "mutual coercion, mutually agreed upon."

Many of our less wholesome behaviors, such as rapid technological change or wars, can disrupt tradition and lead to misuse of the commons. If the commons were presented as a prisoner's dilemma for a community, then individuals would have two options, cooperation or defection. Defection, or private mismanagement of common areas, can lead to economic madness, as Paul Shepard indicated. If there are unavoidable costs to defection, then it would be the less likely choice.

The tragedy of the commons is not that the traditional commons system did not work—it did, because it was controlled ethically in a community—it is that other kinds of commons have no such rules governing their use. The tragedy occurs especially when there is a transition from an ethical community use to a free market system that encourages overuse.

In a larger sense the whole planet is the common, and its limits are obvious, between the sun and the greater vacuum of interstellar space. The photograph of the earth made a logical connection between the common earth and the necessity of conservation. So, if the earth is a commons and humans are threatening the health or existence of the commons, then humans need a common ecological ethic, ort some set of rules that they can respect.

In another sense, most problems are local not global. So, at the same time, we have to take responsibility for our local environments, and respect the cultural carrying capacity, as a term for the whole integrated concept of capacity that includes human luxury as well as minimal eating and nutrition.

4.5.3.1.2.1.2. Goods

People made their own goods in foraging societies. For the Desana, goods from the gardens support people, although small numbers of extra goods are traded for clothing, machetes, soap, salt, aluminum pots, fish hooks, and rarely a gun. With farming came specialization, and more goods that could be made and accumulated. Goods became a form of wealth, especially in groups like the Kwakiutl. In Crete, written records show how goods were directed to the palaces and redistributed from there. The demand for goods between Europe and China resulted in the introduction of the camel in the Trans-Saharan trade and boosted the amount of goods that could be transported. Donkey caravans from the Mediterranean brought obsidian and other goods to Mesopotamia, and ivory combs from the Indus river. The value of goods was determined by supply and demand.

Industrial farming depended on a new kind of market, a large market for cheap goods. The English encouraged trade to increase the number of goods. The industrial revolution decreased contact with the natural world and objectified what was left. As a result of drastic changes in the production of economic goods, other political, social and even psychological changes occurred. Other kinds of order were de-emphasized. Human relationships became based on economic allegiances instead of kinship, and were formed in societies, not communities. Money became a symbolic representation for the value of labor and land. Land and labor became commodities.

Economics has become more and more global. Where peoples used to trade material goods, fish for roots or feathers for leather for instance, now all things have a common symbolic value, most often expressed in yen or dollars. This means that whoever works cheapest, and with the lowest cost of living, sells the most.

As the automated production of goods became more efficient, goods-producing jobs declined. The goal of production moved from the production of goods to efficiency. Needs are almost completely defined in terms of commodities. Even relation needs a commodity, travel, to be fulfilled. Once the need is defined and partly satisfied, it becomes a right. From walking to the right for passenger miles at high speed in a mere fifty years.

The material goods of human societies have been increasing. Goods are privately owned more often. Items that used to be shared in neighborhoods or communities, such as lawnmowers or radios, have many separate individual owners. While this is good for the modern economic system and for the ideal of convenience, it weakens socials bonds, dependencies, and trust. Private goods, like gardens, can be public goods, especially in a city to renew the air. Private goods, like a loaf of bread, can be owned and not shared. Public goods are those that can be enjoyed in common, since one consumption does not subtract the possibility of another consumption. Air is a good example and not yet rare.

Stung by the suffering of people with new unsatisfiable needs, modern governments have offered to produce more, safer goods, instead of ecologically analyzing the relationships of needs to satisfactions. People have to imagine and construct new frameworks so that they can develop satisfactions that are not dependent on commodities.

Surprisingly, once basic needs are met, for food shelter, respect, and confidence, according to Abraham Maslow, then happiness is not increased much by additional material goods or money. The things that make people unhappy are when their higher needs, such as self-esteem, are not met. Lack of love or security, lack of communication or appreciation, lead to unhappiness. People try to balance their lacks by acquiring more money or things. The culture has to be meaningful, as well as secure and equitable, with lower extremes between wealth or status.

4.5.3.1.2.2. To Guarantee Fair Distribution of Power & Rewards

Power in physics is the capacity to move. Socially it is the capacity to act. Power is the capability of making things move or happen. Power is not only making things happen, it is a way of controlling which things happen. Having power determines who gets to decide. Power allows dominance. If power is concentrated in one or few people, then there are fewer opportunities to challenge or limit it. Concentrated power can lock or gridlock patterns of movement; this is not always healthy, since many problems need many different solutions. Absolute power is no longer accountable to lesser power.

In archaic societies everyone has some power. Power is given to those who have better abilities, at hunting, healing or coercing. The real power there is the ability to persuade others. First persuasion yielded power, then strength, then knowledge. Power can come from strength or a connection to another form of power, ancestral or holy. Power can be derived from the permission or weakness of others.

The creation of large dense communities required new forms of power, due to size and organizational problems. No matter how big the bureaucracy, for a while it only controlled human muscle power. That limited their reach, regarding armies or builders (of pyramids). Traditional states had trouble controlling regional potentates or their armies completely. New forms of energy expanded human power and control.

Scale of governments or corporations gives them more power. Management techniques, or technology, can augment power. Power allows more waste. But, power diminishes the ability to see or feel, or suffer, the consequences. Power often reduces the perceptions of those who have it, such that they use power to simplify ecosystems rather than imagine working inside the systems. Having rewards gives people more power. Such power can be expressed in buying patterns as well as bribes.

Corporations increase their power temporarily because of our failure to grasp their nature. Misunderstanding of power or nature can allow a temporary expression of power. The competitive way of life, as it dominates, distorts the meaning of power and demand, and causes imbalances of power and demands.

Governments with concentrated power can be tyrannical about redistributing power or rewards. Inequality in rewards is maintained by the concentration of power. Rewards and power are treated as primary needs, which results in imbalances. Power can lead to detachment from primary needs and natural wealth. Rewards give more power, so that the two form a positive feedback loop that sometimes cannot be controlled, by the user or the less powerful. The cycle becomes self-reinforcing.

If power is spread more evenly, through many leaders and many kinds of leaders, then they can check each other and power is balanced. Under many constitutions power is dispersed by dividing into separate branches. Humans are momentarily powerless to replace or transcend the circuitry of natural or cultural systems. People often have the power to deny using the power, especially when it can destroy places or human values.

4.5.3.1.2.3. To Promote Safety & Security

Without safety and security of the primary needs of food, clothing, and shelter, there will not be a cultivation of secondary needs that promote higher culture. Ultimately security is the availability of food, materials and energy. There are levels of security, starting with a healthy ecosystem, and the ability to use it, protect it, and restore it. Security also requires that resources be available to be used, conserved and substituted. Socioeconomic security requires a healthy culture that can distribute basic needs equitably and efficiently. For people to be secure, all of their needs, such as self-actualization, have to be encouraged.

Safety is a basic human need, physiological and psychological. Safety is increased when people accept limits on their social behavior, when natural disasters are anticipated, and when technological extremes, such as nuclear accidents and wastes, are safeguarded. Safety does not mean wiping out large predators, such as alligators or sharks, to ensure than no one ever gets bitten or eaten; it means reducing exposure to wild animals, by reducing the overlap or reduction of territories. Animals have to balance their own safety with migration and food-getting. Safety does not mean killing every form of bacteria or virus; it means limiting exposure, being healthy, and preventing the spread of pandemics through accelerated travel and sharing. Ecologically, safety means leaving functioning ecosystems outside the circle of human domination.

Safety is increased with higher standards for sanitation as well as for technology. It means minimizing the potential for harm in a work or play situation. This occurs when governments and corporations take higher risks than the communities and cultures, accept a burden of proof for new developments, and accept higher margins of error and certainty. Safety is increased with laws, such as those concerned with cheating, thieving or killing. Laws that limit dangerous things, such as alcohol or guns, expand safety.

4.5.3.1.2.4. To Adapt to Internal Forces of Change

The principle of change indicates that nature is in flux, culture is in flux. Politics is concerned not only with how power and authority are exercised but with how these relationships get transformed. We are interested in the forces that sustain consensus as well as in the forces that bring about change. Intensive change is the development of consciousness or social sophistication; it is characterized by consciousness, connection, and communication. This should apply to cultures also. Extensive change results in the development of cities and technology; it is characterized by conquest, colonialization and consumption.

As humans stay in place, they tried to extract more resources from the same area. This requires new ideas and technology. Which resulted in denser settlements, which resulted in new technology and new social organization. This is extensification. Innovation is influenced by the growth and intensity of population, by the expanding and intensified activities of states or cities, and increasing trade and commercialization. The ease of communication also increases rates of innovation. Perhaps that is why it went east-west in Europe and Asia and north-south in the Americas. Accidental innovation became a culture of innovation, that is, a part of the culture that was encouraged and used.

Civilization is shaped by extensification and intensification, and complication and complexification, although on a local level there were booms and busts of individual civilizations. Why do countries lag behind in industrialization? Why do they need to join the race? Economists argue that their policies or attitudes are at fault, that their environment might be poor. That they are unorganized. That they are fearful of change or exploration. That does not seem likely though. Often the problem can be resources or domination, but sometimes, it is the social predisposition to change, as a result of historical or cultural factors.

4.5.3.1.3. To Represent Citizens

The early Sumerian temple tower, with a hieratically organized city surrounding it, became the model for the Hindu world mountain Sumeru. The king was the central human representative of power. First there were cities and then empires of cities, with kings to represent citizens. Leaders always try to represent members of a group.

In archaic cultures, when people could not present themselves at every meeting, it was useful for someone to represent them. Some degree of household autonomy is sacrificed to some larger order group in return for greater security against attacks by enemies or from starvation. A government promises to judge disputes in values and to protect its citizens from external attack.

The people of a nation-state were first given full sovereignty officially in 1648, with the Peace of Westphalia. Representative governments still use that sovereignty to claim responsibility for their actions, without recognition of any international body. They also represent citizens in matters of economic opportunities and trade, to attempt to guarantee access and fairness.

4.5.3.1.3.1. To Educate People Scientifically & Morally

Thomas Huxley thought that people in nature were Hobbesian, unfit for civilization unless culture educated them. The same essential belief was held by Sigmund Freud. But, the contexts were, and are different. The thought of Huxley was dominated by ideas of competition and fitness, and that of Freud by individuals in society, from hysterical women and conflicted children to selfish businessmen and power-blinded leaders. The context now is drones in an industrial flatscape.

The modern state has an educated bureaucracy to manage information. It requires compulsory education, resulting in mass education and mass literacy. Mass literacy disenchants the cosmos by undermining traditional and magical ways of thinking. Testing traditional knowledge became a habit. Difficulties of dealing with a half-hearted half-educated public, of dealing with a hard-hearted, hard-headed professional elite, only add to the problems of a culture.

Education may be a necessity, but it has to be a freely offered broad education, based on rich philosophies and sciences. It has to be adaptive. Tribes in the Brazilian Amazon, for instance, have stopped using chainsaws and tobacco, to live more traditionally. They also set up educational centers to show others how it is done.

4.5.3.1.3.2. To Protect Their Rights

In a traditional English village, inherited bundled rights provided commoners with rights of grazing and gathering fuel wood non-destructively "by hook or by crook," which indicated the way wood was gathered by shepherds. The form "commons" is plural, and refers to the whole group of commons, subject to these effects.

As rights have expanded, they have also been made more explicit in laws and codes. There is a form of an International Bill of Rights, covering human and ambihuman. This would cover workers, who expect to have equal rights and opportunities.

4.5.3.1.3.3. To Outline Their Duties

In addition to having rights, citizens also have responsibilities to participate in government and to live as wisely as possible, to make good places. Often, duties have to be made known through education and communication from the government.

4.5.3.2. *External Functions of Politics*

External functions are those outside the boundaries of the nation. But, those functions may have dramatic influences on the shape and course of a nation.

4.5.3.2.1. To Coordinate Interactions with Other Nations & Cultures

The possible kinds of interactions of groups or cultures are: To ignore each other; to exist separately with trade or contact; to compete for resources and people; to cooperate with each other; to fight for dominance or territory; or to destroy the other.

Table 4-5321-1. Kinds of Conflicts and Wars

Kind	*Reason*	*Examples*
Personal Conflicts	Insults, broken promises	Palouse, Aborigines
Leaders Conflicts Group conflicts	Social insults, food, territory	Uruk
General's wars	Food, Luxuries	Babylon
Psychological wars	Prestige Glory Idealism Patriotism	Greece, Macedonia
Professional Soldiers wars Army wars	Food, territory, unification	China, Rome
Religious wars Royal wars National wars	Territory conversion	France, Britain
Economic trade wars	Control of resources in colonies	Britain, France, Spain, Portugal, Germany
Chemists war, with poison gases and explosives	Territory expansion	First World War
Physicists war, with radar and nuclear weapons	Territory expansion	WW II
Mathematician's wars, with computer-guided missiles.	Resources Potential threats	Iraq Gulf War 1
Electronic wars, to destroy computers and databases.	Destruction of information or economic structure	WW II
Ecological wars	Destruction of land base Against Nature	Vietnam, Iraq Pesticides, Medicines

For interactions leading to violence, a number of trends that can be seen. These are feedback loops that loop around to the beginning, also. For instance, the failure of neighboring economies can lead to the failure of trade, or the failure of negotiation. The failure of defense, the failure of contextual system, and the failure of a structure for individual participation can lead to the collapse in the meaning of participating. The fragmentation of social responsibility can led to an isolated self-image (unrelated to the external), to lack of self-confidence, and to a decline in participation. Conspiracies for societal control can lead to international conflict, government intervention, guerrilla warfare, and massacres.

On the other hand, alienation can lead to selfishness, gangs, and anarchy, then neighborhood control by criminals, psychological stress of urban environment, substance abuse, family breakdown, lack of control, and violence. Prejudice can lead to segregation, discrimination against indigenous populations, destruction of cultural heritage, ethnic disintegration, an inadequate sense of identity, psychological alienation, and violence.

Conflict starts with people, but extends to animals, natural events, and nations, leading to the unfortunate metaphor of war. What is a definition of war? Does it have to do with the number of battles or dead? Or with having a professional army? Does it have to have a beginning and an end? Does it have to be a certain scale? Does it have to be agreed on by both sides or all parties? The whole process and its effects are complex. As violence and conflict occurred on larger scales, they were called wars.

4.3.5.2.1.1. Personal Conflicts

Individual conflicts often resulted from insults or broken promises. In Mesopotamia, people paid fines for hurting others. For instance, severing the bones of another man with a weapon resulted in a fine of one mina of silver.

When arguments could not be resolved, there was open conflict. Resolution of individual conflicts was usually informal within the confines of the community. For Inupiat and other egalitarian cultures, song duels are a ritual pacific form. The disputants publicly insult one another until the audience laughs down the loser. If individual disputes are not resolved, then they are settled by community consensus.

The predominant value in small cultures, and then large, was harmony. This minimized conflict that might have resulted from inequality. Confucian concepts of ritual and etiquette helped to regulate social conduct and made people feel good about their station. For example: "Inequality is the nature of things" and "seek no happiness that does not pertain to your lot in life."

4.3.5.2.1.2. Group Conflicts

In a Tiwi example, a dispute occurred because the elders of one band reneged on their promise to bestow daughters for marriage to the sons of another band—a violation of norm of reciprocal marital exchange. Two war parties of fifteen warriors each met in adjoining territory. They wore white paint symbolizing anger. Both sides exchanged insults the first day. Then agreed to meet the next day for socializing and renewing acquaintances. The third day the duel resumed; words escalated into wild spear throwing that wounded a few spectators and warriors. Then, the fight ended.

Is warfare limited in band societies? Could it be due to a lack of interpersonal competition for status? Are people taught to be restrained? Are conflicts are resolved by ridicule before getting out of hand?

Population density was controlled by the traditional approaches to resources. In archaic societies, cooperation and consensus, as opposed to competition and individual exaltation, permitted planning to remain informal. Population growth triggered competition and conflict, which led to positive feedback of the thing that caused the stress.

In Mesopotamia, complex systems can be managed locally by farmers, although there may be conflict by upstream users. But, cooperation is necessary, especially in villages and between villages later. As a result large-scale water systems were first managed by religious leaders, then later by secular leaders. Usually irrigation is necessary because the rainfall is unpredictable. Wars started from depressions in food production or storage. Mesopotamian cities had councils of elders to make decisions. They would appoint a leader to lead in war or trade. The word for this leader, lugal, eventually came to mean king; after 2800 BC, gods changed from natural forces to war gods. Agriculture gave a surplus; that and property led to excitement of war. In China, in the Warring States period, the nature of war changed: From an Aristocratic monopoly of soldiers to one with permanent standing armies, professional leaders and peasant soldiers. Conflict was resolved centrally by a chief or king, who held a monopoly on power.

Does every society have war? The anthropologist Carol Ember surveyed band societies and found that sixty-four percent waged some kind of war.

4.3.5.2.1.3. Intercultural Conflict & War

The reasons for war have gone from insults and personal conflicts having to do with bride exchange, broken rituals and personal honor to group and external reasons, such as territory, resources and patriotism.

The nation states are closely related to large-scale violence, usually having to do with trying to consolidate their power (global war). They then have a monopoly on power(monopoly rents). They specialize on political and economic issues (e.g., the Spanish ransacked the planet for gold). They enlarge and consolidate their territoriality, which gives them increased capacity for marshaling resources (maximum global functions with minimum territorial burden).

Wars, like the recent one in Iraq, are based on weak assumptions: That the war can be waged by blasting away any threats, and that it can be contained by using conventional weapons. But, like most actions, it has affects that can get magnified in the larger system. Furthermore, the war breaks out of the barriers that the participants try to create, destroying properties and civilians—there are no longer any safe buildings or noncombatants, and ruining the social and ecological fabrics.

All wars now are ecological wars that destroy the basis of civilization. There are no nonmonotonic effects with war. There are no side effects. There are only effects and they can all be measured. Perhaps the next war will be to directly destroy the ecological basis of a nation, an extreme scorched earth policy by the aggressors. These wars would be social wars that destroy entire generations and traditional social structures. Perhaps, future wars will have less to do with honor and territory and more to do with crises, such as famine, population, and environmental collapse. Perhaps, wars will have to do with symbols and religion, again, as they have in the past.

4.3.5.2.1.3.1. Advantages & Disadvantages of War

War has advantages and disadvantages. One of the advantages of war is its long tradition, being simpler to understand than rights, the attendant macroeconomics or the workings of enantiodromia. Another advantage is that war is cheaper than ever before, especially to attack. Of course, war is also stimulating and fun, at least for the victorious survivors, who are bonded by the danger.

It used to be that it cost more to attack than to defend, with preparations and supply lines. Now, It costs more to defend against a bullet than to attack with one. Cheap bullets destabilized the western U.S., but laws and enforcement, also with bullets, restabilized it. Possibly the same thing will happen with cheap cruise missiles. Maybe not, as the cost of attacking will now continue to be less than defending. This means that the world may become more violent and less stable. Can information warfare be cheaper, or less destabilizing? No, because information attacks may be even easier to perform. Unless everyone has the same weapons.

The disadvantages are greater than the sum of all advantages. Ecosystems in disputed and ravaged territories are always disrupted, damaged or destroyed. Human suffering is always made worse than it was before the war. And, war never solves the original problem.

4.3.5.2.1.3.2. Style & Scale of War

Politics is the management of people through equitable distribution of resources, and the management of relations with neighbors or trading partners, using negotiation or force, war if necessary. Wars have changed in style and scale. From disagreements, wars have become a centralized state function. From religious reasons, where Gods lived with people, it has gone to the secularization of state concerns.

The essence of war is to defeat or destroy an enemy. Governments are efficient killing machines. On the average in the nineteenth century, states killed 3.7% of their subjects. In the twentieth century they killed 7.3% of the world population.

War as related to growth. Often the same factors that allow unsustainable growth allow unsustainable war. Growth is promoted for its advantages, usually without recognition of its difference from development.

The historical rhythm between war and peace inevitably leads back to peace. When does peace occur, when it is won? Unlike many forms of war, peace is a process with a less rigid beginning. It can never be won or kept permanently. It does not have the prestige or honor of war, and perhaps this is why so much less times and effort is devoted to peace.

Is there a way to limit war, within the context of peace? Is there another way to humiliate or embarrass an enemy, and let some other kind of balance be found? Is there a way to limit violence to heroes or to leaders, as if either would agree to have their individual expertise be responsible for a whole population.

4.5.3.2.2. To Decide Matters of Emigration & Immigration

National governments have the right to determine their own immigration policies, even though the policies are influenced by many other factors, from disasters to invasions, which are outside the control of governments. Governments need to balance emigration and immigration in a context of a satisfactory population suited for its environment, based on their own carrying capacity and principles.

4.5.3.2.3. To Adapt to External Forces of Change

Extensive evolution is the horizontal spread of species. Extensive change is the spread of cultures through many ecosystems. Human migration was a form of extensification. That is, when the size of foraging communities made hunting and living problematical, some humans moved away. Only when they could not or would not move, did it become intensification, which required different strategies to live, such as intensive food-gathering strategies. Exploitation of new areas shows the ecological power of the species. Intensive development of a place shows the creativity of the species.

Politics has to adapt a culture to external forces, from climate change to invasions of exotic plants and animals. A culture cannot escape the rhythms of nature. Some events, such as earthquakes or floods can be anticipated with plans and architectural designs.

4.5.4. *Ecological Politics & New Functions*

Ecological, economic, social, and religious phenomena are part of the broad definition of politics. The basic goal of such a politics is the "survival of the community" as William Ophuls identifies it. Politics is the interactive means of providing the basic food and necessities of a community. As survival is survival in nature, politics rests on an ecological foundation. The organization of a community must be in accord with natural laws and limits. Political participation depends on information, much of which can be provided by observers and scientists.

There can be no separation of politics and ecology. Every political act has ecological consequences and every ecological decision is a political demand for control over use of the environment. Ecological consciousness can complement political consciousness.

The ecological, social, and political problems of today do not have simple disciplinary solutions. The problems are cosmological and must be solved on that level. But a single cosmology cannot solve all problems in all places. Where human understanding is still underdeveloped, humanity cannot afford to suppress the diversity of thought necessary for adaptation to the diversity of environments, or to eliminate ecosystems and the societies adapted to them, which explains why archaic cultures are valuable.

Practicing holistic, or metapolitics, is the recognition that humans are part of a larger community, a larger whole that includes all humanity and all the earth, with its species, habitats and resources.

4.5.4.1. Deciding Goals of Ecological Politics

The government of a community is a framework to maintain the lives of people. For the original archaic peoples, tribal teachers were adequate. In our representational republic, representatives are relatively uneducated, except in law, and less capable of institutional change.

The functions of government are to support the functions of politics with specific actions, such as: To make laws; to decide the meaning of the laws, how they are applied and related to a Constitution; to lead the country and to make sure the laws are obeyed. These action specifically include: To protect the nation; to command the military; to manage internal affairs; to coordinate national activities; to manage external affairs; to represent the nation to other nations; to coordinate trade; to coordinate information; and, to coordinate education.

4.5.4.1.1. Expressing the Purpose of Government

Central government has lost sight of its own purpose, which is *not* the sum of special interests or its own desire for perpetuation. Government has always had other reasons for existing. Some of these reasons are to:

1. Hold a vision of the common good, where 'common' means common to all beings in the ecological community as well. Make goals conscious, with some flexibility to enhance the vision over time. Balance public and private interests.
2. Coordinate the means to satisfy the long-term needs of the community, balancing freedom and regulation. Tie rates of consumption to the limits of the system—this means controlling resources and land use, in essence determining the physical shape of the community.
3. Regulate the community. Link it to cultural values. Determine the closure and

openness of the community; rates of increase or decrease, through births, deaths, or immigration. Encourage or discourage some forms of technology or trade. Provide work opportunities to members.

4. Protect the community from internal and external threats: natural disasters, criminal elements, and other communities. Most of these threats are unavoidable. Some are long-term and rare; others are constant and have low intensity. Some are part of the human condition; others the result of historical balances that cannot be restored easily or quickly. Be aware of them and minimize the damage.

4.5.4.1.2. Increasing Participation

John Dewey believed that personal face-to-face communities were necessary to a free and open government. The local communities need not be isolated as they have been in the past; they are more open and active, connected to other larger communities. Government requires trust and goodwill; these arise more easily in communities of acquaintances.

Citizenship is too complex for television or even electronic global villages; it must exist in the community, in person, in place, where individuals can learn about each other in context. Government by local meeting assumes the common sense and wisdom of the common person in an open exchange of belief and need.

Often this kind of involvement takes more time than just voting annually or having one person decide for many. The effect of presenting a problem before a traditional American Indian council was to slow down response by passing it to the entire constituency and getting a consensus. This ensured due consideration.

To encourage the participation necessary for effective democracy, or communism or socialism for that matter, government should solicit public opinion. Land-use agencies do so. Government should offer real power to people—power should devolve to lowest level—by changing the local political institutions to start. Montana or Vermont in the U.S. offer examples of how to change local participation.

4.5.4.1.3. Taking & Yielding Powers

Central government must shrink in order for local government can expand. Some things must be done at a national level, such as the protection of watersheds, rivers, and the atmosphere, to make sure of minimal or median protection. Some protection must be enforced at the international level, also.

Following the federal model, delegated powers go to the highest level, and reserved powers to the community. So, a new national government would coordinate internal and external defense and security; maintain law and order; and, set ground rules on economic exchange to ensure fairness. The most important responsibility of government is to set standards for itself and its institutions. The constitution would instruct the courts to interpret clauses as narrowly as possible.

The new nation would have an administrative department to handle taxation, budgeting, and purchasing. It may coin local money, perhaps on the model of the Local Employment and Trade System—LETS—on Vancouver Island in Canada, which records credits and debits on a computer, which can then only be used locally. There may be departments to protect the civil rights and liberties of the people and a department to protect the environment. Environmental disputes could be resolved by mediation, as was developed in Seattle in the U.S. in 1980s. The nation would also conduct foreign policy, provide technical services to communities, and maintain regulatory offices.

To avoid insularity, being set against the rest of the world, each nation could create an office of global communication, which could set up connections similar to sister cities program. It might be beneficial to join confederations of other similar areas, especially those that could offer complementary crops, or a larger union, such as the United States, for preferred trading.

Spending on education, roads, welfare would be done at local level. There is some risk, especially with education or wilderness restoration, but the breakdowns and errors will not be devastating as with centralized planning. Citizens will need to do some of the work of government as well as make the decisions. They should have total control over some things. The judgments of the people are more important than the efficiency of those judgments. It may not be necessary to have many separate authorities or committees; it might be better to integrate policy-making bodies so they are not too specialized.

With centralization of functions, money has become the primary source of security for most people. Welfare, as giving money to those who do not have it, may reduce homelessness or disease, but it cannot restore family. Family needs a supporting community context—institutions. Decentralization, and the power it would return to local communities, may also help the family as a source of security. Money is an enormous simplifier, but many things cannot be simplified. Decentralizing would make government and economics enough of a small scale to be understandable and manageable.

Decentralized communities fitted to their ecological location are more suitable and livable than urban spreads. Some cities may still be fairly large and dense like those envisioned by Paolo Soleri. Some may be smaller and rural like those suggested by Murray Bookchin. Their relation to support areas would be more explicit and include large amounts of natural and domestic vegetation. As much as possible, the cycles of materials would be closed.

As cities become more sustainable, their forms may change. They may become more compact, with more multiple-use streets, as a focus for human activity, and less involved with cars; buildings could use solar power, efficient heating, perhaps integrate roof-top crops; integrate older buildings into new groupings to integrate services, play, and work, with living; local public spaces and services; recycle waste into cycle; regenerating derelict land, either as green area or new construction.

Preconditions to a sustainable, steady state, economy include pollution control and the redistribution of resources more equally. The redistribution of resources and improvement of environmental quality are more important than increased production by sophisticated technology. This strategy calls for social and educational organization more than technological style. Styles of technology must be determined by culture and context. Such development requires a local authority working with suitable economic and ecological conditions. No authority can be effective without the participation of the people.

4.5.4.1.4. Prescribing an Optimum Size

To restore participation, we have to consider the limits of participation. The current large populations of many places may seem to be too large for direct democracy, as does any of the projected optimal populations.

Twenty five thousand people is large for direct democracy. Many of the cities and towns in an area are approaching this size; thousands have already exceeded it. The size has to be small enough for people to meet and "exercise government" in James Madison's words. Kirkpatrick Sale concludes that the optimum size for direct meeting is five hundred. Participation becomes more difficult as the size increases. Bryan and McClaughry suggest 2,500 as a maximum, since in larger groups people cannot all know one another and the assemblies tend to become a debating forum for a few.

Putting these figures together, we could design neighborhood sizes to be from 500 to 2,500; these neighborhoods would make up communities of 5,000-25,000; and the communities would bring a regional or even national population to 400,000-500,000. Differences in size seem to be close to powers of 10. Each Community legislature would be 40-60 people. Perhaps these sizes are close to the optimum; and, these numbers are similar to Christopher Alexander's *A Pattern Language*. About 3,000-4,000 people are needed to support an elementary school. The Swiss are a good model for government levels—with a national government equivalent to the Swiss national, communities equivalent to Swiss cantons, and neighborhoods to Swiss communes.

The scale of our understanding changes with the scale of our focus (or subject or periphery). Can we relate our population scales to powers of ten? Does it make sense. Every few powers of ten in size results in qualitative change—this may be a universal pattern, for stars or residences. Starting with 15 (and it should probably be a range, from 6-22), the patterns appears in Table 4-5414-1.

Table 4-5414-1. Scale of Human Groupings with Powers of Ten

15 Family (or team, special groups)	
150-	Friends
1500-	Community
15000-	Village or town
150,000-	City
1,500,000-	Large City
15,000,000-	Nation
150,000,000-	Large Nation?
1,500,000,000-	Planet?
15,000,000,000-	?

Groups do not behave like individuals; cities do nor behave like groups. There is a point in critical mass reaction where the mood of the mass becomes indifferent. As long as the size is small enough for recognized identity, people will behave with concern. At a larger size the ideology, which is capable of anything through indifference, can takes over, according to Leopold Kohr.

Small communities are essential to the democratic ideal and other ideals. The uniqueness of place gives belonging and identity. The whole community gives meaning and richness to life. The population density of some places may cause some difficulty, however. People will not be within walking distance, but may have to travel 20-30 miles to meetings or communicate remotely through telephone, computer, paper, or friends.

4.5.4.1.5. Protecting Ways of Life

A government could impose a limit on the intensity of development of the entire area, with transferable development rights (TDRs) assigned to each unit of area. Any project would have a TDR value—25 for an apartment building, for example, or 1 for an individual home; TDRs could be bought or traded from other landowners. All land could be held in communal ownership and leased to farms and businesses, except preservation and conservation lands.

Compartmentalization avoids the need to compromise every ecosystem for human use. Multiple use systems should only be part of the picture. First, the government could ensure a protected environment of mature ecosystems, then productive systems, and then multiple use, and urban areas. This could be done through function (not activity) zoning. The landscape needs to be zoned (compartmentalized) to provide a safe balance between protected ecosystems and used ones. Restrictions on land and water are one means of avoiding overpopulation or overexploitation.

Long before the limits of food or space are reached, or the ecological balance is lost, or a vital minimum is exhausted, phosphorus, for example, the quality of life will sink lower. Regardless of how much protein or energy can be provided to support human life, human happiness will be problematic in large, insecure populations. The question is not how many people agriculture and technology can support in one place at once, but what kind of life is possible for those who have no choice but to live in that place. The limiting factor is that condition in the environment that approaches the limits of tolerance of individuals. The population density may be the limiting factor. It may be living space. It may be wilderness. It may be beauty—aesthetic space.

At a limit, the cost of change accelerates. We seem to understand technological limits, to sailing ships and computer chips for instance, but not to individuals or groups, not environmental or ecological. Calculating these kind of limits is difficult—too much data, too much uncertainty.

Peter Drucker points out that economists from Adam Smith to the conservative F. A. Hayek argued that it is impossible for governments to control or manage the economy, especially in an information age. Recognizing, on the basis of mathematical models of complexity, that detailed management of the biosphere is beyond human capacity, a government should minimize its management, to coordination of communities or larger alliances. The biosphere is dominated by natural communities of which we are largely still ignorant. Detailed planning of complex open systems is not necessary. Planners are not in a position to attempt detailed models of future situations because many relevant parameters remain unidentified, and many of those known cannot be quantified. Plans can be made within the limits of variables, although it is not safe to be limited by lethal variables, as Gregory Bateson recognized. Closeness to limits reduces flexibility, that is, uncommitted potential for change. Vagueness and lack of detail are acceptable in planning, because people will fill in the details. Furthermore, it is almost impossible to plan every detail of a dynamic chaotic system. That does not mean stagnation, that a rice field must always remain a rice field or a town a town. What the government should preserve is the pattern, not the details, is limits, not directions. The limits are to be applied to scale not development.

Therefore, we must limit human intrusion in every system. Government should zone some segments to be free from human activity, and tailor human-made systems to approximate the form of the natural systems replaced. Interference is a broad term for the negative side of human activities. There are numerous forms of human interference: Overexploitation, introduction of exotic species, pollution of air and water, and the subsumption of habitats, in shape and size. Interference is caused by large human population growth, with its requirements of poverty, inadequate metaphors and images, which are too anthropocentric or short-term, uncontrolled change or transformations, as a result of colonialization or revolution, and political or economic failures, from wars or market internationalization.

4.5.4.1.6. Paying Costs & Leveling Extremes

Relative to European communities, many nations have less funding for public services, such as parks and public transportation. Some nations have traded public support for higher levels of private affluence, which has not made people any happier. In fact, they are more insecure; and they can become far poorer, and then second-class and neglected.

Many cultures should try, like Sweden, England and Japan have tried, to weaken the connection of material reward for achievement. Income distribution is too unequal. Full internationalization through trade would bring only greater extremes, which most populations can least afford.

The communities could levy taxes on property. But, there is a discrepancy in the wealth of communities. The nation could collect income taxes, and communities could claim a percentage of taxes collected. The community and nation could both tax the same bases: Income, sales, meals, property, and fuel, as many nations are now. The nation would set a ceiling on each tax; the community rate could be zero on some. Or the communities could do all taxes and give the national government a percentage, although differences in wealth might be maintained; then the state could return a percentage to make up equality in education and environmental protection. The important thing is that taxes are used to direct development and reflect the true costs of the society that people want.

Government could change taxation procedures to reduce growth instead of stimulating it. Talcott Parkinson suggests that taxation beyond a certain point yields declining marginal returns. Government could use a single tax rate flat at some percentage, perhaps from 10 to 25; and then pay everyone a fixed amount of income for basic needs, from 3,000-10,000 USD.

Similarly, property taxes could be appropriately scaled to use. One way to keep farms as farms is to tax land by use. The more important the use, farming for instance, the lower the taxes. Buying farmland for shopping centers would result in discouragingly high taxes.

It is difficult to persuade people to pay more in taxes, to vote to keep less of their income. But, through education or understanding, a culture could expand the understanding of the self and expand self-interest in that way. Some catastrophe might work towards equality, but that might have other high costs to society.

Of course, most of these taxes could be eliminated entirely, if only environmental uses and losses were taxed. Some taxes, such as luxuries or heroic income, might be maintained temporarily until the extremes of ownership and wealth are leveled. This form is expanded in later sections.

4.5.4.1.7. Meeting Limits of Government

When a place has a reputation for being small and livable, it attracts more people, until it is no longer small and livable. But, imposing limits and stopping growth are problematic. Government could impose limits on birth through licenses, perhaps risking rebellion, through limits on housing and public services, possibly causing shantytowns, or through peer pressure, which could contribute to social disorder. Nature is self-organizing, and, society is self-organizing, but we need to recognize some limits and define others, and take responsibility for keeping to those limits in order that the self-organizing process not break down. Limits are fundamental to understanding nature and life.

4.5.4.2. *Planning*

Our local communities are proud to attract more people and larger industries, but do so thoughtlessly, without regard for the limits of population size or the rate of energy use, without sufficient consideration of the effects on the quality of our lives or on the quality of the environment. Although we make plans for people and their activities, the plans are usually reactions to growth and change. The formal development from planning results in a complex of problems, from pollution to ugliness.

We have always tried to exceed the physical and biological limits of places rather than recognize them and be guided by them. This eutopian framework suggests an approach to comprehensive planning based on the biohistory of an ecosystem, the cultural values of the people, and knowledge of the limits for sustainable development. This approach makes the limits explicit and set sustainable goals within those limits. As a synthetic framework, this approach provides for the health of the ecological system, as well as for the health of its human inhabitants.

4.5.4.2.1. Central Planning

At first a means of controlling people through zoning, central planning has expanded its reasons to include sanitation, economics, and aesthetics as well. Although central planning grants some consideration to support areas and aesthetic factors, cultural traditions and the natural environment are not often primary concerns.

Planning in general means deciding on goals to be achieved in specific situations. For central planning, the goals are usually small and not comprehensive, such as the rate of emission of sulfur oxide, and usually end up being a compromise in cost-benefit analysis. Planning tends to neglect or dismiss the distribution of negative, uncertain, or nonmonetary effects. Furthermore, we seem to have no mechanism for developing long-range plans. Certainly, there seems to be no way to deal with long-term, slow catastrophes, from erosion to climactic change.

Most plans address *problems*, from wastewater treatment to air pollution monitoring. Everything else, from employment to pests, is also considered as a problem, and not a direct effect of the cultural implementation of some idea or technology. Most plans seem to be extremely good at compiling area data, from topographic to climactic. These plans are concerned with determining the adequacy of the infrastructure, such as utilities, streets, or sewers, to support actual and projected population growth. Development plans, for water or power for instance, are comprehensive in the sense of seeking to meet all needs of the public, agriculture, and industry. But, they fall prey to all the assumptions of the industrial culture. They tend to be multipurpose with the aim of providing maximum net benefits through management of watersheds, fish and wildlife, and flood control. Both uses of "multipurpose" and "maximum benefits" are based on misunderstandings. Multipurpose in practice means human use, and maximum benefits have proven to be dangerously unstable. Modern resource management strives for maximum sustainable yield, based on partial knowledge of population size and great ignorance of population flows.

Development plans also tend to call for the eventual development of *all* resources in an area; Brazil's Plan 2010, which would develop all of the Amazon with 136 high dams, is a good example. A one-world planned economy is an even greater threat. It is based on unlimited industrial production, unlimited commodity consumption, increased exploitation of nature, and the free flow of resources and labor across cultural borders. This kind of planning requires the abandonment of local controls on development or trade.

All countries are expected to open their markets to outside investment, eliminate tariff barriers, reduce government spending, especially to the poor, convert small-scale, self-sufficient farming to agribusiness, and open all land to resource gathering. Planning is thus characterized by a utilitarian globalism that denies value to the systems that support it.

As a result of central planning, the patterns of life have become the products of market forces and stylish transportation operating in a sterile, abstract order. In the U.S. citizens are criticized for having a "frivolous" culture based on "savage" capitalism. Capitalism increases the pressure for uniformity, a single pattern of existence. Formal development is more concerned with an assembly-line model—simple, isolated, efficient, and easy to maintain. People become remote from, and indifferent to, the system that supports them. They acquire unrealistic images of the world and harmful values and then make bad decisions based upon them. They have not developed qualitative indicators on ecological health or quantitative measures of social health, much less an ecocentric view that would value preserves of nature for themselves.

4.5.4.2.2. Ecological Planning

A number of proposed plans to heal the earth and improve human communities have been presented in popular books. Many of them are too philosophical and general, suggesting that we could change values without showing how or urging us to alleviate some of the symptoms without addressing the disease. Other plans, such as the *Limits to Growth*, are too global. And still others, such as *Design with Nature*, are less concerned with limits than with basic conservation. *Goals for Mankind* offers a similar compendium of global goals that can essentially be summarized to be health and freedom for people in a healthy environment. Many of these plans offer admirable models, but little in the way of specific goals or paths.

A plan should consider the whole system. Communities should be designed for an optimal fit within the limits of the system. Ecological planning considers an optimum or satisfactory population within one ecosystem, although it is connected to others by trade for some necessities or luxuries. This kind of planning is a conscious adaptation of the benefits of technology to the traditional idea of physical, as well as cultural, limits. Direct observation and traditional knowledge yield far more "information" about the societies of animals than autopsies and mathematical models. An outline of a comprehensive plan is presented, to deal with some of the implications, as well as question them.

1. Identify our place within its natural boundaries. Most places exist in a uniquely identifiable ecosystem, with recognizable boundaries and a unique history and character.
2. Calculate the optimum amount of wilderness to preserve the natural cycles indefinitely. If the current area is less than our calculations, restore the difference and set it aside as a reserve.
3. In the remaining area, zone areas for appropriate use, including conservation, preservation, reservation, and artificial areas (with historical, cultural, and functional importance).

4. Identify the resources needed for human use, including raw materials and the productivity of the areas. This productivity can be used to calculate a base line population.
5. Apply cultural modes—in style, values, and technology—to set limits on technology and population. Preserve the cultural values. Renewable resources will sustain a population longer than energy capital like oil or gas.

As part of the formulation of a plan, we have to examine the natural and cultural histories of a place. We need to understand interactions in the ecosystem, as it was with no humans, as it was lightly settled, and as it is now, dominated by humanity.

4.5.4.2.2.1. Limits & Planning

Our modern cosmology, with its basis on machine metaphors and the principles of plenitude gets in trouble because it does not understand how basic the concept of limits is to the physical universe, to life, to ecosystems, and to human constructs, such as cities and economics. Limits are important at all levels, starting with the physical.

The limits of the universe, like the speed of light or quantum of a field, put limits on freedom. Events are limited to localities; size limits function; history limits development. Biological order is built on physical and chemical orders. That is why life is limited to such a narrow range of conditions. And that is why the most complex orders are vulnerable to changes in their substrates; energetic radiation can alter and destroy an individual, a small change in climate can destroy crops and civilizations. Complex orders always depend on simple orders. The success of life depends on miniaturization, where a prodigious number of overlapping mechanisms are packed in a small space. These are persistent by virtue of built in regulation circuits; and open enough for novelty.

Life involves a vast number of interacting structures. Living consists of complex behaviors whose limits are defined by rules of order that can be empirically described. The earth is suitable for life because of three kinds of limits: Solar radiation has stayed within certain limits for 4 billion years; the biogeochemical cycles of oxygen, carbon, nitrogen, phosphorus, sulfur, water have stayed within certain limits; and, the environment has been constant enough for organic evolution, but variable enough for natural selection to be challenged.

The earth is suitable for humanity because of another set of limits: It is covered with a vast array of ecosystems that interact and in which plants and animals create niches; there is material cycling and energy flow within limits; the prolific nature of biological reproduction, which is limited by processes such as predation and by material and energy cycles; and the carrying capacity can sustain a given amount of life.

Cosmologies make the world manageable by limiting it. They are also tuned to the limits of the local ecology, within their knowledge of interactions; the long-range ecological consequences of drainage, irrigation or overexploitation contribute to the deaths of cultures. But many archaic cosmologies are a form of tens and limitation. Most try for adaptation before domination, according to Gerardo Reichel-Dolmatoff.

Human cosmologies themselves are limited and contradictory. All cosmologies cause destruction and waste; all produce the opposite of the good intended. Archaic and modern, occidental and oriental worldviews are complementary but not complete. The very circumstance that makes each cosmology unique—-being in a unique place—-ensures that each is limited. Often, a cosmology is accepted as unquestioningly as a language, technology or place.

Cosmologies have a sense of place, with its beings and features, which is necessary for information on how to live and to get food. The ecological benefits of rootedness are that people will take care of a place if they realize they are going to be there for a thousand years. Having a place means that the inhabitant has stock in it and participates in its unfolding, through planting and caring. Detailed understanding of plants in a locale allow gathering of food and medicine. People in place, being in place as used here means in a human scale in unique surroundings. acquire a sense of community, nonhuman and human; shared set of values and concerns; health and spiritual benefit. An ethic, ecologically, is a limitation on freedom of action in the struggle for existence, according to Aldo Leopold.

4.5.4.2.2.2. Limits: Ecosystem/Global

Ecosystems and organisms depend on a complex set of conditions, which can be limiting factors. Organisms are affected by the quantity and variability of materials, if they require a minimum of them. Organisms are also affected by their own limits of tolerance to those materials. Life is limited by elements and physical factors. Such as light, water, gas, or salt. Life is limited by too little of an element (Liebig's law), as well as by too much of an element(Shelford's law of tolerance).

Ecosystems and organisms are always interacting and changing. Conditions in an ecosystem are limiting factors that determine speciation or extinction. James Lovelock hypothesizes that every element in the system is related in a feedback network to every other element. For example, the biosphere can control the temperature of the surface and the composition of the atmosphere. On the other hand, soil types and the weather can limit vegetation; invasions of vegetation change soil types.

An ecosystem is forced to accept an upper limit, beyond which it cannot grow any further. Further growth results in destruction or disruption of itself and nature. This is an ecological law of the maximum. To survive, an ecosystem depends on the interactions and balance of many variables, most of which are not well understood. In forestry we try to maximize one of those variables. When that happens, the balance or harmony is altered, and although it may take decades or centuries for the consequences to be known, the system is affected. William Ophuls noted that nature abhors a maximum. While it is meaningful to speak of an optimum diversity, as the result of limits and the interaction of many factors, a maximum diversity may never be reached. As Paul Weiss noted, the patterns of organic nature are a combination of order and diversity; order involves constraint while diversity requires freedom for difference. Maximum order would result in a static universe, where a maximum freedom would create a nonordering chaos.

This is why is it not a tragedy of cosmic proportions when species such as sharks or sow bugs fail to respond to the lure of possibility. These species have fitted themselves to relatively unchanging environments and do not need to change for the sake of change or for the sake of diversity. We may not know what is the minimum, optimum or maximum forest cover for a particular watersheds. Science might try to identify minima or maxima but philosophy can aim at optima or satisficia. We are so ignorant of the complexities of ecosystems that it is suicidal to pretend to "maximize" their use for resources. A free market has to be limited by conservative calculations of ecological balance. It is almost impossible to estimate the economic value of natural balance.

The ecological social approach, or a redistributive environmental strategy, to development makes it irrelevant to discuss global limits to growth. Local limits are far more significant to majority of population—regardless of how much food exists, people will starve unless they can get it. Such an approach would also mean limiting humanity and its technological effects, limiting human use to local impacts, and letting other beings live without interference. It is not necessary to dominate or terraform an ecosystem completely to save it. Ecological planning has to consider the limits of ecosystems, as well as the limits variability and stability, in deciding how human activities mesh with domestic or wild ecosystems.

4.5.4.2.2.3. Limits: Human Psychological & Social

As members of cultures and physical beings, humans have very real limits. Physical limitations are well known. We cannot run or jump as fast or far as we want, for instance. We cannot lift too great a weight. We seem to be allocated only a finite number of heartbeats, according to Isaac Asimov, in a single lifetime.

Some of our problems are the outward manifestations of inner causes, as Ervin Laszlo argues. Our perception and thought structures have limits, as do tools based on them. Science has been logically limited, so far, with a predicate logic. Most theories are based on Aristotelian logic that is deductive and bivalued. Analytic science has reached its limits. Data and information developed by hard studies have undercut the paradigms that guided their investigation. The compartmentalization of scientific fields has exposed the complex connections of the subjects. Early science saw the world as mechanism; modern biology is seeing it as resembling an organism. Organismic trends can be seen in sciences, from relativity and Gestalt psychology to ecology.

There seem to be limits to our personal space and levels of tolerance to human intensification, also. Urban intensification leads to the question: Is there a limit to human numbers? Perhaps space, but is there a psychological limit? People in cities seem to do well with high-contact, high-proximity living. What happens when people are crowded or feel crowded? Physical complaints, emotional complaints, sexual dysfunction, or feelings of fear, seem to be expressed often. There may be limits of crowding. Are there social limits, in terms of the number of people one can tolerate? We may have a requirements for personal space, home space, and wild space.

Psychological limits may be the basis for some of the great failures of human life, for instance, the "failure of perception." We cannot see slow change or anticipate it. No one really sees the incredible interdependence of humanity and nature, of diversity and success. We do not seem to be able to see others as feeling human beings. Then, there is the "failure of imagination," that limits our understanding and visions of future. We can explore planets and modify genes, but cannot seem to offer functional education or meaningful jobs, dignity in retirement, or goals for living. We insist on individual rights and freedoms, but neglect the whole framework for individual success. We cannot imagine the importance of difference or challenge. Our international system is going to produce a boring uniformity and a painful collapse. Humans can even create virtual worlds by limiting what could be received; for instance, if a being could see in the x-ray part of the spectrum. Yet, human imagination is limited, as is human knowledge. Many organisms exist of which we know nothing. Their worlds have little meaning in a human world. The "failure of feeling" keeps values and morality as local effects. We do not extend respect or love to distant others. Our personal values and beliefs do not let us. Finally, the "failure of nerve" dooms us to cleave to the familiar, to ignore other alternatives, and to fear change or equality.

4.5.4.2.2.4. Limits to Human Cosmologies

The very circumstance that makes each cosmology unique —being in a unique place— ensures that they have limits. The entire cosmology of a culture is concerned with some form of adaptation. Particularly in agricultural societies, cosmologies are gauged closely to seasons. Cosmologies make the world manageable by limiting it. They can be also tuned to the limits of the local ecology, within their knowledge of interactions.

Some cultures ignore the long-range ecological consequences of drainage, irrigation or overexploitation, and these cultures may decline and die. But many archaic cosmologies are a form of fitness and limitation. Most try for adaptation before domination, according to Ricardo Reichel-Dolmatoff. The Tukano Indians try for adaptation before domination; many of their rituals limit the amount of hunting of fish and game. The ritual cycle of the Tsembaga can be interpreted as a regulating mechanism to maintain limits on fighting and using resources, according to Roy Rappaport. Their rituals also facilitated trade and distributed local food surpluses.

Yet, human cosmologies are limited and contradictory. All cosmologies cause destruction and waste; all produce the opposite of the good intended. Archaic and modern, occidental and oriental world views are complementary but not complete. Often, a cosmology is accepted as unquestioningly as a language, technology or place. Jeremy Rifkin states that cosmologies are a way of hiding the unimaginable: Voids, gaps, and immense size. They relieve apprehension. They make the world manageable by limiting it. They make the world comfortable and small. Rifkin claims that humanity inflates its daily activity into its image of the universe.

The ideology forms part of the personal identity of the members of a society. Even in cultures with alternative cosmological explanations, individual choice is determined more by social and educational levels than by rational comparison and evaluation.

The adaptiveness of religious belief is related indirectly to ecology. Religion is sometimes a responsive to ecological and economic conditions, for instance, when the Catholic Pope decreed fish to be eaten on Friday to help fishermen. Many archaic religions are even more a form of fitness and limitation. However, once powerful cosmological ideas are adopted, they can influence many cultures over centuries. The principle of plenitude, restated in Christian terms, says that an intelligible creator gave an earth of *unlimited* bounty to humanity for their use. This principle seemed to be confirmed in the Renaissance with the discovery of the richness of heaven, of microscopic life, and of unexplored continents. Many modern political ideologies and economics have been shaped by the principle of endless wealth. Adam Smith once calculated that the real price of anything was just the toil acquiring it. Inequality in a world of abundance could only exist through human suppression and exploitation of other humans. The invalidity of this principle came with the recognition of limits.

Many modern cosmologies, modified by advances in technology, pretend that the limits can be exceeded. The new scientific cosmology of biotechnology is still economic in a primitive sense. Only the myths have changed to include greater manipulation. It is still concerned with utility, growth and efficiency, as short-term goals. The problem with efficiency is that it is defined within such narrow limits. True efficiency means continuity over long periods of time, as with natural processes. Long-term exchanges in nature are not efficient in the industrial sense. The large sense of economics is the measurement of nature.

4.5.4.2.2.5. Economic Limits

The focus of economics, however, is rather narrow, in that the concept of resources is very limited and the unlimited wants of human beings are not much discussed. There is no psychology or ethics; there is no ecology or aesthetics (thoughts about beauty). There is no concern with the triviality of the free choice of a worthless 'doodad.' There is no concern with the welfare of the other beings that share the ecological community. Economics has to attempt to understand and predict the behavior of complex, interconnected systems where individual behavior and flows of energy and material are important.

Some theorists, like Paul Samuelson, have concluded that growth is necessary to rid the economy of disparities. The need to grow is intrinsic in this kind of economic system. A large literature has treated perpetual growth as the only conceivable state of affairs. Capitalism depends on growth for stability. There is some analogy with plants, to be elaborated on later. Some stability can be gotten from growth in early stages; later, stability must result from limits and metabolism. Growth in plants can delay the onset of senility by ridding the plant of waste products in more diluted form. However, too much growth produces a strain on tissues and early decay. In fact, one herbicide promotes excess growth as a means to kill weeds.

Development instead of growth would equalize wealth more efficiently—after all, economies have been growing for at least 400 years and increasing the disparities. There is no necessary association between development and growth, as Herman Daly and others have shown. Development means the introduction of an innovation. Economic development will require technology. Ecologically sound technologies will minimize stress to the environment. Economies could be modeled after mature vegetation and not successional vegetation, where diversity in scale is greater. A community is forced to accept an upper limit, beyond which it cannot grow any further. Further growth results in destruction or disruption of itself and its environment. This is the law of the maximum. Production could be stabilized in a steady state economy, a mature economy, like a climax system, where processes and cycles are constant. A steady state economy must be based on natural laws and ethical principles. Natural laws include thermodynamics and ecological theories. Rules of economics, laws of nature, and ethical principles must be related.

There is another distinction between growth and development. The ecological social approach to development makes it irrelevant to discuss global limits to growth. Local limits are far more significant to majority of population. Regardless of how much food exists, people will starve unless they can get it. Redistribution of resources and improvement of environmental quality (home environment) are more important than increased production by sophisticated technology. The natural capacity of regional photosynthesis must be limiting factor in development, especially in tropical and subtropical areas.

The earth is finite economically as well as physically. No matter how technologically expert humanity becomes, the multitudinous beings of the earth cannot be "produced" in the same manner as three billion years of evolution. Nor can we throw away the ambihuman dimension (*ambi* is the Greek word for "surrounding") and ignore at that cost. There is no romance or honor in throwing away every life that we can.

Economic culture defines the means of production and livelihood, the techniques of distribution, and the values and norms underlying economic behavior—and this can be more closely related to kinship. But, economics is transcribed by limits of human comprehension and imagination.

4.5.4.2.2.6. Political Limits

Politics is the art and science of human government. The first goal of politics is to ensure the survival of the human community. Then, it has to maintain the affairs of that community. Politics is the interactive means of providing the basic food and necessities of a community. As survival is survival in nature, politics rests on an ecological foundation. The organization of a community must be in accord with natural laws. Political participation depends on information, much of which can be provided by observation and science.

Robert Bellah and associates make three distinctions of politics, of which politics of the community is the first. It is followed by politics of interest, where different interests are pursued, according to agreed-upon neutral rules; conflict tends to overwhelm consensus. Finally, the politics of the nation addresses the "higher" affairs of the nation. It is more concerned with leadership than citizenship. Like the politics of interest, it accepts the status quo of relations of power or distributions of inequality. Symbolism becomes more important; perhaps because now the citizen diminishes in importance and the symbolism unites and includes them minimally.

Government has become subservient to economic actors, according to John B. Cobb. Partly because the ideology of economics is so positive. It proclaims that continued growth will solve most of the problems of modern civilization, from poverty to conflict, although the promise has not been fulfilled in the past 800 years. The problems have increased: Food shortages, housing shortages, energy shortages, unemployment, inequality of opportunity or goods, environmental deterioration, increase in weapons, and insecurity.

These problems continue due to the limits of politics. For example the size of society is a real limit; if there are too many people, within a limited territory or limited system, politics cannot provide them with an identity or control them. Size limits the distribution of things, such as food, housing, jobs and wealth, also. The time-frame of the political society is also a limit; the short-range visions of national interest are often inimical to the long-range ecological requirements of the support system.

The participation of the members of a society is also a political limit. Communities have always had face-to-face limits, in terms of numbers of face-to-face encounters. In terms of distance, communities have always limits, also. The politics of community is small-scale and local. It is moral consensus is applied to daily operations.

Leadership is another limit; the pool of applicants is usually relatively narrow. The desire to use power is a limit. The will to power does not seem limited to the community, as does participation. The will to power can be found in human communities over 10-13,000years. It seems to have evolved from the simple domination of other community members to all of nature.

Security is a problem for local communities and national governments. It becomes more difficult to protect against most any kind of weapons. Perhaps the solution is a global one, with the coordinated change in national policies and world-wide distribution of excess wealth.

Some facts result in motivation or fear, fear of the future in general, or of the reactions of others, in general. One response is a defensiveness in face of an unpredictable future. Thus, psychological limits intersect with political limits.

4.5.4.2.2.7. Limits of Human Populations

A number of recent studies have suggested that the human population of the earth could be much larger. Several other studies have recommended lower populations based on resource availability. All of the studies mentioned are concerned with finding a maximum human population. This thought experiment suggests ways to calculate an optimum population on a region by region basis, using a deductive, synthetic, conceptual model based on data generated from research on net primary productivity (NPP) and net community productivity (NCP). An NCP model is used to calculate regional goals for populations, considering cultural and economic goals. A deductive approach is necessary because accurate measurements of productivities in most ecosystems are lacking and exactness in some values can be misleading. A synthetic approach is necessary to integrate quantitative and qualitative data. In combining measures of qualitative and quantitative, it is simpler to set aside the first and then to calculate the second. The model must be conceptual because of the inherent fuzziness of the systems.

4.5.4.2.2.7.1. Weak Studies of Populations

Many theorists have estimated maximum or optimum sustainable populations for humanity. DeWit (1967), for example, estimated the maximum human population at 1 trillion, 22 billion (1.022×10^{12}). He then suggested that with 750 square meters added per capita, for forest and recreation, this number would be trimmed down to 146 billion. With consideration for animal protein in diets, the number would drop to 73 billion. These calculations are based on very simple variables—the light scattering coefficient (0.3), leaf area index (5), and a mean synthetic rate. Apparently, there are a number of simple assumptions, also: ideal photosynthetic conditions would always occur; the earth would be covered by a strict monoculture without any natural habitats; humanity would occupy of the first consumer level of all food chains, thus entirely eliminating any competing wild animals; and, humanity would eat all the food available, with no waste.

Colin Clark (1967) put the population of the earth at 47 billion, at a U.S. standard, and 157 billion on a Japanese standard. Who knows how many could have been crowded in with the lowest standard? Earlier, Clark (1958) had foreseen a population of merely 28 billion, using a Dutch standard of productivity and density, at 365 people per square kilometer. Apparently, he neglected to account for the ghost acreage used by the Dutch. Furthermore, he assumed a very large total arable land area—in fact, 63 percent of the total surface area of the planet instead of 24 percent of the land area—by counting the tropics twice.

Darin-Drabkin and Lichfield claimed that with efficient use of land resources, the world population could be 50 billion, and that we have the technological power to fulfill these estimates, but need organizational changes and proper planning to allow it. Burlingh et al. (1975) estimated the absolute maximum food production of the world at about 30 times the production of 1970, and presumably an equally high human population. They arrived at this figure on a basis of quality of soil, climate, and water. It did not seem that any biological factors were considered. Weinberg and Hamilton presented a steady state model of 10 billion at 400,000 Kcal per day for the year 2050, but made no mention of how the population would stabilize at that level or what kind of economics would support it or for how long.

H. R. Hulet (1970) assumed that plant and animal products were equally necessary. By making them equal requirements—animal and plant production at 4.2×10^{15}/Kcal each—he suggested that the earth could support an optimum population of 1.2 billion. Considering the Law of the Minimum, he offered a series of optimum populations as a function of resources (Table 4-542271-1).

Table 4-542271-1. Population Functions (after Hulet)

Wood production (4×10^6 cal/yr/cap)	1 billion
Energy rate	600 million
Fertilizer rate	900 million
Aluminum rate	500 million

He concluded that as technological and agricultural systems expanded, population could also. But the rates of use for the minimum items he provided were not renewable or slow; they were based on U.S. standards of consumption. In the U.S., energy use is increasing at 6 percent per annum; three times faster than the population.

Westing (1980) estimated a global carrying capacity based on five areas of renewable resources (Table 4-542271-2). His estimates range from 1.5 to 3.9 billion people. These numbers are for a maximum population, however, not an optimum. It is important to remember that an optimum is almost always less, much less, than a maximum. Note also that his paper, written ten years after Hulet's, when wood production had increased significantly, gives a doubled figure for a population limited by the availability of wood. In fact, neither rate is sustainable. Annual timber cutting in the United States in 1969, for example, exceeded annual growth by 50 percent. Most cutting since then has exceeded growth by different percentages, despite propaganda to the contrary. Furthermore, wooded areas are still being cleared for agriculture, housing developments, and industrial sites.

Table 4-542271-2. Population Estimates (after Westing)

Total land	2—3.1 billion
Cultivated land	1.5—3.3 billion
Forest land	2—2.9 billion
Cereals	1.7—3.3 billion
Wood	1.9—3.9 billion

Westing recognized that most usable renewable and nonrenewable resources fundamental for human life are used in direct competition with wildlife. He commented that most studies did not consider wildlife and habitats. For a straight energy calculation of an optimum population, assuming the unavailability factor has been subtracted first, refer to Table 4-542271-3.

Table 4-542271-3. Optimum Populations (1980 Per Capita Use)

United States	120×10^6 Kcal	189 million
France	4×10^4 Kcal	567 million
Japan/Argentina		1.13 billion
Portugal		2.84 billion
India/China		5.27 billion

The French Per Capita energy use is approximately the same as the world average. Portuguese energy use is the same as agricultural trade. The main difference between this model and earlier ones is the pattern of utilization and the assumptions of minimal waste and no growth.

Whittaker and Likens (1975) have suggested that an agricultural world, where humanity lived as peasants, could support at least 2 billion, perhaps 5 or 7 billion people. Another study (Wittbecker, 1976) calculates a global population at 1.6 billion, based on the sum of populations from all regions and assuming a switch back to traditional agriculture. A population for Eire, based on traditional methods on current cultivated hectares, is almost 5.5 million—more than the current population of 4.2 million; cultural changes have kept the birth rate low since the potato famines in the mid 1800s. By contrast, a traditional population for Sri Lanka—roughly the same area as Eire—would be 7.6 million people, which is less than the current population of 10.6 million; cultural changes, such as the switch to cash crops such as tea and rubber, have allowed population to increase.

The current high levels of population there and elsewhere, at a large range of standards, can only be maintained through the constant takeover of natural habitats for arable land, or through the drawdown of fossil fuels, and by economically cheating the poor and powerless. Since the quantity of wild lands and fossil fuels is quite limited, either human populations must adjust to renewable resources or technology must provide substitutes to avoid a population crash. Most of these optimistic studies see the limits of world food supply as being determined by the physical limitation of land, water, light, and chemicals, or by the logistics of transportation and conscience. They assume that areas of cultivation can be expanded into planetary biological support systems, without fear of ecosystem collapses. They assume that efficiency can be enhanced indefinitely without the risk of genetic instability. They assume that novel sources of food will be used, ignoring cultural limitations). These requirements may never be met, as others have realized. Many theorists consider the global population explosion to be natural and unavoidable; Eric Jantsch states that population growth is an example of autocatalytic nonlinearity—"it constitutes an essential factor in the creative act of gestalt formation."

The assumptions of these studies are frightening. Humans have already modified animal and plant associations dramatically, simplifying patterns of energy and chemical exchange, and solidifying themselves at the end of many food chains as a dominant species. A dominant is a species with greater influence than any other in its biotic community, changing the lives of other species and the character of the habitat. Humanity is a pandominant species. As such, humanity reclaims, overgrazes, clears, depletes, and wastes at a level that threatens the stability and existence of many systems. One of the ecological consequences of human activity is the degradation of wild habitats for human developments(food, housing, and recreation) and the introduction of novel elements into the biosphere— elements that have not been harmoniously worked in over time. The biomass of the human species (6×10^{14} Kcal) probably far exceeds the biomass of any nondomestic species, and that biomass is supplemented by the tremendous biomass of domestic animals, which is four times greater. This domestic biomass forms an equivalent population that consumes much of the same food, such as milk, fish, and grain. The domination of humanity is related to other characteristics as well, such as a large annual increase (2 percent), a high structural organization (information and matter), and high energy use (globally, 13 times mammal equivalents). This dominance has major effects on ecosystems: transient perturbations in energy relations (from oil spills, burning); chronic changes/shifts of systems (from dams, irrigation, chemical wastes); species manipulation (from the import and export of exotics); and, interference competition with wild species (as opposed to exploitative competition, which can be stabilizing). None of these effects are exclusive to humans as a species, but they are excessive, rapid, compounded, and large-scale.

4.5.4.2.2.7.2. Studies that Place Humans in Ecosystems
Human populations inhabit specific ecosystems and are parts of them. In the past humanity was adapted to and limited by the productivity of ecosystems. Ecosystems result from the interaction of all living and nonliving factors of the environment. These systems are profoundly affected by both random and purposive physical and biological factors. As a result, habitats change and organisms adapt. By modifying their habitats in the process of living, organisms change the characteristics of the system and force further adaptation. More important, organisms are limited by the productivity of the system in varying degrees.

As an ecosystem ages, pressure is put on some populations by other populations. Competition and predation become more complex. Mammals are the best regulated of highly evolved species. Their behavior is controlled and population regulated through the use of space. Most populations, furthermore, regulate their density well below the limits of the food supply, often by as much as 50 to 70 percent. Territoriality can be correlated inversely with trophic levels and productivity. But populations can also be limited by specificity of prey or plant source, size of prey or plant populations, predators, and natural events (catastrophes).

Mammals alter their habitats through chewing, digging, and burrowing. Rodents can dislodge earth at a tremendous rate (18-120 m^3/ha/yr). In many cases these activities improve the conditions for growth of vegetation. Mammalian grazing promotes regrowth and the movement of seeds. Bison and prairie dogs were responsible for much of the character of the American plains. Rodent caches may account for 15 percent of Ponderosa seedlings. Beavers and other rodents create microsystems. Caribou and wolves transfer energy between systems.

Mammals are bound by biological requirements that must be met if a population is to survive. These functional requirements are rather minimal for humans, however, being only food, clothing, shelter, and reproduction. Other requirements, such as respect, comfort, and self-fulfillment, depend on socio-cultural systems. Human beings are mammals— omnivorous, social, bipedal, featherless, symbol and tool-using, game-playing, neotonous, bilateral-hemispheric generalists. As Woodwell (1976) phrased it, humans live as "one species in a biosphere whose essential qualities are determined by other species." Like other mammals, humans change their habitats to suit themselves. But, human beings are pressuring many ecosystems by through simplification and conversion.

Human populations are limited by the real biological constraints of ecosystems and biogeochemical cycles. Eugene Odum was one of the first to consider the implications of such limits. Odum (1970) suggested using land area as a measure of human carrying capacity. The minimum per capita acreage requirements, with a temperate area like Georgia as a model for a quality environment, is 2.02 ha (5 acres). The percentage of areas is broken down in Table 4-542242-1.

Table 4-542272-1. Acreage Per Capita (after Odum)

Food-producing land	30 percent
Fiber-producing land	20 percent
Natural support areas	40 percent
Artificial areas	10 percent

The natural areas are based on minimum space needs for watersheds, as estimated by land use surveys. Food-producing land includes acreage for domestic livestock. Extrapolating this technique to the entire planet, assuming that wilderness area has been considered in the calculation of natural areas, and converting for the differences in productivity of ecosystems, the population calculation comes to 3.97 billion, which is close to the 1970 world population.

Samuel Eyre attempted to describe the wealth of nations in terms of the productivity of ecosystems, specifically Net Primary Productivity (NPP), with nutrition equivalents in NPP for mineral resources. Eyre also devised a common denominator to consider organic and inorganic assets together. Population carrying capacity can be formulated only using both. He assigned a nutrition equivalent unit weights of metal, but this calculation depended on a dollar value for food and minerals.

Gross Primary Production (GPP) is the rate of energy storage by photosynthesis (equal to photosynthetic efficiency) in autotrophs (plants). The maintenance and reproduction of plants is paid for by the energy expenditure of Respiration (R). The amount of energy stored as organic matter after respiration is identified as Net Primary Production (NPP), which equals plant growth efficiency. The NPP accumulates through the history of a system as plant biomass expressed as kilocalories per square meter (Kcal/m^2). The kilocalorie is used as a unit of energy flow and production; it is a useful common denominator for these calculations. The biomass minus the decomposition in a system is the standing crop biomass of that system. The problem of confusing production (amounts) with productivity (rates) is avoided by considering all values per unit area (m^2) over the entire year (m^2/yr). The energy stored in consumers, or heterotrophs, is referred to as secondary production (SP) or assimilation. The storage of energy or organic matter not used by heterotrophs is Net Community Production (NCP). The relationship between productivities is such that autotroph respiration plus heterotroph respiration plus NCP equals GPP, as does the NPP plus autotroph respiration.

In a balanced ecosystem, NPP equals respiration; in an accumulating system, NPP usually exceeds respiration by 1 to 10 percent. Although stable ecosystems tend to produce a maximum GPP, species, biomass, and the production to respiration ratio (P/R) continue to change long after the maximum has been achieved. In fact, as the GPP approaches an asymptote, respiration increases. In a balanced system, tropical rainforests for instance, NCP approaches zero, as adapted heterotrophs become more efficient at using production. In accumulating systems, such as grasslands or young forests, NCP can range from 20 to 70 percent. A balanced system is integrated and self-perpetuating, where production (the photosynthetic fixture of carbon) is balanced by respiration (the oxidation of carbon). As a system becomes balanced, the pressure of selection of organisms shifts; the capacity to live in crowded circumstances with limited resources is favored. Populations that depend on rapid individual turnover (r-selection) are not as successful as populations of large, long-lived individuals (K-selection).

Eyre contended that one must know the productive capabilities of land in its original vegetation to compare with productivities under human management. He found, for instance, that most wild lands are more productive than most agricultural acreage. Rodin et al., and others, have calculated the potential productivity of the wild vegetation of the earth at around 1.19×10^{11} metric tons per year. What is left out of Eyre's considerations is the amount of healthy ecosystem necessary for cycling and renewal. Furthermore, all productivity is treated as economic, to be dispensed with by the nations that occupy the land.

4.5.4.2.2.7.3. Community Models

It is possible to calculate a sustainable population using NCP instead of NPP, however. An NCP model is deductive, synthetic, and conceptual. A deductive approach is necessary because accurate measurements of trophic level productivities in most ecosystems are lacking. A synthetic approach is necessary to integrate quantitative and qualitative data. The model must be conceptual because of the inherent fuzziness of the systems.

For tropical and temperate grasslands, NCP may approach 60 percent of NPP, although 30 percent is much more likely. Temperate forests may also approach 30 percent. The population supported by NCP, while truly sustainable, is much less than that suggested by other models. Furthermore, of the NCP produced, 75 percent is unavailable, 5 percent is eaten by pests, 65 percent of that which is harvested is inedible, 80 percent of that which is edible is lost in processing, and 25 percent of that is wasted during consumption. The food consumed is barely 1 percent of the productivity. The consumed portion of NCP of the tropical rainforest could support only 2.1 million people. Temperate forests could support 25 million. Mediterranean vegetation could support only 375 thousand. Tundra vegetation could only support 62 thousand people worldwide. The oceans could support over 22 million people. Tropical grassland, however, could support over 358 million people. Croplands, based on original grassland and forest vegetation, could support an additional 183 million, even though it has 1.5 times the area as tropical grasslands. The oceans have 23 times the area as croplands, but could support only 1/8th the number of people.

This model is concerned only with the values of natural productivities. Domestic lands were assigned productivities based on those of the original wild vegetation, which in fact is usually higher. Land currently used for agricultural practices was divided into original vegetative states and productivities were added to calculations from wild lands. In effect, crop lands are heavily subsidized by fossil fuels, fertilizers, and pesticides, allowing them to produce 10 or more times the food calories, and thus support a much higher population—how sustainable a population depends on whether renewable energy can replace fossil fuels. When NCP is used to calculate a maximum population, by adding the calculations for each vegetational unit of the earth, the measurements are consistent with previous low figures—1.08 billion—after the same percentages of inedibility and waste are subtracted from a large preliminary figure (82.6 billion).

The population of 1.08 billion is almost identical to a flat 1 percent rate of the NPP, subjected to the same loss percentages (900 million). The use of increased animal protein would reduce the number correspondingly. This availability factor is on the order of 75 percent. The figure, 1.08 billion, is a maximum. An optimum population, arrived at by a 50 percent rule, of 540 million would insure against problems due to fluctuations in productivity. Richard Watson arrived at a figure of 500 million, based on U.S. levels of consumption with the present industrial production. Kozlovsky intuitively estimated 500 million also as an equilibrium population. Others, such as Arne Naess, have suggested low figures based on traditional livelihoods.

There is a maximum carrying capacity for humanity. This capacity is unknown, but could be less than the current population. The carrying capacity is that population which is sustainable on a long-term basis of renewable and nonrenewable resources or energies. For humans, this capacity must include domesticates, as human equivalents, since many domesticates compete for protein consumption. Domestic animals extend the carrying capacity, since many of them consume agricultural wastes or use lands marginal for agriculture, but domestic animals are not as efficient as wild populations. Technology could expand the carrying capacity to some extent, with higher yield crops and resource substitution, but also it could reduce the capacity with unforeseen side-effects, the use of pesticides, for example. War and social disorder reduce the ultimate capacity. Furthermore, the capacity decreases as the per capita use of energy and resources increases. Carrying capacity calculations often just consider food energy, but all needs—clothing, shelter, transportation, information generation, aesthetic satisfaction, wilderness (as places for other species), ecosystem functioning—must be included. Given the current political and technological situation for humanity, it is difficult to imagine how any equilibrium could be reached without a significant reduction of human populations.

There are also minimum viable populations ton be considered. For the human species, it is unlikely that the lower limits will be approached in the foreseeable future. There are several lower limits to keep in mind, however, as shown in Table 4-542243-1. The limits presented are conservative estimates.

Table 4-542273-1. Minimum Limits}

Genetic minimum	5,000
Fertility	50,000
Ideomass	500,000
Social contact minimum	1 million
Evolutionary advantage	10 million

Many theorists worry that a reduction in human natality could cause humanity to stagnate. Natality is the birth of new ideas, also; this is J. B. Calhoun's concept of ideomass to supplement sheer demomass. Ideas and culture can create endless possibilities from a unique beginnings and adaptations to support a stable or declining population without reducing the quality of interactions.

4.5.4.2.2.7.4. Limiting Human Impact (for the Planet as a Whole)

Human populations are not based on simple mathematical calculations. Sometimes the political aspect of population is of utmost importance, for instance, when the minority Tamils in Sri Lanka breed for political control with the Sinhalese. Sometimes population increase is encouraged by tradition or allowed by ignorance of birth control measures. Calculations of population and carrying capacity are made difficult by emigration and drawdown.

Changing population traditions and growth may be difficult. If Leopold Kohr's velocity theory of population—in which aggregates of humans increase their mass through an increase in speed—is correct, then slowing the growth will be more demanding and may require some form of institutional control. Paul Ehrlich has noted that human numbers press against human values, that is, changes in quantity precipitate changes in quality. Quality of life will sink lower long before the limits of food or space are reached. Overpopulation has already blocked many people from desired adventures: pioneering or hunting as a way of life, land for everyone, or an airplane in every garage.

Population involves the entire humanity-technology-environment spectrum; it cannot be approached from just one direction. Lower densities of humans will always be able to harmonize more successfully with biological processes. For the long-term survival of the human species, adaptability to environmental changes is necessary. This requires a wide diversity of gene pools, which is achieved by a relatively large population divided into local, partly isolated groups. And this requires healthy regional ecosystems. The optimum size of the global human population is the sum of optimums for local habitats. The basic question is not how many people can exist at once, but what kind of life is possible for those who do exist. Before an optimum human population for the world can be calculated, other questions must be addressed, such as level of technology or affluence, and systematically investigated.

There are several ways to calculate a population, based on the IPAT model or the Palouse model. The IPAT model, first proposed two decades ago, represented the efforts of population biologists, ecologists and environmental scientists, such as Barry Commoner and Paul Ehrlich, to formalize the relationship between population, human welfare and environmental impacts. The IPAT model postulates that environmental impact (I) is the product of population (P), per capita affluence (A) and technology (T).

But, the model has limitations, such as not providing an adequate framework for discerning the various driving forces of human-driven environmental change. As a consequence, the IPAT model does not produce quantitative data.

The Palouse Model (Equation 2) is more complex, although it has similar problems.

Equation 2. $P = \{([A \bullet N] + E) \bullet T \bullet C\} / U$

Where "A" is the indefinite series of areas, from wilderness to fiber-producing areas, "E" is a series relating the energy of food and other resources, "N" is the net used productivity (the NPP minus the unavailable productivity), "P" is population, "T" is an overall technology modifier summed from used technologies, "C" is a totaled cultural modifier based on habits and images, and "U" is food requirements plus resource requirements, which divides the sum of the productivity and energy available. (This model is expanded in Section 4.5.4.2.3.3.2.)

The model raises more questions that need to be answered before the model can yield more than a very loose number. How much land should be left in its native state? Enough to save one of each kind of ecosystem? Enough to support atmospheric cycles? Enough for exotic ecosystems? Enough for recreation and public service? Ideally, all of those.

At what level of luxury should humans live (in Kcal/yr.)? Just basic amenities (such as clean air and water)? As a function of psychological space? With maximum cultural or material variety? The goal of all animals, including humans, is to live well.

What are the physical limits of resources? Should all nonrenewable resources be used, and, if so, at what rate? Should only renewable resources be used? Nonrenewable resources are being depleted. If for no other reason, their use should be minimized or saved for transitions to renewable resources.

What is an optimum? Laboratory studies with rats show that, with a choice of optimum rat environments, some rats will reject the optimum. Is it meaningful to speak of an optimum for all of humanity? For other species?

How these questions, and others not asked, are answered determines an optimum human population. In calculating an optimum population within ecosystem restraints, few have considered minimum wilderness preservation, air and water quality, genetic minima, nonrenewable resources, appropriate technological innovation, the importance of cultural frameworks, adventure, research, beauty, uniqueness, and other intangible experiences—although these dimensions cannot be quantified, their existence can be associated with attributes that can be quantified. If human civilizations were based on complex ecosystems, they would be more complex. The complexity encountered in nature may serve as a Zen koan-work, to bring us to rational breakdown and to an alternative: humanity as a self-conscious, self-limiting, poetic species.

4.5.4.2.2.8. Adjusting to & Dealing with Limits

There are many ways to deal with limits, whether they are human, economic, or cultural. One obvious way is to ignore them. Ignoring limits usually results in cultural collapses, as happened at Easter Island (Rapa Nui) and other places. Ignoring limits leads to disaster. Recognition of limits, however, does not mean that people are forced to live as bacteria. Limits allow a considerable amount of freedom within their numbers. Limits define locality, local spaces, and local systems, from the global. Limits are not only important for life, but also are implicated in diversity and maturity. But, even these limits are based on the physical.

Another way to deal with limits is to try to expand them. Eric Jantsch states that limits to growth can be overcome by the evolution of dynamic structure; they appear as widened limits. Expansion has happened before as the result of ecological evolution. To some extent, technology can expand limits.

Or, we can respect limits and work within them. This is an ethical way. The limit of Buber's ethical dialogue, I-Thou, is that it is concerned only with the present; posterity is ignored. Human duration is too short. The systems and processes that we deal with are intergenerational. Economic rationality is even more limited. Future values are discounted at current rates. Planting a tree that takes 30-200 years to mature is calculated to be uneconomical (the giant sequoia takes almost 200 years to flower, for instance). Who benefits from it? If our ancestors had planted trees for our benefit, it would be easier to justify the continuation of the effort. But identification through time is related to that with place, or locational stability. People not living in place have no vested interests, hence they have no ethics for the place.

The land ethic Leopold had in mind was a sense of ecological community between man and other species. When we see land as community to which we belong, we will use it with love and respect. Such an ethic would change the human role from master of earth to plain member of it. Predators are members of the community; no special interest has the right to exterminate them for the sake of benefit for itself. This attitude is important for habitat protection. Leopold describes the extension of ethics as "actually a process in ecological evolution. Its sequences may be described in ecological as well as in philosophical terms. An ethic, ecologically, is a limitation on freedom of action in the struggle for existence." An ethic, philosophically, is a differentiation of social from anti-social conduct. These are two different definitions of one thing. The thing has its origin in the tendency of interdependent individuals or groups to evolve modes of cooperation.

Is there a limit to human ethical feeling? How many individuals can a person care for? 100? 50,000? A billion? Or, how many animals or plants can a person care for? How many living beings? Are these fair questions? Obviously, there is some kind of limit—otherwise, everything alive would be cared for and protected. People seem to be limited to their extended families and groups, as well as some kind of larger groups, a race or nation.

Although physical and chemical limits are real limits, they are modified by psychological limits, which are aggravated by socio-cultural limits, which are further increased by political limits. Each "higher" more complex level makes other levels worse. The attempt to exceed limits becomes less efficient. The discussion of limits is expanded in the following sections, especially when applied to wilderness or human populations.

Humanity is part of nature, as valuable and unique as cranes or louseworts, but not more valuable or unique. This attitude would entail using what is necessary, exploiting some ecosystems completely, changing a place to fit human aspirations, and killing plants and animals for sustenance. All animals and plants alter their environments to fit to some extent. But it would also mean limiting humanity and its technological effects, limiting human use to local impacts, and letting other beings live without interference. It is not necessary to dominate or terraform the earth completely. Humanity could contain itself to five percent of the earth's surface and ecosystems and only visit or ignore the remainder. This ideal requires change, even different ways of being.

Humans have made radically different conditions that they must now accommodate. If the mind is exposed to economy of nature, as revealed through living systems, humans may recognize the necessity of balancing values. Total win-lose conflicts are unwise. Value systems concerned with dynamic equilibrium, aesthetics, complementarity, reciprocity, justice, interdependence, reconciliation, and intuition are the language that biology speaks.

Myths, with transformations, and metaphors, with structure of integrated differences, are modes for conveying ecological sense; they are less concerned with survival than the survival value of a good fit between dualisms of life. Equilibrium is needed between self-restraint and self-expression, between self-protection and self-restriction. It is not self-expression or self-restraint, but all in a satisfactory fit. Fitness attunes us to limits.

4.5.4.2.3. *Zoning & Functional Division of the Planet*

Zoning is a way to regulate activities that might conflict. For instance, as a way to minimize settlements in a flood plain to protect homeowners. Or, to separate wilderness or recreational activities from industrial activities.

Michael Soule identifies eight paths for potential biotic survival, including "In situ" (bounded areas with little human disturbance; the problem is that few are large enough for natural processes to operate in or for MVP of large species) or Restoration projects (intensive management of usually small areas to restore sets of species). In situ strategy is the most likely to save entire ecosystems and their processes, as well as biogeographic assemblages, indigenous species, local populations, and overall genetic variation. Right now, inter situ locations have intermediate significance in saving these same targets. Extractive reserves have the potential to save traditional human cultures in place. Zoological parks have an intermediate chance to save populations and assemblages of species. Restoration projects have the potential to save entire ecosystems. Some goals of Conservation Biology or Ecoforestry can be applied to optimum populations, the design of preserves, and zoning.

Design itself has many dimensions. The size and scale of reserves have already been mentioned. The shape of a reserve, the extent and shape of its edges, is also of critical importance. Edge effects started out being considered as centers for diversity. Now, many edge effects are considered to be destructive. In the 1940s, it was thought that edges should be increased to provide bountiful game crops. Edges have proven to be good for both game and weed species. Edge effects, however, are detrimental to populations adapted to forest interiors. Too many edges reduces forest diversity at local and regional scales. Different species perceive edges differently. So, how deeply should edges penetrate forest stands? Overall, within 50 meters—but this may be too general.

The pattern of a design is also important to control patchiness and fragmentation and to allow for future change. For example, for a Palouse grasslands preserve, a spider-shaped area was recommended to incorporate many north-facing slopes and to allow for possible greenhouse effects. A good pattern should also anticipate temporal change.

4.5.4.2.3.1. Functional Division of Earth

Humanity has removed itself to the outer-space support systems of cities. Even an ecological caring society will not cure a rift with nature. For this reason also, wilderness needs to be put away from resources. Our actions are not separate from other living beings. All living beings play a role in life support system; all types of habitats should be maintained. There has to be a limit to human settlements.

The earth must be treated as one whole, although composed of levels of interacting parts. The life cycle of many animals in Africa is associated with gigantic circular migrations to take advantage of seasonal grazing and water supplies. Ocean and air currents ignore national boundaries; birds as well as virus are likewise international.

The areas will not be divided uniformly within national boundaries. If areas are drawn within nations, it will be unfair for England, which has replaced all of its wild areas in becoming rich, to insist that Zaire retain all of its wildlife and remain poor. Justice will not be achieved this way. Empty space on earth is an illusion. Unused (for human purposes) space must be dedicated to preserving the diversity of biomes. It is especially important to preserve ecosystems where humans are not dominant.

E. P. Odum (1975) suggested partitioning ecosystems into 4 compartments according to energy sources: Solar-powered natural ones, like lakes, forests and grasslands; naturally subsidized solar-powered ones, like marshes, streams and coastal zones; man-subsidized agroecosystems; and, fuel-powered industrial systems. The first category provides basic life-support; the second is valuable because of the high energy flow per unit area/time. Both of these have the capacity to assimilate limited wastes from the last two. The third can produce life support; and, the fourth is parasitic on its surroundings for all its functions.

The last category would occupy only 2% or less of world land area; the third would occupy about 15% of land area, with possibility of small increase. Odum recommends a 30% natural forest cover world-wide, with 50-60% in wet climate areas, to balance dry areas. Silviculture should be part of agroecosystems; this would leave hilly, mountainous, thin soiled and other fragile areas for natural forests.

Odum's suggestion has been expanded. The definition of the four areas sets the frame for proper conservation and development policies, which define what we must do and when. Many natural and artificial values can be conserved this way. This also determines the type of sciences that deal with each areas. The four basic areas and their divisions are (roughly based on the work of C. Doxiades):

Table 4-54231-1. Basic Divisions

1. Wilderness Preservation Areas
1.1 absolute preservation; no admission
1.2 scientific investigation only; no machinery
1.3 exploration and inventory; no machinery
2. Common Places Conservation Areas
2.1 home of primitive peoples
2.2 some hunting and gathering; temporary camps
2.3 some exploitation below regeneration rates; limited machinery
2.4 some recreation; limited machinery
3. Domesticated Areas
3.1 herding/nomadism
3.2 traditional agriculture
3.3 managed forests
3.4 modern agriculture (large scale)
3.5 special and general recreation
3.6 rural communities/services
4. Artificial Areas
4.1 roads, air and water lanes
4.2 residential
4.2.1 medium density
4.2.2 high density
4.3 city, commerce/services
4.4 major urban center
4.5 megalopolis
4.6 industrial
4.6.1 light
4.6.2 heavy

Each of these areas would be divided into appropriate areas, depending on uses ranging from virgin land to heavy industrialization. The percentages of surface areas appropriate for each land use are: (1) 50% real wildlife, (2) 30% wildlife used, (3) 16% cultivation, (4) 4% human residence and industry. As many parts of the global ecosystem should be kept as natural as possible. We need examples of all types of natural areas. We cannot have less than 50% until we learn more.

Thoreau wanted a balanced interplay of wilderness and civilization. The opportunity to operate in four basic contrasting environments would contribute to the richness of emotional life; a spirit of each place. Each category and subcategory could be characterized by its limits on human activities.

Area 1.1.—No human activity should be allowed. There is no scientific answer as to how much of the surface should be included in this area. Preserving areas of all types, including polar areas and deserts might add up to over 50% land area.

Area 1.2.—human visitation for scientific data-gathering; machines excluded; all types of environment; scientific studies.

Area 1.3.—some exploration for mapping purposes.

Area 2.1.—partly inhabited by primitive tribes (continuation); scientific observation.

Area 2.2.—hunting and gathering by some cultures; temporary camps.

Area 2.3.—coordinated permanent camps; sports-recreation; no machines; visited systematically.

Area 2.4.—natural wildlife controlled and exploited by humans; commercially exploited, uncultivated forest; transport machines (rail, air or road); greater pressure requiring greater care.

Area 3.1.—herding and nomadism by certain cultures; limited machinery.

Area 3.2.—traditional agriculture and husbandry; nonirrigated; herding; animal shelters.

Area 3.3.—methods of cultivation of forests, oceans and wet areas;

Area 3.4.—modern methods of cultivation and livestock, including factory farming; automation, industrial methods, scale of energy;
almost complete elimination of natural landscape.

Area 3.5.—provision for human recreational need, excluding homes and built-up settlements; hotels, dude ranches.

Area 3.6.—rural communities, with minimal services; organic communes.

Area 4.1.—permanent paved roads, following existing ones where possible; new roads planned to make minimum impact; restricted lanes and volumes of air and water transport.

Area 4.2.—light residential; full services; light commerce, low density (70/ha=28/acre); provision of facilities and services; some employment. Mainly residential with increased commerce, etc. Density of 110/ha (45/acre).

Area 4.3.—high density city dominated by central area functions (communications, services); unsuitable for complete range of human development. Density 300/ha(120/a); Athens had 80/ha.

Area 4.4.—isolated area for heavy industry and waste disposal of noxious waste here; recycling.

Area 4.5.—the tremendous areas for large urbanizations.

Area 4.6.—completely industrial areas, with almost no functioning ecosystems.

Lack of a comprehensive concept concerning these areas has resulted in wrong use, as with Love Canal in the United States. Not so much wrong use of mixing but wrong use of energy and materials, that is, one-way use from resource to unintegrated waste. Each area would be designated to areas best suited for them. Each area must be saved from invasion by others: industry into housing, housing into farm land. Amount of area calculated for regions 3 and 4 could be increased from areas 1-2, but not by more than 5 percent.

We have had no historical experience with many things, from mass communications to the scale of private transportation. We do not understand the implications of mobility, population, economy, and energy. So how do we increase some settlements but not others?

All of these percentages and areas must be regarded as tentative until studies can be carried out. In each continent, region, or nation, land will be assigned appropriate percentages of use, depending on geography, current development and potential. Policies will depend to some extent on political systems and cultural values (private ownership).

We must assess the percentages of land surface occupied by every area classification currently. Global ecological balance depends on better understanding of the differences and relationships in nature and on the application of results of more careful and detailed studies for which this study provides only an outline.

We also need to systematically divide water into areas. Since most pollution comes from land first, these areas need to be coordinated with the four land areas. Water includes the three regimes: atmospheric, underground, and open (oceans, rivers, and lakes). Connected waters to land areas may be considered as having the same classification.

Coastal and Air areas may also be defined according to the same schema. The official figure for the length of coastal areas of the oceans is 261,300 km, but the real figure is undoubtedly much larger. Because of the special human affinity for coasts these should be considered separately. The water and air are ultimately shared by all.

4.5.4.2.3.2. *Wilderness & Preservation*

Wilderness has been the name used for uniformly empty, useless, valueless, and untamed areas. Although what remains is still largely untamed, wilderness is not uniform or empty, useless or valueless. Recently, wilderness has been recognized as holding resources and as being important for recreation. Wilderness is also the source of agricultural and cultural richness. Indeed, these reasons are presented as justification for saving wilderness. But there is a more basic reason to save wilderness: it exists apart from human concern and industry, and it is the only sanctuary for wildlife apart from humanity. It is filled; it has its own values and uses, even if they are nonhuman. It is wilderness itself.

No one of the ancient or modern descriptions and definitions are adequate to define wilderness. An understanding of ecological knowledge shows the importance of wilderness itself and what it entails—other beings for themselves, dynamic system stability, and finally the health of systems upon which humanity is dependent. A set of definitions are recommended to justify and plan the preservation of many kinds of wilderness.

4.5.4.2.3.2.1.Wilderness As Emptiness

The Romans did not really have a word for ultrahuman nature. Words or phrases they applied to wilderness include *vastitas*, meaning emptiness or waste, *solitudo*, referring to loneliness, and *locus desertus*, meaning deserted place. The definition changed slightly over centuries. The modern word wilderness comes from the Middle English word, *wildern*, and the Old English, *wilddeoren*, meaning "of wild beasts." Wilderness was understood as a region uninhabited and uncultivated by humans; an empty region; or a confusing multitude.

In the 17th century, untilled heaths and fens were regarded as a reproach to civilization. John Houghton in 1681 called Hampstead Heath a "barren wilderness." Cultivation was a symbol of civilization, whereas "wild and vacant" lands were thought of as a deformed chaos. Similarly, mountains were thought worthless. Wilderness meant uncontrollable, and was applied to the beasts of inhospitable regions. Something wild was untamed or untilled. Samuel Johnson (1755) defined wilderness as "a desert, a tract of solitude," a wild place inhabited by deer.

As more and more areas were cultivated, a new attitude to wild nature developed. George Hakewell, a clergyman, defended mountains in their "pleasing variety" as God's work. By the late 18th century, the appreciation of wild nature had religious overtones. Nature was beautiful and morally healing. Wilderness had spiritual power. Feelings of awe and terror were transposed to mountains and tropical forests. Many scientists and theologians of the time maintained that all created species had a necessary part to play in the economy of nature.

By the nineteenth century, wild nature was appreciated for itself, without dependence on artistic models. The preference for cultivated landscape had been challenged. Explorers praised the virtues and challenges of wilderness. Yet, much of it was left alone, because it was considered to be useless or inaccessible or too much trouble to cultivate. It was left blank on maps. But even blank spaces may be valuable. Aldo Leopold said that "to those devoid of imagination, a blank space on the map is a useless place, while to others, it is the most valuable part." The blank spaces, least influenced by human activity, are wilderness—the only possibilities for wilderness. We are learning that some of the blank spaces should to be left blank; that they have their own functions and ultrahuman needs; that they can be granted rights in the human system. There should be wolves and bear and martins and lichen somewhere, without human company.

By the twentieth century, the appreciation of wilderness came from those who realized that their lives had been impoverished by being set apart from it. At the same time, now that humanity is competing for protein in every ecosystem, wilderness is regarded as a resource. Scientists since Bacon have considered all of nature as a resource for the creation of human wealth.

Human influence in every ecosystem has led to views of wilderness as playground or the aesthetic source of calendar photographs. The image of wilderness as playground is not historically novel. The entertainment value of wilderness, after all, was what prompted Congress to designate Yellowstone, the first national park, as a "pleasuring ground." Although Muir called the National Parks "temples," he also referred to them as the "people's playground." Even today, wilderness is valued as playground. "Wilderness is for people..." not for flora or fauna, it has been said.

When we can control nature, for whatever purpose, the result is not always good. Our style of efficiency and proclamation of mass values destroys unique places. The environment of few significant places becomes a flatscape. It is turned into uses. John Fowles warns that "We shall never fully understand nature (or ourselves) and certainly never respect it, until we dissociate the wild from the notion of usability—however innocent and harmless the use." This uselessness lies at the root of our hatred and indifference.

Many people in modern societies are learning to be dispassionate (uncaring), unattached (placeless), and objective (uninvolved), without really thinking about the effects on society and the environment. Perhaps, it is a result of confusion or fear. We still fear nature because it is uncontrolled and ultrahuman. Nature is feared as unfathomable and uncontrollable. Natural processes entail death; we fear death. Nature was desacralized by the simple filter of science, the false promise of control. We distance ourselves from what is uncontrolled or unowned. This detachment is the greatest threat to the welfare of nature. The attitude of distance toward beings excludes consideration of necessary information. Science accepts the sentience of plant life, but does not adjust its methods. Human knowledge grants emotions to animals, but uses them badly anyway.

The problem with utilitarian ethics is that it permits the use and exploitation of any natural object, including human beings. Based on a limited science, the ethic failed to value those beings and communities for which no use was known. The majority of the beings in nature have no human uses. Even ecologists cannot think of uses for many large birds and mammals. This makes a coldness in the heart of our coexistence with other species. The vivisection of the world depletes our ability to feel compassion for it. We are destroying the voices of existence, according to Neil Evernden because they are considered unimportant.

Wilderness has been treated as a physical or recreational resource. Wilderness has been appreciated for human intellectual or emotional enrichment, but not for its own sake. Any attempt to save wilderness for its own sake is met with resistance and argument. This paper constitutes one argument for wilderness for itself. An argument based on ecological knowledge, normative knowledge can be used to justify wilderness areas apart from humanity.

4.5.4.2.3.2.2. Wilderness As Filled With Beings

Empty space on earth is an illusion; it is filled with life. Wilderness is rich in species. Estimates of the total number of insect species for the planet range from 300,000 to 3,000,000. In every hectare of forest or grass land, there are billions of worms, mites, and bacteria.

Wilderness is self-making; it predates humanity by several billion years. The word nature is derived from a Greek word, meaning "that which is born." The taoist term for nature (*tzu-jan*) means the spontaneous, that which is so of itself (automotive). It is continually reborn in individuals and systems. It is composed of systems, which renew themselves and regulate the process so that structural integrity is maintained.

Before humans named it, the whole earth was wild. It moved gracefully to constraints; life flowed into available space. Nature is beyond words. Words are tools that disturb and rearrange primordial nature. Fowles holds that tools addict us to purpose. Words (or even film) cannot capture the sounds, scents, temperatures, moods, ambiance of a woods, the levels ascent, or the subtlety of relationships. The wild defeats paper and frame.

The wild earth is older than the human worlds. Its components have been re-combined and recycled over eons. Human bodies contain the ashes of stars, elements of rock and gas. Animal cells share components with trees. Human brains share the patterns with reptiles, birds, and other mammals. Living beings are essentially participants in the world, from the quantum level through the ecological and cultural. All beings participate. There are no lone observers. A world without observers would be impossible. And all these beings interact in a large variety of ways from cooperation to predation and parasitism—not just the last two.

Nature is wrongly seen as a pyramid of death, a slaughter of all by all, even by contemporary philosophers and biologists. The processes of nature are called inhumane. They see nature, in Tennyson's words, "red in tooth and claw." But the metaphor is inadequate. The mycologist Heinrich deBary coined the word 'symbiosis" in 1876 (or 1879) to describe "living together." It defined a constant and intimate relationship between two dissimilar species. Such economy of effort is very common in nature. Living together is convenient for food and protection. Symbiosis has considerable ecological importance. Herbivores, such as cattle or sheep, depend on intestinal microflora for digesting grass—for survival. Two species that live together closely enough are biologically symbiotic. Symbiosis acts as a higher level store of genetic information; a lichen is fungus plus algae, where each is part of environment for the other. It may benefit the individual. There are elements of symbiosis in wolf-deer predation. In one sense, all interactions are positive and symbiotic on the planetary level. Symbiosis is a necessary and beneficial relationship.

Under a broad definition, human activities such as planting trees, growing corn, raising camels, qualify as symbiosis. Without suitable places for livestock, or with cruelty to animals, the symbiotic element disappears, and the human relationship is parasitic. With care, domestic animals benefit from association. The same can be true of pets, bats, owls, coyotes, rats, and cats. Humanity is part of the system. The earth is a loosely formed spherical organism, with all its working parts in symbiosis. Leopold described conservation as an attempt to harmonize civilization with the land towards a "universal symbiosis." Dubos sees humanity and Earth as "constituting a diversity of systems of symbiosis that constantly undergo adaptive changes and thus contribute to a continuous evolutionary process."

Cooperation is a natural rule. It creates communities, living entities in ecological balance. There is a wisdom of life, a "biosophy" in Stephan Lackner's words. In spite of the emphasis on competition and survival from predation, Lackner notes that there are peaceful habitats; the predatorless islands that Darwin knew, African lakes filled with flamingos. Lackner notes that the largest mammals—elephants, hippopotami—are vegetarian; the smallest insects—gnats, mayflies—are too small to be preyed upon. George Schaller estimates that less than ten percent of the grazing animals on the Serengeti die of violence. Karl Popper conjectures that besides the theory of the hostile environment (passive Darwinism), there is a complementary theory of the friendly environment. Many organisms are active explorers, searching for new, friendly environments. Birch and Cobb consider that the richness of the world and the freshness of response are matters of novelty.

Waddington said that the "general anagenesis of evolution is towards what may be crudely called richness of experience." The goal of all creatures is to come into the fullness of being, as a spontaneous flowering. For Whitehead, all living things have three urges: "to live, to live well, and to live better." Living better is being more attuned, stimulated, integrated, receptive, and spontaneous. The goal of each form is fulfillment.

4.5.4.2.3.2.3. The Nonhuman Use of Wilderness

Wilderness is ecologically important, for it accounts for ninety percent of the energy trapped by photosynthesis from the sun; it is crucial in the global energy system. Trees and plants are good sources of energy. Energy and materials can be produced through photosynthesis. In fact, wilderness is the greatest producer of renewable sources of energy and materials. As habitat for incredible number of species, it is insurance against dangers of simplified agricultural systems; it is a depository for genetic types. Biological species are essential to the maintenance of ecosystems. Wilderness is a natural resource that will be increasingly valuable on an overpopulated earth. Wilderness is the source of evolutionary process.

Adolf Portmann shows that every form of life appears as a gestalt with a specific development in space-time. All living forms develop an image of their environments. Genetics provides the proper image choices for some—frogs, for instance, focus most closely on objects that have the same size and trajectory as flies. Others must learn what is valuable using their senses; wolves, for instance, have to learn how to hunt as a group. Animals have their own universes that are strange and fascinating.

Jakob von Uexkull suggests representing the unfamiliar world of animals with bubbles to denote the self-world or phenomenal world of an animal. According to von Uexkull, perceptual and effector worlds form a closed unit, the umwelt. "Figuratively speaking each animal grasps its object with two arms of a forceps: receptor and effector. With the first it invests the object with perceptual meaning, with the second operational meaning." The world—life-image—is what has meaning for an organism. It is a focus. The first principle of the life-image theory is that all animals from the simplest to complex are "fitted to their unique worlds with equal completeness." A simple world corresponds to simple animal; a well-articulated world to a complex animal. Von Uexkull implies that the human world is only one of the many possible. Animals are not suboptimal beings relegated by evolution to second-rate habitats. They are optimally fitted. In a strange and complex earth, other species fit in places that humans cannot.

4.5.4.2.3.2.4. Wilderness has Ultrahuman Value

Humans have stripped the world of qualities and significance and claimed them for themselves. By valuing humans alone, value is made subjective and ends are without value. The human perspective realizes only a small part of the spectrum of rich possibility of experience. Humans use an incomplete source of value for ultrahuman beings; the source is human need. Moral science is not so much concerned with good/bad or right/wrong as with fulfilling needs: survival, reproduction, and prestige. Abraham Maslow established a hierarchy of human needs, beginning with food and continuing through social acceptance to self-actualization. But human needs are based on the health of the earth. Human needs extend to include a foundation of wilderness. Nature, which is self-supporting and self-managing, is a life-support system. Human systems depend on natural ones, for recycling of wastes, water, and air. But human growth is logarithmic. So is human need, and therefore, human destruction of beings and habitats. Human need shapes facts, like resources. But, as Goethe recognized, all fact is theory, a blend of perception, imagination, and needs. Realizing this, it can be argued that the need for wilderness is fact.

Holmes Rolston advances the idea that nature is the source of values, including human. The roots of culture are in wilderness, also. Rich sensory and emotional experience can be derived from contact with the wilds. These values are not often mentioned by economists and planners. But values usually encode information having survival or prestige importance. Perhaps the most valuable thing is living time. This may be why humans value walking in the woods or observing the production of art. The value of wild nature is its independence and wildness. Wilderness is the given primary reality, the matrix for life-images. If we can admit the independence of nature, that things continue in their own complex way, we may feel more reverence for them. Wilderness evolved without human help or interference, in equilibrium with the physico-chemical environment. We create human landscapes out of wilderness, from wilderness. The value of wilderness is that it cannot be reproduced. It can be apprehended only in nameless present by living senses. It can be known and entered in person, not behind glass.

Scott Lehmann stresses that a conviction that 'wilderness is valuable' is not very useful by itself. But, that is exactly the point. Human value has nothing to do with use. Human use is too narrow a perspective. So is science too narrow. Every being has an intrinsic value, before any utilitarian value to humanity. Maimonides said: "It should not be believed that all beings exist for the sake of the existence of man. On the contrary, all the other beings too, have been intended for their own sakes and not for the sake of something else." And, Goethe, in his letter to Heinrich Meyer (1792) stated: "The purpose of life is life itself ..." Associations of plants and animals are just as unique as their components. The value of wild nature is its independence and wildness. Perhaps our astonishment constitutes some of its value. If we can admit the independence of nature, that things continue in their own complex way, we may feel more respect. We can contemplate with admiration, sense as well as manipulate. The emergence of new moral attitudes depends on a more realistic philosophy of nature.

Value refers to the in-itselfness and for-itselfness of the process of realization. Besides having value for itself, it shares value with the rest of the universe. Everything has signification in the universe, says Whitehead. "Remembering the poetic rendering of our concrete experience, we see at once that the element of value, of being valuable, of having value, of being an end in itself, of being something which is for its own sake, must not be omitted in any account..." The basis of all value is being (the verb form). "Everything has some value for itself, for others, and for the whole. This characterizes the meaning of actuality. By reason of this character … the conception of morals arises."

There is a hierarchy of value from simple forms to complex, John Cobb suggests, corresponding to richness of experience. Although there is no hierarchy of being, there may be one of richness or value, depending on the frame of reference. There is value for each, and for the whole. And the whole may invert the value. Ecology can expand the narrow anthropocentric evaluation and see things from the perspective of the whole. Lovejoy's "Great Chain of Being" traces the deductive order in classical nature from Greeks to German idealism. But Lamarck inverted the chain in his theory of transformism; mind is immanent and can determine trans-formations. Although the hypothesis of inherited characteristics was rejected by Darwin, who shared that hypothesis but denied mind as an explanatory principle, both Lamarck and Darwin inverted the value of life. By inverting the great chain of being, Lamarck escaped the directive that the perfect must precede the imperfect. The result of Elton's food chain was the realization that the bottom link—plants—is the most important.

Ecological value starts at the broad base of the pyramid. Even a set of genetic codes is normative; genes show what is valuable in a life-image. So value is morphic on all levels of being. A sort of natural relativity of frames of reference encompasses the smaller human system. Perhaps we should not argue that things have value in the human system. Let us just respect the ultrahuman system. Bees have bee value; wolves have wolf value. Wolves are not efficient at binding nitrogen; neither are humans. Lichen are poor predators, but they break apart rock better than bighorns. From a functional point of view, all beings are equal. In an ecocentric perspective, all beings have intrinsic value and are equally important.

4.5.4.2.3.2.5. Wilderness Definitions

Wilderness has been undefined; it has been defined as emptiness; and it has been defined recently as a place "where man is a visitor who does not remain." The popular wilderness definition, where man is only a visitor, is utilitarian; humans visit for recreation and relaxation. There are flaws in this utilitarian approach to wilderness. Wilderness is considered a resource it is related to human experience exclusively. Wilderness does not have to be destroyed or used or even visited to be beneficial. It is valuable as wilderness, because it is there. Humans are using already a very large percentage.

The definitions have become more complex as more about the functions and values of wilderness becomes known. There are differences between ecological definitions of wilderness (undisturbed climax which change with time) and cultural definitions, which change with time. Wilderness is more than a geographical concept; its boundaries can never be just geographical. It is an abstraction. It is hard to define wilderness exactly or objectively. Any region that can be described is also a state of mind. And because it's a state of mind, it's ambiguous. Consider the biblical attitude toward wilderness, first as a howling wasteland, then as a sublimity greater than the world of "man," that enabled contemplative prophets to see the Divine more clearly. What originally held the disorder of humans, now holds the order of nature. Wilderness is a changing process. And, the English language becomes imprecise and mystical dealing with process thought. Benjamin Whorf pointed out the problem with verbs. A forest is not just there ("is"); it is being or becoming. Many attempts to define wilderness limit it to human experience. Obviously, that cannot be done. Wilderness is empty of most human experience. Therefore, it is undefined.

Reserves of every kind of region need be saved. Shepard concludes that the consequences of ecological conflict in technological, humanistic civilization are either exile or sanctuary; and sanctuary is the only solution available to the humanitarian ideology. The idea of sanctuary recognizes the multiplicity of factors necessary for viable populations. But, although there are sanctuaries for frogs and ferrets, it would be impossible to establish one for every creature. Shepard discounts the humanitarian objective as considering only "worthy" species. Yet, if all species were considered, the whole planet would end up as a sanctuary. He considers sanctuary an unfeasible 19th century political solution, based on a time when space was unlimited. Shepard states that exile (extirpation or extinction) and sanctuary are allopatric choices (life not occurring together). Allopatry (from the base word for fatherland) is seen as being consistent with the tradition of personal property, domination of nature, and model of the nation state.

What Shepard has overlooked is that humanity was never sympatric. It never lived together (the meaning of sympatry) with major carnivores. So allopatry is consistent with the nonexploitation and nondomination of nature as well. Sanctuary as personal property or as ultrahuman property is a moot point if habitats are saved from destruction. Domestication and enslavement are sympatric forms. A conservation program, despite Shepard's argument, cannot be based solely on sympatry, although at the planetary level, all beings are sympatric, parts of planetary cycles (or of a planetary being, Gaia). Not all species occur together in the same place. Sympatry describes only those that do. At the habitat level, allopatry is "intelligent" use of available resources by animal communities. Large herbivores, such as elephants and rhinoceros, may choose poorer quality food and avoid competition with smaller animals and exploit an untapped food source. Many interactions between different species contribute to the mutual benefit of the members of the community, as well as to the community itself. Humanity must be allopatric with most wild species and allow them to develop independently in their own places.

Gary Snyder questions whether complete compartmentalization is healthy. He suggests that some land is saved like a virgin priestess, while other is overworked like a wife, and some is brutally reshaped like a girl declared promiscuous. This argument is certainly good for human lands, but wilderness lands are being compartmentalized for the use of other species. Wilderness can be delineated by human use and nonhuman use, as well as by global function. Eight distinct kinds of wilderness can be identified.

1. Sacred landscapes set aside out of respect for habitats where humans do not belong and for species that cannot coexist with humanity.
2. Foundation areas, that contain active communities where the integrity of life is not challenged by human interference
3. Reservation lands, for the support of nonindustrial native cultures expressing traditional ways
4. Preservation areas, where the system is maintained by human activity, such as burning;
5. Restoration areas, which are set in a pattern by human activity, but may not need further intervention;
6. Neopoetic communities, which have been created through exotic species over time but naturalized;
7. Conservation parks, for multiple use of resources without interfering with the operation of the ecosystems;
8. Wild agriculture, managed wild systems, antelope and fish

By just wilderness, the first three areas are included indiscriminately. They all have similar values, as mirrors of existence, examples of natural, complex processes, expressions of love for nature, and wild, ultrahuman places. The most important definition treats it as a vital organ for the life of the earth, the generator of hydrological, geochemical, and atmospheric cycles. It is where ambihuman species live for their own purposes and benefits. None of these "definitions" is sufficient to identify wilderness exactly, just frame it with words.

4.5.4.2.3.2.5.1. Sacred landscapes
No direct human impact would be permitted; mapping would be possible from satellite or airplane observation. Examples of landscapes are fisher habitat and large areas of tundra.

The image of humanity is the world. The world may be added to haphazardly or through accumulation in a well-defined way. It may be transformed to incorporate new images. As the world becomes more complex, its image becomes more complex. In many archaic cultures, portions of the landscape are regarded as sacred. This may be a mountain at the center of the world, a burial ground, or the portion of a river that belongs to the fish gods. The agricultural societies considered the village, gardens, and fields sacred and surrounded by profane wilderness.

No longer is space just sacred or profane. The ideal landscape became the middle region, the garden between the complete order of the city and complete chaos of the wilderness. The garden is cultivated from the wilderness as a middle landscape. (The city expands at the expense of gardens.) The garden is a human order, but not usually sacred. Not only poets, but politicians and economists created the image of the ideal.

Sacred spaces are highly valued locations, with limited access and usage. Such spaces have great symbolic value to societies they are associated with significant events (or divine manifestations). The assignation of sacredness may result from functional necessity: To prevent flooding, limit population, or balance the economy, as the Catholic church did with fish in Friday. Wilderness is wholly other, ultrahuman. And so, it is sacred, in a literal way. It is distant. And that distance allows appreciation of a different kind. The sacred is contact with the wild self, contact with the larger whole, and the source of spiritual inspiration.

4.5.4.2.3.2.5.2. Foundation Landscapes
Wildernesses have basic value as pools of existence that reflect styles of existence from bacteria to human beings. Their most important value is as a vital organ for the life of the earth, the generator of hydrological, geochemical, and atmospheric cycles. It is also where ultrahuman species live; it is their sanctuary from humanity. Wilderness is ultimately the basis of civilization and freedom. In George Orwell's dystopia, 1984, the rulers of the police state abolish wilderness because it supports freedom of thought and action.

Foundation landscapes protect natural processes. There is no scientific answer as to how much of the earth's surface should be reserved. An areas of the planet that have effectively remained unused are placed in this category (approximately 30 percent). Representative areas of all types, including many polar regions, deserts, and tropical forests may add up to 50 percent of the total land area. Large undeveloped and uninhabited areas, such as Antarctica, are designated foundations. A very large percentage of oceans, lakes, estuaries, and wetlands are included in this category. The category is further subdivided by the impact of human presence.

Foundation landscapes enclose the most sensitive and fragile areas. This would include the majority of wild areas. No human presence would be permitted in the most sensitive parts; data would be obtained from satellite mapping. In less sensitive areas, visitation would be permitted for scientific studies or other acquisition of knowledge; most machinery would be excluded. Foundation Areas house the depth of wilderness. The depth of wilderness is what allows recolonization after disruption or human influence. The reserve function of the wild forest is its most important.

The earth tolerates a diversity of human worlds and animal umwelts. Yet, human imagination is limited, as is human knowledge. Many organisms exist of which we know nothing. Their worlds have little meaning to us. The earth has innumerable modes of being that are not human modes. Our direct intuitions of nature tell us that the earth is infinitely strange it is alien, even when gentle and beautiful. It seems always mysteriously impersonal, unconscious, immoral, hostile, sinister. J.B.S. Haldane recognized the strangeness of nature. "Now my own suspicion is that the universe is not only queerer than we suppose, but queerer than we can suppose." Perhaps the queerness results from sheer complexity. George Perkins Marsh believed that the equation of animal and vegetable life was "too complicated a problem for human intelligence to solve, and we can never know how wide a circle of disturbance we produce in the harmonies of nature..." Nature's complexity should be a source of inspiration. In W.M. Wheeler's words, nature is "an inexhaustible source of spiritual and esthetic delight." He concludes: "Our intellects will never be equal to exhausting biological reality. Why animals and plants are as they are we shall never know, of how they have come to be what they are our knowledge will always be extremely fragmentary, because we are dealing only with the recent phases of an immense and complicated history ... but that the organisms are as they are, that apart from the members of our own species they are our only companions in an infinite and unsympathetic waste of electrons, planets, nebulae, and suns, is a perennial joy and consolation."

Foundation areas could include common lands, that is, one of each kind of ecosystem regardless of aesthetic value. Foundation areas, left by themselves, would require a minimum of management costs. Lehmann and others have said that wilderness costs, in lost opportunities, resource use, and administrative overhead, but without considering free services of wilderness. Similar to slavery, this conclusion is based on the assumption that everything that can be used should. Margalef states that any "human intervention in nature, even presupposing good intentions, can rarely be reconciled with the idea of strict conservation." Wilderness is actively changing, but it maintains itself. Often, when it is suggested that wilderness areas be left without human management, critics conclude that humanity is not being considered as part of the environment. This is an unfortunate conclusion and untrue. Humanity does not have to interfere with every system to be a part of it. Interference competition is far more destructive to systems than exploitative competition.

Foundation areas, even without human presence would benefit humanity, as the source of evolutionary changes, as source of natural wealth. Space unused for human purposes is dedicated to preserving the diversity of biomes. It is especially important to preserve ecosystems where humans are absent or not dominant. The earth must be treated as one whole, composed of levels of interacting parts. The life-cycle of many animals in Africa is associated with gigantic circular migrations to take advantage of seasonal grazing and water supplies. Ocean and air currents ignore national boundaries, as do birds and viruses.

Some visitation is allowed for development of the larger human self, but without permanent residence or machinery. Human impact would be limited to noninterventive scientific pursuits or collections. Examples of foundation areas would be large areas such as the Amazon.

4.5.4.2.3.2.5.3. Reservation Areas
Reservation areas are natural systems in which humans play a limited role, limited by net ecosystem productivity and human images of the place. Humans would be in a symbiotic relationship with wild ecosystems, for example, the Miskito Indians in Central America.

4.5.4.2.3.2.5.4. Preservation Landscapes
These lands are those that have cultural values and are maintained by limited human intervention in that state, such as ruins or many kinds of grasslands. Many of these areas would permit only limited access for cultural or recreational purposes. Preservation Areas are also home for archaic cultures.

Large areas under continuous habitation by archaic cultures (primarily hunting/gathering or swidden agriculture) would fall in this category. Herding and nomadism by archaic cultures would be normal, but only with traditional tools. Other landscapes maintained by human or domestic animal presence would be included, for example, parts of the Mediterranean, some of the English landscape, Japanese shrines, or various central continental grasslands.

Preservation areas Boundary areas, ecotones or lands bordering agricultures, include small patches influenced by human activities, corridors. Human impact could be traditional, experimental science. An example would be central long-grass prairie.

4.5.4.2.3.2.5.5. Restoration Areas
Restoration areas are experimentally restored ecosystems that had been disturbed by human activities or natural events, for example, Lake Shagawa. Another example is the Tall Grass Prairie in central North America.

4.5.4.2.3.2.5.6. Neopoetic Areas
These areas have been started by human activities, but saved as self-regenerating systems and not built up or interfered with. An example is California grasslands, which have been changed with Mediterranean plants, e.g., wild oats, wild mustard, wild radishes, wild fennel.

4.5.4.2.3.2.5.7. Conservation Landscapes
These areas assume the regular exploitation of ecosystems by humanity, but well below natural (nonhuman-subsidized) regeneration rates. Natural wildlife would be controlled and exploited by humans, including the commercial exploitation of uncultivated forest. Commercially exploited areas, natural forests, shorelines, wetlands would be included in this category; exploitation of wild plants and animals would be part of normal activity. Transport machinery (rail, air or road) would be permitted as impacts would be limited to right-of-ways. The greater pressure on these conservation areas would require greater care. The kinds of communities permitted in these areas would include communities that depend on natural ecosystems and permanent camps. Recreation would be allowed, but with limited machinery. Human recreation permitted in temporary camps and with some light machinery. An example of a conservation landscape is the Boundary Waters Canoe area in the United States.

Thousands of local wheats in Turkey and Ethiopia have become extinct in the past decades. The goal of the conservation of genetic resources is to explore and collect vanishing genetic heritage, to evaluate, conserve and document the material for benefit of present and future generations. If wilderness is to be saved indefinitely, knowledge of types and areas is necessary. The national parks and forests should be provided with research scientists as insurance for the preservation and use of the parks. Only such research can answer questions as to whether the park is big enough and shaped correctly to constitute a proper habitat for its inhabitants. Animals do not need to be saved from natural death, a great regulator of life, but from suffering, experimentation and premature extinction.

All beings modify and exploit the earth to some extent. Humans do. For successful exploitation, power must be limited and its density controlled. As the capacity for the earth is approached, conflicts may increase. A human ecology must provide each living human being with a satisfactory environment that must be in equilibrium with the rest of the earth. It may have to accept that there are too many of our species, wanting too much, and that we must learn to share and change our wants and ways of achieving them. Charles Elton has remarked that true conservation means looking for some wise principle of coexistence between man and nature, even if it means a modified kind of man and a modified kind of nature.

4.5.4.2.3.2.5.8. Wild Resource Systems (or Wild Agriculture)

Human impact limited to taking or harvesting species for protein or fiber. Includes minimal cutting or trees from forests and collection of plants for medicinal and cosmetic purposes.

All wild systems have had human influences for thousands of years. Even now, human influence extends into every system through pesticide drift and artificial gases. Yet, over 30 percent of the earth now has no permanent human settlements. In general, none of these wilderness areas will be inhabited (at least by industrial cultures), none of them will have permanent built roads, and management would be limited to temporary intervention.

4.5.4.2.3.2.6. Wilderness & Wildness

The wilderness saved now is all that remains. It cannot be multiplied or increased—without great effort, cost and the investment of hundred or thousands of years, anyway. Saving habitats means saving large areas of land. It means placing wilderness, which is the support for cultivated and industrial areas as well as natural communities, off limits to development and perhaps to any use. Every community should be allotted a place for self-development, a foundation area. Wilderness is not land that is locked up, as several U.S. senators have claimed. It is already interactive. The foundation of a house, to use a metaphor for wilderness, is not available for firewood. We do not have to camp on it or put asphalt on it to live with it. We are already living with it. Foundation areas are not museum pieces, as some scientists seems to think. They are working ecosystems, and only humanity is directly absent.

For the sake of argument, we have been talking about all these areas is if they were the same. They are not. We are suggesting different kinds of zoning, or compartments, to use another term. Gary Snyder questions whether complete compartmentalization is healthy. He suggests that some land is saved like a virgin priestess, while other is overworked like a wife, and some is brutally reshaped like a girl declared promiscuous. This argument is certainly good for human lands, but wilderness lands should be compartmentalized for the use of other species. Almost every wilderness argument has been extremely anthropocentric; conservationists are accused of saving wilderness for their rich, elite, backpacking friends, while use advocates, wise or otherwise, are accused of wanting to convert every kind of ecosystem to human use. But, we are not alone on the planet; the millions of species that evolved before we did have created ecosystems that support humanity. Certainly some wilderness should be set aside for the support of regional and global cycles, although this is a weak anthropocentric reason, but some wilderness should also be set aside for habitat for other species. Wilderness can be delineated by human use and nonhuman use, as well as by regional and global function.

Each of these categories would occupy different percentages of the planetary surface, depending on calculations of minima and maxima, and depending on cultural values and decisions. Estimates place fifty percent of the land area in the first division (eighty percent of ocean and water surfaces); sixteen percent in each of the other three (five for water); leaving two percent for completely artificial landscapes—industrial or city (three for water). These figures are consistent with several earlier proposals. Eugene Odum, suggests 30% forest cover world-wide, with 60% in tropical areas. Paul Shepard offers 75% of the land area left wild in a techno-cynegetic society. C.A. Doxiades suggests 67% of the surface area of the planet left in wilderness.

The current reserves are inadequate; wild species decline on limited reserves. Designing reserves requires a biogeometry, knowledge of the shape and extent of ranges. Left to themselves, in an adequate preserve, a few species may be dropped or generated. Progress, however, deprives many species of existence, simplifying ecosystems of which we are still greatly ignorant. Industry destroys billions of creatures more than unscrupulous hunters or multiplying homesteaders. The rate of extinction of species exceeds our rate of learning about them. Humanity as a species must develop an awareness that it is infinitely enriched by the organic world, and is poorer each time a species is eliminated or natural community degraded. Shepard debunks the model of exploitation. "The historical model we have for being lieutenant of the planet is the shepherd and husbandman, who always destroy the wildness of the animals they seek to protect." The professional form of this is the "wildlife management," where the manipulations are more extensive and include fencing, marking, trapping, breeding, and game farming.

Bertrand de Jouvenel (1968) proposed a Stewardship of the Earth that emphasized harmony with nature, a balanced ecosystem, and a peaceful environment. Rene Dubos claims that reverence for nature is compatible with the responsibility for a creative stewardship of the earth. To this end, he recommends a Benedictine philosophy, which would exercise more control over nature than a Franciscan approach, which would allow more wildness and empathy. Walter Taylor pointed out that the U.S. Secretary of Agriculture Henry Wallace had suggested the need for a Declaration of Interdependence similar to the Declaration of Independence in 1776. Communities and species do have the right to survival. We commonly speak of corporate entities, churches and institutions as having rights and duties; this is not that different from species, which could have an equivalent constitution and rules.

Science assumes that humanity has the knowledge to support a stewardship of the earth. Perhaps we do not have this knowledge or the proper attitude. Humans can never be stewards as long as they dread every intrusion of wild nature into their gardens, neighborhoods and fields, as long as they try to control everything. Stewardship is based on knowledge that we only think we have. Nature is already self-governing; we do not need to be stewards. Nature will heal herself, either through self-healing of the human species or its self-destruction, which would allow alternate forms of life to flower. There are several possible human relationships to the earth or to wild beings: among them, pandominant species, lord-and-master, good steward, or fully conscious and self-limiting beings. Historically, we have preferred lord-and-master.

There is an arrogance, from Plato to J. S. Mill to Rene Dubos, that all human action improves the spontaneous course of nature. Dubos, for example, is disturbed when farms revert to natural vegetation. He dreads forest regrowth in New York as barren and uninteresting, where Leopold felt that it was a welcome prophecy of nature's second-coming. Dubos and others have claimed that "nature knows best is wrong." But ecosystems that seem inefficient and wasteful are many times redundant, and therefore stable and flexible. After all the goal of life is experience, not efficiency; and redundancy promotes experience as well as stability. Natural processes that seem destructive are cyclic and preservative. The humanized landscapes that Dubos praises are the result of a long-term adaptation with limited means. The interplay was slow, so that many species, but not all, could adapt. Beings are shaped by the environment they change.

We have been persuaded than humans are able to transform any habitat on earth and benefit from the transformation. Africans practiced this reclamation, as did Assyrians, Greeks, and American Indians. Occasionally it worked, and long lasting systems were established. But often, the reclamation only contributed to deserts. The United States, Nepal, Brazil, and other countries lose enough soil annually to form a new country.

Human motives appear rational and prudent, but really are often prompted by ignorance, greed, and destruction. The means and scale of the destruction has increased dramatically. Managing other species really means profiting from other species. It means manipulating wild or domestic species to exploit them. The increase of certain species around the world displaces others that may be more adapted. Civilization is both adventurous and stupid. It destroys habitats that are the tissue of its existence.

Humans operations are inefficient for the short-term. They have wasted land and moved on. Wild species have been destroyed and replaced with various cultural domesticated favorites. Then, those domesticates wasted soil and disrupted animal, plant, mineral, and soil cycles. All this happens from ignorance.

The correction for ignorance of ecology is knowledge. The correction for nonacceptance of death is a generalized concept of reverence for life. Death is a part of life, but not a driving force. The reverence for life has to be based on an ecological ethic that understands the necessity of predation as well as altruism. Killing as well as saving. Reverence for life must include awareness of natural laws. Wolves need deer and mice to survive, as much as those need wolves to be healthy. Human intervention into natural communities must be responsible, not sentimental.

Knowledge by itself, however, merely permits a more efficient utility. Knowledge cannot be the sole basis of decision making. It is always incomplete and therefore cannot describe all aspects of the earth that bear on human life or environmental quality. Knowledge must be humane. Abraham Maslow saw the organism as having biological wisdom; it can be trusted as autonomous, self-governing and self-choosing. To treat organisms, and nature in general, we must shift to a taoistic approach, asking rather than telling, observing rather than manipulating; being receptive and passive, not active and forceful; and, being nonintruding, and noncontrolling. It stresses noninterfering observation rather than controlling manipulation. This is part of the paradox of duality; the approach is detached yet concerned; free yet committed; and independent yet responsible. The way to persistence lies in letting be, a form of biological taoism.

The reverence for beings as they are is the law of noninterference, which means "sparing" (Heidegger), "letting alone" (Wilson), and "not killing for pleasure" (Fox). Noninterference is not indifference, which is diffuse. It is caring. Noninterference does not lead to chaos, poverty, and stagnation. The technocratic vision strives for "life under control," but the earth is self-managing, productive, efficient, and orderly. Life does not need to be always under control. Sparing does not value other beings; it means allowing them to be free.

4.5.4.2.3.2.7. Fulfillment of Wilderness
From Rabindranath Tagore, Rene Dubos takes the heroic love-adventure of humankind to be the wooing of the Earth. The wooing of the earth is both sweet and sour; sweet because humans can create enchantment within nature, and sour because they can spoil desirable places—possibly to the point of ruining nature's recovery mechanisms. The wooing of earth implies more than converting wilderness into humanized environments. It also means preserving natural environments in which to experience the wild mysteries transcending daily life and from which to recapture an awareness of the natural forces that have shaped humanity.

Dubos celebrates the human ability to woo the earth by creating new environments that are "ecologically sound, aesthetically satisfying, economically rewarding, and favorable to the continued growth of civilization." While that is noble and just, and realistic, there are two erroneous assumptions contained therein. First, civilization does not have to grow to develop, and it is development that is important, not growth, which is plainly unfeasible. Second, "to woo" means to solicit love or to make love. Dubos assumes we will remake nature in our image, and love, as everyone knows, entails respect or reverence. It does not remake the loved one in one's image, it respects and allows freedom. That is the larger meaning. We should "improve" on nature where we live, but we do not need to improve on all of nature. We do not need to domesticate all beings for human use. Joseph Wood Krutch recognized that ecology without reverence or love is only shrewder exploitation.

4.5.4.2.3.2.7.1. Wildness & Flow
We can create places only by living there; by slowly adjusting all to all. Human modifications of earth can be lastingly successful only if their effects are adapted to the invariants of human and physical nature. Ecological management can be effective only if it takes into consideration the visceral and spiritual values that link us to the earth. Although some of nature can be regarded as a garden to be cultivated, large areas should be preserved as untouched reserves or necessary systems for global balance. Other areas could be studied or urbanized as population centers. Our tendency to redesign nature, rather than tolerate and cherish, is dangerous; it leads to "curing" abnormal people and "solving" weed problems. We do not respect the wild, complex side of human nature.

It is not possible to address the dimensions of wilderness without considering the beings that make it up, without considering it as a unique pattern of lives, without considering animal behavior and human ethics, without considering human evolution and the history of the earth. A true definition of wilderness is not found in human needs and desires. Wilderness is not for people. It is for itself. It makes and defines itself. For humans, it is a land classification. But a unique one, for it is one which is ultrahuman and unmanaged, unknown and unused. Wilderness is not incoherent until human beings give it form. It has form. Wilderness is made by the life-images of the species and beings that inhabit and inform it. Wilderness is not empty. It is full with living beings whose goal is fulfillment.

The primary justification for wilderness as a category is as a sanctuary for other beings that constitute the richness of the earth. Many of these cannot live with humanity; they are allopatric and anthropofugic. If, as many believe, wilderness is necessary for human development, its protection can be justified on that basis also. Its value to humans consists of its diversity, aesthetic nature, and its economic benefits in recycling wastes and trapping energy, for recreation, inspiration and teaching.

Many voices defend wilderness itself. But many more urge its use. We humans are ambivalent; we want to save areas created by the destruction of wilderness. We want areas that bear no record of us, yet we want to leave our mark everywhere. We want ultrahuman places, but we want to live everywhere. We want to use everything, but save everything. There are many alternatives, from indifference to conservation, that is, minimizing the impact of human activity by reducing its scale, to a radical one, creating new social organizations, using high-technology to solve problems and incentives to save natural places.

What is necessary for humans is that they have a human way of being, human worlds. But, these worlds are derived from the wild universe. We need a wild universe to live fully. When we understand our roles and relationships in nature, then we will not be managers or stewards, but participants and sharers in experience. An ecological basis gives human activity dignity. But that in no way negates the life-images of ultrahuman beings. All beings are centers of experience. We are aware of the detrimental effects of human greed on other beings. If we continue that behavior, with awareness, then there can be no hope of any moral human community.

We must emphasizes the value of all ultrahuman beings and try to use as few individuals as possible from as many species as possible—moderate yields, within the limits of the food chain. Many species should not be used at all, whales and wolves for instance. Plants and animals are as much of our heritage as art, history and tools. Ignorance of one is just as sad as ignorance of another. The survival of society depends on an expanded ecological consciousness of the global system in its complexity and connectedness. The spirit of humanity depends on an ecological consciousness that places humanity in a proper relation to the wild places of the earth, taking what it needs, but letting the rest be.

4.5.4.2.3.7.2. Wildness & High Civilization

We must have an appropriate technology to feed and clothe humanity, to support a high human culture. Wilderness will not be saved around the earth without a new economic order that balances the human population, available resources, and equal opportunity. Decisions on these matters, combined with common sense and appropriate technology will determine a course of action for wildlife and human populations, based on the limits of our knowledge and the limits of the productivity and stability of ecosystems.

The current high human populations can only be maintained through theft from other species, theft from future generations, and through the degradation of billions of humans and the ecosystems they depend on. A high civilization has to limit its consumption of natural resources and its place, to account for wilderness and future generations. Wilderness is a limiting factor in the health of human civilization. Although it is not a precise function, we may be close to a minimum for wilderness.

A high culture could limit its consumption of natural resources, its impact, and its population. The human impact on nature can be minimized by minimizing waste.

The surfeit of physical things can be reduced without minimizing the richness of human experience or lowering the quality of life. Some pleasure or comfort, some whim or waste, may be sacrificed. A high culture is not a return to the innocence of foragers, or to the sparseness of the cave. It includes necessary institutions for the arts and sciences limed in transactions with the environment.

We have intense methods of food production, requiring less land. The capability for local agricultures, for distribution. So, let us liberate the land not needed, as suggested by Birch and Cobb, and Rodman. We can give this land back to trees and wildlife to restore diversity and richness, and balance humanity and wilderness. Each species needs to have its own place. Developmental policies need to reflect that. The environmental crisis is a cultural problem, of social character, not a problem of nature.

Perhaps new technologies and new philosophies may permit larger human populations or the expansion to other planets or solar systems, but we ought to be cautious with what we have now—extremely cautious. Conservation and creation must be tempered with preservation. We are too ignorant to tamper with everything. If we cannot live sensibly on earth, the emptiness of space and wilderness of stars will not be home.

4.5.4.2.3.4. *Common Places & Conservation*

A large portion of the planet, probably 25 to 35 percent of the land area, can still be considered still wild, if by *wild* is meant an absence of human habitation and a minimum of human impact—no roads or power lines in particular. Most of these areas are considered unattractive or uninhabitable: too cold (arctic and sub-arctic regions), too dry (the Sahara, central Australia), too remote (southern Pacific islands), or all three (Antarctica); even so, many of these areas have some human habitation.

The most desirable lands were converted to human use long ago. Many rich areas, such as tropical forests and temperate wetlands, whose forbidding characteristics discouraged development for a long time, are being converted rapidly. The cold, dry, remote, often fragile, ecosystems need formal protection before the quest for resources overwhelms them. The rich forests of the tropics need protection before our excessive desire for beef and wood pulp destroys the lungs of the earth. Conservationists, and even much of the public, recognize these needs, but, alas, there has been little support to protect the common areas, the widely known places that we have used and abandoned and reused for thousands of years. Relatively unspoiled common places, like grasslands, are being altered rapidly by industrial agriculture and human population pressures. Grasslands are among the least protected biogeographical provinces on earth. They are being replaced by surplus crops subsidized by energy-intensive industrial agricultural methods.

Wilderness designations in the United States have been inspired by recreational interests, protecting mountain areas good for hiking, camping, and hunting. Most Wilderness Areas are higher elevation areas in the western states. Even in the U.S. western states of Idaho and Washington, other regions—such as basin and range, Snake river lowlands, Owyhee broken lands, and Palouse grasslands—are scarcely represented. Wetlands and forests in the eastern part of the country are poorly represented. Wilderness designation has recently begun to consider potential economic interests and ecological self-preservation—the recognition that the Earth provides free life-support services, as through its biogeochemical cycles. Now, we should plan a global strategy, not only to save wilderness, but to preserve diverse working ecosystems, including semi-natural and artificial landscapes that are not considered wild or special, but are common lands. These lands should be kept at an optimal level, although some percentage should be restored to an ultrahuman condition. The Palouse grassland, extending from eastern Washington to northern Idaho and northeastern Oregon, is an example of a common area needing restoration and protection.

These common areas assume the regular exploitation of ecosystems by humanity, but well below natural (nonhuman-subsidized) regeneration rates. Natural wildlife would be controlled and exploited by humans, including the commercial exploitation of uncultivated forest. Commercially exploited areas, natural forests, shorelines, wetlands would be included in this category; exploitation of wild plants and animals would be part of normal activity. Transport machinery (rail, air or road) would be permitted as impacts would be limited to right-of-ways. The greater pressure on these conservation areas would require greater care. Communities permitted in these areas would include communities that depend on natural ecosystems and permanent camps. Recreation would be allowed, but with limited machinery.

Most conservation strategies are completely anthropocentric, from saving hunting grounds in the middle ages or resources this year. Ecosystem preservation protects entire biotic communities: genes, populations, species, habitats, associated traditional human cultures, and all the processes and interactions. To keep the essential services of nature, from atmospheric cleaning to soil-formation, we need large preserves. Preserves are critical elements in global element cycles; they provide a natural base line for management reference and a unique opportunity for scientific research. Large preserves would increase representation of species and save viable mammalian populations. Such preserves would permit natural processes to occur without human interference. We also need large preserves to derive further benefits from understanding natural processes and direct economic benefits from species. For aesthetic purposes: To see, to participate in nature, and these being the basis of the most popular activities of watching and tourism.

4.5.4.2.3.4.1. Creation & Restoration

There is little information on the restoration or preservation of "near-natural" ecosystems, those that are primarily native, and not subject to major change. The Palouse grassland in the U.S. probably has no natural landscapes and very few near-natural; most are semi-natural (pastures as a consequence of human activity) or artificial (totally humanized). The Palouse could not be restored to its original state, since many species are extinct, but it could be rehabilitated. There are a number of methods (such as direct planting into de-sodded ground) that can be used to restore species-rich grasslands. The inner buffer zone, being rehabilitated, would be the most expensive to create. The outermost buffer would be managed by benign neglect (despite the fact that most ecologists do not list this as a management option). Natural processes, such as wind or species explosions, would be allowed to operate freely, even if they altered the functioning of the communities. In fact, a reserve, in contrast to a preserve, is simply an area placed off-limits to exploitative human activity, where natural processes would be allowed to proceed naturally.

The shapes for the reserves would minimize the dangers from physical and climactic changes. The greenhouse effect could drastically alter the species distributions in reserves, with the loss of many species. Placing the reserves on heterogeneous soil types and topographies increases the chances that a species' temperature and moisture requirements would be met. Simply maximizing the size and number of reserves would enhance long-term survival.

The cost of reserves would be high. Palouse land sells for $2000 per hectare. The cost for restoration of a reserve has been estimated at $20-140,000 per hectare, depending on the density of planting and the area. The costs of buying, rehabilitating, and managing 200,000 ha could cost $2-4 billion. By comparison, this is equivalent to the $3 billion that the U.S. Forest Service wants to spend to add 35,000 miles of roads in roadless areas. Larger areas would reduce management costs; smaller reserves in general require more intensive management and habitat manipulation.

4.5.4.2.3.4.2. Palouse Common Places Example

The Palouse grassland is a geographic region of approximately 6 million hectares (roughly 15 million acres) centered in southeastern Washington state in the United States. Its origin, topography, and soil composition are unique. Wind-deposited loesses form steep, rolling, dune-like hills that overlie the Columbia River basalts. The primeval vegetation was composed of dense stands of perennial bunch-grasses.

Dry-land farming has almost completely replaced the original vegetation, although fragments can be found in fence corners, right-of-ways, cemeteries, and inaccessible slopes. In spite of its uniqueness, there has been no successful attempt to save more than patches of the original vegetation. No large stands of native grasses remain (Wittbecker, 1995).

4.5.4.2.3.4.2.1. Discussion of the Palouse

Palouse vegetation is determined primarily by climate, but sometimes by soil and topography. Unlike most other grasslands, the tall-grass prairies to the east, for example, it developed without significant grazing or regular fires. Fire plays a minor role in these habitats, since the perennial species sprout from underground parts—fire kills sage, a native dominant in some communities. In its natural state, grazing did not significantly affect the composition of the grassland, although white-tailed deer were present. Archeological history indicates small populations of Bison and Pronghorn Antelope as recently as 2000 years ago, but no permanent populations have established themselves since, possibly due to increased snow depths west of the Rocky Mountains.

Human agriculture was the first intensive use of the region. As agricultural technology became more advanced and the demand for crops increased, less desirable segments of the prairie were tilled. Smaller islands of native vegetation, regarded as waste places, were left for livestock grazing, since the native vegetation was palatable and nutritious. However, native vegetation was easily injured by close cropping and unable to compete with introduced exotics, such as cheatgrass or bluegrass, on disturbed sites. Even the few remaining natural stands, on the steepest slopes and boundaries, have been influenced by fertilizer and herbicide drift.

The native grassland is far richer in species than farmland. Ninety-three mammalian species representing 58 genera have been recorded in the present Palouse. Ecological studies of many habitat types are rare and localized to individual areas. Some areas are still unstudied and there has been no treatment of the habitat type as a whole. Many of the plant associations in the Columbia Basin Province are not represented in any of the small research areas in Washington or Idaho, although some of them would be included in proposed areas. Poor representation at present is one reason why large reserves are needed. Existing natural areas are not nearly large enough to save viable mammalian populations.

Large reserves are important for many other reasons, too. They help perpetuate global element cycles; they help maintain the integrity of wild gene pools. But, reserves can be justified by anthropocentric arguments, also; in addition to recreation, resource-banking, and research opportunities, reserves have importance for watershed management and as gene pools. Knowledge of biological history and cultural change over time contributes to the understanding of a place and keeps ecology in touch with geology as well as with human economics and politics. The most important biocentric argument for wilderness is autological—ultrahuman species need ultrahuman places. Wild reserves would permit natural processes to occur without human interference.

4.5.4.2.3.4.2.2. Reserve Proposal

The desired size of the preserve is a complex function of the area's key species, quantity of suitable habitat, and minimum viable numbers of species. Large-bodied vertebrate species tend to have lower population densities, thus a reserve with self-sustaining large-bodied vertebrate populations will likely be adequate for herbivores, insectivores, and primary producers. The key mammal species in the Palouse are coyote, badger, and mice, with white-tailed deer as regular visitors.

Estimating the minimum number of individuals in a population to guarantee a high probability of survival results in widely varying minimum areas, depending on the key species selected. An effective population of at least 500 individuals may be needed to maintain genetic viability of each animal species. Usually, large carnivores are a sensitive indicator of carrying capacity. In the Palouse, however, coyotes can adapt to humanized landscapes, so preserving their entire range is not critical for the survival of that species, though generalist species should be preserved in at least one large range. The minimum area should be large enough for the protection of all endangered or threatened invertebrates, which play key roles in such basic ecological processes as predation, recycling, and pollination.

Although there have been debates over whether a single large reserve is better than several small ones, the shape and size of a Palouse Reserve should be determined by habitat studies of the unique natural history and conditions. The key plant species to be protected are Idaho fescue and Snowberry, with all their ecological relationships to micro-organisms and arthropods. The large carnivores and herbivores (coyotes and deer) can adapt to more artificial conditions, so their needs may not be limiting factors. The size should be large enough so that species will not be vulnerable to "extinction vortices," caused by genetic or environmental stochasticity. In a Palouse reserve, disturbance from farming, grazing, or recreation would probably be the greatest threats.

The recommendation for a Palouse reserve is 1 large area about 200,000 hectares, doubly buffered by rehabilitated fields and then by a greater amount of fallow agricultural land; 3 areas of 3000-10,000 hectares; and 22 satellite areas of 8-25 hectares, which would probably not be buffered. Saving a million hectares would be economically or politically difficult under the current industrial monolith; 1 million hectares is about 17% of a region that profits immensely from growing grains and legumes. In a truly eutopian vision, that is, one concerned with maintaining good places and characterized by rational planning and realigned priorities, 3 million hectares would be set aside, but such a vision can only be created with radical changes in political and economic regimes.

Based on a suggestion that at least five units of each habitat be preserved, the reserve is divided into over 25 unequal areas. The largest portion of the proposed reserve is located on the eastern edge of the bluebunch wheatgrass province and includes five habitat types.

The proposed reserve would be laid out on the SW/NE axis with large fingers extending to the NW and SE, to maximize the number of protected NE slopes. Since the winds are predominantly from the SW, herbicide drift would be minimized. A range of elevations across areas would minimize the effects of climactic change—the possibility of extreme change is rarely considered in wilderness design. Soils, drainage, and land-use history and ownership would also receive similar considerations. This would allow management for diversity on three different scales: community, habitat, and region. The reserve would extend into the ecotones separating grassland and forest provinces; the natural edge effect would benefit many species. The medium sized areas and the satellite areas, with good planning, could be linked by wildlife corridors that could follow property lines or fence rows.

Fence rows could limit human access and might even encourage animal dispersal. For example, small mammals and birds use fence rows for travel and nesting; mice, deer, and coyotes also use fences for partitioning space. Fences also create microsites for different communities. Rex Daubenmire (1942) noted that slight differences in soil depth and moisture permitted different types of associations in the Palouse; this was especially true around fences, which caused greater dust and snow accumulations. Because local soil conditions result in a wide variety of habitat types in small areas, even small preserves might have a rich flora and fauna.

4.5.4.2.3.4.2.3. Management Concerns

Although very few people consider the Palouse a "priority habitat," it does need protection, and it should be a part of the Biosphere Reserve Program. A large area of undisturbed grassland would be a valuable ecological baseline, for comparison with the domesticated landscapes of farming and grazing. Variables such as productivity or the effects of climate could be compared between the natural and artificial systems. Native species would be preserved, both for our sake and their own.

The Palouse is common, in the sense of being taken for granted and treated badly. Being common does not mean being without value, however. The Palouse is attractive and useful because it is so productive and forgiving. The Palouse seems like a simple grassland. Having fewer organisms or simpler patterns does not mean being less unique. The Palouse has a special beauty, more subtle than a rainforest or a desert, that has become rare. If that specialness is not to disappear completely, it must be protected, now, in a reserve.

Saving common lands requires a wide vision that entails making decisions without complete knowledge, ecological or economic, within the human communities that must bear the cost. Because many of the mysteries even of common habitats have not been unraveled, saving large numbers of habitats from human intervention is a hedge against the ultimate price of ignorance—extinction of animals and plants and, eventually, humanity. The Palouse is a unique ecoregion, and there is very little of it left in a natural state. Let us save it and then save samples of common places everywhere.

4.5.4.2.3.5. Domesticated Landscapes & Transformation

These areas are those that have simplified for human needs and must be maintained with human labor. Forests, grasslands, seas, and wet areas would be manipulated for human benefit. Typical activities would include traditional agriculture (nonirrigated) and animal husbandry, with appropriate tools (may be mechanized), herding, and animal shelters. Managed forests, using appropriate methods of cultivation, would be present, as would modern agriculture (large scale), using modern methods of cultivation, and livestock, including some factory farming, automation, industrial methods, and a larger scale of energy, which implies the subversion of the natural landscape. Special and general recreation would occur, with a provision for large-scale human recreational needs, such as hiking, skiing, and boating, using limited mechanical aids, excluding built-up settlements. Rural communities would be evident, with minimal services. This becomes a larger and larger percentage as more areas are converted.

4.5.4.2.3.5.1. Herding & Agricultural Zoning
Throughout their history, humans have used animals and plants for food and clothing. Animals were followed, herded, corralled, tamed, and finally bred. Plants were domesticated later. As technologies developed, human relationships with animals and plants changed.

Many areas have been domesticated for thousands of years. Over time, the ecosystems have adjusted with different species, some domestic. Herding has been used especially in areas that could not support domestic plants. It is the best use of such lands in some senses.

4.5.4.2.3.5.1.1. Wild Herding Zoning
Surprisingly, although humans have had domestic animals for many thousands of years, and herded them, they have rarely attempted to herd and live off wild animals. Wild herding could attempt to keep animals close by, but not domesticate them or try to control their reproduction. This would not be the same as hunting, because people would not hunt individual animals. This would allow native animals, such as antelope, to use plants and animals that humans do not. Furthermore, the land cover would be allowed to develop on its own without human control or interference.

4.5.4.2.3.5.1.2. Natural Agriculture Zones
Natural agriculturalists manipulate habitat to favor the crops they want to grow, working within the laws of ecology to tilt the ecosystem in favor of more desirable plants he wants. The desirable crops virtually invade and grow like weeds. Natural farming methods might be applied to restore some areas where erosion has destroyed the system. Natural farming, however, by sowing seeds and putting up fences to keep out domestic animals for a while, could contribute to restoration, without massive machines and chemical dumps.

4.5.4.2.3.5.1.2.1. Regenerative Zones
Regenerative agriculture is organic, without chemical fertilizers, pesticides or herbicides. It regenerates the soil to a healthy state. Beneficial wildlife, especially birds, can return to prey on insect pests. It allows human health to recover as well. Wildlands, wetlands and the larger environment surrounding are regenerated by organic practices. The community is also regenerated by recycling and reducing pollution.

Regenerative agriculture has become a holistic science over the last twenty years, as regenerative farmers also work to restore local woodlands and streams. Farmers also interact directly with the local community on whole new levels, perhaps by local marketing, organic matter recycling or entertainment farming. They work to teach others this regenerative system of farming.

4.5.4.2.3.5.1.2.2. Permaculture Zones

Permaculture zones classify three dimensional areas according to the amount of human attention needed to maintain sustainably the function of each zone. Permaculture principles can be applied in terms of reducing energy and water needs, harnessing natural resources, and creating a harmonious sustainable environment in which to live, work and relax in.

The Permaculture design model is concerned that useful connections are made between components in the final design. The formal analogy for this is a mature ecosystem. In much the same way as there are useful connections between the sun, plants, insects and soil, there will be useful connections between different plants and their relationship to the landscape and humans. The permaculture design also has multiple outputs. In terms of Holmgren's application of H. T. Odum's work, a useful connection is viewed as one that maximizes the rate of useful energy transformation, that is power.

4.5.4.2.3.5.1.2.3. Natural Systems Agriculture Zones

These zones would be located in natural ecosystems that would be harvested for human use. These perennial polycultures would use mixtures of perennial grains to provide food. Due to the mixture, however, the system would be more labor intensive, that is to say economies of scale may not be possible.

4.5.4.2.3.5.2. Scenic & Touristic

In North America, starting with the gold rush and continuing through the oil rush, hunting rush, and tourist rush, human populations have increased far beyond the carrying capacity of the land to support them. Large permanent supply lines, sometimes inadequate, are needed to support population levels. The Inuit today have been invaded by a consumptive culture, which is so far unable to adjust its urban industrial world view to environmental limits—even refusing to acknowledge many limits.

The carving of crest poles spread north to the Tlingits, but especially as an economic enterprise. Canadian and U.S. schools encouraged natives to carve to be more self-sufficient. In the south, the Nootka and others, who had simple poles topped with birds and sometimes copper or human images. Model poles began to be carved for tourists. Then they were used in ceremonies, for funerals and as teaching devices. Argillite, a slate or carbonaceous shale, for tourist things (soft and damp when mined, it hardens later). In 1926 the Canadian government and the Canadian National Railway decided to restore poles to the western route of the rail to encourage tourists. Whites took, preserved and relocated poles near the rail line. Natives objected, saying that the railway got all the benefit. Tourists wanted the Natives in native costumes and with many totem poles. Villages with only one pole and people in jeans were not interesting. Many things were created especially for tourists, such as basketry items mimicking lampshades, platters, suitcases, dollies.

In Africa, threats to Masai herds include loss of wildlife. Elephants and giraffe maintained grasslands by grazing them. Now, after big game-hunting and fencing, large tracts of grass lands are converting to scrub, which cannot support cattle. The Samburu are switching to camels, which can feed on scrub, have richer milk, and are equally popular with tourists. Masai are sought out by tourists. Some villages can earn income from sightseeing tours. However, many Masai crafts are not made by the Masai nor sold by them. Of the crafts that are, money may not reach them. Tourism accounts for 40% of Kenya's earnings. Should tourism be allowed in archaic cultures? Are parks without people the answer, in Africa or the

Amazon? Maybe it was once true in traditional times, but now, with population growth and lack of rules, how can we? The costs of conservation is high, and dependent on foreign aid. Should wilderness be set aside at expense of archaic cultures, which also need protection from interference?

Scenic vistas have dramatic aspects. Many places are unique in their height or emptiness, in their diversity or expanse. As tourist areas, however, the experience high human traffic, which may interfere with them; and, traffic requires roads, stations, and signs.

4.5.4.2.3.6. Artificial Areas

Some areas have been completely modified and have few remaining natural features. Human dominance would vary from light residential areas, with permanent paved roads (following existing ones where possible or if not new roads planned to make minimum impact; restricted lanes and volumes of air and water transport) to residential areas with full services, including light commerce at low density, with provision of facilities and services, including some cottage employment, increased commerce at medium density, and recreation of any kind, with any machinery, to cityscapes dominated by central area functions (communications, services) at high density, interspersed with manufacturing transitional areas for light industry and services. Although a high biomass of animals and plants may exist, there would be low diversity and few wild species. There would be an isolated area for heavy industry and the disposal of noxious waste.

Artificially created environments are cultural environments. Nature thus acquires characteristics from agriculture and social institutions as well as from geology and climate. The practices of environmental conservation must be complemented by careful policies of environmental creation. Societies have images of the future that influence their policies, and those images can be shaped by knowledge and persuasion. Human beings can and have created new ecological values by collaborating with, or following the laws of, nature.

Many ecosystems in the temperate zone are artificial. The diversity of ecosystems in England result from human intervention. English parks are based on an imperfect understanding of natural systems, and on their potentialities, as well as on a series of happy accidents. Much native vegetation is a social artifact. In Scotland, for instance, forest cover was reduced from 55% of the total area to 5% by primitive stock-keeping and agriculture; moors decreased by half, but meads increased eight-fold. The soil was more fragile, so the accidents were not happy and the forests did not grow back. Some forest plantations have been started. Dubos mentioned Kentucky, Western Europe and Japan as examples of radical conversion from the native fauna without disorganization.

Cultural landscapes must be preserved too. It is as important to preserve areas where humans have generated social, aesthetic or emotional values (different from wilderness) from a partnership with earth as it is to preserve wilderness. Dubos believed that human values can transcend those created by natural forces alone. that potentialities of the earth remain unexpressed until manipulated by human labor and imagination.

Even here, roads, air and water lanes should only be two or three percent of the area. Other artificial areas include medium and high density residential areas, the various densities of cities and their services, and light and heavy industrial areas.

These areas might be most appropriate for technology testing in the short term and the long term, especially new genetic changes, which might require years or decades of testing, and nanotechnology, if it has been modified to interface with human chemical energy processes. Given the chaos and uncertainty, as well as openness and evolution, of natural systems, artificial areas might be able to offer a greater degree of control and insulation from other processes.

4.5.4.2.3.6.1. Traditional Human Places (Cities Suburbs & Industry)
The ideal landscape became the middle region, the garden between the complete order of the city and the complete chaos of the wilderness. The garden is cultivated from the wilderness as a middle landscape. The garden is a human order, but not usually sacred.

The city expands at the expense of gardens, fields, and then wilderness. The whole city, and not just the temple, was conceived as an earthly imitation of the cosmic order, a sociological middle cosmos, established between the macrocosm of the universe and the microcosm of the individual. Through the priesthood, the one essential form of all was made visible.

Greek *kosmos* originally meant order, then later good order and universe. The polis, the classical city-state, was a place where the citizens depended on and maintained the whole, which cared for and outlasted the individual. The stoics declared the cosmos to be the great city "of gods and men." The citizen became related to the cosmos as a whole, in the same way. The Greek idea of nature was based on an analogy with the human body, with thoughts and feelings. The universe was a kind of living organism. Thus, the Greeks also kept an anthropomorphic perspective toward cosmology. Cicero defined *mundus*, as universe, as the common home (*domus*) or city (*urbs*) or state (*civitas*) of gods and men.

The ancient Chinese held a notion that the earth is a living being. In taoist science, feng-shui or geomancy, subtle forces flow through the earth like blood in a body. To build a city, it was necessary to consult the landscape, chart the flow of forces, and place buildings with utmost care to be in harmony with the place. Geomancy is finding the right place and petitioning the spirit of the place to build a wonderful city.

The city becomes a center for intensification and excitement, but it also causes environmental challenges to appear faster and larger. A city changes the local environment around it, pulling in medium size cities, as it did in Mesopotamia, and reducing the number of small satellite cities, which are made continuous with the influence and attraction of the center city.

Cities are where most people live, where most resources and energy are consumed, and where most wastes are produced. To avert further destruction of the earth's life-support systems, cities must be transformed into places where people can live healthier lives while reducing their ecological impacts. Concentration and intensification have opposite effects. The first may reduce the footprint of the disturbance, but it may attract more resources from further away.

Cities designed and lived in before much machinery and cars, had a more on a human scale, and were more livable. Jane Jacobs' book, *The Death and Life of Great American Cities,* is a strong critique of the urban renewal policies of the 1950s, which, Jacobs claimed, destroyed communities and created isolated, unnatural urban spaces. Jacobs advocated dense, mixed-use neighborhoods and frequently cited New York City's Greenwich Village as an example of a vibrant urban community.

In her work Jacobs also tackles the question of economic booms. Great cities with flourishing economies have precipitated these economic booms. She asserts that it is through import replacement that cities have such economic growth. She also asserts that cities are at the root of all economic growth, from agricultural, manufacturing, and technology growth to the information explosion, and therefore import replacement is the cause to all economic growth. This idea challenges one of the fundamental assumptions of Classical and Neoclassical economists, who consider the nation-state to be the main player in macro economics. Jacobs argues that it is not the nation-state, but the city which is true player of this world wide game. She speculates on the further ramifications of considering the city first and the nation second, or not at all.

Cityscapes still contain gardens. Gardens and parks have been designed to express an idealized view of natural and agricultural scenes, since Sumerian times, at least, over 5000years. Many names for the landscape—grove, lawn—are drawn from the imagery of the garden. The complementary aspects of landscape planning are the invariants of a given area and the artistic imagination of a planner. Conrad Aiken recognized that "The language and the landscape are the same, for we ourselves are the landscape and are the land."

Cities still place constraints, include physical, psychological and social, on their inhabitants. Forms include shelters and services. Forms can be monotonous, especially with a larger scale, unless care has been taken to make the designs diverse and stimulating. Industrial zones are basically artificial, requiring only connection to natural inputs and paths to the sinks of the natural environment.

4.5.4.2.3.6.2. Ecodevelopment Example: Ecovillages

Over thousands of years people created good communities in good places, often by accident, sometimes by intent. But, these places developed on a smaller scale, such that people learned to fit their activities to the place. Often their cultures were unconscious ways of ecological adaptation. Myths and taboos limited the taking of rare or overly-desired species. Human technology and tools increased productivity without interfering with ecosystem processes.

We can use such societies as examples for some of our modern communities, with understanding that the scale has changed, our technology is far more effective, and our cosmology has become more mechanical. For a modern community to prosper into the indefinite future, the community has to adapt a conscious ecological plan. It has to learn to mimic a mature ecosystem in terms of diversity and balance, flow and waste. Not only do houses and buildings have to fit the greater environment, from geology to climate, but the economic efforts have to minimize exotic elements and waste and work to fit with other efforts in the equivalent of a food chain.

To some extent since the 1950s, intentional communities have allowed people to consciously address the creation of good places with concern for scale and fitness. Paolo Soleri started to create designs for ecological cities intentionally on a larger scale.

Somewhat later, the ecovillage movement addressed these concerns, of community, ecology, and spirituality, by trying to create small-scale villages, within an urban or rural setting, within the matrix of other communities and places, that would be more sustainable. Many such villages are intentionally created as models of sustainability in action, for the purpose of working against the degradation of the overall cultural and ecological environments. The intent is not only to create good places. connected with other good places, but meaningful ways to live.

Cities, the largest of human structures, resisted conversion to an arcology. The size and investments required for arcologies discouraged investors and builders from creating them. Instead, architects generally remained at the building-size projects. Some architects have suggested altering the design of buildings and combining uses. For example, an apartment building block typically encompasses one square hectare of land. At best, in most cases, the block is 50 percent structure, and 50 percent surrounding yard, lawn, patio, or sidewalks. This footprint has 0.5 hectare enclosed space, and 0.5 hectare open-sky space. Rooftops as designed, however, are rarely usable for additional "live loads" of constant use activity nor deadloads of landscaping and outdoor furniture. By moving to a "stepped pyramid" form, it is possible to have a footprint of 1.0 hectare of enclosed space plus nearly one hectare of open-sky space.

Buildings designed as boxes with "square-shoulders", reduce the open-sky area of the maximum possible solar penetration. This reduces the solar gain for patio gardens as well as reduces the possible locations for solar-power panel positions. In worst cases, blocky buildings cast shadows which not only darken the adjacent streets, but also shade lower portions of buildings at nearby distances. Skyscrapers and high-rise buildings can put large zones in shaded darkness, especially at higher latitude locations in winter, which ensure that the buildings have to use more utilities and can raise heating and lighting expenses for lower levels.

The stepped pyramid form, or modern ziggurat, casts minimal shadows on its own tiered plazas, and does not influence the solar access of neighboring buildings. A structure with twelve floors casts a smaller shadow profile than a six-story block building. The street-level of a pyramid would have one hectare of shops, offices or workshop businesses. The second level would have offices. The higher levels could have apartments with garden areas.

Manhattan Island has a night-time residency density of 1.5 million people. The area of the island is 67.34 square kilometers. A reasonable building density would be 45 buildings per square kilometer, assuming that 20 percent of the area would be reserved for industry and public access. Replacing blocks with pyramids, the potential population of Manhattan would be about 2.23 million people without any building being taller than six levels and no building shading any other. There would be 2,800 hectares of commercial space on the street level, ample for transportation, services, shopping markets, and workshop businesses.

4.5.4.2.3.7. How Much do We Need of Each Area?

How much wilderness, common areas, domestic areas, or artificial areas need to be saved, on a global level, to prevent or reverse increasingly significant annual losses of biodiversity and ecosystem services? In his previous studies (1970, 1972), Eugene Odum suggested that forty percent of ecosystems needed to be set aside as wilderness to provide ecosystem services for human populations. This thought experiment continues that idea.

How do we know how much wilderness the planet needs? Or how much an ecosystem needs? Or how much species or guilds of species need? It is thought that if we save a large percentage of the planet in wilderness that will be enough to automatically save both ecosystems and species. However, it will not save part of each kind of ecosystem. It is thought that we can save minimum viable areas of each kind of ecosystem, and that would save most of even the larger species. But, that will not save at least one population of each species. For that, we need to inventory all the species on the planet. Then we could calculate areas with adequate populations, or Minimum Viable Populations (MVPs).

Planet. Wilderness for the planet is hard to determine, because there are so many human variables, not to mention evolutionary variables. The planet will continue, regardless of how we modify it. It may have fewer simpler ecosystems. What we are really concerned about is richness. How rich a planet do we want? Do we want cathedral-like tropical forest canopies that host millions of different species, each making a home there? Or will we settle for desert ecosystems, blasted by centuries of war and destruction? Of course, these ecosystems have their own stark beauty, even after the many of the adapted species have been destroyed. The moon is beautiful, but many people would not want to live there for long, even with the best technical support. The vast number of species existing now have existed for over a million years. These species have proved resilient to all kinds of natural changes and disturbances, except for human ones, which can accelerate extinctions and interfere with renewal.

Ecosystems. How much area does an ecosystem need to be self-renewing? Of course, that depends on the ecosystem. And, the larger and more complex the ecosystem, the greater the likelihood it has dependencies and exchanges with other ecosystems. We know that, especially bird watchers. who follow the objects of their fascination for thousands of miles. Birds carry information, genes, viruses, and seeds to distant ecosystems.

Species. For species, we can calculate minimal areas to maintain minimal viable populations. But, even then, there are many variables to consider, depending on the species. For example, some species, such as wolves, use nonbreeding individuals to help with education of the young. Each species needs to keep a minimum number to be genetically healthy. Biologists have been attempting to determine minimum viable population size (MVP), which is the smallest number of individuals to maintain a population indefinitely. How big should a population be to avoid extinction? Population viability analysis is a comprehensive analysis of environmental and demographic factors that affect the survival of a population. Soule has suggested 500 breeding individuals as the minimum. The 500 breeding individuals is the effective size; the real size may be quite larger, especially where the generations are shifted to elderly or where not all members mate. Long-lived species are also vulnerable, for example, the Pacific ridley sea turtle or African elephant. So, we could safely use 2500 as the optimum number.

How big is the minimum size of area that will sustain a MVP? What is the minimum viable area? For a forest patch size in forest marsh grassland, a minimum area can be calculated for typical species:

- One to four hectares—amphibians, frogs, a very few common edge birds (such as Downy Woodpecker or Great Crested Flycatcher)
- 10 ha—still dominated by edge species, but may have very small areas of interior which supports numbers of forest interior/edge (FIE) species (such as Hairy Woodpecker or White-breasted Nuthatch)
- 50-75 ha—still predominantly edge, but will support small populations of most birds except those with very large home ranges. Least Bittern may be present in marshes of this size - more bird species (Northern Harrier, Short-eared Owl)
- 100-400 ha—all forest-dependent bird species diving ducks (Redhead, Canvasback, Ruddy Duck)
- 1 000 ha—suitable for almost all forest birds. Some forest-dependent mammals present, but the size may still be inadequate for Sharp-tailed Grouse and Greater Prairie-Chicken
- 10 000 ha—almost fully functional ecosystem, but may be inadequate for a few mammals such as Gray Wolf and Bobcat (100 000 hectares has been suggested as a

minimum)
- 100000 ha—fully functional ecosystem. It may be a minimum size for a totally functional ecosystem but some species may still be near their minimum viable population level.

It might be best to use a strategy that combines all three approaches: a percentage of the planet, in addition to special ecosystems and then special species. Without knowing how much wilderness the planet, ecosystems or species need, it may be possible to calculate an amount that might serve those needs.

4.5.4.2.3.7.1. *Wilderness: A Thought Experiment*

If we don't know how much is needed, how much should we save? Obviously, we should be cautious and save as much as we can of every habitat, and if the area is large enough, then probably the top predators will survive and keep the system diverse and healthy. The current reserves are inadequate; wild species decline on limited reserves. Designing reserves requires a biogeometry, a knowledge of the shape of ranges. Left to themselves, in an adequate preserve, species may be dropped or generated. Human progress deprives many species of existence, simplifying ecosystems of which we are still greatly ignorant. Industry destroys billions of creatures more than unscrupulous hunters or multiplying homesteaders. The rate of extinction of species exceeds our rate of learning about them. Humanity as a species must develop an awareness that it is infinitely enriched by the organic world, and is poorer each time a species is eliminated or natural community degraded.

There are many ways to approach minimum (or optimum) wilderness areas. Some ways are simple, and others are complex, perhaps too complex. Some simple ways are probably unacceptable to the human populations. Some approaches can described as follows:

- The Remainder method (or what's left?)
- The 50 percent rule (or best guess?)
- The Acreage method (after Eugene Odum)
- Minimum System Method (considering processes, after Lynn Margulis)
- Key Species: Top predator home range (using genetics and ecoregions)
- The Mixed Method

These approaches are expanded in the following paragraphs.

4.5.4.2.3.7.1.1. The Remainder method

This method works by calculating how much is left and not critical for human needs. This would result in about 30-36 percent of the planet according to satellite information. Early satellite information revealed that 32 percent of the land area was not being used in a way that compromised the systems self-renewal. A 1987 survey by the Sierra Club revealed that about 34% of the land area is undeveloped wilderness in blocks of 400,000 hectares. Only about 20 percent of that is protected. Including Alaska, wilderness in the U.S. is only 4 percent (and two-thirds of this is in Alaska). The U.S. has 233 distinct ecosystems; very few of them are protected, but, 40 million hectares of public lands could be designated wilderness.

Wilderness remaining by continent has been calculated by J. M. McCloskey (Sierra Club, see Table 4-5423711-1). These calculations could form the basis of saving the remainder.

Table 4-5423711-1. Remaining wilderness by continent

Antarctica	100 percent
North America	36 percent
Africa	30 percent
Oceania	30 percent
Asia	27 percent
South America	20 percent
Europe	7 percent

The Wilderness Society estimates that a wilderness area should be at least 400,000 hectares (1 million acres or 1600 square miles). Presumably, this would permit minimum cycling. In "Wilderness: Earth's Last Wild Places," the authors found that "wilderness areas" still cover close to half the Earth's land, but contain only a tiny percentage of the world's population. Over 200 international scientists identified 37 wilderness areas that represent 46 percent of the Earth's land surface, but are populated by just 2.4 percent of the world's population (excluding urban centers). To qualify as "wilderness," an area must have 70 percent or more of its original vegetation intact, cover at least 10,000 square kilometers (3,861 square miles) and most have fewer than five people per square kilometer. Five of the wildernesses are also biodiversity hotspots, home to thousands of species found nowhere else. These are Amazonia, Papua New Guinea, the Congo, the North American desert and the deciduous woodlands of southeastern Africa.

However, this does not tell us if it is enough to guarantee the functioning of representatives all ecosystems, for minimum function. Nor is there any way to tell if it is enough to support human populations at any level of luxury

4.5.4.2.3.7.1.2. The 50 Percent Rule

The land area of the planet is over 149 million square kilometers (over 57 million square miles). Using the 50 percent rule, we should set aside 74.725 million square kilometers (over 28 million square miles). The oceans would also be protected at the same rate. Water area is over 360 million square kilometers. So, 180 million square kilometers would be preserved from extreme fishing, dumping or other use, especially those areas close to shore and upwelling zones.

It is possible to arrive at this rule as a result of exponential change, 50 percent of lake is covered the day before it is entirely covered, if algae is doubling every day. The problem with this method is that a growing population has to be kept within the remaining limits, and those limits have to be exceeded to provide sustenance for those people. This might be possible with an increase in urban agriculture and with more appropriate technology, but it would wreak havoc on traditional diets and agriculture.

4.5.4.2.3.7.1.3. The Acreage Method

Eugene Odum (1970) suggested using land area as a measure of human carrying capacity. Odum was one of the first to consider the implications of such limits. The minimum per capita acreage requirements, with a temperate area like Georgia as a model for a quality environment, is about 2.02 ha (5 acres). The percentage of areas per capita can be broken down and shown in Table 4-5423713-1.

Table 4-5423713-1. Acreage Per Capita (after Odum)

Food-producing land	30%	.606 ha
Fiber-producing land	20%	.404 ha
Natural support areas	40%	.808 ha
Artificial areas	10%	.202 ha

The state of Georgia has a mean Net Primary Productivity (NPP) of 1200 grams per square meter per year (g/m²/yr). Most of the state is a temperate deciduous forest. By comparison, the mean NPP of the entire planet is 782 g/m²/yr—this includes desert and prairie areas— and this is only two-thirds as productive as temperate forest areas.

Using this figure for wilderness of 40 percent per person, and considering the difference between mean productivity, we can calculate an optimum wilderness area for the planet.

Table 4-5423713-2. Calculation of wilderness percentage

6,301,400,000 (9/2003) x .808 ha x 1.53 =	8,484,204,960 ha
Convert to square kilometers	84,842,049.6 km^2
world land area	149,450,000 km^2
Percentage of world area	56.77%

This calculation results in a figure of just over half of the area of the planet for wilderness. Of course, as the population increases daily, then the amount of wilderness required also increases. This is not happening; wilderness is not being created or restored at the same high rate as population growth. Wilderness needs to be increased to provide the ecosystem services that more people need. At the same time, we might find it prudent to decrease our populations.

If we use Odum's whole figure requirements, not just the natural areas, then the situation is revealed to be more critical.

This current area calculation is greater than the surface area of the planet. The only option in this case is to reduce the human populations over several generations—perhaps by lottery, or reverse incentives, or allowing cultural autonomy.

Table 4-5423713-3. Calculation of percentage required for current population

6.301,400,000 (9/2003) x 2.02 ha x 1.53=	19,532,728,389
Convert to square kilometers	195,327,283.89 km^2
World land area	149,450,000 km^2
Percentage of world area	131%

4.5.4.2.3.7.1.4. The Minimum System Method (Process)

When the ecosystem is the smallest unit capable of recycling the elements of its membership, then for each type of ecosystem, a minimum area can be calculated that would permit the processes to continue to function. For example, organic carbon can be expired, fixed, reacted, or transformed. This method requires intimate knowledge of an ecosystem. For most ecosystems, we do not have that knowledge. Until we do, we should use a more conservative approach to determine a minimum or optimum, perhaps based on keystone species.

4.5.4.2.3.7.1.5. Key Species Method (Top Predator and Genetics)

By calculating the minimum viable area for each kind of ecosystem, using the largest habitat of the top predator, and summing for all ecosystems, a figure can be calculated for a wilderness area for the planet (if wilderness is equated with ecosystems MVAs). Predation is a key to initiating and sustaining the diversity of ecosystems. Humans, by wiping out the predators have made the systems less healthy. The predator serves as a top-down limit for prey species, keeping them healthy by altering their behavior, removing young sick and old individuals. Of course, in the system, there are bottom-up limits also, as plants change their chemistries to attract or avoid predators of their own.

The idea of saving wilderness now is the idea of saving ecosystems complete with predators, most of which cannot compete with humans. Yet the systems are crucial to human survival. The conclusion then is to save large systems, as wilderness, complete with predators, not only to keep the biodiversity strong, but to allow those systems to provide those things that humans cannot do without, such as clean air and ecosystem services. The ecosystems have to have viable populations to be stable.

The size and geometry of reserves can be approached from considerations of large predators such as Grizzly Bears and Mountain Lions. It is possible to determine the minimum home range of such species by means of radio telemetry and to calculate a minimum critical area by making some assumptions about the minimum population size needed for genetic considerations. But such a procedure does not take into consideration the problem of dispersal of the young, who may not be tolerated.

To remedy this situation, some have suggested reserves linked by corridors of similar habitat for dispersal so that there would be essentially islands of habitat for people rather than islands of habitat for wildlife.

Maximum population density scales as mass, therefore the requirements of individuals for space increases more rapidly with size than the space available. There is an overlap in the ranges of large mammals and birds. Large mammals that are grazing herbivores are not territorial. As species decrease in size they tend to be more patchy in their use of the environment (but many carnivores having a healthy population size also exhibit this tendency), which leaves a lot of space unused.

We can calculate the areas for a few keystone species for instance, using their home ranges. Home range is the area around the home of an animal, used for feeding or searching for mates, commonly shared by mated animals with offspring. Defended territory is usually a smaller area that an animal will defend actively. Animals also disperse and migrate, both vertically and horizontally. Keystone species can be top predators, such as the Sea otter, who preys on sea urchins, which graze on algae and kelp, the dominant food source for instance fig trees, a connector such as mycorrhizal fungi, a habitat modifier such as the beaver or African elephant, or a mutualist such as lichen. Co-evolving species appear to alter the structures of landscapes over which they evolve. They then tend to have the highest fitness and survive the longest in their physical landscapes.

Predators have different sizes of home ranges, and the shape of their ranges reflects various habitat needs. Some predators have relatively small ranges that have a linear shape. For instance, the skunk, an omnivore that digs subterranean dens along hedgerows), has a home range of 1 km^2 per animal; the range is linear around edges. The red fox, a nocturnal carnivore that also creates subterranean dens) has a home range of 10 km^2 per animal; that range is also elongated rather than round in shape. River otters, piscovores living along rivers and valley streams, have a home range of 160 km^2 apiece, whose shape is guided by watercourses, but also changes by seasons.

Other predators have larger ranges with other geometric shapes and may use several ecosystems. The Canada Goose requires numerous landscape features for summering and overwintering, roosting on open water, feeding in fields by day, resting on shorelines during day, and staying in protected coves in bad weather. Migratory birds are declining as a whole because not all of their territories are protected. In many cases, wintering grounds are being destroyed by logging and burning. These species need many habitats to avoid extinction.

Owls are also vulnerable to habitat destruction, although they reside in one ecosystem. The Spotted owl requires 1000 hectares (10 km^2) of old-growth forest for each mated couple. Other owls have less stringent needs. The pygmy owl requires half the territory and survives in second-growth forests. As a key species the minimum area for a viable population is 250,000 hectares (2000 km^2).

Coyotes each require a minimum area of 700 ha (7-10 km^2) for a home range; although the range follows natural features, it tends to be much less linear in shape. An effective population of at least 500 individuals may be needed to maintain genetic viability of each animal species. Since not all coyotes in a group breed—some become aunts or uncles and help care for pups—it is necessary to assume 3 coyotes per breeding unit. Using coyotes as the key species, the minimum area for a viable population becomes at least 525,000 hectares (or 5250 km^2).

With wolves, it is necessary to assume at least 3 wolves per breeding unit. Assuming a minimum viable population of 750, and a requirement of about 80 km^2 per wolf, the minimum area for a viable self-sustaining wolf population would be 60,000 km^2. Many species, such as wolf, use clusters of ecosystems, including corridors, such as roads or frozen streams, as well as fields and forests; they avoid wetlands and human cities. Larger carnivores, such as Siberian tigers, require even larger areas (two tigers per 10,360 km^2).

This system is quite a lot of work. But, the calculations can be simplified using three levels of tertiary predators across all identified ecosystems. For example, the US has 233 distinct ecosystems. Wilderness for the US, using three levels of tertiary predators for the 233 ecosystems, can be calculated in Table 4-5423715-1.

Table 4-5423715-1. Wilderness Area to Hold Top Predators

77 x 2000 =	154,000 km^2
78 x 5250 =	409,500 km^2
78 x 60,000 =	4,680,000 km^2
Total=	5,243,500 km^2

The total area of the US is 9,373,000 km^2. By this calculation 55.9 percent of the country should be set aside as wilderness. By extension, that percentage could be extrapolated to the rest of the area of the planet (plus calculated sea area).

The current wilderness in the US, which is the result of conservation efforts, is inadequate. Including Alaska, wilderness in the US is only four percent (374,920 km^2, and two-thirds of this is in Alaska). But an additional area (400,000 km^2) of public lands could be designated wilderness in US. Eight times as much would have to be restored.

There is another way to calculate Minimum Viable Areas. That is to sum the requirements necessary for each species, since each species is genetically unique, not simply address the top predators. This is considerably more difficult, since we do not know how many species exist, where they are, or how much home range they need, nor do we always understand how the areas overlap. This calculation will have to wait on a complete inventory and monitoring program for the planet.

4.5.4.2.3.7.1.6. The Mixed Method

A mixed method might be more appropriate given the differentials in wilderness saved, unused but unprotected areas, and humanized landscapes. First, set aside those areas that are essentially wilderness already. Antarctica is especially important because we do not really know if it is possible to save only half of it, while drilling and utilizing the other half. Because it its influence on the water level of the rest of the planet, as well as political uniqueness as a research continent, it should be preserved intact. Other parts of the Arctic, Sahara desert, and Micronesia, should also be set aside, that is be limited in terms of not increasing human impact. Therefore, we add: Antarctica (14,100,000 km^2); Arctic parts of NA, Europe, Asia (calculate); Amazonia, Central Africa, Asia; Micronesia and Australia. Then we add hotspots of diversity, such as Madagascar. Klamath mountains, and most of Ecuador. Then we calculate areas of wilderness from the remaining areas, in terms of countries: China, Europe, Russia, and the United States.

Then, we calculate the areas to be restored to minimum wilderness areas (in the categories, neopoetic and restoration, from the previous system): Africa, Asia, and Europe. This could be done in terms of ecoregions or ecosystems. As percentages, by Division, it results in 90,061,000 or 60.26 percent (See Table 4-5423716-1).

Table 4-5423716-1. Wilderness Calculation by Regions.

System **Ecodomains**	Size (km^2)	% earth	% save	Save size (km^2)
Polar Ecodomain	38,038.000	26%	88%	33,365,000
110 Ice Cap Division	12,823,000	8.77%	87-100	12,823,000
M110 Ice Cap Mtns	1,346,000	0.92	89	1,300,000
120 Tundra Division	4,123,000	2.82	2.2	3,960,000
M120 Tundra Mtns	1,675,000	1.14	1.1	1,590,000
130 Subarctic Div	12,259,000	8.38	6.0	10,200,000
M130 Subarctic Mtns	5,812,000	3.97	3.5	5,500,000
Humid Temperate	22,455,000	15.36	30	6,737,000
210 Warm continental	2,187,000	1.49	10	218,000
M210 Warm cont. Mtns		0.78	86	885,000
220 Hot continental	1,670,000	1.14		790,000
M220 Hot cont. Mtns	485,000	0.33		300,000
230 Subtropical Div.	3,568,000	2.44		700,000
M230 Subtropical Mtns	1,543,000	1.05		1,100,000
240 Marine Division	1,347,000	0.92	80	1,100,000
M240 Marine Mtns	2,194,000	1.50	1.1	1,200,000
250 Prairie Division	4,419,000	3.02		442,000
M250 Prairie Mtns	1,256,000	0.88		625,000
260 Mediterranean Div	1,090,000	0.75		110,000
M260 Mediterranean Mtns	1,561,000	1.07		365,000
300 Dry Ecodomain	46,806,000	32.00	4	10,127,000
310 Trop/sub steppe div	9,838,000	6.73		2,360,000

M310 Trop steppe Mtns	4,555,000	3.11		3,250,000
320 Trop/sub desert Div	17,267,000	11.80		14,100,000
M320 Trop desert Mtns	3,199,000	2.19		2,750,000
330 Temperate steppe Div	1,790,000	1.22		180,000
M330 Temp. steppe Mtns	1,066,000	0.73		400,000
340 Temperate desert Div	5,488,000	3.75		3,000,000
M349 Temp desert Mtns	613,000	0.42		325,000

400 Humid tropical	38,973,000	26.64		
410 Savanna Division	20,641,000	14.11	35	7,224,000
M410 Savanna Mtns	4,488,000	3.07	75	3,366,000
420 Rainforest Div	10,403,000	7.11	70	7,308,000
M420 Rainforest Mtns	3,440,000	2.35	75	2,580,000
Sum Domains				70,707,000
Sum Divisions				90,061,000

4.5.4.2.3.7.2. Setting Aside

After we decide on wilderness numbers, how do we set it aside and protect it? Does wilderness have to be separate from humans? There are species that cannot coexist with humans. There are habitats that are critical to the functioning of the planet; there are habitats too fragile to bear human interference; there are habitats that are of virtually no use to us. All of these could be set aside and monitored from satellite (even airplanes can have some impact on some systems).

The wilderness that could tolerate small incursions, such as scientific research, would probably tolerate some human presence. Those wilderness areas that have coevolved with human cultures could simply be controlled by limiting human presence to traditional cultures. Not every person in every culture wants a car and television. The allure of simple traditional cultures is such that many people would like to go back to them and live in them.

Some wilderness areas could probably tolerate a higher impact from people. Traditionally, borders (between human and nonhuman habitation) have tolerated human impacts. Most borders are between cities and unused areas. These are permeable and undefended, except for the few reserves. The boundaries are fuzzy anyway, due to the ability of pollution and trash to reach any area.

Even the idea of wilderness has been turned inside out, from the nature outside to the nature preserved from us and our activities. Confucius declared that *fang-wai* (outside the square) was the edge of his interest; now that we understand the interconnections, we need to reverse that attitude as well.

The patterns we impose on all of nature now determine what species thrive and what do not, their numbers and ability to move and evolve. We tend to homogenize nature— homogenize is such an appropriate word—it means made the same, but also perhaps the process of sameness promoted by *homo sapiens*, the homogenizer.

4.5.4.2.3.7.3. Common Domestic & Artificial Areas

Ecosystem conservation offers several advantages over a species-by-species approach for the protection of biodiversity. It directly addresses the primary cause of many species declines (habitat destruction), it offers a meaningful surrogate to surveying every species, and it provides a cost-effective means for simultaneous conservation and recovery of groups of species. The idea of representing examples of all ecosystems in protected areas extends back to the nineteenth century in Europe and in Australia and to the early twentieth century in North America. Ecosystem conservation does not remove the need to understand the autecology and the protection requirements of individual species. It is not a replacement for existing conservation measures such as the Endangered Species Act of 1973 in the United States. Not all species that have gone extinct in the United States since European settlement would have been saved by this kind of coarse filter, which does not protect all presently endangered species. We must save rare and endangered species along with this plan.

Conservation plans for all ecosystems should be developed, starting with those that are at greatest risk of degradation or conversion. Global conservation, in general, has to include key elements:

- Salvage the hotspots, including Madagascar, Hawaii, Klamath, and Ecuador
- Keep the five largest remaining frontier forests in tact: Amazon, Congo, New Guinea, Canada, and Russia
- Cease logging in all old-growth forests
- Save wetland systems, rivers, shores; save marine hotspots
- Restore some of the oldest landscapes, e.g., Mediterranean coasts, African wetlands, and Iraqi grasslands
- Restore areas that are overused or under-represented, such as grasslands and deserts.

Conservation plans have to include monitoring. This requires understanding the rates and locations of land-cover change, as the result of natural processes, natural disturbance and human interference, and understanding the processes associated with biological productivity. Both involve monitoring and assessment of landscapes that are changing rapidly due to extensive and intensified land use and natural disturbance regimes.

These plans must be kept in the context of human populations, cultures, and technologies. Human populations must be balanced with wilderness. If human populations cannot be reduced, then large areas of wilderness have to be restored.

Many wilderness areas and ecosystems are endangered, especially those that have a long history of human settlement, such as the eastern coast of North America, or those that hold desired resources, such as the beech forests at the tip of South America.

Most ecosystems, much less ecoregions, have little or no legal protection. They have been ignored only because they are inhospitable or resource-poor. Plans have to have a strong legal component.

Wilderness is not something that can be preserved in one desired state; it is always changing. But, the processes that are continuously creating wilderness can be preserved by preserving the boundaries that limit wilderness from human landscapes. Wilderness cannot be managed; it is that which is not managed. But, people can be managed.

Partly, things change, partly we reinvent things, as well as ourselves. Now we need to reinvent wilderness, cities, homes, intercultural interactions. We need to keep everything in a complete context. That means political changes and population planning, as well as just designating and protecting wilderness.

4.5.4.3. *Controlling Political Issues & Areas of Concern*

Growth is the first area of concern. Most cities and almost every human area is experiencing growth. Economics requires growth. Mesarovic and Pestel stated that "the issue for the economy is not to grow or not to grow; it is how to grow, and for what purpose." Growth was confused with development. They claim that if a workable world system is to emerge, it must be after the establishment of an organic pattern of growth. They assume continued growth. Due care is devoted to describing such a pattern and contrasting it with other, tragically inapplicable patterns of growth. Their treatment of the world system itself was regionalized and multileveled. They recommended that the establishment of organic growth was necessary with no need for special no-growth policies for populations or economies. They assumed that further industrial growth will continue, that economic growth is good, and that this growth solves human problems as long as it is organic. Exponential growth is said to be bad, and organic growth is said to be good. In fact, although organic growth is better, there is little difference during a world crisis; both reach asymptotes of suffering. One need only regard the population crashes of lemmings to see that organic growth can go wrong. In the organic world, growth is healthy only when the rate of change is decelerative in the long run; cancer and population are constant or accelerative. Mesarovic and Pestle failed to realize that continued economic growth in any form is a threat to the stability of the biosphere. Furthermore, they assumed that we are limited to only those two kinds of growth, but that is palpably untrue.

Three other types of growth can be distinguished: Additive, an accumulation of more; replicative, an accumulation of more through reproduction; and mutualistic, where all agents change structure, as in meiosis in cell development.

Rapid European expansion occurred at rates rarely exceeding a growth of 1% per year, and with unparalleled opportunities for expansion into sparsely settled areas, including North America, Australia, South America, and South Africa. Archaic countries do not have these opportunities; violent population growth has wrecked its hope for development, ravaging every resource. Growth itself requires greater material support for houses, schools, medicine, and jobs. The observation that exponential growth in a finite environment cannot continue is indisputable. And rates of exploitation are even more important than the rates of growth. By definition, the “first world” is the industrial West, with its outposts on major continents— Taiwan, South Africa, Israel, Australia, and others.

Karl Marx conceived of the idea of the ‘second world,’ while studying in the British Museum. Although it was intended to be born from the first world, the nonindustrial, agricultural, feudal Soviet Union formed the basis of the second world, which is now dedicated to the same ideal of material progress; the second mirrors the first. The third world is a new conception; ‘poor’ nations became ‘undeveloped,’ then ‘under-developed,’ then ‘developing,’ then the ‘third world.’ Hardin calls this ballet of terms the ‘Flight of the Euphemisms.’ Any terms, like undeveloped or poor, are distortions to some extent; they classify only according to statistical, economic or technological attainments, not by wisdom or happiness. Although the term ‘third world’ is used in this book for identification, on occasion, individual nations will be referred to as archaic, traditional, agrarian or industrial, or be named by place. The third world is really the mother of the first two. It contains the “wretched of the earth,” in Franz Fanon’s phrase, left behind in sullen poverty and gentle greatness. The first two world victimize the third by stealing its resources. The rich become richer, while the poor struggle to live. Ironically, with the value of scarce resources, some ‘third world’ nations are now among the very richest.

The very poor have been romanticized to have a quiet vision of life on a human scale in peace. But this is just the hope of the frustrated and greedy that somewhere else people are more noble and honest. With the advent of radio and television, poor peoples have learned to want material wealth. And they can see what they are being denied. The economy has been growing almost constantly since it has been studied. We have been trying to force it to grow, rather than to let it contract. Peter Drucker reminded his peers that economics is still the dismal science, that everything has a cost (and therefore a price?), and that nothing can be consumed until it is produced.

The economy may not contract; it may continue to develop without growing. There is no necessary association between development and growth. Growth means an increase in size. Development means the introduction of an innovation.

The relations between labor, capital and resources are changing rapidly. Capital investment is no longer economical. The old capitalist market economy is shifting to a neomercantilist one where some global agency can charter local companies to husband dwindling resources. Resource limits may spur the change from growth to development.

J. S. Mill wrote that beyond the progressive state lies the stationary state; each advance is to approach it. Mill wrote that the stationery state was a metastatic state, after the Greek words meaning 'changing in place.' Mill's vision of a stable state was a response to the goal of industrial society. Technological progress would abridge labor, not necessarily increase production. In a state of equity, persons would have room for solitude and leisure development. But Mill did not believe that 'man' was bound to protect nature.

H.E. Daly concludes that the steady state economy is necessary and desirable. He notes that in the definition of economics, as the study of the allocation of scarce means among competing ends, the entire ends-means spectrum is not considered. Only intermediate ends or means are considered, not ultimate. He anchors ultimate means in physics and ultimate ends in religion. So economics falsely concludes that the middle ranges represent the entire spectrum.

A stationary economy is not synonymous with stagnant; there is always room for developing and increasing scope of mental culture. It could be highly sophisticated, dynamic, and imaginative. It is scarcely necessary to remark that a stationary condition of capital and population implies no stationary state of human improvement. There would be as much scope as ever for all kinds of mental culture and moral and social progress; as much room for improving the art of living and more likelihood of its being improved.

A community is forced to accept an upper limit, beyond which it cannot grow any further. Further growth results in destruction or disruption of itself and nature. Production needs to be stabilized in a steady state economy, a mature economy, like a mature ecosystem, where processes and cycles are constant. A steady state economy must be based on natural laws and ethical principles. Natural laws include thermodynamics and ecological theories. Economic development will require technology. Ecologically sound technologies will minimize stress to the environment. Rules of economics, laws of nature and ethical principles must be related. W. I. Thompson projects a movement from succession economy to a mature economy in a global centripetal reconsolidation of planetary culture.

There is another distinction between growth and development. The ecosocietal approach to development makes it irrelevant to discuss global limits to growth. Local limits are far more significant to majority of population. Regardless of how much food exists, people will starve unless they can get it.

Redistribution of resources and improvement of environmental quality are more important than increased production by sophisticated technology. The natural capacity of regional photosynthesis must be limiting factor in development, especially in tropical and subtropical areas. This development calls for social and educational organization more than technological style. Styles of technology must be determined by culture and context. Such development requires a local authority working with suitable economic and ecological conditions. No authority can be effective without the participation of populace. The diversity of types of agricultural and cultural habits may be analyzed from either adaptation to ecosystems or their transformation of them.

To stop growth, a strict regulation of the productive system is necessary. States have be able to control which products are made, with which technology. The machine with a governor excludes any other states besides a steady state. Corrective action is brought about by difference; it is error activated. Practical implications of the steady state are that nonrenewable resources will be conserved as much as possible, by recycling, while erosion, depletion and pollution are minimized; energy sources may be greatly decentralized and diverse. But a stationary state cannot go on forever; there will be an inevitable decline in resource accessibility. Therefore the most desired state is not stationary, but gradually adjusting to resources and values. The human population may need to steadily decrease over thousands of years. Georgescu-Rogen advised even further that we must expect a steadily decreasing state. Since we do not now know enough about reserves and the rate of renewal of some resources or about technological inventions, a metastatic state is reasonable.

4.5.4.3.1. Economy Energy & Resources

Any economy must consider the needs of its members, as well as the limitations of its resources. Surprisingly, political and psychological factors have been and may always be more important restraints on food supply than physical limitations. Deprived peoples have difficulty striving for better crops and supplies; planners have difficulty envisioning an agricultural system operating on ecological principles in specific areas.

4.5.4.3.1.1. Energy

The problem of food energy mirrors lesser problems. There are shortages of land and energy; there is overpopulation; there is degradation of land and people. Any solution must address the whole complex. At least half the population of the earth is suffering from malnutrition. Although some shortages are caused by transportation and logistical difficulties, food production is still falling behind population growth. Infant mortality in third-world countries ranges from 100 to 200 per 1,000 live births, compared to 25/1,000 in the USA.

Ethanol has been proposed to supplement oils. Corn, however, is an expensive food for humans; it is very expensive as an animal feed; it would be prohibitively expensive as a fuel. While normal waste from corn products could be a fuel, growing corn to convert to energy is inefficient. This is another scale problem; waste reuse, although it can be integrated into an industrial ecosystem, it can never be productive on a large scale. It is also heavily subsidized by the public in some nations. Without those subsidies, it is unlikely that people could afford ethanol as a fuel.

Fossil fuel sources of energy are finite and are becoming more expensive to extract. Other kinds of energy, from the chemical and nuclear, to wind and solar, need more technical sophistication to extend them to large scales. Obviously, we should use the simplest forms of energy. These include the sun, earth, tides and agricultural waste. Obviously, we should use the most abundant resources. These start with hydrogen, the most abundant element in our universe.

4.5.4.3.1.2. Resources

There are basically three types of resources: Fastly accruable or renewable; slowly accruable or basically nonrenewable; and, slowly dispersed, that is, really nonrenewable. The earth is finite. Its resources are limited, interrelated and distributed unevenly. The availability of natural resources, such as minerals, is a common concern. We have had to search over the earth for them. In fact, the word metal is from the Greek meaning 'to search for.'

Edward Goldsmith, in *A Strategy for Resources*, admits that the earth's crust may be far more richly endowed than conventional reserve figures suggest. Mineral reserves may be understated by 5 (Limits to Growth) or by a 1,000,000 (Hudson Institute). But they may be located so far and so deep that it would cost more energy to extract them and move them they it would be worth, unless we used a 'renewable' energy source—the sun—to get them. Oil, coal, peat, and some woods are functionally nonrenewable. Geological time periods are required to produce them.

A coherent body of knowledge waits compilation and evaluation which will demonstrate clearly to cultures what resources they possess. Slow accrual and slow dispersal resources must be equally available to all cultures. It is impossible to sustain any quantitative arguments about resources and population pressure on them without a comprehensive overview. Demands on food, fertilizer, energy, and metals are related inseparably. Organic and inorganic assets need to be assessed together. Population carrying capacity cannot be formulated until both resources have been quantified.

4.5.4.3.1.2.1. Organic Resources

We depend on vegetation for far more than food: Newsprint, construction, furniture, clothing, and packaging. Furthermore, with shortages of minerals, many substitutes are expected to be organic. The measurement of total production of vegetation is difficult and complex. Wild plant communities for experimental controls have been altered over thousands of years. Original wild communities were efficient and flexible, more so than domestic communities.

The International Biological Programme embarked on a large number of projects to assess the biological productivity of all the main types of terrestrial ecosystems. Efforts based on these materials have been made to compute the actual NPP of land vegetation at the present. Samuel Eyre based his own calculations on these and other criteria to calculate the NPP of the original wild vegetation.

The original vegetation cover was extremely diverse. The potential natural productivity of wild vegetation may be 120 billion tons of dry organic matter per year. But, not all of that is available for human exploitation. The proportion of NPP above ground could only be 80 billion tons. These are all rough approximations. This is the theoretical amount that would be produced by wild vegetation with no major environmental changes.

Much of this is not utilizable by humans. For example, in the production of forests, 65% of above ground production is not used by most lumbering operations. Possibly, over all ecosystems, the representative average of unusable material is only 50 percent. Taking most of the material out of the system is not desirable because mineral nutrients are locked up in plants. Complete extraction of plants would result in removal of from one to two thirds of potassium in the whole ecosystem. Although there is plenty locked in rock particles, this is released very slowly.

The land has been "improved" to pastures and crops, from natural vegetation. Some crop productivity, usually at experimental stations, is capable of producing high NPP, in total dry weights. The most productive of crops can approach the NPP of the original cover, but average crops produce less than half of the natural cover. This high figure was achieved in Japan, where substantial dressings of fertilizer were applied. In countries where no fertilizer is applied, the annual productivity is unlikely to be more than one quarter of the native vegetation. With grain of cereals or wood of trees, it might be possible to harvest as much as 30% of the NPP; higher percentages might be possible rarely, lower ones more often.

In tropical climates, the NPP does not appear to increase in proportion to increased sunlight. The tropical plant also uses up far more energy for its own metabolism. The hotter the environment, the higher the metabolic rate; so temperate zones produce higher yields than tropical. Although tropical ecosystems offer less potential for high levels of secondary production, microorganisms (autotrophic and decomposing) and invertebrates could be potential converters to high quality animal food.

Unfortunately, there is a trend to deforestation, with no idea of the rates. The mean annual production of wood may still exceed the annual consumption and destruction by humanity. Present trends in forest destruction and expanding pastoralism run counter to any of rational ordering of terrestrial resources that is necessary to cope with population doubling. Annual timber cutting in the United States in 1963 exceeded the annual growth by 50 percent. This wood was used for heating, industry, and paper products. Some of the areas were cleared for agriculture.

Even in low-use areas, there are effects of human occupation. Interference in woodlands may cause secondary growth. Although the old primary forest ought to be able to grow back out of the young secondary, the transition may take thousands or tens of thousands of years. J. P. Schulz has shown that shifting cultivation may stunt the change for hundreds of years.

4.5.4.3.1.2.2. Oceanic Resources

Marine plant life thrives chiefly on upwelling areas along the west coasts of continents with prevailing off shore winds and strong eastern boundary currents such that surface waters are diverted and replaced by nutrient-rich deeper and cooler waters. These are biologically the richest parts of the ocean. This total area, about the size of California, produces half the world's fish supply. The open sea is practically a biological desert.

The sea has a far smaller NPP than land. The rate of turnover is much more rapid; the standing crop of phytoplankton may be less than 1.5 billion tons (wet), or 1% of annual production. The Net Community Product is much more difficult to calculate since all of the consumers have to be identified. In many cases, there is competition with human consumption. It might be possible to assume that half of the NPP could be applied to human use, far less than Eyre's data.

Feeding efficiencies tend to increase from plants to herbivores and carnivores. Efficiencies in human harvest must be geared to productivity of the system. It is assumed that humans want as much as it is feasible to take without destroying the basis of productivity. Harvest in the form of herbivorous animals should be lower than the trophic level efficiency (under 10%). For aquatic carnivores, harvest efficiencies in relation to primary production should range from 1.0% to 0.1% or 0.01% for primary, secondary and tertiary carnivores, respectively. Only the fraction of the population that represents a surplus above the individuals necessary to maintain reproduction and growth of the population itself can be harvested on a sustained basis.

4.5.4.3.2. *Food Production & Agriculture*

Herding is a traditional way to convert wild vegetation to usable protein, without trying to manage domestic crops under uncertain conditions. Herding, however, makes politics difficult. Herders traditionally follow animals, regardless of political boundaries. Recently, some nations have tried to restrict herding to limited areas, which is detrimental to the herds, as well as to the herders.

Agriculture is still the preferred manner of producing food, on fertile soil, supplemented with fertilizers and biocides. Theodore Roszak theorized that the principles of agriculture, with its emphasis on dull, regular, monotonous work, served as a model for industry. Perhaps the principles are not the same; certainly industry does not recycle its products and wastes as efficiently as a nonfuel-subsidized agriculture does. In one respect the principles of industry seem to be applied to modern agriculture—large capital investment.

4.5.4.3.2.1. Arable Land

Like industry, agriculture is faced with very real limits. The amount of light reaching the earth's surface is a limit that probably cannot be increased safely. Most modern plant varieties have about the same photosynthetic efficiency, anyway, although some varieties have a superior leaf arrangement. Plants would be more efficient if the atmospheric composition were manipulated so that ambient carbon dioxide levels were concentrated; this might be done in greenhouses. Phosphorus is a limiting factor: modern agriculture squanders it, so that much of it ends up in the ocean.

The amount of arable land on the earth totals about 1.5 billion hectares. Burlingh et al., as well as other sources, conclude that 3.4×10^9 ha. are potentially arable, but their figures include 1.5×10^9 hectares in the wet tropics. Since most tropical soils are delicate and tend to erode, most farming techniques ruin the soils after several years. The wild systems in tropical areas contribute inordinately to the cycles and services of the planet. Although proper farming techniques could prevent erosion, it is unlikely that they could on that gargantuan a scale. Since good farmland can be created almost anywhere, with good, laborious techniques, there is little reason to gain several years advantage over other marginal soils, as in New England United States by completely destroying climax tropical forests. Arable land could be increased in cities. Many ecologists regard cities as wastelands, but much agriculture could occur there also. Some organizations, such as the Institute for Local Self-reliance in Washington D. C., are growing food in cities.

The world's cultivated land area exceeds 1.2 billion hectares. But there is much abandoned farmland in the United States, Turkey and the Soviet Union, for example. Other countries—Ireland, Japan, Sweden, and Switzerland—have declining cultivated areas. Possibly India and China may expand their areas slightly, although population and city areas are eating into arable eland rapidly. Other areas are irrigation dependent, such as the Middle East and Northern Africa. Central Africa and the Amazon have significant prospects for expansion, although most tropical soils are extremely difficult to manage.

Trying to double the amount of arable land would require enormous amounts of irrigation, as well as the alteration of whole ecosystems. Ten to 15% of the current arable land is irrigated, presently, and these are the best sites. Yet irrigation is ruining much of this land through salinization and depleted water tables. Furthermore, irrigation is very expensive in terms of materials and energy. If the world's entire petroleum reserves were used only for the purpose of doubling arable land through irrigation, those reserves would be used up within 20 years, not even considering energy for machinery and desalinization. Although irrigation can allow 2-3 annual crops in place of 1, pests often breed in the fallow season. Ironically, it is pests like the tsetse fly that prevent more land in Africa from being ruined by overgrazing.

Instead of expanding arable land, there are two other ways of increasing food production: By enhancing the efficiency of production, which may be easier in small operations that mimic natural ecosystems more closely, and, by using novel food sources or production, such as algae or insects.

4.5.4.3.2.2. Diversity

Agriculture has decreased the diversity of crops, worldwide. Fifteen crops provide 90% of the food for humanity. These crops include: corn (maize), rice, wheat, sorghum, millet, rye, barley, cassava, sweet potato, potato, coconut, banana, bean, soybean, and peanut. These crops also occupy 75% of the tilled land area.

There is an unfortunate development in tropical countries toward commercial crops—cocoa, coffee, sugar—and industrial crops—sisal, palm oil—in direct competition with subsistence farming. The result is that native people are fed by enriched white bread and rice from the spoiled countries, who overgrow wheat and rice. This trend makes the system less stable and increases the likelihood of famine. Famine is not new; it is the 10, 20, or 50-year cycle of agriculture and limits of distribution. But now it is compounded by size, business and waste. Land destruction is progressing at a great rate, everywhere. The genetic green revolution and desert irrigation are prescriptions for disaster. The intensity of the technique is commensurate with the magnitude of potential calamity.

Agricultural intensification causes fallow cycles to be shortened; a shift from mixed crops to monocropping; and a shift from natural fertility to artificial fertilizers. Most overgrazing results from commercial pressures and economic forces that play on a rancher. Most farmers operate in a condition of chronic indebtedness to financial institutions. These rules need to be rethought, also.

4.5.4.3.2.3. Agricultural Energy

Modern agriculture is a squanderer of energy and materials. Population pressures influence agro technology. Fertilizers are against the elementary bioeconomic interests of humans. The green revolution further forces dependence of tropical countries on industrial countries, with the disregard for local customs and ecological vulnerability.

The green revolution violates the ecological system and ignores the minimum needs of people, depriving many of their livelihood. Miracle wheats introduced to Iran in 1968 succumbed to a new strain of wheat rust; entire crops were lost with widespread hunger. The Americanization of Indian agriculture, the green revolution, allowed the rich to get richer through the necessity of chemicals, fertilizers, tractors, and large land consolidations; more people suffer than ever before. Good intentions are not enough. Benevolence wreaks the same level of havoc that malevolence did. The efficiency of agriculture in the U.S. is low already. The U.S. produces three times as much food per hectare as India, but it requires ten times the energy. The U.S. farmer may use 5000 energy slaves to provide food for 50 people;

packaging and preparation reduce the ratio to 1:5.

Energy use is increasing three times faster than the human population. The earth is doubling its human population in 30 years, and its energy consumption in 10 years. Much of this energy is used in food production. The "Green Revolution" requires far more energy than less intensive methods. In the United States, for instance, an estimated 1,250 liters of gasoline equivalents are used to feed one person per year. If the known reserves of fuel were spent for the earth's population at this rate, they would be exhausted within 13 years, assuming no use for any other purposes. For some crops in the U.S., such as Brussels sprouts, the amount of energy used exceeds the energy yield, by over 30%. Other crops in the U.S., such as dry beans and rice, use almost as much energy as they yield. Energy for crop production is of two kinds: That to increase yields, through hybrid seeds and fertilizer, and that to reduce human labor, by using tractors and drying rigs. In the U.S., at least, that to save human labor needs to be reduced. Effective use of human power in the U.S. could produce the same high yields, but using only 25% of the energy employed.

Frantic modern agriculture is not sane, no matter how productive, efficient or economic it seems. And it is not really any of these things. The amount of agricultural produce entering world trade is steadily declining, as it has been for two decades. Food self reliance is the goal, through cooperation. It must be placed beyond the confines of narrow economic rationality.

Increased agricultural production does not always create more food, it pilfers other species, or gathers extra energy. Our actions are leading to the replacement of all wildlife by humanity. Georg Borgstrom alludes to the ghost acreage needed for U.S. standards, that is, trade acreage and fish acreage, that are not usually counted. This drawing on other areas for support is same as having additional land area. Japan and Britain are grossly over populated by this consideration. Eventually, these nations will have to negotiate better trades for crops from ghost acreages.

4.5.4.3.2.4. Conversion of Land with Traditional Styles

The myth of a return to earlier rural times is only a myth; the soils have been destroyed. We must learn to create bioshelters that sustain us and provide us with shelter, etc. The soils are starving for organic matter. Waste and manure need to be returned to the soil. Soils, like animals and plants, can be domesticated. European soils bear little resemblance to wild soils. American soils bear even less. Smaller farms could fertilize by crop rotation and manure. Build up soil, choose better seed stock; make hedgerows so birds can better control pests.

Agriculture works best where it is least perfect: that is, the farmer still hunts and gathers, keeps many kinds of plants and animals, produces little surplus, uses no chemicals, grows adapted crops, and has wild lands. That situation may be ecological and durable. Successful agriculture depends on artificial climax or sustained successional state. A pioneer is self-sufficient in more than material. A subsistence farmer is closer to ecological harmony: with gardens, fields, trees, bees, fish pond, pasture, forest, and natural vegetation. Subsistence farming could use low-energy agricultural methods available, instead of dangerous, unclean technology or nonexistent cheap energy. Large holdings with absentee landlords usually produce less than small plots farmed by owners. Large scale farming has advantages, but it is more efficient when done by the owners.

Traditional rice production, for example, is much more efficient, in terms of output energy to input energy ratio, than U.S. production, which is geared to producing great quantities at great costs, with fairly low efficiency. Per capita production is the only criterion by which giant farms qualify as more efficient than small farms. But since unemployment rates are high this efficiency is foolish. Small farms grow more food per acre, of higher quality, with fewer chemicals and less hardware.

Eric Waddell argues that the return to traditional agriculture is the only means to solve the world food problem. Traditional agricultural bases are the best idea for a maximum population just because people are conservative about their diet, and farmers are conservative about their methods. Much of industrial agriculture may choose to become more labor intensive and multicrop; also antihybridization and pro natural pest control. Land must be cultivated as though it were a garden, with diversified and appropriate flora and fauna. The farmer should have a thorough knowledge of the physiography of the land—soils, crops, forests, pastures, mineral content, microclimate—and study the effects produced by flora and fauna. Land must be cultivated as though it were a garden; the agriculturist should be familiar with the features and subtleties of the terrain and crops.

4.5.4.3.2.5. Intensifying Productivity

Historically, farming allows intense food production in a smaller area. Successful cultivation then intensified trading of cultivars and resources between groups. In other regions, notably S. E. Asia, dense population and very small holdings necessitated intensive cultivation, using people and animals but few machines; here the yield is low in relation to energy expenditure. Production can be intensified in two ways: Technology allows the same labor force to increase the output, which may have freed labor for other endeavors such as building religious monuments; and, different crops or irrigation increases the productivity of the land.

After thousands of years, fortifications were erected. Most of the raiding was on fields and storage complexes. This led to insecurity of rural population, so peasants moved to secure fields around regional centers. This intensified agricultural stress. More intensive fields more easily defended. The solution to stress increases the population which intensifies the stress. In the Middle East now, 200 million people are supported on 10 percent of the land by intensive agriculture and intensive imports from trading oil.

Factory farming is an intensive, large-scale, commercial agriculture based on fossil-fuel energy. Agriculture converts raw energy into food energy and exports it into the smaller area of the city. Cities allow intense dwelling in a minimum area.

The purpose of politics and economics, in a nation, is to protect its members and provide them with social advantages. It has to provide the infrastructure of paths and food. Politically it has to provide defense and justice.

4.5.4.3.3. Intensifying Technology

Technology intensifies. There was an intensification of tool-making in the past 3000 years. Population increased and new areas were exploited. The agricultural revolution was a technological revolution. The industrial revolution, by definition, was a technological revolution which resulted ultimately in men being replaced by machines. Only the Luddites faced this limitation, unsuccessfully. Human labor is increasingly dispensable. Industrial culture confuses mechanical with personal power. According to W. I. Thompson, industrialization is really an intensification of civilization.

Because of the exponential growth of technological exploitation, more primary metals have been consumed since 1940 than in all human history before then. Technology is bound intimately with the exploitation of a convenient source of energy, oil. Unless the trend is changed, reserves will disappear. There is a finite limit on actual numbers of molecules of a given resource in the earth's crust, but the practical limits to exploiting the potential are a function of technological activities. So we are importing materials from the past (oil) and the future (soil), for now. Humans are hunters and gatherers of materials whose renewal times are geological scale.

Recent industrial history reflects new tools, new livelihoods, changes in settlement and behavior patterns. All tools, from the simplest word to the most complex computer, are disturbers and rearrangers of primordial nature and reality; they are implements for working on something. They have addicted us to purpose. We look for purpose in everything; to seek an explanation of nature, and to justify the seeking. Humans are tied to their tools and machines. The basic difficulty with the quality of life resides in machines. Machines pollute air, water and earth. Machines of war threaten human lives. Machines displace forms of life; they take up space. They are a competitive species, whose members die, reproduce, evolve, and sometimes think. If humanity is not to be replaced by autonomous, thinking machines, we had better choose and use them wisely. Not all cultures have chosen wisely. In the Temple of the Moon of the Mochicas (circa 1600 YBP Peru), one wall painting in a ritual room shows axes and tools with arms and legs, apparently rebelling against their human makers.

A tool is extension of individual into new kind of specialized animal; technology lets us mutate into new pseudospecies. Neither the ecosystem nor the animal knows what it ought to do, hence the explosion and chaos. For humans, this is what a new ethic and cosmology is needed for: to tell us what to do. We need direction. The Greek root words from which technology was taken meant art (*techne*) and study (*logos*). Technology must become an art again to make good places.

New technology has been a primary force for change for decades; but some technologies, like computing or genetic engineering, may lead to enantiodromia. The use of such tools may have unexpected effects. For instance, mass-produced computers may lead to individual autonomy. New energy technologies could have the same effect. Technologies have the capability to minimize the use of resources.

4.5.4.3.3.1. Technology & Housing Political Issues of Shape of Cities

At some point, politics has to determine how many houses can be built in an area, by relating that number to the population, standard of living, ecosystem productivity, and the application of technology. Politics, then has to decide where to restrict them by location, such as off a floodplain. Community development issues must also be addressed, including affordable housing, the expansion of suburbia, and homelessness.

On any given night in the U.S., anywhere from 700,000 to two million people are homeless, according to estimates of the National Law Center on Homelessness and Poverty. According to a December, 2000 report of the U.S. Conference of Mayors: Single men comprise 44 percent of the homeless, single women 13 percent, families with children 36 percent, and unaccompanied minors seven percent. The homeless population is about 50 percent African-American, 35 percent white, 12 percent Hispanic, 2 percent Native American and 1 percent Asian.

According to the 1996 National Survey of Homeless Assistance Providers and Clients (NSHAPC): Single homeless individuals in 1996 reported an average monthly income of $348, about 51 percent of the 1996 federal poverty level of $680/month for one person.

Over 28 percent said they sometimes or often do not get enough to eat, compared with 12 percent of poor American adults. And, 44 percent did paid work during the month before the survey; 21 percent received income from family members or friends. The statistics get worse: 66 percent of the homeless have problems with alcohol, drug abuse, or mental illness; 22 percent have been physically assaulted; 7 percent have been sexually assaulted; 38 percent say someone stole money or things directly from them; and, 30 percent have been homeless for more than two years.

Affordable housing would reduce some of the homelessness. Affordable housing does not get built because the profit margin is not desirable or because of zoning difficulties. The technology to cheap homes does exist. Houses can be made by the same industrial processes that make cars. Politics can make house manufacturing more attractive through a combination of laws, incentives and taxes.

4.5.4.3.3.2. Political Issues of Technology & Transportation

Politics has to provide adequate roads and public transportation. Politics has to limit the number and kind of automobiles used by the population. It has to consider the same things for aviation and boating. Regardless of kind of power, from fossil fuel to hydrogen or solar, the density of roads and use has to be part of the equation. Roads have to consider the ecosystems that they cover or divide. Some counties in the U.K. and U.S. design roads to include animal crossing. A few roads have been made permeable, to allow more natural flows of water through them. Politics has consider limits to the idea of personal transportation. It may have control bicycles or other nonmotorized vehicles on public roads. Placing these limits will not infringe on the inventiveness of new forms. The safety of the vehicles and their paths has to be a major consideration. Politics has to provide and monitor standards of safety, especially when the environment has to absorb the waste products in its air or water cycles.

4.5.4.3.3.3. Technology & Substitution

Two technological processes are especially important, substitution and recycling. Substitutability of resources is neglected, as well as new forms of power or price mechanisms. The amount of plastics in packaging in the United States totaled 3.6 million tons in 1978 or 15 kg/person. Plastics are used to substitute for wood, leather and metal in many industries. This substitution may temporarily ease pressures on biological systems, but it depends on the supplies of petroleum.

Goeller and Weinberg, in an article the "Age of Substitutability," said that that this age will be characterized by an asymptotic society that settles into a steady state of substitutionand recycling, using renewable resources and practically infinite resources, such as iron or aluminum. They suggest that society could exist on infinite materials indefinitely. Since phosphorus and fossil hydrocarbons may be in short supply, as exceptions perhaps, a good power source would be required, solar or nuclear.

Sophisticated technology can create huge savings on resources. Fuller noted that a one quarter ton satellite outperforms the transoceanic capabilities of 175,000 tons of copper cable, although he may not have included all the ancillary costs, such as launch or recovery.

4.5.4.3.3.4. Recycling & Waste Technology

When a resource, especially a metal, is in generous supply, it is used for many nonessential and trivial purposes. Thorstein Veblen described the use of waste as a sign of social success as "conspicuous consumption." When the metal becomes scarce, then it becomes more valuable and its uses are curtailed; it is then worth recovering from dumps.

The Club of Rome assumes that metals are consumed like vegetables. Although the mines become exhausted, the metals can be recirculated. Recycling could lower some of the costs. For example, the asbestos fiber content of tailings produced and discarded in the 1930s has been found to exceed the grade of new ore now mined. Some urban refuse is often richer in metal content than natural ores currently mined.

4.5.4.3.3.4.1. Recycling Technology

Wastes can be composted into soil and fertilizer. But recycling cans and bottles is not a good enough answer. It is far better to wash and refill permanent containers and save the energy of processing. Changes in society are reflected by the trend from buckets to throwaways to recyclable containers. Our relationship to the earth has changed and this is only a symptom. Obviously, we need to apply the precautionary principle to our use of new technologies. Furthermore, these technologies should be selected according to goals and plans.

4.5.4.3.3.4.2. Waste Technology

Georgescu-Rogen described the economic process as entropy-producing, where entropy is visible as heat, waste and pollution. Every life process generates waste that is recycled in biological cycles. Nature wastes countless lives and materials. Most young creatures die before maturity. Resources may lie buried for millions of years by landslides or volcanic eruptions. Unfortunately, human waste is different: many materials have no processes or organisms to break them down and recycle them—no solutions have evolved; sometimes the quantity, of a toxin for instance, is immeasurably greater than natural toxins. Even natural disasters, such as volcano eruptions that devastate whole areas, are recolonized from outside. The impact of human eruptions around the globe mean essentially that there no longer is an outside. Humans have the capability to fill the entire atmosphere with radioactivity or organophosphates.

Ecology contains tin cans before, during and after their economic life. We need to understand the capacity of the environment to absorb waste over time. There are no sinks on the earth where waste vanishes. Things are only moved around; eventually they return. Some things are dissipated into the earth's environment, the cosmos, which is an energy source and sink. Even this large environment may not be inexhaustible, as many analysts such as Eric Jantsch claim.

Pollution is a symptom of imbalance and improper resource utilization. It is essential to combine garbage composting with sewage disposal. Advanced communities have to adapt biologically sound processes. A serious problem is our lack of understanding of the extensive, long-term effects of pollution on the atmosphere.

Once it is determined that nuclear and chemical materials and wastes are dangerous and long-lived, the nature of the materials and processes is no longer at issue. Jantsch sees nuclear waste as a call for self-transcendence, much like prokaryotes dying for eukaryotes. No comment. What matters is whether the earth can be safe, and the relevant decisions can be made by informed and thoughtful people.

Some waste problems can be solved by distribution or accommodation. Thermal pollution may be used in conjunction with aquaculture—certain fish and shellfish like warm water. Manatees could be used to clean lakes and streams suffocated with aquatic plants like water hyacinth, lily and lettuce.

A homeostatic system resists measures of control. One method to influence it is carefully proportioned measures. For example, solid wastes. There are three ways to reduce the generation of wastes by acting on the valve (flow variable): Reduce the number of products, and possibly the standard of living; reduce the quantity of waste in each product; and, increase durability and hence the life expectancy of a product. J. Randers shows that best results come from policy mix: a tax of 25 % on extraction of nonrenewable resources, a subsidy of 25% for recycling, and a 50% increase in life of products, a doubling of recyclable portion per product, a reduction in primary raw material per product. It takes more expertise to avoid waste than to create it.

4.5.4.3.3.4.3. Redefinition of Waste

A redefinition of waste is needed. What waste is not: Waste is not the difference between a minimum and a maximum. That is merely the unused flexibility of the system, which is variable, of course, due to changing conditions. To use a mechanical engineering example used by Hardin, waste is not the unused capacity of a bridge to carry trains. That is a margin of safety calculated by the engineers to keep traffic below the calculated physical limit of the bridge. In a similar example, limiting cattle on a range, and leaving some plants uneaten, is not waste. Nonefficiency is not a synonym for waste. Cathedrals are not wasteful, except as office buildings; gourmet dinners are not waste, other than as minimum subsistence. Many things, that seem wasteful, are required as aesthetic needs.

In nature, there is waste in a system, but usually other systems can learn to use the waste of the first system as a resource. The waste of the sun is not a waste for the planet. Waste may be unavoidable in any one system. But, good planning can just design an enveloping system to benefit from the first. Waste cannot be eliminated in one system, but it can be a product in the next larger system. Furthermore, much of what we consider waste, such as unused potential, is not waste at all, but a necessary part of the system itself.

4.5.4.3.3.5. Technology & Energy

Technologies require energy, and many kinds are available: Solar, nuclear, fossil, fire, animal, and human. Each has dominated some historical epoch.

There is a myth of human ingenuity that anything can be solved by hardworking people. Fundamentally, industrialized countries have only substituted mechanical energy for muscular energy. But there is no argument that these countries' populations are more productive, with all that energy. Amory Lovins calculated the number of energy slaves at work: The average U.S. citizen has 250 slaves. These slaves consume 20 times the energy content of food to feed the present world population at 3600 kcal/day.

The industrial revolution increased the quantity of energy, but decreased the variety of energy resources. Carbon products are the largest source of energy. Oil production really means extraction. But that is the only economic means to get it. F. de Chardenedes wrote a scenario of natural technology for producing petroleum; if the amount of energy, as heat and pressure, had to be paid for at a human public utilities rate (circa 1970), the cost would be over a million dollars a gallon.

The three main uses of energy are motor traffic, heat for buildings, and manufacturing. Manufacturing is a heavy consumer; it includes: pulp and paper; primary metals (cars);nonmetallic products (plastics, shoes); and chemicals. Some uses of fuel are incredible. Daly cites the fuel requirements of a B-1 bomber as being between 300 million and 1 billion gallons of fuel per year. By contrast all of the buses in all the cities of the United States in 1974 required only 325 million gallons.

4.5.4.3.3.5.1. Energy & Risk

Energy demand has resulted in the use of high risk sources. With nuclear power, the burden of proof for safety is on the agencies themselves; its use in the absence of complete assurances of safety is the worst transgression against life. G. Hardin is even more adamant in his conclusion: guilty until proven innocent.

Even with inefficiency, there is no need to use high risk energy generation. Buckminster Fuller claims that by using only proven energy resources, only proven technologies, and only at proven rates, within ten years all of humanity could enjoy energy income equivalent to United States in 1960, certainly adequate for a good level of luxury for all inhabitants, and nuclear and fossil fuel energy could be phased out.

An analysis of energy supply possibilities is needed to recognize the consequences of actions in terms of resources. The industrial revolution increased the quantity of energy available, but decreased the variety. There are asymmetries in energy balances. Terrestrial energy is the stock, solar energy is the flow. The future will always have less stock, as energy leaves the earth slightly faster than it stays. There is an astronomical difference between flow of solar energy and the planetary stock. A clever civilization would tap the flow.

Technology could be developed to tap the flow. The second law of thermodynamics is a limit; in all processes in which energy is changed, some of it becomes unusable, diffused and very difficult to harness. We ought to reestablish earlier energy patterns for regions, and use combined systems of wind, water, solar, organic and fossil fuels for energy. Singly, these may be inadequate, but as a mosaic they could meet decentralized needs. The energy pattern should be made organically from potentialities.

The consideration of energy policy takes us back to the soil, back to nature. The establishment of decentralized communities based on ecologically sound organically based agricultural practices, with local technology using local energy sources and recycled nonrenewable raw materials. Some of the energy crisis could be avoided by using less consumptive settlement patterns and natural energy utilization. The U.S. crisis is a crisis of too much energy use. Consumption and production of energy must balance, safely. Energy would have to be regulated, although not necessarily small generation energy—wind or solar. Nature is not a free commodity where the laws of energy work.

4.5.4.3.3.5.2. Flow & Pattern

The organic pattern of energy can be shown to be parallel to the ideas of the flow of the Tao, in the Tao Te Ching. Perfect activity leaves no track behind it; this would be a Taoist approach to energy use; small scale and nonpolluting. The Tao presents a society isolated but satisfied. What is rejected by the Tao is the technology that goes against the nature of things.

Energy production should leave as few tracks behind it as possible. Nuclear fission leaves burning, long-lasting tracks. Burning organic fuels leaves pollution and sickness. Even large scale solar projects, extraterrestrial or earth-based would shift large quantities of energy around with unknown consequences. The sophisticated equipment involved would also cover large areas. Any concentrated energy use for large human populations and manufacturing may be too much. Even wood-burning stoves are causing pollution.

Passive solar heating would be adequate for individual buildings. Photovoltaic cells could provide electricity and power cars. Local energy projects using geothermal sources, winds, tides, or the sun are preferable, since their operation does not introduce new material to local cycles. All of these sources would be characterized by a small scale. Larger scale possibilities include hydrogen fusion, although it is not possible with current technology. This first was proposed by Hans Bethe in 1938; the requirements for production were worked out by John Lawson in 1957. The waste products would be heat and helium. It may, nevertheless, be dangerous. Buckminster Fuller offers the sailing ship as a masterpiece of the technological use of energy: it goes through the sea without damaging it, employing the ceaseless winds without depleting any stores of energy.

The public service function of nature provides free services to humanity that are essential to civilization. But when the free services are overloaded and breakdown, we have to pay the costs of repair. Further increase in flows of energy through technology will significantly reduce the capacity of the earth to support humanity. Woodwell notes that use of nonrenewable resources can destroy renewable ones. For example, if acid rain from burning fossil fuels continues in the Midwest, the primary productivity of New England forests may be reduced a net 10%, due to rising soil acidity. This is the equivalent loss of energy from fifteen 1,000 megawatt reactors. A new ecological balance could help humanity to develop in a state of equilibrium and with the solution of environmental problems. It is almost impossible to estimate the economic value of natural balance.

4.5.4.3.3.6. Industrial Ecosystems

The system could be redesigned so that most waste is a resource in an industrial ecology. Like agriculture, industry draws energy and resources from the earth and dumps its waste back.

But, it draws too many resources too fast, and there is nothing that has coevolved to live on the waste. Unlike a pioneer stage of succession, industry is, in Evan Eisenberg's words, at a "fetal" stage. It relies on the maternal biosphere, or the biospheric matrix, to feed it and lick its wastes. Alas, this fetus is very large and could kill the mother soon, unless the fetus matures or the mother reacts faster.

Others, such as Paul Hawkin have suggested the possibility of an industrial ecology that could mimic a mature ecology in its use of recycling and wildness. Industrial designers and managers need only reduce the capital use of energy and streamline recycling. Despite a few paths of plastics recycling, aluminum or steel, most industry is a one-way path from energy and materials to waste energy and materials.

The Danish have created an example in Kalundborg. Statoil sends most of its byproducts to industrial neighbors. Sulfur to a sulfuric-acid plant, heat to local greenhouses, gypsum to a wallboard plant, gas and wastewater to a coal-fired electric utility, which sends its waste steam to Statoil and fly ash to a cement company. A flowchart with boxes and arrows might make it resemble the charts of mature ecosystems.

4.5.4.3.4. *Ethics Rights Justice & Law as Political Issues*

Humanity is an integral part of food chains and part of an organic cycle of birth and death. Humans need to recognize that they automatically participate in everything, and that they cannot unparticipate by choice. Participation starts at the quantum level, through the ecological and cultural. Human nature does not find meaning in an absurd world, but discovers its structure through interaction with the ultrahuman order. Human identity exists partly in relation to nature; the destruction of one involves the other.

The word ethics is derived from the Greek word meaning 'custom,' which itself came from the Sanskrit word for one's 'own doing.' Since it was used in the plural, it meant 'doing together.' The word 'morality' comes from the Latin word for will of the people; the singular meant the 'will' of a person. It was probably derived from the verb 'to measure,' as to measure one's way, or to go one's way. Morals means the 'way of going together.' Ethics means 'doing together,' which of course one does in living together. And, in an anthropometric universe, this is entirely appropriate.

Ethics are assembled inductively, from experience in living in places. Because of the uncertainty of human actions, ethics has to encompass the far past and distant future. No one knew that when DDT killed mosquitoes, it would concentrate in the food chain to kill birds. Values are time dependent, and ecological time can be very long indeed. The futures we invent are viable only if they are compatible with constraints imposed by an evolutionary past. An ethics that requires a long-range responsibility also requires a new humility, since technological power exceeds the ability to foresee its consequences. An ecological ethic recognizes the moral obligation to leave a habitable for future generations.

Leopold proposed a conservation ethic, dealing with human relationships to land, plants and animals. The land ethic Leopold had in mind was a sense of ecological community between humanity and other species. "When we see land as community to which we belong, we will use it with love and respect." Such an ethic would change the human role from master of earth to plain member of it. Predators are members of the community; and no special interest group has the right to exterminate them for the sake of benefit for itself This attitude is important for habitat protection. Leopold describes the extension of ethics as "actually a process in ecological evolution. Its sequences may be described in ecological as well as in philosophical terms. An ethic, ecologically, is a limitation on freedom of action in the struggle for existence. An ethic, philosophically, is a differentiation of social from anti-social conduct. These are two different definitions of one thing. The thing has its origin in the tendency of interdependent individuals or groups to evolve modes of cooperation."

The extension of ethics to animals and land is an ecological necessity. Extended ethics defines a social conduct that is a mode of cooperation and, ultimately, symbiosis. Leopold argued that ethics are voluntary limitations of freedom, necessary in a complex world of which we remain incredibly ignorant. Ethics are developed in response to problems that arise from increasing knowledge. Science has phenomenally increased our knowledge of physical and biological processes. It has now become the basis of our moral code, but it cannot very long be a science divorced from feeling and art if that code is to help us survive. To do this science requires aesthetic perception as well as disciplined thinking and feeling. As there is a rational component to ethical judgments, so there is an intuitive and emotional one, also. An evolutionary ethic suggests that humans avoid tampering with complex evolved systems, not because they are good, but because they are the basis of life at this stage of development. Ecological ethics is situational because ecology is the study of changing systems. The morality of the act is determined by the current state of the system. Adaptive modes should conform to ecological patterns. An ecological ethics is based on attributes of ecosystems and human compliance with ecological laws. The aim of an ethic must be harmonious to the idea of the world's population of living beings.

Ecological ethics is a series of rules for living together. Most sets of ethics make the rules easy to follow. They emphasize the differences (relativism) or similarities (absolutism) of human beings only, or of the individual or the group. But ethics has to confront the individual, embedded in a community, located in a bioregion, on earth. And the rules really are not as easy as human systems have presented. Albert Schweitzer made them too difficult, with the need for a constant valuing, but neither are they that difficult. An ontological ethics can be detailed only on a local level—even when it uses a global strategy.

Individualism can be tempered with the concept of common good, the good of the whole. The whole is diminished by individual loss, but the individual is crippled by the loss of the whole; wild children, for example, are never really human. Human good cannot be considered apart from the common good. We are in larger communities like an organ in a body. In *The Laws*, Plato has the Athenian say to a youth that all things are ordered with a view to the preservation of the whole, each portion contributes to the whole, and every other creature is for the sake of the whole. Ethics has expanded in wholes, from the family, to the human community, and to the ultrahuman community, on which all depend.

4.5.4.3.4.1. Rights

An ecological ethical model is not distorted by human needs and wants when it argues for the preservation of animals and habitats themselves, because they are, as they are. Paul Shepard says the argument is not new, and that its application is ambiguous because "unlimited rights" will conflict with human interest. But, there are two bad assumptions: That human interests are not ambiguous—they are—and that animals will be granted unlimited rights—they will not. Rights seem to follow the expansion of the sphere of ethics, as formal statements of intuitive knowledge. But, codifying rights is more difficult, especially for philosophers, who tend to limit rights with a series of restrictions. For example, a contractual theory assumes a perfect detachment and a rational debate of rules. Animals and imbeciles are left out. Scott Lehman argues that natural objects are not the subjects of experience, certain animals excepted, and so cannot possess rights. He limits experience to mental states. Perhaps he means nervous system states. Some philosophers maintain that a right is a claim to something; others, that it is an entitlement. Richard Watson takes reciprocity as central to the general concepts of rights and duties; few animals and no natural objects have rights intrinsically. He mentions that some primates and mammals are moral entities because they are self conscious, have free will, understand principles, and intend to act accordingly. But the assumption of self-consciousness would rule out children and feeble-minded adults, as well as most living beings. So, a larger definition of claim or reciprocity is needed, without regard for contracts and mutual duties.

When humanity was divided into citizens and slaves, there was no freedom. When it was divided into governors and governed, freedom was advanced by providing the governed with protection against the tyranny of governors. When people became self-governing, protection was needed against majority opinions, by distinguishing between the individual and society. New contraries—public and private—provide clarification of rights. Rights protect the interests of those holding rights. Natural rights are the rights of an underclass that has not been granted legal rights. These "natural rights" are used by minorities to legitimate their claims against controlling powers.

Rights and obligations were first thought of in a political context consisting of customs and practices within and between states. In the 17th century they were thought of in a constitutional context, where forms of government were established to protect natural rights. Now they are thought of in a human context. Freedoms of—speech, worship—depend on institutional protection and are political rights. Freedoms from—want, fear—are extensions of economic rights. Freedoms for—pleasure, reproduction—are biological rights. Pierre Dansereau categorizes the rights of individuals as:

- Physiological: Light, air, water, food, shelter, procreation
- Psychological: Minimum space, forming attachments, free from shocks
- Social: Choice, work, association
- Economic: Income, dispose of property, use resources
- Political: Education, information, participation in decision
- Religious: Adherence to a creed, opportunity to join in groups.

These rights are only very analogous to those of other beings. The extension of rights to animals and plants does not deny any traditional human rights. Animals should be accorded higher moral regard and legal standing to reflect the intrinsic worth afforded by their existence and sentience. Welfare laws to conserve species and to guarantee humane treatment in research, transportation, and slaughter indicate a growing concern among people. A new ethic can keep animals free from human intervention, prejudice, or overuse. Animals should be preserved because they are as they are; their existence is moral justification. Their intrinsic worth is independent of the instrumental values imposed on them by humanity.

Rights seem to follow the expansion of the sphere of ethics, as formal statements of intuitive knowledge. Humanity has taken its own opportunities. These opportunities have been codified for centuries as rights. Now, we must allow other beings equal opportunities. The interrelatedness of life dictates the interrelatedness of rights. And these rights are necessary to the integrity of the whole planet. Humanity developed in a community of animals and plants, as part of a clade on the same tree of life. The quality of human life has always depended on the quality of animal life. Animals have sensations and feelings, as important to them as ours are to us.

The strongest argument for rights is interrelatedness in communities. It is a basis for assigning rights to ultrahuman nature. Existence implies intrinsic worth. Garrett Hardin considers interrelatedness, but interprets it narrowly. He considers rights as rules of competition; every right is a ploy in the struggle for existence, and every right implies an obligation to furnish it. This is good as far as it goes. Life is more than competition; it involves cooperation and play. Rights are formal rules for living together.

4.5.4.3.4.2. Justice & Law

Socrates argued in the *Republic* of Plato against the conception of justice as giving every man his due, and proposed a definition of justice as every man performing his proper function. This proportion of reward to function in the community was named distributive justice by Aristotle. It describes the right to participate in the benefits of science and culture.

If justice is a proportion set up in the community between men and goods, justice is also the restoration of the relation of men and goods, when disturbed. Aristotle called this justice, rectificatory. This has constituted the business of laws and courts. During the stage of universal rights, world order transfers the criterion from the nature of man to the community of men; both law and justice, obligations and rights are reduced to equity. Thus, new nations demand to participate in a common justice, as opposed to the extension of natural rights. How can the idea of equity be given determinate content and political force? Society should be organized on the basis of functions, not rights. Ecological rights could be based on functions. It is foolish not to assign rights to animals, plants and earth because of contractual formalities. Morality means living together; symbiosis means living together. We are living together with everything on the planet, and in the universe. Therefore we ought to extend rights to all beings. Human life depends on a matrix of life. The matrix is historical. It has duration; it extends from the past into a future of following lives. This gives the past and future rights. They are part of the continuity of relationships.

A principle of justice based on need can be extended to the ultrahuman community. It needs to be altered to account for unconscious, interdependent beings. The right to use nature is a right to share. Current legislation on animal experimentation and protection implicitly recognizes the right to live in a healthy habitat.

One problem with the current legal system is that all ultrahuman beings are given the status of inferior human beings, legal incompetents, thus keeping humans in a guardian role. A new legal category is needed that would respect the existence, competence, and excellence of natural beings. Christopher Stone recognizes that the judicial system has granted rights to a variety of inanimate holders, trusts, corporations, and nations, for instance. The legal system already operates with fictions.

Formal law tends to seek guidance on normative issues from the general population, rather than from legal experts. People care for animals and wilderness. Natural rights are defined by positive laws and by negative restraints on behavior. Laws are needed to protect wilderness, now. Legal or religious action almost always precedes the general shift in conscience. The obligation to treat others equally includes an obligation to change human social patterns in the direction of equality.

4.5.4.3.5. Everything as Ethical

Ethics now considers almost every human being and human interaction. The restriction of ethics to exclusively human modes of existence, however, leads to a troublesome isolation. Human beings are not separate from their social and biological communities and these communities are embedded in ecological contexts with biogeochemical processes. Ethics must be extended to the framework and to the nonhuman communities in the framework, without which there would be no human health or wealth. Through science as well as through mysticism, we understand that communities of other beings have their own values and rules for living together. It remains for us to integrate and codify human rules that recognize the values and rights of other beings.

How do we share things, including power? Michael W. Fox proposes a biospiritual ethic as a unifying set of principles, ethics and values that will bring about a nonconflicting state of one earth, one mind. The ethic is based on the biological fact that all humans and living beings are kin and that life is spiritual—love is stronger than violence. It arises from seeing humanity in an ecological perspective.

4.5.4.4. Cultural Interactions

The final function of politics is to moderate how the culture interacts with others cultures. There are only a few basic was to do this: They can ignore each other. They can trade with the other culture. They can attack it. They can try to conquer or destroy it, or to control it or colonize it. Or, they can cooperate or compete in trade ideas.

Culture provides a filter between humans and other humans. So, culture can serve as another evolutionary filter. It designates what we attend or ignore. It has to screen less valuable information to avoid information overload. In a way, culture is like a trigger. It allows less information to activate the system. And, this is the only way to increase information handling without making the system larger and more complex—of course, this is what stereotypes and metaphors do, also. But, culture traps people in behaviors. Sometimes, violence can be used to escape such a trap. Politics has to direct interactions so that they benefit the culture and allow it to survive and prosper.

4.6. *Believing in Place: Ecodeontics (Religion)*

Religions are attempts to understand or control the world, either by understanding the invisible or by having spiritual beings intercede. The word 'religion' comes from the Latin words to 'bind' or 'read again.' Thus religion is a way of binding power or managing the invisible. This is reflected in the neologism, Ecodeontics, from the Greek word fragments meaning 'bind to the house.'

Religion tends to reinforce the integrity and structure of society, by providing a common image, and reinforcing the belonging and commitment to the group. Religious claims about the other spiritual world tend to be counterfactual, but they cannot be too implausible or people would not pay any attention. The supernatural has to play a part in the world. It has to be associated with living beings.

The stories of religions concern events that are deeply meaningful to the listeners. This helps bind the group, also. Religion may help control disruptive forces, especially things about distribution and power. Religion coerces people into a social contract. Religion and story-telling can reduce the variability of individuals in a group. But, this might increase variability between religious groups.

Religious rituals can also stimulate endorphins to the brain. These rituals may include painful poses, rhythmic movements, singing, or trials of endurance. Endorphins have good effects on the immune system. Endorphins allow people to feel positive about people who share their experiences. Trance states are another feature of religion.

What is the world like? How did the world get the way it is? And what is the role of humanity in the world? All cultures ask and answer these questions. Some of the questions could not be answered from direct observation. And so many of the answers are not limited to observable events. Ideas concerning humanity and the nature of the universe tend to form a coherent system in which ideas are integrated or rejected over days or centuries.

Claude Levi-Strauss emphasized that tribal cultures brought the full power of the human brain to bear in expressing a cosmology; the savage mind is as logical as the modern. Levi-Strauss thought that mythology was explanatory, a product of observation and reflection of reality. Myths often were created, however, out of inherited ideational pieces through fitting together (bricolage). But thought proceeded through understanding, with the aid of distinctions and oppositions. Levy-Bruhl's opinion was that participation was more important than logical thought, but the participation is unavoidable. Cosmology serves to explain appearances, if not in observable events, then in terms of a hidden reality. These two realities make up a totality that is explained. Usually the present is explained as the result of past events. Mythology is constructed as a poetic system. Joseph Campbell states that "Mythology—and therefore civilization—is a poetic, supernormal image." Mythologies and religions are great poems. When recognized as such, they point through things and events to the ubiquity of a presence that is whole in each.

Cosmology forms part of the ideological system of culture. Cosmology is a collective image of the universe. Ezra Pound stated that the image is an emotional and intellectual complex in an instant of time. He used the figure of a Chinese ideogram as an expression of a complex image created by throwing groups of elements together without predication, which is a characteristic of assertion. The cosmology includes beliefs about the origin, structure and destiny of the universe.

As a part of a cosmology, religion shared the same functions: To order, to explain, to justify and to integrate people into the image. Confucian concepts of ritual and etiquette helped regulate social conduct and made people feel good about their station. "Inequality is the nature of things" and "seek no happiness that does not pertain to your lot in life." Taoism also emphasizes the way of Heaven, which seems to be stay low to avoid having your head removed.

Bronislaw Malinowski argues that the primary function of religious mythology is to justify human activities, in order to have those activities continued. Unless a cultural phenomenon satisfied a basic human need, it would not be repeated. The human needs that he listed were all physiological: metabolism, reproduction, comfort, safety, growth, and health. But not all needs are physiological. At the other extreme, Jeremy Rifkin considered only psychological needs were satisfied by a cosmology. But there are also social needs to be considered, such as social status.

Religion cannot be separated from economic and political activities. Technology allowed the same labor force to increase its output from food production, which may have freed labor for other endeavors such as building religious monuments. But, cooperation is necessary, especially in villages and between villages later. As a result large-scale water systems were first managed by religious leaders, then later by secular leaders. In China, the elites gained control of the religious system and used it to increase their own political and material power.

Shared beliefs in a religious community may allow it to out compete strictly secular ones. It permits more sacrifice and commitment. Religion may help control disruptive forces, especially things about distribution and power. Religion coerces people into a social contract, although it may reduce the variability of individuals (perhaps diversity) in a group.

4.6.1. *Sacred Profane Divisions*

"Only the Creator, the Maker, Tepeu, Gucumatz, The Forefathers, were in the water surrounded with light. They were hidden under green and blue feathers, and were therefore called Quetzal Serpent." This is a description of the Mayan cosmos, from about 800 CE. Similarities can be discerned between this and other archaic cosmologies: The distinction between the sacred and the profane, the meaning of the cosmos for humans, their adaptiveness to nature, and their personal aspects.

Human groups have traditionally ordered the universe into two realms: the sacred and the profane. Places without special significance, that do not partake of the sense of the center, that are beyond, are profane. The center, as point, place, locale, or vague area, is sacred. What is known and mythologized is sacred. The lived place is sacred. The familiar is sacred. Sacred space is where sacred experience occurs, that is, the manifestation of a wholly different nonhuman order. Most pre-modern civilizations organized space around a sacred center. The areas outside were considered profane wilderness. Peoples as literate as the Greeks and Chinese structured their worlds this way.

People create an image of the world, in which the most sacred part, a mountain, building or grove, is the most central. The image of the center of the earth, *axis mundi*, running through the middle place, may often give a moral or physical strength. In fact, the image of the center, and its cosmic importance, may be necessary for a culture. The image generates hope and confidence.

Humans have a habit of structuring the world with their own group at the center. This ethnocentrism is evident in small tribes as well as in large empires, such as the Roman or Chinese. The center of a cosmos is usually the place of living, the locality of the group creating the cosmology. Ethnocentrism seems to be universal. Even if the center of the world is relatively featureless, myths are created to give it special significance; a sacred event may give sanction to a rock or tree. The center becomes sacred. Temples may be built on the spot. Sacred places. Grounds around tombs of emperors as natural parks. Grounds around monasteries similar. The holy character of ghosts, but satisfied the need for religion and recreation. Yet nature could impart virtue.

As the world becomes more complex, its image becomes more complex. No longer is space just sacred or profane. The ideal landscape became the middle region, the garden between the complete order of the city and complete chaos of the wilderness. The garden is cultivated from the wilderness as a middle landscape. (The city expands at the expense of gardens.) The garden is a human order, but not usually sacred. Not only poets, but politicians and economists created the image of the rural ideal. In traditional Chinese social classification the farmer was ranked below the scholar, but above the artisan and merchant. Jefferson wished for U.S. citizens to be husbandmen in a like manner.

Joseph Campbell thought that whatever gives power is considered sacred. Food is sacred because it gives the power of life. If custom is essentially sacred, as Rank taught, why should money be different. Money gives power, and all power is sacred. But, money is [389] not sacred because its material power occludes the intrinsic worth of things. Its objectivity reduces uniqueness to the common denominator of economic worth in a closed, limited system. The writings of Emerson, Thoreau, Whitman, R. Jefferies, W. Hudson, and others combined a sacred vision of nature with the observational methods of mechanistic science. Later naturalists like Muir, Burroughs, Seton, and G.B. Grinnell stressed the ethical problems of the mechanistic approach and suggested a science that might include the sacred.

4.6.2. *The Invisible & Visible*

The whole city, and not just the temple, was conceived as an earthly imitation of the cosmic order, a sociological middle cosmos, established between the macrocosm of the universe and the microcosm of the individual. Through the priesthood, the one essential form of all was made visible. A religious attitude is directed toward understanding the invisible part of the world. Ritual is an activity directed toward the hidden properties of the cosmos. In a sense religion penetrates other sections of a culture, economic and political. It explains the world in terms of a necessarily hidden reality; that is, by definition. It makes the universe meaningful by reducing it to human terms, by using the human shape as a metaphor, anthropomorphism. It provides individual solace and hope during catastrophe, and security during danger. It enhances and maintains social order, through participation in ritual.

Heraklitus retained the idea of a basic substance, fire, but subsumed it to the logos, the unity of things in their structure. The visible world was that in which the true essence revealed and hid itself at the same time. "Nature loves hiding," as Heraklitus said. The essence revealed itself in signs. The cosmos was the "same for all" and "one and common" only for those who were awake to read the signs. Part of nature is hidden; part will always be hidden by enfoldment. This is the invisible that will never be named. Maurice Merleau-Ponty thought that the subject and the world are mutually enveloping and inseparable. The mind hides on the other side of the body, invisible. Consciousness brings the invisible over into the visible.

The forest ecologist Chris Maser is fond of saying that most of the cycling in nature is invisible because it is underground or in the air. Many cycles are investigated through ecosystem analysis, where energy and materials are traced through transfers through compartments in an ecosystem. The Gaian ecosystem is a network of coupled smaller ecosystems connected by global patterns of water and air that adapt to each other through feedback loops that regulate physical and chemical environment. All based on great bacterial ecosystems mostly invisible to us.

Much of the operation of culture and cities is also invisible. One original purpose of art was to harness the invisible forces of society, nature, and supernature by making them visible. Humans establish formal relationships with the forces of nature by defining them, as they happen, and then, trying to control or placate them. The visible symbols of invisible forces are interpreted by art into known forms or essences. The invisible side of life is visible through art, rituals, myths, ceremonies, and dreams. Communication is possible through these activities. Rites of passage reflect these beliefs.

4.6.3. *Specialization & Sacrifice*

There were ways that chiefs could keep the underlings happy or at least resigned. They could enlist ideology or religion to justify transfer of wealth to the rich and leader. The shared religion also makes strangers act more peacefully without kinship. It gives people a reason to sacrifice their lives for an institution that may have few genetic links.

Religious conduct specified many daily behaviors. Masai Society requires cattle for sacrifices and ceremonies, as well as payment of fines. In China, people believed in many gods and spirits who ruled the universe. To gain their approval, they offered sacrifices to them. For Albert Schweitzer, nature was maintained only by being in contradiction with itself; the most precious form could be sacrificed to the lowest form in the struggle for survival.

William James suggested that the essence of religious experience is a sense of union with a larger form, a deeper, more meaningful reality. For Aldous Huxley, the essence of religious behavior, distilled from his studies of religious behavior, was the golden rule. Confucius proposed "reciprocity" as a word to serve as a prescription for one's life. And, reciprocity is the meaning of the golden rule for behavior: Do unto others as you would have them do unto you. The golden rule has been extended by others, such as Gandhi, who said: "live more simply, that others can simply live," and Ervin Laszlo, who said: "Live in a way that allows others to live as well."

4.7. ***Expressing the Place: Ecopoetics (or Art & Design)***

People express a place, taking on the characteristics, and reforming them to human ideals. The extension of human identity with places expresses the places. Of course, the characteristics sometimes get expressed through human cultures. More often, places become expressions of human ideas and human patterns. Art is the process of expressing things in materials, or art is the expression of patterns.

The spirit of each place is unique. Place is not just location; it is the total sum of objects in the landscape combined into a unique whole. The identity of place often leads to human identity, thus people call themselves after places. The more unique a place the stronger the emotional attachment of the inhabitants. Every place has certain characteristics that enforce the spirit of place, for instance, a strong definition of place or indicators of great age, or where a place distills the essence of larger landscapes. A sense of wildness and water also contribute greatly to the spirit of place.

Every organism creates an image of its place from what is meaningful to it. This image is what fits the organism to its place. Suckers and caddisworms have simple images; coyotes and humans have more complex ones. Boulding (1956) notes that the image as a cognitive construct of the world has several aspects: Spatial, temporal, personal, relational, value, and affectional (emotional) for each individual. The total sum of individual images is a world. Some of the images we impose on nature result from idealized notions of pastoralism or technological futures. Thus, landscapes abound in nostalgic or consumptive trends on many levels of explication—some are iconic, some invisible. We originally perceive the landscape symbolically, but the landscape has other functional dimensions that increase according to use.

4.7.1. *Speaking Writing & Language*

Grooming in primates releases endorphins that create good feelings. Grooming, or contact in general, is a form of physical communication with high emotional content. Time devoted to social grooming correlates with group size. To achieve social integration at a large size would require more grooming by all animals. This upper limit that can be devoted to grooming would limit the size of the group that could be bonded in that way. If the group exceeds the upper size, it may be less cohesive, fragment, and break up. Time is therefore the limiting factor. It appears to be 20 percent of the day is set aside for grooming. Social interaction is limited by the demands of foraging. Thus, the upper limit of individuals is 70-80.

Are human populations different? They need about the same time for foraging. They seem to use about 20 percent for social interaction. Human groups in foraging societies tend to be 70-80, also, although they can be 150 people, sometimes up to 250. What is the difference? Could it be that our communication is more efficient? As a result of language? That is, language can groom 3-4 people at the same time. More than 4 listeners and the conversation tend to split into two groups or more. It allows us to gossip about others and exchange more information. Language is a social communication that requires the participation of listeners. It may be grooming at a distance, hearing distance anyway.

People speak to communicate with each other about the environment and each other. Benjamin L. Whorf demonstrated that language is a source of information concerning culture. Whorf considered language thought. But, thought is more than language. It includes visual and other sensory forms, different than verbalization.

Whorf also claimed that language is shaped by the world and in return shapes the world. That is, people can only express what language can express. The Hopi language, which Whorf studied in the 1930s, differs dramatically from European languages in its sense of time. The language does not have past present and future tenses, which divide time into measured units before and after now. Instead there are two modes of thought: The objective where things that exist now, and the subjective, where things that can be thought about and are becoming, exist. Each thing comes into being with its own rhythm.

Language is closely identified with place, with the particulars and peculiarities of place. Without place and place-bound cultures many languages are dying. Living in different ecosystems, and having a language suited for that system, may have allowed humans to progress more rapidly. Social evolution happened rapidly as a result of languages. It may have also enforced separation and the isolation of communities, genetic as well as linguistic.

Languages reflect the particularities of place. English-speaking people can name at least six kinds of light that are refracted through a prism. The Bassa language, in Liberia, have only two words for differences in a spectrum. The Hanunoo people have names for ninety-two varieties of rice. The Inupiat have 18 words for kinds of snow.

Papua New Guinea may be the most linguistically diverse place in the planet. Four million people in the size of France speak 860 languages (and the French once spoke at least seventeen languages). This diversity might be reflected by the characteristics of New Guinea: It is an Island; it has many barriers, rivers, mountains and ecological zones (vertical ecology);it has been stable; it has wealthy rainforests; it has the presence of diseases, such as malaria; the human groups are self-sufficient; the cultural environment kept the language viable; and, it has not been taken over by European agriculture, that is, it was considered resource poor.

In the past 500 years, it is estimated that about half of the *known* languages on the planet have died. In the thousands of years before that, probably thousands of languages died, including: Etruscan, Cretan, Sumerian, Egyptian, Meroitic (Sudan, 800 BC-300 AD), and Cumbria (Britain). Recently dead known languages include: Cornish (S. England, 1777 died, with last speaker); Mbarbarum (Queensland, died 1972, with last speaker); Manx (Isle of Man, died 1974, with last speaker); Cupeno (Cal, U.S., died 1987, with last speaker); Wappo (U.S., died 1990, with last speaker); Ubykh (NW Causasus, died 1992, with last speaker); and Catawba Sioux (U.S., died 1996, with last speaker).

Australian aboriginal languages are disappearing perhaps at one every year. By 20,000 they decorated caves with crude paintings near Sydney harbor. Population may have been 300,000-800,000 in 500-600 tribes, speaking 200-300 languages, with 500-600 per tribe.

Irish, although it has the official backing of the state, is used less and less at home and rarely as a business language. Yet, around 1000 AD it was rapidly expanding. Irish literature is the oldest and Europe after Greek and Latin. Perhaps only 9000 speakers can pass it to their children.

In the U.S., only six languages have more than 10,000 speakers, including Navajo, Cherokee, and Mohawk. 51 languages, such as Penobscot, had fewer than 10 speakers; another 35 had between 10 and 50; and 79 languages, such as Pomo or Yuki, had most of their speakers over 50 years old.

Language is an important aspect of culture. Language is a form of communication where arbitrary signs stand for concepts. Speech is not required if there are hand motions. The loss of languages may be related to the loss of places, or their equalization through monotone industrial processes or images.

4.7.2. *Design of Places as Ecosystems*

Nature is self-making and self-designing, but we humans now influence every natural system, taking what we need from some ecosystems, enhancing a few, misusing others, and interfering with the rest. We need designs to restore the balance between human needs and natural processes. Ecological designs focus on whole communities that work in the same self-sustaining and self-limiting ways as nature. By consciously creating meaningful order, we can develop ways of producing widespread community wealth while positioning the community for a long, sustainable future in a healthy environment.

There is no guarantee that nature can provide humans with everything they want. Recognizing the lack of guarantee simply recognizes that nature is wild and we must come to terms with nonhuman beings and processes. It is not enough to arrange trees in rows to maximize future harvests; it is not enough to preserve small areas of old-growth without natural disturbances. We must pay attention to the processes that make up the entire habitat, for example, the role of herbivores on trimming vegetation and diversifying it by predation. The design of the ecosystem and its management must ensure that the processes operate to maintain a dynamic state. Furthermore, the context must be conserved. The ecosystem, however, cannot be considered outside of the context of the entire landscape, including human images and institutions.

A number of ecosystems, everywhere, are still wild. Many others have been impoverished, but have the capacity for regeneration. As afforestation proceeds to reclaim wasted lands, as it has in England and Europe, more attention will be paid to the shape of the forest ecosystem. Design principles can guide our decisions. Although design in Europe has primarily been concerned with artificial forests, many of the ideas can be applied to wild forests. Human design, until now, has been primarily visual. It has emphasized aesthetic reaction to a place, but also the uniqueness of a place. It is not enough to create ragged edged forests to satisfy human eyes; it is not enough to leave beauty strips of real forest to fool travelers. Design is needed to create natural spatial patterns and temporal phases across watersheds and entire landscapes. Ecological design considers the whole context.

4.7.2.1. *Definitions of Ecological Design*

Landscape is a heterogeneous area composed of a mosaic of interacting ecosystems of various sizes; an ecosystem is a community of organisms living in place. Design is a human project in which, as Oliver Lucas says "visual and physical parts are assembled in order to achieve a specific end result." Ecological design is the creative modification of ecosystems to repair or enhance their ability at self-organization and maintenance of their complexity and diversity. Diversity, as in biological diversity, means species richness, different age and size classes in a population, and genetic differences in a species, as well as kinds of habitats present in an ecosystem and the kinds of communities occupying the habitats; and the kinds of ecological processes that maintain habitats; and the variety and richness of the planet's genetic heritage. Ecosystems that do so are healthy. A definition of health is the condition of being sound in body or well-being.

Signs of ecosystem health include the homeorhesis of the system (after Waddington), the stability of the system (that is, its resilience after stress, such as floods), the diversity of its components, the continuous recycling of elements, and flourishing. The ecosystem stocks and flows are determined by stochastic variability due to random factors. Ecosystem health seems to mean the links between human health and ecosystems. Robert Costanza stated that: "an ecosystem is healthy ... if it is stable and sustainable—that is, if it is active and maintains its organization and autonomy over time and is resilient to stress." Health as related to ability of ecosystem to continue to provide services.

Health is related to stress, both good stress and bad stress. Stress may be related to the rate of change for the system, in addition to loss or gain of components or changes instructure. Health is the overall ability of a system to maintain itself under a normal range of environmental conditions. Obviously, a pioneer community may change the conditions to favor a new level of the system with new components.

A ecosystem can be characterized by a number of words: productivity, openness, efficiency, maturity, stability, durability, self-making, flexibility, diversity, richness, wholeness (matrix), and dynamic. This list is not meant to be exhaustive.

Living communities are self-organizing systems with emergent properties; they maintain themselves in a state of flux; species are always coming and going, and changing proportions. In a living system, nothing keeps growing forever; things die and are reborn in cycles. The continuation of the system depends on these cycles. The cycles are bound by limits. Both individuals and communities are usually bound by one or two specific limits. Ecosystem health occurs within those limits.

4.7.2.2. *Characteristics of Ecological Design*

The characteristics of ecological design have to address the characteristics of ecosystems and places. The courses of the design have to imitate the processes of systems and places. The wholeness of the design has to address the ideas of spirit of place, and sensory force.

After the processes are identified, they have to be related to the patterns. Design may have to try shortcuts, unless it can afford to take the ecosystem time, which is often many human lifetimes.

4.7.2.2.1. Imitating Courses

Design has to identify patterns of movement and construction, then it has to imitate the processes and patterns, modifying them to include more things of human interest and need. This is more difficult than copying a shape or a structure. Fortunately, imitation is a human strength, even if the recognition of complex, long-term, moving patterns is not. Courses can be thought about using topological and mathematical models representing a four-dimensional landscape.

4.7.2.2.2. Extending Identity & Wholeness

Ecological design has to capture the integrity of a place, or restore it. Unity is a fundamental objective of landscape design. Unity is the way the elements, including shape and scale, of a landscape are combined.

For instance, visually, a forest ecosystem usually dominates the landscape. From a distance, even-aged forests have much the same impact, in terms of color, shape, and scale, as uneven-aged forests. Diversity becomes more important visually at a smaller scale. Natural forms of the forest are unified with the landscape because the margins are very uneven, and open space in the forest is part of the mosaic caused by birth and death of individual or groups of trees.

4.7.2.2.2.1. Spirit of Place

The spirit of each place is unique. Place is not just location; it is the total sum of objects in the landscape combined into a unique whole. The identity of place often leads to human identity, thus people call themselves by their place names. The more unique a place the stronger the emotional attachment of the inhabitants. Every place has certain characteristics that enforce the spirit of place, for instance, a strong definition of place or indicators of great age (trees or rocks), or where a place distills the essence of larger landscapes. A sense of wildness and water also contribute greatly to the spirit of place.

The spirit of Paradise Creek in the Palouse grassland, in Idaho in the U.S., has changed greatly in the past 200 years, but it is still easily defined with elements of water and wildness. The creek is a more intimate place, less dramatic than Steptoe butte or Palouse falls. The spirit of the place is the best guide to design.

Each place expresses a unique combination of elements, including contrasts, dramatic features, and the presence of water. Design can work to be consistent with the recognized spirit of place. If the design recognizes this aspect of the landscape, it may be stimulated by spirit and it may further enhance it—what it should not do is degrade it. Forest design can emphasize some features above others.

Goals of good designs include: To relink people with genius of their places, to revivify image and identity with places, and to develop and maintain the identity of places.

4.7.2.2.2.2. Sensory Force

All the elements of design can be combined in an image. Every organism creates an image of its place from what is meaningful to it. This image is what fits the organism to its place. Suckers and caddisworms have simple images; coyotes and humans have more complex ones. Kenneth Boulding (1956) notes that the image as a cognitive construct of the world has several aspects: spatial, temporal, personal, relational, value, and affectional (emotional) for each individual. Cognition is an active relationship that is creatively shaped by the participants. Participation is not an option by the way—every scientist or inhabitant becomes part of the system of observation. The total sum of individual images is a world. Some of the images we impose on nature result from idealized notions of pastoralism or technological futures. Thus landscapes abound in nostalgic or consumptive trends on many levels of explication—some are iconic, some invisible. We originally perceive the landscape symbolically, but the landscape has other functional dimensions that increase according to use.

Visual force is a psychological interpretation of perceived power in a landscape. As a principle, it is embodied in psychology, art, graphic design, and architecture. The human mind responds to visual force in predictable and dynamic ways, for instance, visual forces in landscapes draw the eye down convex slopes and up concave ones—the strength depending on the scale and irregularity of the landform.

The effect of a forest landscape is not completely visual, however. Smell, sound, touch, and even taste play a large part of our appreciation of forests. Crawling, which is highly recommended by Gary Snyder, climbing, listening, and tasting things, such as soil, bark, or lichen, can expand our perception of other aspects of the forest. De Tocqueville commented on the ceaseless noises in the forests he encountered in Pennsylvania and Ohio—they kept him awake.

4.7.2.2.3. Enhancing Openness & Flexibility

Because the operation of the universe tends to change systems, the design of a place should be open to the types of processes that could destroy the design. Furthermore, flexibility, defined as the unused capacity for change, can be designed into the system. The parts of a system have to maintain the potential for all possible behaviors that could flow from any other part of the system. Openness and flexibility are characteristics of healthy ecosystems, and they can be considered and enhanced by design.

4.7.2.2.3.1. *Anticipation.* To anticipate means to look forward, to think about possibilities or to expect surprises—in a way it is to create thought experiments about possible future events. Responses to things can be worked out.

4.7.2.2.3.2. *Playfulness.* Playfulness is a fondness for play, for engaging in activity because it feels good. Play is an activity that can lead to relaxation, or practice for more serious things relating to food and mating. In another sense it means to move freely within limits.

4.7.2.2.4. Participating in Co-constrained Construction

Designs are limited by the real biological constraints of ecosystem processes and biogeochemical cycles. We must know the constraints in order to create a healthy design. The design has to work within the constraints of the ecosystem. Rather than emphasize static equilibrium, the design should emphasize heterogeneity and learn to adjust to disturbance.

People have needs, animals and plants have needs, the site does have constraints, and these things can be married into a good pattern. Harmony is a constraint of the whole system. Design is the participation in the process of the ecosystem as a harmonious system, with mutually restrained conflicts and constrained influences. An ecological design is a form of co-constrained construction, where the organisms, environment and designs are co-implicative, co-defining, and co-constructing. They all engage in a process of self-assembly, where the whole is the whole system.

4.7.2.2.4.1. *Participation.* To participate means to take part with others in some activity. Designers should participate in a complete design process, guiding involvement and commitment to the art of living together as a community.

4.7.2.2.4.2. *Adaptation.* Adaptation is a process of making fit by adjusting to circumstances, environmental or cultural. Here it means fitting into an ecosystem, within established cycles and functions. It is adaptation that improves the chances of survival for a living being.

4.7.2.2.5. Investing in Stability & Constancy

People often judge the health or wholeness of ecosystems, or the goodness of designs, by how they look. Traditional design has emphasized visual results above all else. Ecological design, however, achieves the same results by paying attention to the structure and function of the ecosystem first. Design has been concerned for centuries with making domesticated landscapes out of wild ones. Now, design is addressing the opposite problem: How to preserve or provide the conditions for wild ecosystems so that they are stable or constant within the processes shaping and affecting them.

Signs of ecosystem health include the homeorhesis of the system, that is, the stable, directional flow of the system capable of resistance, resilience, and accommodation. The design for an ecosystem describes the system in a comprehensive interdisciplinary approach, using dynamic concepts such as constancy and stability.

4.7.2.2.5.1. *Plurality*. Plurality is the condition of being numerous, the existence within an ecosystem of distinctive living patterns. This conditional applies to cultural and political patterns as well; it allows many values.

4.7.2.2.5.2. *Frugality*. Frugality means being economical, avoiding excess and waste. It is the style of use of things. According to Henryk Skolimowski and others, it is a necessary lifestyle in a crowded world.

4.7.2.2.6. Stimulating Productivity & Health

Most products of an ecosystem are produced and consumed and recycled within the ecosystem. Humans need to minimize the external inputs in the form of energy and exotic substances. The community must be restored to health. This means balancing human needs with bird or fish needs in a sustainable pattern.

Ecological design is the creative modification of ecosystems to repair or enhance their ability at self-organization and maintenance of their complexity and diversity.

Health is the overall ability of a system to maintain itself under a normal range of environmental conditions. Ecosystem health is one of the goals of design. The goal, of course, is not an end point that can be reached once, but is rather a continual striving.

4.7.2.2.6.1. *Respect*. Respect means to show regard for, or to avoid interfering with, others. It means recognizing the "beingness," value and rights of the ecosystem itself.

4.7.2.2.6.2. *Responsibility*. Responsibility is the condition of being accountable for one's actions or obligations. Here it means undertaking all aspects of a design, regardless of the expected level of success.

4.7.2.3. *Applied Design As Lessons from Nature*

Design can imitate nature on many levels, from structure and process to landscapes. We can imitate the structure of mature forests by planting on every level of the forest hierarchy, from canopy to below ground. We can use native species. We can imitate the process of forests by allowing birds, bats, and other animals opportunity to distribute seeds and energy to other areas or prey on "pests". We can create microclimates within the landscape that may shift the landscape in new directions. Planting trees, for instance, allows new species to become established under their protection. Ecosystem health is one of the goals of design.

The landscape provides its own metaphor for design. The landscape is a unique individual, a community, a dynamic system of interacting patterns—the human pattern is a part of it now and should be preserved as part of the whole pattern, but not necessarily as the only pattern or a completely dominant one. Most products of an ecosystem are produced and consumed and recycled within the ecosystem. Humans need to minimize the external inputs in the form of energy and exotic substances. The community must be restored to health. This means balancing human needs with birds or fish needs in a sustainable pattern. Each element in a pattern relates to others and to the whole.

Natural patterns can suggest a number of regularities to understand for the design of ecosystems, according to Michael Soule. Well distributed species are less at risk than concentrated ones. Large blocks of habitat are safer from species extinctions. Blocks close together are more effective than those far apart. Contiguous blocks are better than fragmented blocks. Interconnected blocks are better than isolated blocks. Corridors can make functionally larger blocks functionally. Roadless blocks are better. Human disturbances similar to natural ones are more likely to be resisted or accommodated.

These rules work well with natural processes that operate in ecosystems, such as metalysis (building up and breaking down), animal movement, interelement flows, human interactions, and shifting mosaics. For instance, forest fragmentation can be reduced through the design of forested areas, taking into account the genetic diversity of the trees, catastrophic conditions, minimum viable populations, corridors, and edge effects. The survival of organisms usually depends on one of two factors in the web of relations. These factors can be modified by design.

Wild landscapes are affected by climate, soils, interactions, and disturbances. Domestic landscape is affected by land use as well. The greatest changes have been brought about by the destruction and creation of forests. With the predominance of artificial forests, it is important to consider the qualities of naturalness in the landscape. Forests are expected to meet the needs of society by producing timber, creating wildlife habitats, and providing recreational opportunities for people. But, forests are also expected to look natural.

The English Forestry Commission's guidelines (1994) to principles and practical applications of forest design may be of use. They represent an established standard. They do not cover every aspect of landscape design or details of design techniques, however,; nor do all of them apply to wild forests. The guidelines indicate what to look out for, and which situations may need special attention. Forest landscape design is a complex subject, as are forestry and ecology.

The values of the land and forest must be most carefully assessed. The characteristic qualities must be identified and measured for uniqueness. Comprehensive landscape plans should be required when planting or extensive felling is planned on a large scale. The patterns established at these times may persist for many years or centuries. Good design maybe able to resolve conflicts between characteristic qualities of the landscape and the changes from use. Also, the design should last as long as possible and should be self-sustaining.

4.7.2.4. *Stages of Ecological Design*

Ecological designs can be applied in five stages. These stages are typical of all designs and were used for designing Idaho, Washington and Oregon forests:

• First, review the situation, observing patterns of movement, population change, land use, building and development, boundaries, limits, and life. Conduct ecological and functional analyses.

- Then record all of the resources, from physical resources to cultural resources. Survey the area and create base maps, from geological to zoological maps.
- Next, evaluate the interactions in terms of impacts, needs, goals, and limits. Assess the whole system and create a series of plans, from the site plans to value plans.
- Start to design, which is a community process requiring the participation of all people, including the elderly, handicapped, and poor, as well those ultrahuman beings who cannot voice their concerns. Synthesize simulations and models(conceptual, capability, and suitability). Make another series of plans, from landscape plans to

policy plans, within a master design.

- Finally, implement the design together and maintain it. Use appropriate measures and techniques, emphasizing native species over time to ensure the stable processes of transformation. Provide services for continuity and management.

Forest design must work within the components, structure, and function of the forest. Unless it does, it will not be long-lasting or satisfactory. Because design has to work with a forest, whose aspects are often ambiguous, fuzzy, changing, and general, design has to be able to work with these aspects, as well as within the constraints of the forest.

Forest design takes far more time than graphic or automobile design, due to the complexity, size and longevity of its subject. A number of factors have to be carefully assessed first. Forests require a lot of observation before activities can take place. Forests can be highly reactive to change. Some people value different character of forests than others.

4.7.2.5. *Levels of Ecological Design*

As a design process includes planning of ecosystems, this mean a fourth level of design—notice that after number four, all are properties of normal applications of design: (4) The community, for instance forest, city, or corporation, (3) systems, such as ecosystems, traffic, or industry, (2) products, as habitats, houses, roads, or plant sites, and (1) components, such as trees, fungus, bats, tools, rooms, cars, or land. Many problems occur at level three—more are to be expected at the new level four.

All levels of design need to be addressed, from the conceptual to the political, and are involved in all stages of the process. This involves new challenges for ecological design, which has to:

4. Relate a project to its total context (fourth level of design); be concerned as much with cultural survival, justice, and wilderness preservation as with efficiency and aesthetics.
3. Consider the whole perspective (ecocentric, perhaps); the proper vision is of the whole community in which we dwell. Apply ecological concepts, such as networks and carrying capacity.
2. Make designs are anticipatory, flexible, pluralistic, polyvalent, and polytechnic. Make open guidelines for long-term decisions.
1. Essentially, work backwards from values and goals, and from the bottom up and inside out, drawing designs from the genius of place.
0. Participate in place, care for all inhabitants, and assume responsibility for the designs.

One of the shortcomings of design has been the lack of consideration of the higher levels, which can result in specific problems with safety or recycling.

4.7.2.6. *Principles of Ecological Design*

Principles must be flexible to mirror the flexibility of open systems; flexibility is provided by diversity in fact. Sample principles:

- Work with the duration of the ecosystem. Succession can be assisted, slowed, or speeded up, but not skipped or ignored.
- An ecosystem is designed by its limits, time, scale, complexity.
- Only details and exceptions form the system; only generalities are important—so you have to live with contradictions.

- Make the smallest number of changes.
- Adapt the process of change to the site.
- Seek the best use for products; everything in the forest is a resource for something; many can be directed to human use.
- Extend the life of things through cycling, then return them to ecosystem; things can be recycled indefinitely, as in an old growth cycle.
- Preserve the components, structure, and function.
- Understand the patterns and connections. Make sure they are not broken.

Principles cannot be reduced to other principles, because they emerge from earlier ones and they expand into new forms. Expansion is why we cannot fit them into the old cans.

These design principles are based on basic principles of nature, such as the principle of change. Nature is in flux, culture is in flux, everything is. The climate will change, the shorelines will change. Human understanding and behavior is changing. But, nature is also self-regulating. This is the principle of self-regulation. Nature has evolved to maintain its stability in the face of many kinds of disturbances from planetesimal impacts to changes in atmospheric composition. Nature will continue to regulate itself even as human beings make dramatic changes. The danger is not so much that nature will collapse, as that humanity will lose those things that it values the most, from cool air to wilderness.

Another principle is the principle of flow, which is necessary to the functioning of organisms as well as to the biosphere. Flow can only be realized through structure, however, cell or ecosystem. It is the problem of free flow or division by membranes. At each level of a natural system, from cell to biosphere, the units involved do more exchanging internally than externally with other units at the same level. Flow and division must be in balance

The principle of separation, by walls, barriers or membranes—nonflow or limited flow, is crucial to design. Barriers are necessary to maintain form and integrity of individuals as well as of ecosystems. By removing some barriers, such as releasing carbon that has been locked up for eons, we unbalance natural cycles. Even though landscapes exist within large ecoregions, which exist within the biosphere, they need closure to maintain their integrity. At the level of local ecosystems, there need to be fewer closures, so that there is a flow of genetic information within a species as well as between species. Human activities have the effect of blocking the flow at local ecosystems, with asphalt and wires, yet increasing the flow between large regions with ships and airplanes—mostly the flow of pests and domestic species.

Finally, the principle of error, or play, permits diversity. So many things are thrown at the flow of life. There is not just a little error. Half of everything seems to be error: billions of pollen grains, billions of eggs, millions of species. Transmission is not flawless or efficient, but it is generative of difference and diversity. The diversity of the current ecological world evolved through the breakup of Pangaea, which provided the distances and barriers to isolate species.

4.7.2.7. *Applying Ecological Designs to Places*

All design elements are related psychologically by designers, as focus or frame, as contrast or uniformity, as dominant or recessive, or in a number of other pairs. Good ecological design means not violating any of the aforementioned principles and ideas.

Design can improve the results of bad practices. Bad forest harvesting practices often result in geometric wastelands. Good design can correct reliance on straight lines, parallel lines, right angles, and perfect symmetry. In cutting or planting to improve natural appearance a number of things have to be considered, including the age of the forest, windthrow, the width of corridors, and the minimum size of the habitat.

4.7.2.7.1. Ecological Design of Place

There are basic geometric elements of any design, from three dimensions (volume) to 2 (plane), 1 (line), and 0 (point) dimensions. These elements can vary in numerous ways, by number, position, direction, size, shape, interval, texture, color, and temporal. Furthermore, the elements can be organized into groups by nearness, similarity, and difference (diversity), into structures by rhythm, tension, balance, and scale, and finally into a whole with sensory force and a spirit of place (genius loci). All of the elements interact in complex and unpredictable ways. The spirit of the place is the most important principle to be conserved or enhanced.

Geology, climate, disturbance, and stability all produce diversity. Landscape diversity is linked to ecological diversity, which depends on diversity of the substrate. Different ecosystems introduce diversity into a landscape, but different ecosystems often can look similar. Excessive diversity can lead to confusion in a landscape design. Increased diversity also has the effect of reducing scale, so adding diversity can be used to do reduce the scale. A high level of diversity is acceptable if one element is clearly dominant or if the differences cannot be recognized from a distance.

Psychologists have recognized the need for diversity for people's quality of life and emotional well-being. Ecological diversity in the forest has been reduced by human activities, such as planting or grazing. The overall landscape diversity of many forests has not fared as badly, due to the addition of human artifacts, which increase it. An increase in ecological diversity would lead to an increase in diversity of the landscape, however. It would also tend to reduce the scale, but this would not be a problem in large forests.

Process applied to components yields pattern. Nature is composed of patterns. Organisms have characteristic patterns, such as the branching of trees or the cloud forms of tree crowns. Lichens have lobes, wood grain under stress has spirals. The cracks in tree barks form nets. Patterns are not still. A circular pattern through time can be recognized as a spiral (the earth's orbit for example). The pattern should allow for surprises and discontinuities; it can do this if it is flexible. The design of forests is vulnerable to surprises because nature is chaotic (unpredictable) and science itself is uncertain (by definition) about patterns of change in forests.

4.7.2.7.2. Ecological Design of Buildings

Design cannot be separated from social and political questions. Designs, as Churchill recognized, especially buildings, steer our experiences and actions. According to Langdon Winner, design includes the deliberately chosen, enduring forms of both material and intangible entities that affect human relations. Ecological design is an emerging field that aims to recalibrate what humans do in the world according to how the world works as a biophysical system and a cultural entity. Design in this sense is a large concept having to do as much with politics and ethics as with buildings and technology.

Green buildings are sustainable, energy-efficient buildings. In the United States, buildings account for 36 percent of total energy use, 65 percent of electricity consumption, 30 percent of greenhouse-gas emissions and 30 percent of waste output. Making new buildings that use recycled, nontoxic materials, recycle their own waste and derive multiple benefits from the renewable natural resources around them, such as the sun, wind, rain, not only reduces their environmental impact, but also creates indoor environments that are healthier and more inviting places to live and work.

Adapt existing buildings has advantages. The creative reuse of existing buildings offers the multiple benefits of "harvesting" already built structures, reducing the demand on our overflowing landfills and allowing people to maintain a stronger connection with the history and development of a city.

Changing buildings leads to rethinking transportation. Efficient, convenient public transit can be used to get within and between neighborhoods, reducing congestion, pollution and the use of fossil fuels. Modifying roads, especially in neighborhoods to slow or redirect traffic, could allow more walking and bicycling.

Design could integrate the city into its environment like a living organism in an ecosystem, although in this case the city is an extension of living human organisms. The city could generate its own energy with solar and wind power. The effects of climate could be modified with building shapes and vegetation, Design would influence human connections and diversity, with good psychological shapes and connections. Design could also reduce the scale to a more human scale, requiring less management and rules.

4.7.2.7.3. Ecological Design of Tools and Things

The ecological design of tools considers tools in every aspect as an extension of the human mind. Tools and things as extensions of human ingenuity. The design examines the effects, short-term and long-term, of the tools, and attempts to balance the trade-offs. Tools have environmental effects, as well as physical, biological, and cultural effects on their inventors and users. Tools have different levels of involvement, as well as levels of intensity.

4.7.2.8. *Ecological Design Management*

Ecological design is not finished with the design. The system may need to be managed as a result of the design. Noninterference matrix management is proposed as a technique for design. An ecosystem exists as part of matrix that many interacting elements. Any activity in the matrix can have some effect on these elements. The whole matrix needs to be managed with the ecosystem in mind.

Understanding of the principles of ecology can lead to better management. One critical message of ecology is that if we diminish variety in the natural world, we debase its—and our own—stability and wholeness. Many ecosystems have been simplified and degraded. Perhaps we do not have sufficient knowledge to manage a complex landscape because it is too complex to understand scientifically. But we can understand the pattern and drive it in a healthy direction with minimal intervention. We must do all that we can to restore its richness and the natural processes that created the richness.

4.7.2.8.1. Noninterference Matrix Management

A noninterference approach to ecosystem management, the essence of a Taoist way, is to let the system take its own course. Therefore, once the temporary constructs were in place, whether planting or cutting or any other manipulation, the system would be allowed to develop without further interference.

In nature, noninterference means 'letting be.' Noninterference matrix management (NIMM) is not indifference, which is diffuse. It is caring. Noninterference will not lead to chaos, poverty, and stagnation. The technocratic vision strives for "life under control," but the forest is self-managing, productive, efficient, and orderly. We need to practice the rule of noninterference so that all beings can enhance themselves. Noninterference can be derived from nonviolence, or from taoistic nondoing. This attitude would entail using what is necessary, exploiting parts of some ecosystems, changing a place to fit human aspirations, and killing plants and animals for sustenance. But it would also mean limiting humanity and its technological effects, limiting human use to local impacts, and letting other beings live without interference. It is not necessary to dominate or terraform the ecosystem completely to save it. Noninterference matrix management weaves people back into the fabric that supports them and in a sense makes them subject to the constraints of ecosystem processes.

NIMM would manage the system with minimum subsidies; manage activities that could upset equilibrium; manage sustainable conditions; align human activities with natural processes; work with system instead of attacking it; and, restore context. According to Garrett Hardin, many of the ideas necessary to fitting humanity into the pattern of nature are known but not yet popular. For instance, exponential population growth, or economic growth, cannot be maintained very long. Human communities cannot grow 4 percent per year without disastrous consequences to the infrastructure and the quality of life. Growth cannot be continued because the landscape is limited, in terms of productivity, energy, and resilience. Thus, we need to fit our population into the limits of the landscape, although some limits can be expanded by technology or by lowered expectations. The carrying capacity of the area is not only a function of the limits of the community, it is equal to the number of people multiplied by the level of comfort, the quality of life style. Having more energy and space means having fewer people.

Design may be costly. For example, grassland restoration costs about $1500 per acre per year, based on the first two years. Forest restoration costs more. Design may take a long time—longer than human lifetimes.

The ecosystem may be too complex to design. How do we design an ecosystem, such as a complex, self-making, self-sustaining wild forest? We could identify the parts and functions and try to duplicate them, but that would prohibitively expensive and time-consuming. Management has to recognize the limits of design. Limits of ecological design include the fact that ecosystems are wild, and we have no real control over them. The scale of ecosystems is often too large to manage everything. The longevity of ecosystems is too long, and we will never complete the design in human lifetimes. The costs may be prohibitive—indeed, we have depended on the free goods of ecosystems for economic advantages, that would disappear if we had to rebuild the system. Other human limitations apply to our ability to see and understand any ecosystem.

4.7.2.8.2. Uncertainty & Management

Carl Walters discusses how we recognize and measure uncertainty. He presents material on decision theory. He distinguishes between three kinds of uncertainty in natural systems:

4.7.1.8.2.1. Regular disturbances over time generate unpredictable and uncontrollable changes. Certainly this is true, but we have found that regular disturbances are also responsible for creating diversity in ecosystems; furthermore, regular disturbances "habituate" the ecosystems, so that often the system requires the disturbance to continue. Certainly, as he suggests, monitoring is necessary. The uncertainty of these disturbances is usually just in the exact timing and scale, not whether they will occur or not.

4.7.1.8.2.2. Statistical uncertainty about parameter values of functional responses, e.g., production rates as a function of stock size, can lead to problems with exploitation. He suggests the solution to this is simply assigning probabilities to parameter values. Probability is mathematical way of describing "random events," such as coin tosses. Nature, of course, is not random, even at a quantum level. Probability is also a degree of belief, according to Savage (1954). Probability theory centers on the notion of uncertainty as it applies to independent or conditional propositions. There are at least two kinds of probability: Personalistic, which depends on personal beliefs and confidence, used in Bayesian contexts, and Objective, which deals with repeatable events. Probability, in the face of some history, becomes conditional probability to reflect events that have already occurred. In nature, unfortunately, we are not always sure which events have occurred and which have not.

4.7.1.8.2.3. Basic structural uncertainty about what variables to consider can lead to indecision. Incomplete structural representation implies time-varying parameters and surprises involving changes. Surprises are usually the result of incomplete knowledge, but also can be from the emergent properties of systems.

These three kinds of uncertainty are not that different, being basically levels of ignorance. There are other levels and kinds of uncertainty. There are levels of uncertainty, from the quantum to the human. Elementary processes at the quantum level are not subject to a precise description in time and space. Predictions about location and velocity are just statements of probability—this is Heisenberg's uncertainty principle. The effect of this principle on epistemology is that our exact interpretation has to be abandoned. There are uncertainties at higher levels of organization as well; for instance, the hysteresis of some magnetic solids determines their subsequent behavior—without knowledge of initial conditions and all past events, it is not possible to predict present or future behavior. Furthermore, the higher up in levels of organization, the more kinds of uncertainty there are. There is genetic uncertainty, as well as environmental and social.

There are also many kinds of uncertainty, as suggested by Max Black. There is fundamental uncertainty in the thing/event/pattern, as well as in the channels (noise, meaning) of relationship. At the human level, uncertainty can be distinguished between ignorance and conflicting knowledge.

Ignorance can further be divided into kinds: vagueness (indeterminate knowledge),probability (confidence in partial knowledge), incompleteness (of knowledge, missing elements), irrelevance (place in pattern unknown), and fuzziness (overlapping interpretations).

Conflicting knowledge also can be broken down into kinds: anomaly (incongruity of knowledge, simple error), ambiguity (alternative interpretations of meaning), inconsistency (simultaneous untruth), equivocation (knowledge is constant and inconstant, true and untrue), and belief (confidence in subjective knowledge, taboo).

There are probably other components of ignorance, as well as other ways that they can be combined into a formal typography. There are many different kinds of ignorance, maybe more than kinds of knowledge, and that these determine our approach to the practice of making places. If we think of knowledge as an expanding sphere in a space of ignorance, then as the sphere grows, the surface area in contact with ignorance also grows.

Walters stresses that managers have to live with uncertainty, and then makes the great connection that management decisions are essentially gambles. Gambling is a profession that acknowledges the operation of chance and makes conclusions in the absence of facts—few people are successful at it. This is an important admission, that we do not have facts to base our actions on, that nature is a stochastic process, and that ecosystems always changing. Furthermore, we do not know for sure what effects our actions will have on the forest, which used to live so long. If we do not act responsibly, we are gambling that the cost is more than damaging the planet. If we act responsibly we are gambling also. Successful gambling suggests that the proper attitudes for gambling with nature are awareness, humility and courage, not arrogance, fear and maximum use.

Although he suggests that inaction is an inappropriate alternative to gambling as a result of confusion, in fact, inaction is a very appropriate alternative in the face of confusion— when in doubt about an ecosystem, perhaps we should not interfere.

Walters suggests that managers should embrace uncertainty rather than always strive for "certainty-equivalent" policies. That is, they should try to define a set of possible outcomes, e.g., models, consistent with experience, rather than make a single best prediction. managers have to live with uncertainty, but they can make intelligent gambles, based on their ability to work within real ecological and physical limits. Of course, they should not gamble with all their resources at once.

4.7.2.8.3. Precautionary Principle

We should adopt a precautionary principle, which asserts that, if harm is threatened, and if there is uncertainty about the seriousness of the harm, then precautionary actions must be taken. Since the 1970s, in fact, this principle has been incorporated into Swedish and German environmental laws. This principle means that not doing something, "benign neglect," becomes a valid management option. Carl Walters suggests that inaction is an inappropriate alternative to gambling as a result of confusion, but inaction is a very appropriate alternative in the face of confusion—when in doubt about an ecosystem, we should not interfere. As appropriate alternatives, both inaction and action can be suggested by theory.

Traditional planning asks "what is most likely to happen," where uncertainty planning asks "what has already happened to create the future," in terms of long-term trends and demographics, according to Peter Drucker. Traditional approaches to uncertainty management are probabilistic, focusing on uncertainty as a measure of randomness or of confidence. Due to the degrees of uncertainty, as well as the dangerous aspects of modern technologies, this principle has to be applied strongly and with severe consequences for its violation. Ecological harms are occurring on regional and global scales more often.

4.7.2.9. *Ecological Design Limits*

People often judge the health or wholeness of ecosystems by how they look, as a function of richness, despite the fact that ecosystems have to fit into extreme climates and places. Traditional design has emphasized visual results above all else. Ecological design, however, achieves the same results by paying attention to the structure and function of the system first. Design has been concerned for centuries with making domesticated landscapes out of wild ones. Now, design must address the opposite problem: how to preserve or provide the conditions for wild ecosystems.

Design must address the common good, that is, the good of the entire ambihuman community; it can do so by: Promoting the well-being of all individuals in larger community, deciding what is preferable, attempting to regulate and anticipate all effects, encouraging convivial activity, recognizing links and dependencies, mediating the relation between technology and community, and alleviating some of the problems of industrial society.

Designs provide a framework for natural and artificial processes to work in. The patterns in design are echoes of patterns in nature. Good designs learn to embrace error and failure, so necessary in open systems. Most ecosystem designs will not be restorations, because of the uncertainty about the kinds and associations of native vegetation. Furthermore, humans are now an large part, although not yet an integral part, of the system; therefore it could not be restored to a premodern or prehuman state, even if we knew the proper or historical state. This design is not the biotechnological design of a new ecosystem, either; we cannot accurately control and predict ecological events in most ecosystems. However, we can steer some of the events in a known direction—known because we have historical records of the system, although not complete. We can also reduce those human activities that we know alter the conditions of the forest, such as overcutting and pesticide use.

Although ecological design attempts to restore some kind of balance, the balance does not exclude human activity. Rather, it integrates it into the larger community. A moderate number of human impacts can be absorbed by the system—too many destroy the systems capacity for self-maintenance. The design should be open to evolution and to human technological and social development. The design should be based on a model of ecosystem functions, considering diversity, complexity, and the maintenance of natural process—natural here meaning a self-sustaining system composed of elements now lost through human disturbance.

How do large-scale processes influence design? That is, how can design be flexible and open enough to cope with change? How does design accommodate those processes? The processes provide constraints on the designs, which when implemented force some constraints on the processes themselves. The question is how to limit the latter constraints so that the processes are not fundamental shifted into a new regime. For example, a dam is located depending on waterflow, canyon walls permeability, location, rainfall, and other factors. After being built the dam changes rainfall, erosion, waterflow, species groups, and other factors.

The role of designers is to optimize or satisfy the fitness of people with their environments. To fit cultural goals to ecological characteristics and limits. It is adaptive creativity, not just for the current technology, but because it needs to adapt to the technological and natural environments. An ecological design involves designers and people in reshaping and recreating a self-sustaining community. Individual resources are limited. The relationships to strive for here are community relationships. Furthermore, there are limits for human manipulation of other communities. Total control has limits, also. We should not aim to try to control the forest and its habitats. We have to trust that natural processes are self-correcting and organizing.

An ecological design is the creation of a clear vision of the ecosystem that is aesthetic, useful, and self-sustaining. Ecological design fuses art and ecology from the work on forests and rivers to agriculture and buildings. Some of the relationships can be captured by maps and drawings, but not the dynamic four-dimensional qualities of the system itself, which can only be understood by dwelling there for years. Nevertheless, a simulation of the view from foot or airplane is more compelling than a recital of the statistics or species lists.

The goal of ecological design is not to restore, but to revitalize and reinhabit the forest. We do not want to live in the dead bones of a mechanistic failure. We want to live in a healthy environment with aesthetic appeal—aesthetic appeal is a requirement for human health. Every system has physical, biological, economic, and political characteristics. The design, planning, and management for a forest, for example, describes the system in a comprehensive interdisciplinary approach, using dynamic concepts such as feedback and stability, recognizing limits to change and sustainability with different levels and scales of structure and function in an anticipatory, flexible planning approach, recognizing human and nonhuman goals, and incorporating personal and institutional interests.

Ecological design is the design of communities. We design places as organic wholes to promote the well-being of individuals and the common good. The immediate goals of design are to reverse degradation and reclaim places for communities, but also to work to increase public awareness of the interdependence of communities, to create environmental quality, and to transform public values by generating new metaphors for living. Unlike what Bailey and other say, regional design cannot ignore culture, politics, and economics. Design has to commit to limits, constraints, and optima (physical, biological, psychological, and social).Ecological design can restore the interconnectedness of the systems, especially ecological and human social systems.

Ecological design has to start with the basics, the simple stuff, and the traditional solutions and values, first. This should be easy—these things have been developed and improved for thousands of years. Once we put the constraints in place to preserve human and ecosystem health, then and only then should the design start to employ new technologies and patterns.

4.7.2.9.1. Example of the North Slope of Alaska

An example of ecological design within limits is the 'Ecological Development Plan for the North Slope of Alaska: Goals and Limits For Wilderness and Human Populations.' This thought experiment was originally based on a wolf survey that lost its funding. To read the entire plan, see Wittbecker (1991).

4.7.2.9.1.1. Introduction

This design recommends a series of goals and limits in a comprehensive plan for wilderness based on the biohistory of an ecosystem, the cultural values of the people, and knowledge of the limits for sustainable development. This approach makes the limits explicit and sets equitable goals within the limits. A synthetic framework provides for the health of the ecological system as well as for the health of its human inhabitants.

The plan relates wilderness to resource use by the cultural values of an optimum human population within a home ecosystem, although that system is connected to others by trade for some necessities or luxuries. Cultural goals are keyed to the traditional idea of physical limits. Five general steps are followed from a comprehensive plan.

1. A place is identified within its natural boundaries. The North Slope Region is a uniquely identifiable ecosystem with recognizable boundaries and a unique history and character.
2. An optimum amount of wilderness is calculated to preserve the natural cycles indefinitely. If the current area is less than the calculations, the difference is restored

and set aside as a reserve.

3. The remaining areas are zoned for appropriate use, including conservation, preservation, and artificial areas (with historical, cultural, and functional importance).
4. Resources available for human use are identified, including raw materials and the usable biological productivity of the areas. Productivity is used to calculate a base line population.
5. Cultural modes—where culture is an ecological activity—set limits on the use of technology and resources. Cultural values drive changes.

As part of the formulation of a plan, it is necessary to examine the natural and cultural histories of the North Slope Region.

4.7.2.9.1.2. Description of the North Slope Region

The arctic coastal plain is a physiographic province in the Arctic, covering 200,000 square kilometers, extending 129 kilometers from the foothills of the Brooks Range to the Arctic Ocean. The coastal plain is composed of marine, fluvial, alluvial, and aeolian deposits. Large elliptical lakes cover 40% of the land surface (over 60,000 square kilometers) in the northern part of the plain. Relief is low, with river terraces, polygons, and occasional pingos (single conical hills); the slope is only 1centimeter per kilometer. Coastline (1,800 km) relief is only 2 to 5 meters, although some bluffs along rivers and streams can exceed 20 meters. Only the Inaru river lies within the plain; other rivers originate in the Brooks range or foothills and flow through the plain to the arctic ocean.

Off the arctic slope are the Chukchi sea and the Beaufort sea, divided by Point Barrow. The shallow Chukchi covers part of the continental shelf. The Beaufort covers a narrow shelf dropping to the deep, 2,400 to 3,600 meters, Canada Basin. Water circulation is dominated by a north then east flow. Beaufort coastal waters have a limited amount of nitrogen and phosphorus; with low light levels, there is low annual primary production.

As light increases in the spring, algae blooms provide nutrition for zooplankton and pelagic crustaceans— copepods, amphipods, shrimp, which support whales, seals, and birds. Millions of birds come in the spring—loons, geese, terns, and gulls; puffins and cormorants nest in colonies. By summer, arctic cod and sculpin come close to shore; cisco and other anadromous fish also travel along the shore.

The climatic zone is arctic, with long, dry, cold winters and short, moist, cool summers. Winter lasts from October to late May; summer is cool, cloudy, and short. With only 12.5 to 37.5 centimeters (5 to 15 inches) of rain, the slope is a cold desert. Day length varies from continuous darkness to continuous light. Long-term patterns of climate, especially glaciation, has had a dramatic effect by limiting populations.

Because of the annual low temperatures, the depth of ground that freezes in winter is greater than the depth of ground that thaws in summer, resulting in permanently frozen ground, the permafrost, just below the active layer (freeze-thaw). The permafrost is a complex, dynamically balanced, historical system, with vegetation, substratum, topography, solifluction (soil flow), and animal activity all interacting.

There are 574 taxa of vascular plant flora, only 8% of which occur in the plains, and 90% are found in the foothills. There is low species diversity: Only189 species of birds and 24 species of terrestrial mammals. One basic characteristic of boreal food webs is flexibility in the food habits of animals. Boreal feeding niches are broad and flexible, and without the extreme specializations seen in tropical and even temperate food webs. Boreal food webs, being made up of so few species, are also fragile and undergo dramatic shifts, that is, violent cycles of populations. The tertiary consumers, humans, wolves, and jaegers, depend on three basic primary consumers of vegetation: microtine rodents, ground squirrels, and caribou.

The first boreal hunters were primarily Mongolian in origin. From Carbon 14 dating, they have been there for at least 27,000 years. Hunting on the coast tundra has occurred for at least 6,000 years, probably 12,000. Boreal humans were limited by the seasonal progression of harvests: caribou, moose, seals, fish.

Hunter-gatherers lived in bands; local bands formed part of a larger linguistic and breeding community. Local bands gathered regularly into larger group for ceremonies or food sharing. Trade between Inuit groups had been constant, so that the Inupiat had knives, metal vessels, beads, and tobacco before direct contact with Europeans.

Inuits developed a culture that allowed them to survive in an arctic environment. The arctic small-tool tradition was their technological base. Inuit weapons were fashioned from bone, ivory, and driftwood; all hunting tools were engraved with magic images to encourage the animals to be killed. Inuits killed caribou, bear, wolves, and other animals. They fished, and hunted waterfowl and seals from kayaks. For hunting whales and walrus, crews of Inuit hunted in an open-hulled vessel, umiaks, covered in walrus hide.

Traditionally, production was not separated from the domestic functions of families. Activity was home-based. House types were determined by the composition of a household and the anticipated length of residence. For a nuclear or extended family, a caribou skin tent (*iccellik*), dome-shaped over willow branches, was traditional, requiring 20-23 hides. Snow and moss were packed around the sides. Sod houses were also employed for winter settlements. The snow-block house was built as a one-night shelter. Sleeping bags were made with caribou skin.

The north was first invaded by Europeans in search of right whales, then walrus, bowhead and blue whales. In the 1860s whalers were trading rifles, ammunition, flour, liquor, tobacco, and molasses for caribou meat and furs. Fur traders came later to take their own beaver, wapiti, bear, and musk ox. Explorers also decreased populations of caribou, musk ox, bears, and walrus to feed their expeditions.

Starting with the gold rush and continuing through the oil rush, hunting rush, and tourist rush, human populations have increased far beyond the carrying capacity of the land to support them. Large permanent supply lines, sometimes inadequate, are needed to support population levels. The Inuits today have been invaded by a consumptive culture, which is so far unable to adjust its urban industrial worldview to environmental limits—even refusing to acknowledge many limits.

There are now eight fairly large, year-round villages in the North Slope region: Kaktovik, near the refuge, Barrow with over 3,500 people, Wainwright, 161 km southwest, and Point Hope, another 161 km southwest; also, Anaktuvuk Pass, Atqasuk, Nuiqsut, and Point Lay—most with less than 300 people.

The 12 regional native American claims try to fit native cultures to their ecosystems. They are natural groupings that arose from long-term dwelling. But, they only represent the commercial interests of natives. The North Slope region however, is congruent with an intermediate level of Alaskan government, the North Slope Borough. Often the boundaries of boroughs conform to natural geography and seek to interrelate cultural and economic activities. This encourages reliance on regional resources.

The tundra area, including the National Petroleum Reserve (NPR) and Arctic Wildlife refuge (ANWR) is controlled by the Arctic Slope Regional Corporation (ASRC), which has created subsidiaries: Arctic Slope Alaska General Construction Company, A. S. Consulting Engineers, Eskimos, Inc., which operates gravel pits in Barrow and rents heavy equipment, and Tundra Tours, which owns a hotel and provide bus service to schools.

Humanity is a pandominant species in most regions and ecosystems. This dominance has major effects on the North Slope Region: Transient perturbations in energy relations from oil spills and burning; chronic shifts of systems from trails and chemical wastes; species manipulation from the import and export of exotics; and, interference competition with wild species, as opposed to exploitative competition, which can be stabilizing. None of these effects are exclusive to humans as a species, but they are excessive, rapid, compounded, and large-scale. With a comprehensive, ecological, long-term plan, changes and impacts can be anticipated and directed.

4.7.2.9.1.3. An Ecological Design & Plan for the North Slope Region

In planning for this region, it is necessary to consider an optimum human presence within ecosystem restraints, as well as air and water quality, genetic minima, nonrenewable resources, appropriate technological innovation, the importance of cultural frameworks, adventure, research, beauty, uniqueness, and other intangible experiences. Because it will always be subject to human influence and revision, wilderness cannot be considered apart from human population growth and human activities.

The wilderness of the North Slope Region is a sum of areas related to land area, ecosystem cycles, productivity, technology, culture, and an optimum human presence. An optimum population is calculated by adding the total annual productivity (in Kcal) to the total annual resources (in Kcal), multiplying that sum by technological and cultural modifier fractions, and dividing that by the annual per capita requirements for food and resources. The available area is obtained by subtracting wilderness areas, conservation areas, and other areas to be reserved from the total area. A technological modifier, based on the use of technology in extending or contracting the food or mineral productivity, and a cultural modifier, based on the application of cultural values in determining area and productivity are applied to calculations. For the total figures for all essentials—food, shelter, clothing, transportation— energy is converted to Kilocalories and placed on an annual budget, averaged over 1, 10, or 100 years.

4.7.2.9.1.3.1. Wilderness Reserves

The Inupiat have no word for wilderness (nor was there a word for city); for them the land is peopled with reasoning, speaking beings. At low population levels and high dependence on the resources of the land, it is unlikely that the Inupiat would destabilize the North Slope by their hunting activities. However, with cultural invasions and changes, it is necessary to make land use definitions explicit. Wilderness, however, is not a simple thing; at least eight kinds of wilderness must be distinguished, from sacred landscapes to wild resource systems.

The North Slope Region is already divided into major areas. The National Petroleum Reserve, established in 1923, covers 96,000 square kilometers. None of the Petroleum Reserve has been designated as wilderness. According to the NPR-A Task Force, fewer than 300 tourists visit the Reserve per year, outside of Barrow (as of 1978). The Reserve would not support much recreational activity anyway because of the fragility of the tundra and its long recovery rate from insult. Much should be saved as wilderness itself, as a symbol of natural and cultural heritage. Although tourism is a commercial industry in Alaska, the potential income from North Slope tundra is probably very small. Furthermore, tourist and hunting activities could interfere with migrations and subsistence activities. The current primary use of the Reserve is subsistence. The Arctic National Wildlife Refuge in the far northeast part of the state, comprises 76,000 km^2. The mission of the Arctic National Wildlife Refuge is to safeguard sufficient land to meet people's needs for an area where the entire spectrum of human benefits associated with wild creatures and wild lands are enhanced and "made available." The objectives include protection of the calving grounds of the 100,000 caribou in the porcupine herd.

If both these areas were reclassified as wilderness, then whole ecosystems could be protected. Ecosystem preservation protects entire biotic communities: Genes, populations, species, habitats, associated traditional human cultures, and all the processes and interactions. To keep the essential services of nature, from atmospheric cleaning to soil-formation, large reserves are needed. Reserves are critical elements in global element cycles, also. They provide a natural base line for management reference and a unique opportunity for scientific research. Large reserves would increase representation of species and save viable mammalian populations, that is, maintain the integrity of wild gene pools. Such reserves would permit natural processes to occur without human interference. We also need large reserves to derive further benefits from understanding natural processes and direct economic benefits from species—and, for aesthetic purposes: To see and to participate in nature (these being the basis of watching and tourism).

The desired size of the preserve is a complex function of the area's key species, quantity of suitable habitat, and minimum viable numbers of species. Large-bodied vertebrate species tend to have lower population densities, thus a reserve with self-sustaining large-bodied vertebrate populations will likely be adequate for herbivores, insectivores, and primary producers. The key mammal species in the North Slope Region are caribou, wolves, fox, lemmings, and microtine rodents. Determining the minimum number of individuals in a population to guarantee a high probability of survival results in widely varying minimum areas, depending on the key species selected. Carnivores are usually the least dense of species and have the smallest populations. They are also sensitive indicators of ecological integrity. Protection of the rarest elements should provide security for more dense species in an ecosystem. Usually, large carnivores are a sensitive indicator of the carrying capacity. Minimum habitat protection is necessary for the protection of endangered or threatened invertebrates, which are responsible for maintaining basic ecological processes through predation, recycling, and pollination.

Although there have been debates over whether a single large reserve is better than several small ones, the shape and size of North Slope Region wilderness is determined by habitat studies of the unique natural history and conditions. The size should be large enough so that species will not be vulnerable to "extinction vortices" caused by genetic or environmental stochasticity. In this reserve, disturbance from resource reclamation or recreation would probably be the greatest threat.

The recommendation for a North Slope Region reserve is 1 large areas, about 80,000 km^2, buffered by native cultures. The hunting and gathering lifestyle requires huge expanses; the reserve recommendation for native cultures is 100,000 km^2. The reserve would be managed by benign neglect—despite the fact that many do not list this as a management option. Natural processes, such as fire, wind, or species explosions, would be allowed to operate freely, even if they altered the functioning of the system. The large areas of the proposed reserves would be laid out on top of the petroleum reserve and the wildlife refuge. Larger areas would reduce management costs; smaller reserves in general require more intensive management and habitat manipulation. Saving 190,000 km^2 might be economically or politically difficult under the current industrial monolith, but due to climactic and physical limits northern Alaska is not expected to be a major agricultural or resource contributor. The cost of reserves would be not be high, even in terms of lost potential, when the costs of resource extraction and transportation are considered with the more neglected costs of loss of habitat and local sustenance.

The north-south orientation for the reserves would minimize the dangers from physical and climactic changes. The greenhouse effect could drastically alter the species distributions in reserves, with the loss of many species. Placing the reserves on heterogeneous soil types and topographies increases the chances that the temperature and moisture requirements of species would be met. Simply maximizing the size and number of reserves would enhance long-term survival.

Animal populations vary according to cycles or unknown causes; they move and concentrate or disperse. Because animals fluctuate in such a cyclic manner on the tundra, management by the maximum sustained yield concept is even more difficult and questionable. Furthermore, the ranges of caribou and birds cover more than one ecosystem; wilderness designation for areas outside the North Slope must be coordinated. Simply setting aside large areas of habitat will result in smaller anthropogenic disturbances and fewer ecological surprises

4.7.2.9.1.3.2. Resources

A resource is defined as a form of energy or matter that is essential for the survival of an organism, community, or ecosystem. The North Slope Region is relatively rich in organic resources. Perhaps the most important resource is the living land itself, the ecosystem. The single most important system factor is climate, especially its fluctuation over 10,000 years. From the land, people extract materials for food, shelter, and clothing. They put buildings and roads over that land. They dig up tons of gravel for roads. They depend on the plants, animals, resources, and energy.

The North Slope Region is a tundra. There are no stands of forest, although there are willows around rivers and streams and the imported decorative trees and grasses that people have planted and maintained around their homes and businesses. Productivity is low and it varies wildly, from high, dry ridges with small scattered plants to coastal sedge meadows. The annual above-ground net productivity on the polar desert plateau is low. A large portion of nutrients and primary production is below ground. Most of the productivity of plants is not directly usable by humans, but is important as support for native mammals. Agriculture has never been self-sustaining due to northern latitudes, severe climate, and thin soils. The Matanuska experiment in the 1930s and 40s was not successful. The Delta Barley project in the 1980s stalled on erodible land. Subsistence homesteaders and gardeners, however, grow a limited variety of produce. Agriculture on the slope is not ecologically appropriate. Perhaps a family urban agriculture might be possible in settlements.

The original grassland supported good numbers of mammals, from lemming and to caribou, musk ox, arctic hare, arctic fox, and wolves. During the long, dark winter, animals either migrate away (birds, fish, marine mammals), hibernate (grizzly bears, marmots, ground squirrels), or remain active (caribou, wolves, fox, lemmings, polar bear), if they have physiological adaptations to severe cold.

Caribou diet varies with place and time of year. In the spring and summer they feed on growing buds and leaves of willow, plants uncovered by snow, and new grass and sedge shoots. They feed on the bases of grass and sedge plants in fall and winter, and on fungi and lichens (not their preferred food) in the winter. There are four caribou herds using the north slope. The Caribou population has changed dramatically, from 1 million in 1920, down to 160 thousand by 1950, and climbing to about 400,000 in 1978.

Small mammals—voles, lemmings, ground squirrels—have a higher biomass and faster reproduction than large grazing mammals. Lemmings at Barrow reach peak densities every 3 to 6 years, declining to near zero between times. Although at each irruption, the lemmings destroy tundra grasses and mosses, they provide food for foxes, weasels, owls, and jaegers. Their grazing and burrowing increases decomposition rates, releases nutrients, and in general increases plant growth. Marine mammals, especially seals and whales, live in or migrate to the area.

The arctic slope has sparse duck populations. Although there are shallow, closely-spaced ponds, they have a low productivity of plants and invertebrates. On the coastal tundra, the summer population of wading birds has been estimated (1974) at 5,500,000. In the eastern part of the Arctic refuge, 400,000 snow geese gather in late summer. Invertebrates outweigh birds and mammals.

Despite a shift to a cash economy for some materials, Inuits rely on hunting, whaling, and trapping for much of the food and materials. Adults consumed significant animal prey, although there was a decline in the 1970s. This decline may have been because of the decline of the herds or the increase in availability of imported foods.

Neither domestic animals or crops are likely or perhaps even desirable in the boreal tundra. Although the biomass of a system (its stock) can be relatively large, the rate of production (interest) is low in the Arctic. Human use of the stock is often fast. Humans also try to rechannel the energy to more easily usable forms, reindeer for instance. The introduction of reindeer has caused severe oscillations and overgrazing—reindeer do not migrate—which has reduced the carrying capacity for caribou. Because of the fragility of the tundra system, caribou are the best way of converting minimal solar energy to useful energy. If the caribou could return to half their primeval numbers, they could support the entire native population. Exporting skins or meat out of the area could contribute to destabilizing the ecosystem.

The North Slope Region has sunlight, water, and wind, but not nearly enough to provide energy to residences and industries. Hydroelectric power is not well-developed. Wind and solar energy are not abundant. The big difference between renewable energy resources and energy use is the quantity of oil. There are large amounts of gas and oil, and there are known reserves of coal. Coal estimated at 20% of world resources. Oil (19 billion barrels) and natural gas (2.9 trillion cubic meters).

The oil alone (19 billion barrels) could last a small North Slope population (7,000), at normal rates (7 gallons per day per person) for a long time (over 44,000 years). The same oil used for the entire state might last over 700 years; but, used for the entire U.S. population (250 million people at 3.5 gallons per day per person), it would only last for two years. As a salable resource, therefore, oil could be depleted very rapidly.

Combined systems of wind, water, solar, organic, and fossil fuels for energy could minimize impacts. Singly, these sources may be inadequate in the long-term, but as a mosaic they could meet decentralized needs. Passive solar heating, combined with superinsulation, would reduce energy costs for individual buildings. Photovoltaic cells could provide back-up electricity and power for vehicles. Local energy projects using geothermal sources, winds, or the sun are preferable, since their operation does not introduce new material to local cycles. All of these sources would be characterized by a small scale. The public service function of nature provides free services to humanity that are essential to civilization. But when the free services are overloaded and breakdown, the public has to pay for repair. Further increase in flows of energy through technology will significantly reduce the capacity of the system to support its human population (even large-scale fusion techniques). The overuse of nonrenewable resources can destroy renewable ones.

Because the North Slope Region rests on large deposits of oil, its mineral wealth is considered to be great. It does have large mineral deposits, but also high production and transportation costs. There is gold mining near Juneau, Fairbanks, and Nome. There is some copper, molybdenum, lead, zinc, and silver. Iron, at 10 million metric tons, and zinc, at 24 million metric tons, are the most common.

4.7.2.9.1.3.3. Resource Use & Technology

Inuit technology is ingenious and elaborate. Many implements were made out of bone, ivory, and stone; needles, combs, awls, spoons, cut and shaped with a rotary bow drill. Hunting weapons included bow and arrow, lances, fish spears, and harpoons. Household utensils included lamp bowls, pots, cups, bags, buckets, and storage bags. Like figurines, most all tools were decorated; this attention probably made them better tools. Many tools have been adopted by modern expeditions: dog-team sledges, snow goggles, parkas, fur boots, kayaks, and igloos.

Inuits adopt new tools when they can. Snowmobiles are replacing dog teams. The cost of owning and operating a snowmobile usually exceeds the monetary income derived from the land. But, since the snowmobile reduces the hunting time, more time is available for wage employment in the industrial culture. Sometimes the new tools are inappropriate to the Arctic. Many buildings are industrial temperate zone designs; many uniform commercial products are not durable or useful. Imported peaches or bluegrass lawns are luxuries. The waste is dumped in landfills.

Technology can be used to expand or contract resources. Technologies have the capability to minimize the use of resources, but they also have negative effects. Technology has greatly increased the kind and quality of materials used for buildings and machines, especially aluminum and other light metals and silicon constructs. Yet, the scale of technology produces pollution that reduces the productivity of natural and agricultural systems. Inuit cultures have survived for over 10,000 years. They are durable because their practices have been ecological. They mimic the flexibility of animal and plant populations with their movements and population. The Inupiat are generalists. The generalist strategy combines low investment with high efficiency; it is conservative enough to carry people over periodic drops in available resources. Lower resource consumption rates.

The Inupiat are adaptive and have adapted some modern equipment; but these things require cash, which means selling goods or working in a cash economy. Their culture becomes incorporated into a cash economy. Hunters shift from hunting to trapping, fox, for instance, to participate in the cash economy. Trapping, being much more individualistic, weakens cooperative ties and ceremonies. Having motorized transportation permits hunters and trappers to travel further and faster, and to live in fewer villages, with increased population size and density. As Inuits adapt to new economies, their subsistence and dietary patterns also change.

Energy used to be cycled within the arctic system, with Inuits as top carnivores. Much energy now flows from the south in the form of food and materials. Oil energy flows south. Less energy flows from the north in the form of handicrafts and skins.

Odum points out that the human strategy is the opposite of the ecosystem strategy. Ecosystems increase the ratio of biomass to respiration, whereas humanity reverses it— possibly leading to boom and bust oscillations. Since the arctic already has such oscillations, the superimposition of greater oscillations may sooner lead to extinctions or collapse, of species or human patterns.

A favorite Inupiat pastime was the telling and retelling of stories. Television also has repetition, but not the same depth of involvement. Radio stimulates images in the way books do, because neither impose images, although neither of these forms may be as involving. Improvements in technology are not necessarily improvements in human existence, but technology must be examined constantly, if people are not to lose some of the characteristics, such as empathy, that they value most highly. Some communities may wish to examine vehicles, computers, and television, and eliminate them or modify their use.

Subsistence hunting depends on very-long-term conservation. The ecosystems have to be kept intact to preserve the full diversity on which the hunters depend. Hunting, without a strong cultural tradition and values, does not automatically result in conservation. Inupiat have learned to use snowmobiles and guns, and have access to processed food.

Where traditional ways have been adequate for 10,000 years, the Inupiat compete for resources with an industrial culture that has no sense. If a culture cannot adapt itself to its ecosystem and abide by its limits, then the culture will perish. Disruption of ecosystem processes, as a result of excessive continuous resource demand, can lead to destabilization or collapse of the ecosystem and extinction or emigration of species.

The Inupiat had limited energy sources, but were able to change or move to avoid declining marginal productivity. In tribal societies, people live by harvesting wild food and the energy of the sun; their populations reflect the productivity of systems controlled by weather, landforms, and local cycles; they adjust to the conditions.

Most natives have adapted the infrastructure of the invading culture; they are taught and shown the patterns of consumption and use of an outside culture. They may try to maintain their ancient wisdom formed in a subsistence society that understood snow and permafrost and caribou and whales. They can aim for an optimum use, far below a maximum. In 1982, the council of Arctic Village, an Athapascan Indian community of 120, voted to limit residents to five caribou per person (from the Porcupine herd). Ironically, based on the track record of science and technology in the north, the ancient ways may still prove to be better—capability and flexibility rather than specialization and complexity.

Within unknown limits, all people need to establish a pattern of use that is sustainable and flexible, leaving other possibilities open. The general pattern is: Sustainability within cycles of depression and exuberance, economics fit to the resource base, without external input in energy or money, prevention of unwanted ecological or social costs, and harmonization of cultural differences. The root problem is how to live with technology in a mature manner.

Technological multipliers are efficiencies that allow a limited increase in the carrying capacity of the entire ecological system. The load on the system can be expressed as the resource demand multiplied by number of people. Trade diminishes some resource or food capital, but can be used to avoid some minimums. The Critical Minimum is a factor limiting carrying capacity. It is not known. Assigning a multiplier to the effect of technology is difficult due to the contradictory impacts of technology; here, it is estimated at 2, which is to say that technology has an overall benefit.

4.7.2.9.1.3.4. Cultural Values & Goals

There have been two cultural patterns on the North Slope Region. An archaic one, that offered family egalitarianism and limited impact, and a technological one based on republican democracy and industrial consumption. The cosmology of the latter contains the image of Alaska as resource storehouse. The cosmology of the first has an image of the region as a place to live with other beings.

The Inupiat developed a cosmology appropriate to the place. Most of the universe was addressed in folklore and did not play a significant role in religious behavior. Religious attitudes, however, played a significant part in hunting. Animals were considered to be endowed with many of the same abilities as humans: the ability to reason, to talk, and to react. Although animals allowed themselves to be taken in a hunt, if insulted or mistreated they would withdraw their presence, to the detriment of human survival. Hunters attempted to placate and compel animals with ritual. The world of animals was considered to be superior to that of humans. Every individual had power of some sort. Shamans were more intimate with the supernatural world and had greater power. He or she performed acts necessary for the good of the group.

There were elaborately evolved restrictions and taboos relating to the animal world. For instance, weapons made for caribou could not be used on sea beings. Although there was no limit on the number of caribou to be taken, there were restrictions on how the carcasses were to be treated; the heads of dead caribou on a sled were bent upright so they would not hit the ground. By contrast, only five wolf traps could be set at one time (five being enough for a parka); if more were set, the wolves would be offended.

The essence of tribal experience is distilled into metaphors, into myths and stories. Native knowledge exists in a spiritual framework. The country is sacred to the natives. Folklore stresses the Raven (*tulugaak*) as creator of the world by bringing the land up from water and differentiating night from day. The world itself is considered flat, resting on four wooden pillars.

The Inupiat can not compete against industrial culture, but it has not been completely eliminated either. The monoculture of coke and cars tends to overwhelm the subtle adaptations of the Inupiat. A coherent culture provides a secure platform for exploration and open-mindedness; it provides criteria to judge the 'special effects' like television. A conscious culture, perhaps with elements of native traditions, with appropriate technology and knowledge, could improve and guide the culture of the market place; television, as does telephone and radio, unites people through information, but a strong common culture is indispensable for uniting their hearts.

It may be that the Inupiat of the North Slope Region require 10 times the space as rural settlers; and, it may be that these settlers require 10 times the space as their urban counterparts. Or, it may be that people would prefer to have more discretionary time and less work time, as was possible in archaic societies.

There are numerous advantages of way of life of North Slope Region bands: Fewer working hours, about 4 per day; more leisure to talk, sleep, engage in rituals; a diet adapted to conditions; deliberate underproduction, below the maximum levels; deliberate control of population growth below maximum levels; and deliberate under use of resources, resulting in small ratio of people to resources. Subsistence economics means simply that surpluses are not accumulated. This might make bands more vulnerable to food shortages, although the ratio reduces the possibility; furthermore, industrial cultures have far higher incidences of starvation.

Indicators of the quality of life cannot all be expressed numerically; they are elusive. Quality depends more on the spirit of the society than a count of material possessions. Even orthodox ecological criteria are not adequate to evaluate the quality of a particular environment for human life. Nevertheless, a cultural multiplier can be assigned to aesthetic and support space needs; it is estimated at 0.7, which means that the hybrid industrial culture does not satisfy enough needs or requires more space per person.

Traditional goals for Alaska as a state include: The stability of economic growth, Beluga coal commercially produced, new mines opened, increased oil production, North slope oil shipped to Japan, railroads extended, Susitna dam put on line, becoming a major agricultural exporter, and building a world-scale petrochemical plant.

Alaska's future seems to be dependent on oil. Because of its remote location, harsh climate, rugged terrain, and small population, the Delphi group considers that its future is tied to resource development rather than to manufacturing or service. Thus, a development plan seems implicit, but not as a coherent popular strategy with goals and priorities.

The goals of a native-culture dominated society might include: Self-determination, population limits, paths of cultural renewal, wilderness limits (more nonhuman creatures), limited resource use (renewable especially), and better communication. Many local limits can be transcended, but global limits perhaps cannot be transcended. Therefore, local cultures are related to some extent, as part of a sum.

Ultimately, North Slope civilization depends on wilderness—for recycling, pest control, genetic diversity, soil-making, and water purification, among other things. North Slope civilization depends on native cultures for vitality, and those cultures depend on wilderness for their context and imagery.

The North Slope Region has adequate resources to feed and care for its current population—at a reasonable level. The rates of use of energy and some minerals are far too high. Demands and values are not related closely enough to the particular beauties and limits of the home ecosystem. The explicit goals of the cultures need to be fit within the limits. Local limits are significant. The North Slope Region, for instance, faces very real limits in rainfall, minerals, and fragility of the landscape. Every community is forced to accept some upper limit, beyond which it cannot grow any further.

The size of communities in the region is small enough for people to meet and "exercise government" (in James Madison's words). Kirkpatrick Sale concludes that the optimum size for direct meeting is 500 (and 7 of 8 villages are less than 500). Probably participation becomes more difficult as the size increases. Bryan and McClaughry suggest 2,500 as a maximum, since in larger groups people cannot all know one another and the assemblies become a debating forum for a few. The North Slope is a good candidate for direct democracy. The population density of much of the North Slope Region may cause some difficulty, however. People will not be within walking or sledding distance, but may have to travel long distances to meetings or communicate remotely through telephone, computer, paper, or friends. People will have to impose limits to keep things from growing. They will have to make and enforce limits on immigration and birth, limits on housing and public services.

4.7.2.9.1.3.5. Human Population Goals

The population of the North Slope Region is currently about 5,700, of whom 4,560 are Inupiat. In 1828 the population from Barrow to Point Hope was estimated at 3,000; perhaps another 1,500 people lived east of Barrow. After European contact, the native population dropped to about 1,700. The population has been increasing until it is almost the same as precontact numbers, although concentrations and livelihoods have changed.

Is 5,700 people too many? Or 4,560? How many more could there be? Is this number above or below a maximum carrying capacity? What kind of freedom do the 230,000 people in Anchorage have? Can a large human population function in harmony with the arctic environment? An optimum carrying capacity can be calculated for this region. The carrying capacity is the population sustainable on a long-term basis of renewable and nonrenewable resources. Carrying capacity calculations often just consider food energy, but all needs—clothing, shelter, transportation, information generation, aesthetic satisfaction—must be included. This introduces cultural elements into consideration, so carrying capacity must be considered as cultural carrying capacity.

Calculating an optimum population based on ecosystem productivity is relatively simple. Based on current caribou populations (180,000 and using maximal ratios—1/30—to simplify the calculation), the North Slope Region could support an optimum of 1,338 people (or an maximum of 6,000). Calculating a population based on seal and caribou populations, and assuming diet is 50% seal, 20% caribou, 20% whale, and 10% fish and fowl, the optimum population 1,500 (or maximum of is 6,600 people).

The advantage of community productivity as wealth is that the wealth is sustainable— plants and animals are renewable and minerals can be recycled. The net community production (NCP) takes all of the food chain into account—the hundreds of other species. Technological production (in integrated greenhouses or algae farms) of food in this area has not considered seriously.

A larger population could be supported with more resource use. The population could be expanded possibly to twice as much by technology use. Larger populations could be supported with extra-regional trade. Of course, the cultural use of resources could raise or lower the potential population.

4.7.2.9.2. Wildness & Culture

The Inupiat way of life is still an ecologically viable way. The values of its way keeps the culture knowledgeable of and adapted to the environment. Instead of adopting the capitalist value system with its addiction to resources, which could degrade and destabilize the Arctic Slope region, the Inupiat must find a political way to subordinate the economy to their traditional culture and keep within the limits for ecological viability. Technology could be directed to use local resources instead of providing a means of importing food and clothing rapidly from other regions. Politicians could be directed to supply local needs instead of trading fish, oil, and copper for money. One immediate strategy would be to seek control of traditional lands. Another strategy would be to seek a steady state population until the caribou herds rebuild and then calculate an optimum.

Growth with minimal planning might be tenable for a short time if Alaska had renewable resources that could be traded in a sustainable way. But, oil, gas, and gold damage and destabilize the ecosystems, according to David Klein (1970). Native organizations, such as the Inupiat Paitot, resist too much exploitation, but are not powerful enough to stop the largest exploitation schemes, such as the trans-Alaska pipeline. Regardless of whether petroleum quantities justify lineal development, most of the NPR should be converted to wilderness designation.

A number of concrete recommendations can be made to address the wilderness goals of the North Slope Region: (1) the gradual increase in preservation of wild and restored lands, perhaps up to 90 percent by 1990, to protect ecosystems and resources, (2) educational campaigns for public awareness of the uniqueness of the North Slope Region and of the effects that lifestyles have on it, (3) the offer of tax incentives to private and public corporations to redesign their paths in the community, (4) the promotion of equity in opportunities and rewards for all people, and (5) a geographical/environmental data base for the intelligent management of the land.

This plan recommends two large wilderness areas (wilderness definition 2) coinciding with the two largest caribou herd migration areas, 80,000 for the west Arctic herd and 80,000 for the Porcupine herd; these sizes should be adequate to preserve the associated predator populations as well.

The goal of planning is community success and personal happiness, based on self-reliance in food and shelter, self-sufficiency in food, and self-limitation in size and desires. If human patterns were based on mature ecosystems, civilization would be far more complex; human values would allow for the welfare of humans, animals, plants, and land. The inhabitants have to be wise enough to be disciplined, to leave wilderness for other beings, and yet to make good places for ourselves.

The establishment of the North Slope Region as an independent region would not be unrealistic. The plan recommends 80,000 for a reserve (wilderness definition 3) for the operation of traditional Inupiat culture; this area would overlap about 50% of the two herd wilderness areas.

Calculating an optimum population, based on carrying capacity, but modified by technology and culture, can result in a higher population. One calculated number was 5900 people. Obviously more can live in the region, but they have to be supported by imports from other areas, which reduces the capacity of those areas to some extent.

According to an Inuit myth, the first human beings lived on an island, Mitligjuaq, in the Hudson Strait; no one ever left and eventually there were so many that the island began to sink under their weight. An old woman shouted: “Let it be so ordered that human beings can die, for there will no longer be room for us on earth.” And it came to pass.

4.7.3. *Art & Development—Showing the Invisible*
Art has been a unifying theme of human activity for at least the past 50,000 years. Definitions of art in anthropology and social science are numerous and diverse. Art is part of culture, according to E. B. Tylor and others (1940s). Art is any embellishment of ordinary living that is achieved with competence and has describable form (Herskovits 1951). Art is the process of expressing things in materials or patterns, that others, with similar experiences, can unfold and enjoy (Various, 1950s). Some anthropologists considered art as the expression of patterns on the body or cave walls that increased personal identity and identity with prey. Others suggested that art was an activity that socializes people into their community and gives them insight into their fitness in the cosmos. Art is the creation of public objects or events, by manipulating a medium, to create qualitative experiences (after George Mills, 1971).

Biologists regarded art as a form of play activity, that is, learning to create forms that increased the likelihood of survival. Artistic products triggered effects in observers that increased their consciousness and fitness with nature.

Once artists started defining art, definitions were free from verbal restraints. Marcel Duschamps said: "Man invented art. … Art has no biological source …" The romantic ideal was that art in representing and miniaturizing nature surpasses nature. Where definitions were once limited to styles or topics, now they are concerned about the meaning of art. Art is anything you do not understand, the internal workings of the human body or the cosmos. Art became a reaction to previous art. Daniel Bell argued that the essence of artistic modernism is the desire to violate established norms, defy community standards and question authority. This might explain the outrageousness of modern art., as it becomes harder to find things to undermine. Art is the manipulation of forms. Internationalism in art manipulates nationalism, which manipulate artists, who manipulate nature, only artists manipulate unique details and events, unlike scientists, who manipulate repetitive patterns.

As art became more valued, it became regarded as a commodity. Art is what is defined by the market. A civilization that defines itself in economic terms will have an economic art, like modernism or avant-gardism. Robert Hughes: "What strip mining is to nature, the art market has become to culture." Does an economic value devalue the work of art?

Human beings are embedded in culture, which is embedded in nature. Nature is a process that encourages emergence. Humans create extensions that push culture and nature into new shapes and patterns. Art is an extension. It is not separate from human expression and feelings, although it can be treated as a separate thing. The audience and artist are part of the same process of human extending. Art can be a trigger for change (or explosive decompression, to continue the weapon metaphor). If art is an extension of human behavior and cannot be separated from it then it has to be as moral as any human behavior. But, if art is a form of behavior that examines and changes other behavior then it should not be limited to the prevailing mores.

As a process art extends into and out of civilization and human experience. That makes art difficult to define, without getting into problems with human perceptual limits on containment. Ortega illustrates the dilemma: "Art is a window with a garden behind it: one may focus on the garden or on the window." We think of art as containing or contained. We think of nature in the same way. Nature does not simply reproduce individual beings, like trees and birds, nature creates forests through individuals. Art is not a collection of works; it is a process creating human expression. The dilemma is caused by focusing. If we look with peripheral vision, we can see the whole.

4.7.3.1. Metaphors of Art
Art has also been defined as a new modern supermarket of forms. Does the modern culture accelerate change, collapse time, and disintegrate forms, or rather the temporal and spatial limits of forms, so that all forms are available in a modern supermarket of forms to be used by anyone for any purpose, without understanding, or is it just a misunderstanding? Art forms have become as indeterminate as forms in physics, and just as temporal. Nor one categorize artists who disregard every convention of history, tradition, and space-time. For art, any mode or combination is valid. Art is what the artists or viewers say it is.

But, despite what many say, other metaphors can reveal something constant about art. Art is an amoeba. There are dozens of ways of deconstruction now, each decreating and disintegrating some part of culture, each demystifying and decomposing. MTV mixes aesthetic boundaries at will and accelerate the images pillaged from surrealism as well as horror movies.

Science becomes just another form of socially constructed reality. Art is science. *How does art work?* Like science, each new medium undercuts the old paradigm. Like quantum mechanics, which was the death of objective reality, which is what it was supposed to prove, art undermines the parent idea. Something new flourishes, then it is old. Then the revolution is over, and art is part of the new environment that will require further art to see. The media become the field of the new ecosystem. Unlike science, art uses the rare, unique, individual instance for its lesson. Science uses numbers to relate samenesses and forms. As quantum mechanics showed that participation in the field was not an option, so art requires participation for its consumers to recreate the subject themselves. Does art have a different goal from science? Despite their subjects, art and science both work to expand human consciousness. They both use metaphor. They both make the invisible visible. They both use technology.

Art is a trap. Like hunting and agriculture, and language, art limits what people do and can do. Art is a trap, because creators, participants and viewers are limited by styles or demand.

Art is a mirror. Maybe more like a circus mirror. otherwise art would just reinforce the current environment and become part of it. Art distorts things, then presents the distortion for attention. McLuhan notes that print technology is familiar and soothing, "numbing" actually. If one pays attention to art, however, one is rarely soothed. And, what does a mirror do but copy?

4.7.3.2. Mimesis & Plagiarism
Is art just a kind of copying? Of course, young artists learn by copying. Some mature artists copy natural or industrial forms; other artists copy other artists. Is this plagiarism? Do modern artists plagiarize tribal, chiefdom and extinct styles? Is this economic or artistic colonialism? Obviously, in the past, other images have been absorbed from other cultures, but they are not creatively or imaginatively integrated. Are we too tired to invent our own connections, use our own histories, or reflect the complex consumer technocracy?

Mass media rips forms and images out of context and pursues us with them, accelerating through our consciousness. Many artists also try to erase any trace of cultural specificity, which allows the market to define the value of art. Is this wise? What does it do to culture?

Culture is the nongenetic transmission of information. When anthropologists address ape behavior, which can be more sophisticated in human situations, they claim the sophisticated apes have become enculturated with human culture. But, apes are born imitators and they use that in ape societies. Other words sometimes used for imitation include emulation and facilitation or perhaps fashion, which was used by Wolfgang Kohler, who observed apes inventing new games in the early 1920s.

If copying is a drive, then it may reinforce itself. Copying behavior may have shifting goals depending on context. Chimpanzees learn to crack open oil-palm nuts using tools. This may start with young copying the mother's behavior, to be like her or to please her as part of the mother-infant bond. They may have to practice for three years to be able to coordinate hitting nuts with one stone as an anvil and another as a hammer. Perhaps identity-based learning shifts to reward-based learning. This nut-breaking is most often done at the end of the fruit season, when fruits are not available. They can get nine times the kilocalories that they put in as effort. Tool technology is a critical skill for eating and maintaining health. Stones are often kept at the nut "factory" although mothers may move the stones to familiarize infants with them.

Chimpanzees, and apes, and bonobos are educable because they need to be to be able to survive certain environmental conditions. Often, behavior does nor change for a long time, especially in chimps and apes, but, then culture is conservative and has inertia as well as innovation.

Hand clapping, during mutual grooming in chimps, has been called a social custom. It can be taught or observed. Active teaching is a form of social transmission of behavior. Although not as complex as symbolic communication, it requires understanding of more than one personal perspective.

One popular idea is that biology restricts our freedom and culture frees us from biology. But, the capacity for culture rises from biology. And, culture can restrict freedom as much as biology. Psychologists define culture as problem-solving, but, what does human clothing fashion have to do with problems? Or for that matter, religion, food preferences, art, dancing, or social styles of politeness?

Biological entities replicate using DNA in genes. The genetic complexity of chimps, apes and human allow them to have the capacity for culture, and art. But, how do cultural ideas or artifacts replicate? Richard Dawkins promotes the 'meme' as a new form of replicator, the unit of cultural transmission. According to Dawkins, a replicator has three basic properties: Fidelity—good copies can be made, although occasional errors are made; fecundity—many copies are made; and, longevity—the copies can last for a long time.

The idea of memes is to compare the copying of cultural information with copies of genes between generations. Not all the parallels may be useful. Genes are instructions; memes are more like suggestions. DNA is kind of like a flowing river, but the cultural winds above also make changes in the landscape. Memes are more like the wind, faster and less constrained.

Ideas can arise in one brain and language can act as a carrier wave to other brains. Orcas in Argentina teach their young how to beach themselves to catch seals; the behavior is dangerous and often requires the teachers to help the students back in the water. Changes seem especially noticeable between generations, as regards clothing and decoration. Ideas mutate, get selected, and then spread. Do they mutate like genes? Or, are they created in brains and expressed and interpreted?

How is ecology of ideas different? From genes? Ideas have no size or weight nor requirement for space. No apparent carrying capacity, although there may be human limits to the number and shape of ideas, or fitness. Do they compete with one another? Yes, obviously.

Are transfer and replication the same for ideas? Is one the other? There is no reproduction rate. Does that mean that fitness loses its meaning? No, many ideas are not fit. And they die out. But, do memes die? Can examples show the ideas, like Nazism or medicinal bloodletting, are not dead.

Do ideas change by random mutation? Or by intentional ordering? Ideas are sifted and sorted. Many are selected consciously. Many are improved intentionally. Conscious selection allows estimation, prediction and evaluation. Eutopias is a conscious set of ideas.

There is also complex interplay between memes and genes. This could be coevolution. An example is reactive bear behavior to tourist behavior. People hung food on a rope in a tree; bears figured that out and got the food. People started hanging the food on a rope between two trees. Bears figured that out. People started keeping food in cars. Some bears learned to pop out the windshields of vans; others jumped on the roofs of compact cars to make the doors pop open. Some bears specialized in certain cars. Steel boxes are now recommended.

Perhaps human copying is a form of appropriation, or maybe an example of evolution. Kwakiutl designs have been appropriated for numerous commercial uses; crest poles are sold as parts of sets of toys. A Japanese manufacturer appropriated a Kwakiutl design and sold 250 million in hand-knitted sweaters. Canadian artist Jack Shadbolt produced semi-abstract native-themed paintings for sale. Another clothing company reproduced, on a t-shirt, Aboriginal artist John Bulun-Bulun's work "sacred waterhole"—and a court awarded him $135,000 for that appropriation. So, what is the solution for art? Formal legal protection through patents or copyrights? Being unsigned? Unrewarded? The function of art used to be collective. It was for the entire tribe or band. So, it was not signed. In a global economy, driven by market forces, art can be lifted from its context, changed and transformed easily and quickly. Perhaps simple respect and recognition of the source is enough; perhaps not.

4.7.3.3. Functions of Art

Perhaps at first art was a visible connection to animals that were to be hunted, after the fasting and deprivation of an shaman or animal master led to dreams and the expression of dreams on the hidden walls of a cave. Art is an attempt to restore control of society and environment. Art is a link to nature or the holy. Art is a way to make the invisible visible. Art and the struggle to come to terms with death.

A way to express prestige or status. It was also a luxury afforded by few. Art produces luxury goods. Pretty things. To enhance prestige with rare or beautiful things. What is the esthetic urge to embellish ones life? This urge, whatever, causes technology to follow trade goods, copper, silk, jade, glass, pottery. The price of silk in Rome was equal to gold.

The subjects of art have changed, from animal forms to natural, abstract, and invented forms. The makers of art have changed also, from shamans and individuals to specialists, professionals, and teams. The uses of art have spanned from body decoration and clothing to tools, commercial things, luxury items, and cities. But, the functions of art remain multitudinous and useful. Art has functioned over the past 50,000 years as: Human perception; the physical expression of existence; a tool for the symbolic expression of the processes of nature and the limits of art; imitation of nature; dialectic communication; play; meaning; and, perhaps most importantly, survival technique.

4.7.3.3.1. Art as Human Perception

Art is an emergent quality of life. Art is a way of seeing or viewing new perspectives. Art is a way of revealing the hidden or invisible. Maybe this function of art is the most basic one. As Picasso said: "In art we express our conception of what is not visible to nature." Paul Shepard argues this also, from the perspective of human ecology.

For the Kwakiutl, the purpose of art was to harness the invisible forces of society, nature, and supernature by making them visible and to validate social system through display of status. It was crest art that symbolized the owners mythic origins. Both art and play are metaphorical. The metaphorical aspect goes beyond normal activity. It is "as if" the dream was true, as if the hidden reality was exposed and described. Art distinguishes the extra-ordinary. It makes the ordinary special, by combining it with the hidden. History of art: Pattern and vividness are attractive to humans. Art enables people to act as if they were animals or gods. The environment is the source of opportunities and limits for human life. Humans establish formal relationships with the forces of nature by defining them, as they happen, and then, trying to control or placate them. The visible symbols of invisible forces are interpreted by art into known forms or essences. The NW spirits move in an annual cycle also, coming closest to humans in winter. The arts acknowledge their presence and send them off with honor for another cycle. Artist selects from animals in environment, from mosquito to whale. Some fantastic animals, such as sea wolf or thunderbird. Supernatural. Some combined animals, partly human and partly natural, with human base figure. Whole figures are rarely carved, but features are used and sometimes exaggerated. Art materials were taken from what existed there: Cedar for straight boards and canoes and posts.

Art is a way of making the invisible visible. What is invisible? Everything in the environment, or behind it: Gods, meanings, small beings, the environment itself. For modern print, as a medium, it has been an environmental factor for so long that it is invisible. This is more true of news media. Their invisibility allows them to influence us without our thinking or examining. Even new things, such as technology, that are persuasive and pervasive, are invisible.

The ecological environment, even including news media, is a complex, low-intensity, undemanding process. By making things high-intensity and simpler, art makes them visible. Art miniaturizes it subject, so it can be seen as a whole. It intensifies it so it can be seen as dynamic. It accelerates it, so it can be seen at once (in the short and limited human attention span).

Expression is embodied in perception, as perception is in the body, as the body is in the world. Artistic ideas must first be grasped in their fleshy existence; only if they are first experienced can they be grasped, and then they cannot be transformed into ideality. These ideas, musical and literary among others, cannot be detached from the sensible appearances, as can the ideas of intelligence, which are erected into a second positivity; thus the abstract language of science is freed from the horizon of perceptible reality, while the most artistic seem trapped within it.

Understanding a pattern is also a bodily reaction; creation and re-creation are analogous processes. The feeling is perceived, not conceived; the work of art is contemplated, not cognized; what the artist makes, the audience remakes. The created Pattern becomes a cultural object which alters the complex of the environment, and both organisms must adjust. Most learning is a reflection on experience; art reproduces experience. In the presence of a work of art, its full style is felt; the image presents itself to us, not as a representation, but as a presentation. This presentation is kinesthetic, an expressive dynamic form which appeals more to the kinesthesis of the body than to in tuition. The artist works like his nervous system, selecting highly significant cant details from his environment, and synthesizing them into an excitatory complex. As Whitehead explains: "The human body is an instrument for the production of art in the life of the human soul. It concentrates upon those elements in human experiences selected for conscious perception.... In this way the work of art is a message from the Unseen. It unlooses depths of feeling from behind the frontier where precision of consciousness fails."

Is iconic perception cognitive and aesthetic perception contemplative? Is it the reduction to cognition that poses the problem? Is it solved when all perception is recognized as feeling? Artistic images and symbols are aesthetic and equivocal; signs are linguistic and univocal; since images have such poor semiotic value, it is dangerous to reduce them to cognitive terms. When art works become cognitively generalized, their aesthetic specificity is lost, and they become poor vehicles of knowledge. An aesthetic object should not offer a reassuring vision, which interprets or identifies nature, but a naive vision, which surprises, shocks, fascinates or seduces the senses, which awakens desire and stirs the imagination, and which furnishes a feeling of the invisible.

The function of art in tribal society was to merge people with the cosmos. Art educated the perception to see that the links were visible. The function of art in industrial electronic society has changed somewhat, so that it is to train people's perception of the environment. Rather than just orient people to novelty, art could also embed them in the cosmos.

4.7.3.3.2. Art as Play

Nature and art create in a playful manner, building and destroying in innocence. The innocence of becoming is the unity and inseparability of things we call opposites and contradictions: Being and becoming, nature and man, freedom and necessity. The innocence of becoming is the silence of art—a silence paradoxically dynamic, full of tension and polarity, yet possessing form and balance. Since all experience is comprised of aesthetic events, and since art in human activity is broadened beyond artistic, art is generic. As in Aristotle it involves feeling and making; art is the education of nature (from the Latin '*educare*,' meaning to 'lead forth' or 'draw out'). Life and art merge: experience in art is the dynamic process of life itself, in continuous oscillation between opposite poles; life and culture are both aesthetic phenomena. The similarity uniting diverse cultures is explained by the human condition: the first expressions of the basic ambiguity, of man as a body-mind, shape the living environment. Artistic creation is the spontaneous movement of the human body as it expresses its universe.

The process of recombining and recontexting as a learning and enjoying activity. Like education and modeling, Art is a process of recontexting. Writing for instance takes meaning out of its aural context. Art takes ideas from the unconscious. Using a combination of explanation, symbols, and play. The artist releases ideas from the unconscious, or unexpressed and conscious, for creative purposes. Artist needs a safe place, or security, to allow the vision to emerge to a new context. This creativity is a kind of play. Put the idea in a new context and play with it, to see it in a new way. Artists are considered dangerous because they play with ideas, because they question ordinary ways, and because they play with serious subjects (like thought experiments). Art also creates an explanatory context by allowing the viewer to think or perceive several things at once (just like the operation of a metaphor). In this kind of illuminating moment the viewer can see many things at once, sometimes the entire whole thing. But art can show how things fit together, into a whole.

4.7.3.3.3. Art as Physical Expression of Existence

Cultures have been unique programs, using local materials and ingenuity, for satisfying basic needs, material and spiritual. Cultures become the residue after the flood of goods and industrial ideas. The diversity of solutions dries up. Everything looks and acts the same.

For some cultures, such as the Japanese, art must represent impermanence. The shortness of a bloom is related to its beauty. Art is respect for beauty, as part of ordinary life, and need not be permanent or timeless.

The greater part of human experience is primitive and emotional; ultimately nonverbalizable, literal language is too abstract to deal with it. Anais Nin wrote that "Poetry, which is our relation to the sense, enables us to retain a living relationship with all things." It not only allows expression to savage experience, to feeling, it also is a doorway to mystical experience and transcendental visions; it creates new dimensions. Art is an indirect language, an expression, in which the individual elements need no meaning, but gain meaning by entering particular relations within a context. Art achieves only through indirection like sailing; good sailing requires knowledge of the forces at play, the wind and currents are beyond control, and ability to use them to its fullest advantage. Prose motors into the wind but loses the wind in the noise.

Painting is like a completed ontology of the visible. At the other extreme, music is too far beyond the designatable to depict anything but the turbulent outlines of Being, according to Merleau-Ponty. In a painting the artist touches the two extremities: in the depth of the visible, something moved, which engulfed his body; the invisible. Because every visible thing gives itself as the result of a dehiscence of Being, it functions also as a dimension. Therefore the proper essence of the visible is to have a layer of invisibility, which is noticeable only as a certain absence. The frontal properties of the visible reach the eye directly, but there is also that which reaches from below, and that which reaches it from above (and frees it from the gravity of its origins). In addition to the outer movements, painting tries to capture the secret ciphers. A photograph records the visible instants of time, but it loses the overlapping; painting makes the overlap visible. Yet painting does not present itself to the mind, but to the sight, to offer an imaginary texture of the real. Vision learns by seeing: it sees the world, and its inadequacies; it reaches beyond the "visual givens" to open upon a texture of Being of which discrete sensorial messages are only punctuations.

Painting comes from the eye and addresses itself to the eye, bringing into the world the form of things whose seal has not been broken; it exists in the visible like natural things, but never the less communicates through them. Merleau-Ponty contends that the painter "must affirm ... that vision is a mirror or concentration of the universe or that, in another's words, the *idios kosmos* opens by virtue of vision upon a *koinos kosmos*: in short that the same thing is both out there in the world and here in the heart of vision—the same or, if one prefers, a similar thing, but according to an efficacious similarity which is the parent, the genesis, the metamorphosis of Being in his vision." This is what the painter wishes to capture. To see the visible, it is necessary not to see the play of shadows and lights around it; profane visibility rests on a total visibility which the painter tries to recreate, to liberate the phantom in it.

The history of thought and the history of painting are parallel. History is a unity, with different forms. The artist, or the philosopher, can only transcend the present when he is rooted in the past; the past requires this transcending. Whereas the painter can suspend the world, in his innocence, the philosopher is summoned by the demands of his age. In Aristotle, philosophy is theoretical and art is a productive knowledge, but Merleau-Ponty reverses this conception by having art exist for no other end than its own, while philosophy must speak for a certain end, in a certain way.

Perhaps art maps experience, as an expression of the development and extension of human awareness.

4.7.3.3.4. Art as a Tool

Art is a tool for the symbolic expression of the processes of nature and for the limits of art itself. The human Body and human imagination have limits. Art must share a common code to communicate and so must impoverish the unique; each poet should have his own language, but given conventional speech, such language can only be silence, and uncommunicative. As empty space is a part of modern painting, or silent intervals a part of Webern's compositions, the void places in poems are indispensable to the completion of the poetic act. As every word approaches an idea, it becomes surrounded by space, and the blanks and silences become part of the poem. The object varies with each interpretation, but retains unaltered virtual existence beyond. Silences of this sort insert themselves within the poem and at the close. Audible sound is replaced by a filled silence which is neither tonal echo nor reflected presence. What the poem cannot express, speaks here; it may become the poem. The world surpasses itself in its echoing antithesis and defining negation, silence. Ortega pronounces that: "Language is always bounded by a frontier of ineffability, by that which absolutely cannot be said in any language."

Is art an adequate approach to the ineffable? As a work of art may subtract or add more and more elements in order to reach into the ultimate ground of reality, it becomes more hermetic and incomprehensible, until it can express, or point toward, only freedom or nothingness. Malevich's silent white art—an empty canvas—reveals man's freedom to him. But for fear of limiting freedom, he says nothing at all. Empty canvases mark the limit where art becomes silent and invisible; but are they meaningful? Malevich paints whiteness, Reinhardt paints blackness; Being is empty and full.

Is poetry more adequate than painting? How does the poet transcend the limits of language? By being silent? Ungaretti suggested that the poem should be like a brief tearing of silence. Mallarme argued that ideally a poem should be silent, white. In Holderlin the silent intervals are a sovereign logic in his poems, unfolding with the poem—until he stopped writing. Rimbaud demanded the freedom of silent reference in his poems, until he also turned from writing. Adorno wrote, "No Poetry after Auschwitz." Are poets reduced to silence in the face of ineffability? There are three interpretations of their silence: (1) the paradox of silence as final logic of poetic speech, (2) the existential exaltation of action over verbal statement, and (3) the refusal to write in dehuman times. Is the unspoken word richer, somehow? Is silence the ideal of modern poetry? If art is articulate, it may use, and be silence; but if art eliminated all "speaking," does it not trap itself in a dumb stasis? Does it not reduce it self to one dimension? Silence is not enough to describe Being.

For the Arunta of Australia, the purpose to present existence and dreamtime, but not with personification or idealization. Art is related to nature. In fact, nature and spirit are inseparable. Art is symbolic but also sacred, because it is direct contact with supernatural. People make homes and paths. Often their art diagrams the nodes and lines, as spheres and tubes (things and relations, things and flows). Aborigines express physical and metaphysical events and places with circles and lines. Expresses worldview: Things are always the same. Assent to the terms of life and suffering. Art is a tool to represent the invisible as a model for belief (ecodeontics or religion).

Art is a tool for intensification and miniaturization. Perhaps art is a recycling program for old images, a carnivore that eats raw images, words or ideas and remakes them into flesh, excreting what is not useful.

4.7.3.3.5. Art as Imitation of Nature

Nature is organic, like Aristotle considered a work of art, and whole. Culture is a holistic perspective concerned with the unity of things. This is a perspective of art. In a work of art the whole is never sacrificed to the parts. Excesses and imbalances are permitted only in relation to the whole.

Marjorie Grene finds an organic theme as the source for Aristotle's emphasis on artistic unity. Art is a mimesis of nature, It does not copy nature's products, but, in a like manner, presents unified wholes; works of art are structured like living things. According to Aristotle, when things exhibit unity they have order, when things have a definite size and order, they are beautiful (in art and nature). Only by being organically structured, in fact, can art be mimetic, and only by being mimetic to life can art works be organisms.

Since all experience is comprised of aesthetic events, and since art in human activity is broadened beyond artistic, art is generic. As in Aristotle it involves feeling and making; art is the education of nature (*educare* = 'lead forth,' or 'draw out'). Life and art merge: experience in art is the dynamic process of life itself, in continuous oscillation between opposite poles; life and culture are both aesthetic phenomena. The similarity uniting diverse cultures is explained by the human condition: the first expressions of the basic ambiguity, of man as a body-mind, shape the living environment. Artistic creation is the spontaneous movement of the human body as it expresses its universe.

In China, the wealthy pursued their interests in bronze vases, bowls, plates, carved jade jewelry, glazed pottery, stone sculpture, woodcuts, paintings, opera, architecture. Art was enjoyed for art's sake. But, also as attempt to achieve spiritual meaning. Elements of nature, such as plum blossoms and geese, had symbolic meanings, such as courage and hope, and a happy married life. Landscape art is different. The perspective is ambiguous rather than well-defined as in Europe (never developed mathematical linear perspective). Little emphasis on the horizon. They are more atmospheric than detailed and naturalistic. The axes of painting are mountains and water (which is the first definition of painting). A later sensitivity was more natural and less formal.

Art originates in living beings (not exclusively human); it is organic in the Aristotelian sense. The work of art is simultaneously self-realization and world constitution; it determines the individuality of the artist and the physiognomy of his world. Aesthetic knowledge is explained by perception; perception is explained by expression; since explanation is rooted in experience, and experience is historically progressive, the circularity of explanation (in two dimensions) becomes a spiral (in three).

4.7.3.3.5.1. Art & Poetry

Poetry is a kind of knowledgeable activity with a product for an end. Aristotle defined art at various times as: having to do with creation; using powers; "principles of change in another thing;" and being concerned with the same thing as chance. Poetry was *poeisis*, a fashioning of random data into a significant statement of universal relevance. The whole is structurally unified into a complex thing. This unity is based on the organic theme in Aristotle. Art is a mimesis of nature, but it does not copy nature's products. It presents unified wholes, in a like manner. Works of art are structured like living things; when things exhibit unity, they have order, and when things have a definite size and order, then they are beautiful, in art and nature. In fact, only by being organically structured can art be mimetic; and only by being mimetic to life can art works be organisms. Poetry imitates, not fragmentary reality, but the essential whole. The poet is a maker, as the etymology of the word indicates. Myth and poetry are considered instruments of knowledge.

Poetry intensifies the world; it does not produce a new one from superior authority. The poem is a knot of turbulence, what Pound called a vortex and Yeats a gyre. G. Snyder relates that the Japanese term for song, *bushi* or *fushi*, means a whorl in the grain. The whorl is the world in miniature. The poem is an event in miniature. This miniaturization is an entry to being in the world. If art is like nature, then it must be changing and temporary. It may also share the mosaic of nature.

4.7.3.3.5.2. Hologram as Metaphor for Art

The brain orders its universe in dreams, through language or images. The hologram has been proposed as a metaphor for memory; perhaps it could be offered as a metaphor for art. The two terms of a metaphor, one a pure reference, the other an embodied reflection, interfere with one another to produce the metaphor itself, an interrelated whole. As the hologram captures the phase as well as the amplitude, the metaphor expresses lateral meaning as well as denotative. Literally, the metaphor ('carrying beyond') is a hologram ('whole writing') that is capable of describing the universe, that which is turning into one whole. Art is an ordering process for making of new organic unified wholes.

A hologram may be read out then by shining a copy of the reference beam across the film; the resulting holographic image is like the freezing of light waves from the object in a window. This provides an interesting parallel to Ortega's definition of art: "Art is a window with a garden behind it: one may focus on the garden or on the window." Is the hologram, then, an art form? Does it translate the view more accurately than a painting, or only differently? A hologram is like a window; the object is frozen in light; it is a faithful "reproduction" that mirrors the object, and in which the object can be observed three-dimensionally.

Art may be the same type of condensation on a window as a hologram. Waddington suggests that there may be an analogy between some modern paintings and a hologram; for instance, Mark Tobey's Messengers may seem chaotic and meaningless, under the wrong reference, as would a hologram. Then again, the hologram must frame a garden, whereas a painting may be of the window itself. The hologram could be used as an art form, like photography, or it could be used to produce new possibilities in art. Since a hologram can be produced from a computer printout, it is therefore possible to generate any object on the printout, even one that has never existed, such as a griffin.

4.7.3.3.6. Art as Communication

Inuits developed a culture that allowed them to survive in an arctic environment. The arctic small-tool tradition was their technological base. Inuit weapons were fashioned from bone, ivory, and driftwood; all hunting tools were engraved with magic images to encourage the animals to be killed. Art is more than an object. It is the release of the essence of a thing, it is a help to the expression of hidden form. Art is a form of dialogue between a human and an other. Form and function are determined by feeling. Thus part of an antler that fits the hand well may become a chisel handle with minimum modification.

If it is the form which reaches into stillness, how is it created? Art is a dialectical process that symbolizes the productive activity of nature. In Nietzsche, the two art sponsoring deities, Apollo and Dionysius, are symbols of the metaphysical principles of life and being. Art owes its continuous evolution to this duality. All experience is aesthetic experience; existing is feeling (*aesthetos* = sense perception). Aesthetic experience enters into the depth of being, into the chaos and terrors of existence, to reach fulfillment in the formative activity of art, which contains both suffering and exultation.

On Easter Island, Rapa Nui, art was used as a military and political tool to establish dominance. Obviously this was more important than social or ecological concerns about using the last trees for transport. The statues faced inland. The statues also communicated the status of the builders. The building of the statues communicated the control of the builder. Raising the statues linked the builders to the holy.

For the Maya, art was used to legitimize the reign of a ruler, by proclaiming conquests and ancestral ties. With no standing armies, deterrence was propaganda signaling. The positive feedback fed into monumental architecture and sculpture, as well as painting and art. Massive investment in public displays displayed the strength of the centers (i.e., they were so strong a people, they could squander wealth and labor on art and architecture). Art and sculpture depicted military themes. Rulers were shown judging or standing on captives; captives were shown mistreated, tortured, executed (e.g., at Bonampak). This signaling of strength also recruited more people to live near the center.

Communication is related to media, as extensions of the human body and mind. Art is the creative communication of the environmental changes, including technology, which allows human response and adjustment to the limits of that environment. Reordering the world is one role of an art object; the Romantic poets are a good example of this, as they made “nature” visible to Europeans. Television can involve people if the images are accepted as part of a moral obligation to a larger community, regional or global. Perhaps the internet with television can lead to active dialogues.

As technologies developed, human relationships with animals and plants changed. Hunting, grazing, and agriculture provoked large ecological disturbances. Early domestic animals were revered, but nondomestic animals were considered competitors or nuisances. Now, animals are treated as commodities processed in factories and wildlife is regarded as useless. Technology also promotes land degradation. Plowing causes erosion. Problem with the degradation theory is that time and leisure are needed for technological innovations.

People starving rarely invent their salvations. Then technology adjusts with terracing or contour plowing. Then technology degrades with fertilizers or pesticides. The artist uses new technologies, in a new environment, to perceive the older, "sloughed off" environment. The technology is usually part of the new environment. Can it be perceived? Can it perceive itself? Is art, or technology, reflexive enough to present the new environment? How useful is reflexivity? What happens when art is the environment? Can it perceive itself then? As humans know, new environments can be dangerous. Perhaps that explains our reluctance to move into new environments unless we are pressured by resources or others.

Print technology detaches one, and it is "numbing, according to McLuhan. Electronic technology involves one in depth, according to McLuhan, merging the individual and the environment; perhaps because audio is blended with moving images. But, doesn't print merge one with an environment? It seems that without imagination, the electronic environment is no more enabling. Print can combine the person and environment in an image. Electronics has that capability, also. Electronic involvement can require less imagination and therefore less involvement. There is no touch involved in either media.

Furthermore, archaic norms and value systems crumble under the images and artifacts of industrial culture. Art becomes a form of transcultural communication that eliminates the place specific details. It is not just things, practices and languages that are lost; it is the stories, the skills, the artistic creation of forms and things.

4.7.3.3.7. Art as Meaning Itself

As the artist introduces new meaning into the world, she modifies the world of others, and this meaning sediments as bases for a common world. Art is a historical process which unites many human transcendences into a single people. The painter is an example of a person's relative independence from his present in his particular modes of expression. He is guided by two motivations: what he intends to paint, and his style, as a historical influence. The work of art is also a limit since it is incarnated. The artist creates her own style, a system of equivalences that she establishes and by which she concentrates the meaning which in her perception is still scattered, and makes it exist separately.

The greater part of human experience is primitive and emotional; ultimately nonverbalizable, literal language is too abstract to deal with it. Anais Nin wrote that "Poetry, which is our relation to the sense, enables us to retain a living relationship with all things." It not only allows expression to savage experience, to feeling, it also is a doorway to mystical experience and transcendental visions; it creates new dimensions. Art is an indirect language, an expression, in which the individual elements need no meaning, but gain meaning by entering particular relations within a context. Art achieves only through indirection like sailing; good sailing requires knowledge of the forces at play, the wind and currents are beyond control, and ability to use them to its fullest advantage. Prose motors into the wind but loses the wind in the noise.

Meaning in art is a case of logical meaning and, apparently, an essence. The artist returns to the "things themselves" as they appear contextually in our pretheoretical perceptual experience, as natural meaning structures which are basic to our artificial logics, as the "Logos of the aesthetic world." Nin writes: "There is a form of writing which is like the art of music. It can affect us through the senses directly without first appealing to the intellect, without going through an analytic or conscious process. In this, it acts more like our life experiences, which enter the Body directly before we are able to dissect them." In poetic language, ideas are not produced by words in the ordinary usage of empirical terms, but as a result of the carnal relations of meaning, the halos of signification words owe to their history and uses. Wittgenstein's proposition, "Whereof one cannot speak, thereof one must be silent," perhaps refers to descriptive speech but not injunctive speech, as in various art forms: in music, a composer presents a set of commands, which, if followed by the reader, can lead to a reproduction for him of the composer's experience. There is no attempt to describe the sounds, or the feelings aroused by them. Poetry seems to be a more complex attempt to do the same, using descriptive terms as a set of injunctions. Poetry delivers a message, not by conceptual statements, but by language-formed images for our contemplation. Its functions oriented toward the structure and form of the message itself; the words and sounds, and their sequence, interactions and effects are in a unique order. At an existential level speaking becomes synonymous with thought, as illustrated in poetic language, where words mean as signs to communicate a lived experience. Like mythology, poetry can be analyzed, but not without the loss of meaning. Remember, the idea is a result of the carnal relations of meaning, the halos of signification arise from use.

Art presents itself as pure information, as itself. It works to make the visible abstract, and to make the abstract visible.

4.7.3.4. Questioning Art

These definitions and functions lead to questions about art. How does the permanence of art matter? Should art be consciously ephemeral? Is art a calling or a business? Art used in business shares the same goal, to expand profit, to provide efficient entertainment dollars, as the business. Art as a business simply means an exchange, which may have different purposes, from entertainment to education or edutainment. Is the purpose of art to shock? As a form of education, perhaps.

Art used by politics can be used for expanding acceptance of the status quo in the current environment. Of course, art can show the invisible connection running through the political movement, which may not be flattering.

Does art destroy context? Only if it is the subject. The context is what is changing and what requires art. So, art can ignore context or reform it, but it cannot destroy context. It is like the question: Can one being, the tiger, destroy the field of being? Being? No.

Is art out of context? Yes, by definition. Art selects unique items from the context and uses them to educate. Art is also out of convention. And, out of celebrity—imagine, celebrities as artists. This is a part of art as service to the state and art as commerce. People's attention shifts from the art to the artist, which may not be bad if the life is the art, as the Balinese might contend, and often the life is more evocative and interesting than the art, as often the explanation is more poetic then the figured expression. But, celebrity art is art in service to the old environment, in an uncritical way, rather than the revolution or prefiguring of the new environment. Art can use science or politics as the environment. One has to be careful when science, or politics, or business uses art to scaffold and glamorize the old environment.

Is it possible to identify long-term trends in art? Are the following statements true? The number of art objects increases as the population increases. The number of art objects increases as the mobility of the people decreases. The number of art objects increases as the complexity of the culture increases. The richer the environment and the denser the human population, the grater the wealth of art. Over time, art becomes more abstract. Artists always adapt more modern materials, like witchetty-grubs and acrylics. Art is always concerned with control, from seeing the invisible to decorating, to controlling people, communication, and the level of abstraction.

If these are not true, are there other trends that might be? Is art better in violent times? Is art richer in rich time? Changing times? High-trade times? Can art be related to culture, beliefs, climate, eating? This question is in contrast to theories of technical evolution or limits of physical environment. Is this question a European question? Could it conceivably be asked by a Tierra del Fuegan?

Can the art of one culture be understood by the members of a different culture? Does art have a fundamental role in promoting peace (as Frederick Schiller believed, more effectively than violence or law)? Is ecological design independent of culture, or just another attempt by industrial culture to dominate other traditions?

4.7.3.5. Art as Survival Technique

Art may be a way to accept limits. Art could be a digital or analog program to continue the expression of evolution and diversity. Art could be a politically active way to promote needed things and actions, within limits, with sufficient power to restore balance to human development with the planet, a map to survival.

Gary Snyder explains that maps, histories and lists are just part of the menu of biogeography. Biogeography is political; it destroys a nation's pretensions; it cuts off exploitation and discourages travel for sensation. It is businesslike and playful, and it can nurture or kill, says Snyder. Gary Snyder's later proposal of poetry as an ecological survival technique makes good sense. Poetry is defined as the skilled and inspired use of language and voice to embody rare and powerful states of mind common to the singer and all who listen. But the poet has the power of Orpheus, to call up a world from nowhere to dwell in. Poetry has an ancient, mythic function. Not different from science, but more diffuse; not better than science, but more comprehensive. What is required for survival? Order, diversity, cooperation, knowledge of biogeography, etc. Poetry teaches that these things are valuable.

Of course, archaic world views have been considered survival mechanisms. Poetic ecology is housekeeping on earth. As Bachelard stated, the poet's occupation makes the universe into a habitable cosmic house. Only when we are comfortable can we study the house; that study is called ecology. The poetic imagination, with a primitive awareness and scientific knowledge, may save our world. Gary Snyder discerned an undercurrent in civilization since the late Paleolithic. He considered Buddhist Tantrism to be its finest and most modern statement: "that Mankind's mother is Nature and Nature should be tenderly respected; that man's life and destiny is growth and enlightenment in self-disciplined freedom; that the divine has been made flesh and that flesh is divine; that we not only should but do love one another ... these values seem almost biologically essential to the survival of humanity." Jonas Salk concluded that only wisdom will ensure human survival.

A humanism based on Schiller's ideas must have a whole image of the place of humanity within nature, and not a transcendent view. It must confront the past without the baggage of sentiment and the future without the paralysis of dread. The appreciation of the differences of other cultures (the difference of Horace, for instance) will allow us to transcend our present identities. Responding to any art with pre-established values makes one miss the uniqueness of the experience presented. Art would broaden the mental worlds of observers and encourage tolerance and wonder. Education in aesthetic humanism embraces three concepts: liberation, play and community. Liberation is freedom from contemporary limits of identity. Play is imaginative experience, natural learning entered into freely; education should be more like play than work. Society can gravitate into groups to live, but communication across the barriers is necessary for a world community. The wholeness of humanity needs to be affirmed, but from a firm cultural base. The complete surrender of cultural identity is as dangerous as too little openness. Every culture needs its own local, sacred center, which cannot be broken if the group is not to perish.

It was the belief of Frederich Schiller that human society could be improved by political means. But, after studies on the Thirty Years War in Europe, he became skeptical of the ability of politics. After reading a work on art, Schiller considered it historical proof that art can achieve what violence and law cannot—art educates and liberates the individuals of society in a gradual and peaceful process. In spite of the cultural forces dominant at any moment, an individual can determine a different course. Unlike the classical humanism committed to lessons of the past, the aesthetic humanism of Schiller was open to possibility.

Art can achieve what violence and law cannot—art *educates* the individuals of society in a gradual and peaceful process. In spite of the cultural forces dominant at any moment an individual has the potential to determine a different course.

Art can provide a whole image of the place of humanity within nature and not a transcendent view. It could confront the past without the baggage of sentiment and the future without the paralysis of dread. The appreciation of the differences of other cultures could allow us to transcend our present identities. Art could broaden the mental worlds of observers and encourage tolerance and wonder.

Art could enlarge or alter the perceptions of all human beings on earth with the selection and presentation of relevant information to form an ecological consciousness. The survival of society now depends on a consciousness of the global system in its complexity and connectedness. The spirit of humanity depends on a consciousness of its proper relation to the wild places of the earth. Art, combined with an ecological science and politics, could be adequate to deal with the creation and maintenance of good places on earth.

On the other hand, these archaic norms and value systems crumble under the images and artifacts of industrial culture. It is not just things, practices and languages that are lost; it is the stories, the skills, the artistic creation of forms and things. Art has the potential to restore or save these necessary things.

The true architects of culture have been the historians, artists, propagandists and poets; the poets provided the national archetypes. Nation-building wars may provide heroes for popular imagination; but poets and writers make the identities. The real identity of leaders is provided by propagandists and poets; the leaders only provide the horsepower. Their mask is determined by cartoonists and writers.

But, the mask is by definition subversive. Pablo Picasso stated that: “Art is subversive.” Paul Shepard applied the idea to ecology: “Ecology is a subversive science.” Subversion, from two Latin words combined, simply means to ‘turn under,’ as in to bury wastes or old ideas. Art could rebuild places and the commitment to places, reuse old landscapes and reestablish lost connections. Imagine what art and ecology could do together: Regenerate art as a holistic ecological activity, or ecology as aesthetic design fitting human effort into natural processes.

5.0. **Eutopias: Making Good Places**

When the word 'Eutopias' has been used in literature, on occasion, as a counterpoint to 'utopias,' to describe a dream world of contradictions between rusticity and civilization, as in Bernard Mandeville's verse:

T'enjoy the world's conveniences,
Be famed in War, yet live in Ease
Without (great) Vices, is a vain
Eutopia seated in the brain.

This second meaning of utopia, eutopia, is not used often. It means simply 'good place.' Good places do exist. They can be described, and even expanded. Some of the traits that make them good can be understood and repeated. A formal compilation of general characteristics of good places, a eutopias, can extend the application of utopian thought. Perhaps the number of good places can be increased with understanding of traditional ways and with more effective metaphors. Good places, and good societies, can be partly understood through certain paths of ecology and culture, which can be explored.

The earlier and foreign images of the world, created by archaic and classic cultures, are dismissed as being outmoded, useless, or unrealistic. Utopias have been offered as ideal schemes for social and political development. Sometimes, utopias offer memorable images. But, most utopias are rejected as irrelevant dreams and self-indulgent imaginings. Yet, as Pierre Dansereau has said, the failures of pollution, poverty, and urban decay are failures of the imagination. Rejecting the solutions of imagination, therefore, can only make the suite of human crises worse.

While utopian ideas might be stimulating, other traditional or accidental ideas might be applied more rapidly. Political realism dismisses utopian ideas as naïve and impractical. People are afraid that such ideas would destroy their investments in power and in collections of wealth. Political realism, however, is deficient in imagination and results, as well as deficient in courage and will. Despite great revolutions in technology and intensification, most people are still poor and threatened with displacement. Despite great revolutions in technology and living conditions over the past several thousand years, people are still very much shaped by unique cultures and small groups. These traditional institutions work very well on a local scale in a specific place. The information is available, but there is no framework to make it work, yet.

The word utopia was meant to be ambiguous, either no-place (*outopia*) or good-place(*eutopia*). Perhaps the ideas of no-places have less relevance. The images of good places, however, might be worth considering, with the images of wholeness.

A complete eutopian structure would start from such an image of wholeness. It would be a framework that holds all the other pieces of solutions and makes them part of a whole thing. It would help us to understand the whole thing, the whole set of relations with people, land, and other beings, with other cultures and the ranges of technology. Eutopias would be a framework that holds many centers, beginning with the centers of different human cultures and including the centers of nonhuman communities and guilds. Eutopias has to address the whole, because the health of the whole depends on the health of the whole parts that make it up—the health of wildernesses, the health of civilizations, and the health of ecosystems, that is, inhabited places that are made through living.

Utopias may be impossible, but good places are possible. Good places already exist everywhere. Perhaps by logically working out the limits and adaptability of humanity, good places can be created by almost everyone.

In totalitarian utopias, people must be controlled, the threat being that, without discipline there would be immediate disorder and the utopia would collapse. As we know, it is impossible to avoid disorder, which is a necessary part of any physical and social system. A system with flexibility, such as a eutopian system, could incorporate disorder without being destroyed. Flexibility is an important eutopian characteristics, along with resilience, adaptivity, and surprise—places can offer excitement and danger.

Taking its cue from utopian writings, eutopias asks the broadest questions about society and the environment. Where would you like to live? In urban or rural or mixed areas? What should your place be like? Populated with what kinds of plants and animals? How exciting should it be? How close to nearest large store? How often should your neighbors visit unannounced? Once a week, a month, always, never? What should your neighbors do? Be artists, scientists, laborers, game players, or be multi-faceted?

In that sense, eutopias fits in the tradition of thought. The approach is both descriptive and prescriptive, which allows people to interiorize the system and appreciate the details. Eutopias is a thought experiment to apply real alternatives, derived from the baseline of cultural experiences, to the problem-ridden monolithic applications of industrial civilizations.

A eutopian structure, one that is unambiguously good, could be designed and described. It would outline a science and politics adequate to deal with the creation of good places on earth, where all beings are equal within a framework of high human culture within nature. It would propose an adaptive cosmology to place human values within a global ecology and to balance human development and the preservation of wild places. The framework could protect the integrity of culturally-based nations and to address global issues. Global and local issues would be differentiated; the myths of global communities and one-world people would be explained and discarded. Pretending to be world people is a mistake based on a misunderstanding of human limits. People are the products of places and cultures, of land and nations.

Unlike either the political realism of nations or the ideal designs of utopias, a eutopian plan would be based on traditional human and cultural realities and would propose only modest and reasonable changes at local and international levels. For example, there are 500 million indigenous peoples in fifteen thousand distinct groups, such as the Uighur in China or the Kuna in Panama; and, there are over two billion people in hidden nations within massive political structures, such as the Azerbaijanis in the Soviet Union, the Kurds in Turkey and Iraq, or the Tibetans in China. Furthermore, there are regions in some countries, such as the Pacific Northwest in the United States or Wales in Britain, that may prefer independence to forced membership in a confederation. A Eutopian plan could allow any indigenous people with a traditional culture to become an independent nation without fear of conquest or compromise by existing political states. The benefits would outnumber those of a global monoculture and the negative aspects would be more manageable.

In eutopias, it is important to work with what exists, with what has already worked in place, with history and culture. Eutopias can start in California, Sri Lanka or any place that can support a nationhood aligned with a cultural history, without having a standard army or special currency.

The holistic framework would protect the creation of thousands of good places, or eutopias, based on human cultural realities and on the ecological limitations of places, that is, on homelands. These models already exist—they just need to be better defined and then protected from misguided ideas of growth and progress, as well as from the dominant, consuming industrial culture.

Culturally-based nations can provide a greater spectrum of intricate descriptions and possible relations than any one single thought experiment. That is why they need to be the basis of good-places-in-the-making. The framework is enriched by the ideals and visions of nations. This is consistent with the intuitive knowledge that many styles of life are simultaneously sustainable.

This is where eutopias would work effectively: Accepting the unpredictable nature of natural and social processes, yet managing adaptively on a micro-planning level. Eutopias would incorporate tradition and working past elements, rather than breaking with them. As Edmund Burke, reflecting on the French Revolution, notes that society has gone through a natural process of development, forming traditions and institutions that provide rules of behavior known to the people. We tend to forget how much discipline already exists in every society. People abide by most rules, from driving on one side of the road to preserving their children from molestation.

Eutopias would have to accommodate different styles of living, from agriculture to hunting, fishing and herding. People have been nomads for a longer time than farmers or city dwellers. It might be better to accept that movement and dislocation is part of the harmony of life and try to minimize the negative effects, especially combined with consumerism and energy waste. The civilization of cultures is not a steady state, but it could be accommodated in a flexible framework.

A new model may solve some problems but will definitely create new problems; Nicolai Machiavelli reasoned that this would always happen. This is a good argument for eutopias, since changes would occur mostly on a small scale, easily correctable. That is why eutopias can be a framework, by limiting human evil and good to small places, with minimal control or competition.

There is no single model of government that will fit all cultures, all traditions, all needs or all nations. But, governments can be held to common standards through an international body. Eutopias tries to structure the global commonwealth of nations by proposing specific common goals, regarding weapons disbursement and population limits. It offers a consideration of distributive justice among nations and between the generations of every nation. It offers visions of culturally-based nations in equilibrium with nature, using ideas from ecological design and conservation biology. Such a eutopian process would solve many problems, from problems of scale to political inappropriateness. Not only would more nations exist in the framework, but alliances and networks would form and reform, as they do now. How would it work? A political framework, based on traditional cultures, coordinated by a global 'regulating' body, based on a modified United Nations, would be described and implemented.

The eutopian vision would start by addressing immediate needs for everyone, but then allow people to be as extravagant as they want within the general social limits. Most people like the place they live, and they like their culture, so it is not necessary to force them to stay in place or be what they are. That does not mean that they do not want to live better, just that they do not all want to be Swiss or U.S. citizens.

This framework would also reconcile the well-being of society with the health and continuity of living ecosystems. The international framework would provide paths for policing, international education, and other national needs. This plan would provide a path to independence and formal interdependence.

The eutopian framework is more than is a simple perspective; it is a design for a self-renewing process using proven cultural methods to improve human situations and environments. The frame is not a final goal but a way to allow the many small useful, culturally determined changes that we need for our survival on the planet as we like it.

Eutopias is a cheap solution because it uses the parts already in place. It is within reach of any people in any culture, regardless of how fast or technological, because it limits the ways of big, expensive solutions.

Eutopias is not a big law; but, it is a big story, large enough to allow all other stories. It is changing and open-ended. It is conservative, but that becomes an instrument to create peace and justice everywhere. It keeps human rights connected with species rights and land rights, through an integrative political approach. It anticipates and responds to problems and catastrophes, through an understanding of psychologies.

There is a lure of good places. What does lure mean? Something we desperately want? We have often moved to places that we have only heard about. If we do not live in healthy places, we want to move to healthy places. Sometimes, we remake our places until they have the characteristics of good places.

Good places exist. What does existence mean? Good places are found or made. People groups, with strong ethics and common beliefs and understandings, by respecting limits or keeping well below them, can make good places through their actions.

5.1. *Describing Good Places Someplace*

Good places are a confluence of good human societies in healthy ecosystems. We have described some of the qualities of healthy ecosystems. And, we have described some of the qualities of good human societies. To have good societies in good places will require good images and good actions.

5.1.1. *Characteristics of Good Places in a Eutopian Framework*

From political character studies to technological promises, utopias have kept close to the contemporary forms of society. The possibilities described for the future seem to be circumscribed by the limits of human imagination. The entire literature of utopia, imaginative as it is, cannot match the actual diversity of cultures for richness or the depth of nature for wonder.

Many archaic societies employ a set of principles, different from industrial cultures, that may be more adaptive to place. Instead of regarding the "universe as mechanical, humanity as master, and all persons as equal," the Yaruru consider the "universe static and internal, humans sensible to other's wants, and all beings equal;" by contrast, the Navajo consider the "universe personal and orderly, events primary, and the family first." The ways that people live in place reflect their principles. For instance, the Yaruru are much less likely to overwhelm their home place than any industrial culture.

Other modern metaphors can promise more adaptive behavior for industrial culture. A machine metaphor used by Kenneth Boulding, "the earth is a spaceship" suggests the limits of the earth and the value of its life-support system, but it masks other realities. The metaphor of the spaceship is a closed system model, which leads to inadequate understanding of open, natural systems. The earth is an open system that sustains life. The earth has no single captain with authority. In fact, the image of a spaceship does not fit a large, organic, nonmechanical system. Alas, the earth is not a spaceship. It is far more dangerous and uncontrollable. It is clear that many human behaviors and many human institutions in the past, which may have been appropriate to a large planet, are entirely inappropriate to a small closed spaceship. "We cannot have cowboys and Indians, for instance, in a spaceship, or even a cowboy ethic," Boulding says, "We cannot afford unrestrained conflict, and we almost certainly cannot afford national sovereignty in an unrestricted sense." Another metaphor in popular use, such as "the earth is a garden," is a better model for reintegrating humanity into a balance with nature, because the garden is a small balanced system directed by humanity, and part of the larger environment, and dependent on it. The rule of the garden is empirical and based on observation: If you do something, then something else happens. Even so, the metaphor of the garden has important limits. Humanity does not have adequate knowledge to direct all of the processes of nature.

In naming a new science of ecology, Ernst Haeckel combined two Greek words (eco-logos) meaning "the study of the house." Ecology relates to dwelling, to the frame that contains us. The desire to refine a focus on our problems has allowed the frame of reference to be neglected. This metaphor has turned attention to the whole. But, it too is limited. The house herein is not a construct any more than a spaceship. And, there is not just one house; there are many unique ones with individual characteristics and connections.

Eutopias, as a general description, uses a root metaphor of many places, in different bioregions. This eutopias is a framework for human cultures, to preserve the unique image that a society needs to guide it and to make it different from others. To be effective, in contrast with the ideal characteristics of ideal cities, a eutopian framework embodies attributes that are compatible to the values and norms of living cultures.

By being attentive to the characteristics of place, and those of a good society, a eutopian framework can be described by its own characteristics from groundedness to comprehensiveness. These characteristics are quire different from the characteristics that can be observed with utopias or industrial designs.

As noted (in Section 3.3.2.1.), utopias provide images of ideal societies in abstract settings—literally nowhere. Utopias promise newness, order, happiness, and re-inheritance in an ideal political, technological framework. They banish the irrational, the irreparable, and all conflict. The common characteristics of many utopias, however, make them unworkable and unsatisfactory. Utopias tend to be ungrounded, that is, literally no-place, with no reference points in real places, ecosystems and bioregions. Utopias present a static order that is an unchanging model of perfection that is without history or change. Assuming the possibility of perfection (a teleological goal), human happiness is presented in a final order requiring tremendous discipline on human activities. Utopias aim to be ingenuous, to universalize the best of human society in a gigantic, centralized model of an international community ruled by a special government of the elite. The rules and laws of utopias are generally simple and arbitrary, denying the complexity of physical, environmental, and social problems, as well as of chaos, chance and evolution. Utopias tend to be homogenous, in terms of temperament, race, skills, and the uniformity of clothing and homes. Finally, utopias are incomplete; nonwestern, nonindustrial societies are ignored for the conversion of the sad metaphor of the storehouse of nature being looted to create masses of possessions for an exploding population.

The weaknesses of formal utopias are addressed in a framework of eutopias, good places. Eutopias are grounded in local places. Their realistic properties, based on real cultures living with the ecological knowledge of local places, promise the possibilities of successful implementations.

5.1.1.1. Eutopias are Grounded

Eutopias are grounded in places, and grounding may resemble a knotting in the physical process of existence. Eutopias are realizations of human ideas and designs. The making of places is an ordering of a distinct structure and with an important center. Humans and their communities are embedded in places. The Taureg of the Sahara have created an image, a world, that cannot be relocated to the rain forest of the Campa in Peru. When the symbols of a world lose their meaning, through wrongful application or abstraction, sickness and disintegration result.

Being grounded in place allows us to rediscover a participating consciousness and a symbiotic connection to the living earth. The organization of perception, meaning, and thought is intimately related to specific places. The commitment to a place implies acceptance of its limits. Place is a focus of meaningful events and a platform for ordering a world. The individual image of a place is modified by memory, experience, emotion, imagination, and intention. The social image of a place is influenced by individuals, myths, history, and consensus. The images reinforce each other. Each place and culture is unique.

5.1.1.2. Eutopias are Dynamic

Eutopias are dynamic expressions of the productivity of places. Nature and human nature are not static orders; they are flexible, historical, and irreversible. Worlds have been built by peoples over so many thousands of years that it is not necessary to start from raw sensations for a new image. Societies build images that reflect knowledge of themselves and their environment. The problems of many human societies can be rooted in their anthropocentric images of the universe. But, the solution cannot be a uniform cosmology of the earth. The strengths of cultures lay in the diversity of values and in their fitness to particular places. A holistic eutopian cosmology can preserve the differences in a whole image of the earth. The image cannot be a rigid shell to contain everything, but rather a flexible, organic network holding all human and natural groups. The eutopian cosmology recognizes the value of a whole dynamic biosphere. The eutopian order permits traditional cultures and natural processes to be self-ordering and self-renewing without the imposition of a rigid order from above. They allow cultures to be healthy and harmonious.

5.1.1.3. Eutopias are Adventitious

Societies are part of an unending, imperfect process, without a final state. Furthermore, the attempt at perfectibility through self-improvement causes disharmony, which is part of the same imperfect process. Each eutopia is a practical application to place. It is open to the differentiation and diversity that exists. It accepts confusion and conflict—but constructive, scaled conflict, not insoluble, that can lead to education, understanding, and the abandonment of stupidity. Many utopias imply that society can be remade according to reason. But reason is not large enough. Experience is necessary; the unconscious is necessary, and ecological design is necessary. Utopias pretend that all factors governing a system are known and that their effects can be calculated. Unknown factors determine a large part of the operation of any system. Furthermore, there is chaos in every system; there are plagues and random frenzies. Eutopias recognize and absorb unknown factors. They enhance the richness of place.

5.1.1.4. Eutopias are Sophisticated
All the contents of the human species cannot be captured by a single policy. The eutopian framework protects difference and diversity from a uniform global policy. Within the framework, cultures are decentralized and autonomous; people identify with their local culture. Because the earth is finite, there are physical and biological limits to growth and progress. Eutopias voluntarily limits human influence within ecosystems. This does not mean that humanity cannot modify some ecosystems or become space-faring—just that it should not dominate every ecosystem or transform the entire matrix to human products, in order to luxuriate or to explore space. The process of producing goods results in waste, even at low rates of use. The principle of limited good is respected; desired things exist in nonexpansive quantifies. Eutopias is an approach that respects limits, maturity and subtlety, as well as beauty and elegance.

5.1.1.5. Eutopias are Complex
The eutopian framework is multidimensional and pluralistic. Balanced development, rather than growth, is emphasized. When a culture falls out of balance with its local environment, massive disruption often results; industrial cultures have only avoided disruption by trading advantageously with other locales, using fossil fuels, and promoting institutional inequality. Small cultures have built-in checks; furthermore, their cultural definition of good helps to maintain balance between other species and the use of ecosystem productivities. This allows them more stability.

Gross imbalances of wealth and health are the unforeseen consequences of exploration, colonization, science, and management. These imbalances will become more unmanageable with time and television, or with any communication. The primary obstacles to equity and justice are political and managerial, not to mention inertia and poor priorities.

Regional areas are limited in size, to avoid problems of scale. Historical smallness, even lacking natural resources, has not been an obstacle to wealth for many countries, for instance, the sovereign German states of Hamburg or Bavaria. The merits of urbanization do not require a large population. Local concentrations of artists, philosophers, and scientists are capable of creating a distinct civilization. Cities fifty times as large as classical Athens or Florence have not been fifty times as creative. Eutopias would generate loyalty and complexity within limits of size and impact.

5.1.1.6. Eutopias are Heterogenous
The eutopian frame is unselective. It accounts for all human diversity and variability, for prisoners and misfits, artists and technophiles, and for the insane and the aged. It is pluralistic. It rewards and uses individual differences in constitution and character. Humans are not perfect or interchangeable. It accepts inequities, although biological injustices exist

and can be ameliorated, and social injustices can be rectified. The eutopian frame is flexible enough to incorporate the positive features of

traditional civilizations. Through its respect for the validity of all cultures and understanding of the responsibilities of cultures, it works to define an authentic concept of humanity. It tolerates fluctuation, irregularities, uncertainty, and diversity, which are characteristics of open systems.

5.1.1.7. Eutopias are Comprehensive

The levels of application of human norms are both universal and local, depending on the context. For example, there are some universal human behavioral standards, such as a prohibition against incest or against eating human flesh, but local expectations conform with cultural values—and indeed, cannibalism and incest have been important parts of some societies at times. The eutopian frame tolerates and integrates all cultures. Each culture determines the style and complexity of its individuals. Eutopias strives for concerned noninterference, but offers advice and assistance to all cultures to integrate new attributes or common concerns, such as the equality of women, into the culture.

Eutopias considers the total community as a self-making whole. Human cultures make a place within nature. Culture is an immeasurable complex of material and spiritual achievements inside nature, by modifying and using nature. Nature changes with culture. Nature is the locus of the centers and images of all living beings. Nature is thus an important basis for all cultures. The self-ordering processes of nature must be protected through formal preserves of areas or through limited human impact on other areas. The eutopian framework allows people to identify with their domiture, that is, nature and culture.

5.1.2. *Describing the Dynamic Structure & Framework of Good Places*

The characteristics must be expressed in action. Characteristics are part of a structure of a good place. Characteristics can be described, but pieces could be missing. The structure is necessary to contain and fit the characteristics, regardless of what is missing.

5.2. ***Thinking about Making Good Places in Context***

People live in place. One secret to making good places through living is to have the places be good places, so that people might develop their potentials. There are many potential levels of action, many ways of being, starting with the individual and continuing through the international. However, individuals are part of families, which are part of groups that often include extended families. Larger groups may include clubs or corporations, communities or counties, states or regions. For the sake of simplicity, communities are considered under the heading of individuals. States and regions are considered under the heading of nations.

5.2.1. *Being Individuals in Good Places*

From a cosmological or ecological perspective, living organisms interpenetrate deeply into nonliving forms and the earth. Individual organisms are woven into a complex fabric. Their activities reshape the fabric. Human beings, to make good places, have to consider their individual and social actions, their participation, as responsibilities.

5.2.1.1. *Individual Participation in Place*

To make a good place, individual participation is required. In order for places to continue, participation is necessary. Participation enters the constitution of place; it is not a fusion where things lose their identity, but a mutual folding together where each becomes part of the identity of the other. To exist is to participate in place. Participation is felt, not thought. Participation leads to knowledge and mystical experience.

There is no way to validate the wholly subjective experience. Each human being inhabits her autonomous world. The autonomy of consciousness lets us believe the lies and the truth. As William James and Paul Tillich have argued, beliefs about reality affect real human actions. This can introduce the new into the world and permit respect for difference.

Ervin von Weizsacker is convinced that our actual task is the realization of the human self, that the possibility of world peace stands or falls on that success, although he is skeptical that it will happen. The world transforming idealist can be as neurotic as the rest and works against himself. The Buddha says the cause of suffering is attachment. Perhaps, but attachments grant status to the other. Perhaps suffering is unavoidable. A worse threat to human being and the safety of place is detachment.

The absolute individual reflected the absolute state. The absolute individual becomes more self-important through detachment from the environment and personal nets. The state gave the individual a vote, which many people use to reduce participation. The decay of mutual participation and constraint has led to alienated individuals.

The locus of political sovereignty is individual, who is limited in giving away proxy rights. Politics has to be a participatory process, where the individual has some power over decisions effecting them. Participation is necessary, not only politically, but to establish the existence of perspectives throughout the population as a whole.

Aldous Huxley noted that there are three things we center our existence around: Technological gadgets, political power and morality. Each center can be used as a means to solve problems; sometimes people believe that their problems would all be solved if everyone had a radio or computer, were a communist, or were a Mormon. As the collection of gadgets for some grows, the cost to others creates more problems. With the emphasis of political power, only the mechanism changes—from a dishwasher to the corporation or from the toaster to a political party—but, the problems increase. Moral systems allow automatic decisions also, but to no better effect. The idolatries of gadgetry, celebrity and morality occupy our attention. People struggle and suffer in the resulting turmoil. While these concerns are necessary, it would be better to focus consciousness on the unity of the situation, not on single filters of experience.

A rotation of civic, vocational and professional responsibilities would awaken different senses in an individual and round out self-development. Complete society might create complete humans. A eutopian framework could deepen the sense of personal worth, allow spiritual growth, and augment people's need to participate in governing institutions. People are capable of being the most delicate gauge of the health of a place and ecologically intelligent scale.

5.2.1.2. *Responsibilities of Individuals*

Each individual has responsibilities that cannot be evaded or given away. Some of these responsibilities are: To cultivate the self, that is, to be healthy and fulfilled; to engage in meaningful work; and to practice simplicity.

5.2.1.2.1. To Cultivate the Self. To be Healthy and Fulfilled

Each individual is responsible for her body, for her health, and for the direction of her education. Education is a life-long project, involving the investigation of and respect for other points of view. The idea of a right to health and education is less important than the moral obligation to preserve one's health and educate one's self. The possibility of transformation and peace stands or falls on the success of the realization of the human self. The individual is responsible for the style and simplicity of her life and for its effects on nature and society. The individual is responsible for being tolerant of others, and is free to choose work, mates, or places. An individual can achieve self-realization through participation in place, that is, home.

5.2.1.2.2. To Find Good Meaningful Work

One should choose work that is interesting and positive, that uses proper technology and fits in the system. Work should be respectful of resources, foster a cooperative approach to economic problems, and promote self-help and self-sufficiency. For the individual work should be a form of right livelihood, which was a step on the Buddha's eight-fold path to purification and liberation. Work would be expressed as the needs of the place dictate, regardless of economic incentives to convert a place into a set of commodities. Work would be expressed with an efficiency that comes from experience.

Work is done with other people; in that sense it is moral, also. In her study of the moral underpinnings of work, Jane Jacobs, through a Platonic dialogue of her characters, collected and sorted them to find that they fit two patterns of moral behavior that were mutually exclusive. She calls these two patterns "Moral Syndrome A," which is a commercial moral syndrome, and "Moral Syndrome B," which is a guardian moral syndrome. She claims that the commercial moral syndrome is applicable to business owners, scientists, farmers, and traders. And, she claims that the guardian moral syndrome is applicable to government, charities, foraging societies, and religious institutions. She also claims that these moral syndromes are fixed, and do not fluctuate over time. It seems likely that MSA developed from MSB as a way to cope with the changes as a civilization of strangers developed around agriculture and cities. Jacobs suggests that mixing these two moral syndromes can result in the Mafia, communism, and New York Subway Police. MSA can produce goods that are harmful to people and destructive to the environment.

5.2.1.2.3. To Practice Simplicity

Responsibility for the ecological production of goods is important. Choosing for yourself involves knowing an ecology and knowing a politics, that is, being astute, and protecting what is valued. The people of industrial nations could cultivate voluntary simplicity, living in a way that is outwardly simple and inwardly rich, by consuming less. Simplicity can be carefree and joyous, without being a form of self-punishment and asceticism. To live on earth with a minimum of influence is a noble goal. Each person has the responsibility to change the self and to abandon behavior that is inconsistent with acceptable practice. We have an obligation to make the world livable, not only for ourselves, but for other native species.

One should share with others, practice stewardship of domestic landscapes, and be frugal. Simplicity also involves not interfering with self-governing nature and not imposing one's personal morality on others. Physical enjoyment and cultivation of the inner life are valuable. The desires for superfluous material goods, for praise and power, for moral superiority and fixed opinions, are not as valuable. The manifest goal of all activity is the attainment of felicity, a subjective state. Meaningful satisfaction comes from enjoyment rather than the consumption of commodities, for instance, from walking through a forest, listening to birds, enjoying flowers, and watching a sunset, without destroying any of them. One goal, shared by all living beings, is the richness of experience. People can enhance richness and minimize the destructive results of their participation in the process.

5.2.1.2.3.1. To Limit and Value Possessions

Among Australian aborigines, a sense of responsibility was continually renewed by visions of Dreamtime, through which people were restored to unity with heaven and earth. There is an alternate reality where an individual can be healthy and complete, and be concerned with a healthy and complete system for all life. The Greek ideal can be inferred from qualities of ancient sculpture: Simplicity, proportion, and harmony. Plato's *Republic* was designed to avoid certain evils, most among them, the overcrowding of the economy, the proliferation of needless goods and services, and the infiltration of those who supply them.

People in industrial nations could cultivate voluntary simplicity, living in a way that is outwardly simple and inwardly rich, by consuming less. The main aim would be spiritual, but a lesser aim would be to acquire a minimum wealth to maintain life. This attitude would also require new values: Stewardship, not progress; austerity, not affluence; permanence, not profit; responsibilities, not rights; people, not professions; betterment, not biggerment; and enoughness, not moreness.

Lao Tse recommends cutting out the desire for superfluous goods, then the desire for praise or fear of blame, then desire for power, morality, and opinion. For him, physical enjoyment and cultivation of the inner life are best things of all. The desire for superfluous material goods, for praise and power, for moral superiority and fixed opinions, are expressions of greed and strife.

H. D. Thoreau emphasized the necessity of natural simplicity as a way to avoid the desperate life of those weighed down by strife and greed. A Harris survey in the U.S. in May 1977, showed that people are expressing choices that reflect changing values towards simplicity, although the survey did not indicate how they would face the consequences.

People can exercise personal choice: Choosing to keep an older bicycle or car, buying fewer clothes, and traveling less. The planet is being transformed for human comforts. Freedom has come to mean being 'without limits.' Two reasons for not acquiring material possessions are the ecological responsibility for their production and a goal of personal development. Each person has the responsibility to change the self, to abandon behavior if it is inconsistent with acceptable practice. People have an obligation to make the world livable, not only for ourselves, but for native species; to try to stop destructiveness exactly in proportion to the destructive results of participation in the process. People must realize that the more we are removed from a ruinous process, the less is destroyed, disturbed, or achieved. People need to learn deny things that work against thinking and living as simply as possible. Then, they will have time and the ability to find a responsible ecology, to find meaning in appropriate environments and to discover our potentials.

5.2.1.2.3.2. To Have an Appropriate Diet & Food Production

Although human dietary habits were stable for long periods, they have been changing, for economic, personal, and social reasons. Many people limit their intake of animal products to milk, butter, and eggs. Some are vegetarians or vegans; others concentrate on fruits. Humans have been represented as omnivores, carnivores, or fructivores by different factions. In view of human control over animal and plant populations, a reexamination of the use of animals and plants is critical. There are both advantages and disadvantages to strictly carnivorous or vegetarian approaches.

Humanity has a long history of eating animals, and traditions of eating are important to the integrity of archaic cultures. Besides flesh, animals provide many high-quality materials that cannot be duplicated by an appropriate technology: Wool, leather, lard, tallow, and manure. Most animals eat food that humans cannot; they concentrate protein and convert low-quality protein to high-quality. Where animals are allowed free range, they graze in nonproductive places, such as steep, rocky slopes. There are good reasons for livestock production—ruminants can harvest vegetation on marginal lands not suitable for cultivation; they also consume roughages not suitable for human consumption. Many native grasses yield one and a half times as much energy per acre as potatoes and 2-3 times as much protein.

Nonruminants, such as chickens, are useful as refiners of plants and wastes into processed foods. In periods of abundance, animals could be fed surplus grain, then eaten in lean years. Food animals would serve as shock absorbers. Foods of animal origin provide essential amino acids in which plants are deficient, and which the human body cannot make. For example, plants lack the vitamin B12; plants also may be inadequate for reproducing females. The energy efficiency of eating animals may be low, but there are other factors, including quality of protein or taste.

There are many arguments against eating animals, especially as practiced in industrial cultures. Often, animals compete with humans for the same food, corn and soybeans, for instance. This is especially noticeable in western societies, where food animals are fed corn and soybeans for fattening. In 1980, livestock ate enough food for 14 billion humans. Intensive meat production causes tremendous organic waste and water and air pollution. The intensive production of large numbers of animals for consumption causes great organic waste and pollutes water and air with animals wastes and process byproducts. Much land is not suitable for factory farming or industrial agriculture. These methods often ruin habitats.

Moreover, factory farming methods are inhumane—calves, pigs, and poultry are squeezed into the smallest possible spaces. Currier Holman said that his business at Iowa Beef Processors "is very much like waging war." And, as in war, the innocent suffer. The inhumane treatment of food animals results in lower-quality protein, while drugs and chemical additives, beyond the toxins and saturated fats already in meat, increase the danger to consumers. Circulatory and heart diseases are linked to a diet based on animal foods.

Wild animals are still taken by some cultures. It might be possible for industrial cultures to also take wild animals in limited numbers. Wild animals are already adapted to their range, which is often unsuitable for domestic animals. They are usually more efficient converters than domestic animals. For example, the oryx (antelope) needs only one-third the water of cattle and is immune to many of the diseases that affect domestic cattle. It is twice as efficient at making flesh, and it is immune to most diseases that afflict cattle. And wild animals are adapted to an ecosystem. However, wild animals are unsuitable for domestication, so the take would be limited by health, habitat and reproduction.

Aquaculture has many advantages. Fish are better converters than mammals; fish would have a better distribution and be more acceptable to many cultures. Fish protein is readily assimilated by humans.

Insectaculture would have more advantages. Insects comprise over 80% of all animal species, and they are edible. Their advantages for diets include: fast breeding: many insects breed extraordinarily fast and occur in great numbers. One aphid, if all offspring lived, could produce 1,560 sextillion aphids, equal in weight to all of humanity. Most insects are high in protein and calories. Insects are cleaner than that cleanest of domestic farm animals, the pig. Insects are an abundant, but neglected, source of food.

Limiting the diet to plants avoids the suffering associated with food animals. The practice is more efficient overall; more people can be fed per acre. Vegetarianism could allow vastly more people to be supported on the same acreage. It would encourage humans to explore and use wild plants. Much animal suffering would be avoided. As early as 1961, the Journal of the AMA stated that a vegetarian diet could prevent 90% of thrombo-embolic disease and 97% of coronary occlusions. Plant protein produces lower levels of cholesterol in the body than animal protein. Cholesterol levels in the blood have been linked to heart disease.

By eliminating the cereals fed to animals, the acreage in production could be reduced by 51 percent. Grain is an efficient food, having a high calorie to waste ratio. Furthermore, technology could reduce the area needed to produce edible plants by 95 percent, with greenhouses, hydroponics, and alternate sources, such as algae. Vegetables offer other, untapped, sources of protein, for which appropriate technology currently exists: leaf protein, which is abundant, efficient, inexpensive, and suitable to tropical and subtropical growing areas; algae, which is efficient and protein rich; single-cell protein, from yeast and fungi, which is fast, efficient, waste-free and pest-free, and can be grown on petroleum waste. These new microbial foods could lessen the burden of land use. Furthermore, the use of wild vegetables could encourage the exploration for and use of neglected and unknown plants.

But, there are disadvantages to vegetarianism. Plants are living organisms, also, and many plants are living when they are eaten, although sentience is a more important criterion. Many of the alternative sources, such as single-cell protein and algae, are deficient in amino acids or are of poor nutritive quality. Cereal crops, by themselves, do not provide a good balance of proteins, so many kinds of plants would have to be used. In some areas, this would mean importing protein, at the risk of economic imbalances and threats to local self-sufficiency. Plants also have poisons, to discourage their predators, human or invertebrate; many plants must be cooked before being consumed. Much of the land pressed into production by industrial agriculture is not suitable for domestic plants; habitats are ruined by the development of land for special crops, such as chocolate, tea, coffee, and tobacco.

It is obvious that humans are not pure carnivores, but it is less obvious that they are not pure herbivores or fructivores. Physiological evidence, such as the shape of teeth or length of the intestine, points to an omnivorous existence for many thousands of years. Although some human cultures concentrate on large animals or on roots, most cultures depend on a combination of sources of food.

Consumers cannot afford to be uninterested in nutrition, agriculture, health and ethics. Health is rooted in wholeness. Vegetarianism is not a compartment, separate from agriculture, social living patterns, or wilderness. It is an ethical response to factory farming. If humans eat animals, let them eat the whole animal, and use it completely. We must know the animal we eat, whether it is wild or domestic.

Vegetarianism is an ethical response to the suffering promoted by factory farming. But vegetarianism is not a compartment separate from industrial agriculture, social mores, cultural traditions, the rights of wild beings, and the necessity of sufficient wilderness. Diets are part of the cultural traditions that provide individual identities for all people. Cultures maintain regional differences and emphasize the unique social aspects of consumption; meals often provide important social and psychological benefits.

Many of the problems associated with human patterns of consumption are problems of scale, efficiency of exploitation, and a universal, commercial diet. Our lust for food has resulted in a war against other species, less reported than human conflicts, but waged more constantly, viciously, and mostly out of sight. We cannot eat without killing animals or plants. Human cultures are based on killing. Often, our wants and charities result in deaths. If we have zoos, to save a few species after destroying their habitats, then we must kill for those animals that are carnivores, as we do for our pets.

Killing has been disconnected from necessity. Humans must eat to live. For food, it is necessary to kill. But food should not be wasted. The fruits of nature are not to be destroyed for temporary gains. The consequences of all acts should be examined as thoroughly as possible. When we recognize that animals suffer, we act humanely toward them. When we recognize that their interests might collide with ours we must learn to act justly, also. Are animals interests completely identified with needs?

Knowledge of our carnivorous history should not paralyze us with guilt. Rather, as Raymond Durgnat says, it is "a reminder that what is inevitable may also be spiritually unendurable, that what is justifiable may be atrocious, that the best we can do will always be an organized butchery—and the possible best is itself light-years from fulfillment." Durgnat concludes that when we realize that society is "an organization of deaths as well as of lives, can we become more aware, gentle, alive, sad, free for a Schweitzerian reverence for life ... for all lives, and not just the next one."

Knowledge makes us aware of the costs of eating and, perhaps, inspires us to eat as part of a simpler and more frugal pattern of life. To make choices, we need a way of calculating. Michael Fox presents a scale of suffering of animals: For example, dairy cattle are the least intensively raised while veal calves suffer the most inhumane conditions. A conscientious person minimizes animal suffering by limiting diet.

For understanding the ecology of beings, Aldo Leopold offers the image of a biotic pyramid; individuals at the top, fewer and usually larger, are more complex and capable of feeling. We argue about the extent that each species can suffer, think, or anticipate, but there is evidence of feeling on all the levels of being. As John Cobb implies, the entire pyramid must be healthy and whole. Human health is rooted in the integrity of ecological systems. Physical, mental, and spiritual well-being are dependent on a good diet. Being a vegetarian may be the best option for the urban residents of industrial cultures. For archaic or post-industrial cultures, where human needs are kept simple, limits are respected, and all beings are revered for themselves first, being a conscientious omnivore (the term is from Michael Fox, who is a vegetarian himself) is a middle way that preserves the meaningful rituals of eating, yet uses animals and plants in an humane and optimal way.

5.2.1.2.4. To be Peaceful: Gandhian Nonviolence

Some groups of people, concerned with defending vacant lots, wilderness areas, ecosystems, and the earth as an organic body, have advocated using any means necessary. Earth First!, for instance, showed many of us arm-chair sympathizers that saving the earth requires more active participation than just letter-writing and circulating academic articles. They were among the first to put their bodies where their ideals were, and their stance has made some profiteers and environmental rapists more cautious. But, Earth First! made its wild reputation by monkeywrenching. Although the group has been effective so far, there is a possibility that its tactics may cause long-range difficulties in the form of violent retribution or the overreactive destruction of wilderness, as well as polarization inside other environmental groups. This discussion contrasts the tactics of Earth First! with a Gandhian nonviolence.

Dave Foreman of Earth First! has said, "Monkeywrenching is nonviolent resistance to the destruction of natural diversity and wilderness. It is not directed towards harming human beings or other forms of life." Many Earth First! tactics, however, make this statement questionable. Is Earth First! really nonviolent, at least according to Mohandas K. Gandhi's definition and practice of nonviolence?

To be peaceful, one has to practice nonviolence towards people and towards the ecosphere. One should be vigilant against military intervention, practice conciliation, and resolve differences face to face. Individuals could try to deinstitutionalize legal confrontations.

Gandhi characterized his ethics of group struggle by the Sanskrit word *ahimsa*, meaning "nonhurting" and "nonviolence." Ahimsa is a closely related set of prescriptions and descriptions. Gandhi said: "Ahimsa means avoiding injury to anything on earth in thought, word, or deed." He adopted a wide interpretation of 'injury.' The subject 'anything' included all living beings and perhaps nonliving things. Ahimsa is the absence of *himsa*, meaning hurting from the root *hins*, meaning hurt, a form of the root *han*, with a larger number of meanings, such as strike, kill, destroy, or dispel. Gandhi mostly had living beings in mind, but injury to nature, to natural processes, could come under the general principle of ahimsa.

Indeed, the concept of ahimsa is so wide that it could be interpreted to be an act of violence to abstain from efforts to prevent injurious acts, such as the exploitation of wilderness for profit. But, Gandhi referred specifically to destruction, as part of sabotage, as himsa, even if the things destroyed were not the property of anyone.

Gandhi also said: "Ahimsa really means that you may not offend anybody, you may not harbour an uncharitable thought even in connection with one who may consider himself to be your enemy." Mental forms of injury include hurting people's feelings, their dignity, or their relationships— but the feelings and relationships must be positively valued, that is, it would not be himsa to hurt feelings of hatred, nor to save a victim from wrongdoing. Furthermore, some actions may be in accord with ahimsa if they are performed 'wholly unselfishly', although Gandhi does not accept the postulate of unselfishness as sufficient for the qualification of nonviolence. Most selfless terrorism, however, is still *not* nonviolent.

5.2.1.2.4.1. Injury & Hostages

Mental forms of injury seem to occur in Earth First! campaigns, from name calling and humiliation to suggestive publications. By its verbal and physical stance, Earth First! seems to be violent towards many groups, including loggers and RVers. Foreman tacitly admits this when he makes the distinction between blockades, which are forms of nonviolent civil disobedience, and monkeywrenching, which involves conscious destruction of equipment, while recommending that they not be combined in the same campaign.

By its attitudes Earth First! polarizes most people. The Earth First! slogan is, after all, "No compromise in Defense of Mother Earth!" This really polarizes the opposition. One problem with opposition is its either/ or character: "if you're not with us, you're against us." This of course is a logical fallacy. Positive action does not have to be bilateral or dualistic; it can transcend simple opposition and be positive without being adversarial. Gandhi was always willing to compromise on nonessentials. He characterized himself as a man of compromise because he was never sure that he was right. Compromise is an essential part of the nonviolent person, *satyagraha; ahimsa,* as unselfish love, demands compromise. There are principles that admit no compromise, and furthermore, if the compromise fails, the *satyagraha* is ready for "battle" (in Gandhi's term).

Could Earth First! be truly nonviolent and still be effective? Could Earth First! be more visible and less destructive? Perhaps it could, with the adoption of a Gandhian campaign to defend wilderness. Before this campaign is presented, however, consider how violence may be a dangerous tactic for groups interested in preserving and protecting wilderness against an industry-dominated consumer public.

1. The hostage, wilderness, is very large and vulnerable to counterattacks, as well as to swings in fad (which often determines how people vote to dispose of the object of the fad).
2. The destruction of materials, such as bulldozers, is usually illegal. Property is considered sacred in The U.S. and Norway, as well as in many industrial and consumer countries.
3. Violence tends to polarize opponents and some of the undecided against the long-range goals of a group, regardless of how well supported and argued.
4. Wilderness is symbolic of people's right to earn a living—even if the people in this case are loggers, drillers, or roadbuilders, who may destroy it in the process of using it.
5. Violence leads to escalation by opponents to protect their equipment and property.
6. Violence leads to violent conflict as a *style* of opposition.

Most of the thousands of direct actions on behalf of the environment have been nonviolent in the Gandhian sense and some have been effective. The "hug the trees" movement in India, for instance, physically blocked excessive logging in the Himalayas. The Chipko (meaning "hug" or "cling to") movement started out to preserve trees by embracing them before axes could be used and has resulted in a ten-year ban on tree-felling in over 550 square miles of the Uttarakhand in India, a major source of timber and water power.

The main goal is the "judicious use of trees," according to Chandi Prasad Bhatt, the founder, and not complete preservation. The movement is pressing for a complete remaking of forest policy; they are also responsible for planting trees (over a million so far). One tenet of the movement is that the erosion of human values follows erosion of the land.

Both ecology and human ecology offer the best support for saving wilderness and humanity, which depends on wilderness. The principles from these sciences can inspire a series of norms that are the foundation of a Gandhian nonviolent campaign.

5.2.1.2.4.2. Hypotheses & Campaigns

Certain hypotheses can be suggested and segregated by common themes so that the derivations are more obvious and can be linked more easily. In this instance, hypotheses are presented through the process of induction. This procedure is genetic rather than logical.

The first set of hypothesis notes that humans have the same genetic stock and the same basic interests: Food, welfare, self-actualization, love of place. Human make their worlds of facts from observations and theories colored by needs and wants, within perceptual and imaginative limits. Humans live together in human communities within biological communities in geographic places. Humans can communicate and do so to build cities and make art. Humans are capable of great trust and exhibit this in their use of automobiles, weapons, and money. Humans are capable of violence, for many reasons, from self-preservation to misunderstood symbols. Living in separate locations with differences in languages and cultures can lead to misunderstandings. Humans are autonomous beings and can choose their behavior. Humans can change their behavior. Nature is self-making and self-maintaining. Where nature is left alone, by definition wilderness, animals and plants reach their own balance. Animals and plants live in communities in geographic place. Animals and plants are autonomous beings.

The second set suggests that all beings (human and ultra) have long-range interests, especially in the continuation of their ecosystems. The presentation of facts, as uncolored by needs and wants as possible and within the limits of perception and imagination, increases the probability of understanding. Living together in the same community fosters cooperation and understanding on immediate, common goals. Working together on common projects increases cooperation. Constructive work is more binding; a constructive program is more meaningful and generates trust and communication. Work abstracted from the community can cause its opposite (enantiodromia), that is, destruction instead of good. Misunderstandings can be corrected in peaceful confrontation. Constructive confrontation concentrates on faulty understanding in a situation and not on the opponent's action or personality. The focus on misunderstanding reduces violence. Humans are responsible for their own actions. Humans enjoy being with animals and plants in general and being in wilderness and using it for living and recreation. Human economies are based on wilderness, ultimately. Wilderness should be saved for many reasons.

The campaign to save wilderness should be constructive and positive and be addressed to issues and relationships, using unbiased facts. It should be well-advertised and simple, sticking to goals. The campaign should be nonviolent and given to the possibility of agreement, including compromise on some aspects. Violence against transgressors of wilderness may result in further violence against wilderness or perhaps further violence on the part of defenders to guarantee wilderness. The campaign should apply to a local area and address those who are knowledgeable in the local community. Responsible persons, acting in a group with an appropriate lead time, concerned with their own community, present an undeniable case.

From these hypotheses can be derived norms that regulate personal and group behavior in a nonviolent campaign. Here are some examples of norms. Live in the community with your opponents. Formulate the essential, shared interests and try to cooperate on the basis of these interests. Refrain from provoking or humiliating your opponent. Seek personal contact with your opponent and his group and be available for meetings. Trust your opponent. Learn about wilderness.

Act as an autonomous, responsible person. Choose attitudes and actions that reduce conditions that lead to violence. Find common interests to build on. Present unbiased facts. Be flexible and ready to compromise. Be constructive. Suggest alternatives. Trust people, at least until that trust is betrayed, then respond in kind.

Act nonviolently, peacefully, and responsibly. Defend wilderness. Do not stop, even if it seems to be saved.

5.2.1.2.4. Tactics & Norms

The Marsh Institute, for instance, uses an educational tactic by offering to performing a consulting service for free on a small scale; the farmer and logger then can compare directly the results and the methods, such as pesticides versus integrated pest management to reduce aphids on a barley crop, or sanitation cuts versus single-tree trimming to eliminate dwarf mistletoe infestations. Then, there is defensive monkeywrenching, such as tree spiking—with signs notifying cedar thieves, and passive resistance, such as not informing hunters or biologists of the whereabouts bears or other wildlife. In general, logging and farming prices are unrealistically subsidized by free goods from nature, as well as by government subsidies. Prices should be geared to costs, and tariffs and taxes used to balance trade discrepancies. The answer is not to fight over trees in wilderness areas or to give up wilderness as a one-time boost to timber interests. It is to attack the old, unreal ways of cutting, and to attack unrealistic economic policies. The long-term strategy is to save wilderness as completely as possible and to include the cost-benefit analyses of alternate plans. Sawmill workers have been complaining that wilderness removes real jobs from their grasps; this misperception needs to be exposed as the short-coming of a problem economy.

Anti-ecological decisions usually come after years of planning by government or industry bureaucracies, at extravagant costs. Therefore, it is best to seek out and address plans at the earliest stages, before momentum can carry the plan. This is hard to do when the bureaucracies resort to secrecy. One of the goals of Earth First! is exposure of destruction and the increase in public awareness. It might be more effective to team with reporters and send news flashes on destruction and illegal use than to destroy equipment.

These norms provide a consistent guide to reacting to dishonest or violent opponents in the struggle. For instance, if your opponent uses biased reporting, merely provide a factual presentation, with evidence, without resorting to provocation or name-calling. Although the stupidity or badness of opponents is not an issue, some people are stupid and bad, so it is a factor to be considered. The stupid and bad must be neutralized some way, by simple diversion through relatively harmless assignments.

Nonviolence is easily misunderstood. If your opponent respects violence in defense of property, and you misjudge your opponent and offer nonviolence, which is perceived as weakness, prompting him to violence, what should you do? Perhaps, you should stay in the center of the issue and be active, but do not respond violently yourself. Most such opponents eventually recognize perseverance as an indicator of strength.

It is important to formulate one very clear, concrete, easily understandable goal for an action and alert the opponent to that goal as soon as possible. This is very hard to do when it is difficult to know who the opponent is and how to reach them. Any action can be part of a larger campaign, however, and this may be important for psychological reasons, since many actions are unsuccessful in reaching their goals, although that does not reduce their importance. The success of the campaign does not depend on the success of each single action. One effect of actions is to attract the attention of the public, who might act rightly with knowledge.

A campaign may be part of a larger movement, such as one to eventually put 40 to 50 percent of the landscape (of North America or the planet) into special preserves. Such a movement may take many campaigns and a hundred years. But, the goal is important.

Wilderness campaigns are not really constructive in the sense of architectural or educational campaigns, because they are defensive. Furthermore, what is defended is, first, in the process of change, so it can not be saved as it is; second, it will always be vulnerable; third, wilderness is an ambiguous concept misperceived by opponents; and fourth, it is ultrahuman. What is to be saved is the potential for the evolution of uninhibited development of species and ecosystems.

Violence polarizes opposition. A stance of "no compromise" polarizes opposition. To turn your opponent into a supporter, compromise is far more effective than violence or coercion. We have to compromise. We don't know enough not to—we are not sure enough of the consequences of our actions, which often have the opposite effect of the one intended. Furthermore, compromise can be such that it satisfies the opponent's ego without giving up much, because it creates a state of cognitive dissonance, where a little token is enough to convince people that something is owed in return, a much bigger concession in this case, regardless of whether they know about or care about how cognitive dissonance works.

The code of nonviolence, as presented by Gandhi, is not a rigid system. Exceptions are possible and, under some situations, even desirable. Arne Naess suggests that a small piece of a technical installation, a dam, for instance, could be destroyed in order to avoid the greater destruction of an area. Nevertheless, this violence is an exception and not a norm. Earth First!, by compromising on nonessential issues and using violence only as a warning, might increase its effectiveness. Either way, Earth First! already exemplifies an important, and neglected, aspect of Gandhi's philosophy: *you should follow your inner voice whatever the consequences.*

5.2.1.2.5. To Share in Governing Process

One should get involved in government and change, focus on political effort, and work to decentralize and debureaucratize institutions. We need to work to make laws to equalize representation, encouraging the poor as well as the highly privileged, who are less likely to accept far-reaching changes, to participate. We have to work to create new goals and purposes for society, especially those related to survival and happiness. We need to offer service to others, to volunteer for civic groups, and to challenge discrimination and prejudice. Involvement means to recognize and participate on at least three levels, from the local and regional to the global.

5.2.1.3. *Individuals in Communities*

There may be a genetic tendency for individuals to live in groups and to behave altruistically towards their kin. Individuals rarely exist outside of families and larger groups. Living in groups can be related to Maslow's list of human needs, especially to the satisfaction of needs such as self-actualization. The groups inculcate values, behaviors and language into individuals, so it is difficult to consider an individual without the presence of their group culture. Culture is everything created as a group, tribe or nation, physical or ideal, in the past or present. This embraces cookware, arrows, steam engines; artworks, books, legal codes; symbols, values, social structures. A cultural system surrounds the network of human interactions with raw materials, forms of life and other humans. Culture includes all of the expectations, understandings, beliefs, and commitments that influence the behavior of human groups. Many social ceremonies reinforced the cohesion of a group. Gatherings were regular, to celebrate the first fish or last crop of berries, as well as social ties. People of different groups visited, traded, played sports, and gambled. The band was the economic group. Food-sharing, division of labor, diet with variety. Most sets of ethics make the rules easy to follow. They emphasize the differences (relativism) or similarities (absolutism) of human beings only; or of the individual or the group; or of good feeling, reason, or desire. But ethics has to confront the individual, embedded in a community, located in a bioregion, on earth.

The fundamental organizing principle of human communities was kinship. The first social networks in human history were symbolized by reciprocity; relations were more important than the gifts. Reciprocity becomes established in a group with the internalization of norms. Internalization of norms may be a gene-cultural co-evolution. So, altruistic behavior can be internalized, like any other trait, as long as the costs are not excessive. Cooperation could become stable under Darwinian conditions in large social groups, provided that defaulters were punished. Reciprocity, it seems, can be positive or negative. It seems that people are willing to punish "free-riders" at a cost to themselves and no benefit; this is sometimes called altruistic punishment. People may be motivated to punish by strong feelings of resentment; even if they personally have not been directly cheated, they may want the punishment out of fairness. Traditional theories of reciprocal altruism cannot account for this kind of punishment. Maybe groups select for the traits, since free-riders could destroy the trust and harmony of a group. It shows that individuals will cooperate in a reciprocal fashion without regard for future rewards. Strong reciprocators that punish others may help the group survive, especially under crisis conditions such as famines or earthquakes. Is this a cultural form of punishment, as opposed to an individual one? Perhaps it can be understood in terms of group dynamics rather than selective advantage or cost/benefit.

Humans have a habit of structuring the world with their own group at the center. This ethnocentrism is evident in small tribes as well as in large empires, such as the Roman or Chinese empires. The center of a cosmos is usually the place of living, the locality of the group creating the cosmology. It is the center for individuals and groups.

5.2.1.3.1. Kinds & Importance of Groups

There are several basic types of kinship groups: The nuclear family, the expanded family, and various types of descent groups, that is, a group of people who claim common ancestry, such as lineages or clans. Each type can be characterized by its size, formal political system, marriage forms, and leaders. Nuclear families last only as long as parents and children live together; descent groups are corporate groups in that they are permanent units that continue to exist even though their membership changes. Membership is usually determined at birth and last for one's lifetime.

Other groups, such as tribes and states, are not based on kinship, although component groups may be. A tribe is a group of nominally independent communities occupying a specific region, sharing a common language and culture, which are integrated by some unifying factor. A chiefdom is a regional polity in which two or more local groups are organized under a single chief, who is at the head of a ranked hierarchy of people. An individual's status is determined by closeness of one's relationship to the chief. The state is the most formal of political organizations with political power centralized in a government which may legitimately use force to regulate the affairs of its citizens, as well as its relations with other states. States maintain civil order and socioeconomic contrasts through a central government and specialized subsystems. The populations are divided into socioeconomic classes, or strata, and states draw a line between elites and masses with the former clearly separated from the latter in activities, privileges, rights, and obligations. The major concerns of government officials are to defend hierarchy, property, and the power of the law.

One function of a community is to free people to follow occupations of their own choice in which they may cultivate their gifts, and grow in strength, skill, understanding and achievement. Another function is to give them the opportunity to make their contributions worthwhile to the group.

5.2.1.3.2. Responsibilities of Communities

The responsibilities of a community are: To educate people, in schools, corporations, libraries, and museums; to protect people, with public health programs, sanitation, hospitals and fire departments; to keep the community heritage, through cultural events, shared customs, festivals, art displays and museums; and, to create community wealth, in the form of parks, wilderness areas, monuments, and public buildings.

5.2.1.3.2.1. To Educate Individuals

The community teaches an individual its language, behaviors and values. Sometimes we forget that people want to live their traditional ways of life. They are willing to make sacrifices to preserve it or to reestablish it. The Inuit of Chesterfield Inlet in Northern Canada, for example, have established a community like that of their ancestors, rather than embrace mainstream Canadian industrial values. Language, for instance, is an emergent property of a group of human beings.

5.2.1.3.2.2. To Ensure Health & Safety

Groups must ensure the health and safety of the group, of individuals and of the local environment. People live in a mixed community of beings. In archaic cultures, other beings were considered equals, with their own homes and territories.

5.2.1.3.2.3. To Ensure Resources & Food

Communities used enforce rules about common resources, from hunting grounds to sheep pasture. Although an individual hunter can choose to hunt and live outside the territory of the group, the group has to be sure that the territory it claims can support all the people in the group. Communities are the matrix of inspiration for the use of resources for tools and the use of tools for resources.

5.2.1.3.2.4. To Provide Ways of Transport

Paths are structures that enable people move easily, knowledgably and efficiently from and to special areas. A walking community is highly permeable. As paths become wider more permanent structures that can accommodate wheeled vehicles, their edges interact with human places. Streets are public places. Other public ways are developed for towns and cities, from escalators and conveyer belts to buses and mag-lev trains.

5.2.2. *Being Nations in Good Places*

A nation is an independent and self-guiding conglomeration of people—sometimes with more than one culture. Nations often have special properties, such as a national currency, flag, and armies. Formerly, nations were as self-reliant as communities.

There is an operation of metalysis with nations as well. Through isolation and identity, people divided into cultures and nations. Nations have unified and disunified. The reason for nations is the desire for autonomy, to practice unique cultural beliefs and traditions. For nations, sovereignty is limited to culture (or subculture?). Isolation and commerce have to balance. Free movement would cause an imbalance.

Nations usually form during violent conflicts, sometimes as a result of settlement patterns, language and geography, sometimes from ethnic cleansing, war or partitioning. National formation has never been entirely rational and planned. Nation states are closely related to large scale violence, usually having to do with trying to consolidate their power. They then have a monopoly on power. Often, a nation will specialize on political and economic issues, for instance, when the Spanish ransacked the planet for gold. They enlarge and consolidate their territoriality, which gives them increased capacity for marshaling resources, that is trying to maximize global functions with minimum territorial burden. But they have not been stable; there are violent fluctuations. They are part of long cycles.

People are parts of community, but they learn to become part of a nation. They have to learn how to interact as members of a nation.

5.2.2.1. The Import of Recognizing Nations

A nation can claim a measure of truth and reality, but, it does not need to contradict the truth and reality of other nations. A nation often sees the source of its identity in people, and more, people in place, unique peoples in unique places, with unique histories, stories, traditions, and values.

Preservation of a nation appeals to qualities inherent in established ways and to people's moral right to maintain their distinctive customs against change. There are also esthetic reasons for preservation: to preserve styles, merit and achievement. Ethnic identity and consciousness of gender make finer grids of groups.

Groups progress when inspired by a common vision and supported by common values. Cultures need to be reminded of the values already woven in fabric of existence. Reinterpret old beliefs, base a vision of the worlds in places on earth, to be.

Devotion to nation, community, groups, clubs, corporations, teams, and co-ops, can provide a better personal identity. But, humans require more than what is local. They require broader horizons or larger identities. The larger culture or the nation provides this. Forced cultural integration, however, breeds tensions, as it has in Zanzibar or Rwanda.

Humans also require less than a global identity; they require a measure of provincialism. The provinces would be defined by culture, by region and language. Individuality requires distinctive features. When these are stripped away by global coca-colonization, people fight to preserve local culture; the Irish fight to use Gaelic, the Welsh for self-rule. The real feelings of innumerable groups of people center on much smaller regions of the world than nations. For example, Britain is composed of Scotland, North and South Wales, Northern Ireland, Anglia, and Saxony. It may mean more to be Quebecois than Canadian, Kurd than Iraqi, Mongolian than Chinese.

5.2.2.1.1. Regional Differences between Nations

After the differences in environment, and different historical directions, people have different ways of imaging their places, of distributing their wealth. Variations of custom, languages, religion, organization, philosophy, and perception between communities are what enrich the totality of national experience, as the talents and skills of individuals enrich communities. People need to realize potentials as member of the human community and nature.

Herbert Read presents differentiation as a measure of social progress. "Progress is measured by the degree of differentiation within a society. If the individual is a unit in a corporate mass, his life will be limited, dull, and mechanical." But if a unit on his own group, with space and potentiality for action, then he can develop. People need to work on their own solutions in their own places; this would allow maximum diversity and satisfaction.

Heterogenization is beneficial to all nations; it enriches the cultural resources, provides niches for individuals, and supplies many patterns that may be adapted. It also increases the speed of cultural evolution or the richness of exchange. The process of heterogenization may proceed by localization or interweaving; this is interlocality differentiation. Yet, some homogenization is equally important. This is intralocality differentiation, where each locality heterogeneity increases while differences between them decrease.

5.2.2.1.2. Regional Politics of Nations

The polis in ancient Athens was made for the amateur. Its ideal was that every citizen should play a part in all of the many activities of the polis. We must have right politics as part of right livelihood. We might explore a convergence of thought of Jefferson, Gandhi and Mao, according to W. I. Thompson.

What matters is not perfection, but the basic value orientation of the polity. The developed countries may prefer Jefferson's republican simplicity to Hamilton's national power; Jefferson's vision of republican simplicity would be more manageable than Hamilton's vision of national power and commercial complexity. The latter has been tried and the need for it may be gone; it is time to try the former. "That government is best that governs least," according to Jefferson. Although the smallness of cultural units would not guarantee freedom, with a smaller scale tyranny, at least the tyrant is visible, flight would be possible, and revolt more likely. The best disinfectant against the danger of a selfish demagogue is wise and skeptical education. Lawmakers and executives would be chosen by lot, rotation, election or a combination of procedures. Similar procedures used in Athens, Florence and other places. Thoreau's Walden is an extended sermon on the necessity of natural simplicity as the only way to avoid desperation of power and possessions. People must be free to choose. If people are left alone to do what they please, to follow their nature, a new social order will emerge of itself. Nature is not a forced order.

Cultures cannot be planned, nor are they logical. As long as they are coherent, contradictions may abound. The form of rearrangement is immaterial. With a cultural unit approach, groups may pursue anarchistic isolationism or disinterested internationalism. It would be against any kind of morality to prescribe forms of government and administration for peoples.

Other nations may prefer Gandhi's vision or Mao's original vision of small communities. Other visions may be invented. In the eutopian approach, groups may pursue anarchistic isolationism or disinterested interrepublicanism. What matters is not perfection, but the basic value orientation of the polity.

Privatism or socialism could work, depending on population size, resource distribution, traditions, or other factors. Privatism works well within a system, with intrinsic responsibility. The problem with the socialism is its contrived responsibility; who shall watch the watchers? Commonism, a systems of cultural commons, or altruism, might not work as large systems.

Anarchism is not only a stateless society but also a harmonized society that exposes humans to the stimuli provided by both agrarian and urban life, physical and mental activity, communal solidarity and individual development, spontaneity and self-discipline, elimination of toil and promotion of craftsmanship, regional uniqueness and world brotherhood. An anarchistic society would establish harmony of human and nature and human and human.

5.2.2.2. Responsibilities of Nations

The planet is experienced on a smaller frame of reference than global unity or nations; people live on the local level. Local knowledge is knowledge in place, earned in place by generations of inhabitants, through visions and trials, experience, and stories. Thus, individuals are preserved in societies that are preserved in places that are preserved by individuals and societies. Laws, politics, architecture, sports are things of place. They are shaped with local knowledge. A local area is limited by the limits of vision, a horizon. As protectors of place, Nations have explicit functions.

5.2.2.2.1. To Conserve Ecosystemic Boundaries

Nations have the responsibility of keeping their environment healthy, of conserving local ecosystems and places. Human activities cannot be isolated from societies or ecosystems. Cultures must adapt to ecosystems or biological regions to survive. All beings modify and exploit the earth to some extent. Humans do. For successful exploitation, power must be limited and density controlled at a national level, which could balance communities and places. A human ecology must provide each living human being with a satisfactory environment that must be in equilibrium with the rest of the system.

5.2.2.2.2. To Manage Resources: Distribution

Every nation has the power to use its local sources in any manner, within the limits of damage and pollution set by an international association. Every nation has the duty to conduct activities in a manner respectful of their effects. With the information available now, from a more extended resource inventory and with optimal ideas about renewal, climatic conditions, traditional land-use patterns, local cycles, and ecological requirements (limits), conservation is more effective. Its goal is to support a steady state economy within optimum ranges based on natural and human limits.

Technology can be used to increase carrying capacity to some extent, but not infinitely. Although, technologies tend to homogenize people and places, the same technologies may be used in different ways, especially adapted to local requirements, for instance, the use of tin cans as cases for radios. Heterogeneity is beneficial to all countries; it enriches the cultural resources, provides niches for individuals, and supplies many patterns that may he adapted. It also increases the speed of cultural evolution or the richness of exchange.

A nation could promote appropriate technology to manage resources for its region. Dangerous technologies would he reduced through wholesale substitution, if not of materials, than by labor-intensive solutions. Traditional housing, for instance would be preferred; its form and design are integrated into the culture, it is adapted to the local climate and is usually less expensive, due to use of local materials. Much traditional architecture is authentic and unselfconscious; its forms fit the context of place and develop in response to place. With arcologies, the urban ecologies designed by Paolo Soleri, the city can change its relationship with nature; an arcology is a good solution for an urban culture, as it solves the problems of waste, resource-use, scale, obsolescence, and segregation.

One necessary condition for the preservation of finite resources is sovereign power. To share resources without the discipline of power invites the tragedy of the commons. The limit of sharing has to coincide with the limit of sovereignty; otherwise runaway destruction could result. Every nation has the responsibility to conduct its economy without causing damage to its ecological base or to other nations. Environmental risks and damages must be identified and solutions found before economic processes can be implemented.

5.2.2.2.3. To Maintain Population

A nation is responsible for maintaining the health of cultures and individuals. Traditional cultures provide personal security, respect for the individual, responsibility for actions (self-discipline), social integration, concern for others, and reverence for nature. Traditional social structures, with networks of marital relations, inheritance, and rationalizing myths, are closely adapted to the local environment.

Preservation of identity appeals to qualities inherent in established ways and to people's desire to maintain their distinctive customs against change. There are also esthetic reasons for preservation: to preserve styles, merit, and achievement. Ethnic identity and consciousness of gender make finer grids of groups. Devotion to groups, clubs, corporations, teams, co-ops, can provide a better individual identity. Cultures allow different frames of accomplishment. The sum of cultures provides a greater sum of accomplishments. A global reference for a competition or cooperation is restrictive; instead of one winner at a global level, there are 2,000 winners in 2,000 cultures. Limits allow different forms of expression and knowledge.

Cultural identity is necessary to the benefit of places; for, if everyone is a citizen of the world, who will protect small places of little economic or scientific importance? In the one world system, dependence on free conscience to produce conformity is vulnerable to the smallest minority of nonconformists. And there will always be some minority.

Nations educate their members; they are responsible for the ecolacy, numeracy, literacy—and now imagacy—necessary for individual survival and actualization within the culture. The emphasis would be local: Local culture, history, geology, botany, and economics. Cultural education may not emphasize competition or excellence for economic or political purposes, as is so often done in industrial cultures.

Cultures thrive when inspired by a common vision and supported by common values. Variations of custom, languages, religion, organization, philosophy, and perception between communities are what enrich the totality of experience, as the talents and skills of individuals enrich communities. Progress is measured by the degree of differentiation within a society. People need to work on their own solutions in their own places, resulting in maximum diversity and satisfaction.

Every nation that claims sovereignty accepts responsibility for keeping its population within the ecological limits of its place—that is, the biological and cultural carrying capacity. The carrying capacity is the maximum population supported indefinitely in a given habitat. Solar energy is limited; reserves of fossil fuels are limited; ecological productivity is limited. The carrying capacity of human population includes consumption and production impacts without damage to the integrity of the ecosystem. Carrying capacity is not a rigid number, however; it can be increased or decreased by numerous factors.

A single population policy for every nation is unfair, since nations are at different stages of development. There are dilemmas posed by necessity to equate global balances and national needs. Every state needs a comprehensive population policy, closely related to environmental and technological policies, and within the constraints of their agriculture. Every nation that claims sovereignty must accept responsibility for keeping its population within carrying capacity (as a fuzzy set). Carrying capacity can be expanded to include the notion of self-reliance. This offers more flexibility for trade. Mao committed China to a policy of *tsu li kong sheng*, "regeneration through one's own efforts"; this is self-reliance.

5.2.2.2.4. To Provide Paths of Power to Individuals

For Aristotle, politics was the science of the possible. The city (*polis* in Greek) was a human artifact whose structure could be modified by reason; it was potentially a work of art, limited only the capability of the artist. Nations present the possibilities of power for individuals.

The function of politics is to ensure that decisions are taken at the right level. A nation protects individual freedoms, guards regional culture (values and identity), and holds groups accountable for the use of power. Regional politics limits the scope of institutions by showing that the institutions can be destructive to society as a whole.

The restriction of freedom, either through tyranny, cultural uniformity, or crowding, results in a decrease of variety, which is created by spontaneous play, which is necessary to flexible and enduring social systems. It is not necessary to prescribe forms of government and administration for peoples. Many forms of government that are size-specific, such as democracy and communism, could be possible in scaled nations. Although the smallness of cultural units would not guarantee freedom, with a smaller scale tyranny the tyrant is visible and corrective action possible and more likely.

Social and personal advantage can be combined in secure social orders. In societies where non-aggression is conspicuous, an individual serves her own advantage as well as that of the group with the same act. Ruth Benedict used the term high synergy to describe such secure societies. The institution insures mutual advantage; the acts are mutually reinforcing. High synergy institutions transcend the polarities of selfishness and altruism. Virtue pays because the rewards for selfishness coincide with benefit for the society. The social structure of low synergy cultures insures opposition and counteraction; the advantage of one individual is a victory over another, as in a zero-sum game. Wealth may be distributed or concentrated, depending on factors, such as synergy, generosity, reciprocity, and cooperation.

New politics can start at the community, over community issues, like housing, transportation, or pollution. People need to save their own identities and places first from corruption and degradation. Power can be shifted to local levels through self-reliance and participation.

The size of nations would be defined by place and culture. Such a limit would increase power to individuals. There can be no separation of politics and ecology. Every political act has ecological consequences and every ecological decision is a political demand for control over use of the environment.

5.2.2.2.5. To Provide Opportunities for Needs of People

Every nation strives for self-reliance. Communities can be self-reliant by producing enough food and shelter, by limiting their population to what can be produced, by sharing tools, by recycling and repairing, by using handicrafts rather than manufactures (shoes, furniture), by using local products and raw materials (soil, minerals, plants), by using general and not specialized machines, by having multipurpose factories, by networking with other communities, and by doing without things that are not needed (bombs, food additives, plastic bottles). Food would be produced and available within local groups, so there would be no reliance on large-scale food production and distribution. Local production would eliminate transportation cost and waste and diminish dependency.

Many leaders of nonindustrial areas have attempted to reduce links with overindustrialized areas. They have attempted to balance population and resources on a local level; they have placed local values before international ones. Julius Nyerere in Tanzania promoted an African socialism. Mao Tse-Tung placed the peasant before the urban dweller in China. For India, Gandhi envisioned a familiarization of society, where property had a common ownership. Each village was a complete nation, independent of its neighbors for vital needs, characterized by self-rule and self-restraint. Later, Schumacher addressed himself to India, urging officials to try to keep poverty rural, and develop village systems to solve it, as worked out by Gandhi, instead of urban complexes.

National control would mean the regionalization of transportation and communication, the administration of social services by a smaller and more responsive bureaucracy, and the resurgence of more direct forms of economic interaction, such as labor-gift and barter exchange in local neighborhoods. A free market system involving each nation could work; one major difference is that supply would be regulated by an international association. Nations could coordinate worker controlled industries and producers cooperatives, forged from kinship or cooperation. Credit unions and mutual insurances could replace big banks and insurance companies. The nature of work could change with a change in scale; it would be self determining and nonexploitative, resulting in greater harmony between worker and employer,

Traditional cultures often have wealth-leveling properties, absolute property ceilings, fixed wants, and production coupled with need; this results in a stable economy. Efficiency and productivity are less important than use and appropriateness. Advertising, the creation of desires and needs, is less important in face-to-face economies. In a small state, people can decide how to use scarce resources and how to distribute them. They can decide whether to be conservatively sustainable or to grow and gamble on innovation and substitution. Local communities have different economic attitudes. Canada, for instance, may cut and sell Canadian timber at a loss, but British Columbia may decide not do that with its timber.

A rational approach to economics is not adequate because of differences in wealth; whether saving, spending, or investing, what is rational to a pauper is not rational to an industrialist or the reverse. But, a rational response to an irrational system increases the wobble of the system. An irrational response hurts the individual. International equalization is useful to put a rational approach in perspective. Economics is connected to ecologies, at any level. Agriculture needs to be connected to optimum rates of production, which are far below the maximum. Society must start paying the true costs.

The populist governance model gives responsibility to communities. Food would be produced and available within own groups, so there would be no reliance on large scale food production and distribution. Local producing would eliminate transportation cost and waste and diminish dependency. A lower population fed by organic agriculture will result in greater flexibility.

How fast can a country become self-reliant? What political or social damage would change do? How would international problems be solved; for instance, when cutting trees in Nepal causes floods in Bangladesh, and deaths because crowding caused the poorest to live on flood plains? The poor in the highlands everywhere effect those in lowlands, often adversely. How should these be treated? depopulated, emigrated? Every nation can be self-reliant; almost none are self-sufficient.

Self-reliance, through decentralization, is freedom. Decentralization is a structure most appropriate to the U.S. idea of freedom; it is also the most cost-effective. The task is to make new one more visible. What is at stake is the right of nonhuman beings to exist. Freedom (in decentralization) is opportunity to define local norms for behavior; a mandate for different traditions in unique communities. Equilibriums can be reestablished through decentralization. Different kinds of constraints on a system are possible. Variety permits a wider range of responses to change in environment. Variety also produces the unexpected.

5.2.2.2.6. To Provide Paths for Resolution of Conflicts

Given differences between cultures or human groups, from images to languages, there is potential for misunderstandings and conflicts between them. A nation has to provide ways to resolve those conflicts. Conflicts can be resolved through communication. If that does not work, arbitration can be made available. Courts would be used as a another resort. Without these paths, and others, retreat or violence are more likely.

5.2.3. *Being in an International Framework*

A global coordinating body could help with communications between nations. It would help different cultures adapt to each other and learn from each other. Global interaction has been a tendency for over five hundred years. How should an individual or a culture be in an international context? What kinds of visions should we follow or goals to make?

5.2.3.1. Necessity of the Framework (UN Currently)

Founded in 1945, to replace the League of nations of 1919, the United Nations (UN) describes itself as a "global association of governments facilitating cooperation in international law, international security, economic development, and social equity." From 51 countries, the UN has expanded to 191 member states in 2006, which the UN considers to be virtually all internationally recognized independent nations, except for Taiwan and several others.

The UN, with its system of 30 affiliated organizations, attempts to solve global problems, from disease and poverty to the environment and war. The UN agencies define the standards for air travel, telecommunications and consumer products. It has developed international campaigns against drug trafficking and terrorism. Its agencies try to assist refugees, to clear landmines, to expand food production, and to reduce disease.

The United Nations has six main organs: The General Assembly, the Security Council, the Economic and Social Council, the Trusteeship Council, the Secretariat, and the International Court of Justice. When nations become Members of the UN, they agree to accept the obligations of the UN Charter, an international treaty that sets out basic principles of international relations. The Charter expresses four purposes: To maintain international peace and security; to develop friendly relations between nations; to cooperate in solving international problems; and to harmonize the actions of nations. In September 2000, the members of the UN met to set an international agenda for the new century. The Millennium Declaration lists measurable goals in seven areas: Peace, security and disarmament; development and poverty eradication; protecting the common environment; human rights and good governance; protecting the vulnerable; meeting the special needs of Africa; and strengthening the UN itself.

The UN has been successful in encouraging dependent people to become independent and then incorporating people into its system. In 1960 the General Assembly adopted the Declaration on the Granting of Independence to Colonial Countries and Peoples, which resulted in 60 former colonial Territories attaining independence and joining the UN as sovereign Members. When the UN was formed, 750 million people lived in non-self-governing territories; by 2006, that number was reduced to about 1 million.

The UN has been successful in affirming the fundamental equality of all people and to counter racism. A long UN campaign in South Africa contributed to ending the system of racial segregation known as apartheid. In 1994, a UN observer mission observed that country's first all-race elections. A World Conference in 2001 examined ways to combat racism, racial discrimination, xenophobia and intolerance.

An important mandate of the UN is the promotion of higher standards of living, full employment, and conditions of economic and social progress and development. The UN web site notes that as much as 70 per cent of the work of the UN system is devoted to accomplishing this mandate, based on the belief that eradicating poverty and improving the well-being of people everywhere are necessary steps to create conditions for lasting peace.

5.2.3.1.1. Structure of the UN Main Organs

The structure of the UN is geared give a voice and a vote to all Member nations, to formulate policies on the goals of the system and to resolve international conflicts.

5.2.3.1.1.1. General Assembly

The General Assembly is a "parliament of nations" in which member nations consider the world's most pressing problems. The membership includes every defined nation and each member has one vote. Decisions on key issues, such as admission of new members, the UN budget, and international peace and security, are decided by two-thirds majority. Other matters are decided by a simple majority. Recently, efforts have been made to reach decisions through consensus, rather than through formal vote.

The General Assembly holds an annual regular session from September to December, although it may resume later or hold a special or emergency session, if necessary, to address subjects of particular importance. At its 2001 session, for instance, the Assembly considered over 180 topics, including globalization, AIDS, conflict in Africa, protection of the environment, and consolidation of new democracies. The Assembly does not have the authority or power to force any member to act on UN decisions, but the UN considers that its recommendations “are an important indication of world opinion and represent the moral authority of the community of nations.”

5.2.3.1.1.2. Security Council

The Security Council, under the UN Charter, has primary responsibility for maintaining international peace and security. The Council may convene at any time that peace is threatened. Under the Charter, all member nations are obligated to carry out the decisions of the Council.

The Council has 15 members. Currently, five—China, France, the Russian Federation, the United Kingdom and the United States—are considered to be permanent members. The remaining 10 are elected by the General Assembly for two-year terms. Decisions of the Council require nine ‘yes’ votes. A decision cannot be made if there is a veto by a permanent member.

When the Council identifies a dispute that may become volatile, it explores ways to settle the dispute peacefully. It may suggest principles for settlement or encourage mediation. If fighting has begun, the Council tries to secure a ceasefire. It can send a peacekeeping mission to separate the forces until a truce can be set up.

The Council can take other measures, such as economic sanctions or an arms embargo, to enforce its decisions. It can, in extreme situations, authorize member nations to use “all necessary means,” including collective military action, to see that its decisions are carried out. The Security Council has established well over 50 peacekeeping operations. The Council can also make recommendations to the General Assembly on the appointment of a new Secretary-General and on the admission of new members.

5.2.3.1.1.3. Economic & Social Council

Under the overall authority of the General Assembly, the Economic and Social Council coordinates the economic and social work of the UN system. As a central forum for considering international economic and social issues, and for formulating policy recommendations, the Council fosters international cooperation for development. It consults with non-governmental organizations (NGOs) to maintain a vital link between the UN and civil societies of member nations.

This Council has 54 members, who are elected by the General Assembly for three-year terms. The Council meets throughout the year and holds its major session in July, when a special meeting of Ministers discusses major economic, social and humanitarian issues.

The Council’s subsidiary bodies meet regularly and report back to it. The other bodies focus on issues such as social development, the status of women, crime prevention, narcotic drugs, and environmental protection.

There are five regional commissions to promote economic development and cooperation in their regions.

5.2.3.1.1.4. Trusteeship Council

The Trusteeship Council, considering its work complete, is composed of the five permanent members of the Security Council. The rules of procedure have been changed to allow it to meet if required. The Trusteeship Council had been established to provide international supervision for 11 Trust Territories that were administered by seven member nations, and to ensure that adequate steps were taken to prepare the Territories for independence. By 1994, the Trust Territories had attained independence as separate nations, or they had become part of neighboring independent countries. If the mandate of the Council is not changed, then it will probably be abolished.

5.2.3.1.1.5. Secretariat

The Secretariat carries out the administrative work of the UN, as directed by the General Assembly, the Security Council and the other organs. The Secretary-General, as head of the Secretariat, provides overall administrative guidance. The Secretariat consists of several departments and offices. Its staff of about 7,500 is drawn from 170 countries.

5.2.3.1.1.6. International Court of Justice

The International Court of Justice is the main judicial organ of the UN. The Court consists of 15 judges elected jointly by the General Assembly and the Security Council; its purpose is to settle disputes between countries. Although participation in a proceeding is voluntary, if a nation agrees to participate, then it is obligated to comply with the Court's decision. The Court also provides advisory opinions to the General Assembly and the Security Council.

In 1998 the General Assembly called a conference in Rome to establish an International Criminal Court (ICC). The ICC Court formed in 2002 and heard its first case in 2006. It is the first permanent international court charged with trying those who commit the most serious crimes under international law, including war crimes and genocide.

5.2.3.1.1.7. Special Agencies of the United Nations

Specialized agencies, as part of the UN system, address almost every area of economic and social endeavor. The agencies provide technical and practical assistance to countries around the world. In cooperation with the UN, they help formulate policies, set standards and guidelines, foster support, and mobilize funds. These organizations have their own governing bodies, budgets and secretariats. They report to the General Assembly or the Economic and Social Council.

Close coordination between the UN and the specialized agencies is ensured through the UN System Chief Executives Board for Coordination (CEB), which includes the Secretary-General and the heads of the specialized agencies, funds and programs, the International Atomic Energy Agency, and the World Trade Organization.

The International Monetary Fund, the World Bank, and 12 other independent organizations are linked to the UN through cooperative agreements. These agencies, among them the World Health Organization and the International Civil Aviation Organization, are autonomous bodies created by intergovernmental agreement. They have wide international responsibilities in the economic, social, cultural, educational, health, and related fields. Some, such as the International Labor Organization and the Universal Postal Union, predate the UN itself.

The Office of the UN High Commissioner for Refugees (UNHCR), the UN Development Program (UNDP), and the UN Children's Fund (UNICEF), work to improve the economic and social condition of all people.

The World Bank provides loans and technical assistance to developing countries to reduce poverty and to advance "sustainable economic growth." The World Bank, for example, provided more than $17 billion USD in development loans for the fiscal year 2001 to more than 100 developing countries. The ILO (International Labor Organization) formulates policies and programs to improve working conditions and employment opportunities, and sets labor standards for all countries. The FAO (Food and Agriculture Organization of the UN) works to improve agricultural productivity and food security, and to better the living standards of rural populations. The IMF (International Monetary Fund) facilitates international monetary cooperation and financial stability and provides a permanent forum for consultation, advice and assistance on financial issues. The ITU (International Telecommunication Union) fosters international cooperation to improve telecommunications, coordinates usage of radio and TV frequencies, promotes safety measures, and conducts research. The WMO (World Meteorological Organization) promotes scientific research on the Earth's atmosphere and on climate change, and facilitates the global exchange of meteorological data. The IMO (International Maritime Organization) tries to improve international shipping procedures, raise standards in marine safety, and reduce marine pollution by ships. The WIPO (World Intellectual Property Organization) arranges international protection of intellectual property and fosters cooperation on copyrights, trademarks, industrial designs and patents. UNIDO (UN Industrial Development Organization) promotes the industrial advancement of developing countries through technical assistance, advisory services and training. And, the IAEA (International Atomic Energy Agency) is an autonomous organization that emphasizes safe uses of atomic energy.

5.2.3.1.2. Purposes Goals & Actions of the UN

Like any human institution, one important goal of the UN is to strengthen itself. Programs that involve every nation and its people reinforce the image of the UN as an international body, despite some failures and criticisms. Funding, however, is used by many nations, as a stick to drive the UN in different directions.

5.2.3.1.2.1.To Maintain Peace & Security

The primary purpose of the UN is to preserve world peace. Under the UN Charter, member nations agree to settle disputes by peaceful means and refrain from threatening or using force against other states. UN peace efforts have produced some dramatic results. The UN helped to defuse the Cuban missile crisis in 1962 and the Middle East crisis in 1973. In 1988, the UN sponsored an offer of a peace settlement ended the Iran-Iraq war, and in 1989, UN-sponsored negotiations led to the withdrawal of Soviet troops from Afghanistan. In the 1990s, the UN was instrumental in restoring sovereignty to Kuwait, and it played a major role in ending civil wars in Cambodia, El Salvador, Guatemala, and Mozambique. UN efforts helped to restore the democratically elected government in Haiti. The UN helped to contain conflict in other countries through disarmament and peacemaking.

5.2.3.1.2.1.1. Arms Control & Disarmament

The 1945 UN Charter envisioned a system of regulations that would ensure the least diversion of the world's human and economic resources towards armaments. The use of nuclear weapons weeks after the signing of the Charter highlighted the necessity of arms limitation and disarmament. In fact, the first resolution of the first meeting of the General Assembly, on January 24, 1946, was to establish a Commission to deal with problems raised by atomic energy and to make specific proposals for the elimination of atomic weapons and other weapons of mass destruction from national armaments.

Halting the spread of arms and reducing and eventually eliminating all weapons of mass destruction are major goals of the United Nations. The UN has established forums to address multilateral disarmament, including the First Committee of the General Assembly and the UN Disarmament Commission. Items on their agenda include consideration of a nuclear test ban, outer-space arms control, a ban on chemical weapons, nuclear and conventional disarmament, nuclear-weapon-free zones, the reduction of military budgets, and measures to strengthen international security. The UN supports multilateral negotiations in the Conference on Disarmament and in other international bodies. These negotiations have produced agreements such as the Nuclear Non-Proliferation Treaty (1968), the Comprehensive Nuclear-Test-Ban Treaty (1996), and treaties establishing nuclear-free zones.

The Conference on Disarmament has 66 members representing all areas of the world, including the first five major nuclear-weapon states (the People's Republic of China, France, Russia, U.K. and U.S.). This independent Conference is linked to the UN Secretary-General through a personal representative, who serves as the secretary-general of the conference. Resolutions adopted by the General Assembly often request the conference to consider specific disarmament matters. The conference annually reports its activities to the Assembly.

Other treaties brokered by the UN prohibit the development, production and stockpiling of chemical weapons (1992) and bacteriological weapons (1972); ban nuclear weapons from the seabed, ocean floor (1971) and outer space (1967); and ban or restrict other types of weapons, such as landmines. By 2001, over 120 countries had become parties to the 1997 Ottawa Convention outlawing landmines. The UN encourages all nations to adhere to treaties banning weapons of war. The UN is also supporting efforts to prevent, combat and eradicate the illicit trade in small arms and light weapons, the weapons of choice in most, 46 of 49, major conflicts since 1990. The UN Register of Conventional Arms and the system for standardized reporting of military expenditures help promote greater transparency in military matters.

The International Atomic Energy Agency, through a system of safeguard agreements, attempts to ensure that nuclear materials and equipment intended for peaceful uses are not diverted for military purposes. The Organization for the Prohibition of Chemical Weapons collects information on chemical facilities worldwide and conducts routine inspections to ensure adherence to the chemical weapons convention.

5.2.3.1.2.1.2. Peacemaking

UN peacemaking attempts to bring hostile groups to agreement through diplomatic means. The Security Council may recommend ways to avoid conflict or secure peace through negotiation or through recourse to the International Court of Justice. The Secretary-General also can play an important role in peacemaking, by bringing any threat to peace and security to the attention of the Security Council; the S-G can use the offices of the UN to carry out mediation or to exercise quiet diplomacy, personally or through special envoys, to resolve disputes before they escalate.

5.2.3.1.2.1.3. Peacebuilding

The UN has started to address the underlying causes of conflict, such as health, wealth, and education. Development assistance is a key element of peace-building. In cooperation with UN agencies, donor countries, host governments and local and international NGOs, the

UN works to support good governance, civil law and order, elections and human rights in countries struggling to deal with the aftermath of conflict. At the same time, it helps these countries rebuild administrative, health, educational and other services that have been disrupted by war. UN-supervised elections in East Timor in August 2001, allowed people to cast their ballots for a democratically elected assembly.

Some of these activities, such as the UN's supervision of the 1989 elections in Namibia, mine-clearance programs in Mozambique and police training in Haiti, take place within the framework of a UN peacekeeping operation, and may continue when the operation withdraws. Others are requested by governments, as is the case in Cambodia, where the UN maintains a human rights office, or in Guatemala, where the UN helps to implement peace agreements.

In Africa, UN field missions continue peace-building activities in Guinea-Bissau and Liberia; and they remain in Angola and Burundi to support various initiatives aimed at promoting reconciliation. At the request of the Security Council, the Secretary-General has provided a comprehensive analysis of conflicts in Africa along with recommendations on how to promote durable peace.

5.2.3.1.2.1.4. Peacekeeping

All UN peacekeeping operations must be approved by the Security Council, which sets up UN peacekeeping operations and defines their scope and mandate in its efforts to maintain peace and international security.

Most operations involve military duties, such as observing a ceasefire or establishing a buffer zone while negotiators seek a long term solution. Others may require civilian police or other civilian personnel to help organize elections or to monitor human rights. Operations have also been deployed to monitor peace agreements in cooperation with the peacekeeping forces of regional organizations. These forces are provided by member states of the UN, which does not maintain an independent military.

Since the UN deployed peacekeepers in 1948, 123 countries have voluntarily provided more than 750,000 military and civilian police personnel to engage in 54 peacekeeping operations. UN peacekeeping is a vital instrument for peace. Currently, 47,650 UN military and civilian personnel, provided by 87 countries, are engaged in 15 operations around the world.

Peacekeeping operations may last for a few months or continue for many years. The UN's operation at the ceasefire line between India and Pakistan in the State of Jammu and Kashmir, for example, was established in 1949, and still continues. UN peacekeepers have been in Cyprus since 1964. By contrast, the UN was able to complete its 1994 mission in the Aouzou Strip between Libya and Chad in about a month.

Total UN peacekeeping expenses peaked by the end of 1995, when the total cost was just over $3.5 billion. Total UN peacekeeping costs for 2000, including operations funded from the UN regular budget as well as the peacekeeping budget, were $2.2 billion. UN peace operations are funded by assessments, using a formula derived from the regular scale, but including a surcharge for the five permanent Security Council members.

5.2.3.1.2.2.To Develop Friendly Relations between Nations

Through its activities, the UN tries to increase the participation of developing countries in the global economy. The UN Conference on Trade and Development (UNCTAD) promoted international trade. UNCTAD also works with the World Trade Organization (WTO), in assisting exports from developing countries through the International Trade Centre.

The UN provides the means to settle disputes peacefully. The UN has played a major role in helping defuse international crises and in resolving protracted conflicts. It has undertaken complex operations involving peacemaking and humanitarian assistance, and it has worked to prevent conflicts from breaking out. After a conflict, it has increasingly undertaken action to address the causes of war and to lay a foundation for durable peace.

5.2.3.1.2.3.To Cooperate in Solving Problems

Many nations have problems as a result of historical paths and economic inequities. Many problems are geological or meteorological. Other problems arise from conflict.

The UN offers help with resettlement, after a crisis subsides. The UN helped to repatriate refugees to Mozambique, provided humanitarian assistance in Somalia and Sudan, and undertook diplomatic efforts to restore peace in the Great Lakes region. It has helped prevent new unrest in the Central African Republic, and it is helping to prepare for a referendum on the future of Western Sahara.

The UN offers help with reconstruction. In Kosovo, the UN is rebuilding schools and providing student supplies, as part of a wide-ranging assistance effort. The Security Council established an interim international administration there in 1999, following the end of NATO air bombings and the withdrawal of Yugoslav forces. Under the umbrella of the UN, the European Union and the Organization for Security and Cooperation in Europe are working with the people of Kosovo to create a functioning, democratic society with substantial autonomy. Municipal elections in October 2000, and the casting of a Constitutional Framework for Provisional Self-Government, paved the way for Kosovo-wide elections for a legislative assembly on 17 November 2001.

5.2.3.1.2.3.1. Promote Human Rights

World War II atrocities and genocides led to a consensus that the new organization, the UN, must work to prevent such tragedies in the future. So, the pursuit of human rights was central to the creation of the UN. An early objective was to create a legal framework for considering and acting on complaints about human rights violations. The UN Charter obligates all member nations to promote universal "respect for, and observance of, human rights" and to take joint and separate actions to that end.

The Universal Declaration of Human Rights, proclaimed by the General Assembly in 1948, sets out the basic rights and freedoms to which all human beings are entitled: The rights to life, liberty and nationality; the rights to freedom of thought, conscience and religion; the rights to work and to be educated; the rights to food and housing; and the right to take part in government. The UN states that the Declaration is not legally binding to nations. Then, it states that the rights are legally binding due to two International Covenants, to which most nations are parties. One Covenant deals with economic, social and cultural rights, and the other addresses civil and political rights. With the Declaration, they constitute the International Bill of Human Rights.

The Declaration laid the groundwork for more than 80 conventions and declarations on human rights, including conventions to eliminate racial discrimination and discrimination against women; conventions on the rights of the child, against torture and degrading punishment, the status of refugees and the prevention and punishment of the crime of genocide; and declarations on the rights of persons belonging to national, ethnic, religious or linguistic minorities, the right to development, and the rights of human rights defenders. The UN Commission on Human Rights (UNCHR) is the primary UN body charged with promoting human rights, through investigations and offers of technical assistance.

The United Nations and its various agencies are central in upholding and implementing the principles enshrined in the Universal Declaration of Human Rights. A case in point is support by the UN for countries in transition to democracy. Technical assistance in providing free and fair elections, improving judicial structures, drafting constitutions, training human rights officials, and transforming armed movements into political parties have contributed significantly to democratization worldwide.

UN human rights field activities are currently being carried out in nearly 30 countries or territories. They help strengthen national capacities in human rights legislation, administration and education. They investigate reported violations and assist governments in taking corrective measures when needed.

Promoting respect for human rights is increasingly central to UN development assistance. In particular, the right to development is seen as part of a dynamic process which integrates civil, cultural, economic, political and social rights, where the well-being of all individuals in a society is improved. The eradication of poverty is a key to the right to development.

The UN is also a forum to support the rights of women to participate fully in the political, economic, and social life of their countries. The UN contributes to raising consciousness of the concept of human rights through its covenants and its attention to specific abuses through its General Assembly resolutions or Court rulings.

5.2.3.1.2.3.2. Promote Health & Human Development

The UN programs and funds for health work under the authority of the General Assembly and the Economic and Social Council to carry out their economic and social mandates. The World Health Organization (WHO) coordinates programs aimed at solving health problems and at the attainment by all people of the highest possible level of health. It works in such areas as immunization, health education and the provision of essential drugs.

The UN has special health programs for children. Every year, 3 million children are saved by immunization, but almost 3 million more die from preventable diseases. UNICEF, WHO, the World Bank Group, several private foundations, most of the pharmaceutical industry, and many governments have joined hands in a new initiative (the Global Alliance for Vaccines and Immunization) to reduce deaths from diseases to zero. Other UN agencies work with local officials and NGOs to meet the health needs of children in conflict situations, such as this UNICEF-led immunization campaign in Afghanistan. The UN Children's Fund (UNICEF) is the lead UN organization working for the long-term survival, protection and development of children. Active in some 160 countries, areas and territories, its programs focus on primary health care, immunization, nutrition, and basic health education.

UN programs try to eradicate diseases. UNICEF, UNDP, the World Bank and WHO joined forces in 1998 to launch a new campaign to fight malaria, which kills more than 1 million people a year. Joint initiatives to expand immunization and develop new vaccines have enlisted the support of business leaders, philanthropic foundations, non-governmental organizations and governments. The UN supports programs on HIV/AIDS, including grass-roots education campaigns, in 155 countries. Many other UN programs work for development, in partnership with governments and NGOs. Smallpox was eradicated from the world through a global campaign coordinated by WHO. Another WHO campaign has eliminated polio from the Americas, and aims at eradicating it globally by 2005.

The UN Human Settlements Program (UN-Habitat) assists people living in health-threatening housing conditions. UNHCR's assistance for Pakistan's Afghan refugees focuses on education and health. To pay for this assistance, the UN has raised billions of dollars from international donors. In 2001, the Office for the Coordination of Humanitarian Affairs launched 19 interagency appeals, raising more than $1.4 billion to assist 44 million people in 19 countries and regions.

5.2.3.1.2.3.3. Promote Education

UNESCO (the UN Educational, Scientific and Cultural Organization) promotes education for all, cultural development, protection of the world's natural and cultural heritage, international cooperation in science, and freedom and communication of the press.

5.2.3.1.2.3.4. Encourage Economic Development

The UN is in a unique position to promote development; it has a global presence and a comprehensive mandate to address social, economic and emergency needs. The UN tries to be neutral and not represent any particular national or commercial interest. And, it offers a voice for every country, regardless of wealth or power, on major policy decisions.

The UN has played a role in building international consensus on action for development. Beginning in 1960, the General Assembly has helped set priorities and goals through a series of 10-year International Development Strategies. While focusing on issues of particular concern, the Decade strategies have consistently stressed the need for progress on all aspects of social and economic development. The UN continues to formulate new development objectives in such key areas as sustainable development, the advancement of women, human rights, environmental protection and good governance. The UN continues to build new programs and actions to fulfill the objectives.

At the Millennium Summit in September 2000, leaders of nations adopted a set of Millennium Development Goals aimed at supporting development; at eradicating extreme poverty and hunger; at achieving universal primary education; at promoting gender equality and empowering women; at reducing child mortality; at improving maternal health; at combating HIV/AIDS, malaria and other diseases; and at ensuring environmental sustainability. These goals have specific measurable targets to be achieved by the year 2015, such as: Reducing by half the proportion of those who earn less than a dollar a day; achieving universal primary education; eliminating gender disparity at all levels of education; and dramatically reducing child mortality while increasing maternal health.

The UN and its agencies, including the World Bank and the UN Development Program (UNDP), are the premier vehicle for furthering development in poorer countries, providing assistance worth more than $30 billion a year. The UNDP is the largest multilateral source of grant technical assistance in the world. The UN also publishes annually the Human Development Index (HDI), a comparative measure ranking countries by poverty, literacy, education, life expectancy, and other factors.

5.2.3.1.2.3.5. Offer Poverty Relief

Other UN social service programs address poverty and relief, especially in vulnerable Africa. A UN System-wide Special Initiative on Africa, a 10-year, $25 billion endeavor launched in 1996, combines all UN efforts into a common program to ensure basic education, health services and food security in Africa.

Relief work for Palestine refugees has been carried out since 1949 by the UN Relief and Works Agency for Palestine Refugees in the Near East (UNRWA). As of 2006, the Agency provides essential health, education, relief and social services, as well as implements income-generation programs for more than 4 million Palestine refugees in the region. A UN Coordinator oversees all development assistance provided by the UN system to the Palestinian people in Gaza and the West Bank.

5.2.3.1.2.3.6. Provide Human Assistance

Disasters can occur anywhere, at any time, from flood, drought, earthquake or conflict. The cost to human communities is lost lives, displaced populations, communities incapable of sustaining themselves, and great suffering.

After a disaster, or famine or war, the UN system provides emergency assistance, in the form of supplies, food, shelter, medicines and logistical support to the victims, many of whom are children, women and the elderly. When disasters occur, the UNDP for instance, coordinates relief work at the local level, while promoting recovery and long-term development. In 2001, for example, following a devastating earthquake in India, the agency helped local communities, while working to reduce long-term vulnerability.

In providing humanitarian assistance, the UN has to overcome major logistical and security constraints in the field. Reaching the affected areas can be a major obstacle, getting adequate supplies another. Recently, many crises have been aggravated by an erosion of respect for human rights. Humanitarian workers have been denied access to people in need. Warring factions have deliberately targeted civilians and aid workers. Since 1992, over 200 UN civilian staff members have been killed and 265 workers have been taken hostage while serving in humanitarian operations. In the effort to prevent human rights violations in the midst of crisis, the UN High Commissioner for Human Rights has taken an active role in the UN response to emergencies.

The UN coordinates its response to humanitarian crises through a committee its humanitarian bodies, chaired by the UN Emergency Relief Coordinator. Members include UNICEF, UNDP, WFP, and the UN High Commissioner for Refugees (UNHCR). Major non-governmental and intergovernmental humanitarian organizations, such as the International Committee of the Red Cross, are represented as well. The UN Emergency Relief Coordinator is responsible for developing policy for humanitarian action and for promoting humanitarian issues, such as helping to raise awareness of the consequences of the proliferation of small arms or the negative effects of sanctions.

People who have fled war, persecution or human rights abuse, that is, refugees and displaced persons, are assisted by UNHCR. At the start of 2001, there were 22 million people of concern to UNHCR in 120 countries, including 5.4 million who are internally displaced within a nation. About 3.6 million Afghans accounted for 30 per cent of refugees worldwide, followed by 568,000 refugees from Burundi and 512,800 from Iraq.

War and civil strife have separated an estimated 1 million children from their parents since 1994, made 12 million more homeless, and left 10 million severely traumatized. UNICEF seeks to meet the needs of these children by supplying food, safe water, medicine and shelter. UNICEF has also promoted the concepts of 'children as zones of peace.' It also created 'days of tranquility' and 'corridors of peace' to help protect children in war and provide them with essential services. And, in countries undergoing extended emergencies or recovering from conflict, humanitarian assistance is increasingly seen as part of an overall peace-building effort, along with developmental, political and financial assistance.

5.2.3.1.2.4.To Harmonize the Actions of Nations

There is a musical analogy with the health and operation of ecosystems, and with the health and harmony of nations; health equals harmony. The four essential elements of music are rhythm, melody, harmony, and tone color. Their combined effects form a web of sound. The combination can lead to the idea of physical motion. Rhythm can lead to monotony. Melody is associated with emotion. A good melody should be of satisfying proportions.

Unlike rhythm and melody, which come naturally, harmony gradually evolved from an intellectual conception that was unknown until the ninth century A.D. The earliest form of harmony was called organum, when you harmonized in intervals of thirds or sixths above the melody. Harmony is the study of chords and their relationships (chords are sounding together of separate tones, a full chord is made of three or more tones).

What can music tell ecology? Harmony continues over time. If a forest has harmony it has to be seen over time, long periods of time. Furthermore, harmony is related to wholeness. The word "whole" comes from the Indo-European root *kailo*, which is also the root for the words health and holy. The concept of the whole forest is relevant. A forest that has very complete complement of interacting beings. A whole forest can renew itself without replanting and pesticides.

David Bohm, in his theory of the implicate universe, proposes that health is a result of a harmonious interaction of all the analyzable parts that comprise the extricate order—cells, tissues, organs, the body—with the surrounding larger environment. Health is a quality that is grounded in the total order of the environment (or implicate order). Health is a dynamic quality of the entire movement of the environment (holoverse) as it flows. As organisms sometimes interfere with others or with the flow of change, the harmony breaks down—we call that disease. Health is the dance of bodies that interpenetrate (in Paul Shepard's image).

None of the bodies are completely independent or completely bounded; they are interdependent and open systems. A body is only maintained by a flow of energy and materials from its environment—much of this flow is in the form of other entities, usually much smaller, such as prey, insects, bacteria, viruses.

What can music suggest to a global political framework? The essence of harmony is allowing all elements to have a voice or sound. That is one purpose of the UN, to allow each nations to have a voice. Harmony is more likely if no one nation can dominate the others all the time at every level.

5.2.3.1.2.4.1. Governance of Nations

The UN has helped run elections in countries with little democratic history, including recently in Afghanistan and East Timor. In East Timor, UN-brokered talks between Indonesia and Portugal culminated in a May 1999 agreement that opened the way for a popular consultation on the status of the territory. Under the agreement, a UN mission supervised voter registration and an August 1999 ballot, in which 78 per cent of East Timorese voted for independence over their regional autonomy within Indonesia. In August 2001, a major step was taken in that direction, with the election of a Constituent Assembly which drafted the constitution for an independent and democratic East Timor. The people of East Timor, in August 2001, cast their ballots for a democratically elected assembly, under UN-supervised elections.

For countries in transition to more open government or democracy, the UN provides technical assistance for holding free and fair elections, improving judicial structures, drafting constitutions, training human rights officials, and transforming armed movements into political parties. In 1994, a UN mission observed South Africa's first racially-open elections. The UN has supervised elections in Namibia and elsewhere.

The United Nations works to support good governance, through civil law and order, elections and human rights, in countries struggling with the aftermath of conflict. The UN works to promote dialogue between parties, to establish a broad-based, inclusive government, with an appropriate political framework and leadership. In the Pacific, the UN helped the government of Papua New Guinea and the Bougainville parties reach a comprehensive agreement covering issues of autonomy, referendum and weapons disposal. In Haiti, following international action to restore the democratically elected government, the UN offered a comprehensive program that emphasized human rights, consensus-building and conflict-reduction, with the strong participation of civil society. The UN supports good governance with its programs.

5.2.3.1.2.4.2. Environment of Nations

Environmental conventions sponsored by the UN have helped to reduce acid rain in Europe and North America, to cut marine pollution worldwide, and to phase out production of gases destroying the ozone layer of the planet. International diplomacy, as a result of other conferences, such as UNCED, may recognize that nature is a finite source of resources, as well as a finite sink for wastes and a finite regenerator of cycles. The UN provides some oversight on overuse of resources, such as forests and fishing grounds.

5.2.3.1.3. Activities of the UN

The UN has developed many public forums, so that people from many nations can address the challenges and problems that are either world-wide or global.

5.2.3.1.3.1. Conferences

When an issue is considered particularly important, the General Assembly may convene an international conference to focus global attention on the issue and to build a consensus for consolidated action. A recent example is the UN Conference on Environment and Development, also called the Earth Summit, in June 1992, which led to the creation of the UN Commission on Sustainable Development to advance the conclusions reached in Agenda 21, the final text of agreements negotiated by governments at UNCED.

There are other examples. The International Conference on Population and Development, in September 1994, approved a program of action to address the critical challenges and interrelationships between population and sustainable development during the next 20 years. The World Summit on Trade Efficiency, held in October 1994, focused on the use of modern information technology to expand international trade. The World Summit for Social Development, held in March 1995, underscored responsibilities of nations for sustainable development and found commitment to plans that invest in basic education, health care, and economic opportunity for all, including women and girls. The Fourth World Conference on Women, in September 1995, sought to accelerate implementation of historic agreements reached at the previous World Conference on Women. And, the Second UN Conference on Human Settlements (Habitat II), in June 1996, addressed the challenges of human settlement, development and management in the twenty-first century.

5.2.3.1.3.2. International Years

The UN declares and coordinates "International Years" in order to focus world attention on important issues. Using the symbolism of the UN, a specially designed logo for the year, and the infrastructure of the UN system to coordinate events worldwide, the various years have helped advance key issues on a global scale.

5.2.3.1.3.3. Agreements Treaties & International law

The United Nations Charter specifically requires the UN to undertake the progressive codification and development of international law. Over 500 conventions, treaties and standards have resulted from this work, and they have provided a framework for promoting international peace and security and for spurring economic and social development. Nations that ratify these conventions are legally bound by them.

The International Law Commission prepares drafts on topics of international law which can then be incorporated into conventions and opened for ratification by States. Some of these conventions form the basis for law governing relations among States, such as the convention on diplomatic relations or the convention regulating the use of international watercourses.

The UN Commission on International Trade Law develops rules and guidelines designed to harmonize and facilitate laws regulating international trade. The UN has also pioneered the development of international environmental law. Agreements such as the convention to combat desertification, the convention on the ozone layer, and the convention on the transborder movement of hazardous wastes are administered by the UN Environment Programme.

The UN negotiates treaties such as the Convention on the Law of the Sea, to avoid potential international disputes. Disputes over use of the oceans also may be adjudicated by a special court. The Convention on the Law of the Sea seeks to ensure equitable access by all countries to the riches of the oceans, to protect the oceans from pollution and to facilitate freedom of navigation and research. The Convention against Illicit Traffic in Narcotic Drugs is the key international treaty against drug trafficking.

The United Nations fosters international efforts to create a legal framework against terrorism. Twelve global conventions on the issue have been negotiated under the auspices of the United Nations, including the 1979 Convention against the Taking of Hostages, the 1997 Convention for the Suppression of Terrorist Bombings, and the 1999 Convention for the Suppression of the Financing of Terrorism.

The International Court of Justice (ICJ) is the main court of the UN. Its purpose is to adjudicate disputes among states. The work of the ICJ continues from 1946. The IJC has heard cases where the Democratic Republic of Congo accused France of illegally detaining former heads of state accused of war crimes and where Nicaragua accused the United States of illegally arming the Contras—the source of the Iran-Contra affair.

5.2.3.1.3.4. UN Courts

The UN also runs international criminal tribunals, including the International Criminal Tribunal for Rwanda (ICTR), as well as ones for the former Yugoslavia (ICTY), the Special Court for Sierra Leone, and the Ad-Hoc Court for East Timor.

5.2.3.1.4. Strengths of the UN

The UN has increased the involvement of nations with global issues, from disease to health and population. The UN has involved schools in its programs, through programs such as a "Model UN."

The efforts of the UN have resulted in upsurges of activism and declines in conflicts. A report produced by the Human Security Centre at the University of British Columbia, with support from several governments and foundations, documented a dramatic, but largely unknown, decline in the number of wars, genocides and human rights abuses over the past decade, after the end of the Cold War. The report, published by Oxford University Press, argued that the single most compelling explanation for these changes is found in the activities of the UN.

The report singles out several specific investments that have been particularly effective. There has been a six-fold increase in the number of UN missions mounted to prevent wars, from 1990 to 2002. There has been a four-fold increase in efforts to stop existing conflicts, from 1990 to 2002. There have also been: A seven-fold increase in the number of groups and other government-initiated mechanisms to support peacemaking and peacebuilding missions, from 1990 to 2003; an eleven-fold increase in the number of economic sanctions against regimes around the world, from 1989 to 2001; and, a four-fold increase in the number of UN peacekeeping operations, from 1987, to 1999. These efforts were both more numerous and often substantially larger and more complex than those of the Cold War era.

The UN has been able to broker independence for some nations. In East Timor, UN-sponsored talks between Indonesia and Portugal culminated in a May 1999 agreement which paved the way for a popular consultation on the status of the territory and eventually for an independent and democratic East Timor.

The UN has tried to consider nationless people. The UN has extended its efforts to groups of people within nations. For example, in 1977 the UN organized a meeting to discuss the creation of indigenous rights under international law. Indigenous peoples usually do not have armies or national currencies, perhaps not enough to define them as independent nations.

5.2.3.1.5. Inadequacies of the UN

The founders of the UN expected that the organization could prevent conflicts between nations and make future wars impossible, by fostering the ideal of collective security. Those expectations have not been completely realized. During the Cold War, about 1947 to 1991, the division of the world into hostile camps made peacekeeping extremely difficult. Following the end of the Cold War, the UN was expected to become the agency for achieving world peace and co-operation, as military conflicts continued to increase. The breakup of the Soviet Union, however, left the U.S. in a unique position of global dominance as self-anointed peacekeepers. This has created a variety of challenges for the UN. Even where it has been successful, the UN has been unable to create peaceful conditions lasting enough for its peacekeepers to withdraw.

Perhaps because of its nature, or its relative powerlessness, the UN has had failures, problems, and issues. One reason might be because it assumes that nations are rational players, certain of wants and communications, existing on a benign, stable planet. The intergovernmental nature of the UN means that it must reach consensus; it is an association of 191 member states and not an independent organization. Even when actions are mandated by the Security Council, the Secretariat is rarely given the full resources needed to carry them out.

The UN fails at times. In some cases, UN member nations have shown reluctance to enforce Security Council resolutions. Iraq may have broken 17 Security Council resolutions dating back to June 28, 1991 as well as trying to bypass the UN economic sanctions. The U.S. violated international law when it invaded Iraq. The UN did not respond; nor has it responded to other violations of international law by the U.S., especially related to the treatment of prisoners. For nearly a decade, Israel defied resolutions calling for the dismantling of settlements in the West Bank and Gaza.

The UN failed to prevent the 1994 Rwandan genocide, which resulted in the death of nearly a million people, due to the refusal of the security council members to approve any military action. The UN failed to intervene during the Second Congo War, 1998-2002, which claimed nearly five million people in the Democratic Republic of Congo, and failed to carry out and distribute humanitarian aid. The UN failed to intervene in the 1995 Srebrenica massacre, despite the fact that it had designated Srebrenica a "safe haven" for refugees and assigned 600 Dutch peacekeepers to protect it. The UN failed to successfully deliver food to starving people in Somalia; the food had been seized by local warlords, and a U.S./UN attempt to apprehend the warlords seizing these shipments resulted in the 1993 Battle of Mogadishu.

The UN has been unable to control sexual abuse by UN peacekeepers— men from several nations have been repatriated from UN operations for sexually abusing and exploiting girls as young as 12 in a number of peacekeeping missions. A 2005 internal UN investigation found that sexual exploitation and abuse has been reported in at least five of 16 countries where UN peacekeepers have been deployed, including the Democratic Republic of the Congo, Haiti, Burundi, Cote d'Ivoire, and Liberia. This abuse seems to be widespread and continuing, despite revelations and investigations by the UN Office of Internal Oversight Services.

The UN has scandals, problems and security issues. The inclusion of nations such as Libya and Sudan, whose leaders have weak records on human rights, on the United Nations Commission on Human Rights of nations, is an issue of concern. These countries argue, perhaps with some justification, that Western countries, with their history of colonial aggression and brutality, have no right to argue about membership of the Commission. One solution would be to qualify members by their current records on rights.

The Oil-for-Food Program was established by the UN in 1996 to allow Iraq to sell oil on the world market in exchange for food, medicine, and the other needs of ordinary Iraqi citizens who were affected by international economic sanctions, without allowing the Iraqi government to rebuild its military after the first Gulf War. The program was discontinued in late 2003 amidst allegations of widespread abuse and corruption; several people were implicated in bribery. Under UN auspices, over $65 billion USD worth of Iraqi oil was sold on the world market. Officially, about $46 billion was used for humanitarian needs, and additional revenue was used to pay for Gulf War reparations through a Compensation Fund, the UN administrative and operational costs for the Program (2.2%), and the weapons inspection program (0.8%).

The UN is limited by its lack of power. Although the UN has been effective at times dealing with limited kinds of conflicts, it has had problems dealing with unresolved, long-term conflicts, such as the Basques and the Spanish or Israel and Arab countries. UN concern over the Arab-Israeli conflict spans five decades and five full-fledged wars. The UN has defined principles for a just and lasting peace, including two benchmark Security Council resolutions in 1967) and 1973, which remain the basis for an overall settlement.

Sometimes the UN seems powerless with its own agencies. How should the UN deal with the World Trade Organization? Should it make it a nonprofit? How should the UN regulate goods and people? Is the free movement of goods or people across any boundary a good idea? The World Trade Organization needs to be made sustainable. Perhaps, it could be incorporated into the UN. But, the WTO must have environmental assessments, which are more important than WTO rules. New rules would give preferential treatment to trading partners with strong environmental policies and labor practices, and human rights.

The UN depends on voluntary dues from its member nations. Its funding is often too little. Expenditures of the UN system on operational activities for development, mostly for economic and social programs to help the world's poorest countries, amount to approximately $6 billion a year, excluding the World Bank, International Monetary Fund and International Fund for Agricultural Development. This amount is roughly equal to 0.75 per cent of world military expenditures of over $800 billion.

5.2.3.1.6. *Proposals for a New Framework*

In 2004, allegations of mismanagement and corruption regarding the Oil-for-Food Program for Iraq led to calls to reform the UN. There have been many calls for the reform of the UN, but there is little clarity or consensus about how to reform it. Some nations want the UN to play a greater or more effective role in world affairs, while other nations want its role reduced to symbolic humanitarian work.

An earlier, official reform was initiated by UN Secretary-General Kofi Annan shortly after starting his first term on January 1, 1997. Reforms mentioned included changing the permanent membership of the Security Council, which still reflects the power relations of the victors in the 1945 war; making the bureaucracy more transparent, accountable and efficient; making the UN more democratic; and, imposing an international tariff on arms manufacturers worldwide.

In September 2005, the UN convened a World Summit that brought together the heads of most member states, in a plenary session of the General Assembly's 60th session. The UN called the summit "a once-in-a-generation opportunity to take bold decisions in the areas of development, security, human rights and reform of the United Nations." Secretary General Annan had proposed that the summit agree upon a global "grand bargain" to reform the UN, revamping international systems for peace and security, and human rights and development, to make them capable of addressing the extraordinary challenges facing the UN in this century. World leaders agreed on a compromise text with such notable items as: The creation of a Peacebuilding Commission to provide a central mechanism to help countries emerging from conflict; the agreement that the international community has the right to step in when national governments fail to fulfill their responsibility to protect their own citizens from atrocity crimes; a Human Rights Council, since created and operational; an agreement to devote more resources to UN's internal oversight agency; several agreements to spend billions more on achieving Millennium Development Goals; a clear and unambiguous condemnation of terrorism "in all its forms and manifestations"; a Democracy Fund; and, an agreement to wind up the Trusteeship Council due to the completion of its mission. The UN is a recognition of our human limits—also that otherness needs to exist. It is too unsatisfactory, but too important to abandon.

A global unity cannot govern itself; it requires an external controlling agent. But, small nations could govern themselves, and the UN would be not a supergovernment, but a global center for small-scale business and small-scale politics. The UN should not be a union or an association, but a weak federal government with a few more powers than the largest nation. Perhaps the UN should also have a division of power similar to nations, that is, an executive branch, a representative legislature, with equitable representation, and a judiciary.

The traditional way of governing, from the Medes and the Persians, to the Swiss and the U.S., has been to reduce the size of the governed unit, not the size of the governing unit. The Roman approach was to divide and rule. The Duke of Sully and Henry IV of France planned to limit the size and number of European states to fifteen of equal size. Hitler applied the same strategy to Prussia and Austria. Kohr alleges that all successful empires share this small-cell pattern.

One way to divide a great power is to host a war to deunify it. Another way would be to give them a gift, according to Kohr, of proportional representation in the global federal union, such as the UN. A conventional federal principle of government grants each sovereign unit an equal number of votes irrespective of size. International law does not distinguish between degrees of sovereignty. Otherwise those with more population, territory, or wealth might be considered better. In theory, regions of nations could represent their regions, regardless of what other federal union they might belong to. Centralized systems would have to be decentralized for membership. For checks and balances to work the UN has to be larger than the largest of nations.

How would the UN limit size, though? By territory, population? What would a maximum be? 20 million? How can we have equality of nations? Should that be decided by equal numbers? Equal territory or places? A global organization, such as the UN, can be a suboptimal solution with plenty of flexibility. It can be satisficing rather than optimal.

The UN would be an agent of disunity, of political anarchies. Its concern would be with global things and coordination (rather than strict order). All nations would have to be dissolved through representation to allow voting. But have a maximum size. The natural list of nations already exists, within the unions formed violently in the past 200 years. Aragon, Valencia, Catalonia, Castile, Galicia, Warsaw, Bohemia, Moravia, Slovakia, Ruthenia, Salvonia, Slovenia, Croatia, Serbia, Macedonia, Transylvania, Moldavia, Walachia, Bessarabia, Macedonia, Sicily, Basque, Catalania, Scotland, Bavaria, and Wales, among many.

A central governing body for the earth has two very important functions: To insure a diverse biosphere, on which all humanity depends, and to equalize the opportunity of humans to live in health. There is no organization that addresses either of these functions. A global organization is necessary to coordinate the system. Either individual nations or a partially-responsible global institution is inadequate. The global organization must have the regulatory powers to maintain a healthy environment and to coordinate the constituent nations. It might have a new name as well, with new powers and responsibilities. In his pamphlet Common Sense, Thomas Paine encourage people to revolution. Later, he proposed that the U.S. help form an "Association of Nations." As a working title, the Association of Nations can replace the United Nations, which are not really united and which do not really represent all nations.

5.2.3.1.6. *Strengths and Weaknesses of a New International Body*
A revitalized UN or a replacement body would correct some of the weaknesses of the old pattern and perhaps offer new strengths. However, it would also develop strengths and weakness of its own, that would have to be addressed during its operation through some kind of adaptive process.

5.2.3.1.6.1. Potential Strengths of the Frame of an International Body
A holocultural framework, such as the Association of Nations, can identify and attempt to solve global problems, such as the greenhouse effect or acid rain, that cannot be solved at the level of a culture or nation. It can address the working of opposites in human affairs, where the solution to one global problem may cause another.

By being a global framework, it can adjust international economics. Local communities are based on traditional cultures, which have long-term lasting power. Traditional cultures often have wealth-leveling properties, absolute property ceilings, fixed wants, and production coupled with need— all of which results in a stable economy. Efficiency and productivity are less important than use and appropriateness. The framework can promote limited and rational economic development and coordinate international economic exchanges, protecting those cultures that choose to remain outside networks. It can put restraints on the current international community, from large corporations to large federations.

The framework can provide a holistic education of all cultures, besides that of the local culture. It can archive knowledge of other cultures. The experiences of many lives are encoded in myths, along with natural phenomena, supernatural beliefs, moral values, and features of the culture. All interpretation and recounting of the past is mythmaking. Mythic symbols store information concisely, which makes it possible for a person to assimilate the collective experiences of a culture. That is why myths reflect the detail of a culture.

The framework may be capable of realigning social boundaries to ecological realities; the boundaries of a watershed or ecotone would be more appropriate than geometric lines. A natural region supports a great deal of life without human intervention; it produces enough life to support a reasonable number of humans. We need to know natural associations and limitations because these determine the harmony of development.

A framework can justify a wide diversity in nature and accommodation to natural laws. It can recognize the value of the total biosphere and respect all forms of life, past, present, and future. It can do so, because, unlike traditional or industrial cultures, it is conscious of itself and its purpose.

5.2.3.1.6.1.1. Being Conscious

Creating a holocultural image requires changing the gestalt of images of self, nature, and society. That effort is revitalization. Unlike classic cultural change, revitalization requires the explicit intent of the members of society; it depends on restructuring elements already in use or known. Where the culture remains responsible for the performance of ritual or the preservation of doctrine, the images are preserved. When the images are anticipatory, they lead to development and social change. Attractiveness reinforces the movement towards them. We are dependent now on our consciousness of the entire system of nature and humanity. Undertaking a conscious orderly change in our living habits, before it is forced on us by an unbalanced environment, gives us more options.

5.2.3.1.6.1.2. Recognizing Context

Cultures change as the result of human interactions in nature. Nature and cultures are in a constant state of flux; cultures have much in parallel with biological species. Our thoughts and ideas, tools and cultures, are as much a part of nature as other species or peat bogs. To preserve our cultures and natural environments, we must understand that they are examples of a dynamic order brought forth by the earth in its history. We are physically dependent on nature. We are psychologically dependent as well; without signals from nature, our minds become closed and dead. We also are physically and psychologically dependent on culture. Yet, the diversity of habitats and cultures is allowed to erode.

A global framework for cultures depends on important principles drawn from ecology. One role of ecology could be to urge the toleration of fluctuation, irregularity, uncertainty, and diversity. As adaptive systems, cultures change as ecosystems change. And sometimes ecosystem change is a result of cultural change. They are linked together.

If humans adapted more closely to the complexities of natural ecosystems, then human cultures would be more diverse and stable. If humans adapted to the complexities of natural ecosystems, then human societies would be more complex. The proper attitude of an ecological framework is care, a positive spontaneity, but also a "letting be," a reverence toward the wild alienness of nature, a willingness to comply with the limitations of natural systems, and a willingness to reduce human dominance.

What this means practically is that the local environment would determine the extent of a culture. The framework would recommend wilderness areas sufficient for a culture, although the shape and expression of these areas would depend on the kind of culture. Eugene Odum has calculated that for the temperate southern United States it takes two acres of wilderness to support each human being.

Cultures can determine the minimum, or optimum, amount of wilderness areas to support local ecosystems; for instance, very large areas are required in tropical ecosystems or deserts, relatively little ones in grasslands and temperate forests. They can determine the natural productivity and the percentage to be used by humans, as well as artificial productivity and costs. Cultures can key their population to natural productivity for long-term sustainable existence. And, they can multiply any increase by trade-offs, such as a reduced standard of living or exchange with another group.

The framework could recommend an optimum size for each human population. A nation must have a population large enough for economic advantages in food production, education, and entertainment, and for political tools. As Leopold Kohr has noted, the size of a culture is determined by the function it fulfills. The function of a state is to provide its members with protection and other advantages that they do not have as independents.

When a state becomes too large, it cannot offer protection—it cannot offer even clean air or water. The country of Andorra, with about ten thousand people is stable, sovereign, and healthy; the Greek, Italian, and German city-states that furnished much of Western civilization often numbered less than twenty thousand individuals. As the size of a nation increases, the negative factors of civilization, such as overcrowding and breakdowns, increase. Technology has the capacity to allow some expansion, but not an infinite amount. A comprehensive population policy must be created for larger cultures, and it must fit into the context of wilderness and other cultures.

5.2.3.1.6.1.3. Being Comprehensive

A holocultural framework includes all human cultures without judgment. The framework provides a higher resolution image of the whole, since it incorporates all human cultures. It includes all its members, recognizing that each says something worthwhile. It is not details or knowledge of the operation that is critical, but an understanding of the wholeness of order.

The framework can interact with nature much like the mythic, but understand the rational and mechanical sides of thought. It would not be a conglomerate of sciences; it would not be limited by the facts of any science, even ecology. The insights of people of every culture must be considered. Each person tells of a way the world is; together, these ways make a holistic framework.

The framework includes ultrahuman cultures in its consideration. It can create wilderness zones that would have various limitations for conversion or use. It can reserve large areas of wilderness for ultrahuman beings and biogeochemical processes. The framework can attempt to combine the best single elements of industrial culture with the superior components of primary cultures, in parallel with Gordon Taylor's paraprimitive solution. High technology can offer immense benefits, with restraint and appropriate limits. Primary cultures can satisfy the human needs for belonging and status.

5.2.3.1.6.1.4. Making Authentic Images

Individual cultures, in their unique cosmologies, create images of the human place in nature, in terms of mastery, community, or participants. The holocultural framework offers a holistic value of human worth, outside of any one local perspective. It promotes and protects universally accepted values: Reciprocity—the repayment of obligations; territorial integrity for cultures; legitimacy—the value of children born in wedlock; and the working of opposites—life and death, sacred and profane. It can promote basic rights human rights: The right to land, food, shelter; to equal opportunity to develop, regardless of race or sex; to participate in global affairs as desired; and to live without excessive discrimination or conflict.

5.2.3.1.6.1.5. Protecting the Diversity of Cultures

Each culture is a response to a unique place, and so there is a diversity of cultures. Industrial culture condemns to backwardness any culture that is not part of its global electronic neural system. This definition of backwardness means only a lack of fast things or professional enslavement. Primary cultures do not lack art or play, or food, tradition, freedom, or happiness.

It might be good for cultures to be uncoupled economically; it might be a sound option for traditional societies unwilling to make the same mistakes as industrial ones. The framework would keep cultures separate and coordinate any exchanges between them. It would resolve disputes that arise from territorial expansion, the past movements of people and borders, or the unequal expansion of cultures in the same territory, such as the Sinhalese and Tamil in Sri Lanka.

5.2.3.1.6.1.6. Providing Order

The AN would have to be stronger than the largest nation, through arms control, disarmament, and its own weapons program. It would have to disarm the nations substantially. Police operations would be quite similar to the current United Nations, although there would be differences. The AN would have a permanent police force, kept in regional divisions. The funding for the police force would be as a result of the income of the AN, not dependent on voluntary membership fees.

The first force of Police would always be unarmed. If a second force is necessary, it would be well armed. In fact, AN police forces would be better armed than any single nation. A maximum residence time would be set, and an exit strategy required.

Of course, the AN would work to prevent conflicts. Although many conflicts would be reduced by the changes in the Eutopian framework, there will always be conflicts between cultures. AN forces would try to solve the problems through understanding, compromise and consensus.

5.2.3.1.6.2. Potential Weaknesses of an International Framework

History does not show a progressive unfolding of human betterment; loss and defeat are much of the texture of daily life. A framework will not be able to solve all problems, especially ubiquitous ones like hunger. No human construct is perfect and completely comprehensive. No human framework can expect to solve every problem to everyone's satisfaction. The framework can be expected to exhibit a number of weaknesses.

5.2.3.1.6.2.1. Being Abstract

A holocultural framework is a general human construct, which may not be implemented. There is no working model of global unity. Our experience with international cooperation on an immense scale is minimal. Our ability to plan our cultures and foresee our impacts is minimal. Other abstract ideals, including democracy and communism, have been disappointing and severely modified in practice.

Kinship is more rigidly localized than other dimensions of culture, which can be more rapidly disseminated and assimilated. The transition from kinship to a simultaneous abstract global citizenship may slow. Kinship loyalty sometimes clashes with global perspectives. A framework trying to lessen the conflict and resolve contradictions may be faced with more conflict.

5.2.3.1.6.2.2. Being Uncritical

Such a framework, by definition, accepts any human culture, even bad ones. It cannot make judgments about use. In avoiding ethnocentrism, it must accept failure. It may not be able to deal fairly with cultures that are dying out, because they are unfit or because they are victims of a large coercive culture. Yet, it cannot artificially support bad images. It must preserve the process of making and sustaining a way of living, not every individual culture.

This framework does not reject or judge cultures, but incorporates all the practicality and paradox. It makes no distinctions between right and wrong or good and bad; these polarities are more like positive stimuli useful to development. Hence, there is no evil, as considered in many cultures, only suffering that results from lack of wisdom. Many customs, like sacred cows in India, at first glance, seem to be dysfunctional. But, even sacred cows provide dung for cooking fires.

5.2.3.1.6.2.3. Being Contradictory

It used to be, as Karl Marx said, that village life enslaved the human mind with traditional rules and subjugated it. No more—too much communication is a greater threat. Our excess communication tends to wear out our ability to feel empathy and react to suffering. Some cultures may overcommunicate and others may undercommunicate. Undercommunication may result in ignorance and suffering; overcommunication may result in passiveness and insignificance. The framework will have to abide more than one contradiction.

5.2.3.1.6.2.4. Being Weak

The framework may not have the power or authority to make agreeable boundaries. It may be unable to set aside large enough areas for natural processes. It may not be able to dictate population restrictions for some cultures without seeming to be genocidal or prejudiced. It may not be able to achieve an agreeable redistribution of some kinds of wealth. Any action may result in some dislocation and suffering. It may be impossible to limit the interdependence of nations.

The framework may not be able to deal with incompatible cultures or the divisive forces of large industrial cultures. Some cultures may refuse to participate. It may not be able to handle large differences or to limit the influence of powerful corporations, which have no local accountability. Some traditional cultures may have trouble incorporating new ideas, such as the equality of women.

5.2.3.1.6.2.5. Being Fallible

A holocultural frame may try to address global problems that may be insoluble within its range. Some of its actions may have negative consequences for some cultures. For instance, in mediating boundaries that have changed over centuries, it may be difficult to rectify imbalance, theft, or suppression. Cultures have dominated, displaced, merged, or destroyed other cultures for millennia. No one knows how far back to trace a wrong. The dividing line might always seem arbitrary. It may be appropriate to return lands to the Pawnee, but not to the people that the Pawnee displaced.

5.2.3.1.6.2.6. Being Naïve

What does it mean to be naïve? Artless? Ingenuous, which we already consider a weakness of utopias? What is naïve? That people will accept inequity as it worsens? That people will always accept cheating and discrimination? Is it naïve to think that corporate greed will benefit starving children? Is it naïve to think that people will give up heroic luxuries to help people somewhere else on the planet? That nations will find it in their interest to break apart along ethnic or economic lines? A eutopian framework may always seem naïve by some definition or example, but by contrast with hard "realities" it will always seem less naïve.

5.2.3.2. *Responsibilities of a Global Framework*

There is already a world system. But it is not, and should not be, a stagnant, monolithic industrial system—to say that there is a human body is not to say that all organs have decided to become kidneys. A global order is necessary to govern this world system. The Associated Nations (AN), an elected body, should have the regulatory powers necessary to maintain an healthy global environment. It should have regulatory and advisory powers to maintain the independence and integrity of its constituent nations. It should have regulatory and punitive powers to rectify resource and human rights infringements; only this body would have police powers and large impersonal weapons. Various advisory bodies would recommend policies and actions to nations. The Associated Nations has six basic functions: To ensure a diverse biosphere; to manage resources; to protect unique cultures; to coordinate representation; to provide services to nations; and, to create peaceful conditions. For instance, to ensure a Diverse Biosphere, the AN has to identify, zone, conserve and preserve landscapes. Then it has to monitor, protect and restore the landscapes as necessary.

5.2.3.2.1. To Ensure a Diverse Biosphere

To ensure a diverse biosphere, on which all humanity depends. The Associated Nations works to conserve genetic resources and ecosystems. Preservation of entire systems is addressed on a global scale. The Associated Nations is responsible for planetary monitoring of all major biomes and their ecosystems. The destruction of basic landscapes sets the frame for proper conservation and development policies. Many natural and artificial values are conserved this way.

In ecological ignorance, our ancestors cut down the cedars of Lebanon, ruined the Mediterranean, created dust bowls and deserts. History supports the notion that no civilization has ever recovered after ruining its environment. Some, like Egypt or Rome were replaced from the outside. Others, such as Ur or Rapa Nui, were rebuilt by wealthier invading peoples. A few, like the Mayans, settled for a lower level of complexity.

Ray Dasmann distinguishes between ecosystem people and biosphere people. Indigenous traditional societies are examples of the former, and technological societies are in the latter. The former live within a single ecosystem usually and are dependent on it for survival. If ecological rules, such as "do not overkill," are violated, they perish. Island people do not tolerate overpopulation. But biosphere people can draw support from any ecosystem on earth; if one place is ruined by exploitative pressure, then another place can be drawn from. But with absolute increase in human numbers, they cannot be completely insulated from ecosystem failures. Knowledge of local collapse in distant states, which are part of the resource commons, could precipitate an ultimate collapse.

A natural way of extermination is by stimulating overgrowth. The natural way of preserving things and increasing the base of life is to contain growth; instead of expanding form, it duplicates it. Julian Huxley calls this adaptive radiation. Speciation uses a wider range of noncompetitive food sources than the expansion of a single species. This type of adaptation occurs in humans also. The relationship between groups can be characterized as symbiosis, living together. Biologically symbiosis increases the chances of its organisms for survival. Humans should choose goals that are symbiotic from the alternate paths.

But problems arise when societies become larger and inflexible, when the variety that insures tribal and small communities yields to nation states. Nation states have no value in a global order. The designation of cultural units does not involve a major revolution. Revolution is a false dilemma; it does not reflect the possibility of thousands of microrevolutions on farms, factories, and families, all at local levels.

5.2.3.2.1.1. Basic Landscapes

The basic landscapes and their divisions have been described (Section 4.5.4.2.3.2.5). Sacred landscapes are critical for keeping many cultures healthy. Preservation landscapes, also of high cultural value, are maintained by human exploitation using traditional ways, including herding and nomadism. Conservation landscapes allow a higher level of exploitation, but without the addition of subsidized energy or the impacts of heavy equipment. Domestic landscapes have been simplified and manipulated for higher levels of use; fields and forests are managed by industrial methods and tend to have the characteristic of domesticated, controlled lands. Artificial landscapes have been almost completely modified and covered, for travel, urbanization, and industry, although a few wild species may be present. Foundational landscapes protect the processes that create wilderness (often the most fragile and sensitive areas).

By keeping these divisions separate, and limiting the kinds of activities in them, the landscapes themselves will be healthier. Not all activities are compatible; furthermore, living systems have developed in relative isolation, and function best with a limited amount of interference from other species or other systems.

5.2.3.2.1.2. Calculate Global Ratios

Each land or ocean system would be classified and put into an appropriate category, ranging from pristine to heavily industrialized land. Each category would occupy a different percentage of the planetary surface, depending on calculations of minima and maxima and depending on cultural values and decisions. At a minimum, approximately fifty percent of the land area would occupy the first division (eighty percent of ocean and water surfaces); sixteen percent in each of the other three (four for water); leaving two percent for completely artificial landscapes—industrial or city (two for water). These figures are consistent with several earlier proposals. Eugene Odum suggests thirty percent forest cover worldwide, with sixty percent in tropical areas. Paul Shepard offers seventy-five percent of the total land area left wild in a techno-cynegetic society. Constantin Doxiades suggests fifty percent of the surface area in wilderness. We cannot preserve less until we learn more about the requirements for large cycles.

5.2.3.2.2. To Manage Common Resources

The Associated Nations would have the power to designate areas for conservation, including the oceans and atmosphere. It would regulate all industrial and residential use of common resources. Furthermore, it would form new institutions, both regulatory and advisory, such as a Associated Nations Environmental Agency, to deal with resource availability and alternate technologies, create global scientific bodies to study global ecological balance, collect data on global systems, explore remote areas, and maintain a central library of all information on sciences, technologies and cultures. It would maintain reserves of food and minerals for emergencies and catastrophes. The Associated Nations could perform a resource function for all nations, maintaining large crop margins, for instance for seven years, to secure survival. Future survival should depend on systems sufficiently flexible and elastic to sustain moderate failures in parts of the world without causing catastrophes to a connected food system.

This attitude applies to the entire technology and survival controversies: Irrigation, tankers, nuclear power, pesticides, population, deforestation, and genetic engineering.

Finally, the AN would recommend optimum populations for nations, although it would not enforce those figures. Optimal sizes would be calculated, based on social and ecological limits, as well as on traditional values. Every nation needs a comprehensive population policy, closely related to environmental and technological policies, and within the constraints of their agriculture. A single population policy for the world is unfair, since cultures are at different stages of development with different values. There are dilemmas posed by necessity to equate global balances and republican needs. Since the allocation of resources to a nation and the representation of a nation would be determined by area and not population, there would be no reason for a nation to exceed the optimum figure. If a nation wanted to expand its population, it could do so through a number of means: Trading off with other nations or through the development of new food technologies, such as attached greenhouses for every building. If growth exceeded a safety margin established by the Associated Nations, demographic policies would be strongly recommended by the Associated Nations, without prejudice or malice.

Humans have modified their surroundings as much as possible within their power to improve their lives. Recently, they have done so to improve nature. Believing that nature was incomplete, they added plants and animals, then added fields, structures, canals and dams to feed the plants and animals. In *The Origin of the Species*, Darwin observed that insular biotas were glaringly depauperate in general, until supplemented by human culture: "man has stocked ... far more fully and perfectly than has nature." Unfortunately, filling up nature and perfecting the earth did not proceed with ease; there were setbacks. Exotic animals and plants ran wild and became pests. Fire control caused raging fires. Dams became the sources of diseases. Canals introduced pests. Plantations ruined soils. Irrigation projects salted up the soil. Peter Matthiessen noted that where great, wild creatures ranged, vermin prosper. If the influx of new organisms of all kinds continues unabated, all life on earth will eventually become homogenous or drastically changed; then it will also be strained through the filter of adaptability to humanity and their managed crops. This flora and fauna will undergo change as our living habits change. On the other hand, if we plan our future, we can include squirrels instead of rats, butterflies instead of cockroaches.

The induced instability of ecosystems is an important cause of economic, political, and social disturbances throughout the world. The disturbances are passed on to humanity. And where our intervention has unbalanced nature, we need to repair. Each biota has developed only once in the history of the world. And once lost can never be regained. Some environmental degradation can be reversed, but not biome or species extinction. We do not know how many whole systems that we have destroyed. Nor do we know which element of a system is a more crucial one. Very little is known of degradative synergies from noise, heat and pollution.

5.2.3.2.2.1. Renewal of Resources

Ecosystems damaged by social activities may reacquire lost ecological qualities by natural processes. Ecosystems dependent upon periodic natural disturbances, for instance floods or fires, may be markedly changed if these disturbances are controlled. Damaged ecosystems may also be rehabilitated to a condition that includes some of their original characteristics and some beneficial to humanity. Much of our manipulation of nature is highly desirable for us, perhaps even for some natural systems. Some possibly may be enhanced by management techniques to an improved condition different from the original. Some may remain degraded.

Ecosystems have enormous powers of recovery from traumatic damage. They can overcome the effects of outside disturbances by progressively reestablishing ecological equilibrium, even if not exactly to the original state. Frequently, other potentials are activated by the outside disturbance. Unfortunately, the key word is outside. No disturbances are outside, anymore. If all ecosystems are disturbed, no improvement can come from outside either.

The rate of healing of injured systems is often more rapid than expected. Good management and human commitment can aid the process. The recycling of degraded environments is one of the urgent tasks of our age. Marsh envisioned man as a coworker with nature in the reconstruction of "the damaged fabric which the negligence and wantonness of former lodgers has rendered untenable." Under loving care, even very degraded ecosystems can be made productive and satisfactory for humanity, although not the same as the original. "Even the most successful programs of reclamation and the best artificial environments cannot of course duplicate the subtleties and complexities of natural environments; but most of them will improve in time," states Rene Dubos.

Of course, they might, as they become less under control and less invaded. Perhaps humans have a secret desire for simple environments. Dubos seems entranced by the bleached islands of Greece, as well as by deserts. Do we want all earth to be those abstractions? Dubos has spoken eloquently of the humanized landscapes, but there is no one to praise what has been lost. There are inherent values in wilderness as much as in the humanized landscapes. The impoverishment of Southern Europe may be aesthetic, but it leads to human impoverishment.

Should ecosystems be restored to a close approximation of the predisturbance condition? It is always feasible to establish more than one type of ecosystem on a disturbed site. Due to climactic change, faithful restoration may be impossible. Certain ecosystems are perturbation dependent. Untouched reference areas could be preserved as models.

5.2.3.2.2.2. Conservation of Resources

The Emperor Asoka, in Third century BC India, took a positive stand on wildlife conservation. The same Asoka earlier had caused roads to be built with periodic rest stops, for the care of animals. In most literatures there is little mention of the long history of rural conservation. The rural economy is the result of centuries of careful cultivation. Wise people have always treated resources with care. Many countries in Europe and Middle East had strict regulations for management of water, wood and farmlands. Vergil's *Georgics* were written in support of government policy to remedy the decay of rural lands.

Carrying on from the UN, the AN would work to conserve genetic resources and necessary wetlands and watershed forests. The Biosphere Programme of UNESCO is a worldwide monitoring of all major biomes and their ecosystems based on international agreements. Conservation efforts are concerned with: The dynamics of a system; the interactions within the system and effects of climactic and system changes, including human involvement; the varying time scales of concern; and, the preservation of natural communities with a high degree of integrity, subject to intrinsic processes.

Conservation is basically a problem in ecology and must be addressed on a regional and global scale. Conservation is practiced as an affluent token; even poorer societies are conserving with the expectation of tourist income. Volunteer financing is inadequate. Long-term AN financing is necessary to extend the effort beyond the lifetimes of politics. The cost of saving the wild for a common heritage must be borne equally. If Brazilians cannot extract minerals from the Amazon basin, if an Indian peasant loses a bullock to a Bengal tiger, there must be some balance.

Richard Allen contends that the way to save the world is to invent and apply patterns of development that conserve living resources essential for human well being and survival. Although resource conservation is thought of as specialized and limited, it cuts across all human activities, and should be incorporated.

5.2.3.2.2.3. Management of Resources

History records the debris of some civilizations that tried to manage their resources and failed; they existed in the Americas, the Middle East, Africa, Asia, the Pacific, and Europe. Natural resources were originally defined as objects provided by nature for human use. This concept has been expanded over thousands of years to include minerals, wildlife and people. Eric Jantsch claims that humanity now acts as a systems manager at all levels, where management is an activity that aids evolution, acting with it, recognizing and applying an ethics that transcend an individual level.

Even most of the noncultivated land surface of the earth is being managed; elephants, giraffes, crocodiles, wolves, caribou, snail darters, redwoods, prairies flowers are managed or else destroyed. Many are done in by human recreation, with its attendant necessary vehicles. Even a modern, balanced exploitation may destroy forests and fisheries. Currently, many resource managers espouse the ideas of equilibrium maintenance and maximum sustainable yield. These ideas are poor guides to management. By trying to maintain habitats in equilibrium, we often set them up for catastrophic decline, for instance, in fire-climax pine forests, or destroy resident species, such as the condor.

The use of maximum sustainable yield in wildlife management has resulted in the degradation of the populations involved, whales and salmon, for instance. A carrying capacity is not constant; species that live near the limit of capacity cannot be killed at a maximum. Even small numbers, for the Sandhill crane, less that six percent, hunted could result in extinction. This may be true of wolves, bears, mountain lions, and other species.

Some managers, like whalers, are far worse. They do not try to manage for a continued maximum yield; they try to maximize the economic value of a resource, in spite of an awareness of extinction—the rape of one "resource" provides the capital for the rape of the next.

The idea that everything should be managed is based on an extreme belief that nature is a resource to be processed. Furthermore, management is self-perpetuating and self-justifying. The objective of resource management is to increase the measure of quality of life for affluent people in overdeveloped countries.

Management, even conservation management, has been based on economic objectives. And, as Aldo Leopold pointed out, the weakness of relying on economic motives is that most members of the earth's community, such as wildflowers and song-birds, have no economic value. Yet all the members of the community contribute to the integrity of the whole, which is vital to maintaining what we do consider important. Those beings with no economic value are ignored, or worse, labeled as weeds or vermin and destroyed so that crops and animals with short-term advantages for human ends can be substituted. The goal of this institute is that kind of temporary control.

The impulse to manage nature is an expression of the judgment that we know how the world should be run. But we are finding that we do not know at all. We did not know about the effects of DDT or radiation or chemical dumps or special drugs. The whole approach of the conservative position ignores the physical and ecological dimensions of resources. The vandal position is even more basically ignorant. Laws of ecology must be obeyed for these laws determine our existence and that of "resources." There can be no "balance" between obeying some laws and disobeying others.

We are unbalanced. Our whole industrial world view is unbalanced. Balanced resource management will still unbalance nature, though perhaps at a slower rate. The balance of nature has to come before the balance of resources. We will continue to be unbalanced until we enlarge our understanding of nature and let ecological limits suggest new technologies and techniques. A balanced relationship between humanity and environment is necessary.

Such a balance must be based on conservation, if it is to avoid harmful 'side-effects' and provide benefits. We must invent patterns of development that also conserve living resources essential for survival and well-being. This kind of conservation is not a special and limited activity; it is a process that affects all human activities. Conservation must be integrated with development to ensure that vital parts of the biosphere are protected or modified only in ways it can sustain. A conservation strategy has to identify the most significant objectives, according to criteria of biological importance and urgency of need. Damage to life support systems that are becoming irreversible—extinctions, habitat destruction—these need the highest priority. Every independent group should prepare proposals for cooperative programs concentrating on biomes that cross boundaries, that is, tropical forests, rivers, and global common areas, such as oceans and atmospheres.

Common resources can be managed, if the system is managed so as to minimize fluctuation and interference, so as not to impact the stability of the process, so as to harvest an appropriate production, but so as to be aware that even a stable sustained yield of a renewable resource might change deterministic conditions so that resilience is lost and a chance event could trigger sudden change and the loss of integrity of the system. Some resources can be restored or renewed, although intervention may be inappropriate if natural cycles, including catastrophic events, are not understood. Natural processes of recovery work slowly, but good management can accelerate them.

5.2.3.2.3. To Protect Unique Human Cultures

Cultural patterns relate human communities to the ecological areas in which they are embedded. Any culture is only one of many possibilities. There is no single or correct way. By 1900, humanity had spread through 1,000 different cultures and 3,000 languages—-roughly equivalent to the number of natural biogeographical provinces and subprovinces on earth. Whenever groups were geographically separate, there was differentiation, which enforced separate cultural identities.

The real feelings of innumerable groups of people center on much smaller regions of the world than nations. For example, Britain is composed of Scotland, North and South Wales, Northern Ireland, Anglia, and Saxony. It may mean more to be Welsh or Irish than British, or Quebecois than Canadian, Kurd than Iraqi, Mongolian than Chinese. Forced cultural integration breeds tensions; in the USSR by 1988, the tensions exceeded the force and advantages of integration. Rwanda, and Tanzania are additional cases.

Many cultures in established countries, like Scotland in Britain or the Nyiha in Tanzania are organized cultural communities, but are not permitted to join the UN because they do not possess armies. A new world order would permit autonomous groups to join the Associated Nations according to cultural or linguistic affinities and not merely force of arms. Nations could break up into preferred "natural" units. Every cultural group, or nation, is considered equal, regardless of size or sophistication. These things would contribute to their protection from forces that fragment and destroy cultures.

5.2.3.2.4. To Coordinate Representation of Cultures

The Associated Nations, would function as a global coordinating body with powers and limits. In order to coordinate representation, the AN would create a representative body. Each nation would provide a set number of representatives to the governing body of the Associated Nations, which would provide a forum for designing the governance of the earth, one in which everyone can participate.

5.2.3.2.5. To Provide Services to Nations Groups & Individuals

Garret Hardin noted that the Marshall Plan, which channeled $12 billion for rebuilding Europe after 1945, was not entirely altruistic; it was meant to keep the USSR out of Europe. After the Marshall plan succeeded in Europe, Mr. B. Hardy convinced Truman to extend the plan to the world; the result was the beginning of foreign aid in 1949. But, the plan did not work as well. Over $80 billion and 25 years later, most aided countries are still poor. What was the difference? The plan in Europe recreated an industrial civilization. Foreign aid expected illiterate, fatalistic, poor populations to try something new to their experience. Although it was funded at about the same level ($3 billion per year), it was expected to help 20 times as many people—2 billion instead of 100 million. Furthermore, the expectations were wrong. A. van Dam suggested that the Marshall Plan extended to the world in the same spirit and at the same level would prove more rewarding and stimulating. That amount of investment is only equivalent to the production of two day's goods and services in North America, Europe and Japan. Linear thinking, combined with economic altruism, can lead to great short-term successes, but to long-term problems.

Strictly, human altruism occurs only on a tribal level. Biological heredity makes kin altruism possible. Money makes a flexible and reciprocal altruism possible between unrelated members of the species. Money also makes people less materialistic; it is a symbol. It frees people from the calculation of gift-barter. Telescopic philanthropy—the term is from Charles Dickens—is not coupled with responsibility. Our gain is unrelated to any effect on the recipients. That irresponsibility may be part of appeal. Hardin wrote that human actions should be guided by charity, but he applied the older understanding of the virtue, which he noted confers benefits and refrains from injuring, but does not shrink from inflicting suffering to achieve real good. Amiability is a weakness not to be confused with charity. It is the source of foreign aid, which ruins the longer prospects of a self-reliant life.

Any kind of aid is a distribution from a commons, as Hardin notes. Dealing with a global commons is more effective through laws that end destructive behavior. Furthermore, laws are a second-order altruism; an individual does not risk acting as a lone altruist, in violation of Hardin's Cardinal Rule of Policy, which is to never ask a person to act against his own self-interest. If it can be demonstrated that the long term effects of egoistic actions are harmful, people may be persuaded to forgo short-term gains. To prevent a tragic end to commons, the system must be changed. People will be reluctant to change because of the uncertainty of technological forecasting and political stability. Political change may be similar to lifting one's self by the bootstraps.

The tragedy of the commons could happen in mass economy, not necessarily in an information economy. Hardin believes that coercion, centrally controlled by majority rule, is required for survival. Self-interest and knowledge of capacity could also avoid tragedy. J. Martino claims that private property can eliminate the tragedy also; self-interest of the owner dictates. In this case, a global socialism would work because it would be responsible to national cultural units, whose interests would be represented. Economic cooperation in world of scarcity will not solve environmental problems.

The Associated Nations, by comparison, would furnish only temporary aid in the form of help and education. The kinds of aid would be threefold.

5.2.3.2.5.1. Rescue and assistance. Responses by the UN to earthquakes in Peru (1970) and Nicaragua (1972), droughts in Africa, floods in Bangladesh, and other disasters, are indicative of the promise of cooperation, although many efforts have been minimal or diverted into administrative mazes.

5.2.3.2.5.2. Civic action would be concerned with representation and voting. Mass media would be available to every culture for referenda. Civic action would include technical projects, such as farming or reforestation. Civic action might also address inequity and population or immigration concerns. Civic action groups, similar to the U.S. Civilian Conservation Corps (1930s) or the U.S. Peace Corps (1960s-), could apply appropriate technology on request. Richard St. Barbe Baker suggested armies for reforesting the Sahara; the technical feasibility has been demonstrated in places, but any whole effort is crippled by politics.

Special education would be a priority. Education would be especially important at this common level, since cultures would dispense local specific information. All scientific and technological knowledge shall be available to all states, as regulated by the Associated Nations. Scientific research and development, especially on environmental problems, will be promoted in all states. The AN will support a free flow of scientific information and experience. The AN will also ensure that appropriate technologies are available to all countries. The AN will award basic educational and research grants to all humans, for whatever use desired. An earth university might be established. For public health, the AN would create indicators of social and biological health, as well as monitor health, trade, and social quality. It would establish centers on epidemic and disease control, recognizing that health in general depends on healthy global cycles and ecosystems. For financial health, the AN would standardize exchange rates and provide banking facilities. It would work to stabilize the prices of commodities and materials and set common business standards in terms of work and pay units and wade values—the unit of wage shall be a human work unit, which shall have equal value for all. The Associated Nations would provide laws and courts to address problems of justice.

5.2.3.2.5.3. National armed forces could be replaced by an unarmed Associated Nations (AN) police force; AN enforcement would consist of a persuasive presence for the observation of law and order. The UN force in Cyprus (1960s) performed this function admirably. An AN charter would ensure the inviolability of personnel and their right to intervene in any conflict when asked by any group. With world consciousness, the weak and disadvantaged can get food and shelter by appeal or right, without plundering, without war. War would be allowed to evaporate with the protection of all people by a central governing body. Fighting would occur, between individuals and small groups as interests conflict and communications falter. Nonviolence is possible only between rational individual human beings. Sometimes force is necessary. Therefore, some armed security would be necessary, and part of the police force would be armed. A shift from military to police forces could provide that security. This police force could provide humanitarian intervention.

5.2.3.2.5.4. To Manage Populations. Human populations tend to expand, especially after agriculture, to fill up the human niche, to use every possible resource and to reduce every kind of flexibility. People then become excited by the intensity of people and addicted to the luxuries of trade. For humans not to destroy the planet, we have to plan an a satisfactory or optimum population rationally or irrationally.

At low population densities, management only needs to apply to interpersonal behavior. Management in this sense is at the limits. A global framework could proscribe behavior at the limits, and prescribe on a cultural level. It could balance the equation of access, rights, and distribution.

5.2.3.2.5.5. To Adjust Human Equity & Species Equity. The AN would be responsible for distributing a portion of the wealth of the earth among humanity. In this sense it would be socialistic. It would regulate resources for a common good that included all living beings. A redistribution of the imbalance of wealth would require an attitude like altruism.

5.2.3.2.6. To Create Concord (Peace) & Accommodate Discord

Like the UN before it, a global framework would try to create the conditions for peace and eliminate war. What is war, again? War means fighting or hostility, campaign or invasion. But, more than just being a kind of conflict, it has the connotation of physical harm or complete destruction. Long ago, war changed the level and scale of violence, by involving more people and by making the fighting more detached. Now, war is guaranteeing its own continuity. The dangers of the military-industrial complex, that Eisenhower warned of, have increased exponentially. The original axis of military and industry now includes, in the U.S., the enthusiastic complicity of the entire legislative branch of the government, which is supported by the military and industries, and the inbred ideas of the new forms of think-tanks, which are supported by the military and industries. All four are staffed by escalating revolving-door policies. The profit-driven American way of war, where the U.S. presents itself as the arsenal of democracy and the reluctant guardian of freedom, works to ensure the coca-colonialization, that is, the economic colonialism, of weaker countries and the physical subjugation of disagreeable countries.

Wars are advertised and sold to the citizens of a nation using the same dishonest and fallacious, but effective, techniques. Supposedly, one can consider war as the imposition of order through the death and destruction of people causing disorder. Is war simply contention or conflict? Contention means verbal strife or dispute. Dissension means difference of opinion or opposing groups. Conflict means strife, struggle, collision, hostility, fight or battle. These words may have been adequate at one time, but war, as institutional destruction, has to be eliminated; the word has to be refer to behaviors that are no longer exhibited. So, what will remain? All those smaller pieces, such as fighting, conflict, dispute, and strife. These will not go away. But, they can be kept small and handled by different kinds of treatment, and maybe rarely, force. Discord is a good word to collect the human behaviors. Discord means a disagreement, from the words meaning 'hearts apart.'

What then is peace? Is it dull and uninspired? Why is it desirable, if it is dull and impossible? The word 'peace' is from the Latin 'pax,' which is from IE 'to fasten.' Peace can also mean concord, harmony, amity, friendship, or just 'quiet.' Popular definitions of peace start with 'freedom from war' and go on to 'freedom from public disorder,' 'freedom from disturbance' and 'freedom from disagreement.' This is a negative way to define freedoms or peace. Peace must therefore be order or agreement.

Agreement is a going together without conflict, which is what morals are, a going together. Accord is the fitness of things considered together. Concord means of the same heart and mind. It also means agreement or harmony. Concord can mean the friendly relationship between nations or a treaty. Concord is a better word; it avoids the negative connotations of peace.

The problems of war, aggression, nationalism, disarmament, degradation, goodness and peace are problems of human nature, that is, of symbols, cultures, politics, and ecology. They are not simple problems and not easily solved. They are interrelated as human groups and natural ecosystems are interdependent ecologically, politically, economically, and technologically.

Human interactions are dominated by symbols. The most powerful set of symbols, embodied as nationalism, has a direct relationship to war. The intellectual rationalization for the continuous preparation for war is the old Roman adage: "If you want peace, prepare for war." This adage has been so completely taken into the modern heart that most of the larger nations have spent half of every century in war, according to Pitirim Sorokin. Preparation for war has always led easily into war. There seems to be no reason that the present preparation will lead anywhere else.

Starting peace means ending preparations for war. Modern communications, radio and television, could reveal the concern for peace virtually everywhere. But, they also aggravate the images of inequality. It is unrealistic to expect cooperation without some fair redistribution of resources and manufactures. Equalization would allow trust. Trust would allow many customs and prejudice barriers to fade away.

Communication through art could have a fundamental role in promoting peace. Frederich Schiller believed that there was historical proof that art can achieve what violence and law cannot—art could educate and liberate the individuals of society in a gradual and peaceful process. In spite of the cultural forces dominant at any moment, an individual has the potential to determine a different and peaceful course.

5.3. ***Getting to Good Places***

Having analyzed, or at least regarded, the whole of human history, it is possible to see long-term trends or problems. Having discussed and dissected what the human relationship to the earth actually should be and imagined goals and responsibilities, it is time to outline the physical steps that could be used to create a eutopian situation. There is always some path to a destination. There is always some way to proceed to goals. Having documented the catalogs of losses and suffering, it is necessary to act now. The first step is to strengthen the identities and boundaries of nations. We can start by offering the status of nation, with a voice and one vote, to any culture willing to participate in a new international government.

5.3.1. *Finding Workable Divisions & Sizes*

Before the advent of large-scale civilization, according to A. Keith, the habitable earth formed a mosaic of separate territories and peoples; and this grouping favored rapid evolutionary change. The size of the group depended on the fertility of the territory. In archaic peoples the group varies from 50 to 150 individuals; he calls these local breeding, competitive groups, units of evolution.

The tribe was the appropriate form of social organization for humans at the subsistence and pastoral stages. Tribes are a second step in production of evolutionary units; local bands are federated into larger units, tribes. Tribal groupings can reach a third stage, the national level, as exemplified by the Iroquois confederation or the Aztec state. Perhaps cities were followed by leagues of cities or city-states. Keith concluded that from an evolutionary point of view, city-states carry a weakness which eventually proves mortal. All go the way of Nineveh. He suggests that Egypt illustrates the weakness. Then nations followed. The national feeling requires larger loyalties and consciousness of common destiny. The territorial sense, conscious ownership of homeland, is important for human evolution; it is certainly not less important than kinship.

5.3.1.1. The Stage of Nations

The nations are a new stage. Each new nation enlarged the size of ecosystem and increased the number of interrelationships. Is the nation system maturing? Nations are primary, secondary or tertiary producers depending on whether they trade with raw materials, processed goods or manufactured goods.

A nation, by definition, has a single, central government. A national people occupy a continuous country, have consciousness of the larger unit, as well as a consciousness of samenesses and differences, and of national security. The state is a type of strong central government with a professional ruling class divorced from kinship bonds. It is stratified and internally diversified. It may have developed in a number of ways: Irrigation, warfare, population growth, trade, or integration by religions.

The people of a nation-state were given full sovereignty officially in 1648, with the Peace of Westphalia. Representative governments still use that sovereignty to claim responsibility for their actions, without recognition of any international body. Perhaps this is why so few are willing to transfer any perceived sovereignty to an international body. Nations insist on sovereignty. But the sovereignty is concerned with rights more than responsibility. Sovereign governments have the relation of anarchy between them, that is, they are without a way of governing the community of nations.

The whole land surface almost of the globe is divided into centrally governed states. Currently, human affairs are managed within the framework of autonomous national units. Individuals live within boundaries they did not choose, whose boundaries were drawn with straight lines. States often impose one life-style on their populations and reeducate citizens to accept the regime.

5.3.1.2. Limits of Nations

Lord Acton observed that nationalism aimed solely at making a nation, the abstract idea of the political state, and not at encouraging liberty or prosperity. He also predicted that the result would be moral and material ruin. Perhaps abstraction is the ruin of humanity, more than nationalism.

Many large nations are basically exploitative: The United States concentrates on Latin America; Europe on Africa; and Japan over Southeast Asia; Russia over Eastern Europe; and China over Tibet. Some of the reasons for the existence of great powers are: The rapaciousness of society; the acceptance of war; and the economic advantages of large scale operation.

The cultures of industrial nations are based on unethical accumulations of materials. The success of large nations may be due to accumulations of power, the wrong kinds of power. Power is capacity to carry out reasoned intentions, even if the reasons are irrational. Power is centrifugal now, as people have more influence over future. The power has shifted to interference with nature. Reason can live in natural landscapes without radically altering them, but irrational use destroys them. Locke declared that the negation of nature was the way to happiness. The purpose of government was to allow people freedom to produce wealth for themselves. Many nations have usurped that freedom for their leaders. All economies are upset by power relations, but the dislocations still make the wealthy wealthier.

Nations use war as a tool of politics. War has allowed the seizure of the resources of other cultures. War has helped nations grow larger through conquest and incorporation.

Large scale does confer temporary economic advantages, especially if the economic system is only short-term. The scale of nations is a problem. Some nations are too big to cope with some issues and too small to cope with others. For employment, law, ands liberties, decision-making has to be close to the personal level. However, Ervin Laszlo thinks that nations are too small for effective transborder decision-making, for labor, resources, and marketing. He thinks nations may be too small for national and environmental security. Many territorial and environmental issues are unmanageable at local or national levels. Nations have regional and global responsibilities that they seem too small to execute, especially related to regional peacekeeping. But, nations are already too big, which is why Canada sells all the forests in British Columbia, even if BC does not want to sell them, and BC sells Golden's forests, even if Golden does not wish to sell them. Nations are too big to balance the needs of communities. The perspective of great nations is ethnocentric and it fails to distinguish scale and size, which matter.

5.3.1.3. Discord Conflict & Nations

The population explosion and technicalization of human life is occurring in a framework of nationalism. The world order of nations commits those who believe in the theology to war with one another. The war ethos has been expanded and reduced to absurdity, as so many know and say. War has become so big that there can be no victories or victors, and possibly no survivors. The only remaining purpose can be the total destruction of the combatants, defined as nations, as well as most of civilization and large parts of the earth.

Strangely, the only people who do not know this—or admit it—are those in decision-making positions, who sadly are compelled to prepare for what they subconsciously must know would be a terrible disaster. Their power has made them the most secure prisoners of their nations, afraid to be caught in any criticism. Yet, they direct the money, skill and knowledge of their citizens to projects leading to misery, servitude and hideous death, and not to life, liberty and happiness.

The intellectual rationalization for the continual preparation for war is the old Roman adage: "If you want peace, prepare for war—*si vis pacem para bellum*. This adage has been so completely taken into the modern heart that most of the larger nations have spent over half of every century in war, according to Pitirim Sorokin. Preparation for war seems to lead easily into it. There seems to be no reason that the present arms race will lead anywhere else.

The economic rationalization is even more crucial. Armament piling has become a vital part of the U.S., Russian, Iranian, and Guatemalan economies. The recovery from the U.S. depressions in the 1930s was not complete until the rearmament surge to combat the Axis powers. The Korean and Vietnamese wars also spurred the U.S. economy. U.S. prosperity has its basis in the preparation for death. The fear of Russian competition ensured government expenditures of billions of dollars. The fear of terrorism has spurred spending at a greater rate. The vested interests in this system of death are almost insurmountable. The concrete companies and bomb makers put up their own puppet politicians to guarantee their part of the spoils.

Perhaps it is the fate of humanity to die a radiant death, taking most other living beings with them. Who could resist the glory of complete annihilation? But what is fate? The hand of God, or the idea of Tolstoy—that historical events are determined by the summation of innumerable decisions by the anonymous masses of humans that add up to a tendency? In particular terms, the small decisions to buy fur coats, poison coyotes, gouge oil, acquire new cars or vacation homes, donate money for the 30 million (in Robert McNamara's 1970s estimate) children who die of starvation every year, add up to fate. Modern fate certainly concentrates power in corporate, military and political hands, which change positions with startling ease. Nevertheless, this power is still in the hands of human beings, who cannot find consolation in death or total ruin. We need to stop pushing toward catastrophe. We cannot accept the existence of "absolute" weapons and genocide; we cannot accept complete destruction.

5.3.1.4. Nations & Disarmament

There is an alternative. People can ask their national leaders to stop making and selling weapons. They can ask manufacturers to stop making toys of grotesque weapons. Someone has to heal the schizophrenic breach in our moral lives, between human welfare and total destruction. Our technology provides wonderful opportunities for communication. One can see the concern for concord (or peace) in Russia, Germany, China, and South Africa on television, as well as traveling through the countries.

Furthermore, disarmament could be accomplished within a week. Taking this first step would add to the prestige of the country bold enough to do it. This is not a new idea. Earl Osborn, founder of the Institute for World Order, proposed the concept of sudden disarmament in response to the tedious phase-out envisioned by most plans. An agreement would not involve much negotiation. The United Nations could post a police force to disable all military ordinance (even hammers and welding equipment are effective on quiescent weapons). In fact, "rapid disarmament would not be difficult. A 1,000 planes each carrying 100 trained inspectors ... could distribute ... these men at all major centers in Russia and the United States within 24 hours," according to Osborn. This police force could be staggered by nationality to avoid cheating. And, the same force could be used to settle international disputes. There is precedence for using unarmed peacekeeping units to mediate between hostile groups, in Cyprus, Kashmir, and Lebanon, among others.

How could disarmament work? Humans are one of the most pacific of large animal species; they have a tremendous capacity for kindness and decency. Violence results from fear and ignorance. Violence is learned and so can be reeducated. Morals can change according to conditions, and not be defined for all time. A slow gigatrend of disarmament has already removed weapons from households and cities. The atavistic obsession with violence as a solution to problems is inimical to an ecological outlook. The image of society can change from competition and aggression to cooperation and mutual respect.

If we fear for our safety, we need only remember the success of nonviolence in India or the success of guerrilla actions in Southeast Asia and Central America. This alternative certainly may be regarded as utopian, but then the arms race is incredibly utopian, with its policy of complete destruction that no one really intends to implement. After such a war, there will be no place on earth, and that is the original meaning of utopia. Working for disarmament is working towards eutopia, the making of good places.

5.3.1.5. Decline & Breakup of Large Artificial Nations

A new nationalistic fervor helped Napoleon set up the framework for a new Holy Roman Empire, but he did not take the new nationalism seriously, and the same nationalistic feelings contributed to his downfall. Karl Marx also underestimated nationalism, in predicting that class patriotism would replace national patriotism. But a new tribal or cultural patriotism, with its smaller and more manageable focus, may be stronger than nationalism, if it is not crushed by a worker patriotism, with its purchased citizenship in multi-national corporations.

Large nations built up through a process of fusion, as smaller cultures were forced into larger national networks. As with physical elements, the fusion of too many units becomes unstable. There seems to be a process of national fission, where new nations are created out of old ones. Perhaps it is more accurate to say that old nations are left from the disintegration of new large nations. Lithuania and Latvia were old and legitimate nations before becoming part of the USSR. Kenneth Boulding states that once people have tasted sovereignty, it is hard to return to being provincial within a large nation.

Union or reunion, in the case of the many divided countries, like Korea or Germany, would be difficult because it would involve the sacrifice of power by many who enjoy it; the correction of inequities would also require sacrifice. Smaller states might have a better chance for good government. India could break into ten states and develop independently; this separatism started with Pakistan and Bangladesh.

5.3.2. *Recognizing Global & Local Difference*

Recognizing local difference is quite easy. People simply recognize differences, although not necessarily all the differences. Recognizing global differences and the difference between global and local systems is more difficult.

5.3.3. *Transitioning to Global Local Politics*

People have been predicting catastrophes and global shifts for the past 50 years. They have identified discontinuities ranging from weather patterns to disease patterns to political upheavals and collapses of alliances. The trends they identify are unsustainable. The responses are unknown or uncertain. The predictions are uncertain. So, we need some kind of direction, as well as an understanding of catastrophes.

Many kinds of catastrophes are possible and that raises questions. What are the contributing causes? How can they be changed? Other questions include how can catastrophes be diverted or lessened or stopped? Disasters regularly arise from natural and economic systems. They are part of the process. Recently, we have a new catastrophe, called the U.S. problem, the U.S. emergency, or the U.S. crisis, even though it is a human crisis or a global crisis. Other catastrophes that befell other cultures, from the Mesopotamians to the Greeks to Rapa Nuins, were generally local and did not effect the remainder of humanity. Of course, the Mediterranean was changed, as was the Fertile Crescent and many islands. Now, we are more connected. Regional and local problems can become global. Perhaps we should act as if we were wise and make changes immediately.

5.3.3.1. *Adopting a Psychology of Catastrophe*

The word catastrophe means 'down turning,' from the Greek word fragments (*kata trope*). A catastrophe is a down-turning, literally. Most catastrophes are assumed to be fast, sudden, and manageable. But, catastrophes come in all ranges of speed, size, temporality, visibility, and combinations. Our language is poor in its terms for catastrophe. A slow catastrophe could be called a bradycatastrophe. A long-term catastrophe could be called a chronocatastrophe. A large catastrophe could be called a megacatastrophe, a global catastrophe perhaps. An unseen catastrophe could be called a cryptocatastrophe. A multi-pronged catastrophe could be called a polycatastrophe. Unfortunately, catastrophes can and do occur in many combinations. Thus the loss of species by the planet might be called a polycryptobradochronomegacatastrophe.

We know of many catastrophes that have happened in the past or distance. Such catastrophes may have included an asteroid strike that ruined the dinosaur dominance or the ecological collapse of freshwater systems, but until they happen to us, knowledge of them will not be adequate to inspire change and preparation.

Widespread poverty from gross inequities may cause current, close catastrophes, such as nuclear war or biological warfare. Richer countries will need to recognize that the poverty of others is not in their interest, especially as potential markets. Inequity may never be erased. Perhaps some inequity is good and stimulating, but gross inequity needs to be limited. Famine may come home with a vengeance, with a crisis of industrialized agriculture. But famine is not the real problem. The real problem is the inability of people to perform on unbalanced diets—to behave in a human way. This covers mental retardation, laziness and other incapacities.

5.3.3.1.1. Bad Circumstances

Humankind possesses incredible scientific evidence of environmental wobble, biological imbalances, and the unfitness of entire domestic species, but knowledge moves few to action. Probably nothing will be done until catastrophes become common experiences.

Perhaps good can come from catastrophes. Eric Jantsch wonders if major catastrophes on earth mean only a weeding of a garden by evolution; the loss of many lives that may permit more beautiful flowers. On the other hand, humans made the garden they way they like it. Many catastrophes would not improve our situation through chance.

Catastrophes concentrate attention on a landscape and its people and that is their benefit on human affairs. Ideally, it should not take catastrophes to precipitate corrective measures. Instead, we might resent the necessity to change. William Catton recommends that we do not indulge in resentment. Our bad present circumstances result from the innocence and hope of our ancestors; they are the result of decisions to have babies, fires, televisions, tractors, and status. The understanding of catastrophe may let us avoid it or at least ameliorate it. Catton makes the biological analogy that die-off is a signal to overshoot, and overshoot leads to habitat damage. The agent of a post-irruption crash may be starvation, war or just behavioral stress.

If civilization collapses, the struggle back to a technological society will have greater limitations. Accessible minerals have been scattered; the gene pool has been greatly reduced. Then it may be too late. Our species may die. It is hard to image all life on earth dying. Even the worst of catastrophes would leave a simplified ecology of mosses and slugs. Weeds and invader species would prosper. Habitats would be ruled by the natural laws of ecology again. Perhaps herbivorous animals would build up large populations again. Fortunately, catastrophes do not occur in completely destructive patterns. Limited starvation occurs before total starvation. There will be uncomfortable smogs before acid rains destroy all crops. But these things could signal the necessity of immediate corrective actions.

Eric Eckholm described how economic and political pressures, which are derived ultimately from population pressures, forced farmers to intensify their efforts to increase crop production. This seems to instigate an utterly dismal cycle of population expansion, environmental deterioration and poverty: As the population expands, arable lands are used to capacity, and sometimes beyond; as the soil deteriorates, it requires more fertilizers that cause more hazardous conditions that decrease agricultural capacity; people starve, but the population increases, and marginal lands are used to meet increased demands, or food is imported from other lands.

Eckholm describes how the usage of such marginal lands can result in a dust bowl phenomena, when climatic conditions revert. Clouds of topsoil rolled east into cities from the dust bowl in the central United States in the 1930s. Afterwards, national conservation programs were able to restore some of the mythical fecundity, through pasteurizing, strip cropping, terracing, and contour plowing. However, current production efforts are causing greater losses of topsoil, and farmers are abandoning some of the conservation methods for economic reasons. Eckholm concludes that free market conditions encourage dangerous trends. The lesson of the latest dust bowl may be forgotten until the next one.

The world food supply is in danger. Almost all the continents, except North America, now have grain deficits. And as energy and capital become dearer, the likelihood of adequately feeding the world grows remote. As topsoil loss contributes to a decline of world productivity, the higher expectations on the remaining acres may impair their productivity.

This deterioration is most severe in poor countries, but its effects should concern everyone. Eckholm concludes that the United Nations must identify, analyze and marshal world resources against these trends. A scientific method will take a long time, however, and poor countries cannot wait. They must attempt a rural regeneration to stop urban drift. He interprets these trends as indicating the sinking of marginal peoples on marginal lands into a quiet helpless poverty, leading to later urban deterioration—perhaps less quiet.

The serious firewood shortage over most of the earth is also charted by Eckholm; it is considered due to population pressures on the remaining woodlands. Pressures on British woodlands in the 1300s forced people to turn to coal as a fuel source (a source then regarded as inferior). The timber famine reached Europe in the 1700s. It had existed in China and India over a thousand years before. There have been crises in clothing also, as animal hides became scarce, and then wool came into short supply as farming was expanded into grazing lands; cotton was imported from colonies, where it ruined fertile lands; now, synthetic fibers are made of petroleum products and chemicals. Each substitute required more energy to produce. Humanity has provoked a global crisis of local crises.

5.3.3.1.2. Fast Change

Most scientific studies have stated or implied that change cannot be fast, that people could not adjust, that social disruption would result, and that chaos would finish what ignorance and technology could not. The most serious drawback is the time of implementation. The Club of Rome claimed a 20-year feedback lag. *The Ecologist* cited a social inability to adapt to rapid change. Everyone assumes the time scale remaining before collapse will be long enough for their plans to be implemented. But, these studies also propose slow, long-range plans, while warning at the same time that the earth is facing imminent, drastic change. Certainly, if their plans are implemented too slowly, and if the population or pollution doubles again, surpassing some unrecognized critical level, then there will be worse disruption. Their beliefs depend on the truth of the limits to the rate of social change. As Plato recognized, it is never too late to reverse the fatal tendency towards decay, however late the hour.

It is questionable whether these time limits are absolute. For instance, war produces relatively fast, far-sighted—if wrong-headed—policies. War is actually stimulating to many individuals and often produces a national determination and purpose that peacetime does not. War is said to unite whole peoples in a common cause. But, war is a catastrophe for most living beings and for the equity of wealth.

There are other times when human beings face rapid and catastrophic change without chaos. Social systems can adjust if there are popular reasons. When a dam breaks, there is an emergency, and millions of people can mobilize to meet it. When the earth quakes or volcanoes erupt, when people flee from rival nations, there are emergencies, and they are met quickly.

There are millions of people starving every month in India and Africa, and others in Europe, the Pacific and the Americas. This is a catastrophe also, though it is slow, constant and quiet. These are emergencies and should be treated as such. Furthermore, whole species are disappearing, and whole ecosystems are wobbling, from exploitation, desertification and pollution. These are great catastrophes and much more final in character than any local one. In fact, many of the previously mentioned problems are only symptoms of these. Especially since we are part of the web of life and may be working toward our own extinction. Miners used to take canaries into the mines with them, because they are sensitive to carbon monoxide poisoning. The death of a canary was a warning. The extinction of so many species now may be such a warning to us.

The trouble with complex, self-regulating systems is that very small changes have large consequences—Reid Bryson points out that shifted rainfall patterns caused whole cultures to disappear. In some cases, where conditions, like drought, are cyclic, in the Sahelia region of Africa, humans expand during the good times, only to perish when the drought returns. In other cases, human activities, such as deforestation or overgrazing of herds, can cause weather changes. We tend to approach the limits of use of some environments. We also tend to overdependence on modern high-energy methods of agriculture and on some key resources, like water.

We need to adapt consciously to slow catastrophes. The environment is changing too fast for genetic adaptation, so our change will have to be psychological and social. Social changes can occur very rapidly when the time is right for them. Oil-producing nations, for instance, became financial equals of industrial countries within months. Pressures are building for radical change.

Change can intensify and accelerate. Small changes in different weather patterns can lead to dramatic changes in climate. Changes in climate can lead to dramatic changes in the species composition of ecosystems, as well as to human food production and housing costs.

5.3.3.1.3. Applying a Disaster Psychology

Catastrophe has its own psychology. Humankind possesses incredible scientific evidence of environmental wobble, biological imbalances, and the unfitness of many domestic species, but knowledge moves few to action. Catastrophes, on the other hand, concentrate attention forcefully. When a dam breaks, millions of people mobilize to meet the emergency. When the earthquakes or volcanoes erupt, the emergencies are met quickly. The psychological effects of a hurricane—fear of suffering and dread of loss, accompanied by exhilaration—have admirable 'side-effects;' the definiteness of danger and the immediacy arouse people to great heights of cooperation. People react admirably to catastrophe. They choose sensible directions and agree to practical expedients. War also produces relatively fast, far-sighted—if wrong-headed-policies. War is actually stimulating to many individuals and often produces a national determination and purpose that peacetime does not. War unites whole peoples in a common cause. There are times when human beings face rapid and catastrophic change without chaos. They can adjust if there are popular reasons.

Immediate foreknowledge provokes a greater response than indefinite expectations: hurricanes occur periodically, but not always in the next three hours. Probably nothing will be done until catastrophes become common experiences. As choices become more important, more urgent, errors are more disastrous. It should not take catastrophes to precipitate corrective measures. There are millions of people starving every month in India and Africa, and others in Europe, the Pacific, and the Americas. This is a catastrophe, although it is slow, constant, and distant, and perhaps because of that, it is neglected. Perhaps distance limits the ability of people to react to problems.

Furthermore, whole species are disappearing, and whole ecosystems are wobbling, from exploitation, desertification, and pollution. Environmental deterioration proceeds so slowly that the change is invisible, that is, until a catastrophic threshold is crossed. Those catastrophes, in places from the Tigris and Euphrates to northern Africa, seem to be long-term. The trouble with complex, self-regulating systems is that very small changes have large consequences—shifted rainfall patterns caused whole cultures to disappear. In some cases, where conditions, like drought, are cyclic, in the Sahel region of Africa, humans expand during the good times, only to perish when the drought returns. In other cases, human activities, such as deforestation or overgrazing of herds, can cause climatic changes. The scale and rate makes our situation seem natural, but that is because it has been a slow catastrophe, just now approaching the threshold.

But, humans should not adapt to these catastrophes. As Rene Dubos has pointed out, humanity is enormously adaptable and resilient. We could probably survive almost any physical or social conditions by adjusting to them; for example, overcrowding and smog are the norm in some areas. Hans Selye suggested that an organism is in more danger from its adaptive reactions than from external agencies. Adaptation has its own dangers. We might become less humane, less creative, and less concerned with starvation, suffering, crowding, or destruction. Our goal should not be to survive under any conditions, however difficult and unpleasant. Our goal should be to create an optimum life in an optimum environment. Perhaps this goal needs to be framed first in a humanistic or religious framework.

We do not know where to look for meaning in a cultural world, plagued with dispossessed people. Nothing is meaningful if civilization goes mad changing whole ecosystems, as Paul Shepard suggests it is. The adaptational environment is too fast for genetic adaptation. We have adapted from arboreal animal to nomadic hunter to agriculturalist to urban chemist in a short time. Little is understood about the adaptiveness of human social behavior.

If we could precipitate a disaster psychology for slow environmental or cultural catastrophes, then the priorities and motives of people might be changed. But how should everyone be convinced that there is a crisis? The change is so slow: Fewer eagles, fewer salmon, more people, and more beverage cans. The causes are so complex, and responsibility is so difficult to assign.

Unfortunately, a history or theory of catastrophes does not engage us. People need to feel situations before they act, and people will need to feel themselves as part of a delicate web of relationships before they act with ecological wisdom, as once they had to feel that the earth was round by going around it, as lately they had to see that the earth was an oasis in space by leaving it.

5.3.3.2. *Taking Immediate Action*

A transformation of world order is necessary before the next large catastrophic event. But perhaps, as many have said, the dreadful has already happened—in Hiroshima, Vietnam, Africa, and lesser known places—and the cultural transformation is already awakening or has already finished. Most cultural transformations are invisible.

Maybe the gloomy forecasts are right. But, it is the function of a Cassandra to be always wrong. To be successful, Cassandra always has to be willing to be proved wrong. If she is disbelieved and her warnings go unheeded, then her prognostications may come true. But, if she is given credence, steps are taken and policies are changed to falsify her warnings.

Political response is not enough to change the system. Many problems are social or cultural; many are personal. As people change their attitudes, social and political changes can be made. This could take generations. A violent revolution could occur within weeks, but generations could be wasted imposing patterns on personalities. Social change can spark personal change.

The future is said to be in our hands, but our hands are in the cookie jar or on our genitals, and we are afraid to let go to touch something different. Perhaps we are just insecure. Disaster is not inevitable, unless we refuse to change our ways, in which case disaster is inevitable. So, we need to change. The change is going to discomfort many, but not as many who are being discomforted by not changing directions.

How long do we have before we have to change? A decade? A hundred years? When should we start? Last minute? Last week? Next month? How should we acquire urgency? Or the political will to change or the wisdom to change in better ways? What are our long-term interests? Is it necessary to declare a state of emergency? Another alternative? Not a war against nature or injustice, which is a tired metaphor, but just a rational response to a long, slow catastrophe. The response has to be immediate action.

5.3.4. *Taking Coordinated Action on Three Levels*

After adopting an attitude towards catastrophe and starting to take action immediately, people, nations and a global framework will continue to act on those three levels. All three are necessary, but perhaps the individual is the most critical and important, if not to make changes, then to stimulate change for others and at higher levels.

This action has to include a survey of the state of the world, of all ecological and social systems. We have to make radical changes in the economic and political systems that arrange, produce and distribute wealth. We have to create a framework for world order without making the mistake of a gigantic global dictatorship, even if it is for the good of equality and environmental health. An order has to be based on bioregional models where nations are based on ethic populations in specific regions, where the operation is of interlocking hierarchical systems.

5.3.4.1. *Acting as an Individual/Community*

The redesignation of cultural nations does not involve a major revolution for most people. Most people prefer to remain in their native culture, with a geography they fit into, an economic system that reflects their values, and a religion that explains their image of place. Revolution is a false dilemma, it does not reflect possibility of thousands of individual actions on farms, factories, and families, all at local levels. Individuals can control their lives. They can make choices to be self-reliant or to limit their impact on their supporting environment. Margaret Mead noted that this is how anything has ever been done, by the actions of committed individuals.

5.3.4.1.1. Individual Steps

Of course, small steps can work. Little things matter, if the scale of the repetition is large enough, that is, if others do it also. Individuals can try to question things that seem to be automatic behaviors, such as wanting children or requiring personal transportation.

5.3.4.1.1.1. Live Frugally & Eat Well

A first step for an individual can be to live frugally, as suggested by Henryk Skolimowski. You can reduce your role as a consumer by not seeking meaning in the acquisition of things. That would result in fewer clothes, fewer tools, fewer toys, and fewer luxuries, which would force the economy to change from obsolescence and growth to development and permanence.

That does not imply being impoverished, just fewer luxuries.

Simplicity can continue with the diet, by eating lower on the food chain, with more vegetables and fruits, and by getting local food, which keeps the money circulating in your neighborhood. Reducing the consumption of meat will reduce the demand for meat and cattle, which will reduce the use of rangelands and resources for these animals.

An individual can live without many personally-owned things, such as a computer, table saw, car, or lawnmower. Many of these things can be rented for short periods of time, if they cannot be shared among the neighborhood. If you live in a city, use public transportation or cars for hire.

5.3.4.1.1.2. Make Homes, Personal Possessions & Things

In many cultures, individuals assembled their own houses. That may not be possible if the house is complex, but any individual can participate in finishing and personalizing their house. A house is often the largest item people will build or buy. There are many ways to reduce the impacts of homes, for instance, build in a place where it will not cover fertile soil or interfere with water cycles. New houses are a major source of habitat loss.

Reduce the immediate impact on its setting. A big, new tract house with a lawn that that demands water, pesticides and a fuel-consuming power-mower, can be avoided. Individuals can make their houses more efficient and naturalize their settings, with native plants in a sensible arrangement. If you have a yard, use natural plantings that require only natural water from rainfall. Arrange them sensibly around your house to cool or protect the house. Make things that you need to keep your shelter useful. Compost organic waste to renew the soil. Reduce the size of the house. Houses have been ballooning in size. All that extra space needs to be heated and illuminated, mostly by burning fossil fuels, and all that extra material needs to be extracted from the earth, then manufactured and transported.

5.3.4.1.1.3. Reduce Consumption of Materials & Space

Avoid buying things that cannot be recycled or reused easily. Avoid buying things that you do not need. Avoid paying for them with credit. Use less paper or energy at work. Reuse paper and envelopes.

Other needs, such as pets, can be shared, also. Michael Fox has suggested that shared pets would benefit many pets and reduce the number of unwanted animals—the millions of unwanted animals that are abandoned and then often disrupt wild or conservation areas.

5.3.4.1.1.3.1. Reduce Consumption of Fossil Fuels & Reduce Emissions

Walk or bicycle if you can. Get a hybrid or hydrogen-powered car if you can; otherwise try to buy a more efficient combustion engine, some of which can get 50-70 miles per gallon. Share your car with others in a carpool or pooled shopping trips, or share their cars. Take care of the car by getting it tuned and keeping tires inflated. Drive it more slowly.

Individuals can work to make changes, such as reducing the number of cars and roads, especially in roadless areas. Perhaps the number of cars could be restricted by a modified lottery system that allowed emergency vehicles and special needs vehicles and then randomized the remainder below an absolute number, which would be related to the number of roads and their condition, including crowding. Individuals could encourage a road-building moratorium; fix up the old infrastructure or replan what roads to be discontinued or changed into areas for housing or commerce.

5.3.4.1.1.3.2. Reduce Material Waste

Reduce unnecessary waste in your home. Do not heat it or cool it beyond a reasonable amount; if you want to be really cool, for instance, consider moving closer to a planetary pole. Use natural kinds of change such as open windows or evaporative coolers. Use natural light or mechanical nonelectric devices to open cans.

Reduce toxic substances inside the house. There are many potentially hazardous chemicals and chemical combinations, even with substances that are benign on their own but toxic in combination. Recycle things like aluminum, if you cannot avoid buying aluminum containers; recycle glass, batteries, and newspapers, if you cannot avoid them by reading on a computer or discussing things with neighbors.

5.3.4.1.1.4. Be Healthy

An individual can eat well and exercise regularly. Exercise can be a part of everyday living if you can walk to work or bicycle to visit friends. Reduce the amount of time sitting, especially in front of computers or televisions. Televisions and computers can be important parts of our education and communication, but they should not be the only part or the most important part.

5.3.4.1.1.4.1. Develop Your Self

As an individual, you can learn something new about your location, for instance about many of the invisible parts of your environment, from insects to elemental cycles. Develop yourself, before having children or caring for children.

Insist in pursuing a form of livelihood that is interesting and rewarding. The most important loss, according to Robert Reich, is the loss of the individual's power over their livelihood. Fewer people have the right or opportunity to choose their work, to make a living, to participate materially and meaningfully in society. They are constrained to taking opportunities present in the system, which is indifferent to people, other than as disposable, replaceable or surplus cogs.

5.3.4.1.1.4.2. Play

Individuals can engage in meaningless, unproductive activity. Do something for fun, without regard to its usefulness or rewards. Make something and take it apart. Play. Play is imaginative experience. Play is not limited by arbitrary rules and economic goals.

Play is the method of learning for most juvenile animals and a means of enjoyment for many adult animals. For humans, play is imaginative experience, entered into freely. Much human activity is play, in place in a community. Even science and philosophy are forms of play, as attempts to solve the puzzles of existence. The state of play is whole and simultaneous. The rich flowering of human nature is possible only when the constraints of need are replaced by leisure and abundance.

5.3.4.1.1.5. *Participate in Community & Culture*

Individuals need to participate in their communities. They can join conservation groups and volunteer to help with specific projects, such as saving bird or bat habitat. Work with the county or state on long-term plans for beautification or change. Planting trees works well for carbon sequestration, as well as beautifying an area and filtering pollution.

5.3.4.1.1.5.1. Work to Guide Nation & Corporations

Try to limit the power of corporations; work to revoke or amend their charters. Redefining corporations from fictional individuals to real, responsible collectives would make them more responsive to their effects and damages. Work to equalize things. For instance, try to tie politicians' salaries to the average working wage, which would certainly be an incentive for them to increase the average; and, try to tie their generous retirements to standard programs, such as Social Security in the U.S.

Make sure that political representation is more equal, with less gerrymandering by politicians for their own benefit. The Electoral College, for instance, was a good idea for a time when news was slow and people were less informed. There is no excuse for ignorance of the voters, now, so direct elections should be more effective. Try to get the constitution amended to reflect the value of wild nature and its automatic services that are incredibly expensive for us to duplicate.

5.3.4.1.1.5.2. Seek to Limit the Domination of Government & Corporations

Individuals can try to make sure that the size of corporations and governments fits the size of your community. Get commitments to invest locally to keep symbolic wealth cycling locally. Take community action to describe limits and restraints on corporations. Take legal action to correct corporate irresponsibility. Take personal action to educate others about elegant simplicity. Take political action to get representatives to address these issues, from local safety to international coordination.

Protest unhealthy actions. Write letters or make conversations about concerns. Start a dialogue with your neighbors. Take direct actions against harmful behaviors, which may be personal, community, or corporate. Get your community to be weapon free, not only from large weapons of national defense but also from personal weapons larger than knives. Get the community to observe itself.

5.3.4.1.1.6. *Reduce Conflicts with Consensus Conversation & Compromise*

Individual steps are incredibly simple. Start by listening. Accept what other people say and try to understand a different perspective. Many conflicts result from misunderstanding, and many misunderstandings can be resolved through communication.

5.3.4.1.1.6.1. Dualisms Conversations & Consensus

Many archaic cultures had formal paths to resolve conflicts, through conversations and consensus mediation. The principles of small cultures are similar: All decisions were made by consensus in which everyone participated; chiefs were not coercive rulers, but teachers and leaders with specific duties limited to their realm—medicine, war, or ceremonies for example. Cooperation and consensus, as opposed to competition and individual exaltation, permitted planning to remain informal.

Government by local meeting assumes the common sense and wisdom of the common person in an open exchange of belief and need. It requires trust and esteem. Often this kind of involvement takes more time than just voting annually or having one person make most decisions. The effect of presenting a problem before an American Indian council was to slow down response by passing it to the entire constituency and getting a consensus. This ensured due consideration. Living generations are responsible for limiting their actions within a reasonable framework of cost and irreversible change. The standard requires conversation and some consensus about the limits, which are never exact. Ecological value has to be balanced with socially optimal resource allocations that consider past and future generations as well.

5.3.4.1.1.6.2. Communicate Common Sense

Individuals can try to get other people to agree to be use more common sense. Try to get legislation changed to favor conservative measures. For instance, reducing the number of cars may be difficult. After all, the entire might and weight of advertising dollars is given to showing large vehicles ravaging the natural environment for fun—and it is fun to drive large bush-ripping, flower-crushing vehicles across the streams and the entire landscape, make no mistake—it just results in the suffering of other people or living beings.

Seek to have federal laws requiring recycling, as part of a larger use program. Saving humanity, or the planet, is ultimately a political task, and people do have power. Citizens who were concerned about the removal of ozone-depleting chemicals from cosmetic products, or the efforts to ban unnecessary herbicides and pesticides, convinced others that these were important topics, requiring legislation and marketing awareness. As the marketplace has taken over a larger percentage of the public arena, people have more clout than ever, because the marketplace is accessible to everyone, and it is vulnerable to fads as well as to safety issues. Businesses are vulnerable to much smaller trigger effects. To change business emphasis, it may take only a change in profit margin of a percent or two. Mobilization, through communities, television, or the internet, has been made easier. Everyone wants to live in a healthy environment, and to provide one for the following generations.

5.3.4.1.1.7. *Volunteer for Your Community*

Volunteering is just the formal recognition of the required unavoidable participation in one's native society. It is formal because many people are estranged from the requirements of a modern culture or have forgotten how it works, as a result of alienation. Archaic peoples knew when to help others on projects. Modern people often have the impulse, which is reflected in the many volunteer organizations, but the effort is often disconnected, uneven or inadequate.

5.3.4.1.1.7.1. Serve Your Nation

Everyone should agree to serve their nation for a two-year period, by working on educational, security, maintenance, or emergency projects. The infrastructure of a nation, its roads, bridges, and canals, often need more maintenance than can be hired out. Special amenities, such as parks or low-cost housing, are often not considered profitable enough to design or build. Although emergencies rarely visit a community more often than once every twenty years, across a nation, emergencies are daily events and critical enough to need extra labor to resolve.

Volunteering would also reduce the need for standing armies or drafts for military conflicts. This kind of service can be put to work on restoration projects, from houses and neighborhoods to conservation areas and wilderness areas.

Many other activities can serve a nation. You can also serve the nation by sponsoring laws that raise a minimum wage or issue coupons for minimum support for food, education, and health-care. Writing books and creating designs is also a service.

5.3.4.1.1.7.2. Volunteer Service for the Community
This service is common sense. You learn from the community that supports you with help and love. Each person should spend from two to four hours per week on community projects, from holding office to building centers or being a companion to the young or elderly.

5.3.4.1.1.8. *Assess Your Self-Performance*
One can always improve one's behavior. Sometimes people will consume too much or eat unhealthy foods or drink too much. The act of self-assessment can help people reduce self-destructive or harmful behavior.

It is important to keep things in perspective. Whether or not we buy a second car makes a big difference; washing out plastic grocery bags makes a smaller difference.

5.3.4.1.1.8.1. Question Things
The Union of Concerned Scientists has a handy rule-of-thumb for consuming: Ask yourself, "How big is it?" Buying an energy-efficient refrigerator reflects more on conservation than replacing the light-bulb inside it. Buying juices in cardboard containers has less impact than recycling aluminum cans. Another good test is to ask more questions: "What is this purchase supporting? Is it efficient production? Does it work towards my personal needs or goals?" Driving a fuel-efficient car or choosing nonbleached toilet paper is a decision that will be noted by the manufacturer, the distributor, the retailer and the market.

5..3.4.1.1.8.2. Act on the Answers to the Questions to Explore Changes
Asking questions is a good start, but then you should take actions based on the facts that the questions have uncovered. Actions have some effects elsewhere, even if they are not immediately apparent. Explore changes in yourself as a result of your actions. Adjust your behavior based on what you see. Maybe the actions will not have exactly the effects that you want, but you will be more likely to understand how these actions mesh with others.

5.3.4.1.2. ***Community Steps***
People live in communities. The community is responsible for the health and activities of individuals. Common values are reflected on the community level, since they are imbibed by individuals in the community. The community has distinct actions that can be taken for human culture to get to good places.

5.3.4.1.2.1. *Defining Itself as a Community or State*
Human community organization is a cultural universal. Communities are self-making. A community has a special image and set of special beliefs related to place and history, such that they are unique and cannot simply accept many of the beliefs of other communities.

A community implies that the experiencers share ways of experiencing or the same experiences. This sharing enables an individual to go beyond a finite view, to see the embedded community as one of many ways of relating the self to the universe.

5.3.4.1.2.1.1. Setting the Goals of a Community
Unconsciously or consciously, a community sets goals related to its style and size. The science of ecology to identify those goals, especially regarding impacts on wild ecosystems. Ecological

knowledge forms the basis of the normative aspects of living together, that is, ethics, and the maintenance of the affairs of a community, that is, economics and politics. There can be no separation of politics and ecology. Every economic or political act has ecological consequences and every ecological decision is an economic or political demand for control over use of the environment.

5.3.4.1.2.1.2. Implementing Ecological & Cultural Goals
Local goals are appropriate for watersheds and habitats. Educational and social goals have to originate at the local level. The following goals, are not exhaustive:

- Zone wild ecosystems first
- Measure the productivity of all ecosystems to contribute to a regional or global inventory, from which to make intelligent decisions
- Preserve all special areas, such as heritage land or old-growth areas
- Protect the long-term health, integrity, and ecological balance of ecosystems, that is, the ecological and evolutionary processes that make ecosystems
- Work with indigenous peoples to get control for them of their lands, which are often an important part of their culture
- Protect local water and air sheds
- Provide habitat for all species, since all species contribute to the functioning system, even agents of disease
- Manage commercial land for permanence and sustainability
- Provide exclusive areas for recreation and aesthetic appreciation
- Achieve true multiple use of land by strict regulation and rationing
- Set up land trusts for intergenerational protection
- Protect the health of human communities
- Open communications with all groups working in the region
- Develop paths for public participation in land use decisions
- Broaden local economies from resource extraction to invention and specialization
- Set up cooperatives to refer and share work
- Combine forest and agricultural crops where appropriate (through tree crops or permaculture)
- Diversify local institutions that deal with ecosystems
- Encourage survival of small, ecosystem-based communities
- Increase manufacturing efficiencies above 75 percent (compared to a 50% world average, 67% in Japan, and 40% in Thailand)
- Promote the full use of ecosystem products; support small-scale businesses that produce new, high-quality products
- Educate all people to feel their connections to their place, because, until they feel them, they will not act ethically or ecologically
- Educate people to realize that long-term sustainability requires healthy places, and that protecting places protects jobs
- Educate people in the science, art, and techniques of place (literacy, numeracy, ecolacy, and imagacy)
- Develop criteria for the ecologically responsible use of places, as well as standards for certifying practices and products
- Work to incorporate findings from practical and theoretical sciences into a unified

knowledge of places
- Resacralize places—desacralization is a defense mechanism against loss of meaning—by emphasizing the meaning of wild places
- Base the size and kind of a culture on the limits of the ecological place.

These goals, and others, need to be discussed and modified in a community context, through discussions and meetings.

5.3.4.1.2.1.3. Convening a Constitutional Meeting

Starting is simple; a community or group of communities in an area, for example in Eastern Washington and Northern Idaho in the U.S., could convene a constitutional meeting. The constitutional convention would work out a new government, possibly more radical than the 1972 Montana constitution, which required a local government review process, which mandated real change in the form of government, although few changes took place. The convention may suggest new boundaries of areas, according to watersheds or other boundaries. An example is in the Washington State statute of 1967.

Because of the differences in size, and the need for even the smallest community to be enfranchised, it may be useful to adopt a floterial district system as in Idaho. A representative could represent other communities.

The convention could draw on the cultural depth of the region. Goals that have been only thoughts or hopes could become explicit expressions. This kind of change would depend on the flexibility of the federal government to accept the reformation of some of its states or regions. If the communities decided, they could create a new nation.

The new nation could join the Unrepresented Nations and Peoples Organization (UNPO), which promotes the respect of human rights of all peoples through nonviolent tolerance and self-determination. UNPO offers an international forum for nations and peoples whose causes are not addressed by the existing UN or by existing bodies. UNPO provides assistance to 26 groups, representing 50 million people, including the peoples of Mari and Tibet. The new nation could give recognition by other unrepresented nations, such as Scotland or the Karen. The nation could apply for recognition as an independent nation through the United Nations or other new international organization (perhaps Associated Nations).

5.3.4.1.2.2. *Balancing Budgets for Community or State*

A community needs a budget to limit its wealth to actual quantities. A budget generally refers to a list of all planned expenses and revenues. A budget is an important concept in microeconomics, which uses a budget line to illustrate the trade-offs between two or more directions of development.

The budget of a government is a summary or plan of the intended revenues and expenditures of that government. A budget is usually prepared by a separate department and then submitted to the legislature for consideration, changes and approval. Any budget should be required to be balanced in the long run.

As the technology of money and banking continues to make spending easier, individuals, families, companies, and governments need to adopt more effective budgeting methods, strategies, and tools to balance their outflows to their inflows and to avoid deficit, or debt-based, spending.

Annual or multi-year budgets could be replaced or complemented by monthly forecasts or rolling forecasts. Monthly forecasts provide more up-to-date financial plans. Many large businesses reforecast their original budgets on a quarterly basis. As months pass, the actual income achieved and expenses incurred can be compared to the budget and forecasts. Variances between these financial plans and actual delivery can then be analyzed to provide information that can improve performance.

5.3.4.1.2.2.1. Income for Community or State

Income is the form of usable or tradable wealth that is used by the community to pay for public expenses and amenities. As Herman Daly and John Cobb point out, the means of raising community funds would also be the means of attaining personal and community goals. Taxes could be ways of internalizing costs.

5.3.4.1.2.2.1.1. Taxes for Community or State

This term 'tax' in its most extended sense includes all contributions imposed by the government upon individuals for the services of the state, by whatever name taxes are called, whether it is tribute, tithe, talliage, impost, duty, gabel, custom, subsidy, aid, supply, excise, or other name. Taxes are any charge of money or property imposed by a government upon individuals or entities that are within the government's authority to assess such charges. This term, however, generally does not include charges imposed in exchange for the provision of specific goods or services, such as bridge tolls or sanitation fees. Although most modern taxes are levied on the basis of economic measurements such as income, consumption, property, and wealth, some governments also impose excise taxes or other taxes.

5.3.4.1.2.2.1.1.0.1. *Traditional Kinds of Tax*

Many traditional taxes require complex schemes to avoid cascading taxes or the regressive unfairness of some taxes. Many taxes were *ad hoc* additions to a tax code to add income from new or overlooked profitable activities.

State of Being Taxes, or Title Taxes. An example of a 'state of being' tax is an *ad valorem* property tax, which is not an excise tax, but which may be imposed on the property or the person who owns that property at a certain moment of time, for example, on July 1st of each year based on the state of title at that given moment. The 'state of title,' state of ownership, or property by reason of its ownership, is what is being taxed. The next year, on July 1st, another such tax is imposed again in the same way on the same property and person, even though there has been no 'change' or intervening 'event.' The amount of the tax may change from year to year, based on the change in the value of the property or a change in the tax rate, or both, but those are separate issues governing how the tax is computed. What is being taxed, fundamentally, is the state of title, which is not an 'event' but instead a 'state of being.'

Event Taxes. For purposes of the U.S. constitution, for example, an excise is essentially any indirect tax, or 'event' tax. In the constitutional sense, an excise includes gift taxes, estate taxes, payroll taxes, sales taxes, miscellaneous excise taxes, and income taxes on any income other than income from property—in short, any tax that is not a direct tax. Excise taxes are taxes paid when purchases are made on a specific good, such as gasoline, which is often one of the major components of an excise program. Excise taxes are usually included in the price of the product. There are also excise taxes on activities, such as on wagering or on highway usage by trucks. An excise tax can have several general excise tax programs.

Excise duties usually have one of two purposes: To raise revenue or to discourage a specific behavior. Taxes such as those on sales of fuel, alcohol and tobacco are often justified on both grounds. Some economists suggest that the optimal revenue-raising taxes should be levied on the sales of items having an inelastic demand, while behavior-altering taxes should be levied where the demand is elastic.

Capitation Tax. The capitation tax or "head" tax is an assessment levied by the government upon a person at a fixed rate regardless of income or worth. A poll tax is an example of a capitation tax. Originally used as a duty on the importation of convicts and slaves, it was later used to discourage the poor from voting.

Severance Taxes. A severance tax, on items that are 'severed' from their context, would have the effect of limiting the use of nonrenewable resources, such as coal or oil, as well as slowly renewable resources, such as forests.

Income Taxes. In 1913, the Sixteenth Amendment to the U.S. Constitution was ratified. It empowered Congress to tax "incomes, from whatever source derived, without apportionment among the several States, and without regard to any census or enumeration." The Internal Revenue Code is today embodied as Title 26 of the United States Code and is a direct descendant of the income tax act passed in 1913. While some U.S. states, such as Nevada or Florida, do not have an income tax, all residents and citizens of the United States are subject to a federal income tax, which requires a minimum level of income.

Sales Taxes. A sales tax is a tax on the consumption of goods or services. Normally, it is a certain percentage that is added onto the price of a good or service when it is purchased. To avoid double taxation, or 'cascading' taxes, where an item is taxed from design and production to retail sale and resale, a sales tax should charged ideally only once on any one item. A retail sales tax is charged only on retail transactions, not on businesses buying raw materials for production or on businesses reselling finished goods.

A related type of sales tax is the *value-added tax* (VAT). All sales, retail and wholesale, are taxed in this scheme. To avoid multiple or cascading tax, every business is refunded the amount of VAT remitted by their suppliers. Most nations, especially European nations such as Norway or Ireland, have sales taxes or VATs at the national or local level.

5.3.4.1.2.2.1.1.0.2. Effects of Traditional Taxes.

The combination of these taxes often makes them regressive. In terms of fairness, sales taxes are generally regressive, that is, poorer people tend to pay a greater percentage of their income on sales taxes than richer people, because they generally tend to spend a far higher percentage of their income for food, clothing, shelter, and medical care. In some locations, items such as food, clothing, or prescription drugs are exempt from sales taxes, ostensibly to alleviate the burden on the poor. Some of these exemptions, such as exemptions for clothing or prescription drugs, actually may actually have the opposite effect and make the tax more regressive, since poorer individuals may spend a smaller percentage of their incomes on these items than do richer individuals. Some of these taxes discourage the efforts that they tax. For instance, a tax on value added by labor to commodities can discourage labor.

5.3.4.1.2.2.1.1.0.3. New Kinds of Taxes for Community or State

When Arthur C. Pigou introduced the concept of 'taxing bads' in the early 1900s, economists and politicians discussed with the idea, but its use was limited. Environmental taxes are functionally nonexistent in the current tax code of the United States. A few such taxes have been proposed, most recently a BTU tax during the Clinton presidency, but the defeat seems to have discouraged any kind of environmental tax since then.

Taxing 'bads' is essentially a 'tax shift.' Taxes are reduced on things that should be to be encouraged, such as work or savings. The loss of revenue is compensated by taxes on 'bads,' that is, on things to be discouraged, such as pollution and waste. A tax shift could help mitigate the impact of pollution charges on some businesses; it would remove taxes that discriminate against low-income families. A tax shift might also improve general fiscal health, as well as over-all environmental health.

Rather than use old labels, which were the result of *ad hoc* additions to many tax codes, it might be simpler to present them as what they are taxing, that is what kinds of income we want and what kinds of behavior we want to encourage. These kinds of new taxes can be put under four categories: Use, Loss, Adjustment—or the misplacement of resources, such as pollution—and Distribution. Several of these might be regarded as temporary, until a permanent conservative system is developed.

5.3.4.1.2.2.1.1.1. *Use Taxes for Community or State*

A use tax is similar to many consumption taxes or a few severance taxes. These taxes take a percentage for a service for any resource. A use tax would have the effect of limiting the use of nonrenewable resources, such as coal or oil, as well as the use of slowly renewable resources, such as forests. The rate of the tax would be related to the scale of the economy, as well as to the carrying capacity of the ecological support system.

Tax collectors would monitor points of entry, from wellheads to forests, to ensure that the tax would be fair, and that it would be paid. The tax would be easier to collect and harder to avoid than income taxes. It could be, and probably would be, included in the cost of any commodity that used the resource. Few commodities do not require some resource.

5.3.4.1.2.2.1.1.1.1. *Air Use Tax for Community or State*

Taxes on output or on polluting inputs are called presumptive because their target is the pollution presumed to be associated with the activity. A combination of the two types, those that reduce output and those that reduce emissions per unit of output, could replace emission fees and reduce monitoring requirements and their prohibitive costs. The taxation of fuel use can be a powerful, indirect instrument for controlling air pollution because of the direct connection between fuel use and emissions. This is not quite the same as taxing the pollution; in this case the goods are taxed to address the pollution.

Indirect instruments, such as a tax on air, are designed to reduce the scale of output, as important complementary measures in a program of cost-effective pollution control. Air is used in almost every operation of an industry or bureaucracy. Clean air is used for cooling and to provide oxygen for the occupants of buildings.

5.3.4.1.2.2.1.1.1.2. *Water Use Tax for Community*

Water is such a good solvent, it can be used to enhance many kinds of processes. Water is used in almost every operation of an industry or bureaucracy. It also is used for cooling and for drinking.

Bottled water consumption, which has more than doubled globally in the last six years since 1999, is a natural resource that is greatly affecting the world's ecosystem, according to a new U.S. study. "Even in areas where tap water is safe to drink, demand for bottled water is increasing, producing unnecessary garbage and consuming vast quantities of energy," according to Emily Arnold, author of the study published by the Earth Policy Institute, a Washington-based environmental group.

A water use tax would encourage efficient use of water by making waste too expensive.

5.3.4.1.2.2.1.1.1.3. *Land Use Tax For Community*
Land use tax is a base tax on land, which would encourage efficient use of the land, as opposed to, for instance, building taxes, which if higher than tax on land allows land to lie idle or buildings to deteriorate. It would reduce runaway and harmful speculation on land. Users fees would have to be high enough to cover the costs of the services provided, according to Daly and Cobb. Cities that increased land taxes and decreased building taxes have encouraged better building programs.

This tax would tax the use of soils by agriculture or forestry, especially rapid-use forestry with short rotations (or perhaps on less than 500-year cycles). Agricultural land should also be taxed on its unimproved value, according to Daly and Cobb. The taxes would be local rather than national. Improved agriculture, such as regenerative, would prove to be more efficient as a result of the higher initial cost and the lower taxes on improvements or profits.

5.3.4.1.2.2.1.1.1.4. *Element Use Tax for Community*
Under the Clean Air Act, the U.S. Environmental Protection Agency (EPA) sets national standards for ambient air quality designed to protect public health and welfare. The EPA defines acceptable levels for six 'criteria' elements that are air pollutants: Sulfur dioxide (SO_2), nitrogen oxides (NO_x), ozone, particulate matter, carbon monoxide (CO), and lead. Along with emissions from natural sources, emissions of air pollutants from stationary sources, such as industrial facilities and commercial operations, and mobile sources, such as automobiles, trains, and airplanes, contribute to the ambient levels of those pollutants.

Some U.S. states levy a 'severance tax' on mineral production because these assets are nonrenewable. Once the minerals are produced, or 'severed,' from the ground, they are gone forever. Since many of these assets, such as coal and oil, come from federal lands in the U.S., which are owned by all the citizens of the states and the nation, citizens are entitled to a fair rate of return on the assets. An element tax could capture the value of these assets for current and future generations of citizens.

5.3.4.1.2.2.1.1.1.4.1. *Carbon Tax for Community*
Carbon appears in every living cell; life on earth is carbon based. However, out of place, carbon can poison life or cause changes to the climate of the planet. Carbon dioxide is recognized as a greenhouse gas, and increasing levels in the atmosphere are linked to global warming and climate change.

A carbon footprint is a measure of the amount of carbon dioxide emitted through the combustion of fossil fuels. In the case of an organization, business or enterprise, a carbon footprint is part of their everyday operations, as it is for an individual or household. A carbon footprint is often expressed as tons of carbon dioxide or tons of carbon emitted, usually on an annual basis.

A carbon tax would link the effects of burning carbon-containing fuels to the cost of repairing the damages from burning. A carbon tax could greatly offset other tax rates, as well as reduce energy consumption, especially fossil fuel consumption, and address climate change. Although there would probably be a small reduction in measures of wealth, such as the GNP or ISEW, there would be offsetting tax reductions elsewhere in the economy; and some of the money would go toward improving efficiency.

Displacing these old taxes with use taxes would improve the productivity of the economy. A high carbon tax, coupled with reduced tax rates on income and profits could generate a significant gain for each dollar of tax shifted. The gain would come not only from improved economic efficiency, but from reduced investment in infrastructure, in reduced operating costs due to higher energy efficiency, and in reduced environmental damage.

5.3.4.1.2.2.1.1.1.4.2. *Nitrogen Tax for Community*
Nitrogen is a critical element of life; it is a major part of the atmosphere. In the form of nitrite, nitrate and ammonia, nitrogen enters watershed, oceans and streams from fertilizers, animal wastes and decomposing organic matter. If nitrogen levels increase, algae increases dramatically. Nitrogen oxides (NO_x) usually enter the air as the result of high-temperature combustion processes such as those in automobiles and power plants.

One way to help control NO_x would be to tax emissions from stationary sources. For example, firms might adopt currently available abatement techniques whose capitalized costs are lower than the tax they would otherwise pay. If a regional allowance trading program was put into place, the community could tax only the stationary sources of NO_x that do not participate in the program.

5.3.4.1.2.2.1.1.1.4.3. *Sulfur Tax for Community*
Sulfur is also an element required by living beings, and it also appears in the atmosphere. Sulfur dioxide belongs to the family of sulfur oxide gases formed during the burning of fuel, mainly coal and oil, containing sulfur and during the operation of metal smelting and other industrial processes. Exposure to high concentrations of SO_2 may promote respiratory illnesses or aggravate cardiovascular disease. In addition, SO_2 and NO_x emissions are considered the main contributors to acid rain, which can degrade surface waters, damage forests and crops, and accelerate the corrosion of buildings.

One option is to tax emissions of SO_2 from stationary sources not already covered under an acid rain program. Marketplace strategies can be coupled with regulations.

5.3.4.1.2.2.1.1.1.4.4. *Other Element Taxes for Community*
Phosphorus is a relatively rare element. Isaac Asimov calls it the bottleneck of life, a good example of Liebig's law of the minimum, which states essentially that something has to be in least supply. In the case of life, it is phosphorus. Phosphorus is a necessary constituent of protoplasm, but it is sometimes the element in least supply. Most phosphorus is locked up in phosphates in rock, which is weathered and 'escapes' to the oceans, which act as a sink. Sea birds play a large part in returning it to land. In fact, guano deposits are mined for fertilizers. Harvesting fish for food returns some to land, but little effort is made to recycle it. A community tax on phosphorus would encourage more of it to be recycled.

Oxygen is a major component of the earth's crust. It is a significant component of the atmosphere. For living beings, it allows energy to be released as part of their respiration. In the atmosphere, oxygen can be converted by natural processes, such as sunlight or lightning, to ozone. Ozone is not emitted directly into the air but is formed by the reaction of volatile organic compounds (VOCs) and nitrogen oxides (NO_x) in the presence of heat and sunlight. Ozone occurs naturally in the upper atmosphere and provides a protective layer there. At ground level, however, ozone is a prime ingredient of smog. Short-term exposures, from one to three hours, to ambient ozone concentrations have been linked to increased hospital admissions and emergency room visits for respiratory ailments. Ground-level ozone has remained a pervasive pollution problem in many areas. Rather than directly tax oxygen use, communities might want to tax the VOCs or chemicals that interfere with ozone.

The advantage of a broad-based tax on VOCs is that it would affect large and small sources of the compounds. A disadvantage of such a broad-based tax is that it may be regressive. To the extent that the tax raises the prices of consumer goods, it may take up a larger share of household income for low-income consumers than for higher-income consumers.

5.3.4.1.2.2.1.1.1.4.5. *Species Use Taxes for Community*

Tax would be charged on the taking of members of any species; this tax is not the same as a hunting or fishing license, which is required to test knowledge and dictate that only a certain number can be taken. Many of these species are slowly renewable or functionally nonrenewable. This tax would work to encourage preservation, take-limits, or take-efficiency. This would tax the use of virgin species for commerce.

Species, as well as fresh air and water, are the interest of an ecosystem; the ecosystem itself, on life-bearing land, is the capital. If the capital is allowed to be used, the taxes on the land should be very high. Using the interest is a more sustainable practice.

5.3.4.1.2.2.1.1.2. *Loss Taxes for Community*

A loss tax is a tax on losses from the capital base, that is, it is a tax on the destruction of resources, not just their use or on their negative impacts. It is similar to a dispossession tax or a capital depletion tax.

Resources can be sorted into one of three types of resources: (1) fastly accruable, that is, renewable in economic terms over a short time horizon; (2) slowly accruable, which are basically nonrenewable within a human lifetime; and (3) and slowly dispersed, which are really nonrenewable in most economic frames, but are actually renewable in geological time, which is rarely considered. Most of the wealth used by modern economies is nonrenewable. These resources are limited, interrelated and distributed unevenly. Forests are a special problem; although trees can grow to a good size in 30-40 years, forest ecosystems may take 300-600 years to develop and then last for thousands of years.

Oil, coal, peat, and some woods are functionally nonrenewable. Geological time periods are required to produce them. Mineral reserves may be understated, but, they may be located so far and so deep that it would cost more energy to extract them and move them than they are worth, unless we used a 'renewable' energy source, such annual plants, or a very slowly depletable resource, such as the sun.

Slow accrual and slow dispersal resources should be equally available to all cultures. It is impossible to sustain any quantitative arguments about resources and population pressure on them without a comprehensive overview. Demands on food, fertilizer, energy, and metals are related inseparably. Organic and inorganic assets need to be assessed together. Population carrying capacity cannot be formulated until both resources have been quantified.

This kind of loss tax is on things or processes that interfere with other things and processes, things that cause runaway feedback or the destruction of cycles, things in other words that reduce our continued use of and enjoyment of the earth. This tax would have the effect of internalizing both ecological and social costs; since all consumers would be paying the real costs, no consumers would be protected. It should also have the effect of reducing pollution.

Many of these things have been subsidized for many decades as a result of special interests. The purpose of this kind of tax is to change behavior that depletes resources and discourages labor. It also can pay for the damage caused by misplaced or used resources.

5.3.4.1.2.2.1.1.2.1. *Land Conversion Loss Taxes for Community*
This tax would be on the conversion of complex systems to simpler, and more expensive to manage, systems, for instance, the conversion of forests to agricultural fields, or the conversion of fields to parking lots. The rate of taxation would be directly related to the cost of restoration plus the loss of productivity over the period of time before restoration.

5.3.4.1.2.2.1.1.2.2. *Nonrenewables Loss Taxes for Community*
Nonrenewable means functionally nonrenewable in a human lifetime, although technically such resources are renewable in the very long term as a result of very slow processes, such as continental drift and folding. Nonrenewables includes geothermal energy, as well as fossil fuels, such as oil and coal. The rate of taxation would be related to the cost of human production of those things.

5.3.4.1.2.2.1.1.2.2.1. *Geothermal Loss Taxes for Community*
Geothermal energy is the long-time flow of the energy of formation of the planet to space; it is not inexhaustible, although it may not be used up within the lifetime of the human species. It should be taxed at the community level because it is inequitably distributed, and to provide support for other energy sources that require more research and development

5.3.4.1.2.2.1.1.2.2.2. *Fossil Fuel Loss Taxes for Community*
In a way, energy does not cycle because it is lost in equal quantities from the planetary system; however, energy may be trapped by cycles on the earth for millions of years—in coal or oil, for instance. Oil, coal, peat, and some woods are functionally nonrenewable. Geological time periods are required to produce them. Assuming 1×10^{16} grams of carbon are fixed each year by photosynthesis, averaged over a billion years, the total mass recycled of carbon alone exceeds 10^{25} grams. The total coal and oil deposits in lithosphere are estimated at 10^{19} grams. However, a small amount of detritus does not get recycled; these ordered compounds form deposits. Marine organisms produced large deposits of oil, possibly in a reducing atmosphere; forests left ranges of coal.

Oil and tar may have been first extracted near the Zagros Mountains to be used for binding tools. Since then, oil production really has meant extraction. That is the only economic means to get it. It would be expensive to synthesize. F. de Chardenedes wrote a scenario of natural technology for producing petroleum; if the amount of energy, as heat and pressure, had to be paid for at a human public utilities rate (circa 1970), the cost would be over a million dollars a gallon. If we based the cost on human labor and technology, so that 1 gallon of oil is worth a million dollars, as Buckminster Fuller also calculated, then it would be used with more circumspection.

As technology improved, wood fires also shaped metals and provided steam for power; coal fires provided electricity; oil fires provide electricity and motion. Furthermore, oil and gas can damage and destabilize ecosystems.[7] Oil use places extra carbon and pollutants in the air. A tax on oil would reduce its extraction and use, as well as encourage renewable, cheaper alternatives.

5.3.4.1.2.2.1.1.2.2.3. *Natural Gas Loss Taxes for Community*
Natural gas could be taxed at a much higher rate, a tax of 25 percent instead of 5 percent of the market value of the natural gas. This higher rate would discourage the use of natural gas, and to encourage the development of alternative and relatively renewable sources, such as solar energy, although the sun has a finite lifetime also. Natural gas reserves could be used as a transitional energy source or for special needs.

5.3.4.1.2.2.1.1.2.2.4. *Coal Loss Taxes for Community*
Coal would also be taxed at a higher rate, a 25 percent tax based on the sales price per ton, instead of a one-cent-per-ton or a dollar-a-ton or a small percentage of the sales price. This should cover, not only the cost of restoring the landscapes or mines, but also the medical costs, such as black lung, associated with the mining of coal.

5.3.4.1.2.2.1.1.2.3. *Slow Renewables Loss Taxes for Community*
These resources are essentially nonrenewable in terms of a human lifetime. Although some trees can mature in forty-year cycles, many require eighty, a hundred years or over three hundred. Forests themselves may take many hundreds of years to mature.

Although individual fish can mature in one to four years, fish populations require much longer time periods. Fish populations are also vulnerable with relatively low percentages as takes.

5.3.4.1.2.2.1.1.2.3.1. *Trees & Forests Loss Taxes for Community*
Taxation can be a controversial field. Many states or provinces have special private forest tax laws that exempt timber from taxation or defer the tax until harvest. Economists fight over whether the bare land is the capital and timber a 'goods-in-process,' or whether the timber is the accumulated capital. One could even state that the forest ecosystem is the capital, not just the timber. Trees, as well as other species, fresh air and water, are the interest on a forest; the whole forest itself is the capital.

In the U.S., the Oregon Legislature intended to tax timber like any other crop, while forest land would be taxed at reduced values, under an overhaul of timber taxation that cleared both houses. Constanza and Daly suggest a natural capital depletion tax could be applied to forest use that is not biologically sustainable. Sustainability here is used to mean no extrinsic costs, that is, we cannot sustain forest use at the expense of clean air and water, at the disruption of processes, and with the destruction of plant and animal communities. This would tax the use of virgin materials for commerce. Both ideas would improve forest use, but could be extended to other ecosystems.

Forest land should be taxed on its land value. Since the forest is the capital, and trees characterize the forest, the depletion tax on trees needs to be high. If enough trees are removed the forest dies. The tax should be applied so that the benefits stay local.

5.3.4.1.2.2.1.1.2.3.2. *Fish & Fisheries Loss Taxes for Community*
Wildlife has commercial value, that is, fish can be caught without being produced. The constancy of the environment, even with its changes, has allowed many kinds of animals to live in large populations. Fish have quite explicit needs for temperature, stream bed character, and cleanliness. Taxing fish takes should allow populations and habitats to recover. The community must be restored to health. This means balancing human needs with fish needs

in a sustainable pattern. Some fish might never be hunted or taken, especially those with high ecosystem or cultural value.

5.3.4.1.2.2.1.1.2.4. *Fast Renewables Loss Taxes for Community*
Fast renewables are things that are usually part of daily, monthly or annual cycles, such as solar power, hydropower or wind power. They can be taxed at a relatively low rate or even a negative rate to encourage development.

Large-scale use of any of these methods of production could result in other problems. Other than the aesthetics of giant wind or solar 'farms,' concentration could cause problems with ecosystem interference and stability, as well as with human safety and aesthetics.

5.3.4.1.2.2.1.1.2.4.1 *Wind Power Loss Taxes*
Wind is a term for the movement of air, usually the result of heat differences in the air and planetary rotation. Although wind energy is not infinite, it will continue for many millions of years. The purpose of taxing it is recognize its values and limits, as well as to shift to more benign energy sources. Wind power could have a negative tax (or tax credit) of 2 cents per kilowatt-hour for electricity produced from a wind farm.

5.3.4.1.2.2.1.1.2.4.2. *Solar Power Loss Taxes*
Energy from the sun drives most cycles on earth. Even though solar energy is finite, it will continue, and increase slowly, for billions of years. Solar power could have a negative tax of 3 cents per kilowatt-hour for electricity produced from arrays of solar collectors. There would be tax credits for solar domestic or business water heating.

5.3.4.1.2.2.1.1.2.4.3. *Hydropower Loss Taxes*
The earth-moon system generates tides on both bodies. The water cycle on earth pumps water into the atmosphere and its descent can also generate power. Hydropower would have a negative tax of 1 cent per kilowatt-hour for electricity produced from large or micro generators.

5.3.4.1.2.2.1.1.2.5. *Waste Loss Taxes for Community*
Some waste can be considered a measure of loss from the system, that is, resources that are not captured for reuse. Many wastes come from industrial, agricultural, or mining operations and include toxic substances. A lot of waste comes from private households, in the form of paint, pesticides, and poisons. Many wastes, from solid wastes to energy, would be taxed by weight or neutralization costs and at amounts necessary to discourage their use or to encourage their incorporation into industrial cycles.

5.3.4.1.2.2.1.1.3. *Adjustment Taxes for Community*
Adjustment taxes are regulation or correction taxes to regulate or harmonize cures and causes. They are not applied to loss but to meet the cost of recovery. For example, water returned to the watershed that does not meet purity standards would be taxed. Many forms of solid waste, such as garbage, refuse, or sludge, could be taxed by volume, at a rate of 50 cents per cubic foot. Other forms of solid waste, such as hazardous waste, tires, batteries, or nuclear by-products would be taxed sufficiently to pay for their isolation or break-down. Recyclable solid waste, such as animal excrement or rock, would not be taxed if it was recycled properly.

5.3.4.1.2.2.1.1.3.1. *Sin Adjustment Taxes for Community*
If people do dangerous things, and expect their government to take care of them, then government has to tax those things that result in illness. This tax would benefit people who suffer from bad habits. Due to its relatively high rate, it might reduce the consumption of addictive substances. The money from these taxes would address human consumption requiring health care, especially for health care beyond what the individual can afford.

5.3.4.1.2.2.1.1.3.1.1. *Cigarette Adjustment Tax for Community*
Inhaling the smoke from burning in plants in moderation may be stimulating, especially in ceremonial settings. However, overuse has serious health consequences. An industry exists that makes profits by selling such things to users and addicts.

Like an excise tax, this tax is on sales of cigarettes, usually a fixed fee on each pack of cigarettes sold. This tax can double or even triple the retail cost of cigarettes in some U.S. states.

5.3.4.1.2.2.1.1.3.1.2. *Alcohol Adjustment Tax for Community*
Drinking the products of fermented plants may also be stimulating. It certainly has a long, rich tradition, from native peoples eating fermented berries to the large-scale production of ale in Mesopotamia after 9,000 years ago. This relatively high tax would offset the increased medical and social expenses from its use.

5.3.4.1.2.2.1.1.3.1.3. *Drug Adjustment Tax for Community*
Many communities try to control the kinds of drugs used. Many drugs, that may have greater or lesser consequences than those categorized as legal, may be illegal. Legal drugs would be taxed according to their purpose or perceived social benefit. Illegal taxes would be taxed at a much higher rate.

However, if all substances were legalized, then taxes could be used to control their use. There could be additional social benefits, such as decriminalization and deimprisonment, and all the costs that that entails.

5.3.4.1.2.2.1.1.3.2. *Pollution Adjustment Taxes for Community*
All forms of pollution would be taxed, especially those related to regional and global cycles of the elements necessary for metabolism of living beings. The amount of pollution that a firm or product releases into the air, water, or soil would be taxed. A pollution tax would address the 'market failures' that arise when businesses and consumers are not held responsible for the full health and environmental costs associated with their activities. Pollution taxes make polluters pay for their damages and incorporate these costs into their decisions and product prices (Specific elements of pollution, such as carbon and sulfur, are addressed in Section 5.3.4.2.15.1.1.3.2.1.).

Water pollutants, for example, could be taxed on the basis of Biological Oxygen Demand (BOD). Dissolved oxygen is necessary to sustain fish and other aquatic life. Generally, firms that are subject to water pollution standards do not pay taxes or fees based on effluents that regulations still allow. Most of the high-volume BOD dischargers, referred to as point sources, are publicly owned treatment works, paper and pulp mills, food processors, metal producers, and chemical plants.

5.3.4.1.2.2.1.1.3.2.1. *Industrial Adjustment Taxes for Community*
Particulate matter is the general term used for a mixture of solid particles and liquid droplets found in the air. Those particles come in a wide range of sizes: Fine particles are less than 2.5 micrometers in diameter, and coarse particles are larger than that. The particles originate from many different stationary and mobile sources, as well as from natural sources. Fine particles result from the combustion of fuel in motor vehicles, power generation, and industrial facilities as well as from residential fireplaces and wood stoves. Coarse particles are generally emitted from power plants, factories and sources such as vehicles traveling on unpaved roads, materials handling, crushing and grinding operations, and windblown dust. Some particles are emitted directly from their sources such as smokestacks and cars. In other cases, sulfur dioxide (SO_2), nitrogen oxides (NO_x), and volatile organic compounds (VOCs) interact with other compounds to form particles.

A tax on coarse particles could force some electric utilities and manufacturing plants to install improved electrostatic precipitators, wet scrubbers, or other equipment to reduce their emissions and lower their tax burden. Reductions in emissions caused by the tax would be economically efficient if the additional abatement costs were less than the social benefits from reduced pollution.

Opponents of this kind of tax argue that it would impose an excessive burden on firms that already incur costs to comply with current standards. Firms have escaped their justified burdens of safety and efficiency, so this is not a good reason. To the extent that the tax would raise the price of energy generated in this way, it might be regressive. But, once the price of energy reflects its true costs and effects, and once many unfair taxes are removed, then some disadvantaged people and businesses can afford to pay a larger percentage for energy costs.

5.3.4.1.2.2.1.1.3.2.2. *Agricultural Adjustment Taxes*
In order to make its impressive gains, agriculture tries to control more of the growing process of crops; it does this by adding fertilizers and pesticides, as well as by using special, energy-intensive machinery. These things would be taxed to offset their costs and effects.

5.3.4.1.2.2.1.1.3.2.2.1. *Pesticides Adjustment Tax*
Pesticides can have a number of adverse impacts on human health and the environment, imposing substantial costs on society. These include direct financial costs, such as the treatment of water, and wider environmental costs, such as loss of biodiversity, which are much harder to value.

Pesticide impact reduction is a prime candidate for the use of economic instruments. A pesticide tax would greatly benefit farmland birds. The tax could invoke the 'Polluter Pays' principle and reduce the inappropriate use of pesticides, as well as raise revenue to pay for solutions to the environmental impacts of pesticides. Preliminary economic analysis shows that a well designed tax would not significantly damage farming incomes or competitiveness.

5.3.4.1.2.2.1.1.3.2.2.2. *Fertilizer Adjustment Tax*
Fertilizer is used to increase the productivity of crops, although rarely is that productivity equal to the that of the original land cover. Fertilizers rob plants of root mass and foods of nutritional values. This means that more of the crops have to be planted, and people that have to eat more of the plants to get the same nutrition. Fertilizer pollution causes other

problems, especially with nitrogen and phosphorus.

The application of environmental tax shifting to pesticide and fertilizer use could provide an incentive to minimize pesticide and chemical fertilizer use, and to generate revenue to allow for tax relief in other areas.

5.3.4.1.2.2.1.1.3.2.2.3. *Water & Other Adjustment Taxes*

To encourage crops to grow in areas that have low rainfall, such as central Washington state in the U.S., farmers irrigate their crops, even crops that are not native to the area, with water drawn from an aquifer. Since water from aquifers is often renewed at very low rates, it would be taxed at a rate designed to reduce use to replacement rates.

5.3.4.1.2.2.1.1.3.2.3. *Personal Pollution Adjustment Taxes for Community*

A tax would be charged on personal pollution, from solid waste to discharges from machines. A fuel tax would be imposed on the sale of fuel to cover the calculated pollution from its use. In the United States, the funds are often dedicated to transportation or roads, so the fuel tax is considered to be a user fee. In other countries, the fuel tax is a source of general revenue. Sometimes the fuel tax is not imposed on fuel that is not intended for transportation, such as fuel used to power agricultural vehicles or for home heating oil. This creates an economic incentive for the illegal use of fuel. The solution is to tax all fuel regardless of use.

Because of the inelastic nature of the demand for fuel, in the short run the tax would be an effective source of revenue. In the long run, however, people adjust their consumption; over a period of years, people would consume less as the price increases, by driving less or by buying more fuel-efficient cars or furnaces. A fuel tax could be a way to reduce reliance on environment-damaging fossil fuels.

5.3.4.1.2.2.1.1.3.3. *Sale of Heritage Items Adjustment Taxes*

A heritage tax would be applied to any thing considered part of the natural or cultural heritage, such as special buildings or landscapes. It would also include unique art works produced by the culture. It would include a tax on the export of such things without a value-added component, such a raw logs or unprocessed resources.

5.3.4.1.2.2.1.1.3.4. *Financial Speculation Adjustment Taxes*

The speed and detachment of money may artificially change the values of things that should remain uninflated or undepressed. A speculation tax, like a Tobin tax, is a small tax on each international financial transaction.

There are many potential benefits from a modest tax on financial transactions, such as the buying and selling of shares of stock or blocks of foreign currencies. Such a tax would have the effect of reducing short-term speculation in these markets, thereby making them somewhat less volatile. It would slow capital movements. It would also cut back some of the economic resources that are wasted in these transactions, since if the number of trades declines, the money spent on these trades would decline as well. In addition, it would make the tax code fairer, since most financial speculation is conducted either directly or indirectly by wealthy people. Just as poor and moderate income people pay taxes when they gamble at a casino or buy a state lottery ticket, a speculation tax would simply be applying a comparable tax to gambling in financial markets. Finally, a speculation tax could raise an enormous amount of revenue for a community or state.

5.3.4.1.2.2.1.1.4. *Distribution Taxes for Community*
Distribution taxes are the same as reapportionment for luxuries and large incomes. Many large incomes are so large that they are heroic. Combined with heroic inheritance and profits, these incomes essentially remove their receivers from any sense of local community by concentrating wealth.

5.3.4.1.2.2.1.1.4.1. *Heroic Possessions Distribution Taxes for Community*
A heroic possessions tax is a tax on products that are not considered essential, in other words, luxuries. A luxury tax is similar to a sales tax or VAT, except that it mainly affects the wealthy because the wealthy are the most likely to spend heavily on luxuries such as expensive cars, boats, houses, or jewelry. One might even consider this tax as a 'nonuse' tax, since many luxuries are not often used and are taken out of circulation for others.

Although a luxury tax of some sort may be justified on many things, the regulations would have clarify implementation in order to achieve at least some degree of equity. Although the tax would be collected at the community or state level, it might be coordinated at the national level, considering the multiple residences of those who can afford other luxuries. Regulations have to broaden the definition of luxury. It is hard to argue that a luxury tax is unjust and counter-productive, considering the massive suffering and starvation outside the gated compounds.

5.3.4.1.2.2.1.1.4.2. *Heroic Income Distribution Taxes for Community*
A heroic income tax would be applied to incomes over a certain amount, as determined by a ratio of one to ten. Herman Daly and John Cobb quote a range of the acceptable inequality of income at ten to one, although some corporations, like Ben & Jerry's ice cream, used to limit it to a one to seven ratio. As Cobb and Daly point out, the idea of unlimited inequality works against the notion of community. As they also wisely point out the goal of a community is not some perfect ideal of equality, but a limited inequality that allows individual differences to show, and individual rewards for luck or skill, but also allows others to catch up. And, this idea, rather than being new, is explicit in many biblical accounts of Hebrew laws governing landholding and usury, among other things. So, this tax rate would be quite high for those with heroic incomes. Paying such a tax would make these people heroes and in fact they would likely acquire heroic status for paying.

5.3.4.1.2.2.1.1.4.3. *Heroic Transfer Distribution Taxes for Community*
A transfer tax is a tax on any transfer of wealth, including inheritance, death duty, estate, or bequest. It has similarities to death taxes, inheritance taxes, and estate taxes, but it is much broader. It is applied to any transfer of wealth, to spouses, children, relatives, and even bequests to trusts, foundations and charities, and not on the total wealth. This would include payment of life insurance benefits or financial account sums to beneficiaries.

This tax would require relatively heavy regulation. This tax would serve to prevent the perpetuation of wealth, free of tax, within wealthy families. It would work towards reducing the accumulation between generations, thus leveling the field more for the next generation.

5.3.4.1.2.2.1.1.4.4. *Heroic Profit Distribution Taxes for Community*
A community has to be responsible to keep the economic playing field level, to use a popular sports metaphor. In the event that a person or corporation records a profit that could distort

the community, then the community has to limit that profit through taxes to not more than ten times the mean profit in the community (not the average or median).

5.3.4.1.2.2.1.1.5. *Discussion of Transitional & Missing Taxes for Community*
Under this new tax scheme, there would be no taxes on buildings, equipment or inventories. There would be no corporate income tax, although, as Daly and Cobb suggest profits would have to be distributed to all shareholders as income. There would be no personal income tax, no property tax, and no sales tax. Why? Because things people want or work for should be encouraged. Once the real price of oil and other goods has settled out, no other taxes are needed for value-added things.

In order to reach this position, however, there may be a series of transitional taxes, such as an added tax on all new vehicles.

5.3.4.1.2.2.1.2. *License Privileges for Community*
A license is a formal authorization by law to do something, such as to marry, hunt, or practice law or medicine.

5.3.4.1.2.2.1.2.1. *Marriage License for Community*
A marriage license is permission from a legal authority for the marriage of two people to be performed. The requirements differ depending on the time and place: Licenses to marry have been granted since the Middle Ages. Valid marriages can occur without a license, for example, by obtaining pardon for having married without license, or by cohabitation and representation as husband and wife in jurisdictions permitting common law marriage.

Every state in the United States issues marriage licenses. After the marriage ceremony, both spouses and the official sign the marriage license; some communities or states also require a witness. The official or couple then files for a certified copy of the marriage license and a marriage certificate with the government.

What is the purpose of such a license? Health? Tracking? People who wish to live together with the formal approval of a government need to meet certain requirements of residence or health. Some groups believe that needing to obtain a marriage license from the State in order to be married is unnecessary or immoral. Combined with other rights or privileges, such as having children, a marriage license would be required to indicate preparedness and intent, as well as to collect benefits.

5.3.4.1.2.2.1.2.2. *Reproduction License for Community*
A reproduction license would be permission from a legal authority for two people to have offspring. The requirements would differ depending on the time and place. The purposes of such a license would be several, starting with qualifying the parents, through a reading program or knowledge of health issues, such as smoking or drinking, and continuing with the health and safety of children, and with consideration of the effects on the health and fitness of the entire gene pool and on society.

Parenting skills would be tested and improved, much as certification and licensing improves day-care and educational providers. Poor parenting can be linked to for juvenile delinquency, crime, and sexual abuse. Children with poor personal and social abilities can burden the social system, especially the educational system; as adults they can continue to burden the system and produce another maladapted generation.

The penalty for not having the license might range from a fine to reallocation of the child to another family or institution, although that alternative would have to have numerous safeguards for the health and safety of the child.

One incentive for having the license might be a monetary rebate. Another incentive could be an extra coupon towards education or luxuries. This license would not limit the number of children or help with the planning of institutions, such as schools—that would be the purpose of tradable vouchers to have children (see Section 5.3.4.1.2.2.2.1.5.).

5.3.4.1.2.2.1.2.3. *Driving & Vehicles Licenses for Community*
A driver's license is an official document which states that a person has the qualifications to operate a motorized vehicle, such as a motorcycle, car, truck, camper, trailer, or a bus. Driver's licenses are generally issued after the recipient has passed a successful driving test and proven that they meet the age, education and understanding requirements. Different categories of licenses may exist for different types of motor vehicles. The difficulty of the driving test may vary considerably between jurisdictions, but perhaps be above minimum national standards.

Vehicles themselves must also be licensed for use and safety. The term 'motor vehicle' includes automobiles, trucks, buses, campers, trailers, and motorcycles. Motor vehicles are valued by year, make, and model in accordance with a formal vehicle valuation manual. Values are based on the retail level of trade for property tax purposes.

5.3.4.1.2.2.1.2.4. *Voting License for Community*
A license to vote would be required to demonstrate that the voter is living, lives in the community, speaks the language, and understands the basic issues. This would prevent some kinds of fraud.

5.3.4.1.2.2.1.2.5. *Wavelength License for Community*
Any use, for broadcast or communication, of common frequencies of the spectrum of wavelengths, would be licensed by the agency responsible for regulating interstate and international communications by radio, television, wire, satellite, or cable.

5.3.4.1.2.2.1.2.6. *Business License for Community*
Virtually every U.S. state licenses businesses. A business is defined as any commercial enterprise, trade, occupation, calling, profession, vocation, or activity engaged in, conducted, carried on, advertised, or held out to the public to be a business by any person, agent, or employee for the purpose of gain, benefit or advantage, either direct or indirect, with the principal objective of livelihood or profit through repetitive means.

The business license certificate is evidence of having met community requirements and that a fee has been paid for doing business within the limits of a community or state. Most codes requires that people obtain a license when conducting any business activity within the community, even if the business is located outside the area and has a license from another area. Other permits may be required to operate a business.

The fee imposed is solely for the purpose of obtaining general revenue. Business fees could help pay for community services like roads, fire, police and other community services that benefit businesses, business owners and the general public.

5.3.4.1.2.2.1.2.7. *Collecting License for Community*

Anyone who collected plants, animals, fish, or any living substance for its body, compounds or DNA, would be required to have a license. The license would certify that they have the knowledge to collect, and it would limit the number of people collecting. Issuing licenses would help to protect commons, such as parks and wilderness areas, as well as to coordinate the timing of collection and the limits of collecting.

5.3.4.1.2.2.1.2.8. *Weapons License for Community*

A weapons license would verify that knowledge and experience requirements are met, and reviews and researches of criminal history have been conducted for information that might preclude the issuance of a license, such as a criminal record or history of violence.

The community could create a bureau responsible for the issuance and denial of licenses. This bureau would receive and examine licensing applications for statutory compliance and verify the applicant's eligibility for licensing through former employers, educational facilities and examination of criminal history records. The kinds of weapons could be limited to knives, arrows, single-shot firearms and multi-shot fire arms; certain weapons, such as automatic guns, tanks, bazookas, and missiles, would not be licensable for individuals. Applying for a license to carry a weapon for self-defense is a right of the law-abiding members of many communities in many nations. Small individual weapons can be carried responsibly, properly and safely.

5.3.4.1.2.2.1.3. *Fees & Tolls for Community*

Fees could be charged by the community for use of community space or renewal cycles, for example, visiting fees or sanitation fees. Tolls could be charged by the community to cover extra expenses for transportation, for instance, bridge tolls or road tolls. Special fees could be charged for tourist attractions and for heritage sites.

5.3.4.1.2.2.1.4. *Labor Use for Community*

Many forms of volunteer work take place in a private setting or for an NGO dealing with social problems or environmental conservation. People volunteer for hospitals or libraries, food programs or the humane society. Many volunteers are coordinated by large networks for specific services. A community could encourage this or support it, with many kinds of recognition and reward.

5.3.4.1.2.2.2. *Payout for Community*

The community has to pay for its existence, as well as the expenses, amenities, and services that it trades for its existence. The payout is available from income, and should balance income, although savings accounts and debt-load can expand or contract temporarily.

In economics, business, and accounting, a cost is the value of inputs that have been used up to produce something, and hence are not available for use anymore. In business, the cost may be one of acquisition, in which case the amount of money expended to acquire it is counted as cost. In this case, money is the input that is gone in order to acquire the thing. This acquisition cost may be the sum of the cost of production as incurred by the original producer, and further costs of transaction as incurred by the acquirer over and above the price paid to the producer. Usually, the price also includes a mark-up for profit over the cost of production. Costs are often further described based on their timing or their applicability.

When a transaction takes place, it typically involves both private costs and external costs. Private costs are the costs that the buyer of a good or service pays the seller. External costs, or externalities, in contrast, are the costs that people other than the buyer are forced to pay as a result of the transaction. The bearers of such costs can be either particular individuals or society at large. External costs are often nonmonetary, and as a result difficult to quantify for comparison. They include things like pollution, things that society will likely have to pay for in some way at some time, but that are not included in transaction prices.

Social costs are the sum of private costs and external costs, that is, both the costs internal to the firm's production and external costs not included in the firm's production. For example, the purchase price of a car reflects the private cost experienced by the manufacturer. The air pollution created in the production of the car is an external cost. The manufacturer does not pay for these costs and does not include them in the price of the car, so they are external to the market pricing mechanism. The air pollution from driving the car is also an externality. The driver does not pay for the environmental damage caused by using the car. A psychic cost is a subset of social costs, for example from endangering pedestrians, that specifically represent the costs of added stresses or losses to the quality of life. A community has to consider all costs for as long as possible into the future.

5.3.4.1.2.2.2.1. *Issue vouchers for Community*

A voucher is an economic warrant and guarantee for goods and services. A voucher is a certificate which is worth a certain monetary value and which may only be spent for specific reasons or on specific goods. Examples include, but are not limited to, education, children, and medicine. The community would issue certain vouchers for all citizens. Vouchers would be an effective way of dispersing limited resources and reducing inequity.

5.3.4.1.2.2.2.1.1. *Basic Income Vouchers for Community*

A basic income voucher would be issued every month for community residents. This voucher would cover minimum clothing, housing, and food (see Section 5.3.4.2.1.5.2.3.1.).

5.3.4.1.2.2.2.1.2. *Medical Vouchers for Community*

A basic voucher could be issued by the community to allow the consumer to choose basic medical care. Extra care could be acquired through insurance companies that use their own voucher system. In some respects vouchers resemble credit cards, although they would be more specialized. Vouchers may vary in value over time, and may be specialized for use in the separate phases of medical care.

5.3.4.1.2.2.2.1.3. *Productivity Resource Vouchers for Community*

Within a calculated optimum (see Section 5.3.4.2.1.5.2.3.3.), every community within a nation would receive a number of vouchers for its population for the optimum use of resources and processes. Each individual would receive a set of vouchers for specific resources, such as productivity or carbon. Each community could make decisions on population size and wealth depending on other values or trade-offs.

5.3.4.1.2.2.2.1.3.1. *Carbon & Other Element Vouchers for Community*

Every community could be issued a number of vouchers equal to the carbon-bearing products of it population and their standard of living, based on a world standard. These would determine how much was used in each national system, regardless of its population or the intensity of its exploitation. Each community could be given vouchers for each element or compound, such as phosphorus or iron. These would be distributed among the population, according to a simple formula.

5.3.4.1.2.2.2.1.3.2. *Vouchers for Water, Air & Other Compounds*
Everyone has basic requirements for breathing, drinking, or getting materials. The requirements of living, such as breathing or drinking, would not require vouchers, except under extreme circumstances of overpopulation or overconsumption. Trying to issue vouchers for these things, on a personal level, might be too expensive and too invasive.

5.3.4.1.2.2.2.1.4. *Education Vouchers for Community*
An education voucher, commonly called a school voucher, is a certificate by which parents are given the ability to pay for the education of their children at a school of their choice, rather than the closest public school to which they were assigned. This kind of voucher could be accepted at a publicly supported school, although it might not pay for a complete private education. The school itself could decide to accept or not accept a student voucher, depending on qualifications and load limits.

5.3.4.1.2.2.2.1.5. *Child Vouchers for Community*
Several years ago, Kenneth Boulding recommended issuing women a marketable license permitting them to have a limited number of children, saying that the right to have children should be a marketable commodity, bought and traded by individuals, but limited by the state. For instance, every person would be issued a 0.5 replacement voucher share at birth. This number would change every year depending on population and support. For instance, for 2007-2008 the value might be 0 (or 0.1); and for 2034-2035, it might be 0.55.

These vouchers would be tradable and combinable. They could be traded so that, when a couple had a full voucher (1.0 or over) they could have a child. The voucher could not be used until the holder reached a majority age. This would result in a slightly declining population, even with a small percentage of cheating. The size could be adjusted once the population was at the desired target, perhaps at an optimum cultural carrying capacity.

5.3.4.1.2.2.2.1.6. *Others Kinds of Vouchers for Community*
What other kinds of things could be distributed using vouchers? Should there be vouchers for luxury? In a random lotto system for fairness? Or should it be a reward for generosity or some other virtue?

5.3.4.1.2.2.2.2. *Community Costs for Health & Maintenance*
The community has to take care of itself and to maintain itself as a separate entity. The community has to be healthy and vibrant. It has to build and maintain the infrastructure used by its citizens, from heroic architecture and dramatic boulevards to parks, theaters, and public areas. It has to pay the costs of these necessities.

5.3.4.1.2.2.2.2.1. *Promote Self-reliance of Community*
Self-reliance is the basis of security for a community. It is indication that the community has adjusted to its environment. And, that it is strong enough to resist some of the blandishments

of other cultures, as well as recover from environmental disturbances. Communities can be self-reliant by producing enough food and shelter, by limiting their population to what can be produced, by using local products and raw materials, from soil and minerals to plants, by using general and not specialized machines, by having multipurpose factories, by networking with other communities, and by doing without things that are not needed, including bombs, food additives, or aluminum bottles.

5.3.4.1.2.2.2.2.2. *Operate Community Government*
A community has to operate its local government. Leaders and representatives have to be elected and paid. It has to determine and fund the level of local infrastructure, from roads to schools. It has to establish and fund special departments for things, such as joblessness or old-age difficulties. It has to control communications and trade exchanges with other communities.

5.3.4.1.2.2.2.2.3. *Encourage New Business*
The community has to encourage the formation of new businesses and provide ways to extend the useful lives of businesses beyond the first year. The community can attract entrepreneurs, then set up programs to develop new businesses through incubators or special economic programs. New businesses often contribute disproportionately to innovation and efficiency.

5.3.4.1.2.3. *Assess Community Performance*
Adam Smith developed the argument that the only fundamental way of assessing the wealth of a nation is by examining the manner in which it uses its labor. This is true in a very basic way, since food, clothing and housing are products of labor. And, those countries with the highest standards have the largest output per person-hour. However, his thesis is not complete. It attributes no inherent value to the raw materials upon which agriculture and industry are based; these are treated as free gifts of nature, available in infinite amounts. Labor is an incomplete means by which the total economic well-being of nations could be assessed.

The assessment of personal or cultural wealth, for instance, is mostly psychological; wealth may be measured by how many valuables one has, which may be physical, like feathers, cattle, gold, or land, or by how by much status one enjoys, which may be behavioral, such as enjoying deference or a good reputation. Every assessment has degrees of subjectivity. A community has to relate its goals and images to its values and products, as well as to the satisfaction and happiness of its citizens.

5.3.4.1.3. *Future & Leadership*
Making these changes occur all at once would seem unrealistically disruptive, but would it be? Would it be more disruptive than an earthquake or a war, than runaway inflation or economic collapse? Would it fracture the cultures or communities? Would people's lives change that much? Would it change worse than having a lost job, a business fail or death in the family?

These things could all be done if people were convinced by their respected leaders that this was the best way to proceed. Given the fact that the characteristics that make people lust for leadership often detracts from their abilities to lead, perhaps we need a form of draft to get leaders, or perhaps we need to consider persons who have been groomed for decades or from birth, like the Dalai Lama.

We need to examine different models of power and authority: Autocrat, guru, revolutionary, or transformer, as W.I. Thompson suggests. The transformer has authority but no power; the autocrat has both; the revolutionary has power but no authority; and the guru has neither. Perhaps we should consider the guru as leader.

Until we can do things naturally and spontaneously, we must try to act our way into right thinking, while thinking our way into right action. It is a form of intelligence—understanding principles and trends of natural affairs so that the least energy is used. This is also the unconscious intelligence of a whole organism and the whole community.

5.3.4.2. *Contributing at the National Level*

Individual and community efforts may be most important, but action at the national level can coordinate changes or make them a matter of policy. Making the policy can feed back positively into community and individual actions as well.

The national level itself is quite fuzzy. Although people group into families and communities, there are many possible groupings between the individual and the national. The responsibilities of each group may be determined at the national level, or certainly through a process involving the national, regional, local and personal levels. Thus, some things may be more appropriately controlled at a county, state or province level, such as collecting taxes for roads. Other things, such as health insurance, may be best directed at a national level, to ensure the equity of distribution.

5.3.4.2.1. *Catastrophic & Normal Steps for a Nation*

Leadership at the national level could balance the dramatic changes suggested here—perhaps one leader could act like a shaman in Desana society, who is responsible for acknowledging the ecological limits and suggesting the behavior required to stay within those limits. Some of the following steps may seem revolutionary. Unfortunately, revolutions have the connotations of violence and overthrow. Revolutions can be as quiet and regular, and unthreatening, as the turnover of an axle on a wagon or car. Thomas Jefferson suggested that little revolutions, every couple of decades in the U.S., could make the experiment fresh, as well as break up unproductive hierarchies of power. We could start these little revolutions with many small steps, as long as they did not contradict cultural norms.

The first revolution has to be to adopt a new attitude. Adopting a catastrophic psychology for the nation, to address current and imminent losses, is less revolutionary and a more appropriate response to larger scale catastrophes, such as the national loses of biological and cultural heritages. Poverty and inequity are growing problems in every nation. These are reasons to adopt an extreme attitude towards survival.

5.3.4.2.1.1. Changing Nation to a Larger Self

As humanity is extended throughout cultures and nature, the human self interpenetrates so much that it becomes a larger self. The nation is at a level of the stereotype of the people of a culture, the sum of their behaviors, as it were. It is the larger self within the enveloping self of the environment. The nation has to change itself, and to recognize the extension of itself in nature, before it can make other changes. National behavior emerges from the individual behaviors of the participants.

5.3.4.2.1.1.1. Define & Secure Borders for a Nation

Borders are a crucial part of a nation, to keep its identity and define its territory. Borders form the skin that acts as an organ to keep vital processes and cycles inside and allow needed energy and materials to come inside. A nation has to decide what kind of borders to maintain, with what degree of openness. The amount of openness allows the renewal and protection of some areas. Leopold Kohr points out that much violence between nations is the result of bad divisions between them. Two alternatives to bad division are unification or good division.

5.3.4.2.1.1.2. Balance Isolation & Connectivity for a Nation

A nation has to control its relative degree of isolation from other nations. Isolation offers many advantages and a few disadvantages. It allows differentiation, of language and invention, that can reinforce the identity of a culture. It can offer safety from conquest by overwhelming force. Too much isolation, however, can lead to lack of stimulation and to ingrown customs.

A nation has to regulate the amount and kinds of trade. It has to decide how to participate with other nations, to ignore them or to compete with them, and how. It has to play a part in regional and global processes.

5.3.4.2.1.1.3. Balance Immigration with Emigration for a Nation

Some people will always be uncomfortable with the dominant culture of a nation and wish to leave it. Others will always be attracted by the unique qualities of another culture and want to join it. Nations may also try to balance certain skills and subpopulations with its overall goals, depending on their trade specializations.

5.3.4.2.1.1.4. Secure Treaties with Neighbors for a Nation

In addition to having secure borders, a nation has to normalize relationships with its close or regional neighbors. Such official agreements would spell out the extents of exchanges.

5.3.4.2.1.1.5. Set Standards for a Nation

Standards are models or examples of quality or value established by authority or consent that can be repeated as procedures. Standards can be used to certify practices for harvesting or manufacturing. A nation has to set standards for requirements, tariffs, protections, and for emissions and other things. A nation should also set the official standards for those things necessary for political interactions or economic production.

5.3.4.2.1.1.5.1. *Ecosystem Standards for a Nation*

Ecosystem standards would start with zoning. In a typical meso-ecosystem, a minimum area, comprising at least fifty percent, should be left wild. A smaller area, perhaps thirty percent, should be set aside for conservation of forests and rangelands. Based on the planned population, at least fifteen percent should be set aside for agricultural areas to feed the population. City and artificial areas should be restricted to the remaining small percentages, of the least productivity. In cases where the city area exceeded that maximum percentage, an equal area of rooftops or pavement areas would have to be dedicated to agricultural activities. Thus, there would be no limit to house density, only coverage area, and that would be directly related to natural primary productive areas.

5.3.4.2.1.1.5.2. *Social Standards for a Nation*
A nation has to have standards for the health of its people. It has to describe the minimum kinds of housing, utilities, and cleanliness that are acceptable. It also has to decide how much looseness and cheating are allowed, and how much punishment to apply to those who are caught.

5.3.4.2.1.1.5.3. *Economic Trade Standards for a Nation*
Tariffs may be needed for the protection of the manufacturing of special trade items. Nations should have the right to reject trade items that go against their laws or moral codes, or that may be hazardous to people or their environment.

5.3.4.2.1.1.5.4. *Technology Standards for a Nation*
Standards need to be set for those things related to transportation, such as the extent and quality of roads, as well as the safety and efficiency of vehicles. These standards might try to reverse trends towards large-size, inefficient trucks and sports utility vehicles (SUVs); the average SUV emits two to three times as much greenhouse gas as the average compact car. Even if a van is necessary to transport the whole family, there are ranges of choices in terms of quality and gas-mileage. The choice can make a ton of difference in $C0_2$ emissions every year, literally. Gasoline-powered vans and cars can be converted to hydrogen or waste oils.

Unlike those for aviation, most countries do not have a feedback process to make automobiles or boats safer. Every nation could create a department for analyzing every accident and making requirements for standards and behaviors and infrastructure changes. Few nations have national building standards, either.

5.3.4.2.1.1.6. *Establishing Rights for a Nation*
Rights, as an extension of ethics, are simply rules for living together. A nation has to codify those rights so that they will respected by all the residents, not just those who share the dominant culture.

5.3.4.2.1.1.6.1. *Rights for Nature in a Nation*
Rights seem to follow the expansion of the sphere of ethics, as formal statements of intuitive knowledge. Thus, rights for nature are being considered by many people. Paul Shepard says the argument is not new, and that its application is ambiguous because 'unlimited rights' will conflict with human interest. But, there are two bad assumptions: That human interests are not ambiguous—they are—and that animals will be granted unlimited rights—they will not.

The strongest argument for rights is interrelatedness in communities, which is the basis for assigning rights to nature. Garret Hardin considers interrelatedness, but interprets it narrowly. He considers rights as rules of competition; every right is a ploy in the struggle for existence, and every right implies an obligation to furnish it. This is good as far as it goes. However, life is more than competition; it involves cooperation and play. Rights are formal rules for living together. It would be foolish not to assign rights to animals, plants, and the earth because of contractual formalities.

Humanity has taken its own opportunities. These opportunities have been codified for centuries as rights. Now, we must allow other beings equal opportunities. The interrelatedness of life dictates the interrelatedness of rights. And these rights are necessary to the integrity of the whole planet. Humanity developed in a community of animals and plants, as part of a clade on the same tree of life. The quality of human life has always depended on the quality of animal life. Animals have sensations and feelings, as important to them as ours are to us.

Furthermore, the extension of rights to animals and plants does not deny any traditional human rights. Animals should be accorded higher moral regard and legal standing to reflect the intrinsic worth afforded by their existence and sentience. Welfare laws to conserve species and to guarantee humane treatment in research, transportation, and slaughter indicate a growing concern among people. A new ethic can keep animals free from human intervention, prejudice, or overuse. The intrinsic worth of animals is independent of instrumental values imposed on them by us.

One problem with the current legal system is that all nonhuman beings are given the status of inferior human beings, legal incompetents, thus keeping humans in a guardian role. A new legal category is needed that would respect the existence, competence, and excellence of natural beings. Christopher Stone recognizes that the judicial system has granted rights to a variety of inanimate holders, trusts, corporations, and nations, for instance. The legal system already operates with fictions, so the extension to natural entities should not present an insurmountable problem.

5.3.4.2.1.1.6.1.1. *Space to Exist & Opportunity to Flourish*

Every species has to be allowed the opportunity to live, even species that we fear or dislike, such as sharks or viruses. We do not know how these species contribute to the whole process of nature. Giving other species opportunities does not mean sacrificing any human needs, just limiting human influence and interference to a percentage of the earth, perhaps 40 or 45 percent.

In our control of conservation or artificial areas, which include many wild species, we can imitate the process of ecosystems by allowing birds, bats, and other animals opportunity to distribute seeds and energy to other areas or to access their prey, which may be our 'pests.'

5.3.4.2.1.1.6.1.2. *Freedom from Premature Death Extinction & Suffering*

Animals do not need to be saved from natural death, which is a great regulator of life, but from unnecessary suffering, experimentation, and premature extinction. The world would not be a better place without sharks, silverfish, rats, cockroaches, or hyenas. They need their own places, where they can take their own opportunities, live or die. The places, entire ecosystems, need to be saved. If we diminish variety in nature, we debase its stability and wholeness, which we need to survive.

5.3.4.2.1.1.6.2. *Establish Rights for People in a Nation*

Humanity has the right to coexist in healthy diverse, sustainable conditions. The basic things expected by people of their political and economic systems are simple. They are: Equality of opportunity for anyone; jobs for those who can work; security for those who need it; the ending of special privilege for the few; the preservation of civil liberties for all; and, the enjoyment of the fruits of scientific progress in a constantly rising standard of living.

Human rights refers to the concept of human beings as having universal rights, or status, regardless of legal jurisdiction or other localizing factors, such as ethnicity and nationality. The existence, validity and the content of human rights continue to be the subject to debate in philosophy and political science. Legally, human rights are defined in international law and covenants, and further, in the domestic laws of many states. However, for many people the doctrine of human rights goes beyond law and forms a fundamental moral basis for regulating the contemporary political order.

Human rights are taken to be inherent, universal, indivisible and inalienable. This means that everyone has them, they are the same for everyone, all are equally important, and they can not be taken away. Although a right can not be taken away, it can be violated.

Human rights can be divided into seven categories: Civil Rights (equality before the law), Political Rights (participation in government, life, liberty), Economic Rights (right to work for a living wage), Social Rights (having children, education, health care), Cultural Rights (preserve a cultural identity, language, practices), Environmental Rights (a healthy environment, clean drinking water, unpolluted air), and Developmental Rights (the rights of nations to control their own resources).

Civil and political rights are sometimes divided into negative and positive rights. Negative rights, which follow mainly from the Anglo-American legal tradition, denote actions that a government should not take. These include right to life and security of person; freedom from slavery; equality before the law and due process under the rule of law; freedom of movement; and freedoms of speech, religion and assembly. Positive rights follow mainly from the Continental European legal tradition, which denotes rights that the state is obliged to protect and provide. Examples include: The rights to education, to a livelihood, and to legal equality. Positive rights have been codified in the Universal Declaration of Human Rights and in many 20th-century national constitutions.

Human rights can also be based on the 'natural' moral order described by religious precepts. Religious societies justify human rights through religious arguments. For example, liberal movements within Islam have tried to use the Qur'an to support human rights in a Muslim context.

5.3.4.2.1.1.6.2.1. *Basic Rights in a Nation*

Nations have to decide the kinds of rights and guarantee them to their citizens. Some of these basic rights should always include the right to a healthy environment or to be secure.

5.3.4.2.1.1.6.2.1.1. *Right to Healthy Environment Air Water*

Principle 1 of the Rio Declaration of the UN states that human beings are "at the centre of concerns for sustainable development. They are entitled to a healthy and productive life in harmony with nature." While this statement fell short of recognizing a healthy environment as a basic human right, it points in that direction. People need to be assured of having clean air, clean water, and healthy ecosystems.

5.3.4.2.1.1.6.2.1.2. *Right to be Secure*

The right to be secure can mean many things. It can mean being free from invasion of your home. It can mean access to wilderness or land where you can provide for yourself. Insecurity is a way of life for people living in poverty; it means that they might survive, if things improve and if nothing goes wrong, but neither scenario is likely without eutopian change.

5.3.4.2.1.1.6.2.1.3. *Right to Opportunity for a Minimum Home*

If you cannot afford a home, you can ask for help. If you cannot get help, you can build your own. If you cannot build one, you can live in a group that shares one. The right is in the opportunity, not in the ownership. Most people can make or buy homes, if the circumstances are promising.

5.3.4.2.1.1.6.2.1.4. *Right to Opportunity to Minimum Work*

Every person should have the right to work and to receive a living wage for their work. Some nations are unwilling to support nonworking adults and deny them government assistance. Other people who do sacrifice and work very hard, may not earn enough to lift themselves and their children out of poverty. When current economic and legal arrangements hurt individuals, families, and communities, then something needs to change. A change in the laws, or in the Constitution, could provide every citizen with the right to an opportunity to work for a living wage.

A nation must guarantee everyone an opportunity to work at a living wage. Most people want work, meaningful work. People want to contribute to their own well-being, as well as to that of their family, community and nation. It is in the common interest of the community and nation that people who work should not be poor or dependent on others for support.

Millions of people are seeking work. Organizations, such as Oxfam, try to help people and communities to achieve 'sustainable livelihoods,' that is, a means of living that can maintain itself over time, and can cope with and recover from shocks. Oxfam has found that even small shocks, such as a broken washing machine or an unexpectedly large fuel bill, can trigger an imbalance of income and expenditure, and can lead to a downward spiral into debt and despair. There are many factors, such as lack of access to affordable credit, which force people to borrow from moneylenders at extortionary interest rates. However, there are many factors, such as social networks, education, personal skills, and accessible transportation, that contribute to the sustainability of livelihood.

5.3.4.2.1.1.6.2.2. *Tradable Rights*

Many rights, such as the right to reproduce, are partial social rights. As such they must be traded, through a voucher system, to acquire the right to an entire child or for a certain level of luxury. Other rights, such as the right to a healthy environment, can be traded for luxuries or larger populations.

5.3.4.2.1.1.6.2.2.1. *Right to Reproduce*

People have the right to have children, but this right changes in the context of overpopulation, as well as child mistreatment and abandonment. The right is limited when natural or social services are limited.

5.3.4.2.1.1.6.2.2.2. *Right to Share in Luxury*

What is luxury? How can we deal with luxury goods provided by a free market intent on maximizing profits? Luxury, a measure of those things wanted to improve happiness, is being hoarded by the rich or by the middle classes in rich nations. Runaway growth is exacerbating inequalities. If the trends continue without change—not redistributing income from higher to lower-income consumers, not shifting from polluting to cleaner goods and production technologies, not promoting goods that empower poor producers, or not shifting priority from conspicuous consumption to meeting basic needs—then the problems will get worse.

An equally critical issue is not gross consumption itself, but its patterns and effects.

Inequalities in consumption are stark. Globally, the 20% of the people in the highest-income countries account for 86% of total private consumption expenditures—the poorest 20% of people get a minuscule 1.3%. More specifically, the richest fifth: Consume 45% of all meat and fish (the poorest fifth only 5%); consume 58% of total energy (the poorest fifth less than 4%); have 74% of all telephone lines (the poorest fifth 1.5%); consume 84% of all paper (the poorest fifth 1.1%); and, own 87% of the world's vehicle fleet (the poorest fifth under 1%).

And, consider the following expenditures that reflect the priorities of wealthy nations: $8 billion for cosmetics in the United States; $11 billion for ice cream in Europe; $12 billion for perfumes in Europe and the United States; $17 billion for pet foods in Europe and the United States; $35 billion for business entertainment in Japan; $50 billion for cigarettes in Europe; $105 billion for alcoholic drinks in Europe; $400 billion for narcotic drugs in the world; and, $700 billion dollars for military spending in the world.

Compare that to what was estimated as additional costs to achieve universal access to basic social services in all developing countries: $6 billion for basic education for all; $9 billion for water and sanitation for all; $12 billion for reproductive health for all women; and, $13 billion for basic health and nutrition for all.[8] These numbers enforce the stark inequalities in consumption.

What would happen if everyone tried to exercise their right to some luxury? Would those spoiled by, and now needing, high levels of luxury refuse to share it?

5.3.4.2.1.1.6.2.2.3. *Other Rights*

Other rights, such as for basic ecological services, could be controlled by issuing tradable shares for those services. The number of shares would depend on calculations for resources and optimum populations.

5.3.4.2.1.1.6.3. *Work to Establish Justice*

Although many new rights are being extended to cover all of humanity and much of nature, new nations can demand to participate in a common justice. Rights and their obligations, after being reduced to principles of equity, can be addressed by justice, through standards and laws.

A principle of justice based on need can be extended to the ultrahuman community. To be sure, it needs to be altered to account for unconscious, interdependent beings.

5.3.4.2.1.1.7. *Represent All People of a Nation*

A nation has to give equal representation to everyone in its borders. Although distinctions will be made between citizens, noncitizens, visitors, and tourists, each will be represented to some degree.

5.3.4.2.1.1.7.1. *Obligation to Protect Rights & Privileges*

A nation is obligated to protect the rights and privileges that have been defined and agreed upon. Citizens will have more rights and privileges than noncitizens or visitors, because they pay for them with taxes and participation.

5.3.4.2.1.1.7.2. *Obligation to Meet Basic Needs of People of Nation*

A nation is obligated to provide for the basic needs of its people, and, for this reason, has some decision in how many people there are and how they can sustain themselves.

5.3.4.2.1.1.7.3. *Integrate Everyone into Society of a Nation*

A nation needs to be able to integrate people into its society. This often happens with a dominant culture and language, but it has to happen if there are equal cultures or several minority cultures. The nation has be decide how people can communicate, and how they can minimize conflict or stress.

5.3.4.2.1.2. *Amend Constitution to Control Corporations & Special Groups*

Every nation creates a formal or informal constitution. The Constitution guarantees a series of rights, such as protection, ownership of property, and a healthy environment. It also guarantees a series of freedoms, such as expression, religion, choice, assembly, association, and the petition of grievances. It prescribes some duties, such as voting or respect for others. And, it has a number of goals: Stability of government through transitions; predictability of the laws; and, common rules shared by all. In many nations, it may be necessary to amend the constitution, not just to reflect new international standards and agreements, but to reflect the new responsibilities of nations themselves, and of their constituents, especially powerful groups or large corporations with dominating influences.

For individuals, leaders, and corporations, both local and international, the constitution has to define limits to behaviors. Laws have to apply to large organizations, which are by no means necessarily private, or individuals or responsible. They must be liable for their behavior and effects. The nation has to reinstate the system of balances and checks, and government by law, with no exceptions.

5.3.4.2.1.2.1. *Encourage Small Businesses for a Nation*

Small business allows local employment, quick adjustments, and innovation. Small businesses built the United States in that country's first century; even today U.S. citizens hold high regard for small businesses, whose flexibility has provided lessons for big businesses. Nearly 98 percent of all Canadian businesses are small businesses. Small businesses contribute significantly to the economy of a nation, in innovation, in adaptability, and in job creation for women and minorities, as well as to distressed or depressed areas.

With most of the world's business being conducted by small entrepreneurs, it makes good economic sense for governments to implement policies that encourage small-business growth. The five ways in which government can have the most positive effect are by making capital more accessible, facilitating business education, promoting entrepreneurship, reducing regulatory burdens, and protecting intellectual property.

There are many things that go into creating a successful small-business economy, but the first might be the encouragement of entrepreneurs willing to start new businesses. For that to occur, citizens must be able to learn business skills. There are several ways in which governments can assist citizens. The community can create business incubators, as many universities in the U.S. do regularly.

Aside from lowering taxes to encourage business formation, it is important to reduce or eliminate those government regulations that slow business development or encourage business overgrowth. The simpler and faster the regulatory process, the greater the likelihood of small-business expansion. Any government that wants to encourage small business needs to produce laws that protect the innovations of entrepreneurs. Innovation is at the very heart of small-business growth, but if innovations are not legally protected, entrepreneurs may be unlikely to engage in the risks necessary to invent new solutions to social problems.

Accordingly, policies that protect patents, copyrights, and trademarks help small businesses to flourish.

5.3.4.2.1.2.2. *Protect Citizenship for a Nation*
The rights of citizens used to be straight-forward, perhaps because the definition of a citizen was more straight-forward. As nations are becoming increasing complex and societies are becoming more socially and geographically mobile, the idea of citizenship needs reworking.

Nations have the right to identify their citizens, and to offer them benefits. The citizens of a nation have made many investments and sacrifices to promote their own livelihood or the health of the nation. But, there are problems. There is increasing tension between the rights and obligations of citizenship. The differentiation of equal citizenship into group-related rights and special legal statuses, for multiple citizens, refugees, resident aliens, and transient foreigners, puts a strain on the membership. There is growing ambiguity about the collective identity of a people in a nation with open government, and on the significance of citizenship as membership in a national political community.

Some rights are available to all people in a nation. The basic principle for this inclusion is stated by the 14th amendment of the U.S. constitution: "No State ... shall deny to any person within its jurisdiction the equal protection of the laws." Protection is equal for citizens and foreigners; and it derives from being inside the territorial jurisdiction of a nation. Stated in this way, this is a universal human right whose corresponding obligations happen to fall upon a particular nation, in this case the United States.

Although the nation has the power to limit citizenship, and the power to deny citizenship to settled first and second generations of immigrants, using this power will require constraint. Citizenship itself is not a right to anyone from outside. Although foreigners have access to basic human rights, they should not have the same access to citizenship.

The rights of citizens might be better protected, and the power of public security authorities more restricted, under a law on identity cards. A law on ID cards for residents could maintain a balance between the function of public security and the protection of citizens' rights.

5.3.4.2.1.3. *Create Long-term Ecological Planning for a Nation*
For nations that intend to have a long-term existence, it is necessary to have long-term plans, and these plans should reflect the importance of an ecological perspective. It is crucial to know what habitats and resources are within a nation, to be able to delineate them, and to evaluate their health. A national survey is needed. The nation must begin the survey—perhaps the first that any nation would have—immediately. And, it would need to be a survey of every species, including viruses and bacteria, of habitats and cycles, including freshwater resources.

When the survey is complete, monitoring must begin to verify the kinds and extents of changes. When the ecological environment is evaluated and understood, the nation should link population to the carrying capacity of the land. It should link the consumption levels of the population to productivity of the land and to its ability to take and recycle the waste.

5.3.4.2.1.3.1. *Survey All Resources: Measuring & Mapping Inventory*
The first survey should start with the geological: Heat from the earth, thermal vents and volcanoes; geological formations, as well as specific elements. Then, an assessment should be

made, of those elements which should be left in the ground, and how, of those to be taken, they should be taken, and how the land should be replaced to a desired condition.

5.3.4.2.1.3.1.1. *Survey Air Water & Soil Resources*

The air would be surveyed for circulation and quality, not just of the local system, but of the entire airshed, which requires information from neighboring nations. The integrity, productivity, and sustainability of natural ecosystems are intimately linked to air quality. Injury and death of terrestrial animals from airborne pollutants, such as metals and gases, have been observed since the 1870s.

Soil ecosystems are a sink for many air pollutants; wildlife that inhabit the soil environment are sensitive to soil contamination. Air emissions can cause reductions in soil organisms and shifts in trophic structures, such as insectivorous bird species. A reduction or change in decomposers can result in a decrease in litter decomposition and nutrient cycling. The distribution and abundance of salamanders may be influenced by soil acidity. In the United States, approximately 50 percent of the species of frogs and toads and 30 percent of the species of salamanders use ephemeral forest ponds for reproduction. These small pools and ponds can become acidic when they receive snow melt and spring rains that have little contact with the soil buffering system. The problem is that atmospheric monitoring data is almost exclusively urban. Local communities, schools, and individuals can join in the gathering of this important, wild data.

Water would be surveyed, not just amounts and the purity of aquifers and watersheds, but its regional cycles also.

Health is related to the fertility of the soil. Most every culture that has inhabited the planet has come to such an awareness—many after it was too late. We need scientifically-based data on soil resiliency, as well as on the relationships of soil characteristics and conditions to gains or losses in productivity and hydrologic activity. The maintenance of soil quality is one of the most important requirements for long term sustainability of the productive capacity of forest ecosystems. We must use tools and methods that provide early warning signals of impaired soil productivity.

Social and economic health requires us to regulate and monitor those management practices that have the potential to reduce soil productivity. Soil, along with climate, landscape morphology, species diversity and other factors, sets the limits on productivity within a biome through the flow of nutrients, moisture, and air supply to the soil builders. Soil condition reflects a wide variety of other variables, and is therefore a direct and key indicator of site carrying capacity and landscape productivity.

5.3.4.2.1.3.1.2. *Measure Productivity of Resources*

The productivity of the land would be measured. Ecosystems are dynamic: Trees and plants are the food source for all other organisms. Almost every food web is ultimately dependent on the amount of plant tissue or biomass available for consumption. The rate of growth of trees varies greatly in response to a range of environmental conditions, but most notably to climatic factors, such as solar radiation or moisture. Changes in growth rate are affected seasonally and are dependent on the stage in the life cycle of a particular species. Most methods for measuring productivity are based on the repetition of biomass measurements at several points in time, with the increase in biomass representing the net primary productivity.

Productivity studies are very important for a number of reasons: They indicate a great deal about the dynamics of natural ecosystems; are of great value in agriculture and forestry, with domesticated and wild crops and resources; and, are useful in the study of plant-environment relationships through the application of bioassay techniques.

Productivity rates have kept the atmosphere functioning the way comfortable to human beings. Carbon dioxide accumulates in the atmosphere, although deforestation and fossil fuel consumption increase the amounts. Reforestation could remove carbon from the atmosphere. We need about 7×10^6 square kilometers of new forest to store 4×10^6 tons of carbon annually, according to George Woodwell's estimates.

Estimates of the minimum vegetative area for the planet are more difficult to arrive at. Houghton et al. (1990) suggest that the minimum should be about what remained in 1990: About 5.3×10^9 hectares, or 40 percent of the land area, although the area remaining that year is not definitely known.

5.3.4.2.1.3.2. *Monitor All Resource Use for a Nation*

Monitoring should be a national effort, involving scientists as well as citizens groups and special organizations. I.F. Spellerberg defines monitoring as the "systematic measurement of variables and processes over time for a specific reason," such as ensuring that standards are met. Monitoring is the measurement of the parameters that define patterns that indicate health or change in an ecosystem or place.

There are different levels of monitoring, including environmental, biological, and ecological. Environmental monitoring is an umbrella for many activities, including climatic variables and geological processes; for example, the systematic recording of soil and air temperatures, humidity, air pressure are measured to predict long-term climatic change.

Biological monitoring is the regular, systematic use of organisms to determine environmental quality; that is, the state of the environment can be analyzed by how individuals react to pollutants. Biological monitoring has numerous subcategories, such as the biochemical, microbiological, epidemiological, and biotelemetry. Ecological monitoring is the observation of communities to understand long-term ecological processes, such as succession and maturity.

Before monitoring can begin, the objectives and data collection methods have to be nailed down. The purpose of the monitoring program has to be stated; the objectives have to be identified, for instance, the health of ecosystems. Health will probably be the first objective, followed by production or aesthetics. Health can be defined by a set of indicators of health, such as species, or patterns of health, such as stability or productivity. None of the indices that are measured are really adequate to define the health of an ecosystem, because they cannot account for the complexity, richness, and cycling that goes on in the ecosystem. For that reason, health indices are data that need to be resolved on the ground in person by someone who knows the history of the place and has a feel for it.

The basic medical definition of health used to be freedom from disease. Part of a new definition is resilience to stress—of course, there is good stress (eustress) as well as bad stress. Ecosystems respond to stresses in different ways, but usually through a decrease in the indices mentioned. The symptoms of ecosystem stress are major items to be monitored.

5.3.4.2.1.3.2.1. *Increasing the Scale of Monitoring*

We have been monitoring stands and patches. Ecosystems are larger than patches; forests are larger than stands. We should be measuring at larger scales: Watershed, landscape, or biome.

Although some of the technical tools, such as satellite imaging and GIS, are being used, many conceptual tools, such as the Gaia Hypothesis or global design, have been neglected.

Furthermore, we are measuring over short periods of time, a year or two, only to establish a growth rate or productivity. We should be measuring over centuries. We cannot use a short-term industrial approach to measure a few parameters and then pretend we know enough about an ecosystem to take a large percentage of it. Forests ecosystems, for instance, are created by slow processes that take hundreds or thousands of years. In their 36-year study of a 450-year old conifer forest in Washington, Franklin and DeBell (1988) projected that it would take the shade-intolerant Douglas fir 750 years to drop out of the forest. Maser points out how long it takes for coarse woody debris to decay (200-460 years). Soil formation takes millennia; rates can range from 50-100 years per centimeter. We need very long-term studies.

We need to be monitoring every part of the web of interdependence. Monitoring everything will give us a better grasp of the health and normal changes in a forest—especially compared to interference or degradation through improper use. Monitoring should be as comprehensive as possible.

Monitoring shows how an ecosystem changes and at what rates. It shows what areas of the system are critical for the functioning and ecological integrity of the system. It may give an indication of how to rank values in the system. It shows how use of the system can be balanced and how the use can be zoned according to degrees of protection and use.

There are a number of problems or limitations with monitoring. The processes of many ecosystems have not been researched, so there is no baseline to compare measurements with (it is difficult to measure change without a baseline). Anthropogenic (human -caused) disturbances have long-term, synergistic, cumulative effects that are hard to trace. Furthermore, with the complexity of the measurements and the limitations of laboratory facilities or labor; the costs for analysis for some chemicals and metals, may be prohibitive. The cost of investigating over a large area for a long time may be very high.

To some extent, these problems are the reason that monitoring is not a regular part of management plans. Monitoring is crucial to understanding ecosystems. Until we understand how systems change and move around the landscapes, we will not know which changes are important and inevitable and which are the unhealthy result of human interference. Until we understand the changes, we will not be able to adjust our needs to limits.

5.3.4.2.1.3.2.2. *Establishing a Context for Ecosystem Monitoring*

The actual substance of the environment consists of patterns as well as things or individual species. The environment is generated by a patterning of the ecological ebb and flow of energy, substances, individuals, and species across a suitable landscape. The distinction between growing and declining patterns is not arbitrary, and can be arrived at objectively. And, the ecosystem environment is constituted of a large set of events that are objectively definable by specific outcomes.

The procedural consequences of these facts involve practical changes in the relationship: Between the earth and its species over space and time; between the earth and the collective ability of species to respond flexibly to situations beyond their normal patterning in the processes of adaptation and repair; in the flow of energy through the environment and its biotic community; and in the ecosystem's collective patterning of shared needs and governance which we will define as its health or soundness.

Taken as a whole, the process of observing patterns in these relationships leads directly to a fundamentally different way of perceiving ecosystem management. As Richard Hart mentions, patterns are the key to understanding the nature of a forest. In some ways, patterns are prior to things, in helixes, light, fields, and ecology. Paul Shepard and others have written that relationships are as real as the objects that result from them. Ecology attends the overall pattern of relationships, beyond the details.

The challenge to measuring and monitoring ecosystems is to address the patterns. But, the tools will have to be used in new ways in a new framework, perhaps with topology and holograms as metaphors (topology provides a mathematical model for processes, and a hologram provides a model for connected wholeness).

Based on a broader metaphysical foundation, with more comprehensive values, measuring (mensuration) and monitoring need to address patterns of being in an ecosystem and not just a few commodities dictated by a short-sighted economics supported by single-visioned science and technology based in a savage culture. One challenge is to identify the patterns and set up long-term programs to study them and relate them to sustainable use of ecosystems.

5.3.4.2.1.3.3. *Protect Resources for a Nation*

Once the extent of the resources is known, and once they have been categorized for use, they need to be protected. Protection may be temporary, for future resources or use, or permanent, for areas critical to regional and global cycles.

5.3.4.2.1.3.4. *Save Important Ecological Areas for a Nation*

Saving used to mean isolating the areas entirely from human use and interference. But, human aesthetic use and human basic use, as done by archaic cultures, is often part of the process itself. Saving has to mean regulating use, but this is more concerned with reducing consumption below certain levels and not allowing large-scale use for external requirements.

Our species has been shaped by the earth. The desire to save forests, wetlands—all natural ecosystems—is an expression of deep human values (or perhaps a more basic survival instinct). Experience of wildness lets us capture some of our own wildness and authenticity. Our emotional response to the unfathomability of the ocean or luminosity of the desert is an expression of aspects of our fundamental being that are still in resonance with these forces.

We have learned that we cannot just save big trees or waterfalls or geological formations. We have to save the system and process that produces big trees or spectacular forms that we like. Size is important. Thomas Lovejoy and others have made parallel arguments on species loss, leading to the idea of a minimum critical size for ecosystems. The minimum critical size is greater than the areas suggested by the species-area curve because of three reasons, Lovejoy says: Species are identified by individuals, not breeding populations; the species-area relationships are derived from extensive habitats rather than the fragmented ones; and, higher trophic levels, such as tertiary predators, may be excluded from smaller areas—that is, all species are not equal. As an ecosystem is first reduced in size, it can still maintain its characteristic composition and species diversity as a self-sufficient, functioning whole. However, as it is fragmented and impoverished, more sensitive species drop out and its function is impaired. At some point, as the size is reduced below the minimum, the integrity of the system is compromised and the system collapses.

We have learned that shape is important, also. We have to protect interior species and conditions as well as exterior ones. Edge effects started out as being centers for diversity.

Now, many edge effects are considered to be destructive. In the 1940s, it was thought that edges should be increased to provide bountiful game crops. Edges have proven to be good for both game and wed species. Edge effects, however, are detrimental to populations adapted to ecosystem interiors. Too many edges can reduce diversity at local and regional scales. We have to anticipate changes at global scales.

5.3.4.2.1.3.5. *Restore Natural & Wild Areas as Necessary for a Nation*
Restoration is one of the major ways to ensure the survival of species, habitats, territories, and ecological systems. Restoration projects have the potential to save entire ecosystems. Agricultural systems might save domesticated species. Restoration ecology, as a new discipline, tries to address the difficulties of deciding the goals and means of restoration, considering the lack of information about original systems and the loses of component species during the degradation or destruction of the wild ecosystems. The discipline attempts to create self-maintaining neopoetic systems characterized by complexity and diversity.

A holistic science such as landscape ecology addresses the overall patterns of large-scale ecosystems, considering the biogeochemical, atmospheric, and hydrological cycles in relation to the shape and extent of individual landscapes. Landscape ecology can identify candidate ecosystems for restoration, as well as for preservation, conservation, or reservation; it can identify patterns of forestry to preserve larger functional islands.

A crisis science like conservation biology can make recommendations which would preserve diversity and complexity in forests, and would avoid numerous extinctions. The first recommendation is to stop logging old growth and mature natural forests. Then promote cutting practices that respect the productivity and complexity, leaving snags, logs, and many-aged forests. Grant timber leases that are contingent on the maintenance of the productivity and diversity of the land. Reduce fragmentation through the design of forested areas, taking into account the genetic diversity of the trees, catastrophic conditions, minimum viable populations, corridors, and edge effects. Other recommendations are to: Stop constructing new roads; close and revegetate old roads; restore clearcut areas; replant with native species; restore damaged streams and wetlands; restore natural connections, such as corridors and canals; and recommend that reserves be made large enough for minimum viable populations and minimum viable ecosystem areas. Restoration areas, which are set in a pattern by human activity, but may not need further intervention. All areas need plans.

5.3.4.2.1.3.6. *Conserve Use of Resources for a Nation*
Natural resources were originally defined as objects provided by nature for human use. This concept has been expanded to include minerals, wildlife and people. The idea that everything should be managed is based on an extreme belief that nature is a resource to be processed. Furthermore, management is self-perpetuating and self-justifying. The objective of resource management is not to strengthen its defenses or funding, or to increase quality of life for affluent people in overdeveloped countries—it is to adjust the use of resources to address the needs of current and future generations.

In short, conservation management is based on economic objectives. And, as Leopold pointed out, the weakness of relying on economic motives is that most members of the earth's community, such as wildflowers and songbirds, have no economic value. Yet, all the members of the community contribute to the integrity of the whole, which is vital to maintaining what we do consider important. Those beings with no economic value are ignored, or worse, labeled as weeds or vermin and destroyed so that crops and animals with short-term advantages for human ends can be substituted.

Most conservation strategies are completely anthropocentric, from saving hunting grounds in the middle ages to allocating resources this year. This plan proposes ecosystem conservation, which protects entire biotic communities: Genes, populations, species, habitats, associated traditional human cultures, and all the processes and interactions.

Conservation parks are a way to keep resources. Conservation parks are areas set aside for multiple use of resources without interfering with the operation of the ecosystems. Research may be conducted to answer questions as to whether the park is big enough and shaped correctly to constitute a proper habitat for its inhabitants.

5.3.4.2.1.3.6.1. *Limit Use of Common Resources for a Nation*

Common areas need to be preserved. Common areas can be used so that they are preserved in character and function. This can be done through the following limits: Limit access; limit use by large vehicles or houses or buildings; limit harvest; and, limit grazing. This may mean limiting the size of flocks or herds to allow other wild grazing animals. Hunting, grazing, and agriculture provoke large ecological disturbances.

In general, mammalian grazing promotes regrowth and the movement of seeds. Bison and prairie dogs were responsible for much of the character of the American plains, but cattle have led to fencing and the eradication of pests, which leaves the ground overeaten and muddy. The values of keeping healthy ecosystems, such as temperate or desert grasslands, are far more than the perceived losses from limiting domestic animals.

5.3.4.2.1.3.6.2. *Limit Use of Slowly Renewable Resources for a Nation*

Slowly renewable resources can still be used, but with respect to their extents and rates of renewal. In general, we need to: Limit takes to the renewal rate; limit the styles of taking; require replanting; and, charge the cost of use by replacement value, which may be substantial.

5.3.4.2.1.3.6.3. *Limit Aquifer Use to Recharge Rates for a Nation*

First, we need to determine or survey aquifers, then determine recharge rates, then determine the value of water and the cost according to value. To supplement the use of aquifers, plan for the collection of rain or wastewater to reduce demand. Monitor all water use. Treat the aquifer as the principle and water flow as interest.

5.3.4.2.1.3.6.4. *Use & Limit Wild Harvest for a Nation*

If large areas are to be allowed to operate with natural processes, then large areas have to be limited regarding human use or conversion. Perhaps, the human use of some wild lands, within limits, can continue, but it must be part of a total change, unlike permaculture and other strategies which can be practiced as part of an industrial lifestyle. Paul Shepard argues that 75% of the land area should be left wild in a techno-cynegetic society, so that human beings can return to the hunting and foraging lifestyles that shaped our species.

Even if a return to this style of life is not possible for most people in most nations, some ideas can be taken from it. Wild animals could be harvested for wild food, if it can be done with minimum disruption to their breeding and movement.

5.3.4.2.1.3.7. *Create Large Conservation and Wilderness Areas for a Nation*
Create large conservation areas, such as the Wildlands Project recommendations for the Rocky Mountains and Great Plains in the U.S. The goal of the Wildlands Project (Dave Foreman et al.) is to set aside approximately half (50%) of the North American continent as 'wild land' for the preservation of biological diversity, by creating 'reserve networks' across the continent, which would be composed of cores, buffers and corridors. The primary characteristics of core areas are that they are large (from 100,000 to 25 million acres), and allow for little human use. The primary characteristics of buffers are that they allow for limited human use when they are managed with native biodiversity as a preeminent concern.

5.3.4.2.1.3.7.1. *Conserve Places and Species for a Nation*
We have always tried to exceed the physical and biological limits of places rather than recognize them and be guided by them. Most places exist in a uniquely identifiable ecosystem, with recognizable boundaries and a unique history and character. Modifying such places can change or diminish their uniqueness. We can conserve places and species by respecting their limits and limiting our exploitation.

5.3.4.2.1.3.7.2. *Privatize Communitize or Nationalize Resources*
Some resources might be saved through a program of privatization. Perhaps we should privatize where applicable or reasonable. When something is privatized, however, the profit or loss accrues wholly to an individual or group, such as a corporation. Privatization is wonderful for individuals, farmers, and small businesses. It inspires people to work hard, save, and keep their efforts healthy, especially if a forest or farm is the basis for their efforts. Privatizations sandwiched Chinese efforts to grow crops efficiently and were found to be preferred by farmers. Privatization is limited by scale; it works at small scales where the individual knows the details of an operation and profits from their own efforts and restraints.

However, the privatization of large holdings to individuals and corporations does not seem to have worked well. With no sense of stewardship, with a focus on short-term profit, and with little understanding of ecology, land gets exhausted quickly, then used or sold for less profitable ventures, such as running cattle or building houses, which further degrade it. When land is put away from use, or rather for the exclusive use of one person or group, who does not always use it, then it may be conserved. Thus, kings or groups of monks quite often saved trees and forests by putting them behind walls and limiting access. It worked by accident. Larger patterns of land and things should be communitized, that is owned by whole communities, who would enforce rules for use and restrict access. Communities, like monasteries, are managed for long-term, sometimes with better ethics.

For whole regional systems, with watersheds, airsheds, and landscapes of ecosystems, nationalization is a better strategy. The nation can set aside whole systems. The nation can better balance resources with wilderness areas. The nation can take a larger view than the community, although their perspectives often are longer than that of an individual or business group.

5.3.4.2.1.3.8. *Implement National or Regional Ecological Goals*
Once goals are identified and agreed upon, they can be implemented. National governments have been comfortable with short-term economic goals and a few ethical goals, but have neglected goals dealing with the ecosystems and climates.

5.3.4.2.1.3.8.1. *Anticipate Climate Change*

The climate is changing. The changes have been documented. It is just that politicians, and most of their constituents, are slow to react to these average changes, which after all seem slight, compared to the extremes of weather. To a few the changes are threatening, not just to low-lying islands and a few corals, butterflies and trees, but to civilization. Global warming could lead to reversal of ocean currents and other unpleasant, deadly effects. What is the solution for climate change? Use alternative energy sources? Possibly, energy use could benefit from a high-tech solution. A massive development program, like that for the atom bomb, and with the same urgency, could develop and implement alternative energy technologies.

There are many less technical actions that could combat climate change. Planning could always consider greenhouse warming. CFCs and greenhouse gases could be phased out. Contributing agricultural problems could be corrected. Nations could consider full social cost pricing of energy, where the polluter pays. Deforestation could be halted; reforestation projects could be started. Biodiversity losses could be slowed or reversed. Efficient use of water could be increased. Aquifers should be protected. The easiest solution might be to reduce consumption and populations. Nations could share data and participate in international projects.

5.3.4.2.1.3.8.2. *Ecological Goals for a Nation*

Regional or national goals are appropriate for bioregions and isotopes (literally meaning similar places). The number of goals decreases as the scale gets larger.

- Zone ecosystems at the landscape level for preservation, conservation, or selection use
- Maintain ecological and evolutionary processes in healthy landscapes
- Diversify the institutions that deal with ecosystems; relate people diversity to ecosystem diversity
- Reduce fragmentation of landscape patterns with responsible use
- Accept and plan for natural change, disturbance, uncertainty, ambiguity—it's part of the process
- Restore forest cover in North America to pre-European levels
- For the U.S., replant 142 million hectares of forest lands—up to 438 million hectares (the approximate level of cover in 1600)
- For Canada, replant over 80 million hectares of forest, up to 530 million hectares
- For the Northwest (Pacific Northwest coast forest, replant up to 47 million hectares
- For the Northeast (northern hardwoods), replant up to 11 million ha
- For the Southeast (Oak-pine forest), replant up to 129 million ha
- For other ecoregions, replant to a high percentage of 2000 BC levels, especially around the Mediterranean
- Relate (or limit) population and consumption to the productivity of ecosystems—without such limits, forests will eventually be destroyed

5.3.4.2.1.3.8.3. *Cultural Goals for a Nation*

Cultural goals are more complex and challenging. Stabilization would be a good first step.

- Stabilize the cultural environment
- Stabilize population
- Encourage cultural education and history

- Encourage teaching of the local language
- Allow renewal of critical customs
- Work to establish equity between groups
- Allow more opportunities for fulfilling needs and generating wealth.

5.3.4.2.1.4. *Long-term Economic & Political Planning for a Nation*

Human interactions can become more violent as a result of competition, inequity, and limits, requiring more planning and social control. Local planning ignores limits and carrying capacity, long-term deficits and problems, other species, and ecosystems. Local planning is limited to a two to four-year cycle. Regional and global planning are virtually nonexistent. Crises sciences, such as conservation biology and ecoforestry, plan at the landscape level, and economic and political planning need to plan at this level also.

Planning in general means deciding on goals to be achieved in specific situations. For central planning by a federal government, the goals are usually small and not comprehensive, such as a forest cutting level or a single species preservation, which usually end up being a compromise in cost-benefit analysis. Central and most other types of planning tends to neglect or dismiss the distribution of negative, uncertain, or nonmonetary effects, which are characteristic of much of nature. Furthermore, we seem to have no mechanism for developing long-range plans. Certainly, there seems to be no way to deal with long-term, slow catastrophes, such as deforestation.

Most plans address *problems*, such as building roads. Everything else, from employment to pests, is also considered as a problem, and not a direct effect of the cultural implementation of some pattern. Most plans are also development plans that are comprehensive in the sense of seeking to meet all needs of the public, agriculture, and industry. Development plans tend to call for the eventual development of *all* resources in an area. Development is more concerned with an assembly-line model—simple, isolated, efficient, and easy to maintain. We become remote from, and indifferent to, the system that supports us. We acquire unrealistic images of the world and harmful values and then make bad decisions based upon them.

A one-world planned economy is an even greater threat, being based on unlimited industrial production, unlimited commodity consumption, increased exploitation of nature, and the free flow of resources and labor across cultural borders. This kind of planning requires the abandonment of local controls on development, trade, or lifestyles. Planning is thus characterized by a utilitarian globalism that denies value to the systems that support it. As a result of central and global planning, the patterns of life have become the products of market forces operating in a sterile abstract order.

We have not developed qualitative indicators of ecological health or quantitative measures of social health, much less an ecocentric view that would value preserves of nature for themselves. One solution to many of these problems is a reduction in scale for everything from ecosystem use to management units, with local controls and local use primary. Management costs increase with the size of management units; more levels of human hierarchy are required to deal with problems; decisions are slower, and the people who make them are more remote from the site.

Obviously, a plan should consider the whole system. Human needs should be designed for an optimal fit within the limits of the system. Ecological planning considers the health of the system, which is based on intimate knowledge of the system. Direct observation and traditional knowledge yield far more 'information' about the societies of plants and animals than autopsies or mathematical models. A whole-system, comprehensive plan would proceed in stages (as previously shown):

1. Identify the place within its natural boundaries. Most places exist in a uniquely identifiable ecosystem, with recognizable boundaries and a unique history and character.
2. Calculate the optimum amount of wilderness to preserve the natural cycles indefinitely. If the current wild area is less than the calculations, restore the difference and set it aside as a preserve.
3. In the remaining area, zone areas for appropriate use, including conservation and artificial areas (roads or cabins).
4. Identify the resources needed for human use, including raw materials and the productivity of the areas. This productivity can be used to calculate rational exploitation.
5. Apply cultural modes—in style, values, and technology—to set limits on technology and population in the area to be supported.

We would examine the natural and cultural histories of a place, as part of our comprehensive plan, which is actually a deductive, synthetic, conceptual model based on data generated from research on biological productivity, the rates of resource use, cultural valuation, minimum wilderness preservation, air and water quality, genetic minima, nonrenewable resources, appropriate technological innovation, the importance of cultural frameworks, adventure, research, beauty, uniqueness, and other intangible experiences.

Planning is not meant to be a finished work of art—it has to be a changing, adaptive pattern that reflects our participation, understanding and use of the system. Each activity needs to be fed back into the process of updating the plan. Implementing the plan should result in improvements.

5.3.4.2.1.4.1. *National Planning to Survey Human & Cultural Needs*

Abraham Maslow listed a full hierarchy of human needs: Physical needs, such as food, shelter, and clothing; 'safety' needs, including law, order, and security; 'psychological' needs characterized by belongingness and love; 'esteem' needs, such as strength, self-sufficiency, competence, freedom, attention, and prestige; and 'self-actualization' needs, like self-actualization, achievement, and creativity, which develop after other needs are satisfied. Of course, human needs are based on the health of the earth. Human needs extend to include a foundation of wilderness. Human systems depend on natural ones, for recycling of wastes, water, and air. But, as human growth tends to be logarithmic, so does human need, and need shapes facts, like the kind and quality of resources. Granting this, the need for wilderness is as much a fact as the need for food.

Surprising things can be discerned about living a good life: Once basic needs are met, for food shelter, respect, and confidence, then happiness is not increased much by more material goods or money. The things that make people unhappy are when the higher needs are not met. Lack of love or security, lack of communication or appreciation, lead to unhappiness. People try to balance their needs by acquiring more money or things. The effort to balance may result in distortion or violence.

The society and the culture have to be meaningful, as well as secure and equitable, with low extremes between wealth or status. Similarly, there is a hierarchy of needs of a culture: To be grounded in place, to be secure or partly isolated; to have a dynamic order, with human health; to be complex and sophisticated, with checks and balances; to be comprehensive, to allow change and diversity; and, to have and manage adequate resources. Alas, sometimes cultural needs can be perverted, and the needs are almost completely defined in terms of commodities.

It is human values that ultimately determine what, where, and how we monitor management activities. Therefore, it is imperative that cultural values and natural processes be determined and monitored simultaneously. In this way, we can begin to understand which natural processes we use, or favor or discourage, in order to benefit our current value system. Monitoring and evaluation of indicators provides necessary information for understanding the human dimension as it relates to ecosystem management. What makes something valuable is not in its own properties but in its relation to the personal preferences of its perceivers. However, the self-value or usefulness of a thing, being or species does not rest entirely in its appearance but rather in its existence.

5.3.4.2.1.4.2. *National Planning to Monitor Human & Social Resource Use*

The Earth Summit's "Agenda For Change," states that a stable human dependence on natural resources is key to the protection of forest and range lands. Social and economic data is often compiled by standard political boundaries, not on provinces, sub-provinces, landscapes or project sites. Thus, socioeconomic data is not strictly comparable or usable at the sub-province, landscape or project site levels. However, the indicators listed can be either described or mapped onto geographic information systems (GIS), providing an opportunity for integration with other physical, chemical and biological indicators.

For instance, the indicator 'demographics' provides information on where people settle, how many there are, and what they do. This has a profound effect on the environment and its sustainable use and management. Information gathered from non-national sources would assist us in making sound decisions that integrate the diversity of the local and national human population. Data may include age distribution, in-migration, percent of population using resources directly (including recreation visitor days and employment), and where people settle—their distribution, and number including rural interface zoning, roads, type of industry, public services, communication lines, and community activities.

For cultural influences, information gathered through archaeological sources and studies provides information on the historic uses of the surrounding landscapes, and of associated human activities. This data may include ethnographic information and historic use patterns.

5.3.4.2.1.4.3. *National Planning to Promote Efficiency of Resource Use*

Data on local and regional economic health would help us to predict long range trends and landscape use patterns. Data may include real unit costs and price changes, income tied to resource activities, profitability of management, payments to government, percentages of products recycled or reused, changes in the distribution of employment and income, employment patterns tied to resource activities, rural interface land values, types of commerce, welfare payments into the community, and the economic base. Data may include various land-use patterns such as percent of land base set aside in protected areas, dispersed and concentrated recreation areas, grazing allotments, mining activities, vegetation or animal usage (game management areas, firewood, timber, special forest products, fishing spots, water withdrawals), and community contacts (retirees, business people, school students faculties and staff, community clubs, and tribes).

5.3.4.2.1.4.4. *National Planning to Anticipate Technology Integration*
Technology can be broadly defined as the material entities created by the application of mental and physical effort to nature in order to achieve some value. In its most common use, technology refers to tools and machines that may be used to help solve problems. Technology is a technique that lets us use resources to produce products and solve problems; this technique feeds back into culture. Due to the increasingly widespread use of ever more complex technologies and their frequently unintended consequences, problems may arise in their use that are unknown or only partially addressed.

As tools increase in complexity, from knives and levers to computers and space stations, so does the knowledge needed to support them. Complex modern tools require libraries of information that has to be continually increased and improved, then spread and understood. Technology first simplifies life, then complicates it. The combination of cheap goods and complex tasks can lead to sweatshop slavery and unsolved wastes. Some problems are solved, but new ones are created by the unconsidered use of the technology—problems such as toxic waste or radioactive waste.

The use of technology has a great many effects; these may be separated into intended effects and unintended effects. Unintended effects are usually also unanticipated, and often unknown before the arrival of a new technology. Nevertheless, they are often as important as the intended effect. The most subtle 'side effects' of technology are often sociological. They are subtle because the effects may go unnoticed unless carefully observed and studied over large areas and long periods of time. These may involve gradually occurring changes in the behavior of individuals, groups, institutions, and even entire societies. A nation needs to address the effects, all of the effects including unwanted ones, of technology on business, culture, management, and the environment.

The implementation of technology influences the values of a society by changing expectations and realities. The implementation of technology is also influenced by values. There are major, interrelated values that inform, and are informed by, technological innovations. These realities and expectations may alter the world view of a culture, especially as regards efficiency, bureaucracy, or progress.

A nation has to define the goals of technology, from simple technique to the simplification of chores, as well as to consider the concept of nonharm. Then, it has to analyze the results of technology, not just the mass production of things, but the effects on human health and behavior. A nation has to discern what is missing from an application of technology, whether it is human scale or appropriateness. Then, a nation has to decide how to manage the technology.

Many new technologies are not managed for their perceived benefits or losses. Nanotechnology, for example, has the potential to clean chemicals and viruses from the human body, but some minute nanos could pass through various barriers in the body and interfere with brain processes or environmental cycles. Technology can be used to develop renewable energy and restore damaged environments. Technology can be used to promote sustainable localized energy industries, with solar, wind, hydro, tidal, or biofuels. But, technology has to fit the scale and tempo of the environment. Technology has to be sustainable and replaceable.

Technology needs to be integrated into society. It needs to be made appropriate to the goals and desires of a culture. The notion of appropriate technology, however, was developed in the twentieth century to describe situations where it was not desirable to use every new technology or those that required access to some centralized infrastructure or parts or skills imported from elsewhere. The eco-village movement emerged in part due to this concern.

Technology needs to be integrated into the entire environment. Some technologies have negative environmental effects, such as pollution and lack of sustainability. Some technologies are designed specifically with the environment in mind, but most are designed first for economic or ergonomic effects. The effects of technology on the environment are both obvious and subtle. The more obvious effects include the depletion of nonrenewable natural resources, such as petroleum, coal, ores, and the added pollution of air, water, and land. The more subtle effects include debates over long-term changes, such as global warming, deforestation, natural habitat destruction, and costal wetland loss.

5.3.4.2.1.4.5. *National Planning to Encourage Vernacular Design*

Ivan Illich points out that the term 'vernacular' is an old technical word used for Roman law, having to do with 'free' access from a commons or free ownership, such as getting offspring from a donkey already owned. As Illich further points out, the vernacular is the opposite of a commodity, that is, something for which one has to pay. Illich suggests, in the tradition of Schumacher, to use the term 'vernacular value' for those things made by individuals and groups that are *not* destined for the market. These values imply two kinds of technologies, however, one for mass-producing commodities for the market, and the other for the production of personal or community goods that remain in the community. Illich questions the use of the word production, but it is used here in the large sense, as in the production of edible matter by plants.

In a commodity-centered society, ethics, politics and justice are reduced to equitable distribution of commodities, according to Illich. But, the access to resources for vernacular and commodity productions, with lessened economic domination of the latter, can be equitable, without reducing politics to simple access. Illich suggests that there are two choices for educating in a technological society, regardless of whether the energy is hard or soft: (1) People need further education to build solar collectors, or (2) education needs to be made transparent by the engineering of tools. Illich's idea of the vernacular describes a great divide, not just of production and education, but of language, with an every-day vernacular language and a technical, formal language.

Vernacular architecture is a way that cultures express a shared heritage in patterns of construction of shelter. It is a term used by the academic architectural culture to categorize structures built by nonprofessional or untrained builders. Although modernity should not be cause for exclusion, true vernacular is most apparent in the archaic world where indigenous populations produce their own shelter based on traditions of using locally available materials. This architecture can include a wide variety of structures, though domestic and agricultural buildings are the most common.

Another distinguishing feature of vernacular architecture is that design and construction are often done simultaneously, onsite, with nonmanmade materials. And, many of those who eventually use the building are involved in its construction, or at least have direct input in its form. Vernacular building shapes, construction techniques, and other characteristics are often generated from centuries-old local patterns. These patterns continually change, incorporate new technologies and values that perpetuate cultural norms. Vernacular buildings can reflect sophisticated adaptation to the environment and needs.

We have forgotten that the whole purpose of architecture is not efficiency or luxury, but is time and space for living well. For living well, efficiency is unimportant. Most importantly, vernacular design and building allows people to participate in meeting their own needs and expressing their cultural values.

5.3.4.2.1.4.6. *Sponsor New Economies from Biodiversity*
A nation can discover new economies from biodiversity, such as new crops or pharmaceutical plants. The tax use of such things related to biodiversity could generate income for the nation, as well as limit and protect the wild sources.

5.3.4.2.1.4.7. *Specialize in Trade*
In the past, specialization has allowed human groups to successfully fit the requirements of their environment. Successful cultivation, for instance, intensified the trading of cultivars and resources between groups and permitted further specialization. Village specialization may have been a great adaptation. As a result of surplus food and larger population, specialized people can create a flow of specialized objects.

Complex societies depend on production from resources. Increased complexity requires more information processing and more integration of disparate parts. The costs of communication increase. Complex societies need control and specialization. Yet, investment in complexity yields declining marginal returns because of the increasing size of bureaucracies, increasing taxation, and costs of internal control.

Overspecialization reduces flexibility and ability to change, but underspecialization reduces efficiency. Specialization is a way of limiting problems. If the numbers of specialties were reduced, there would be less overhead and higher returns. According to economist David Ricardo, the patchy distribution of resources is not the only reason trade is profitable—trading allows peoples to produce limited items more efficiently, allowing a better payback on their efforts.

Through specialization, now, each nation could become an expert in one area and could reduce taxes in that area. In a global economy, economies of scale are not as important nationally (and therefore, nations do not have to be large at all to have scale advantages, and therefore ethnic groups could form their own nations without giving up some economic advantage). The higher standards of living can be gotten by clever trade and specialization, after self-reliance is achieved.

5.3.4.2.1.5. *Balance Total Budgets for a Nation*
Nations have traditionally promised security and betterment in return for loyalty and service, but the costs of making good on the promises has forced many nations to increase revenues or reduce their obligations. Reconsider. A nation has to first identify the components, all the components, of the budget, that is, all incoming and outgoing energy and materials, as well as symbolic wealth. Then balance the flows over the desired survival time.

When countries started taxing to raise revenue, it made sense to tax what they could. But, after centuries of dramatic growth and development, these old forms of income need to be shifted to forms that can shape self-reliant and constructive behavior. Perhaps tax shifts can put the budgets of nations back in balance.

5.3.4.2.1.5.1. *Income for a Nation*

Income is the form of usable or tradable wealth that is used by the nation to meet its goals and pay its bills. There are many ways of raising funds. A nation could rent its land, sell the services of its citizens, or ask for donations from its citizens. It could sell its resources. Or, it could tax the success of its production and invention.

5.3.4.2.1.5.1.1. *Taxes for a Nation*

This term tax, in its most extended sense, includes all contributions imposed by the government upon individuals for the service of the state, by whatever name they are called: Tribute, tithe, talliage, impost, duty, gabel, custom, subsidy, aid, supply, excise, or other name. A tax is any charge of money or property that imposed by a government upon individuals or entities that are within the government's authority to assess such charges. This term, however, generally does not include charges imposed in exchange for the provision of specific goods or services, such as bridge tolls or sanitation fees.

Although most modern taxes are levied on the basis of economic measurements such as income, consumption, property, and wealth, some governments also impose excise taxes or other taxes. Taxes are usually divided into two great classes, those which are direct, and those which are indirect. The former include taxes on land or real property, and the latter taxes on articles of consumption.

Regarding national-level taxes in the U.S., for instance, the 8th section of Article I, of the Constitution provides that "Congress shall have power to lay and collect taxes, duties, imposts, and excises, to pay" its costs, but "all duties, imposts and excises shall be uniform throughout the United States."

Taxes can have effects other than raising money for a national government. Tax can used as a tool for the equalization of use; many people are luckier or greedier than others. Taxes could be a way of internalizing costs. Taxes can encourage or discourage activities. Today our tax system tends to tax and therefore raise the price of those activities we want to encourage, like investment, employment and property ownership, and tends not to tax or tax at very modest rates those activities that we want to discourage, like pollution and resource exhaustion.

The World Resources Institute argues that current taxes on capital and labor undermine economic efficiency. A tax on capital raises the cost of capital and thus discriminates against technological innovation. A tax on labor raises the cost of labor and thus reduces employment. Displacing these taxes with use and loss taxes would improve the productivity of the economy. For instance, Dower and Repetto suggest that: "Unlike many other sources of federal revenue, a carbon tax would generate overall economic efficiency gains, regardless of how the revenues from the tax are used." A high carbon tax coupled with reduced tax rates on income and profits, according to the World Resources Institute, could generate large gains, not only from improved economic efficiency but from reduced investment in infrastructure and in reduced operating costs due to higher energy efficiency and in reduced environmental damage.

Several European countries, such as Denmark and The Netherlands, are discussing how to achieve the greatest economic efficiency and equity from tax shifting. Sophisticated models have shown that reducing the personal income tax might stimulate short term spending but has modest long term benefits. A better result might occur by directing the environmental tax revenues to expanding investment tax credits. However, that lowers the cost of new capital investments relative to labor and thus could increase unemployment. Also, higher growth could actually increase carbon dioxide emissions. A reduction in payroll taxes could help labor and an increase in investments in energy efficiency and renewables could reduce the linkages between growth and pollution. To balance these changes will be a challenge.

The concept of replacing taxes from production to the environment is a promising shift. It would radically shift the targets of taxation into a more ecological direction. Ecological tax shifts could work in several ways, but all ways would probably have similar key elements. Revenue neutrality would mean that all or virtually all the money raised by increasing taxes on pollution is returned to a community of workers and consumers by lowering or eliminating other taxes on income, payroll, property, or other things. These taxes would not add to the total tax burden, and would even contribute to tax reduction. Another element is a partial exemption for energy-intensive industries to keep them competitive.

An ecological tax shift would not harm the economy of a nation. At the same time it could guide economic behavior towards a more sustainable pattern. For a community, the shift might be transparent or trivial. A similar amount of money given to the government would be returned to the community in services or assistance. For individuals, an ecological tax shift would allow them to have more control over the amount of taxes they pay. In the past, one could only avoid income, payroll or property taxes by going broke, losing a job, or living in a cardboard box. But, one can avoid or reduce ecological taxes by reducing resource use or pollution-causing activities.

Ecological tax policies could be pro-business, but they would discourage the accumulation of corporate or individual wealth in great quantities. Assuming that the desire for wealth is more related to status than to the sickness of "misplaced concreteness," this should not be undesirable to most wealthy people—they would still have the status of having and displaying more than others.

At the same time as the shift, most exemptions, deductions and loopholes, especially those that are environmentally or fiscally damaging, would be eliminated; these things would include provisions for welfare for mature industries such as oil, mining, timber and automobiles. Government handouts to corporations work against common-sense notions of free markets, innovation and fiscal responsibility. The handouts, billions of dollars per year in the U.S., are paid for by taxpayers. Worse, they contribute to mounting environmental and health costs that U.S. society and the environment must bear.

5.3.4.2.1.5.1.1.1. *Use Taxes for a Nation*

A use tax for a nation would take a percentage for a service for any resource. It would be applied to those resources held at the national level, although the nation may impose an additional tax on community resources, for the purpose of balancing communities (see Section 5.3.4.1.2.2.1.1.). A use tax would have the effect of limiting the use of nonrenewable resources, such as coal or oil, as well as the use of slowly renewable resources, such as forests. The rate of the tax would be related to the scale of the economy, as well as to the carrying capacity of the ecological support system.

5.3.4.2.1.5.1.1.1.1. *Air Use Taxes for a Nation*

The misuse of air at the local level causes local pollution (see Section 5.3.4.1.2.2.1.1.1.). Due to its nature, however, air is also a regional and global thing. Taxing air at a national level reflects its regional aspect. Indirect instruments, such as a tax on air, are designed to reduce the scale of output, as important complementary measures in a program of cost-effective pollution control.

5.3.4.2.1.5.1.1.1.2. *Water Use Taxes for a Nation*

Water, like air, is part of local, regional and global cycles. Water is used in almost every operation of an industry or bureaucracy. Although people can use water from precipitation, rivers and ponds, much water use comes from aquifers that underlie communities and provinces. This kind of water use has to be taxed at a national level.

5.3.4.2.1.5.1.1.1.3. *Land Use Taxes for a Nation*

Where land use is large-scale and very long-term, as in agriculture or forestry, some of the taxes would be national. Agricultural land should also be taxed on its unimproved value, according to Daly and Cobb. National taxes would be collected to pay for regional planning and restoration.

5.3.4.2.1.5.1.1.1.4. *Element Use Taxes for a Nation*

Element use taxes for a nation would be applied to every element used for economic purposes, especially those elements which are present in air and water pollution. Many of these elements, especially minerals, can only be used once, although many of them, such as aluminum and iron, can be recycled. Since many of these assets, such as coal and oil, come from federal lands, which are owned by all the citizens of the nation, citizens are entitled to a fair rate of return on the assets. An element tax could capture the value of these assets for current and future generations of citizens.

5.3.4.2.1.5.1.1.1.4.1. *Carbon Use Tax for a Nation*

Carbon is a component of the earth's crust and the atmosphere. In combination with oxygen, carbon forms carbon dioxide, which is recognized as greenhouse gas and has been linked to the greenhouse effect and global warming. William Nordhaus suggested a growing carbon tax to internalize the externality of damage to the climate. This would increase the price of fuels in proportion to how much carbon was released.

The Danish government examined the impacts on their economy if they were to increase their carbon dioxide tax by about $25 per ton. They concluded that if the revenue generated were returned through income tax reductions, there would be a loss in production and a rise in unemployment, but if the revenue generated were returned by reducing the social security obligations of a business, employment and production would both rise.

5.3.4.2.1.5.1.1.1.4.2. *Nitrogen Use Tax for a Nation*

Nitrogen is also a critical element of life; it is a major part of the atmosphere. Nitrogen oxides play an important role in the atmospheric reactions that generate ground-level ozone, which contributes to smog, and acid rain. The U.S. Environmental Protection Agency (EPA) believes that nitrogen oxides can irritate the lungs and lower resistance to respiratory infections like influenza. NOx and its pollutants can be transported over long distances, so problems associated with the pollutant are not confined to areas where it is emitted.

Nitrogen gas is emitted into the atmosphere by cars and industrial processes. Some of it then returns the planet in rain, which can harm plants by changing the nutrient content of soil. Emissions from industrialized nations seem to be stabilizing and nitrogen deposition is even declining in some regions. Rapid population growth and industrialization means developing countries are becoming major emitters of nitrogen.

Since the early 1990s, Sweden has imposed an environmental charge on NOx emissions from large combustion plants. The fee is $2.18 per pound, measured as NO_2. The revenue from the charge is returned to the plants in proportion to their energy production. This refund policy allowed the tax to gain business support, yet the refund policy encourages NOx reductions.

5.3.4.2.1.5.1.1.1.4.3. *Sulfur Use Tax for a Nation*

Sulfur dioxide and NOx emissions are considered the main contributors to acid rain, which the U.S. EPA believes degrades surface waters, damages forests and crops, and accelerates the corrosion of buildings. The Clean Air Act Amendments of 1990 adopted a program to control acid rain; it introduced a market-based system for emission allowances to reduce SO_2 emissions. An emission allowance is a limited authorization to emit a ton of SO_2. The EPA allots tradable allowances to affected electric utilities according to their past fuel use and statutory limits on emissions. Once the allowances are allotted, the act requires that annual SO_2 emissions not exceed the number of allowances held by each utility plant. Firms may trade allowances, bank them for future use, or purchase them through periodic auctions held by the EPA. Firms with relatively low abatement costs have an economic incentive to reduce emissions and sell surplus allowances to firms that have relatively high abatement costs.

Another option is to tax emissions of SO_2 from stationary sources not already covered under the acid rain program. Imposing a tax of $210 per ton of SO_2 emissions from those sources would raise more than $6 billion over the 2001-2010 period, according to researchers. Most firms do not pay taxes or fees on emissions that regulations still allow, although major stationary sources must pay fees annually to cover program costs of operation permits under the U.S. Clean Air Act Amendments of 1990. Basing the tax on the terms granted in those air pollution permits would minimize the cost of administering the tax for the Internal Revenue Service. Opponents to the tax argue that it would impose an additional burden on many firms that already incur costs to comply with current regulations on emissions.

There are market incentives and disincentives. Sometimes market-place strategies have been coupled with regulatory strategies. For example, the nation could impose a cap on sulfur emissions from power plants and then create an emission offset trading system to lower the overall cost of sulfur reductions. One might argue that if the tax is stiff enough, there is less reason for an accompanying prohibition. Sweden's tax on sulfur emissions is about four times the cost of sulfur reduction. As a result, after the imposition of the tax, sulfur emissions fell by 40 percent from oil refineries.

5.3.4.2.1.5.1.1.1.4.4. *Other Element Use Taxes for a Nation*

There is a U.S. Federal Tax on Ozone Depleting Chemicals. In 1989 the U.S. Congress enacted a tax on eight ozone depleting chemicals as part of its Omnibus Budget Reconciliation Act. It extended this tax to 12 additional chemicals and raised the tax on the original 8 chemicals in the National Energy Policy Act of 1992. The Clean Air Act established caps on most chlorofluorocarbons (CFCs), with a phase out occurring around the year 2000. The tax on CFCs was $1.37 a pound in 1990 and 1991, about twice the then current product price. Recycled CFCs were exempted from the tax. The tax was raised in 1990 and again in 1992. The tax rises to $3.10 per pound in 1995 and then rises by 45 cents per pound per year thereafter. The tax is proportional to the chemical's potential for depleting the ozone layer.

In another instance, ozone at ground-level has remained a pervasive pollution problem in many areas of the United States. To control ozone pollution, the EPA has traditionally focused on reducing emissions of VOCs (and, more recently, NO_x). VOCs include chemicals such as benzene, toluene, methylene chloride, and methyl chloroform. VOCs are released from burning fuel, from fossil fuels or wood, or from using solvents, paints, glues, and other products. The EPA can tax emissions of VOCs from stationary sources, such as industrial facilities, such as chemical plants, petroleum refineries, and coke ovens, to small sources, such as bakeries and dry cleaners. The vast number and diversity of stationary sources make it difficult to estimate emissions and the cost of abatement. A tax of $2,300 per ton on all VOC emissions from stationary sources might promote some abatement and could generate slightly over $111 billion in revenues from 2001 through 2010, according to researchers. It could apply to mobile sources.

5.3.4.2.1.5.1.1.1.5. *Species Use Taxes for a Nation*

Tax would be charged on the taking of members of any species. Many species are slowly renewable or functionally nonrenewable. Many species move between local human communities. This national tax would work to encourage preservation, take-limits, or take-efficiency at the national level.

5.3.4.2.1.5.1.1.2. *Loss Taxes for a Nation*

A loss tax is a tax on losses from the capital base, that is, it is a tax on the destruction of resources, not just their use or on their negative impacts. It is similar to a dispossession tax or a capital depletion tax. For a nation, it is a tax on losses from the national base (for comparison, see Community Loss taxes). Functionally nonrenewable resources are often distributed across local boundaries, and should be managed at a national level.

This kind of loss tax is on things or processes that interfere with other things and processes, things that cause runaway feedback or the destruction of cycles, things in other words that reduce our continued use of and enjoyment of the earth. Many of these things have been subsidized for many decades as a result of the power of special interests. The purpose of this kind of tax is to change behavior that depletes resources and discourages labor. This tax would have the effect of internalizing both ecological and social costs; since all consumers would be paying the real costs, no consumers would be protected.

5.3.4.2.1.5.1.1.2.1. *Land Conversion Loss Taxes for a Nation*

This tax would be on the conversion of complex systems to simpler, and more expensive to manage, systems, for instance, the conversion of forests to agricultural fields, or the conversion of fields to asphalt-covered parking lots. This tax on a national level would pay for the required planning to keep land in optimum national or human coverage.

5.3.4.2.1.5.1.1.2.2. *Nonrenewables Loss Taxes for a Nation*

Nonrenewables includes geothermal energy, as well as fossil fuels, such as oil and coal. These resources represent uniquely rich resources and have to be taxed at a national level due to their uniqueness. Furthermore, they should only be used as transitional resources, although they have been treated as limitless resources in the past.

5.3.4.2.1.5.1.1.2.2.1. *Geothermal Loss Taxes for a Nation*
Geothermal energy would be taxed because it is not permanent; it is the long-time flow of the energy of formation of the planet to space. The U.S. Geothermal Energy Act of 2004 (H.R. 4094) would reduce that nation's reliance on oil, gas, and coal-fired electrical power plants by authorizing a new assessment of that nation's geothermal resources and expanding geothermal energy investment by offering tax incentives to develop these resources. However, the use of that nonrenewable source would be taxed at a relatively low rate.

5.3.4.2.1.5.1.1.2.2.2. *Fossil Fuel Oil Loss Taxes for a Nation*
Oil and coal are often located under national territories and may extend between borders. Nations should tax them due to their nonrenewability. And, they should be used as transitional energy sources, now that their maximum economic extents have been basically mapped and their economic cost calculated. Oil and gas should also be taxed because their production and use can damage and destabilize ecosystems.

5.3.4.2.1.5.1.1.2.2.3. *Natural Gas Loss Taxes for a Nation*
Because of its distribution, natural gas may be taxed more effectively at a national level. This would allow better planning and transition to renewable fuels.

5.3.4.2.1.5.1.1.2.2.4. *Coal Loss Taxes for a Nation*
Many environmental and medical costs would have to be addressed at the national level. This additional tax would pay those costs, as well as additional costs for planning at the regional level.

5.3.4.2.1.5.1.1.2.3. *Slow Renewables Loss Taxes for a Nation*
These resources, such as forests and fisheries, are essentially nonrenewable in terms of a human lifetime. They are also not distributed uniformly and cross community boundaries.

5.3.4.2.1.5.1.1.2.3.1. *Forests & Trees Loss Taxes for a Nation*
Nations need to tax the losses of forests and trees, especially considering the status of the earth as a forest planet. Forest land should be taxed on its land value. Since the forest is the capital, and trees characterize the forest, the depletion tax on trees needs to be high. If enough trees are removed the forest dies and the capital can be lost. This tax might be specific to national lands, forests and parks.

5.3.4.2.1.5.1.1.2.3.2. *Fisheries & Fish Loss Taxes for a Nation*
Fish can be caught without being produced. The constancy of the ocean, lake and stream environments, even with tides and shifts, has allowed many kinds of animals to live in large populations. Water tends to cross ecosystem boundaries; fish also move between ecosystems, quite dramatically in the case of salmon. A national tax on fisheries (the fish-bearing environments) and species takes could allow some populations to recover.

Populations can shift to different norms with exploitation. Some populations are stable, but not resilient, such as U.S. Great Lakes fish. C.S. Holling notes the pattern of fishing pressure in the history of the Great Lakes: There is a period of intense exploitation, during which there was a prolonged high-level harvest, followed by sudden and precipitous drops in population. Although not entirely unexpected for sturgeon, with slow growth and late maturity, it was unexpected for herring and whitefish. Fishing pressure shifts the age structure of populations toward younger ages. Apparently, fishing progressively reduced the resilience of the system, so that when an inevitable unexpected event took place, the populations collapsed, with a ripple effect on fish-dependent species, including birds and parasites.

5.3.4.2.1.5.1.1.2.4. *Fast Renewables Loss Taxes for a Nation*
Fast renewables are things that are usually part of daily, monthly or annual cycles, such as solar power, hydropower or wind power. They can be taxed at a relatively low rate or even a negative rate to encourage development.

5.3.4.2.1.5.1.1.2.4.1 Wind Power Loss Tax
Wind power would have a negative tax (or tax credit) of 1 cent per kilowatt-hour for electricity produced from or for national interests.

5.3.4.2.1.5.1.1.2.4.2. Solar Power Loss Tax
Solar power would have a negative tax of 2 cents per kilowatt-hour for electricity produced from or for national interests. There would be tax credits for solar domestic or cross-community business water heating.

5.3.4.2.1.5.1.1.2.4.3. Hydropower Loss Tax
Hydropower would have a negative tax of 1 cent per kilowatt-hour for electricity produced from or for national interests.

5.3.4.2.1.5.1.1.2.5. *Waste Loss Taxes for a Nation*
Various wastes, from solid wastes to water and energy, would be taxed at amounts necessary to discourage their use or encourage their incorporation in industrial cycles. A national tax on waste is an additional tax designed to catch oversights or adjust equalizations in a process.

5.3.4.2.1.5.1.1.3. *Adjustment Taxes for a Nation*
Adjustment taxes are regulation or correction taxes to regulate or harmonize cures and causes. Many forms of liquid or solid waste would be taxed by volume, for example, in the U.S. at a rate of 50 cents per cubic foot. Other forms of solid waste, such as hazardous waste, tires, batteries, or nuclear by-products would be taxed sufficiently to pay for their isolation or break-down. Recyclable solid waste, such as animal excrement or rock, would not be taxed if it was recycled properly. A national tax would allow the nation to recover specific costs for special wastes that a community might not be able to.

5.3.4.2.1.5.1.1.3.1. *Sin Adjustment Taxes for a Nation*
A sin tax is collected to offset the consumption of activities that would require eventual health care. If people do dangerous things, and expect their government to take care of them, then government has to tax those things that result in illness. This tax would benefit people who suffer from bad habits and offset their health care. Due to its relatively high rate, it

might reduce the consumption of addictive substances. The national tax would be used to coordinate and equalize health care costs.

5.3.4.2.1.5.1.1.3.1.1. Cigarette Adjustment Taxes for a Nation

Inhaling the smoke from burning in plants in moderation may be stimulating, especially in ceremonial settings. However, overuse has serious health consequences. An industry exists that makes profits by selling such things to users and addicts.

Like an excise tax, an adjustment tax on sales of cigarettes is a fixed fee on each pack of cigarettes sold. The cigarette tax for the nation would level out community taxes. It varies by state in the U.S. and ranges from $0.07 per pack in South Carolina to $2.46 per pack in Rhode Island. The tax doubles or even triples the retail cost of cigarettes in some states.

5.3.4.2.1.5.1.1.3.1.2. Alcohol Adjustment Taxes for a Nation

This relatively high tax would offset the increased medical and social expenses from its use. Taxes at a national level would fill community gaps and ensure coverage for those harmed by the personal and social effects of excessive use.

5.3.4.2.1.5.1.1.3.1.3. Drug Adjustment Taxes for a Nation

Many nations try to control the kinds of drugs used. Legal drugs would be taxed according to their purpose or perceived social benefit. Illegal taxes would be taxed at a higher rate. If all substances were legalized, then taxes could be used to control their use. There could be additional social benefits, such as decriminalization and deprisonment, and all the costs that that entails.

5.3.4.2.1.5.1.1.3.2. *Pollution Adjustment Taxes for a Nation*

All forms of pollution would be taxed, especially those related to regional and global cycles of the elements necessary for metabolism of living beings.

Pollution charges would be collected on the amount of pollution that a national firm or its product releases into the air, water, or soil. A pollution tax would be a means of tackling the 'market failures' that arise when businesses and consumers are not confronted with, or responsible for, the full health and environmental costs associated with their activities.

There is no doubt that introducing new pollution charges would be challenging in any political system. As a result of declining income and growing budget deficits, however, this approach may become more attractive. Revenue from pollution charges could be utilized to dispense with distortionary taxes as part of an innovative, revenue-neutral tax reform. For instance, they could offset payroll or income taxes. Pollution charges should not even involve additional disruption; they are being used in many countries and several U.S. states. China is using charges to address some of its environmental problems, including water pollution.

5.3.4.2.1.5.1.1.3.2.1. *Industrial Pollution Adjustment Taxes for a Nation*

As a direct result of many industrial processes, solid particles and liquid droplets can be found in the air, ground or water, in a variety of sizes and concentrations. According to U.S. Environmental Protection Agency (EPA) studies, the emissions of particles, especially if combined with other air pollutants, are linked to some adverse health effects. For example, particulate matter can carry heavy metals and cancer-causing organic compounds into the lungs, increasing the incidence and severity of respiratory diseases. Other health effects may include chronic bronchitis.

Since monitoring systems and a permit system are already in place for coarse particle emissions, emissions from stationary sources could be taxed. That tax could be administered similarly to the taxes on sulfur and nitrogen oxides. A relatively small tax per ton of coarse particulate matter could raise over half a billion dollars per year. The national tax on would address mobile pollutions and even out disparities at the community level.

5.3.4.2.1.5.1.1.3.2.1.1. *Carbon Pollution Adjustment Taxes for a Nation*
A carbon pollution tax would link the effects of burning these fuels to the cost of repairing the damages from burning. A carbon tax could greatly offset other tax rates, as well as reduce energy consumption, especially fossil fuel consumption, and address climate change. A number of studies have evaluated the macroeconomic impact of a tax on carbon equal to $100 (USD) a ton. Although there would probably be a small reduction in GNP, which is itself a flawed measure, there would be offsetting tax reductions elsewhere in the economy; and, some of the tax money would go toward improving efficiency.

Displacing these taxes with carbon and other pollution taxes would improve the productivity of the economy. A high carbon tax, coupled with reduced tax rates on income and profits, according to the World Resources Institute, could generate a possible gain of 45-80 cents per one dollar of tax shifted. The gain would come not only from improved economic efficiency but from reduced investment in infrastructure, in reduced operating costs due to higher energy efficiency, and in reduced environmental damage.

5.3.4.2.1.5.1.1.3.2.1.2. *Phosphates Pollution Adjustment Taxes*
Phosphorus is an element. Phosphate is a salt of phosphoric acid. The tax is for the mineral phosphorus content, which is used in animal feed and can cause pollution; phosphorus may also be related to red tide. A tax might reduce excess phosphorous used in agriculture by limiting the use of phosphate in animal feed. This would require knowing what stock owners have, and letting them know how much and when to pay.

5.3.4.2.1.5.1.1.3.2.1.3. *Nitrogen Pollution Adjustment Taxes for a Nation*
Nitrogen is a significant part of the earth's atmosphere. Nitrogen gas emitted into the atmosphere by cars and industrial processes can be a pollutant, however. Some of it then returns the planet in rain, which can harm plants by changing the nutrient content of soil.

Researchers have warned that rising nitrogen emissions from developing nations will soon threaten plant life in some of the most biodiverse parts of the planet. A team led by Gareth Phoenix of the University of Sheffield has shown that, in the mid-1990s, the average amount of nitrogen deposited on the planet's 34 biodiversity 'hotspots' was more than 50 per cent higher than the global average. They say this figure could more than double by 2050, at which time nitrogen levels in 17 of the 34 hotspots will exceed critical levels that European nations have set to protect their sensitive ecosystems. As a result, many of the hotspots will soon be in danger of being damaged by high levels of nitrogen, say the researchers. They add that this may already be true for some areas. The researchers point out that rapid population growth and industrialization means developing countries are becoming major emitters of nitrogen. Emissions from industrialized nations, on the other hand, are stabilizing and nitrogen deposition is even declining in some regions. This suggests developing countries are responsible for the damage suffered by biodiversity hotspots.

Since January 1, 1992 Sweden has imposed an environmental charge on NO_x emissions from large combustion plants. The fee is $2.18 per pound, measured as NO_2. The revenue from the charge is returned to the plants in proportion to their energy production. This refund policy allowed the tax to gain business support, yet the refund policy still encourages NO_x reductions. The average cost of reducing one kilogram of NO_x is $1.20 while the tax is $2.18 per pound.

5.3.4.2.1.5.1.1.3.2.1.4. *Sulfur Pollution Adjustment Taxes for a Nation*
Sulfur is an element necessary for life. It is also used for processes in refineries, smelters, mills, and chemical plants, often producing sulfur oxides as pollutants that contribute to smog and respiratory diseases.

Some nations impose a cap on sulfur emissions from power plants and a tax on sulfur emissions. Sweden's tax on sulfur emissions is about four times the cost of sulfur reduction. As a result, after the imposition of the tax, sulfur emissions fell by 40 percent from oil refineries.

5.3.4.2.1.5.1.1.3.2.1.5. *Elements & Compounds Pollution Adjustment Taxes*
In 1989 the U.S. Congress enacted a tax on eight ozone depleting chemicals as part of its Omnibus Budget Reconciliation Act. It extended this tax to 12 additional chemicals and raised the tax on the original 8 chemicals in the National Energy Policy Act of 1992.

The Clean Air Act established caps on most chlorofluorocarbons (CFCs), with a phase-out occurring around the year 2000. The tax on CFCs was $1.37 a pound in 1990 and 1991, about twice the then current product price. Recycled CFCs were exempted from the tax. The tax was raised in 1990 and again in 1992. The tax went to $3.10 per pound in 1995 and then rose by 45 cents per pound per year thereafter. The tax is proportional to the chemical's potential for depleting the ozone layer.

Many other elements and compounds are toxic, and are used in cleaners and biocides. Taxing them at a national level would coordinate use and research, as well as limits.

5.3.4.2.1.5.1.1.3.2.2. *Agricultural Pollution Adjustment Taxes*
The use of fertilizers and pesticides, as well as specialty and energy-intensive machinery would be taxed at a national level to offset their costs and effects.

5.3.4.2.1.5.1.1.3.2.2.1. *Pesticides Adjustment Taxes for a Nation*
Because of the adverse impacts on human health and the environment, as a result of pesticide use, and the direct financial costs, such as the treatment of water, and wider environmental costs, such as loss of biodiversity, which are much harder to identify and value, there would be national taxes on pesticides (perhaps under an umbrella term, like biocides).

5.3.4.2.1.5.1.1.3.2.2.2. *Fertilizer Adjustment Taxes for a Nation*
Fertilizer is used to increase the productivity of crops, but it causes problems. Even natural fertilizers, such as dung or plant matter, can cause pollution and damages at large scales. The application of national environmental tax shifts to pesticide and fertilizer use could provide an incentive to minimize pesticide and chemical fertilizer use and could generate revenue for tax relief in other areas.

5.3.4.2.1.5.1.1.3.2.2.3. *Water Adjustment Taxes for a Nation*
The pollution of water from water courses or aquifers would be taxed by a national government. The rate would be equal to or greater than the cost of cleaning or purifying the water.

5.3.4.2.1.5.1.1.3.2.3. *Personal Pollution Taxes for a Nation*
A personal pollution tax would be imposed on the use of things that create pollution through personal use, such as packaging or fuel. A fuel tax, for instance, would be imposed on the sale of all fuel regardless of use.

5.3.4.2.1.5.1.1.3.3. *Heritage Items Sale Adjustment Taxes for a Nation*
A heritage tax would be applied to any thing considered to be part of a natural or cultural heritage, such as natural bridges or famous bridges. It would include unique art works produced by people in a culture. It would also include a tax on the export of anything without a value-added component, such a raw logs or raw minerals.

The Heritage tax would be similar to a tariff. One function of a tariff would be to protect special accomplishments or special resources; another would be to raise money for the government and to protect home industries. Too high a tariff would prohibit all imports. Too low a tariff would not protect home industries from those who had less environmental protection or paid fewer employee benefits. Tariffs could be used to encourage economic self-sufficiency and give a government more control of its economy.

5.3.4.2.1.5.1.1.3.4. *Financial Speculation Adjustment Taxes*
The speed and detachment of money as a medium of exchange may artificially change the values of things that should remain uninflated or undepressed. A speculation tax, like a Tobin tax, is a small tax on each international financial transaction.

There are many potential benefits from a modest tax on financial transactions, such as the buying and selling of shares of stock or blocks of foreign currencies. Such a tax would have the effect of reducing short-term speculation in these markets, thereby making them somewhat less volatile. It would slow capital movements. It would also cut back some of the economic resources that are wasted in these transactions, since if the number of trades declines, the money spent on these trades would decline as well. In addition, it would make the tax code fairer, since most financial speculation is conducted either directly or indirectly by wealthy people.

But, it would have to be collected in every country. Otherwise some countries would not collect it to increase their advantages. One or two countries could not afford to impose such a tax, unless all countries were to cooperate. If one country imposed a speculation tax unilaterally, then traders would simply move their business to a nation that did not tax their transactions; the internet, for instance, has given traders an unprecedented mobility. Unless mobility were controlled, the main effect of the tax would be to deprive the markets of the pioneer country of financial business. If large trading blocs were to adopt it, like the European Union, and it was shown to be manageable, then an international agreement on speculation tax, binding on all countries, could be arranged at the global level.

As with all taxes, there will be opportunities for evasion. The incentives for evasion are much greater in the case of money laundering, but there are some effective regulations on money laundering. The demands on financial institutions under those regulations are comparable to the demands that would be imposed with the imposition of speculation taxes. The incentives for evasion in case of copyright laws is also great. The fact that governments can protect copyrights for publications and internet suggests that a speculation tax can still be enforceable. Collection of the tax at the national level would help with coordination and enforcement.

5.3.4.2.1.5.1.1.4. *Distribution Taxes for a Nation*

Distribution taxes are the same as reapportionment for luxuries and big incomes. Combined with heroic inheritance and profits, these incomes concentrate wealth and essentially remove their receivers from any sense of participation in the community or nation. Although such taxes would take more from those who have more, they should not discourage people from working to have more, as well as getting higher status from paying more.

5.3.4.2.1.5.1.1.4.1. *Heroic Possessions Distribution Taxes for a Nation*

A heroic possessions tax is a tax on products that are not considered essential, in other words, luxuries. A luxury tax is similar to a sales tax or VAT, except that it mainly affects the wealthy because the wealthy are the most likely to spend heavily on luxuries such as expensive cars or jewelry.

The provisions of a luxury excise tax in the U.S. were contained in the Omnibus Budget Reconciliation Act of 1990, which was intended to reduce the federal deficit by enforcing a 10 percent surcharge on high- priced products of the automobile, boat, aircraft, jewelry, and fur industries. Effective on January 1, 1991, the tax applied to the 'first retail sale' of luxury goods with a sales price above the following thresholds: automobiles $500,000; boats $100,000; aircraft $25,000; and jewelry and furs exceeding $10,000. The tax was projected to provide revenues of $9 billion in the first five years. It did not.

There was controversy from ambiguities in the determination of tax liabilities. There was concern that the tax was unjustly levied on only those five industries. There were difficulties in the administration of the rule in terms of compliance of the five industries. There was criticism of subjective loopholes of the regulations, evident in the cases of the boat and automobile industry. It was argued that the imposition of the tax aggravated the plight of luxury industries already depressed by current economic conditions and losing jobs and profits. Legislators levied the tax on only five industries, to make the tax easier to administer. Electronics was not used a category because of different types of dealers and equipment. Houses and furnishings were not considered.

For the consumer, the tax seemed too complex and the thresholds difficult to understand or apply. For opponents, the tax was worrisome because, once the statute was in place, it would be simple to lower the threshold amounts or to increase the rate. Whereas autos, planes, and boats require licenses and leave a paper trail, jewelry and furs can be purchased and hidden, making it difficult to collect a tax.

Although a luxury tax of some sort may be justified on many things, the regulations have to clarify implementation in order to achieve at least some degree of equity. This tax would be coordinated at the national level, considering the multiple residences of those who can afford other luxuries. Regulations have to broaden the definition of luxury. It is hard to argue that a luxury tax is unjust and counter-productive, considering the massive suffering and starvation outside the gated compounds.

5.3.4.2.1.5.1.1.4.2. *Heroic Income Distribution Taxes for a Nation*
A heroic tax would be applied to income over an acceptable ratio. Herman Daly and John Cobb quote a range of the acceptable inequality of income at ten to one, although some corporations, like Ben & Jerry's ice cream used to limit it to a one to seven ratio. As Cobb and Daly point out, the idea of unlimited inequality works against the notion of community. As they also wisely point out the goal of a community is not some perfect ideal of equality, but a limited inequality that allows individual differences to show, and individual rewards for luck or skill, but also allows other individuals to catch up within a decade or generation.

A heroic income tax would not prevent some people from getting richer, but it would slow the rate and limit the vast differences. The poverty line in the U.S. in 2005 for one person is $9570; for a family of four, it is $19,350 USD. There is no maximum income, although surprisingly many people receive many millions per year. That would mean that if the minimum for a family of four in the United States was set to $20,000, the maximum income for the family would be $200,000. Although limiting wealth to a ratio might seem like unfair and confiscatory taxation, it would not be dissimilar to the way that many people achieved great wealth—by confiscating free goods of nature or cheating the less powerful. So, this tax rate would be quite high for those with heroic incomes. Paying such a tax would make these people heroes and in fact they would likely get heroic status for paying.

5.3.4.2.1.5.1.1.4.3. *Heroic Transfer Distribution Taxes for a Nation*
A transfer tax is a tax on any transfer of wealth, including inheritance, death duty, estate or bequest. It is applied to any transfer of wealth, to spouses, children, relatives, and even bequests to trusts, foundations and charities, and not on the total wealth.

Due to a unique set of circumstances in the past, including luck, theft, hard work, and inheritance, many large fortunes do not represent taxed income or savings, but they do represent unrealized capital gains which would never be taxed as capital gains under the U.S. federal income tax, for example. Those most affected would be estates of considerable size, after applicable credits. The effect on family-owned farms would be minimal. According to the Center on Budget and Policy Priorities, the American Farm Bureau Federation acknowledged to the *New York Times* that it could not cite a single example of a farm having to be sold to pay estate taxes.

This tax would require heavier regulation. This tax would serve to prevent the perpetuation of wealth, free of tax, in wealthy families. It would work towards reducing the accumulation between generations, thus leveling the field more for the next generation.

5.3.4.2.1.5.1.1.4.4. *Heroic Profit Distribution Taxes for a Nation*
Two words: ExxonMobil. One acronym: BP.

5.3.4.2.1.5.1.1.5. *Discussion of Missing or Transitional Taxes for a Nation*
Under this scheme, there would be no national taxes on buildings, equipment or inventories. There would be no national corporate income tax, although, as Daly and Cobb suggest profits would have to be distributed to all shareholders as income. There would be no personal income tax, no property tax, and no sales tax, so that things people want or work for should be encouraged. Once the real price of oil and other resources has settled out, no other taxes are needed for value-added efforts or goods. In order to reach this position, however, there may be a series of transitional taxes, such as an added tax on all new vehicles.

5.3.4.2.1.5.1.2. *License privileges for a Nation*
A license is a formal authorization by law to do something, such as marry, hunt, or practice law or medicine. The granting of licenses has a long history, in many nations and cultures.

5.3.4.2.1.5.1.2.1. *Marriage License for a Nation*
A marriage license is permission from a legal authority for the marriage of two people to be performed. The requirements differ depending on the time and place. Combined with other rights or privileges, such as having children, a marriage license would be required to indicate preparedness and intent. Although this license is required at a community level, it would be available at a national level, also.

5.3.4.2.1.5.1.2.2. *Reproduction License for a Nation*
In the early 1970s, U.S. Senators Bob Packwood and Jacob Javits seriously discussed having the government license parents to have children. A reproduction license would be permission from a legal authority for two people to have offspring. The requirements would differ depending on the time and place. The purposes of such a license would be several, starting with qualifying the parents, through a reading program or knowledge of health issues, such as smoking or drinking, and continuing with the health and safety of children, and with consideration of the effects on the health and fitness of the entire gene pool and on society.

Although this license would be granted at a community level, it might be coordinated at a national level. The penalty for not having the license might range from a fine to reallocation of the child to another family or institution, although there would have to be numerous safeguards for the health and safety of the child. One incentive for having the license might be a monetary rebate. Another could be a grant towards education or luxuries. This license would not limit the number of children or help with the planning of institutions, such as schools—that would be the purpose of tradable vouchers to have children (see later).

5.3.4.2.1.5.1.2.3. *Driving License for a Nation*
A driver's license is an official document which states that a person has the necessary qualifications to operate a motorized vehicle, such as a motorcycle, car, truck, camper, trailer, or a bus. Driver's licenses are generally issued after the recipient has passed a successful driving test and proven that they meet the age, education and understanding requirements. Although licenses would be awarded at a community level, they might be coordinated or duplicated at a national level, especially for national service or representation.

5.3.4.2.1.5.1.2.4. *Voting License for a Nation*
A license to vote would be required to demonstrate that the voter is living, lives in the community, speaks the language, and understands the basic issues. This could prevent some kinds of fraud. Although this licenses would be awarded at a community level, it would be duplicated at a national level for national elections.

5.3.4.2.1.5.1.2.5. *Media License for a Nation*
Any use, for broadcast or communication, of common frequencies of the spectrum of waves, would be licensed by the agency responsible for regulating interstate and international communications by radio, television, wire, satellite, or cable. This license would be coordinated with community-issued licenses.

5.3.4.2.1.5.1.2.6. *Business License for a Nation*
The nation would only need to issue a separate business license if the business operated across communities, provinces or states.

5.3.4.2.1.5.1.2.7. *Collecting License for a Nation*
Anyone who collects plants, animals, fish, or any living substance for its compounds or DNA, would be required to have a national license for collecting in national parks or preserves. The license would certify that they had the knowledge to collect, and it would limit the number of people collecting. Issuing licenses would help protect national commons, such as parks and wilderness areas, as well as coordinate the timing and limits of collection. The nation would only need to issue a separate license if the business operated across provinces or states or on federal lands.

5.3.4.2.1.5.1.2.8. *Weapons License for a Nation*
A weapons license would verify that knowledge and experience requirements are met and reviews and researches of criminal history have been conducted for information that might preclude the issuance of a license, such as a criminal record.

The nation would have a bureau responsible for the issuance and denial of licenses. This bureau would receive and examine licensing applications for statutory compliance and verify the applicant's eligibility for licensing through former employers, educational facilities and examination of criminal history records. The kinds of weapons could be limited to knives, arrows, single-shot firearms and multi-shot fire arms; certain weapons, such as automatic guns, tanks, bazookas, and missiles, would not be licensable. Applying for a license to carry a weapon for self-defense is a right of the law-abiding citizens. Appropriate personal weapons can be carried responsibly, properly and safely.

A nation needs to stop the spread of assault rifles, pistols, hand grenades, mortars, and other so-called light weapons. There is a growing realization among governments that in the post-cold-war world that the high-powered automatic small arms, more than cannons or missiles, are the prime contributors to regional instability, fueling ethnic or nationalistic wars with uncountable casualties. Governments have to decide whether to concentrate on illicit trafficking or to address the legal trade as well. Many governments consider controlling the legal trade crucial to keeping weapons out of regions of conflict and the hands of dictators.

5.3.4.2.1.5.1.3. *Users fees & Tolls for a Nation*
Fees could be charged by the nation for use of national space or renewal cycles, for example, visiting fees or sanitation fees. Tolls could be charged by the nation to cover extra expenses for transportation, for instance, bridge tolls or road tolls on national highways.

5.3.4.2.1.5.1.3.1. *Conservation & Tourism Fees for a Nation*
Historically, many areas get a substantial part of the money used to market themselves as a destination from a room tax or surcharge on hotels and motels. This is commonly referred to as a 'tourism tax' because the vast majority of it is paid by tourists, as opposed to local residents, and much of the income is used to promote tourism. Although there are always exceptions to the general rule, typically, if this tax is expanded to other areas, it is generally tourist specific, such as tickets for attractions or for food and souvenirs.

5.3.4.2.1.5.1.3.2. *Cultural Treasures Fees for a Nation*
Although much of the heritage of the nation would be free to observe or visit, fees could be charged to limit access. Fees would also be used for special costs such as cleanup.

5.3.4.2.1.5.1.4. *Labor Use for a Nation*
Historically, rulers have exchanged protection and coordination for the labor services of residents or citizens, especially for large public or national projects. Although the use of labor has become much more formal, volunteers are a major source of service.

5.3.4.2.1.5.1.4.1. *Require Service to Nation for Two Years*
Nations need the services of citizens to carry out many of the plans and operations of the government. A national service provides a pool of young or skilled people who are committed to working for a certain length of time. For the people themselves, service provides experience, wages, and a sense of need. National service has been usually understood to mean a form of military service in which all citizens, or all male citizens, can participate, either voluntarily or involuntarily. The involuntary form of service has been called conscription or draft.

Conscription is a general term for involuntary labor demanded by some established authority, but it is most often used in the specific sense of government policies that require citizens to serve in their armed forces. Many nations rely on a volunteer or professional military most of the time, although they reserve the possibility of conscription for wartime and 'crises' of supply. Conscription has also sometimes been used as a general term for nonmilitary involuntary labor demanded by some established authority; for example, in Japan during World War II, Japanese women and children were conscripted to work in factories. This kind of service may last for an indefinite time, such as the length of a war.

National Service is still used to describe the compulsory military service that is still implemented in some countries, including Singapore, Greece, Germany, South Korea, Israel, Republic of China (Taiwan), and Russian Federation. According to the U.S. CIA, comprehensive national service incorporates the Army, Navy and Marine Corps, Air Force, and Coast Guard, and all these branches are controlled by the president. This service usually has a fixed period of time, such as one to two years.

Every nation should require a two-year period of service to the nation, for civil work, police work, or overseas work (as with the U.S. or Switzerland service corps). Every person would have the choice of area of service. The nation would provide training in each area.

5.3.4.2.1.5.1.4.2. *Encourage Volunteers for a Nation*
Some people may wish to voluntarily extend their conscription for a length of time, depending on the project. Many forms of volunteer work take place in a private setting or for an NGO dealing with social problems or environmental conservation. Thus, people volunteer for hospitals or libraries, food programs or the humane society. A nation could encourage this or support it, through recognition or reward. A nation could also develop its own specific programs, for national parks or political activism.

5.3.4.2.1.5.1.5. *Discussion of Income for a Nation*
Most income would come from taxes on inefficiency and waste, as well as from taxes on heroic profit and ownership. A tax on these activities should not dampen the competitive

spirit displayed by many citizens.

Some people worry that these taxes would undermine the competitiveness of domestic industry. This concern has been raised with regard to national green taxes in Europe, where countries operate under the Single Europe Treaty, which forbids taxing imports within Europe, and with regard to state taxes in the U.S., where the Constitution prohibits states from imposing taxes on interstate commerce. Countries and states may prefer to introduce shifted or green taxes cooperatively, rather than unilaterally.

Shifted or green taxes should not impose a competitive disadvantage if imposed on the vernacular economy, on the household sector, or on that portion of the business sector that does not export its products or services. The most significant impact would probably fall on energy-intensive, exporting industries. For a carbon tax, for instance, the greatest burden would fall on a energy-intensive industry, such as Portland cement manufacturers, which is rarely an exporting or interstate industry, because of its high transportation costs relative to its modest value. Nitrogenous fertilizers and a few primary metal industries, such as aluminum and steel, would experience price increases in excess of 0.5 percent, but those industries have other difficulties being competitive in some nations.

5.3.4.2.1.5.2. *Payouts for a Nation*

The nation has to pay for its existence, as well as the expenses, amenities, and services that it trades for its existence. In economics, business, and accounting, a cost is the value of inputs that have been used up to produce something, and hence are not available for use anymore. In business, the cost may be one of acquisition, in which case the amount of money expended to acquire something is counted as cost. In this case, money is the input that is gone in order to acquire the thing. This acquisition cost may be the sum of the cost of production as incurred by the original producer, with the further costs of transaction incurred by the buyer beyond the price paid to the producer. Usually, the price also includes a mark-up for profit. Costs are often further described based on their timing or their applicability.

When a transaction takes place, it typically involves both private costs and external costs. Its private costs are the costs that the buyer of a good or service pays the seller. Its external costs, also called externalities, are the costs that people other than the buyer are forced to pay as a result of the transaction. The bearers of such costs can be either individuals, communities or the nation (or the ecosystem or planet). External costs are often nonmonetary, and as a result are difficult to quantify for comparison with monetary values. They include things like pollution, things that a nation will probably have to pay for in some way at some time, but that are not included in transaction prices.

Social costs are the sum of private costs and external costs, that is, both the costs internal to the firm's production and external costs not included in the firm's production. For example, the purchase price of a car reflects the private cost experienced by the manufacturer. The air pollution created in the production of the car is an external cost. The manufacturer does not pay for these costs and does not include them in the price of the car, so they are external to the market pricing mechanism. The air pollution from driving the car is also an externality. The driver does not pay for the environmental damage caused by using the car. A psychological cost is a subset of social costs, for example as a result of endangering pedestrians, that specifically represents the costs of stresses or losses to quality of life. A community has to consider all costs for as long as possible into the future.

5.3.4.2.1.5.2.1. *Reduce Legacy Debt of National Government*
Debt reduction, especially in countries like the United States, is necessary to reduce the burden on future generations who did not receive the direct, initial benefit. Unfortunately, this may mean some sacrifice on the part of many who also did not benefit from the incurrence of the debt, although in some cases, it was themselves or their parents who benefited.

5.3.4.2.1.5.2.2. *Operate National Government*
The government itself has expenses for its maintenance and activities. The government has to pay for information services, as well as for collection and administration costs.

5.3.4.2.1.5.2.2.1. *Elect Leaders & Representatives at Government Cost*
The government has to pay for its leaders and representatives, as well as for staff to provide services. A government could consider aligning salaries to a national mean, and making it illegal to accept any campaign contributions from special interests that would expect to have special favors done for them. Tying benefits, medical and retirement to a national mean would also work to normalize government service related to the public.

5.3.4.2.1.5.2.2.2. *Open Communications Paths for Nation*
The government has to ensure that there is equal access to media. In the U.S., for instance, it has to allow more fair use of common waves in the national atmosphere.

5.3.4.2.1.5.2.2.3. *Fund Special Departments of Nation*
Research benefits a nation. Research funding is a term generally covering any monies for scientific research, in the areas of 'hard' science and technology, social sciences and humanities. The term often connotes funding obtained through a competitive process, in which potential research projects are evaluated and only the most promising receive funding. Such processes, which are run by government, corporations or foundations, allocate scarce funds. Total research funding in the richest countries is between 1.5% and 3% of GDP; Sweden is the only country to exceed 4%.

Most research funding comes from two major sources, corporations, through special research and development departments, and the community or national government, whose work is primarily carried out through universities and specialized government agencies. Some small amounts of scientific research are carried out (or funded) by charitable foundations, especially in relation to developing cures for diseases such as cancer, malaria and AIDS, or by individuals.

5.3.4.2.1.5.2.2.4. *Ensure Equal Opportunity for Education in Nation*
Equal opportunity is a descriptive term for an approach intended to give equal access to an environment or benefits, such as education, employment, health care, or social welfare, to all, often with emphasis on members of various social groups who have historically suffered from discrimination. Some protected groups include gender, race, or religion. Equal opportunity practices include an organizations' implementation of personnel practices to ensure equality in the employment process. Equal opportunity also applies to equality in housing and public accommodations. In developing countries, equal opportunity is provided by creating jobs accessible to poor people. Economic studies suggest that this is achieved by reducing barriers

to entry and allowing entrepreneurial activity.

In the U.S., the federal government and various state and local governments sometimes institute affirmative action in terms of hiring and government contracting with the attempt to overcome alleged historical discrimination against protected groups. There are three different forms of affirmative action, (1) in education; (2) in government contracting and (3) in personnel practices.

5.3.4.2.1.5.2.2.5. *Allocate Money to Critical Jobs*

In many nations, people are attracted to easy, glamorous, well-paying jobs, such as playing games, singing, or pretending to be other people. These jobs usually pay well above the mean. Many other jobs, that are equally important, such as teaching or police work, pay far less. Other necessary jobs, such as trash collector or store stocker, pay far less than those and have even less status. Perhaps less popular jobs should have higher rates of pay, to ensure that they will get done.

The nation has to try, for itself as an employer and for all employers, to make sure than applicants have equal opportunities to jobs. In many employment contexts, including government entities and federal contractors, not all employers are required to implement an affirmative action program. Employers should compare the incumbency of females and minorities performing specific jobs to the availability of those groups in the recruitment area. Where the employment incumbency in that job group is less than 80% of the availability of that group, the organization should set a better goal. Employers should widely advertise jobs with organizations who might be able to refer qualified minorities or females applicants, although once applicants apply, the employer should choose the most qualified candidate and cannot make employment decisions based on race or sex even where there are affirmative action goals.

Ways of providing equal opportunity are often subject to controversies, since selection or opportunity are difficult to measure accurately. Equal opportunity is said to exist when the outcomes are equal, when people with similar abilities reach similar results after doing a similar amount of work. As long as inequalities can be passed from one generation to another through gifts or inheritance, it is unclear that equality of opportunity for children can be achieved without greater equality of outcome for parents. All applicants should be treated in a nondiscriminatory manner in compliance with established laws, regulations and executive orders. A government should promote a culturally diverse workforce, and a work environment that allows all employees equal opportunity to achieve their full potential, without discrimination.

A government may need to undertake job creation programs in order to assist its people in seeking employment, especially during times of high unemployment or when private sector jobs are lacking or unbalanced—that is, the government may need to create new job categories or emphasize critical jobs, such as teaching or police work. The government can concentrate on making macroeconomic policies to create a supply of jobs or create more efficient ways to match employees with prospective employers.

5.3.4.2.1.5.2.3. *Issue National Vouchers at Birth*

A voucher is an economic warrant and guarantee for goods and services. A voucher is a certificate, a document, evidence or proof, of a promise or commitment, which is worth certain rights, materials or a monetary value, and which may only be spent for specific reasons or on specific goods. Examples include, but are not limited to, education, children, and medicine. The nation would either issue these vouchers for all citizens or oversee the issuance by an assigned authority.

Different sets of vouchers could be issued at birth or later in adulthood. Some vouchers might be claimed as wanted or needed.

5.3.4.2.1.5.2.3.1. *Basic Income Vouchers for People of a Nation*

A guaranteed minimum income is a proposed system of income redistribution that would give each citizen a certain sum of money independent of whether they work or not. In the U.S. this is sometimes known as a 'Basic Income Guarantee (BIG),' 'universal basic income,' or 'guaranteed annual income,' but these systems also often include a method of paying for the income as well.

The system would be a government-administered one that would allot every citizen a sum of money large enough to live on. One amount proposed is 20% of per capita GDP. A basic income voucher could be issued every month, for example, to $7000 a year for U.S. residents. This voucher would cover minimum clothing, housing, and food, although some amount might have to be paid back for carbon taxes or other taxes. Retirement would not be necessary, since payments would continue until death. This program would replace U.S. Social Security. The wealthiest as well as the poorest citizens would receive this. Salaries from employment would be a supplement to this government income.

An often proposed way of paying for this system is through a negative income tax where a government flat tax would be charged to all citizens. The current model of progressive income taxes used throughout the western world could be eliminated, but the system would still be progressive, since those at the lower end of the wage scale would pay less in taxes than they would receive in guaranteed income. For the most wealthy members of society the few thousand dollars of the guaranteed income would only make a small dent in the taxes they would have to pay.

In economics, a negative income tax (sometimes abbreviated NIT) is a method of tax reform that is popular among economists, but has never been fully implemented. It was developed by Juliet Rhys-Williams in the 1940s and later by economist Milton Friedman in 1962. Negative income taxes can implement or supplement a guaranteed minimum income system.

Proponents of a guaranteed minimum income argue that the system has numerous advantages. First, it would simplify the welfare state. The introduction of a guaranteed minimum income could also see the elimination of traditional welfare, the minimum wage, much of unemployment insurance, government pensions, and benefits for the disabled and ill. This would eliminate large amounts of government bureaucracy. It could also see the elimination of the progressive income tax with no adverse effects on the poor, as explained above.

Next, it would prevent any citizen from falling into abject poverty. With a guaranteed minimum income, *involuntary* starvation and homelessness would be eliminated. And, it might eliminate a few of the major problems of the modern welfare state, such as the welfare trap, which is presumed to discourage people from working.

A minimum income could change the character of work. People would have the opportunity to select other kinds of jobs. It would allow them to do work that is productive but does not normally provide income, such as caring for children or the elderly within one's own family, providing public goods, or working for NGOs. There may be other effects, ranging from increases in employment to depression of wages.

5.3.4.2.1.5.2.3.2. *Medical Vouchers for Nation*
The American College of Physicians has begun the process of formulating a new voucher system of financing medical care that could ensure universal access to health care in the United States, based on the strengths of existing health care financing mechanisms.

Vouchers could vary in value over time, and may be specialized for use in the separate phases of medical care. They could be individualized for use by a single patient or a single physician. For instance, American Express's new card, Quattro, will be used by 500 of its workers and those of John Hancock Financial Services in the Boston area.

A double voucher system for patients and physicians could provide mechanisms to assure that the care decisions reached between patient and the physician are not subject to prior review or micro-management by the insurance carrier, which has been a cause of the increasing frustration and anger felt by physicians and patients over denials of care. The voucher solution will go a long way to restoring the beneficence needed for everyone in the medical interaction.

There are a number of ways a voucher could work; it could be good for 24 months of treatment, or for a lifetime. It could also be traded. Birth and death services might be free.

5.3.4.2.1.5.2.3.3. *Productivity & Resources Vouchers for Nation*
Here is one possible way the numbers could be determined: After calculating an optimum population for the planet (based on totals of ecosystem productivity and resources), that number is divided into the estimated quantities for each resource on the planet. Based proportionately on territory, each nation is given a number of vouchers for the optimum use of resources and processes. Nations may then choose to set their populations accordingly, or to make decisions based on other values.

People would have access to a share of the total natural plant productivity in an area. This would replace an idea of straight calculation of 2.5 hectares per person. And, that would be after other land uses were subtracted.

5.3.4.2.1.5.2.3.3.1. *Carbon & Other Elements Vouchers for Nation*
Every nation would be issued a number of vouchers equal to the carbon-bearing products of it population and their standard of living, based on a world standard. These would determine how much was used in each national system, regardless of its population or the intensity of its exploitation. Other elements, such as copper and sulfur, would be treated the same way.

5.3.4.2.1.5.2.3.3.2. *Vouchers for Water, Air & Other Compounds*
The requirements of living, such as breathing or drinking, should not require vouchers, except under extreme circumstances of overpopulation or overconsumption. Trying to issue vouchers for these things, on a personal level, might be too expensive and too invasive, although assigning the vouchers might protect basic needs.

5.3.4.2.1.5.2.3.4. *Education Vouchers for Nation*
National educational vouchers could be issued for skills that are in short-supply or demand. Education would not be required per se, but it would be required to have a driving or voting

license. The nation would reserve the option of requiring the first six to twelve years of education, but any individual could claim the education later. A nation could issue other vouchers for post-secondary education, especially to emphasize areas where skills were needed.

5.3.4.2.1.5.2.3.5. *Reproduction Vouchers for Nation*
A nation might have to issue child or reproduction vouchers, depending on the self-reliance or choices of its communities. If communities were balanced at the national level, then the vouchers would be controlled at the national level. This would allow some communities to have more or fewer children, based on their suite of values. China has something like this in place now, that is, there is a limit of one or two children per family, although children can be placed with childless relatives. Furthermore, the most of the Chinese people recognize that places are overcrowded and think this is fair.

5.3.4.2.1.5.2.3.6. *Other Vouchers for Nation*
All the previous vouchers may be traded. This may be necessary in the case of partial vouchers. Should there be vouchers for luxury? In a random lotto system for fairness? Or should a voucher be a reward for generosity?

5.3.4.2.1.5.2.4. *Special Accounts for a Nation*
In common usage, saving generally means putting money aside, for example, by putting money in the bank or investing in a pension plan. In a broader sense, saving is used to refer to cost-cutting costs or to rescue. In terms of personal finance, saving means to preserve money for future use, by putting it in a safe place, such as a mattress or banking institution. Saving refers to an activity occurring over time, a flow variable, whereas savings refers to something that exists at any one time, a stock variable. The flow refers to an increase in assets, whereas the stock refers to a designated part of the entire assets.

For a nation, special savings accounts would be set up to lessen the extent of catastrophes or emergencies. One of those would be to keep stores of food to feed each community for as long as possible, or at least for seven years. There would also be accounts for helping other nations and for paying dues to the Associated Nations organization.

5.3.4.2.1.5.2.4.1. *Start National Savings Account*
A nation needs to start savings accounts for lean years. Each nation should establish at least one account for food and one for nonrenewables. Perhaps each account should hold a seven-year supply. Genesis in the Christian Bible provides the model: 36 "And the food shall be for a store to the land against the seven years of famine, which shall be in the land of Egypt; that the land perish not through the famine." Some nations may prefer to set aside money, although food and supplies may be more certain than the value of money during a world-wide catastrophic famine.

5.3.4.2.1.5.2.4.2. *Start National Account for International Charity*
A nation should also keep accounts to help the wildlife and people of other nations. This would not only provide safety nets for other nations beset by temporary problems or catastrophes, it would promote trust between nations.

5.3.4.2.1.5.2.4.3. *Pay National Share of Global Costs*
Every nation would be expected to pay a share to support the Association of Nations (AN). As the AN would be self-supporting, with its own forms of income, this would be a form of membership dues.

5.3.4.2.1.5.2.5. *Discussion of National Payouts*
A nation could afford to make these kinds of payouts because of the restructuring of income. In fact, just suspending many subsidies and handouts would free billions of dollars for many nations.

5.3.4.2.1.5.2.5.1. *Remove Legacy & Current National Subsidies*
Many governments subsidize activities that threaten their own sustainability. Lester Brown points out that fishing fleets are subsidized at approximately $54 billion per year (2000), despite the fact that fishing capacity exceeds the capacity of the ocean to yield sustainable fish. Norman Myers and Jennifer Kent placed U.S. agricultural subsidies at $390-520 billion (USD); $110 billion to fossil fuels; and $220 billion for water. All subsidies are over $2 trillion. By comparison if all U.S. citizens received $7,000 per year in negative taxes that would cost under $2 trillion.

Another example is the costs of modern industrial farming in England: $2 billion (USD) for removing pesticides from drinking water, damage from soil erosion, medical costs of poisoning and mad cow disease (which is 90% of what farmers earn); $4 billion for subsidies to farmers (180%); $1 billion for health care costs for poor choices (45%); and $3 billion for loss of productive land.

One action that policy makers could take to meet tax reform or deficit-reduction goals is to eliminate a number of existing tax exemptions, deductions and loopholes that are both fiscally and environmentally damaging. Most notable among these are provisions for mature industries such as oil, mining, timber and automobiles. Not only do these government handouts to corporations work against common-sense notions of free markets, innovation and fiscal responsibility, but they also encourage mounting environmental and health costs that society must bear. The U.S. public is paying billions of dollars a year to help corporations make profits on things that harm the public and the environment.

Nations should remove most subsidies, especially for fossil fuel, farming, and roads. Temporary subsidies might be used to encourage solar power or wind power, or for hydrogen use to replace fossil fuels in transportation.

5.3.4.2.1.5.2.5.2. *Disallow National Tax Cuts*
The bulk of tax cuts, under the graduated system in the U.S., for instance, disproportionately benefit the wealthy and are meaninglessly small for everyone else. With the increases in use and loss taxes, and the elimination of income taxes, there are no reasons for tax cuts (although the small minority taxed for heroic incomes may disagree).

5.3.4.2.1.6. *Organize & Manage Complete Economy for Nature and People*
An economic system is a mechanism (or social institution) which deals with the production, distribution and consumption of goods and services in a particular nation. The economic system is composed of people, institutions and their relationships. It addresses the problems of economics, such as the allocation and scarcity of resources.

5.3.4.2.1.6.1. *Encourage Efficiency*
A nation can encourage efficiency. Allocative efficiency is the market condition whereby resources are allocated in a way that optimizes the net benefit attained through their use. Allocative efficiency is also defined as the production of the quantity that is most beneficial to society.

The term eco-efficiency, identified by the World Business Council for Sustainable Development (WBCSD) in its 1992 publication 'Changing Course,' is based on the concept of creating more goods and services while using fewer resources and creating less waste and pollution. The 1992 Earth Summit endorsed eco-efficiency as a means for companies to implement Agenda 21 in the private sector, and the term has become synonymous with a management philosophy geared towards sustainability. According to the WBCSD definition, eco-efficiency is achieved through the delivery of "competitively priced goods and services that satisfy human needs and bring quality of life while progressively reducing environmental impacts of goods and resource intensity throughout the entire life-cycle to a level at least in line with the Earth's estimated carrying capacity." This concept describes a vision for the production of economically valuable goods and services while reducing the ecological impacts of production. In other words eco-efficiency means producing more with less.

According to the WBCSD, critical aspects of eco-efficiency are: A reduction in the material intensity of goods or services; a reduction in the energy intensity of goods or services; reduced dispersion of toxic materials; improved recyclability; maximum use of renewable resources; greater durability of products; and, increased service intensity of goods and services. The reduction in ecological impacts translates into an increase in resource productivity, which in turn can create competitive advantage. In 2002, Michael Braungart and William McDonough expanded the ideas of eco-efficiency and its practical applications.

5.3.4.2.1.6.2. *Certify National Standards*
Consumer pressures have often influenced the extent of exploitation in the national landscape. Consumers have indicated, for instance, that they want forests for wood products, but also for habitat for other species, aesthetics, recreation, and other values. Certification itself is a demand. As consumer demands change, however, the uses of landscapes may change. Certification may bring about other far-reaching changes.

Certification is one way for consumers to discriminate against products that do not enhance other values. Some may see certification as a way to protect ecosystems and their nonmarket benefits, such as cleansing water and air. Others may see it as a way to sell items for more money, or as a way to avoid the social conflicts over limited resources. Certification is a way for all stakeholders, including consumers, to define and refine demand. But, it cannot be isolated from the large-scale economic and political trends that shape human culture. Problems are part of a matrix of industrial social practices and policies, but there are now pressing economic and ecological reasons to revamp them.

Certification will require more knowledge about ecosystems, as well as improved inventory and monitoring levels and techniques. Certification implies that a complete inventory must be taken of the system to provide a baseline for further assessment. It also implies that monitoring must be continuous. Certification requires much better scientific research to determine minimum, optimum or maximum numbers of features for ecosystem health.

Certification will address concerns of human equity and social justice. Land tenure is one barrier to meaningful certification in many parts of the world. Poverty is another such barrier; as long as the compensation of workers depends on the style of the ruling regime, there may not be enough stability for certification. Certification can contribute to a needful awareness among political representatives, who have been hypnotized by the flow of money. Certification has the potential to increased desperately needed revenue for regeneration, management, and research. It has the potential to increase revenue to local communities, rather than contributing to the flood of cheap raw materials exported elsewhere. It can encourage community cooperation to create millions of jobs and generate expenditures.

Living generations are responsible for limiting their actions within a reasonable framework of cost and irreversible change. The standard requires conversation and some consensus about the limits, which are never exact. Ecological value has to be balanced with optimal resource allocations.

Certification can identify issues that must be examined to bring about sustainable societies exploiting sustainable resources. The concept of sustainability is being incorporated into every level of planning and management. The concept can be used to develop practices that decrease depletion and waste, and reduce the threats to future generations.

5.3.4.2.1.7. *Represent All People in a Nation*
A nation is obligated to represent all people with its borders, even those who do not have formal citizenship. A nation is obligated to guarantee resources first for its authentic citizens.

5.3.4.2.1.7.1. *Obligations to Protect Rights & Privileges in a Nation*
The rights that have been established need to be protected, not only from impingement by other nations or a global association, but from changes within the nation itself.

5.3.4.2.1.7.2. *Obligation to Meet Basic Personal Needs in a Nation*
A nation has an obligation to meet the needs of its people by planning for enough food, through growing or trading, and by planning the development of the structure or the national population.

5.3.4.2.1.7.3. *Obligation to Integrate Everyone into a Nation*
The nation has to be able to integrate people into a national population, through acceptance, education, and shared values and actions. It has to make sure than no conflict between groups or individuals becomes out of control or involves large numbers of other people.

5.3.4.2.1.8. *Work with Other Nations*
A nation has to acknowledge the existence and rights of other nations. It is no longer possible to ignore another nation, although every nation has the right to limit the number of connections and exchanges with other nations. So much, in terms of global patterns or trades, is shared among nations that a nation is obligated to work with another nation if there are common problems that need to be resolved.

5.3.4.2.1.9. *Assess National Performance through Surveys*
A nation has to receive feedback on its performance, from distributing resources to protecting rights. This feedback can come from other nations, its citizens or its representatives. A nation can monitor its own activities, to ensure that standards are followed for the duration of the

activities. Monitoring also allows the standards for the activities to be revised to increase efficiency. The results of monitoring need to be evaluated and fed back into the process of planning and acting. The skills and knowledge of those who do the evaluation determines the success or failure of the principles and standards developed in the plan for monitoring and assessment. Interactions need to be evaluated in terms of impacts, needs, goals, and limits. The whole system needs to be evaluated.

5.3.4.2.2. *Future & Leadership*

Isolated problems stir only mild attempts at reform. But, few problems are isolated. Modern government is the assumption of responsibility for problems without adequate knowledge. Modern citizenship is the abandonment of responsibility, on the assumption that others have adequate knowledge to manage problems. But, our leaders refuse to learn or understand the simplest facts of science or technology. Leaders have to acquire an ecological consciousness.

Many adjustments may be necessary at a national level, such as civil service tests to determine candidacy, as well as board review for basic minimum moral and social qualifications. Making these changes all at once might seem unrealistically disruptive, but would it be? Would it be more disruptive than an earthquake or a war? Would it fracture the cultures or communities? Would people's lives change that much? Would the change be worse than a lost job, a business failure or death in the family?

These things should always be done if people were convinced by their respected leaders that this was the best way to proceed. Given the fact that the characteristics that make people lust for leadership often detract from their abilities to lead, perhaps we need a form of draft to get leaders, or perhaps we need to consider some who have been groomed for decades or at birth, like the Dalai Lama.

We need to examine different models of power and authority: Autocrat, guru, revolutionary, or transformer. The transformer has authority but no power; the autocrat has both; the revolutionary has power but no authority; and the guru has neither. Perhaps we should consider the guru as leader. The guru who would transform the world may choose to use the metaphor of water, and follow the flow of things.

Until we can do things naturally and spontaneously, we must try to act our way into right thinking, while thinking our way into right action. It is a form of intelligence, to know the principles and trends of natural affairs so that the least energy is used in dealing with them. This is also the unconscious intelligence of whole organism and the whole community.

Alan Watts suggests that Wu-wei in ruling is the delegation of authority and of administrative duties—the noninvolvement in fussy details. Wu-wei, not forcing, is the life style of one who follows the Tao. Watts counts great power as a worry; total power is boredom, so that even God renounces it and pretends she is plants and fish and people. Nature can be a model for a culture nourished on nature. Logic can recognize soil fertility and animism as continuums of mind and nature.

5.3.4.3. ***Contributing to the International Level***

A eutopian framework could be implemented immediately. Most global studies, such as those from the Club of Rome and *The Ecologist*, state or imply that change cannot be fast, that people cannot adjust, that social disruption would result, and that chaos would finish what ignorance and technology could not. These studies propose slow, long-range plans, while warning at the same time that the earth is facing imminent, drastic change. If their plans are implemented too slowly, and if the population or pollution doubles again, surpassing some unrecognized critical level, there would be worse disruption. The Eutopian proposal is for immediate action, scaled on a week.

Sudden change is already a hallmark of industrial progress. Industrial cultures have replaced older patterns with great suddenness. Eutopias cannot seem more sudden than the loss of a home or place. Industrial cultures have reduced people's control over the means of production and power. Eutopias does not offer less control. Whole communities have been destroyed by industrial scale. Our social structures are already changing rapidly and impractically.

Let us take immediate action to make the changes conscious and more practical. Eutopias offers movement towards common, achievable goals. Eutopias would be a framework for cultures, where different human experiments are tried. Its variability would insure that we could reject any of the local visions that fail. People may object to giving up too much or not gaining enough. Eutopias may be called anti-human, anti-progress, anti-scientific, anti-technological, or anti-educational, but it is merely a new framework for conducting traditional human activities.

Natural environments and human societies are wobbling. Many contradictory impulses are leading to unbalance; some countries want to consolidate into economic powers and others want to secede into independent units. Human civilization will tear itself apart if we let it. We can slow it down and direct it.

The need to maintain our comfortable status for as long as possible, fatalism that nothing can be done or it is too late, prejudice, ignorance—all are keeping us from moving. There are other reasons not to move: Failure of knowledge, failure of communication, failure of imagination, and failure of nerve. Much human suffering is caused by self-deception, which leads to isolation and then anger, reaction, and more suffering. Real change is difficult in this state, but change is more difficult for people who are starving or oppressed.

For most people in agrarian countries, even freedom from hunger and sickness is utopian. For most people in industrial countries, the choice of a fulfilling profession is utopian. Grinding poverty, economic dislocation, homelessness, are more painful than a transformation to Eutopias. Already most cultures have been transformed by cash crops, mining, tourists, highways, high-rise housing, and condominiums. Physical disruption has been more extensive than the transition to Eutopias could cause.

A long view seems meaningless when so much suffering already exists. An immediate, realistic, coordinated program of action is needed, capable of being implemented by communities and global agencies. We must face our responsibilities directly, declaring that there is no place in a eutopian society for monopolistic and multinational corporations, for the maniacal religion of merchandise, for genocidal military establishments, for urban explosion, for state socialism, for overbearing bureaucracies, or for technocratic politics, we must act to end them. The declaration must be political, through cooperative networks or leaderless consensus, by persuasion and example. The problem of human existence on the planet must be approached without deference to artificial boundaries of states, races, or castes. Poverty, pollution, repression, are concerns of every human community. We must stand and state that nature has limits, that we cannot have all we want.

The application must be immediate. The crisis of exponential growth and destruction cannot be solved just after some final limit is approached or passed. The crisis of ignorance cannot be solved by hurrying ahead and creating more problems. Paradoxically, the best thing to do is stop—stop growing, stop producing, stop running; suspend the race and contemplate a direction. We have been asking how the earth will survive its human populations and how they can be lowered. Let us just freeze growth and see what happens. Let us just freeze the populations—a year or decade of no births. We know that whole countries have lost a generation and continued. We know they have rebuilt again from ruins. We could build from recycled materials alone, so there is nothing to fear from stopping. Immediate social reforms, the reallocation of resources, and the preservation of wilderness are necessary, because of the nature of the problem; we cannot predict global climatic or ecosystemic catastrophes. Substantive change and research cannot be delayed until academic controversies are resolved.

The transformation must be complete; it cannot be done partially. Global political and economic institutions must all be changed. The Associated Nations (AN), more empowered than the United Nations, must have authority for the preservation of nature and human cultures. Holistic change will permit the reorientation and balance of local institutions. For example, air pollution is not independent of industrial processes, transportation, and employment patterns. Communities must be of a size that their members can feel responsible for them. These changes are demanded by new s, ecological balance primarily among them. New institutions must be compatible with these new values.

The approach must be pragmatic and flexible. By its nature, the eutopian frame could reduce some of the stresses of transition, the uncertainty, ambivalence, or reversion. The readjustment to the realities of our new intricate involvement in the whole order of nature and her ecological balance will cause social swains. Some capital of energy and materials may be wasted. Population will be matched to solar budgets or net ecosystem productivities. Production will be redirected to communal needs in transportation, housing, food, and recreation.

There will be problems regarding the breakup into more natural cultural divisions. Some will want to decide boundaries by ecosystem; others through culture, watershed, or political power. The Associated Nations will have to decide when two groups claim the same place or when cultures combine through unions and conspiracies.

There will continue to be other problems. Cutting trees in Nepal causes floods in Bangladesh, and floods cause deaths because overcrowding has forced the poorest people to live on flood plains. The poor in the highlands everywhere effect those in the lowlands, often adversely. The quest for ecological balance means that some ecosystems must be maintained by systems managers, who often overmanage. The larger the human impact, the more control is necessary. Eutopias seeks to improve people's circumstances by enlisting them to save their environment and their way of life.

People cannot be given material equality instantly. But things can be leveled within a culture; cultures with excess may be taxed by the Associated Nations. Providing work for everyone is one way to narrow income differences. The AN, nations, communities, and families must provide it. Worthwhile work requires imagination. The large work force employed by military contracts in industrial countries will be dislocated at first, but that employment is supported by taxes, which could be reallocated for construction and deconstruction—so many highways, manufacturing plants and abandoned buildings.

Crime and civic unrest will not disappear. The Associated Nations and nations could reduce many kinds of global and victimless crimes with new policies. Because most cultures have strong policies regarding drugs, abortion, and prostitution, among other things, the AN would not impose rules on every crime. Dangerous weapons, from automatic guns to tanks, and dangerous products, including nuclear reactors and biocides, would be strictly regulated.

People will still make mistakes and bad choices in a Eutopian framework. But, if a form of government is bad or ineffective, then it can be altered more easily in a smaller, more flexible framework. In the eutopian framework, people can learn from mistakes or unintended side-effects—as when doing good that causes evil. The scale is small, so the catastrophe is small. There will always be some injustice, inadequacy, and unpredictability. Large political and economic institutions have only made it worse. If Eutopias turns out not to be the proper framework to solve these problems, it might lead to a better way.

5.3.4.3.1. *Implementing Immediate Steps for an International Body*

The AN can be based on the UN, but with immediate responsibilities and powers, for protection and preservation, as well as some temporary powers, such as taxation. Too many things have happened in the past 100-500 years. The solutions need to be started now. Immediate steps are necessary to address catastrophic changes.

5.3.4.3.1.1. *Creating Framework for International Body*

The original charter of the United Nations (UN) restricted its activities to peace-keeping and human rights. While these are still important, other things need to be addressed, especially as they directly relate to concord and harmony (peace and health). Many things need to be done. New organizational bodies need to be created. These bodies would operate inside the new Association of Nations (AN) and be harmonized with existing bodies. Other bodies, such as the World Bank or International Monetary Fund, that operate parallel with the AN, must be reintegrated as part of the AN Financing. They must be governed by the AN executive branch, rather than by their own independent governing bodies, which are unduly influenced by wealthy countries. The AN needs to control it agencies and departments, as well as its financing.

The AN must have the power to define itself, to rewrite its charter so that it encompasses more than security and health, or expand those things to include all of human activities. Several studies of the UN, such as the Jackson and Pierson reports, have already focused on organizational efficiency—and efficiency has been the dominant factor in all bureaucracies for the past one hundred years—but the organization needs to be powerfully relevant to address the challenges that modern global trade has issued.

The AN can accept new nations as members. Membership in the AN needs to be opened to those states now represented by UNPO, that is, cultures, without recognition of their land, as well as other cultures that may not be landed at all. Recognition may also be granted to the regional association of cultures or nations.

Independent cultural areas within nations shall have the status of independent nations within the AN. Any culture would be given legal recognition, protection, and full autonomy over their boundaries by application to the AN, which would determine priority of claims (by archaic peoples, agriculturalists, pastoralists, or industrialists). No action would be taken to disband existing nations. Nations could still remain allied with old, larger nations as independent or dependent regions, although they would have only one vote at the highest federation. The nations would determine the use of allocated resources. Local economics and technology would provide for populations. Traditional religions and customs are maintained or permitted to develop.

The AN may consider whether it should extend recognition to voluntary economic associations, that by reason of size or power, such as some international corporations or regional industry associations, have great influence of capital and resources. Once membership is nonculturally or nonnationally based, however, it must be open to NonGovernmental Organizations, that may be dedicated to special interests or to preservations. And, then, perhaps the interests may be religious or spiritual. However, for voting purposes, nations or organizations may have to give up privileges as one or the other. On the other hand, the AN might require corporations and organizations to be based at the community or national level and participate in them materially.

The framework would allow a loose unity. For example, consider this lesson from fifteenth-century China: Too much unity stifles a creative advance; or this lesson from Europe: Too much disunity wrecks a creative advance. Europe had enough disunity to prevent unification, but not enough barriers to prevent the spread of technology and competition for technology. The role of the AN would be to provide just enough unity for nations to advance and develop. A eutopian framework is a form of 'balkanization' with limited barriers and limited unity. The barriers serve to protect the culture, its resources, its people and standard of living.

The AN would create a different structure than the UN. The AN could be modeled as a federal government, although it would be composed of independent nations. This would be necessary due to the responsibility of the UN for global issues and to coordinate nations. The AN would also guarantee fundamental environmental nonhuman rights, as well as human rights based on common things.

The AN would rework its constitution as a formal process of expressions and expectations. Traditionally a constitution is an agreement to divide into rulers and ruled. For the AN it would claim responsibility for all global affairs, including the environment and large weaponry.

The executive branch, the Secretariat, would be elected from the General Assembly. Seven to twelve members would be elected. They would elect a coordinator with the title of Secretary or Coordinator. It would have authority over all departments and agencies. It would mange the organization as an organic whole. The General Assembly, as a legislative branch, would make global and international laws. It would strengthen the Declaration of Human Rights (1945). It may make the planet into an incorporated entity. The judicial branch would be the International Court

Other branches may be separate and equal. Nations would nominate others for these branches. A Commission on the Earth would have the responsibility to consider all events in the environment from earthquakes to biodiversity. Other important Commissions would be the Commission on Beliefs and Religions and the Commission of the Heritage of Cultures.

The Office of Ombudsman would be expanded and strengthened. Although not traditionally a fourth branch of government, as first proposed in Sweden, its function would be to improve the effectiveness and fairness of the global government while protecting basic human and ambihuman rights. One strength would be its power to initiate legal proceedings in behalf of any human and ambihuman constituents or nations. It would be an independent branch with separate responsibilities: To investigate complaints against the UN or its officials for wrongdoing; to investigate complaints against nations; and, to investigate the operation of the UN.

5.3.4.3.1.2. ***Start Catastrophic Measures by the International Body***

Because the challenges are immediate and because the consequences of not meeting them could be catastrophic, these things have to be started immediately. Of course, they are part of a process that may never be complete, like a one-time fix, but they have to be started immediately.

5.3.4.3.1.2.1. *Transfer Powers to the International Body*

The major military powers would grant their powers to the Associated Nations and relinquish their efforts towards global leadership; they also resign from the security council, cease propaganda activities, renounce foreign policy objectives, call back all soldiers from foreign countries, and stop giving away produce, factories, or weapons. They put their technical and educational surpluses at the disposal of the AN. If the USA or Russia is to be a world leader, let her lead in tolerance or in trust. Let her be the first to give allegiance to a world organizing body, the AN, the first to divest themselves of nuclear weapons. If they fear for safety, they need only remember the success of nonviolence in India or of guerrilla actions in Southeast Asia, Central America, or the Middle East.

5.3.4.3.1.2.2. *Disarm Nations by International Body*

In a separate essay entitled "Toward a World Social Contract," Kenneth Boulding examined possible global mechanisms to abate conflict within "Spaceship Earth," including "universal policed disarmament down to internal police levels" and "the organizational union of the armed forces of the world under a limited world government" — both of which are key elements of unfolding official U.S. government disarmament policies envisioning a world "effectively controlled" by the United Nations.

The 1945 UN Charter envisaged a system of regulation that would ensure "the least diversion for armaments of the world's human and economic resources." The advent of nuclear weapons came only weeks after the signing of the Charter and provided immediate impetus to concepts of arms limitation and disarmament. In fact, the first resolution of the first meeting of the General Assembly was entitled "The Establishment of a Commission to Deal with the Problems Raised by the Discovery of Atomic Energy" and called upon the commission to make specific proposals for "the elimination from national armaments of atomic weapons and of all other major weapons adaptable to mass destruction."

The UN has established several forums to address multilateral disarmament issues. The principal ones are the First Committee of the General Assembly and the UN Disarmament Commission. Items on the agenda include consideration of the possible merits of a nuclear test ban, outer-space arms control, efforts to ban chemical weapons, nuclear and conventional disarmament, nuclear-weapon-free zones, reduction of military budgets, and measures to strengthen international security.

The Conference on Disarmament is a forum established by the international community for the negotiation of multilateral arms control and disarmament agreements. It has 66 members representing all areas of the world, including the five major nuclear-weapon states, the People's Republic of China, France, Russia, UK and U.S. While the conference is not formally a UN organization, it is linked to the UN through a personal representative of the Secretary-General; this representative serves as the secretary general of the conference. Resolutions adopted by the General Assembly often request the conference to consider specific disarmament matters. In turn, the conference annually reports its activities to the

Assembly. The conference indicates that disarmament ideas are considered.

The U.S. must keep its own promise to give up nuclear weapons, which it made in 1970 when it signed the Non-Proliferation Treaty. The U.S. must terminate its $8 billion annual program to develop new weapons and agree to Russian President Putin's offer to reduce the mutual nuclear arsenals of about 10,000 weapons down to 1,000, or down the ante to zero.

The U.S. and Russia need find parity with the arsenals of the other nuclear weapons states, China, UK, France, and Israel, with stockpiles in the hundreds, and India, Pakistan, and North Korea each with less than one hundred bombs, should find reduction easier. China has offered to negotiate a treaty to eliminate all nuclear weapons and call every nuclear weapons state to the table. There already exists a plan for a model treaty prepared by scientists, lawyers, and policy makers, which was submitted to the UN as a discussion document. It lays out all the steps for dismantlement, verification, guarding, and monitoring the disassembled arsenals to insure that we will all be secure from nuclear break-out.

Russia and China have repeatedly offered a proposal in the UN General Assembly to ban weapons in space. That is the only way that those two nations will agree to join the U.S. in abolishing nuclear weapons. They do not want U.S. control and domination of space, which indeed is the aggressive mission statement of the U.S. Space Command, to gain military superiority over the whole planet. We must also replace the Non-Proliferation Treaty's guarantee of an 'inalienable right' to so-called 'peaceful' nuclear technology. This is the provision on which Iran is now lawfully relying. It could be nullified by establishing an International Sustainable Energy Agency and phasing out nuclear power. Every nuclear power plant is a potential bomb factory, so it would be impossible to eliminate nuclear weapons eliminating nuclear power.

An international body could disarm all nations; it could fund its work through the redistribution of the $200 billion in tax breaks and subsidies given to the nuclear and fossil fuel industries world-wide. The treaty negotiations and actual dismantlement of the nuclear arsenals could be done within weeks, although years or decades will probably be requested. If nuclear weapons were illegal, Korea, Iran and other nations might be willing to give up new programs, without recourse to attack or bribery.

All other nuclear weapons nations have indicated that they would be willing to give up their nuclear weapons, numbering in the 100s or 10s, if the larger nations do. The U.S. has been the biggest block to the efforts to stop proliferation. It is naive to believe that anything less than the elimination of nuclear weapons will reduce the possibilities of nuclear war and the unimaginable catastrophes that would follow a nuclear winter.

Elimination of nuclear weapons will not stop wars, although it is less likely that wars could be as destructive or final. Other weapons of 'mass destruction' also need to be reduced and controlled, especially long-range guns or automatic missiles. These weapons are designed to kill as many people as possible as quickly and cheaply as possible, regardless of their status as combatants or civilians. The International AN could work to bring every arsenal to parity.

5.3.4.3.1.2.2.1. *Take or Destroy Nuclear or Large-scale Weapons*

Complete disarmament could be accomplished within a week. Earl Osborn proposes this concept of sudden disarmament in response to the tedious phase-out envisioned by most plans. An agreement would not involve much negotiation. Taking this first step would add to the prestige of the country bold enough to do it. The AN could post a police force to disable all military ordinance. A thousand planes each carrying one hundred trained inspectors could be distributed at all major centers in the nuclear countries within 24 hours (1 day).

5.3.4.3.1.2.2.2. *Allow Personal Weapons in International Body*
For Nations, the AN would make available cruise missiles at some agreed-upon level. For individuals, it would allow personal weapons from hands and knives to arrows and single-shot guns, depending on national choices and limits.

5.3.4.3.1.2.3. *Start Year of Consideration through International Body*
The Associated Nations promotes a year of consideration. Starting with population growth, all economic growth, or expansion of claims, would be suspended for a year. Age grouping would be discontinuous for a while, if further years of consideration were necessary. The use of certain poisons or chemicals, especially greenhouse gases and other chemical additives that destroy ozone, would be discontinued.

5.3.4.3.1.2.4. *Start Equity Measures*
The global environment needs to be equalized also. If every nation or person has certain rights to land and resources, then those have to be respected and adjusted. Some AN taxes would be designed to repair the inequity from hundreds of years of unfair trading or accumulation.

5.3.4.3.1.2.5. *Implement Global Ecological Goals*
Global goals apply to the planet as a whole, for Gaia as a metaphor for the living planet. These goals are not simply the sum of local and regional goals. The AN would:

- Reimplement international initiatives to slow deforestation—the UN notes that previous initiatives *accelerated* deforestation, as in Cameroon, where log production is planned to double in the forest, home to 50,000 Pygmies with a unique and valuable cosmology and life-style
- Plant and maintain forests sufficient to guarantee indefinite support of known and unknown global biogeochemical cycles
- Protect fragile ecosystems with global importance
- Reduce threats to ecosystems from acid rain and other nonpoint-source pollutions to less than 5 percent of present values
- Plant 9 million ha of trees each year to meet current demands; for soil and water conservation, plant another 6 million hectares (at an estimated cost of $6 billion dollars); and plant 110 million hectares just to catch up with cutting
- For the planet, reforest 1.4 *billion* hectares to restore the 30-40% forest cover removed in the past 3000 years.

The AN would monitor ecological goals for nations and communities.

5.3.4.3.1.2.6. *Implement Global Cultural Goals*
Global cultural goals apply to every human culture.

- Protect the health of human communities; set standards for health and provide information and assistance to nations.
- Provide educational assistance to nations. Encourage cultural education and history.

Encourage teaching of local languages. Allow renewal of critical customs. Educate all people to feel their connections to their place, because, until they feel them, they will not act ethically or ecologically. Educate people to realize that long-term sustainability requires healthy places, and that protecting places protects jobs and values.

- Coordinate and support economic activities. Broaden local economies from resource extraction to invention and specialization. Promote the full use of ecosystem products; support small-scale businesses that produce new, high-quality products. Increase manufacturing efficiencies.
- Guarantee sovereignty for nations. Offer a platform for communication with other nations. Open communications with all groups working in the region. Work to establish equity.
- Help nations stabilize their cultural environments, as well as their populations and resources, which can be related to the limits of place.

The AN would be responsible for defining the goals that are common to human cultures and coordinating goals on the global level.

5.3.4.3.2. *Managing Nations*

The AN would take responsibility to manage the interrelationships of nations, to try to coordinate trade, reduce conflicts, and foster respect for other nations and cultures.

5.3.4.3.2.1. *Create a Charter & Constitution for Associated Nations*

The Charter of a global association has to expand the original charter for the United Nations. It needs a formal constitution that spells out additional responsibilities. The constitution would indicate how the Association would relate to nations, corporations, alliances, and NGOs.

Global organization should be in the form of a heterarchy, a multi-level structure integrated at regional and national levels. Community levels would self-reliant, but linked with other communities into nations and regional alliances, The regional alliances would not be allowed voting rights separate from their components in the AN. The regions would be too large to be manageable. Corporations currently that did not have to have a cultural or territorial base might be given temporary status until they were integrated.

5.3.4.3.2.2. *Create New Structure & Branches of Associated Nations*

Because a global association will have more responsibilities than the United Nations, it must create a new structure, with new branches and agencies.

5.3.4.3.2.2.1. *General Assembly (Legislative) for All Nations*

The General Assembly would include old and new nations. Conceivably, the number of nations could exceed 3000. The General Assembly would make all laws and rules.

5.3.4.3.2.2.1.0.1. *Define Limits of Global Law.* Global law would be directed at global interactions relating to nations and corporations. It would also address global resources and problems that follow from the cycles and uses of resources.

5.3.4.3.2.2.1.0.2. *Make Laws.* The UN made laws against war. The AN must expand those laws and be prepared to enforce them. Laws also need to passed to address other human concerns, such as slavery or internal violence. Laws have to be proposed to control international spread of disease. A whole set of laws is needed to control global resources.

5.3.4.3.2.2.1.1. *Office of Normalization & Nationalization*
The AN has to issue its challenge to allow cultures or provinces join the organization and to have one vote. Any culture or province would have the right to apply for membership. The size and form of nations would begin to decentralize and normalize. The requirements for nationhood would be basic: A traditional culture, traditional territory, or the size and uniformity of people. Although new nations would vote independently, they could maintain a regional association with other small nations.

5.3.4.3.2.2.1.2. *Office of Standards for Ecosystems Nations & Trade*
Through a eutopian program the Association of Nations (AN) could set up and enforce international ecological rules of trade, for including environmental costs into the prices of things. Economists, especially global economists, have not much considered these 'externalities' which are really free internalities, free so far, because they have been big and plentiful. To compete in an open arena with fixed economic policies and no ecological policies, nations often squander their ecological capital to be competitive in the short-term. There has been no reason political or economic to save resources to be sustainable in the long-run.

5.3.4.3.2.2.1.3. *Office of Budget: Secure & Balance Budget*
This office would be concerned with funding the AN. Where will money come from? What do we have now? Footloose food and crazed capital? In the form of dues, contributions and the sales of bumper stickers? The new office would have to coordinate the income from taxes on global resources and rents and fees on common lands, such as Antarctica.

5.3.4.3.2.2.1.3.1. *Income for International Framework*
Right now, the income from member nations is related to their GNP, a partial indication of relative gross national product. With the capability of taxation, the AN would not rely on voluntary national gifts. Although the AN would tax global resources, such as land and air, it could also tax weapons quite heavily. In fact, Ervin Laszlo suggests an international body could tax military expenditures of nations instead of the simple GNP. This would ensure that nations that produced the most weapons would pay the most for the effects of those weapons. Income from gifts and charity could still be used to combat special problems.

5.3.4.3.2.2.1.3.1.1. *Rent for International Framework*
The AN could charge rent for global lands and oceans. Ownership of much of the planet would be through a corporation or department. The use of these areas or resources would be rented for the long-term. But, they would also be monitored by the AN.

5.3.4.3.2.2.1.3.1.2. *Membership Dues for International Framework*
For the UN, member nations now have to pay a percentage or a flat rate. The United States has the maximum assessed contribution to the UN regular budget—22%. In 2005 the assessed amount is $439,611,612. Actual U.S. contributions to the UN in 2005 totaled $1,959,053,000. This included the regular budget, peacekeeping operations, international tribunals, specialized agencies and subsidiary organizations. The minimum assessed contribution is 0.001%. The scale of assessments for each UN member for the required contributions to the regular budget is determined every 3 years on the basis of Gross

National Product (GNP).

Only nine countries (starting with the largest contributor: United States, Japan, Germany, United Kingdom, France, Italy, Canada, Spain, China) contribute 75% of the entire regular budget. Cuba contributes 0.043% of the regular budget. One of the richest countries, Saudi Arabia, contributes 0.713 percent. The size of contributions, however, does not seem to be related to need, importance or dominance.

In addition to their contributions to the UN regular budget, member states contribute to the peacekeeping operations budget and the cost of international courts and tribunals. The level of these contributions is based on their assessed contributions to the regular budget plus variations which take account of permanent membership on the Security Council. UN members also make voluntary contributions to UN specialized agencies and subsidiary organizations. The administrative costs of such bodies, though, are met from the regular budget.

The UN could acquire much more money if the assessment were related to the defense budget of a country instead of its GNP. This would reduce military research and spending. The U.S. defense budget, for instance, was $343 billion a year, almost 200 times its UN membership fee. That kind of assessment would discourage military build-up and perhaps reduce military spending.

Membership dues would be phased out over five years, until tax, rent and other income programs are operating.

5.3.4.3.2.2.1.3.1.3. *Taxes on Global Things*

This term in its most extended sense includes all contributions imposed by the global association upon nations, cultures, or individuals for the service of the association, by whatever name they are called: Tribute, tithe, talliage, impost, duty, gabel, custom, subsidy, aid, supply, excise, or other. Taxes are any charge of money or property imposed by the association upon individuals or entities that are within the authority to assess such charges. Most modern taxes are levied on the basis of economic measurements such as income, consumption, property, and wealth. These taxes would be uniquely applied to global things or processes or to nations.

5.3.4.3.2.2.1.3.1.3.1. *Use Taxes on Global Elements Cycles*

Use taxes would have the effect of limiting the use of nonrenewable resources, such as coal or oil, or the use of slowly renewables resources, such as forest products or fish, located in global commons, such as the atmosphere or oceans. This would discourage overexploitation of previously 'free' resources.

5.3.4.3.2.2.1.3.1.3.2. *Loss Taxes on Global Resources*

The loss tax is on things or processes that interfere with other things and processes, things that cause runaway feedback or the destruction of cycles, things in other words that reduce our continued use of and enjoyment of the earth. This tax would have the effect of internalizing both ecological and social costs; since all consumers would be paying the real costs, no consumers would be protected. It should have the effect of reducing pollution. The purpose of this tax is to change behavior that depletes resources and discourages labor. It also can pay for the damage caused by misplaced or misused resources, such as the destruction of ozone.

5.3.4.3.2.2.1.3.1.3.3. *Adjustment Taxes on Global Sin & Pollution*
There would be no necessity for sin taxes, unless they were necessary to support explicit AN institutions on health, or for use of such things in International areas. International pollution would be taxed, however. Especially industrial and agricultural pollution.

5.3.4.3.2.2.1.3.1.3.4. *Distribution Taxes on Global Things*
Distribution taxes have the function of reapportioning wealth. These taxes would be assessed in international waters and lands. They would be on luxuries or incomes that have evaded national and community taxes.

5.3.4.3.2.2.1.3.1.4. *Licensing for a Global Association*
A license is a formal authorization by law to do something, such as marry, hunt, or practice medicine. The AN may issue international licenses for some purposes.

5.3.4.3.2.2.1.3.1.4.1. *Licensing International Corporations*
Corporations that have no land base, or that have evaded local charters would be subject to a license with the AN.

5.3.4.3.2.2.1.3.1.4.2. *Licensing Satellites & Media*
Satellites, carriers (or spacecraft), and any items in sublunar or solar system space would be licensed by the AN. These fees would be used for tracking and cleanup. Licenses would be required for wave communications in international areas. Due to the global aspect of waves, it might be necessary to coordinate all media among nations.

5.3.4.3.2.2.1.3.1.4.3. *Licensing National Weapons*
All weapons of a certain kind, from automatic weapons to cruise missiles, would be licensed by the AN. These weapons would be for national police forces and the police force of the AN. Their possession or use would be banned for individuals.

5.3.4.3.2.2.1.3.1.4.4. *Other Licenses*
The Associated Nations might offer international drivers licenses. Licenses would be required also for collecting in international places.

5.3.4.3.2.2.1.3.1.5. *Fees & Tolls for a Global Association*
Fees would be charged to save, restore, and maintain the common wealth of humanity. Such fees would be charged to nations.

5.3.4.3.2.2.1.3.1.5.1. *Global Fees on Common Human Wealth & Great Art*
Although much of the common heritage of the humanity on the planet would be free to observe or visit, fees would be charged to limit access. Fees would also be used for special costs such as cleanup. This fee might be applied to special areas of the planet or for works of art that are considered the heritage of humanity, such as many cave paintings.

5.3.4.3.2.2.1.3.1.5.2. *Global Fees on Wilderness*
Global wilderness areas, such as Antarctica or the North Pacific Sea, would have limited access. The fees would also support research in these areas.

5.3.4.3.2.2.1.3.1.6. *Labor for Global Projects*
The AN could use scientists from other countries. It would also use volunteers from many nations to work on international projects. It could have a formal two-year volunteer program, similar to the programs of nations, but without conscription.

5.3.4.3.2.2.1.3.2. *Payout for a Global Association*
At this level all costs, including environmental and social, have been internalized. The AN has to be able to guarantee that all unintended costs are being paid, as a result of the economic activities of nations.

5.3.4.3.2.2.1.3.2.1. *Operate Organization & Agencies*
The AN has as large a bureaucracy as any national government, in fact, larger than the largest national government. However, since there is an overlap of responsibilities and so many things are duplicated on lower levels, the AN will essentially be a coordinator. Nevertheless, it will have a large number of agencies, as well as the separate governing functions and commissions.

5.3.4.3.2.2.1.3.2.1.1. Health Supplement
Traditionally, an international body has been concerned with the levels of health of the people of all nations. One function is to make sure that the lower levels of health are improving faster.

5.3.4.3.2.2.1.3.2.1.2. Education Supplement
Traditionally, an international body has also been concerned with equalizing educational opportunities for people in every nation. The AN would be responsible to providing education on the common heritage of humanity and on the global environment and history.

5.3.4.3.2.2.1.3.2.1.3. Global Planning
Planning is a large part of a global order. The AN must plan for many contingencies that nations may not have to face independently.

5.3.4.3.2.2.1.3.2.1.4. Global Resource Surveying & Monitoring
Monitoring is crucial for nations and a global order. No coordinated effort has ever been made to inventory, assess and monitor the ecological systems of the planet. The AN would do this.

5.3.4.3.2.2.1.3.2.2. *Restore Global Cycles or Places*
Many places have been impacted by the international problems of pollution and simplification of land, and disruption of global ecological cycles. The AN would develop programs for restoration.

5.3.4.3.2.2.1.3.2.3. *Create Set-aside Accounts for Global Catastrophes*
A large part of the finances and efforts of a global association have to be concerned with global changes in climate and geology. These sometimes catastrophic changes are an integral part of the change and aging of the planet. They have to be anticipated, prepared for, and responded to.

5.3.4.3.2.2.2. *Executive Branch of Global Association*

The executive branch of the global association makes sure that the global laws of the AN are obeyed. Other functions of the branch are: To execute policies; to control policies; to appoint officials; to command the police force; and, to veto legislation. The executive branch has to be formed from the assembly of national representatives. It has to be relatively large, and it has to have salaries, support monies, and travel monies.

In addition to global laws, the Executive Branch would have to oversee the enforcement of laws by nations, especially where the two levels overlap.

5.3.4.3.2.2.2.1. *Coordinate Global Association*

The Secretary, equivalent to the former Secretary-General of the UN, would be the head of the Executive Branch. The Secretary would be elected from the General Assembly and serve a term of six years. The Secretary help from the Assistant Secretary, Police Chief, department heads, and heads of independent agencies. The Assistant Secretary would be head of the legislature and next in line to head the branch. The Police Chief would be head of all AN police forces. The Department heads would advise the Secretary on issues and help carry out policies. Independent Agencies would help carry out policy or provide special services.

The executive branch could be headed by an Executive Council of the department and agency heads; these offices would elect the Assistant Secretary.

5.3.4.3.2.2.2.2. *Security Council of Global Association*

The Security Council would be composed of twelve people elected by the General Assembly every three years. There would be no permanent members. The Council would elect its own Secretary to speak for the Council. The veto power would be replaced with consensus.

5.3.4.3.2.2.2.2.1. *Assess Threats (Internal & External)*

The Security council is concerned with any kind of threat to the global order and the orders of nations and nature. Many of these threats are internal, as nations enter into conflicts with other nations. These threats can be handled by delegations or police actions. But, many threats are external to the planet. They come from solar or interstellar space and tend to be physical.

5.3.4.3.2.2.2.2.1.1. *International Conflict*

Widespread poverty may cause catastrophes. Richer countries will need to recognize that the poverty of others is not in their interest, especially as potential markets. Inequity may never be erased. Perhaps some inequity is good and stimulating, but gross inequity needs to be limited.

Each culture develops rules for living together. A common culture provides an ideal framework for public and private decision making. The Sami in northern Scandinavia have institutionalized ways of avoiding conflict, for instance, by shaming those who would impose their will. The Fipa of Tanzania use cooperative exchange rather than competition to keep the peace. The Akawaio of Guyana believe that community disharmony upsets the spirit world, resulting in illness and misfortune.

Conflicts, territorial or symbolic, are symptoms of insecurity. Many of our wasteful conflicts could be more easily resolved through a neutral international power. Conflicts would still occur. Conflicts over prestige or power, as much as for various crusades as for a true state, still lead to human and environmental destruction. The Security Council would be charged with the responsibility to avoid massively destructive forms of conflict, such as biological war or nuclear war.

5.3.4.3.2.2.2.2.1.2. *Global Threats*
The planet is still a very active planet. Some threats to humanity rise from the normal activities of the planet, such as volcano-building or continental drift. Other changes, such as greenhouse gases, are causing possible threats.

5.3.4.3.2.2.2.2.1.2.1. *Geological Threats*
Typical geological threats include earthquakes volcanoes, and mudslides; rare threats may be comets or sunspot activity. While technology may be able to retard some of these threats or counter them early, the best solutions are design. Cities can be moved from floodplains; cities can be required to have quake-resistant buildings.

5.3.4.3.2.2.2.2.1.2.1.1. *Climatic Threats*
Someone said that civilizations exist through the consent of geology; however, the consent of the 'first empire of climate' might also be needed. Climate is a constant and immediate challenge. Many variables affect climate. One variation, with a long cycle of perhaps 100 million years, is continental drift. When Panama closed continents, it forced the gulf stream north. The earth's orbit around the sun, a 100,000 cycle, also changes climate. This is close to the spacing of ice ages. Other smaller cycles, at 10,000 years or 6,100 years are minor harmonics perhaps. Other activities that influence climate are sunspots, comets, and volcanoes. The most stable periods seem to be the coldest or warmest weathers. For instance, 400,000 years ago, a warm period lasted 25,000 years. If we are in a 10,000-year warm period, it may be almost up. This means that predictions are relevant.

Cooling can lead to disease and depopulation, which can lead to cooling. Farmland is abandoned to trees in times of collapse. Because trees take CO_2 out of the atmosphere, the regrowth of forests after the plagues in 1322-1351 in Europe and China, would have allowed drops in CO_2, which would have allowed climate to cool some.

The years 1997 to 2004 broke most heat and storm records, especially as some of the warmest years on record. We experienced 500-year floods, droughts of the century, and other extremes. Some of these events have been related to greenhouse effects.

5.3.4.3.2.2.2.2.1.2.1.2. *Oceanic Threats*
Ocean currents affect not only islands and coasts, but they affect the entire climates of continents. For example, the effects of the Pacific El Nino current can be linked to droughts in India and China in the past 150 years that caused three times the deaths of the Black Death, and more than the 60 million who died in WWII.

There does not seem to be a single trigger for El Nino. One trigger has to do with water overflow and then back flow from the western pacific. Others have to do with sunspots, which would reduce radiation, or volcanic eruptions. Many of these atmospheric, oceanic and geophysical triggers may converge. Nineteenth-century famines may be correlated with ENSO events that influenced China, Indonesia, Brazil, and southeast Africa.

The adherence to political colonialization and 'free market' economics, along with climate changes, made the suffering in famines worse. Millions died within the market system of the golden age of liberal capitalism. Thus, the famines were political and economic failures. Even at the height of the famines in 1877-78 and the 1890, Britain continued to export grain from India to England, which had its own agricultural downturn. The invisible hand did not lift those starving in India or China. There was starvation before the British, especially during El Nino events in 1596 and 1630, but many droughts did not result in such widespread and deep famines. Market economics and politics magnified the effects. The market economy can spread risk, through insurance companies, when crises are local and intermittent, but they may not be able to respond when the crises are global and ubiquitous. As the risks increase everywhere, fewer things can be done.

Other movements of water, such as tsunamis or floods, can be linked to earth movements, such as landslides or earthquakes.

5.3.4.3.2.2.2.2.1.2.2. *Solar System Threats*

Solar energy is not quite constant; over millennia it has been increasing; over the next billion years it will start decreasing. Collisions with asteroids or comets will always be a threat due to the nature of the solar system. The passage of the solar system through dust clouds will effect the climate of the planet and the solar output. The AN could plan to deal with asteroids and comets.

5.3.4.3.2.2.2.2.2. *Provide Security using a Police Force*

Countries scramble to identify and claim resources that they need, fighting for them if necessary. Resources in short supply include water, timber, and fossil fuels. Countries strive for resource security, but this leads to further fighting and instability. Traditional ownership is stretched, leading to new disputes. All of these conflicts are the kind that could be resolved by the AN. Cooperative solutions are more durable and effective. Violence only leads to resentment and further violence. The AN has to have either the most weapons in the largest police force or the best moral stance.

5.3.4.3.2.2.2.2.2.1. *Expand Police Force*

Peacekeeping, as defined by the Association of Nations, is a way to help countries torn by conflict create conditions for a sustainable situation of concord. AN peacekeepers—police, rather than soldiers and military officers—civilian police officers and civilian personnel from many countries would monitor and observe peace processes that emerge in post-conflict situations and assist ex-combatants in implementing the peace agreements they have signed. Such assistance comes in many forms, including confidence-building measures, power-sharing arrangements, electoral support, strengthening the rule of law, and economic and social development. All operations must include the resolution of conflicts through the use of force to be considered valid under the charter of the Association of Nations.

5.3.4.3.2.2.2.2.2.2. *Use Police Force*

The Charter of the Association of Nations would give the AN Security Council the power and responsibility to take collective action to maintain international peace and security. For this reason, the international community should look to the Security Council to authorize peacekeeping operations. Most of these operations would established and implemented by the AN itself with police serving under AN operational command. In other cases, where direct AN involvement is not considered appropriate or feasible, the Council would authorize regional organizations such as the North Atlantic Treaty Organization, the Economic Community of West African States or coalitions of willing countries to implement certain peacekeeping or peace enforcement functions. In modern times, peacekeeping operations have evolved into many different functions, including diplomatic relations with other countries, international bodies of justice, such as the International Criminal Court, and eliminating problems, such as landmines, that can lead to new incidents of fighting.

The AN would be expected to have many operations around the world, similar to the peace operations of the UN. Recent operations of the UN in Africa include the Burundi Civil War (2004), the Civil war in Côte d'Ivoire (2004), Second Congo War (1999, United Nations Organization Mission in the Democratic Republic of the Congo, Second Liberian Civil War (2003), the Eritrean-Ethiopian War (2000), the North/South Civil War and Darfur conflict (2005), and the Moroccan occupation of Western Sahara. In the Americas, the UN monitored the 2004 Haiti rebellion. In Asia, the UN was involved in: The 1949 Indo-Pakistani Wars. In Europe, the UN put itself in: The 1964 Cyprus dispute, the 1993 Abkhazian War, and the 1999 Kosovo War. In the Middle East, the UN participated in the 1974 Agreed withdrawal by Syrian and Israeli forces following the Yom Kippur War, the 1978 withdrawal of Israeli forces from Lebanon, and the 1948 various cease-fires and assists.

5.3.4.3.2.2.2.2.2.2.1. *Deal with International Conflicts*

The police force would be entrusted to try nonviolence as a first response. In some cases police would be unarmed. If not that did not work, police would use weapons.

Conflict cannot be separated from other things, such as environmental destruction or inequities. For that reason, the Security Council and its force has to consider the broadest meaning of security: It is for people to have the resources and opportunities to provide themselves with their needs. Thus, security has to be addressed on many levels. For example, what do forests have to do with peace? Of the fourteen nations requiring UN peace-keeping operations since 1990, twelve of them have lost over 90 percent of their forests. Perhaps peace-keeping should be abetted by tree-planting. As ecosystems are destabilized, nations become less stable, and must be helped with more than social conflicts.

5.3.4.3.2.2.2.2.2.2.2. *Protect against Interference Events or Disasters*

Police would be expected to be prepared for natural disasters, including the extraterrestrial. The police would focus on the prevention of man-made disasters, from watersheds that were compromised to chemical spills. Collisions with asteroids will always be a threat due to the nature of the solar system.

The flexibility of a natural ecosystem to respond to change has been reduced, so either we have to take over control, which could cost quite a lot, or we have to adapt to the diminishment after some climatic episode.

5.3.4.3.2.2.2.3. *Agencies for a Global Association*

Agencies of the Association of Nations are necessary organs to present information and research to the main branches of the Association.

5.3.4.3.2.2.2.3.1. *Food & Shelter Agency*

This agency is devoted to monitoring the food and shelter requirements and deficiencies of groups in every nation. It would provide information for traditional or ecological ways to build food supplies or shelters. Through international volunteers, it may provide help with expertise and labor.

5.3.4.3.2.2.2.3.2. *Comprehensive Energy Agency*
This energy agency would consider all possibilities: Human energy, animals, fire or combustion, fossil fuel, solar energy, geothermal energy, wind energy, and nuclear energy from fission or fusion. It may recommend and sponsor less hazardous kinds of energy use, such as solar energy and wind energy. It would also work to make sure that energy is not wasted; this may mean building or converting more efficient houses. Without the conservation of energy, giant wind farms or solar farms will dominate the landscape, ruin the scale and be possibly as hazardous as fossil fuels or nuclear power.

5.3.4.3.2.2.2.3.3. *Transportation Agency*
Transportation is a global phenomenon. Many kinds of transportation use the atmosphere and oceans—in fact many new proposals have to do with the undersea environment. The AN would monitor or control traffic in the global commons.

5.3.4.3.2.2.2.3.4. *Economics Support Agency*
Economies are tied together now in global patterns. They might be overconnected. This agency would provide information on economic weaknesses and problems to nations. It would act to provide alternatives to some kinds of economic practices and suggest practices that would eliminate the worst excesses.

5.3.4.3.2.2.2.3.5. *Education Support Agency*
Education needs to be stressed. The goal of education is for people to choose within limits—limits on wealth, waste, and freedoms that might endanger others. Education allows people to choose without being conditioned by brainwashing or dishonest advertising. The UN supported programs to enhance education, and it established the United Nations University for Peace, Costa Rica (a nation that has abolished its army), as an institution of higher learning for education for peace. The AN would continue these kind of efforts.

5.3.4.3.2.2.2.3.6. *Health Agency*
Health needs to be stressed, in every nation. The AN needs to monitor global disease and threats, to reduce chronic disease. Chronic diseases are neglected conditions. Chronic diseases represent a huge proportion of human illness. They include cardiovascular disease (30% of projected total worldwide deaths in 2005), cancer (13%), chronic respiratory diseases (7%), and diabetes (2%). Two risk factors underlying these conditions are key to any population-wide strategy of control: Tobacco use and obesity. These risks and the diseases they engender are not the exclusive preserve of rich nations. Chronic diseases are a larger problem in low-income settings. Research into chronic diseases in resource-poor nations indicates that it is critical to intervene early in the course of any epidemic. Fast intervention could save many millions of lives.

5.3.4.3.2.2.2.3.7. *Communications & Technology Agency*
There is a need for the development and delivery of a strategy for schools and other learning and skills institutions. This agency could provide strategic leadership in the innovative and

effective use of communications and technology to enable the transformation of learning, teaching and educational organizations for the benefit of every learner.

The agency would be charged with regulating interstate and international communications by radio, television, wire, satellite, and cable. Its jurisdiction would cover in fact every nation and international areas.

5.3.4.3.2.2.2.3.8. *Research Agency*

This Agency is the central research and development organization for the planet. It manages and directs selected basic and applied research and development projects for the international body and member nations. It pursues research and technology where risk and payoff are both very high and where success may provide dramatic advances for issues of global importance.

The reductionist path taken by science has yielded tremendous results about how the world is built up out of particles and molecules. Now that we have uncovered the complexity, we need to address relationships. This is where the synthetic path can help, by identifying emergent principles and operations. Science is an open, self-referential, self-correcting system capable of using analytic and synthetic methods.

Advances in science have been quite remarkable. How can it continue being remarkable, but applicable to whole systems? While pure research continues to reveal unimaginable details of ecosystems, and while applied research continues to support sophisticated use, more ecological and landscape research is needed. Research is expensive, time-consuming, labor-intensive, and uncertain, however.

Long-term research is especially important in forest ecosystems, since many of the components live hundreds and thousands of years, and the forest itself can live far longer. Long-term research requires different levels of monitoring, including environmental, biological, and ecological. Environmental monitoring is an umbrella for many activities, including climatic variables and geological processes; for example, the systematic recording of soil and air temperatures, humidity, air pressure are measured by meteorological organizations to predict long-term climatic change. Long-term research also depends on a stable cultural base and shared values between generations.

Other kinds of research needed equally are: Historical research, fire research, productivity research, mortality patterns, mature ecosystem research, key species, artifact interaction, and genetic research.

5.3.4.3.2.2.2.3.9. *Business & Corporations Agency*

Businesses and corporations sometimes have more people, money and power than nations. This agency would be concerned with monitoring and assisting these entities, to ensure that they do not violate the norms of nations or take advantage of cultural limits.

5.3.4.3.2.2.2.3.9.1. *Earth Incorporation*

Political systems are impotent to stop the massive interference in ecosystems by international corporations. The simplest and most direct way to give the earth a voice in the development of the earth by humanity is to incorporate the earth following international law. The entire planet, with its biochemical cycles and nonhuman communities, would become one legal body. Since corporations are human constructs, however, humans would have to represent ecosystems and their wealth of living organisms.

In early civilizations, the advancement of the state was expected to contribute to the welfare of its people. Corporations are recent devices created by states for public purposes. Most early U.S. corporations, for example, were concerned with travel (turnpikes and inland waterways) or safety (fire insurance)—they resembled public agencies more than profit-seeking associations. The exclusive privileges and political power granted to corporations were based on the implicit promise of social services.

The association of economic development with national wealth allowed incorporation laws to be broadened. The corporation was given the constitutional rights of an individual. A corporation is a legal entity, independent from its founders, with its own rights, privileges, and liabilities. It is, however, required to obey laws and pay taxes; and it is accountable for its deeds in courts of law.

Unfortunately, as private good became identified with public good, corporations became larger, more acquisitive, and less concerned with social services. The quest for profit now has the effect of violating social amenities, such as clean air and clean water, instead of ensuring them. No responsibility is taken for environmental degradation since no right of contract or fair use of property has been breached.

Changes in societies, from rural to urban, from sparsely to densely populated, from culturally diverse to monotonic, have transformed corporations and the societies themselves. Business corporations now provide the bulk of goods and services in many states. The scale of these corporations, the processes of production, and the size and needs of human populations, have altered and degraded many ecosystems and biogeochemical cycles.

Successful modern corporations create an identity based on their purpose in providing goods or services; they define their business in terms of profitability, growth rate, cash flow, and competitive position; they develop a corporate vision, with specific objectives and strategies, including long-term vision, collection of ideas and creative implementation, aggressive manufacturing, and reliable finance.

The purpose of a corporation often transcends simple financial gain—the corporation seeks to maintain its own existence, before profit. Financial objectives (sales, assets, profits) exist to sustain its existence. The goals that most motivate corporate managers are survival, independence, self-sufficiency, and self-fulfillment. Yet, these motives are consistent with the financial objectives of the corporation: to maximize corporate wealth. The responsibility of managers is to maximize the value of the company. Furthermore, because corporations are long-lived, that value should last a long time—a good reason for looking beyond the ten-year monetary horizon and the lives of its managers.

Although current wisdom holds that a corporation's only responsibility is to its stockholders, corporations are being pushed to include social purpose in their strategies, again. Alas, they are doing poorly at it. They do not know how much responsibility to take, or where to put limits, or whether to pursue policies that diminish their profits. Corporations have proved spotty in doing social and environmental good. It would be more appropriate to have them deal with the environment as a corporate entity concerned with maximizing its own values. Of course, that would mean no more 'free' resources or environmental services.

The important advantages to incorporating the earth are the same as for incorporating a business. There are at least six advantages: (1) Managerial flexibility; the stockholders are separate from managers; responsibilities are assigned by needs of the corporation. (2) Limited liability; the corporation borrows and repays. It shields its members from hazards to which they would otherwise be exposed. (3) Financial advantage; the ownership of assets can benefit stockholders and the corporation. (4) Tax advantage; investments in the good of the corporation may not be taxed by nations. (5) Estate planning and longevity; the corporation exists indefinitely beyond the lives of its participants. (6) Central management and representation; a large and complex business needs operational and managerial efficiency. Many of the participants have no direct voice in the operation—they must be represented.

The earth incorporated would focus on a core business: to ensure the integrity and continuity of life and all its connections and to secure the opportunity for development free from undue interference. It would operate to optimize values, like any good corporation, but the values would be ecosystem values, such as fungus values and earthworm values, as well as human values.

A temporary Board of Directors would adopt bylaws, elect working officers, approve stock certificates, open accounts, and arrange a stockholders meeting. The stockholders would elect new directors, possibly from Association of Nations representatives or directly from elections, and decide on dividend declarations.

Stockholders, as citizens of independent nations, would turn over common and national property to the Earth Corporation, which would issue stock certificates to the stockholders. The corporation would allocate the purchase price of stock to capital at par value. Most of the shares—the percentage to be determined by the board as necessary to the operation of ecosystems—would be treasury shares. Anything more than par value would go to capital surplus, and only capital surplus could be distributed as dividends. Stockholders have the right to receive these dividends equitably, without resort to traditional distributions of wealth.

Stock certificates denote ownership of the corporation. Although the stockholders own the corporation, they do not own the property of the corporation, the earth, which is owned by the corporation itself. Stockholders, as individuals, groups, or nations, could make agreements about how business would be conducted, about what resources would be used or traded.

The elected board of directors would make decisions of distribution and limitation. Percentages would be deducted from the interest for the operation of the corporation and for equitable distribution to nations less favored by chance with biological or geological wealth. Furthermore, since the dividends would be distributed among people according to net ecosystem productivity and resource availability, no advantage would be gained by nations having large populations.

The basic functioning system would be considered capital, thus limiting the amount of human use of resources and probably the size of human populations. Interest would accrue in the form of net ecosystem productivity and diverted percentages of materials, such as gold or water.

The earth incorporated would solve the problem of having to value ecosystems in monetary or quantifiable terms; its systems would be untouchable capital. The human value of resources like copper, air, or water would be equated to the technological cost of recycling or producing them.

Raw material and energy are only two facets of the capital of a corporation—another is human ingenuity. Thus, human wealth would not be limited by restrictions on the availability of resources, but rather by a shortage of ingenuity.

An incorporated earth would be instrumental in conditioning international corporations to their social responsibility and in internalizing all costs. This corporation and governments could use traditional means, such as credit access, low interest rates, and setting priorities on equity issues, to evoke public interest in smaller and healthier human endeavors. The corporation would keep rights to global territories and resources, issue leases to nations, and monitor scarce resources.

The suggested articles of incorporation are:

FIRST: The name of the corporation shall be The Earth, Incorporated.

SECOND: The purposes for which the corporation is formed shall include: The protection of functioning ecosystems and their living beings from destructive interference.

The conduct of inquiry into the operation of such systems and the role of humanity therein for scientific and educational purposes.

The taking of appropriate legal steps to carry out these purposes.

The maintenance of all real common property, including all lands, seas, and atmosphere, subject to the restrictions and limitations hereinafter set forth, to use only the interest from income therein, reserving the principal thereof exclusively for the aforesaid purposes, it being intended that the corporation be organized and operated for preservational purposes and not for pecuniary profit.

The corporation is organized as a voice for nonhuman beings and systems. No part of the income of the corporation, if any, shall inure to the benefit of any trustee or officer of the corporation or to any private individual having an interest in the corporation (except for reasonable compensation) and no trustee or officer of the corporation or any private individual shall be entitled to share in the distribution of any of the assets of the corporation.

The corporation shall not be authorized to carry on propaganda, influence legislation, participate in any political campaigns, or discriminate against human cultures.

In furtherance of the foregoing purposes, the corporation shall have the following powers:

To accept and hold by gift or judicial order any real or personal property of whatever kind, nature, or description, wherever situated.

To sell, transfer, or dispose of the interest from any such property, but not the principal or any part thereof.

To make, accept, endorse, execute, and issue bonds, promissory notes, bills of exchange, and other obligations of the corporation for monies borrowed for the purposes of the corporation.

To invest and reinvest its funds in stock, bonds, or in such other securities and property as its trustees shall deem advisable, subject to the limitations and conditions contained in any grant or gift.

In general, and subject to such limitations and conditions as are or may be subscribed by international law, to exercise such other powers which now are or hereafter may be conferred by international law upon a corporation organized for the purposes hereinabove set forth.

THIRD: The operations of the corporation are to be conducted on the surface of the earth but the operations of the corporation shall not be limited to such territory.

FOURTH: The principal office of the corporation is to be located temporarily in the Association of Nations (AN), currently in the City of New York, State of New York, United States of America (and possibly relocated to Antarctica or elsewhere).

FIFTH: The number of directors, who shall be known as trustees, of the corporation shall be not less than 30 (a minimum number associated with major ecosystems), nor more than 3,300 (the number of independent cultures associated with biogeographical provinces and subprovinces).

An earth corporation would have a special charter to protect the planet and provide opportunities and services for all the inhabitants.

5.3.4.3.2.2.2.3.9.2. *Recharter International (Landless) Corporations*
Nations are people in place, whereas corporations are people in profit, often out of place. Corporations are like the feudal domains that evolved into nations. They are new kinds of political organizations, rich and arrogant, but they are currently free to leave behind any social or economic mess as they search for freer or better markets.

Nations need corporations more than corporations need nations, Lester Thurow notes. But, corporations are legal individuals. Nations simply need to make laws requiring that every corporation have a national base, preferably where most of the employees work, and must meet their obligations and duties as national citizens.

Laws for corporations, such as Antitrust laws, need to be consistent globally, says Thurow. Laws, like morals, have to react to new technology and new social situations, regarding consumer protection or company protection.

The AN could make a binding treaty on transnational corporate accountability. The AN would also collect data on Transnational corporations. Local corporations would have national accountability.

The question is who should have the power and how much. The AN would have the power to charter transnational or global corporations. The AN could tie corporations to place and force them to be responsible by identifying the corporation as a legal community in a place, rather than a fictitious artificial individual, with no address. Members, or staff and stockholders, would share equal responsibility and liability for the activities of the real corporate community. The AN would recognize that the corporate community has moral obligations which can be spelled out. The AN would recognize that the community is not permanent or abstract. It must have a place and participate in that place as a responsible community.

The AN would create a comprehensive model of requirements, such as stability, justice, and balance for transnational corporations. It would make sure that the charter requires serving public interests as well as private gains, to increase the meaningful work of employees, who want to serve the public good of the place as well. The AN would make sure that profits and losses, gains and costs are all privatized equally. The AN would make sure that all costs are internalized, that the capital of the ecosystem has to be paid for use. The AN would require fair labor practices in terms of reimbursement, where the minimum wage would be above a standard poverty level for a nation and the maximum could be no more than ten 10 times the minimum.

The AN would require a corporation to address environmental problems, using ecological performance standards, and address social problems, using social performance standards. The AN would require plans showing dependencies on environment and society community, then monitor corporations for their economic health, to make sure that they develop and mature correctly. The AN would also make sure they have plans for unforeseeable contingencies.

The AN would monitor corporate responsibility, the natural home communities of corporations for their ecological health, and the health of the home human communities. The AN would set standards for corporations to internalize the loops of production and waste, at least within the system of sharing corporations. The AN would track all impacts. The AN would encourage corporations to promote ecological design, and to guarantee the use and safety of their products and services.

The AN would expect loyalty from corporations, but develop loyalty both ways.

5.3.4.3.2.2.2.3.10. *Office of Personnel & Service*

This office keeps track of permanent and temporary employees of the global association. The association would use a civil service examination, much like that developed in China, to qualify and individuals for positions.

5.3.4.3.2.2.2.3.10.1. *Permanent & Part-time Staff*

The permanent staff would support the functions of the body. All positions, even the assembly, would be paid by the organization, with the understanding that the first allegiance professionally would be to the global body, before any national or community interests.

5.3.4.3.2.2.2.3.10.2. *Volunteers Service to AN or Nations*

The AN would create a separate volunteer program, for those who wanted to volunteer to work in global areas, such as mid-oceans or Antarctica. Some of the volunteers could serve on peacekeeping or emergency response missions in specific nations. The specific commitment could be two years.

5.3.4.3.2.2.2.3.11. *Office of Self-Assessment & Future*

Self-Assessment in an organizational setting, according to definition, refers to a comprehensive, systematic and regular review of an organization's activities and results referenced against a model organization. The Self-Assessment process allows the organization to discern clearly its strengths and areas in which improvements can be made and culminates in planned improvement actions which are then monitored for progress.

Self-assessment can be extremely valuable in helping an organizations to critique its own activities, and form judgments about its strengths and weaknesses. For obvious reasons, self-assessment is more usually used as part of a formative assessment process, rather than a summative one, where it requires certification by others.

The UN needs to assess the performance of independent governments. So, that in addition to a self-analysis at a national level, there will be a AN assessment. Assessment could be based on two-year cycles, although planning could extend to 500 or more years.

5.3.4.3.2.2.3. *Judicial Branch & International Courts*

The judicial branch of government is made up of the system of courts and offices. The Interests Court would be the highest court on the planet. The Constitution would establish this Court and all other courts. Courts decide arguments about the meaning of laws, how they are applied, and whether they break the rules of the Constitution. The functions of the judicial branch are: To maintain the integrity of the Constitution, to interpret laws, and to check the executive and legislative branches and special commissions, by monitoring their activities.

5.3.4.3.2.2.3.0.1. *Monitor Other Branches & All Laws.* The judicial branch monitors the Executive Branch, the Legislative Branch, and all other departments and agencies for the relevance of their behavior to the constitution of the organization.

5.3.4.3.2.2.3.0.2. *Interpret Laws.* The judicial branch interprets the laws made by the legislative branch and their enforcement by the Executive Branch for their closeness to the constitution and the intents of the peoples of all nations.

5.3.4.3.2.2.3.1. *Conflicts & Interests Court*

The Judicial Branch would set up a court to resolve interests and conflicts of the member nations, especially as related to global resources. The Court would decide matters relating to use of global resources or shared resources.

5.3.4.3.2.2.3.2. *Environment Court*

The UN Environmental Programme is relatively weak, from lack of capacity, staffing and funding. It should be made an agency that could coordinate environmental policies for the planet. It should have enforcement powers, maybe taxing powers directly. The environmental agency would conduct environmental surveys and monitoring.

The AN Environmental Court could address all global properties, including the atmosphere, deep continents, oceans, moon, space, bioregions, watersheds, and wild habitats. Many of its issues would be raised by the environmental agency.

5.3.4.3.2.2.3.3. *Office of Rights (Human & Ambihuman)*

Should some rights be made universal? We know that there are universal rules for human behavior, most of which relate to families. We know some things should be universal by law.

An economy has traditionally been seen as a morally neutral body, but even if it has only to conform to the nominal rules of society, it is already a moral agent. Economies are no more neutral than other organisms. Many areas of moral concern already are recognized: Worker safety; affirmative action; advertising truth; foreign investments; and harm to the consumer, public, and environment.

Responsibility occurs wherever the interests or rights of a person, society, or ecosystem are significantly affected by the actions of economic actors. Responsibility can be understood in terms of costs and benefits, that is, through operations and their consequences rather than abstract behavior. Every action entails a gain and a cost (or profit and loss). Profits and losses are distributed privately, socially, or environmentally.

Economies need to work cooperatively to make sure the costs and benefits are extended equally throughout the system. They could start by sponsoring the rational use of rare resources through taxation. Influence the government to determine priorities for wilderness areas or special landscapes. Beautiful, fragile, unique, or endangered ecosystems and species must be protected at the expense of commercial activity. Assigning rights to Nonhuman patterns would support many kinds of preservation.

5.3.4.3.2.2.3.4. *Tax Court*

The Tax Court would ensure that the taxes of the global association are legal and being collected. It may also ensure that the taxes of nations are not out of line with those of other nations.

5.3.4.3.2.2.3.5. *Police Enforcement Court*

This Court would judge cases related to police actions of the global association, in handling problems between nations or corporations.

5.3.4.3.2.2.3.6. *Office of International Trade & Equalization*

The resources of the planet are spread unequally around the globe. Local trade allowed some resources to be acquired through trade, especially things used for ceremonial activities, such as ochre or gold. Trade also exposed people to new ideas, and to distant people and their products. Trade also forced the collection of surplus in excess of needs and perhaps in excess of the system. Many effects of trade are long-term problems and do not become evident for several generations. They are also very difficult to reverse. For a society that needs surpluses to continue, with growing dependents and growing numbers of people, there is little flexibility to change. Other cultures, such as the English, encouraged trade to increase wealth.

Regional trade started to emerge a few thousand years ago, with Roman trade between 100 BC and 400 AD and Mongol Trade with China and Europe, 1250-1350. The Atlantic slave trade in the 1700-1800s linked three continents. The first wave of global trade destroyed many of the traditional societies of the Americas. A later wave of globalization destroyed traditional economic systems. European exports, especially in textiles, undermined regional livelihoods. These processes enriched the Atlantic shores. It also widened the inequities within nations.

Economics has become more and more global. Where peoples used to trade material goods, fish for roots or feathers for leather, for instance, now all things have a common symbolic value, most often expressed in yen or dollars. This means that whoever works the cheapest sells the most.

The purpose of this office is to promote the equalization of trade opportunities. It has to make sure that trade partners are not overconnected. The office would also help nations that are underconnected by trade. The office would safeguard the global aspects of the system. It would create rules for the international regulation of trade.

5.3.4.3.2.2.4. *Commission on the Earth*

The Commission would have trusteeship over global commons. Boundary issues need to be resolved first. The Commission would limit national jurisdiction of oceans to 20 miles from the border, rather than 50 or 200 miles, to protect ocean resources. Furthermore, it would have a say on continental shelves and shore fisheries. The other 65-70 percent of the surface of the planet has to be regulated as a common area.

5.3.4.3.2.2.4.1. *Environmental Surveys & Monitoring*

Surveys need to be made of every kind of ecosystem. Humanity needs to have an inventory of kinds of systems and kinds of changes. How is conserving terrestrial animals part of conserving ecosystem health? We do not know whether animal declines were caused by disease or some other factor, such as competition or predation. We need to find that out.

Monitoring is the key to understanding changes. Disease needs to be monitored as an important indicator of integrity. Other indicators are surveys of key species, habitat mapping and human impacts monitoring. Complex interactions have to be monitored, using a range of indicators at levels from behavioral to ecological. There may be limitations of the bioindicators of ecosystem health. Perhaps we need to find common and endangered indigenous species and monitor them, hoping that would reflect the health of the entire system.

5.3.4.3.2.2.4.1.1. *Create Inventories*

The Commission would inventory every ecosystem on the globe. An inventory is a complete list of all components of an ecosystem, from the geological to the ecological. A complete inventory of elements starts with the shapes of the features in the area, the geomorphology. The large volumes are rounded and natural hills—even the agricultural evidence is almost natural, that is, from the roadside not the air, the fields appear not to be squares, triangles, or circles; a small number of geometric shapes exist in the buildings by the road, but because of their scale are not too intrusive. A complete physical and ecological inventory would then be integrated with economic and cultural values.

With the information available now, from a more extended resource inventory and with optimal ideas about renewal, climatic conditions, traditional land-use patterns, local cycles, and ecological requirements (limits), conservation is more effective. Its goal is to support a steady state economy within optimum ranges based on natural and human limits.

5.3.4.3.2.2.4.1.2. *Create Monitoring of Global Properties*

The Commission would address real global problems, such as global warming, which has resulted in grain harvest shortfalls in recent years. The climate in general would receive renewed and detailed examination.

A global system is the sum of localities and may have unique characteristics of its own, that is, the universe has characteristics that local frames of reference do not have. An ecosystem is directly connected with global cycles and other ecosystems. The system is embedded in larger systems and global cycles, cyclicity. There has to be a good substrate with energy and materials flowing into the system.

The global problems include: Global warming; ozone depletion (chemical caused); disruption of global cycles; contaminations (nitrates, mercury). These threats cause ecosystem collapse. Ecosystem breakdown happens as a result of stresses, singly or grouped, that relate to interference patterns in the system, most of which are caused by the human species now, although the potential for asteroids or volcanic eruptions remains.

Our actions on the planet are experiments, whether we want them to be or not. Ignorance, denial, or cupidity cannot unmake this experimental course, which may be global and irreversible. This Commission would make the experiment conscious and cautious.

5.3.4.3.2.2.4.1.2.1. *Atmosphere Monitoring*

The biosphere can exert control on the temperature of the surface and the composition of the atmosphere. On the other hand, soil types and the weather can limit vegetation; invasions of vegetation change soil types. An ecosystem is a topographic unit, a volume of land, occupied by organic beings, extended over an area and through time, with connections to larger mineral, chemical, water and air cycles. This means they are geographical units that intersect with atmospheric units. Ecosystems have a vertical structure, that includes the levels of climate from micro to topoclimate and macroclimate, to soils, water structures, and bedrock, as well as a horizontal structure. An ecosystem is a process, or a set of interlinked, differentially-scaled processes that may be diffuse in space but are easily defined in turnover times.

Processes encounter each other in a functioning web of an ecosystem, with tangible and diffuse surfaces. Lynn Margulis qualifies her definition of an ecosystem: The smallest unit capable of recycling the elements of its membership. For example, organic carbon can be expired, fixed, reacted, or transformed. This is done through the physiological activities of the members of the system, through breathing, enzymes, or some other way. Margulis states that elements recycle faster within ecosystems than between them. Forests, for instance, act as sinks for carbon. The rapid release of sinks can affect other atmospheric or terrestrial cycles. The biota of the planet appear to regulate the surface temperature, atmospheric composition, and ocean chemistry, for a start, perhaps like the human body regulates its temperature, blood chemistry, and other vital signs. As it achieves a new balance, with human inputs, the atmosphere may cause problems with agriculture and other human activities. Atmospheric monitoring is tied with water and terrestrial monitoring. Parts of the atmosphere could be designated as reserves.

5.3.4.3.2.2.4.1.2.2. *Oceans Monitoring*
Life had once been limited to the oceans. The evolution of living forms expanded those limits. Life, over time, has colonized deep oceanic vents, as well as Antarctic gravel fields. Oceans are not as productive as most land-based ecosystems. Worldwide, oceans could only support about twenty two million people, even though its area is over twenty three times that of grasslands. The ocean bottom acts as a sink for phosphorus. The rapid release of sinks can affect other atmospheric or terrestrial cycles.

This Commission would protect the integrity of the oceans. It could designate ocean wilderness or conservation areas. Perhaps eighty percent of the ocean surface could be designated as a reserve.

5.3.4.3.2.2.4.1.2.3. *Deep Continents Monitoring*
Continents are formed by the movement of tectonic plates. As each continent forms, it develops its own combination of topology and water and climate patterns. Australia for instance has become the driest continent now. As continents rise or subside, in addition to their movement and combination, life has to adapt to the changes. The purpose of such monitoring would be to predict long-term changes as a result of continental change.

5.3.4.3.2.2.4.1.2.4. *Bioregions Watersheds & Habitats Monitoring*
Monitoring bioregions and watersheds involves a larger spatial scale than ecosystems. Regional goals are appropriate for bioregions. Evaluation of data must occur in an integrated manner that spans biological and physical scales, watersheds, administrative boundaries, as well as functional areas. To understand how ecological processes are connected we need to relate information across disciplines and agencies, and collectively perceive the effect of our actions on the environment. This approach follows ecosystem theory (the hypothesis that cycles in nature integrate the physical, chemical, and biological components of ecosystems), and the hierarchical organization of ecosystem functions throughout the landscape. Hierarchy theory can be described as the development and organization of landscape patterns, e.g., vegetation communities, through time and space.

This can be accomplished by incorporating the three primary attributes of biodiversity, as described by Jerry Franklin—composition, structure and function—into four levels of organization—province, subprovince, watershed, and site. Indicators incorporating composition, structure and function at the appropriate levels of organization have been identified for many ecosystems; they range from landscape morphology to human demographics and cultural influences.

For watersheds and habitats one has to consider the impact of any kind of vegetation removal. What is a minimum, optimum or maximum vegetative cover for various watersheds? Science might identify minima or maxima but philosophy and conservation can aim at optima.

5.3.4.3.2.2.4.1.2.5. *Antarctica Moon & Space Monitoring*

The moon has been such a constant for the earth, it might be hard to imagine how things would develop or change if the moon changed or were destroyed. The moon is related to stability of earth system. Because of its relatively large size and closeness, the moon forms the other half of a double planet with the earth. The moon revolves around the earth and both orbit the sun, so the entire lunar cycle takes almost 30 days. The moon exerts a gravitational pull on the earth that is stronger on the closer side. This creates a tidal variation in the heights of the oceans; these vary monthly. For many shallow water creatures, amphibians and mammals, it is good to adapt to these tidal variations. Of course, the earth exerts a pull on the moon, also, but it is less dramatic.

Due to its rotation around the sun, and to imperfections in balance, the earth tilts on its axis. This obliquity of the ecliptic creates seasonal variations, to which most animals and plants have adapted. Any changes, even relatively small ones, could be catastrophic for climate—a one-degree change could account for some ice ages. Jacques Laskar and others have documented the importance of the moon on the habitats of the earth. A stable climate needs the influence of the moon; otherwise, there would be immense variations in the solar heating of the earth's surface. The moon provides energy pulses, stabilizes axial tilt, and causes tides and variations.

Space is an equally important part of the system environment. It is not only the source of the sun, but it is the sink for energy from the earth as well. Stars and the sink of space provide many elements and their proportions. Galactic and solar system dust influences long cycles of the earth's climate.

Life is also challenged by energy and gravity, as well as the moon's behavior. The moon provides daily variations in tides, that provide energy to organisms, although the organisms have to adjust for the different levels of water.

5.3.4.3.2.2.4.2. *Ecological Design & Planning for Commission on Earth*

The Commission is entrusted with finding out what is on the earth, as well as designing forms for the continuity and enhancement of human life on the earth.

5.3.4.3.2.2.4.2.1. *Create Long-term Ecological Plans*

Despite valid arguments against centralized and global planning, the most important thing people can do with civilizations and ecosystems is plan. That is, plan for the ecosystem needs and for human needs, plan for landscape, watershed, preservation, site, alternative use, and social objectives.

The goal of the Commission is to create a practical plan that fits cultures into nature in ways that protect all aspects of domiture. We can break the planning process into various stages, each with accompanying tasks and subplans. This plan, with all its partial plans, is necessary to protect the scale of landscapes that are too large to see, except perhaps by satellite, the parts that are too small to see, such as fungi and viruses, the parts that are too-long-lived for us to observe, such as long successional changes or evolutions, and the parts that we are ignorant about. Without special effort, we are aware only of what we see working in the system during a very short time. We trust that our plans will ensure that the system will remain as a healthy entity for a very long time so that many generations of us can gather our needs from it.

Planning is not meant to be a finished work of art—it has to reflect our understanding and use of nature. Each activity needs to be fed back into the process of updating the plan. Implementing the plan should result in improvements to it.

5.3.4.3.2.2.4.2.2. *Protect Hotspots (such as Madagascar Hawaii & Ecuador)*
The Critical Ecosystem Partnership Fund (CEPF) is a major endeavor to preserve Earth's most critically endangered and biologically richest regions. The biodiversity hotspots are in a state of emergency, according to Jorgen Thomsen, CEPF's Executive Director and Senior Vice President at Conservation International, the managing partner of the fund, who stated, "By engaging local people in biodiversity conservation, we ensure the best chance of success at protecting the environment for future generations." The CEPF, a joint initiative of Conservation International, the Global Environment Facility, The John D. and Catherine T. MacArthur Foundation, the World Bank, and the Japanese government, aims to invest at least $150 million over five years in biodiversity hotspots—highly threatened regions where more than 60 percent of terrestrial species diversity is found on only 1.4 percent of the Earth's surface. The Commission would work to refine and coordinate protection of all identified hotspots. Areas identified here are representative samples, not an exhaustive list.

5.3.4.3.2.2.4.2.3. *Keep Critical Areas Intact (such as Amazon Congo & New Guinea)*
Many areas are critical because of their size, as well as their uniqueness, and due to their out of scale effects on the global system. These areas should be kept in tact, as functioning systems, although they could be used by archaic cultures and by industrial systems, if precautions were taken. Their use by nations would be limited. Earth parks, in the Antarctic, Amazon, Arctic, Northern Canada, Congo, New Guinea, Oceanic areas, and Eastern Russia would be declared immediately.

5.3.4.3.2.2.4.2.4. *Restore Large Areas (such as Mesopotamia & Western China)*
Ecosystem restoration would be begun; massive planting efforts are undertaken. No further expansions would be permitted for development in wetlands or other sensitive areas. Destructive searches for resources would be suspended, in favor of substitution and recycling. No new building would be encouraged until uninhabited ones are restored.
Many special areas, now degraded or destroyed, could be restored. Candidates for restoration would include: the Zagros mountain area between Iraq and Iran, the Shat-el-Arab river system in Iraq, the forests of Lebanon, the central tall-grass prairie in the U.S. for buffalo as a Wildlands Project, and Northeastern China forests.

5.3.4.3.2.2.4.2.5. *Anticipate Climate Change*
What is the solution for climate change? Restore forests and grasslands, and use alternative energy. Should we try a high-tech solution? A massive program, like the atom bomb, only

on alternative energy technologies, might work. Continued global warming could lead to reversal of ocean currents. One way to anticipate change for preservation is in the design of protected areas, using a north-south axis and including many different elevations. Siting cities, or most of a city, over twenty feet above sea-level might be prudent.

5.3.4.3.2.2.5. *Commission on Cultures & Religions*
This commission addresses how to keep cultures healthy and active by understanding the characteristics of a healthy culture and meeting its needs, from being grounded to being sophisticated. Cultures change over time; some develop, some collapse. Many cultures are transformed when they adapt to changes. Some cultures were transformed by domestication or agriculture.

Culture operates like nature, with rhythms of dissolution and reformation. Often the elements of a culture will simply be rearranged by a succeeding culture. A new culture can only be made from the heritage of the old. Our survival depends on the capacity to remake the image of the world from within, phoenix-like.

Like biological species, cultures do not fit perfectly into an integrated whole; there are discontinuities and contradictions. The culture is a loose-fitting patchwork of ideas, things and relationships. Humans can tolerate inconsistency and operate with contradictory beliefs: Soldiers fight for peace; ministers save the unborn for starvation. If the contradictions become too great and maladaptive, then the culture can collapse or disappear.

In the face of a change a culture can either embrace change or resist. Resistance to change is normal as a cultural process. Groups like the pygmies have specialized to fit the requirements of the environment, successfully. This makes it difficult to adopt other cultural arrangements.

On the other hand, resistance to change itself is an adaptive mechanism. According to Betty Meggers, it works as a successful "cultural isolating mechanism." Isolation remember is what allows a culture to develop in the first place. But, then does it force a culture to become stagnant? This Commission will examine the history and features of cultures and religions.

5.3.4.3.2.2.5.1. *Preserve Cultures & Languages*
The spread of the Bantu people in Africa caused a destruction and then a homogenization of languages. But then, because the societies were still relatively small-scale, they soon started to fragment in local mosaic environments. Bantu, for example, spread south as far as it could and people adopted that language, but then dialects started to diversify. Bantu now has produced 500 daughter languages in the past 200 years. Things tended again to a linguistic equilibrium. Of course, other peoples, such as the Hadza, Sandawe, or San Bushmen were pushed to the margins.

Language reflects places. Knowledge of the environment is coded in a language. Local peoples have knowledge from thousands of years of successes and failures. In Palau, they know the 1000 approximately fish species. The Kapingamarangi islanders in Micronesia spread their catch over 200 species without threatening their numbers. Western science has not had time to identify the species, much less create an effective marine management system. The vast undocumented traditional environmental knowledge is kept in indigenous languages by those people.

Languages can disappear for many reasons, normally, because people stop speaking them, especially if the young abandon the tongue and the elderly die. Occasionally, it can happen that a drought, famine, disease, war, flood, earthquake, tsunami, volcano eruption, acts of genocide, or other catastrophe wipes out a people. For example, the Paulohi language speakers in Maluku, Indonesia, experienced a severe earthquake and tsunami several years ago which killed all but about 50 of them.

Some people have said that they do not want to bring children into a world where their society, language, and people have no place. Some have turned to negative behavior like alcoholism, drugs, crime, or killing. The Waorani in Ecuador, the Carabayo in Colombia, and other groups in South America turned to killing, and for that reason some groups have still not had peaceful contact with the outside world.

This commission would work with groups that have diminishing number of speakers. It could help publish dictionaries, audio tapes, and news programs. It would promote groups that are successfully trying to recover their languages, such as Hebrew and Hawaiian.

5.3.4.3.2.2.5.2. *Allow Religions to Act & Flourish*

Religion is a part of culture that binds people to their ancestors as well as to the invisible powers of place. It focuses on the changeless aspects of natural and human processes. Religion concerns itself with an image of the world, that explains what the world is like. It also explains how we can influence it and why we would want to influence it. A shared religion affirms family and ancestral ties, but it also allows strangers without kinship ties to act more peacefully. Of course religion has also served more mundane human purposes, such as to justify the transfer of wealth to a leader or to the rich, or the sacrifice of lives for an ideal nongenetic reason. Religions, and myths, according to Joseph Campbell, are great poems, pointing through things to the ubiquity of a presence in each. Cultural inheritance seems to work. The heritability is high. Children tend to adopt their parent's religion, political views, and leisure interests.

Religions are attempts to understand or control the world, either by understanding the invisible or by having spiritual beings intercede. Religion tends to reinforce the integrity and structure of society, by providing a common image, and reinforce the belonging and commitment to the group. Religious claims about the other spiritual world tend to be counterfactual, but they cannot be too implausible. The supernatural has to play a part in the world. It has to be associated with living beings.

Religious rituals can also stimulate endorphins to the brain. These rituals may include painful poses, rhythmic movements, singing, or trials of endurance. Endorphins have good effects on the immune system. Trance states are another feature of religion. Endorphins allow people to feel positive about people who share their experiences.

Each religion is an attempt at transcendence with its own truths, certainties and stories. Their differences allow common beliefs, such as the golden rule, but also inhumanity. Some social virtues, such as trust, truth, restraint and obligation are grounded in religious beliefs. A contractual economy needs these virtues to survive, but at the same time, it is undermining religion with secularization.

The stories of religions concern events that are deeply meaningful to the listeners. This helps bind the group, also. Religion may help control disruptive forces, especially things about distribution and power. Religion coerces people into a social contract. Religion and story-telling may reduce variability of individuals in a group. But, this might increase variability between religious groups. Shared beliefs in a religious community may allow it to outcompete a strictly secular ones. It permits more sacrifice and commitment. Religion also allows ecological balance for many groups.

This commission would work to preserve the cultural capital built up by religions of thousands of years and regenerate other capital.

5.3.4.3.2.2.6. *Ombudsman*

Although not traditionally a fourth branch of government, as first proposed in Sweden, the function of the Ombudsman would be to improve the effectiveness and fairness of the global government while protecting basic human and ambihuman rights. Through the Ombudsman, nations or cultures could present cases of dispute for resolution. One strength of the office would be its power to initiate legal proceedings in behalf of any human and ambihuman constituents or nations. It would be an independent branch with separate responsibilities: To investigate complaints against the AN or its officials for wrongdoing; to investigate complaints against nations; and, to investigate the operation of the AN.

5.3.4.3.3. *The Future: Will a Eutopian Framework Happen?*

Yes.

If.

But, 'if' always has so many complications. Taking these steps would solve many of the problems addressed earlier. The satisfaction of physical and cultural needs, as a result of living in stable and small societies, would contribute to the health of people. Fitting economic costs and needs to the limits of ecosystems and monitoring the economic process would reduce wastes and pressures on natural processes. The coupling of agricultural productivity to a solar budget, and the conscious restoration of degraded systems, would contribute to the health of ecosystems. Sufficient wilderness would allow the self-maintenance of global cycles. With the increase in security, wealth, and self-esteem, human populations could be dependent on ecosystem productivities and still be diverse and unique.

With the removal of war capabilities and the equalization of wealth, the remaining issues are not the kind to incite violent passions. Disagreements over the best way to raise wheat or maintain a forest may be more easily resolved than deciding the best nation or truest religion. The death of large-scale dogmatic ideology and national idolatry could also mean the end of organized slaughter. We have to perfect the art of resolving conflict. Mastering it through social debate would free unprecedented resources to satisfy social needs. Perhaps a planetary electronic referendum would open communication. In designing the world, everyone can participate. We can reduce the violence to nature and ourselves and transmute it to debate. That which has been hitherto left unsaid-what we want to become, what we could become-could become explicit.

Changes would be made after a period of adjustment. The process must be sustainable and equitable. The population would be adjusted for carrying capacity.

The real answer is still 'if.' If things get worse, if there are catastrophes and collapses, and if people try to take charge of the global course, then a eutopian framework may be instrumental in creating a rich, equitable and peaceful future. But, if things really deteriorate, then a eutopian course is not likely to be attempted.

5.3.5. *The Dance of Art Money and Ethics: Advertising Good Places*

Advertising creates the mythic images of our industrial cosmology—Marshall McLuhan called advertising the "cave art of the twentieth century." The myths are powerful, but trivial, and memorable, but inadequate to convey the meaning people need to live. Perhaps the myths are restricted by their content. If so, then ecologists and artists, and urban planners, historians, and politicians need to use the strengths of advertising to convey ecological sense and traditional wisdom, the feelings of balance and the dreams of nature.

Our dream of nature, and it is still a dream, in modern Western culture, is the dream of order and beauty. But, as Aldous Huxley noted, the dream of order begets growth and tyranny, the dream of beauty, monsters and violence. Our dreams are nightmares because they are not complete. The nightmares are symptoms that reflect unbalanced and immature cosmologies, that is, images of the earth. A traditional cosmology evolves with people's needs, fears, and knowledge. But, if it is incomplete, or if it does not fit environmental conditions, it may fail. Many early cosmologies, primitive or advanced, failed to fit the earth.

The modern industrial image of nature as a resource has resulted in pollution, material shortages, and environmental degradation. A culture that degrades its ecosystem risks its own extinction.

Industrial cultures, however, are not the only cultures in existence. There are hundreds others, although at one time, around 1900, our species had over 1,000 different cultures and 3,000 languages, roughly equivalent to the number of natural biogeographical provinces and subprovinces on earth. Each culture exists in a particular location with a unique history. Later developments are not more adaptive than earlier; nor do they replace them. Ethnic groups are not evolutionary stages culminating in The U.S., but are equally valid ways of life. Each culture is only one of many possibilities, a way. There is no single or correct way.

Each culture has a root metaphor. In the West, it is the machine. The advent of the machine made processes of order more amenable to description. Although only a closed system itself, the machine was a fruitful metaphor for living systems. The theory of the living organism as a mechanical contrivance explained biological phenomena from the physiology of an organism to the processes of cells. The cybernetic machine metaphor was successful at explaining detailed processes without answering fundamental questions of meaning.

Science makes extended use of the metaphorical process to construct its models. For example: "Man is a system" according to Erwin Laszlo or "Man is a computer" according to Michael Arbib. Kenneth Boulding offered the perfect machine metaphor for the operation of the earth: as a spaceship. As a metaphor, a spaceship suggests the limits of earth and the value of a limited life-support system—unfortunately, it also implies something of human creation that can be controlled and fixed by human intention.

The use of the word "ecology" by Ernst Haeckel implied that the natural world was a place to live, a house, rather than a machine to control. Making the earth into a house is fundamentally a poetic activity, according to Gaston Bachelard. Poetry also is a way of understanding the universe through metaphor, a literary device that transfers the characteristics of one term to another. As Picasso said of art, poetry also is a `lie that tells the truth.' For example, William Shakespeare said "The body is a garden; and William Harvey said "the body is a machine." The body is not a garden or a machine, but the metaphors extend our understanding of the body.

Poetry is communicative of the quality of things. Like science, it discriminates the unsuspected in the commonplace. It is not different from science, but more diffuse; not better than science, but more comprehensive. It accepts ontological parity, the equality of beings; aspects of the world are not negated or reduced by one another. As metaphorical knowledge, which may be prerational or metarational, poetry can avail itself still of scientific references. Poetry can measure a whole qualitatively and mimetically, a germ or the cosmos with its imagery. Poetry is a tool for comprehending partially what cannot be known totally. A poetic language could include a view of the interrelatedness of all existence in a sublime ecology.

People need to be made aware of the power of self-determination. People need to feel things, like the immensity and uniqueness of nature or the strangeness of a biting tick, before they can act. Poetry can help people feel themselves as part of the web of life or on an oasis in space. That feeling, more than laws or injunctions, can justify preserving the ecological systems of the earth on which we live. Humanity is a poetic species, as Richard Rorty noted, "one which can change its behavior by the words it uses." We need desperately to change our behavior.

Mythology can join science with feeling to help us change. Mythology is not limited by method. Mythic symbols store information concisely, which makes it possible for a person to assimilate the collective experiences of a culture. Myth combines us with other beings. Mythologies are in fact great poems that function to awaken the experience of awe and humility before mystery, create a cosmology, validate and maintain an established order, and bring the individual into harmony with the whole.

5.3.5.1. Monetary Lies & Pecuniary Pseudotruth

Unfortunately, the myths of the predominate industrial cosmology are inadequate. The myths are powerful, but trivial and misdirected. Poetry and art are undervalued as forms of communication, not to mention as ways of shaping and making. Business has transformed much of art and poetry into advertising, to match the style and attention span of the people in industrial cultures. Advertising, quite literally from the *Wall Street Journal* to college textbooks, refers to its activities as "shaping the American dream." Like art, advertising creates an image of a way of experiencing. Unlike art, it limits its focus for a specific goal—profit. Like art, it mirrors us. Unlike art, it intensifies and glorifies only the positive aspects of culture, ignoring the dark, negative aspects that are equally valid.

Its simplicity is irresistible. Our environment deteriorates according to ecologists, but gets better according to economists. And their pictures are prettier. People want to hear that it is getting better. Advertising tells them it is. People want to act stupid, greedy, and selfish, and spend the inheritance of their children on themselves. Advertising tells them their actions are rewarded. The real issues of life and death, destruction and hope, make people feel helpless and anxious, so advertising draws their consciousness to comfortable trivia.

Despite the ugliness of the dreams of progress and growth, of waste and stylistic frenzy, advertising, using sophisticated techniques and narrowing the focus out of context, makes the dreams desirable and irresistible. People in agricultural and hunting cultures interiorize the abstract industrial vision. African farmers are convinced to buy inorganic fertilizers, even though it degrades the soil; women to buy powdered milk for their children, even if it kills them. Tractors replace draft animals in the paddies in the Philippines, even though they are costly and less energy-efficient; French winter fashions are found desirable in tropical Brazil, even if they can only be worn in air-conditioned villas. People in industrial societies are convinced that their children will be ruined without personal computers. Disposability is offered as a fix to a wanting in the temperament. Advertising fuels the acceleration of conspicuous and compulsive consumption.

5.3.5.2. Ecological Persuasion

Yet, advertising may be the most effective means to reshape desires and reform buying habits. Advertising presents the symbols of modern experience, even if they are just the trivial ones. It could present healthy symbols equally well. Advertising does incorporate traditional values, like family, friendship, and love, although to sell beer and cereal and, sometimes, churches and hospitals. And, like art, advertising lies, although Jules Henry thought it was instead a new kind of truth—"pecuniary pseudo-truth"—not intended to be believed, or certainly, proved.

Advertising is beginning to support more informational functions, such as the dangers of drug abuse and smoking. Advertising creates values—fur coats, fast cars, dark beer, slim cigarettes are certainly recent and artificial values—but it could be used to create positive ecological values and new identities that show that our needs for prestige, esteem, and belonging can be met without stylistic waste at mindless speeds. Advertising could promote new attitudes about appropriate technology, the rights of other cultures, and the place of people in nature. Good advertising could be as subversive and conservative as ecology. It could avoid confrontation with people's values; emphasize positive aspects without negative ones. A good ad could capture and carry the most self-indulgent viewer; for the most part, ads don't require effort, literacy, or consciousness, just attention.

Advertising has been serving the dream of progress, but progress is leading to catastrophe, a long, slow, global catastrophe. When people experience local, sudden catastrophe, they usually respond immediately, with heroism and sacrifice, aiding the victims of earthquakes or floods, sometimes famine. Advertising could bring to consciousness the slow catastrophes of erosion and population overshoot, and, perhaps, invoke the same altruistic and effective responses to them.

To work towards this service, conservation groups could define and promote an integrative mythology as the basis for the framework of diverse efforts to protect life and the environment. Conservation groups could provide a meaningful philosophical foundation, as well as coordination for other humane, social, and conservation programs. But, the approach must be egalitarian: Respect for life cannot neglect human life and suffering. The approach must be eutopian: A new cosmology cannot ignore adaptive cultural traditions that arose in place over centuries. Furthermore, in addition to formal education, they could provide re-education through the most effective means, such as advertising. Conservation groups could spend money advertising ``humane consciousness," moderation, and the joy of living, instead of just consuming or winning. Ecological ads would be unique and compelling, simple and effective. They would advertise not a product, but a way; not for a profit, but for a dream.

The other day, tired of scribbling, I went to visit friends, who were watching a car race. It occurred to me that only with entertainment industries is there so much technical fizz and coordinated enthusiastic teamwork. Imagine all that energy and enthusiasm directed to appropriate technology for reforestation or the proper use of forests. Imagine television coverage of forest work with the same amount of attention and detail. Why not a competition for the most beautiful or productive forest or teams working to restore devastated areas—broadcast by a major network as an important event. It also occurred to me that this remorseless entertainment is an anesthetic against the fear or emptiness or self-searching or death. Continuous entertainment is a kind of guarantee of health, riches, and long-life. Everything that is pleasurable, thought George Orwell, seems to be an attempt to destroy consciousness. Ecoforestry cannot ever compete with entertainment if it raises troubling questions or difficult expectations. As long as the industry can guarantee many forests through the arithmetic of fantasy, we will always seem to be complainers and false prophets—until it's too late, then we will be blamed for not avoiding the catastrophes.

Maybe the situation is not that bad. Maybe we can present images that rival the industry images. Maybe we too can speak the languages of euphemism that large corporations use to conduct their businesses of larceny and fraud. Positive images and pleasing language skills are everything these days; no one really looks for substance. The devotion to money, beauty and youth is our focus. I think one way to compete might be to present conservation biology or ecoforestry as a medical discipline, aimed at restoring forests to health—and advertise it that way. As with any medicine, the patient actually does most of the work to become healthy, although the doctor gets the credit and the payment. This would also lead to more respect for the practitioners, but also to more responsibility and more rules. The first rule, which we might take to be basic, is identical to the first vow of the Hippocratic oath, "Do no harm."

5.4. ***Creating & Maintaining Eutopias Now***

To avoid fanaticism and violence, Karl Popper has suggested that utopians should try to build an open, progressive, partially-planned society, instead of a finished, closed, completely-planned society. Indeed, this is how general systems theory would describe a working, successful society. Such a utopia would have to accept the imperfect nature of man and the changing ambiguity of nature. Utopias is the dream of reason. Eutopias is the dream of small traditions and cultures, reasonable or not. Where an imagined utopia offers revelations promising a desired future, Eutopias offers references from selves and cultures for producing good places on earth now. There is no mechanical prescription for making good places; there is no blueprint or timetable. The current institutions cannot create good places; the market has not been able to create health and equity; even radical ecologists have not been able to create a way—Eutopias is a fourth way. It is not an institution that benefits only the rich; nor is it a schedule of temporary handouts. It is a plan for a framework for local self-reliance and global exchange, that is respectful of traditional cultures and ecological networks. So far, there is only the idea or poetic image. Human will to power might be found in the will to imagine, and then to speak and become. What is our moral responsibility for this power? We can choose to alter our world with new images moderated by new ideals, such as good places, eutopias. Eutopias should offer knowledge and power with charity.

The criteria for eutopias include: Its benefit for humanity; the inadequacies of the present system; a drastic system change as a result of catastrophic awareness; and a low, but not too low, political feasibility. The benefits must be worthwhile to justify the costs. Benefits cannot be vague and unsatisfying when the costs are immediate and painful. Poetry and education must prove the benefits, so that the eutopian alternative can be begun. This code emphasizes its flexibility.

The best thing to do is stop—stop growing, stop producing, and stop running. Suspend the race and contemplate a direction. We know that whole countries have built again from ruins. So there is nothing to fear from stopping—if we know there would be no problem going again. What are the dangers of fast social transformation? Lack of justice? Lack of order? That is what the Eutopian framework is for.

The steady state would provide a period of rest, and time to explore human values and quality considerations. This strategy would avoid eventual hardening of the choices. But it must be instituted at once. The crisis caused by exponential growth and destruction cannot be solved just after some final limit is passed and the ultimate catastrophe begun. The crisis of ignorance cannot be solved by hurrying ahead and creating more problems.

Eutopias exists in the extended present, incorporating past traditions and future values. It would concentrate earthward (down) and inward. Heaven is a perfect home, eutopias is here and now. Eutopias is a new comprehensive philosophy to make sense of the world. Eutopias is comprehensive and global. A broader frame of reference is assumed. It is concerned with evolution unfolding, evolving, and producing new emergent forms, not just a static description. Eutopias is grounded in environmental concerns. Its values must be highest cultural and natural values. It must develop from existing social and political forces. Eutopias is vitally concerned with the well being of society. It regards society as a *sui generis* entity, not just an aggregate for analysis. Conservation is a means to an end, which is human fulfillment in harmony with nature. Human happiness depends on a balance between needs and commodities. In a throughput process it is not possible to economize all inputs simultaneously. There are many criteria for inputs to be preserved. Options must be site and culture specific. Eutopias recognizes and preserves slow cultural knowledge. A social base may be partly developed through ecological education. Social diversity may have cross-cultural appeal. Eutopias would retain the capacity to change and innovate, with changes in environments and human values.

Eutopias would detoxify national rivalries. Racism, sexism and ageism would lose their importance in a cooperative society of advanced communication, automation, equality, humane scale, and meaningful preservation. Eutopias is politically aware. We make political statements by the way we live. Every tradition is only one tradition among many. The higher sanity requires of philosophies and therapies is open, planetary dialogue between modern experience and sacred tradition. Eutopias requires a planning process that bridges all cultures and sciences. It must be a participatory political movement. It must appeal to a large segment of the total human society. Since not all interests will be satisfied, there must be opportunities for transformations or alternate paths. Eutopias would be a framework for microeutopias, where different human experiments are tried. Its variability would insure that we could reject any of the local visions that fail.

The eutopian frame can be justified. It is not just one kind of global society in one place. If we try to make one society, it will change over time, because people are in different places. Solutions come from living in place.

How can we form society within an ecological perspective? By following the principles of ecology and applying them to the characteristics of good places. According to David Orr, certain design principles work with ecosystems and nations: Small units dispersed in space, redundancy, short linkages between modules, simplicity, diversity of components, self-reliance, decentralized control, large margins, quick feedback. A megaframe like Eutopias would allow this.

Eutopias requires a change of attitude. We have to change the framework so that we can change to new minds. We no longer have an external point of view. We are inseparable from the environment and each other, but we differentiate. Eutopias is a nostrum, really, a description of good places for all beings. Its connotation is as a panacea or questionable remedy. This is appropriate since a panacea is a cure-all, a remedy for disease, and a solution to catastrophe.

5.4.1. *What Eutopias Can Do*

Eutopias is a self-conscious panacea. It requires an understanding of the anatomy, physiology, and diseases of the body in question, now the entire human and wild planet.

5.4.1.1. *Eutopias Can Eliminate Bad Approaches & Actions*

Human approaches to the challenges of nature have resulted in many losses. The lure of size and simplicity has resulted in many failures that become traps difficult to avoid or leave. The stresses from these things have resulted thefts, as attempts to balance or correct the situations. The very size and impact of humanity has made theft the only easy option, much easier in the short run than planning or self-restraint.

5.4.1.1.1. Eutopias Can Reduce Losses

Eutopias can reduce the losses of nature and culture by creating a framework to protect them. Eutopias can reduce the losses of health, fitness and accord, by emphasizing them and creating circumstances for their continuity. Eutopias can reduce the losses of equity, renewal and design by offering new designs that allow for a normalization of equity and for the normal processes of renewal.

Losses from accidents and diseases can be reduced by preparedness. Losses from earth and climate changes can be reduced, also, with preparedness for 'normal' events, such as hurricanes, earthquakes, and droughts. Design can also be used to reduce impacts from these events; for instance, by denying building permits on floodplains. The losses from some events, such as droughts resulting from El Nino, can be ameliorated, by having surplus food and supplies stockpiled.

5.4.1.1.2. Eutopias Can Reduce Thefts

Stopping a theft can be as simple as stopping a thief. But, theft has become such a complicated thing, many steps removed from the people who make the decisions and from those who carry them out. Eutopias would address the processes and trails of theft.

5.4.1.1.2.1. Reduce Theft of Life from Ecocides and Democides

Animal and plant lives are stolen for the profits of a few; ecological and human systems collapse as a result of biocide and ecocide. Hundreds of millions of human lives are stolen every century at the behest of a few. Millions die from starvation when regional crops fail; millions more die when the distribution system fails or is perverted for a few. This democide is unacceptable. The plague of power is responsible for the dialogues of death, and the absoluteness of some power is responsible for the massive scale of deaths in Russia, China, and some other nations.

Three reasons for these deaths seem more contributory than others: Inequity, runaway cultural antagonisms, and the use of absolute power. Inequity is simple to understand; some people have more valuables than others. Often, equalization does not happen until during or after a collapse, as it may have happened with the Mayans in 795 AD (or 1211 YBP). Cultural antagonisms seem to worsen when the different cultures are forcibly combined in nations as the result of colonial wars. Totalitarian regimes, especially with great power and in secrecy, behind walls or threats, kill almost as many people as famines—more, when one realizes that many famine deaths come from denial of distribution. Famine itself is the regional failure of food production or distribution. However, the denial of distribution, as when food is saved for trade or the elite, such as the English did to the people of Ireland and India in the 1800s, can contribute to the severity and extent of a famine.

Eutopian responses to these three reasons is: To equalize wealth as much as possible, to separate cultures into separate nations, and to make laws to control ecocides and democides. Laws without supervision and enforcement would not be effective. The UN outlawed war, but when the U.S., U.K., Iraq, Korea, or others decide to violate it, no one enforces it. The AN needs to have the power of enforcement, backed by all member nations. The AN needs to have the moral force to keep national governments open and responsible. The workings of government should be transparent to the people, whether the government is democratic, royal or charismatic. The AN must be sure that people can bear witness to every regime in the world, from Korea and Cambodia to the U.S. and Russia. Bearing witness should reduce the number of secret pogroms. Restricting and checking the power of leaders, in a eutopian framework, overseen by the AN, should greatly reduce democides, including genocides and mass murders. There are no permanent solutions. Eutopias can reduce the theft of life by making it more valuable and more visible.

5.4.1.1.2.2. Reduce Theft of Common Sense

Eutopias can reduce the theft of common sense by fostering and respecting common sense in communities and nations, as well as increasing its value and appreciation.

5.4..1.1.2.3. Reduce Theft of Choice

Eutopias can reduce the theft of choice by creating a framework that offers more choices for people in different communities and nations. It can also address the economic and political processes that reduce choice.

5.4.1.1.3. Eutopias Can Reduce Failures

Eutopias can reduce failures through education and opportunity. Perception, intelligence, and imagination can be taught. Integrity, will and charity can be shown by example. And, if there are enough teachings and examples, these human capabilities will be developed and applied to domiture, that is, civilization and nature. With will and imagination, people can design and build good places.

5.4.1.2. *Eutopias Can Integrate Tools & Designs*

Tools and designs are important extensions of the human mind. Their purpose is to foster and assist survival, not to make it more difficult. Tools and designs can be made appropriate to environmental limits and cultural preferences, both of which are often ignored by industrial approaches.

5.4.1.2.1. Eutopias Can Use Tools Effectively
Eutopias can illustrate how people can use tools with awareness of their effects and impacts. The principle of caution is followed.

5.4.1.2.2. Eutopias Can Thread Characteristics with Plans
Eutopias can thread the characteristics of nature, that is fields, ecosystems, and places, with the characteristics of cultures and good societies to make good places.

5.4.1.2.3. Eutopias Can Avoid the Traps of Noplaces
Eutopias can avoid the lure of or the accidental assembly of no-places. By exposing the lures of no-places, and showing the connections that make no-places into traps, eutopias can neutralize the plague of placelessness.

5.4.1.2.4. Eutopias Can Suspend the Designing of Noplaces
By promoting the understanding of the inadequacies of bad characteristics and bad designs, eutopias can stop the plague of uniformity and paucity. Through an understanding of the consequences of human ambitions and actions, eutopias can avoid many of the evils that result from a civilization on technical autopilot.

5.4.1.3. *Eutopias Can Increase Understanding*
Understanding is a powerful thing. Understanding why an animal bites can dissipate the desire for revenge or punishment. Understanding why things break down can result in an examination of the context and effects of tools, and maybe a simplification.

5.4.1.3.1. Eutopias Can Help Understand Ways of Knowing
The ways that human beings know things is part of the human adventure. There are many ways of knowing, from traditional ecological knowledge to the most abstract science. None of these ways is the only way. None should supplant the others entirely. This is the importance of education, that it be applicable to local place, yet broad enough to put that place in a larger context, of the environment or planet or universe. These ways of knowing allow people to learn the operation of nature and fit human activities in it. This understanding points to the need for basic sciences and crisis sciences.

5.4.1.3.2. Eutopias Can Help Understand How Things Go Together
Ecology is a science of relationships and patterns. Understanding the components of an ecosystem, and the characteristics of an ecosystem, can be applied to the growth, change and development of human systems.

5.4.1.3.3. Eutopias Can Help Understand How Things Happened
History can allow understanding of the regular patterns of human life, as well as the dramatic changes, such as agriculture or urbanization, and how these changes have affected the patterns. History is a record of the cumulative human impacts on living systems, as well as of a few famous people or battles.

5.4.1.3.4. Eutopias Can Help Understand How Things Renew

Systems automatically renew themselves, especially living systems. Systems that do not renew themselves very well, such as agriculture and cities, can be put on track for renewal by linking them with the surrounding natural systems. Eutopias can foster the renewal of human systems by integrating them into natural self-renewing systems.

5.4.1.3.5. Eutopias Can Help Understand How Nature & Culture Work

Nature and Culture are systems. Domiture is the combination of those two systems. Culture was once called a "Second Nature," but human culture has expanded so dramatically that the two systems are better identified as one developed system, now. The fitness of human systems are intimately related to the fitness of species and natural ecosystems. The human attachment to place is critical to understanding why people live where they do.

5.4.1.3.6. Eutopias Can Help Understand How to Live in Place

Eutopias can offer understanding of how people live in places, not only how they adjust themselves to a place, but how they adjust the places to their needs and desires. This mutual adjustment can be ruinous or beneficial. Emotional investment in a place, even love for that place, is crucial to the preservation of the genius of place.

5.4.1.4. *Eutopias Can Start Making Good Places*

Eutopias can start making good places by addressing the economics and politics of human cultures. Economics and politics are large-scale human programs to relate human needs to resources and distributions of resources and goods.

5.4.1.4.1. Eutopias Can Show How to Preserve & Restore, Design & Plan

Eutopias can provide an ecological planning process that offers a structure of limits and divisions for the planetary system that would permit the preservation and restoration of natural cycles and places. An ecological design process would be applied to ecosystems as well as to cities and fields.

5.4.1.4.2. Eutopias Can Illustrate Ways of Making a Living

Eutopias, through a holistic examination of how people make their livings in place, can show how changes can make better places. Economies can be as diverse as tropical or desert places; there is no evolution to one economic style, such as capitalism. Eutopias can show how to integrate individuals, communities and corporations into place.

The current, dominant economic and financial order is unfair to many groups and nations. It needs to be radically reshaped. It is better to do this as part of a directed plan than after some kind of collapse. Power and wealth must be more equitably distributed. A Eutopian plan would try to ensure that the underrepresented would be allocated more resources. Which would increase the demand for basic services. But, would that let them disconnect from any dependence on a world market? Wouldn't that be the goal for every nation? To be self-reliant in terms of food and basic production? Then, nations could reject technologies and products that did not fit their cultures or that would affect their own resources. These are all transitional processes, backed by the educative process and political pressure. A specific political process could be worked out later. There will be a painful readjustment to the realities of our new intricate involvement in the whole order of nature and her ecological balance. Social strains at this time will be unavoidable. A great amount of capital of energy and metals will be wasted. Sophisticated technology may allow a rebirth. A new world will have to be based on a gradually decreasing population. Production will have to be redirected to communal needs in transportation, housing, food, and recreation. It will be a much more humane world, moving from a materialist society based on industrial production and consumption to a contemplative culture based on ecological consciousness and symbiosis. But, people and cultures have to learn how to harmonize human culture with a deeper understanding of the ecology of earth, to become partners, rather than stewards or bosses.

This is the time to define goals in terms of population, quality of life, and preservation of biomes. Goals are not some final state reached once and for all time, but a horizon to be strived for, never reaching. Order is the highest ideal of mind; but chaos is necessary to shape and change it. Natural order is dynamic, creative, logical, and temporal. We live in natural order. To preserve what we are, we must admire the matrix out of which we arose. What is desirable is a relationship with a certain amount of conflict.

Global peace could be dull sometimes, but global war would be deadly. We need competition as well as cooperation. We could make a golden age in the present, not in the past (where Plato put it) or in the future (where O'Neill puts it). Life in the past was sometimes good regardless of environment. There are enclaves of well-being in all parts of the world, in Asia and Africa, in Australia and South America. Why not such good places for all? There is no technological reason why gracious living can not be available for all.

Economies can emphasize different things, from producing family needs (reciprocity), to distribution and redistribution of luxuries, to trade (mercantilism), to unbounded capital and to bureaucratic efficiency. It may be time to emphasize a form of aesthetic efficiency, that is the shared production of what is wanted without as much regard for cheapness and mass production.

5.4.1.4.3. Eutopias Can Explore Ways of Governing

Eutopias can examine how people govern themselves. Political styles can be equally diverse; there is no evolutionary path to one political style, not democracy, socialism or community anarchy. Eutopias can suggest ways to fit governing to culture and place. Politics can use different types of leadership or rules. The rule of law is fine. Perhaps a rule of religious tolerance would allow people to live together, something on the order of the golden rule. Does it always come down to one person, president, pope, king, or dictator? Representing all? In small groups or communities, anarchy could work fine. Rather than rule, perhaps understanding. Would a rule of knowledge work? Or would people have to know too much? When it comes to politics, as with mythology and religion, different people have different levels of understanding. So, government should be simple enough for everyone to participate knowledgeably in.

5.4.1.4.4. Eutopias Can Try Ways of Integrating Religion & Art

Religion is a part of culture that binds people to their ancestors as well as the invisible powers of place. It focuses on the changeless aspects of natural and human processes. Religion concerns itself with an image of the world, that explains what the world is like. It also explains how we can influence it and why we would want to influence it. Art is a part of culture that expresses the invisible parts of society and the environment.

Religion can lead to understanding of the world; it can lead to ecological balance. Art can lead to peaceful ways of interacting with nature and other human beings. Art ruthlessly examines society. Religion reinforces the integrity of society. Art is a survival technique for humans on a wild planet. Religion relates the human to the wild.

5.4.1.5. *Eutopias Can Start Making a Framework for Revolution*

One premise of eutopias is that many things have to be changed simultaneously; some things have to be eliminated, and other things have to be invented. This revolution in thinking and acting, especially on a global and national level, will have to be governed by the consent of people in those nations. The basic everyday experience of human life, will basically remain the same, but the superstructure, that is concerned with global trading, distribution and taxes—that will change.

5.4.1.5.1. Creating an Associated Nations Framework

The police force would be used for positive nonviolent interventions to help people with problems or disagreements. Such interventions would be cooperative efforts by neighboring nations, coordinated through a revitalized Association. All states now have armed forces, whose primary duty is killing their enemies, internal or external, in unending conflicts. The unarmed police force would have different goals: Rescue from catastrophes such as earthquakes; civic assistance, such as vote getting or monitoring; and simple police action, being a persuasive presence in areas of conflict. The AN would insure the inviolability of the police to go anywhere on assignment and to intervene in any conflict when asked by any party. If the AN had most large-scale weapons, it might short-circuit the vicious cycle of armament races, and it is conceivable that the AN would need to use weapons in some circumstances. The UN has successfully used police for the observation of peace and for enforcement, in Cyprus for instance. The UN has used police in response to natural catastrophes, such as earthquakes, in Peru and Italy for instance.

A global association would also coordinate the distribution and use of common resources, which would also be owned by the association as representative of all nations, rather than of individual nations; resources across the planet are uneven and have precipitated numerous disputes for thousands of years. The agency would address real global problems, such as global warming, which has resulted in grain harvest shortfalls in recent years. Climate itself would be a concern.

For any nation, the association could advise on topics ranging from justice to wilderness. Wilderness has an important role in human freedom, as well as in providing ecosystem services. The association could insist on self-reliance, by connecting human population to ecosystem productivity. It could also make sure that local air, soil, and water resources are stabilized.

For all nations, the association could provide education on health and appropriate technology. It could work to provide basic needs for food and health. It could insist on the truth of the ecological situation, on the real costs of economic decisions and growth, especially those that destroy ecological capital, in the form of wilderness ecosystems. It could recommend changing the system to allow taxes on destructive activities, such as excessive carbon emissions, and to normalize the values of resources and wilderness.

5.4.1.5.2. Encouraging & Fitting New Nations

The AN will offer any culture the opportunity to have a vote in the global management of human cultures and natural processes. Those cultures may choose to remain in their current national framework and share one vote, or become independent and exercise a whole vote.

Three levels of responsibility—individual, national, and global—are identified and discussed; each has responsibilities for specific attributes, such as population or health. This does not mean that only nations will exist within a global framework; alliances and networks will form and reform.

The eutopian code divides the earth into zones for preservation, conservation, domestication, and human communities. Human activities are limited to specific zones and, within those, the global authority controls all air, water and land use under complete sovereignty. Political units are formed from existing cultural units; an optimum human population is of each nation is based on a calculation of net community productivity on arable land through traditional agriculture. Common planetary resources are assigned according to the optimum population figure. The development of the nations is regulated by the Associated Nations through charters. Self-reliant nations would decide their own appropriate technology, crops and institutions based on traditional values and are responsible for the ecological education of their constituents. Residents of nations have equal rights and work opportunities, and have the responsibility to participate in government and to live as wisely as possible, to make good places.

5.4.1.5.2.1. Connection & Size

When small societies start to grow, they become successful in different ways. But, that success leads to increases in size, which can lead to a tragedy of scale. Each major technological breakthrough permitted a step increase in size. The size of a local population increased the likelihood of its success. For cultures, size was important. More successful cultures (as measured by size and continuity) were larger and more aggressive. Humans naturally increase the size of societies, but do not know how to stop or limit it. Ambition contends with common sense? Is the problem growth or growth without development for the scale? Henry Simons designated the great powers as "monsters of nationalism and mercantilism" and suggested that they are the obstacles to world peace, and must be dismantled for us to survive.

Leopold Kohr relates cancer to size. But, is it largeness or sameness without function? Cancer allows the mass of one unit to become too big. The cells outgrow their limits. Political cancer is a matter of proportion between large and small units. Cancer is a small-cell phenomenon, but it converts all cells to itself. Would a theory of social gravitation work? At one scale we have fusion and energy, but if the scale is too large that becomes crushing and instability.

The processes of social development can create traps that then determine the direction of development. Changes in scale, such as population size or sedentary living, can decrease the number of options possible, while providing different attractive options, such as the accumulation of luxuries. In this sense a trap is a sudden reduction in options or flexibility due to changes in scale or repetitive pattern.

The sheer size of a united China meant that tributes were enormous and that any commercial income could not compete well with the mass of tribute. The commercialization of states seemed to benefit from proximity to rich routes, small state size and rivalries.

The size of a cultures might not be optimal, since they grow unplanned and wild. Kohr quotes Arnold Toynbee as linking the rise of universal states to the downfall of civilization. Toynbee suggests that one solution might be the return to the Greek ideal of a self-regulating balance of small city-states, *Homonoia*, rather than further macropolitical solutions. Leopold Kohr sees 'gigantomania' as the economic problem of systems. Kohr notes that our choice is not between crime and virtue but between big crime and small crime, not between war and peace, but between great wars and small wars.

Local equalities are easier to acquire first, more than global equalities. We live local lives in home places, with local limits and local pressures. We can calculate an optimum size population of a culture or community, by relating it to the carrying capacity of the land and society. We can establish optimum scale of populations through limits of carrying capacity. We may also calculate an optimum size for the planet, based on the sum of local cultures. An optimum global population might be of the order: Between 0.5 billion and 1.5 billion people, the sum of local optima.

For the Chipewyan people in Canada, for example, the commitment to caribou hunting is ecologically inefficient, since people could spend more energy on secondary sources of food. For the Chipewyan, a deliberate "underutilization" of moose, rabbit, grouse, birds, and fish, is the result of cultural values: The willingness to live below the carrying capacity of the local environment—a characteristic of most hunting/gathering societies—the complex practice of drying caribou meat, and the reciprocity of the kinship system, that is, the act of giving as the basis for future relationships. The cultural decision to hunt caribou as the primary item of subsistence has produced a unique pattern in the utilization of land and in the formation and distribution of social groups. Many foraging groups, such as the Chipewyan, have been successful by staying below the maximum carrying capacity. Culturally, an optimum population has to be large enough to allow variety.

We may not know what is the minimum, optimum or maximum use of an ecosystem. Science might try to identify minima or maxima but management can aim at optima or satisficia. Francisco Varela analyzes the evolutionary process as satisficing rather than optimizing; a suboptimal solution is adequate. A free market has to be limited by conservative calculations of ecological balance. It is almost impossible to estimate the economic value of natural balance.

Kohr's reasons for the greatness of small states: There is a cultural diversion of aggressive energies, or artists are cheaper than soldiers; there is a relief from social servitude, as a result of time and leisure; there is the variety of human experience; and there is the testimony of history. Kohr concludes that it was always "the small state, not the empire, that survived. That is why small states do not have to be created artificially. They need only be freed."

5.4.1.5.2.2. Connection & Speed

The speed of our economies might not be optimal, either. Alvin Toffler has described the fast economies that are forming and concludes that slow economies will have to speed up their responses or risk becoming uncoupled from the fast lane. Yet, it might be good for countries to be uncoupled. Uncoupling economically might be a sound option for traditional societies unwilling to make the same mistakes as industrial ones. Local communities are based on traditional cultures, which have long-term lasting power.

Traditional cultures often have wealth-leveling properties, absolute property ceilings, fixed wants, and production coupled with need—all of which results in a stable economy. Then, efficiency and productivity are less important than use and appropriateness.

Furthermore, Toffler says that the nonindustrial countries are faced with a shortage of economically-relevant knowledge. Are they? What kind? The knowledge of how to find or grow edible and medicinal plants? The knowledge of how to make appropriate houses and cooking utensils? Toffler touts knowledge-based agriculture as a cutting edge of economic advance; how knowledgeable can it be, if it ignores the erosion of soil and the destruction of beneficial insects? Traditional communities have lost more knowledge than we will have in the near future. What happened to our rich biological knowledge of animals and plants, to our rich mythical knowledge of animals and plants?

The path to economic power is through the application of the human mind, according to Toffler, and he urges that "revolutionary" forms of education are necessary. What is more revolutionary than traditional education? Learning about plants, animals, families, and cultures is more relevant than theoretical knowledge; computers and economics can be learned after adolescence. We have more than enough information and secondary knowledge.

Economic success is secondary, as is money, the accumulation of goods, and prestige. We are accomplished in the secondary meanings of life. The satisfactions from being in a culture in place, from planting trees, growing apples, watching birds, playing with children, and making love are primary. And, they are not speed-dependent. Some things have proper speeds. Music is not made more efficient or better by increasing the revolutions of a disk. Food or relaxation require a human speed. We lack the wisdom to act as if we believed this. Is fast technology a necessary part of happiness? Those who are uncomfortable with primary meanings tend to become addicted to power, speed, and possession, as a frantic way to avoid awareness, silence, or responsibility, as a replacement for being grounded in nature.

Nature provides the source, of wonder, of the sacred, of otherness, and of the wild. By submitting ourselves to positive accelerating feedback loops in economics, we distance ourselves from such primary meanings. Nature possesses power that is not speed dependent. Human consciousness has already had a "revolution"—from the wild to the tame—and many of us regret it; a further revolution to remoteness sounds more depressing. Animals used to be directly experienced; now, they are humanized and domesticated. Humanizing the world has made it tedious, uniform, and dull. Economics is dull! Toffler's assumptions are dull! The needs he describes are transitive wants, and their only measurement is quantitative. For fertile nature, we have substituted a sterile model of production and economy. The model is reductive: trees become resources, people become labor. More is more, faster is better. Although speed is our normal response to dullness, the celebration of speed for itself is ultimately unsatisfying.

What is the result of our fascination with speed in everything? Dismissing nature in disgust, we attempt transcendence through speed. We speed away from nature, from our own bodies, and base our civilization on that momentum, praying, requiring, that it never stops. People's souls die, but secure in their power, they manage the things of civilization and inhabit the treeless flatscapes of the malls of commerce, comforted by the banishment of wilderness and the capture of animals in zoos and of free people in reservations, satisfied that their young are mercilessly tied to televisions and computers, acquiring information without touch and speed without grace.

Nations and communities do not all have to follow the same path and the same rules at the same time and at the same rate. Cultural success is not the "survival of the fastest" any more than it is of the biggest or shallowest or newest. Perhaps if we remain unconscious, there will be a power shift to the fastest that will homogenize and level all human cultures. But, we can consciously imagine alternatives and work to preserve cultural and natural diversity and the richness of existence.

We have the knowledge to save cultures, to restore places, to participate in the cycles of the earth, but extra speed and power are not required. The pace of nature is generally balanced and well-established; we violate it at our risk. If we adjust to the pace of the growth of trees and to the movements of animals, we would not be risking catastrophic extinctions and famines, shortages of water and fuel wood, and the death of humaneness.

We do not need to give our power to faster economies. We need to shift power to local communities through self-reliance and participation. A community protects individual freedoms, guards regional culture (values and identity), and holds groups accountable for their use of power. In communities, people can decide to be conservatively sustainable or to grow and gamble on innovation. Communities can have different economic attitudes, paces, and goals. A community that is balanced and flexible, in tune with natural cycles, based on traditional values—in which industrial production is limited to appropriate goods—can absorb the shocks of change far better than an immensely big, powerful, accelerating, postindustrial, national vehicle.

5.4.1.5.2.3. Change & Exchange

In an article on good forestry, Hugh Williams links the maximization of the moral good of creativity to the maximization of public good, then to forests specifically. Williams also proposes maximal creativity as an ethical imperative. Maximizing creativity, however, would lead to chaos, both in one's self and in ecosystems. There has to be a balance between creativity and stability, between innovation and habit. We should not being trying to maximize creativity in human beings or forests, but rather seeking stability and relative harmony. Maximizing creativity may be a meaningless exercise.

In a system of ethics, we might consider maximizing a value, but then we have to decide if it is being maximized for one species or the system, or if it is being maximized for the present or the future, or whether it can be maximized at all. John B. Cobb Jr. suggests that we should act to maximize value in general, at least for every entity with intrinsic value, rather than maximize value for one human or all humans, that is, the greatest good for the greatest number, in the present or the future. Perhaps though, we should aim for an optimum or satisficium here also.

An economic maximum is a monotone value, which only increases or decreases. There are no monotone values in ecology. Desired substances have an optimum value—more calcium is not always better.

Before anything cam be maximized or optimized, we have to be able to measure accurately. We can measure wealth with modified Index of Sustainable Welfare to avoid distortion by size or by the combination of repair and maintenance with production.

5.4.1.5.3. Providing Paths for Individuals & Communities

Isolation can dangerous, whether it is isolated theoretical knowledge or an isolated culture. A maximum isolation is bad. A minimum is bad. An optimum is good. Isolation is dangerous, especially isolated theoretical knowledge. We strive for optimal solutions and control, but should settle for suboptimal and partial control, a satisficing solution.

Eutopias suggests small solutions to big problems. Protecting and restoring ecosystems is a local effort. Reducing gases, that contribute to the instability of the climate, reducing consumption in general, reducing the human population, and reducing conflicts, which contribute to the escalation of wars, are local efforts. Integrating food into ecosystems, to regenerate soils and repair ecosystems, integrating technology into a culture, integrating economies into ecosystems, and equalizing wealth, are local efforts.

Although not every global problem has a local solution, people will have to address them in small ways, too. Climate would be very difficult to change, much less control on a global level.

5.4.1.5.4. *Promoting Health at All Levels*
Eutopias is healthful. Positive health results from being on good terms with cosmos. The idea of right to health should be replaced by moral obligation to preserve ones health. We need to become attuned to the earth, to commit our fate to nature, and not just say that we have faith in modern technology to save us with an artificial environment. We must be flexible, not detached or noncommittal. We must commit ourselves and be able to adjust to necessary changes—to be in a state of risk. The harmonious interplay between humans and environment results in adaptive fitness, which requires a constant expenditure of effort to maintain. We must maintain an environment for plants, animals and humans that is healthy for all. Conservation and creation must be tempered with preservation. We are too ignorant to tamper with everything. More searching and researching are necessary.

Ecological health requires a single system of environment combined with high human culture, with a matching flexibility, to create an ongoing, open, complex system, characterized by a slow change in its basic characteristics. High culture is not a return to the innocence of the Inuit, or the sparseness of the cave. It includes necessary institutions for the arts and sciences limed in transactions with the environment. Flexibility is needed; within limits, a variable can move to achieve adaptation.

The ecologist must create flexibility and then prevent civilization from immediately expanding into it. Flexibility is uncommitted potentiality for change. Flexibility must be distributed among many variables of a system. Freedom and flexibility in regard to most variables is necessary during the process of learning and creating a new system by social change. There are still many possible futures for the earth and humanity, but they become fewer as we burn or destroy the earth's flexibility and our options. Recommendations to reserve flexibility must be tyrannical.

The ecological health of a civilization depends on a single system of environment combined with high culture in which the flexibility of the civilization must match that of the environment to create an ongoing complex system, open ended for the slow change of even basic characteristics. Health is the capacity of the land and water for self-renewal. Conservation is the effort to preserve this capacity.

Although the nature of the biosphere is largely determined by evolution, by organisms adapted to specific parameters and to each other, the anthroposphere tends to be artificial and managed, with only human needs considered. We need to keep as much of the natural world as possible in the anthroposphere; there is a human need for variety, individuality, and the challenge to understand the nonhuman. Emersion in trees and bees is necessary to nourish human attributes that are in short supply: Awe, compassion, reflectiveness, and brotherhood. As humans move from concrete to trees, there may be a profound transformation in a scientific return to animism. The metaindustrial culture is one in which the trees are counted in a census of members of a community. In the shamanistic tradition, people are not viewed as individuals, their history and experience is seen as result of being part of the group.

One significant book in the Hippocratic corpus was *Airs, Waters and Places*, which showed how well-being is influenced by the quality of air, food, land and general habits. It is as important to know from whence your body atoms and molecules came as it is to know the history of a used car.

Atoms that came from stars and rocks make up molecules of seeds, flowers, defecation, and rotting leaves, which are cycled through our bodies. Bodies are open systems exchanging materials with the whole environment. It is therefore important to choose carefully what is put into the body. Good food comes from healthy plants and animals, unprocessed and unpoisoned. Cells and enzymes react poorly to poisons and preservatives. Physical, mental and spiritual well-being are dependent on a healthy diet. As much as possible, one should know the origins of one's food— the soil, the plants—and be able to determine what becomes you.

Illness can only be understood in context, in relation to network of interactions in which person is imbedded; the health matrix. Ill health may be a natural stage of growth and interaction. Temporary illness as part of dynamic balance (especially childhood diseases?). The observation and study of balanced relationships in complex systems should allow us to recapture an experience of harmony and intimation of divine from scientific knowledge of processes. "A truly ecological view of world has religious overtones," according to Rene Dubos.

5.4.1.5.5. *Integrating Acts with Poetic Wisdom*

A planet that is mindless is not entitled to moral or ethical consideration. The earth has a mind but ecocrises are driving it to madness. The alternative to ecological insanity is wisdom. Wisdom is the functioning of a mind that is respectful of its own boundary and processes, according to Gregory Bateson. Evolution is trial and error process of learning; all learning contributes to evolution of global mind. So the cure for ecocrises is the education of minds.

We need reeducation in the demands and recompenses of a sane, realistic world. Peace of mind, security and self-respect must arise out of being someone in a real community, and these will be valued more than conspicuous possessions and idleness. The process would be labor-intensive rather than capital intensive, and intimate rather than pretentious.

A high culture must respect the wisdom of its experience, using necessary technological devices (computers, televisions); be diverse enough to accommodate the genetic and experiential diversity of people; shall limit transactions with the environment, consuming natural resources (capital) only to make necessary changes. High cultures must depend on renewable resources from photosynthesis to wind, tide, and sun to continue "making," which is the source word for poetry, good places.

Poetry expresses the image of human potential, of what better circumstances may have formed. Poetry tells of a goal, even if it is the moral superiority of suffering in the 'third world.' By presenting a goal, poets can become the "unacknowledged legislators of mankind," as Shelley defined them. Poetry creates a fourth world, of unique groups sharing part of the wealth of the earth in a global community. This fourth world is where the past is reconciled with the present and the terrible beauty of the future is born. The terror of beauty, as Rilke recognized, results from its power to shake humans from the refuge of a small identity into an immense strange world.

The lives of humans and all beings has become a collective responsibility. Humanity has to learn to live again on a finite and varied earth. Learning is a transforming experience. Poetry objectifies conscious experience and makes it easier to communicate. Nationalism (1930s) used poetry in service to the state. Used for each nation in this way, poetry shows the diversity of human experience. The essential unity of the earth can be discovered through its infinite diversity. Poetry gives groups and individuals their identity; it articulates societies and authenticates forms of exchange. By transcending the limits of single cultures, it draws all cultures together. The tradition of poetry does not belong to just three worlds; it encompasses them and links them together in a fourth. Poetry is wise language.

5.4.2. *Moving Forward Backward Inward & Outward*

Human ills cannot be cured by a return to idyllic hunting and gathering groups or to a quasi-agricultural, ecologically-caring society. There is no possibility of complete return. Most industrial nations are urban, and are becoming more so, as agricultural countries pack their surplus peoples in cities. Nor can there be a return to 4th century B.C. Greece, or to 17th century China, or to 1910 France, or to any time. Many traditional cultures no longer exist; others are disintegrating under pressure from industrial cultures. Nor can there be a jump to a complete technological future, where technology transforms hydrogen into wealth for everyone. Eutopias works with traditional cultures and realistic planning.

5.4.2.1. Uncertainty & Incomplete Knowledge

The detailed planning of complex open systems is not necessary. Planners are not in a position to attempt detailed models of future situations because many relevant parameters remain unidentified, and many of those known cannot be quantified. Plans can be made within the limits of variables, although it is not safe to be limited by lethal variables, as Gregory Bateson recognized. Closeness to limits reduces flexibility, that is, the uncommitted potential for change. To minimize untested conclusions, Eutopias is based on the values and forms of traditional cultures. This could allow time for rational planning to catch up.

We have to invest and cultivate our inheritance. We must enlarge our human identity, to include other beings and the earth, to include our own posterity and its image of the future, without which we lose the will and capacity to solve problems. Creating the future is necessary to maintain the present. It is meaningful to construct a world that we will never live to see, to plant trees that take two hundred years to mature, to save some of the forests and soils—not for the oil and timber elite or even for the backpacking elite, not for social abstractions or for personal profit, but for our heirs, for them to see and decide to save or to use.

Eutopias addresses the inadequacies of the present system; it offers a drastic system change from the institutional gridlock of elitism, but the change is not so drastic that the feasibility of acceptance is too low. The benefits must be worthwhile to justify the costs. The benefits cannot be vague and unsatisfying when the costs are immediate and painful. Communication and education must prove that the benefits exist, so that the eutopian alternative can be exercised. It must be a participatory movement, and it must appeal to a large segment of the total human society. Since not all interests will be satisfied, there must be opportunities for transformations or for alternate paths.

The eutopian framework is an open, flexible, and partially-planned global relation, instead of a finished, closed, completely-planned society, as imagined in utopias. Eutopias accepts the imperfect nature of humans and the changing ambiguity of nature. Eutopias detoxifies cultural rivalries. Racism, sexism, ageism, and speciesism lose their importance in a cooperative society of advanced communication, automation, equality, humane scale, and meaningful preservation.

5.4.2.2. Action Responsibility & Wisdom

Now is the time to define goals in terms of population, quality of life, and preservation of biomes. Resolving conflict through social debate would free unprecedented resources to satisfy social needs. That which has been hitherto left unsaid—the goals of humanity-could become explicit. Goals are not some final state reached once and for all time, but a horizon. Eutopias offers continuity towards a horizon.

Solutions to uncertain futures can be found in the characteristics of good places and in the principles applied to them. The proper action come out of common sense. We need to be sure that we allow ecosystems to regenerate healthy conditions while we take our needs from some of them. We need to plan for at least seven generations ahead, being flexible and keeping some options open, and being as self-reliant as possible. We need to be frugal with most resources and keep seven years of food and supplies in reserve. We need to identify an optimum population, over a minimum and maybe fifty percent below a maximum. We need to be as playful and joyful as any previous generation.

Science presents us with too many facts, yet we crave to have more. Philosophy presents us with too many values, but we attend to too few. Technology presents us with too many things, but we do not know what we need. We do not need more information or rules, but we do need meaningful ideas. Our attitudes and feelings toward nature need to be revitalized with evocative metaphors that let us accept responsibility for that part of the earth that we build, namely human culture and human landscapes. In order to know what is important and what is valuable, we need wisdom that we may not have.

The words 'view,' 'vision,' 'witness,' 'wise,' and 'idea' are all derived from the Indo-European *weid* or *wid*, meaning to see, understand, or know. Wisdom is knowledge of the larger interactive system, which if disturbed, can generate exponential curves of change. Greed is unwise. Wisdom is recognition of and guidance by a knowledge of the total system. The system punishes any species unwise enough to quarrel with its ecology. Greed, size and pride are unwise. Any course of action, like that just discussed, that ignores ecological stability and intentionality, i.e., the logic of nature, is unwise.

Hans Vaihinger (1911) in *The Philosophy of As If* suggested imitation as a solution to our lack of wisdom. Humans have no choice but to live by fictions; as if this world is the ultimate reality, as if there were free will. Humanity must plan for its future as if its days were not counted (or at least for several billion years). Jonas Salk urges us to behave "as if" we were wise, by using good sense. Wisdom is a new kind of fitness. To survive, we must accommodate ourselves to the new conditions of a radically different life. Survival in this sense is not a win or lose proposition, but a double win.

Wisdom is knowledge of the larger interactive system, which if disturbed, can generate exponential curves of change. Wisdom is the recognition of and guidance by a knowledge of the total system. Lacking knowledge, lacking wisdom, we must behave "as if" we were wise, as if we had good sense. Humans have no choice but to live by fictions, as if this world is the ultimate reality, as if we are responsible for our actions. Humanity must plan for its future as if its days were not counted (or at least for several thousand years). Wisdom is a new kind of fitness. To survive, we must accommodate ourselves to the conditions of the earth.

Wisdom is the disciplined use of the imagination with respect to alternatives, exercised at the right time and in the right measure. But we need practical wisdom, prudence, and intellectual control in virtue, in place of the theoretical wisdom taught by schools. The truths of our unique cultures and the wild earth are apprehended through myths. The poetic language of mythology can fit all the facts and values, things and images, into our hearts so that we can feel them and act upon them—so that we can make good places.

Related to wisdom, there are many corrective factors of human action: Contact, art, love, and religion. Love is the formation of Martin Buber's I-Thou relationships between human and society and environment. Socrates stated that Eros is midway between wisdom and ignorance. He who has no sense of his own deficiency will have no love of wisdom. Love is the desire that good be one's own for as long as it can. Arts and the activities of the mind can correct the excesses of pride. Contact between man and nature and animals can correct the problems of abstraction. And, ecodeontics (or religion), the binding of humans to the invisible powers of place, can correct the effects of detachment.

5.4.2.3. *Can We Make Eutopias? Starting Now?*

Will eutopias work? Yes, if—

and that word makes a difference. We can try to be wise, and act "as if" it might work. Will we survive? Has some limit been exceeded? If not, do we have time to correct our actions? Time has become a real problem, especially since calendars were invented to keep track of big events. If the climate or our lives were more regular, we might not be so concerned with time. The idea of a regular or eternal return might be satisfying enough. If a fundamental limit has been exceeded, what should we do? If it is too late, and it is very difficult to know this definitely or absolutely, then we could party. Or we could act "as if" we were wise. We need to start now. It is an emergency!

6.0. **Ecodex** (Coda)

Eutopias is a total reconsideration of the current pattern of technologies, cultures, value systems, and behaviors, evolving into a low-profile technological ethic suitable for a renewal of ecological understanding. It is a code for preserving those parts of the earth that are needed for renewing the holecosystem and for habitats for the billions of animals, plants and living beings that are part of the earth. It is a code for allowing fair use of that part of the earth that is human. It is a code for human equality in opportunity and worth. It is the demand for a margin from catastrophe, so that if humanity is unable to live peaceably, the rest of the earth will not become extinct as well.

The new theme for people's minds may begin in prose, but it should culminate in poetry. The human mind, under pressure from the dialectic process, grows into more subtle noetic experiences, until ecstatic insight blossoms. We must learn to be an individual in a human society in an ambihuman ecology with amphibian grace.

Poetry expresses the image of human potential, of what better circumstances may have formed. Poetry tells of a goal, even if it is the moral superiority of suffering in the 'third world.' By presenting a goal, poets can become the "unacknowledged legislators of mankind," as Shelley defined them. Poetry creates a fourth world, of unique groups sharing part of the wealth of the earth in a global community. This fourth world is where the past is reconciled with the present and the terrible beauty of the future is born. The terror of beauty, as Rilke recognized, results from its power to shake humans from the refuge of a small identity into an immense strange world.

The lives of humans and all beings has become a collective responsibility. Humanity has to learn to live again on a finite and varied earth. Learning is a transforming experience, but difficult. Poetry objectifies conscious experience and makes it easier to communicate. Romantic nationalism (1930s) used poetry in service to the state. Used for each nation in this way, poetry shows the diversity of human experience. The essential unity of the earth can only be discovered through its infinite diversity. Poetry gives groups and individuals their identity; it articulates societies and authenticates forms of exchange. By transcending the limits of single cultures, it draws all cultures together. The tradition of poetry does not belong to just three worlds; it encompasses them and links them together in a fourth. Poetry is wise language.

What mysteries of the universe we cannot understand, we must accept in faith. One secret is that all things are secret. The working out of the cosmic process is effected by the actions of human beings, by hate or love. Love is reverence for the experience of all beings. Through love, humans can make good places on earth. Eutopias is a presentation of properties, principles, standards, and practices that can be combined to address challenges and problems facing our species and to make good places on earth.

Properties (Characteristics) are characteristic qualities of a thing that distinguish unique individuals, systems, or patterns; Gregory Bateson calls them differences that make a difference. Properties are also essential or accidental. An essential property is one that a thing possesses regardless of relationships with other things. If an essential property is lost, then the thing becomes another species; an accidental is one that makes less difference to essential properties.

Principles are fundamental rules or laws, derived from and based on the properties of the surrounding systems, that we can use to create images or models to meet stated objectives, that is, the goals towards which our action is directed, e.g., a healthy ecosystem or a balanced productive society. Principles unify our images. Standards are models or examples of quality or value, established by authority or mutual consent, that can be repeated as procedures. Standards are established from principles. Practices, as actions or procedures, are regular operations based on standards and principles. Principles match properties to our standards and practices to maintain the system favorable to us.

For example, one property of the planet is its wildness; we live on a wild planet. The corresponding principle is that the planet is self-making and self-ordering, without requiring human control and management. Our objective for the planet is to allow the wild process to continue. We can set standards that are likely to keep the planet wild, stable and enjoyable: Limit our exploitation to less than fifty percent of the surface area; use appropriate techniques to minimize damage to wild systems; or, calculate optimum regional populations based on ecological and cultural carrying capacity. We can match our practices to those standards. For example, we could restore areas that have been damaged; rework or concentrate urban areas rather than converting wild systems; or, limit our population. The samples presented are not meant to be complete.

6.1. ***Properties***

- Global biogeochemical systems depend primarily on wild ecosystems and cycles.
- Archaic peoples lived within the limits of their ecosystems.
- Modern agriculture, forestry and urbanization require great energy subsidies from fossil fuels or nuclear energy to survive. They exceed the limits of forests and land.
- Changes in scale put pressures on human and natural systems.
- We ignore very long-term trends in nature and human history that have very dramatic influences on our activities and management styles.
- Our history of use of surrounding ecosystems has been to exploit them to collapse and then to move on.
- We justify our temporarily successful behavior with myths that allow us to continue the behavior without being responsible.

6.2. ***Principles***

The Eutopias framework recognizes and incorporates into its application a number of basic principles. The Eutopias framework possesses three levels of authority, each with its own area of responsibility. There is a global authority to protect both the planet and human cultures. This authority, the Associated Nations (AN), is responsible for all land, air, and water utilization, for global cycles, and for interactions between nations. The AN gives equal opportunity to nonwestern, nonindustrial cultures to flourish.

Nations are based on traditional cultures, which have long-term, lasting power. Nations are responsible for protecting local environments and for providing a context for individuals. Both globalism and the simple community are necessary, if the community is not to be diseased and the globe impersonal.

The locus of political sovereignty is the individual, who is limited in giving away proxy rights. Politics has to be a participatory process, where an individual has some power over decisions affecting him or her. Participation is necessary, not only politically, but to establish the existence of common values throughout the population as a whole. Individuals have responsibilities for themselves and to their cultural governments.

6.1.1. *General Principles*

- The principle of lawfulness. Every unit of information can be analyzed into lawful components. The discovery of new patterns can confirm, but not prove, the principle.
- Preserve the earth, the heritage of all living beings. It has been modified and maintained by living processes over billions of years, and is required for the continuity of life.
- Keep humanity participating in the entire ecology of the planet. As with all beings, both the whole ecology and the species-specific are necessary for survival.
- The natural resources of the earth—the air, water, soil, plants, and animals—are to be preserved, and where possible, restored or improved.
- Allow all beings the equal opportunity for existence; this does not mean that they cannot be exploited or separated, just that they are not eliminated from the system.
- Humanity must build its civilizations within the framework of a planetary ecology. The human-created environment is necessary for human well-being.
- Economic development and social progress are necessary for the welfare of humanity. But must be conducted within environmental policies.
- The goal of economic and social development is to provide favorable and meaningful human habitations and activities.
- Human settlements must be planned and constructed within environmental constraints and according to ecological priorities.
- Technological processes must be brought into balance with the cycles of the earth. They must not damage or degrade natural cycles. Processes that cannot be controlled or rendered harmless, such as nuclear waste or inorganic pesticides, must be banned.
- Nonrenewable resources must be used sparingly, to avoid exhaustion before a steady-state economy has been achieved. The benefits from their exploitation are to be shared by all.
- All communities and citizens are responsible for the environments. They cooperate through a united body to preserve and promote the whole environment for the well being of all species and posterity.
- All humans have equal right to their share of the resources of the earth and to the amenities of civilization.

6.1.2. *Global Principles*

- No one group has sovereignty over the earth. The Association of Nations (AN) has regulatory powers necessary to maintain a healthy environment.
- The AN will establish a body for managing the earth's resources. And encourage states to establish local institutions.
- The AN will manage the earth through a fund from all states.
- The AN has regulatory and punitive power over all waste and pollution.
- The AN has regulatory power over all mineral resources in the earth, and distributes them equally, according to the principles herein.

6.1.2.1. *Regulation*

- The AN will be the intergovernmental body for the world to guide common policy.
- World-wide and interstate matters will be handled by all states working through the AN. Cooperation is essential to preserve, control and improve environmental and human conditions.
- The AN must ensure that rational planning considers the irrational, that human development considers the nonhuman. Therefore it has the power to refuse state plans.
- Only the AN shall have large-scale police capabilities. Only the AN shall have appropriate weapons. All mass destruction weapons, such as nuclear bombs, killer satellites and biological bombs, shall be reduced and recycled.
- Where population growth exceeds the safety margin established by the AN, on the basis of organic carrying capacity, demographic policies may be imposed by the AN, without prejudice or malice.
- The AN shall create a world bank for the exchange of all nations.
- The unit of wage shall be a human work unit, which shall have equal value for all.
- The AN shall stabilize prices of commodities and materials, and earnings from employment of labor.

6.1.2.2. *Resources*

- Limit use of other species within the limits of the dynamics of those species.
- Respect the use of materials through limited quantities and recycling.
- Respect the cycles of a system that renew or circulate elements.
- All scientific and technological knowledge shall be available to all states, as regulated by the AN.
- Scientific research and development, especially on environmental problems, will be promoted in all states. The AN will support a free flow of scientific information and experience. The AN will also ensure that appropriate technologies are available to all countries.
- The AN will award basic educational and research grants to all humans, for whatever use desired.
- The AN shall maintain quantities of food and currency to provide for disasters of natural and human cause. And shall be responsible for transfer of financial and technical assistance as required.
- The AN shall provide social security to those unable to work because of age or inability.

6.1.3. *Local Community or National Principles*

- The community is the unit of survival and must be kept healthy.
- A community is composed of many interacting species and must be kept diverse.
- A communities embodies the rhythmic changes in the activities of individuals and groups, from which emerges periodicity, including genetic or system changes.
- A mature community is self-perpetuating and homeorhetic, with a dynamic balanced energy-matter budget.
- A community can change state and replace another community, as a result of orderly processes, such as succession or intentional development.

6.1.3.1. *Management*

- Preserve all components, structures, and functions of system, that is, the health of a system providing services to the community or nation.
- Protect and maintain diversity; preserve the overall natural patterns of the system.
- Utilize natural processes for regeneration and protection.
- Current knowledge of the supporting ecosystems of a place is necessary.
- Acceptance of ecosystem limits of a place is necessary.
- Reduction of human demands on a place is necessary.
- Use of a place cannot be independent of human equity.
- The scale of human infrastructure and building should reflect the scale of the landscape.
- Human designs should follow the sensory force of a place and enhance the spirit of place.
- An optimum diversity in a landscape should be valued above a maximum or minimum.

6.1.3.2. *Political*

- All states shall be equal, politically, regardless of size or sophistication. Every state has the right to its own integrity.
- States shall have sovereignty over their cultures, people, and organic and inorganic goods, within the limits of human and environmental rights and health.
- In accord with the AN charter, states have the right to use their assigned resources in any manner, within the limits of damage and pollution previously set.
- Nations have duties to ensure that activities are in conformity with principles set forth in the Association of Nations.
- Nations have duties to conduct their activities in a manner respectful of their complete effects.
- Nations must adopt an integrated and coordinated approach to planning and design, for the benefit of the environment and populations.
- Every nation has the responsibility to conduct its economy without causing damage to any other state or to the whole environment; each state must compensate for damage.
- Environmental risks and damages must be identified and solutions proposed.
- Through the AN, states shall develop laws on human and environmental rights, and activities.
- Education of the people on local culture, ecology and science is the responsibility of the state. The AN will provide aid for promoting enlightened opinion and responsible conduct. Mass media will disseminate educational information to protect and improve the whole ecology.
- Mass media will also be available in state centers for referendum needs.
- If population growth does not adjust to social and economic changes, it must be controlled by state policies.

6.1.4. *Individual*

- The individual is the unit of experience and reproduction and must be kept healthy.
- An individual identifies with her home (habitat) and must keep it healthy.

- An individual depends on his niche (work or way of producing food).
- An individual acts for her self-preservation and self-interest first.
- An individual engages in ethical behavior, as part of a cultural group.
- Individuals have the right to ecological freedom and equality.
- The AN or any state will not have the right to determine values for any other state, within the basic limits of human and environmental rights and equality.
- Individuals are responsible for protecting the environment and improving damaged areas.

6.2. ***Standards***

- Reserve at least fifty percent of the planet in wild areas.
- Calculate optimum regional human populations.
- Allow wild ecosystems to continue to be self-creating and self-managing.
- Modify our myths to provide appropriate images of place.
- Apply known ecological ideas to our exploitation of ecosystems.
- Minimize the effects of articifial corridors such as roads or flight lanes.

6.3. ***Actions***

- Set aside blocks of wild areas first
- Exploit only a limited amount of wild areas that will not affect wild patterns
- Use only appropriate equipment or technology to minimize accidental damage or takes.
- Reduce number of incursions or roads
- Do not take all of one kind (or health or size or shape)
- Use labor-intensive methods (humanity our most valuable resource)
- Be sloppy, leave waste, debris or parts for natural renewal.

7.0. End Matter

This section contains specific notes from the text, as well as definitions of terms, a bibliography, and an index of subjects and names.

7.1. *End Notes*

[1] Called by P. Shepard 'philopatry'
[2] See also C.H. Waddington 1968.
[3] P. Soleri 1966.
[4] F. Sarfatti, 1975.
[5] for a full description, refer to Heinz Pagels
[6] From Herman Daly 1977.
[7] see Wittbecker articles
[8] After G. Hardin, 1987.
[9] in *Instinct and Experience 1912.*
[10] in his book, *Emergent Evolution 1923.*
[11] in *Holism and Evolution 1926.*
[12] From Zeigler, 1976.
[13] Aristotle in Generation and Corruption, 316a5
[14] Bolwig 1963.
[15] Hall 1968.
[16] Eibl-Eibesfeldt 1970.
[17] E. Wilson 1971.
[18] Fagen 1985.
[19] Wittbecker. Courses taught at the Community Free University 1973-74 include: Radical Philosophy, Radical Ecology, and Radical Psychology. Certain distance learning courses and tutorials through the Ecoforestry Institute 1994-99 include: Forest Design, Matrix Management, and Restoration.
[20] Wittbecker, 1983.
[21] After N. Evernden.
[22] See also C.H. Waddington 1968.
[23] Lewontin 1983.
[24] Ortega in his "Meditations on Hunting"
[25] in *Farming without Fields.*
[26] Borgstrom, 1975.
[27] In M. Heidegger's word.
[28] In G. Weiss, 1975.
[29] In *The Subversive Science.*
[30] Quoted in M. W. Fox, 1980.
[31] From Kaplan, 1983.
[32] Fromme is quoted in Hampden-Turner.
[33] The word is from Y-F. Tuan, 1974.
[34] M. Heidegger, *Being and Time.* Tubingen: Niemeyer, 1960, p. 192.
[35] After Berry & Annis, 1974.
[36] Bormann and Likens, 1967.
[37] See Walsh, Loomis, and Gillman in *Land Economics*
[38] Welch 1966.
[39] As P. Singer 1981, points out.
[40] Whitehead, SMW, p. 136.

[41] Recent translations say "Thou shall not murder."
[42] Kaplan 1973.
[43] Jacobs. P. 82.
[44] After K. C. Chang.
[45] Wittbecker, 1970.
[46] Pirie, 1976.
[47] Moore and Lappe, 1975; Gabel, 1975.
[48] Tansley, 1935
[49] Commoner 1972, 1992; Ehrlich and Ehrlich 1990; Ehrlich and Holdren 1971, 1972; Holdren and Ehrlich 1974 for IPAT
[50] Wittbecker, ICE 1986, *Wild Earth* 1994
[51] Based on Eyre.
[52] In *Revelations*
[53] By Lamoureux et al.
[54] Frankel and Soule, 1981.
[55] Gilpin and Soule, 1986.
[56] Odum, Eugene P. 1970. "Optimum population and environment: A Georgian microcosm." *Current History* 58:355-366.
[57] Soule, Michael. 1986. *Conservation Biology: The Science of Scarcity and Diversity.* Sunderland: Sinauer Associates.
[58] Conservation International and Agrupación Sierra Madre. "Wilderness: Earth's Last Wild Places."
[59] Frankel and Soule, 1981.
[60] Bailey, Robert. 1999. *Ecoregions.*
[61] Borgstrom, 1973.
[62] FAO, 1973.
[63] From *Modern Plastics.*
[64] Hayes (1976), cited in Daly, Steady State Economics, p. 169.
[65] M. Feinberg et al.
[66] Maya, From The Popol Vuh: Beginnings (J. Rothenberg and G. Quasha, eds., America A Prophecy. (New York: Vintage Books, 1974), p. 13.
[67] Especially in the works Norberg-Schulz 1980.
[68] Kaplan 1973.
[69] Wittbecker, 1992.
[70] Sahlins, 1968.
[71] Wittbecker 1985.
[72] Klein, 1978.
[73] Hemming, 1971.
[74] Eschenback and Geistauts, 1986.
[75] Jose Ortega y Gasset, 1956.
[76] P. Shepard, p. 140-141.
[77] In C. Otten, *Anthropology and Art.* Garden City: American Museum of Natural History, 1971, p. 46.
[78] Anais Nin, p. 199.

7.2. Definitions

Accommodate. To settle, 'conform with right measure.'

Acculturation. Cultural change that results from contact between different cultures; usually referring to loss of traditional culture after it adopts elements from more aggressive or larger culture.

Adaptability. The ability of an organism to change itself to the limits of the system to increase potential for survival in the environment. The ability to change the system makes for coconstrained construction.

Ambihuman. Everything that is not human; nonhuman; ultrahuman; 'surrounding the human'

Analogy. Correspondence based on similarity; in computers, based on similar electric charges.

Autarchy. Similar to anarchy, but meaning 'self-building,' rather than 'without a ruler.'

Autopoesis. Self-making, as used by F. Varela (autopoeisis).

Biophilia. Generally, the love for life. As coined by E. Wilson, the "connection" between humans and other living forms.

Biotopology. A neologism meaning the study of living in place (what sociology does).

Boundedness. Openness. The ecosystem has to be open to flows of energy and materials

Care. This meaning is based on Martin Heidegger's term for 'letting be,' similar to Goethe's nonintervention.

Carrying capacity. The number of people who can be supported indefinitely in a territory, with a given culture, with a given style, and using a given technology (may refer to any organism, as well as to population or use).

Catastrophe. A down-turning or disruptive event.

Cause. A relationship between events, such that the occurrence of an earlier event must be sufficient or necessary for a subsequent event to happen.

Characteristics. Qualities that distinguish unique individuals, systems, or patterns. Gregory Bateson calls them differences that make a difference.

City. A high-order, multiple-loop, feedback system, characterized by large buildings, created by large groups of humans living in place.

Coevolution is mutual adaptation of species in a system.

Collapse is the rapid significant loss of an established level of sociopolitical complexity. For a culture, collapse is a rapid, significant reduction in an established level of sociopolitical complexity, such that there is less specialization and stratification.

Commotion. Moving with, moving together, turbulence, confusion.

Complementarity. Bohr's term for the idea that two descriptions are needed to understand the whole reality; mutual completion.

Complexity. The quality of being complex. The word complex is from the Latin fragments meaning to "weave with." It means having interrelated parts or ideas.

Compliance. The acceptance of partial freedom and partial conflict, rather than all of either. It is cooperation enlarged to involve an aspect of play, within the framework of freedom and determination.

Conscension. Climbing with.

Contension. Stretching together, struggle.

Continuity, The system is stable and persistent in time, self-maintenance, mature, hysteresis. The process is the same as acquiring information.

Conversation. The communication between two or more beings; 'turning with'.

Conversion. Turning with, transforming, exchanging value.
Conviviality. Social, jovial (from the Latin *convivium*, living together).
Cosmology: An ideological system that explains the order and meaning of the universe and the place of people in it (a collective image of the universe).
Culture. Culture … is that complex whole which includes knowledge, belief, art, morals, law, custom, and any other capabilities and habits acquired by man as a member of society (Tylor). It is an adaptive human expression, developed through materials, ideas and symbols.
Deep Ecology. A movement, as formulated by Arne Naess, that goes beyond a concern with pollution and resource use, with its man-in-the-environment image, to consider humanity in a relational, total-field image.
Design. From the Latin meaning to "mark out." The verb form means to make a pattern or plan. To intend for a purpose.
Difference. The condition of being unlike or dissimilar; 'carrying apart'.
Disturbance. Regular but unpredictable events that destroy a percentage of organisms and their relations.
Diversity (as in biological diversity). Species richness. Different age and size classes in a population. Genetic differences in a species. Kinds of habitats present in an ecosystem.
Domination. In politics, the achievement of power within a group of states (economic, political, religious, cultural or military). Ecologically, of one species by another.
Domiture. The system of culture embedded in nature (has to be a larger term to enclose the previous nature/culture dualism. It has to include human reason and human emotion, rather than simply putting them on opposite columns., as with order and chaos, higher and lower, linear and cyclic), as well as agriculture and wilderness.
Earth. The entire planet, including all life forms.
Ecocybernics. A neologism meaning 'controlling the house' or governing the house (politics is a subcategory, a way of distributing power and luxuries in the cityscape).
Ecodeontics. A neologism meaning 'binding to the house.' (this is similar to what the word religion means).
Ecohygiology. The science of the 'health of the house.' Ecosystem medicine.
Ecology. The science that deals with the relations of organisms to their environment (which includes other organisms).
Economics. Literally, "the management of the house" or law of the house. The formal study of how people use and share resources to make a living.
Ecophilia. Generally, the love of the house (where house is metaphor for ecosystem or earth) expressing the need for meaning and significance in a place.
Ecopoetics. A neologism meaning 'making the house,; or expressing the house. This is how the invisible or power is represented in art, which is a subcategory.
Ecosystem. An adaptively organized system of living beings and elements that recycle important elements within the system (after Lynn Margulis); the smallest unit that recycles biologically important elements.
Ecotopology. A neologism meaning the study of living in place (what sociology does).
Ektropy. the fundamental ordering process of the universe; Hirth's term 'turning in'. Similar to negentropy or syntropy.
Emergence. The unpredictable appearance of new characteristics in a process.
Enmotion. Turning out (cf. Emotion, feeling, course of feeling).

Energy. Energy is inherent power or capacity for action or movement. Or the capacity for doing work.
Entropy. the fundamental disordering process of the universe; Clausius' term 'turning out'.
Escension. Climbing out, emerging.
Ethics. A description of human conduct, or set of rules, determined by custom and culture. Ecologically, a limit on one's freedom of action.
Eutopia. Good place (as used by Thomas More).
Eutopias. Good places that can be made culturally and ecologically.
Event. A change of state, an initial state and a subsequent state.
Eversion. Exversion or ektropy, the spontaneous generation of order.
Evolve. Roll out, unfold.
Evolution is the process by which species reorganize their structures to adapt to the environment. Evolution is an integrated, partly open process that selects whole individuals in whole environments. Evolution flows upwards and outwards as well as inwards and downwards, from the simple to the complex, but also back.
Exploitation: The normal use of a resource by a species. It may also create diversity in ecological systems.
Extension. Stretching towards.
Facts. As Kenneth Boulding notes, "For any organism or organization, there are no such things as 'facts.' There are only messages filtered through a changeable value system." Goethe: composed of emotion, feelings, beliefs.
Feedback. Information fed back into system that causes a change. Feedback may be negative or positive. Negative counteracts the effects of change to maintain the system in a steady state or homeorhetic state. Positive is a change in the system that causes it to move in the same direction, which can destabilize the system.
Flexibility. The potential for changer within a system. It is not being over connected or too rigid or efficient. Able to slough off species or living forms and incorporate new. A measure of unused potential (once the potential is used, the system becomes less flexible) or the capacity to adapt to larger systems.
Form. The contour or structure of something; mode; essence; 'shape'
Frame. Something that encloses or encircles; equal to vehicle or ground
Gaia. The hypothesis of the living earth; Homer's term; James Lovelock's theory.
Gestalt. A unified configuration having properties that cannot be derived from the parts.
Gigatrend. A long-lasting pattern in a human culture, such as accumulation of goods, or the height of early agriculturalists.
Harmony. The simultaneous occurrence of tones; agreement in feeling.
Heterarchy. A hierarchy of values; ambiguous order.
Hierarchy. A sacred order; a vertical arborizing process.
History. A sequence of states. In complex systems, the earlier state, also called the 'past,' contributes to or constrains a subsequent state, such as the magnetic hysteresis of ferromagnets or the behavior of organisms having a form of memory.
Holarchy. A whole order.
Holeconomics. Ecological study of the entire exchange system.
Holocosmology. The whole framework of human cosmologies.
Hologram. A pattern produced on photosensitive medium; a three-dimensional image of an object; equal to holograph; 'whole writing'

Holon. A Janus-faced entity, simultaneously a part and a whole; Koestler's term
Holopoetic. Referring to making wholes.
Home. A place for one or more persons, invested with emotion and time.
Identity. Identity is that persistent quality that is. A system's identity can be described apart from its performance in interactions, but not isolated. Integrity.
Ideology. A body of ideas reflecting the needs or aspirations of a culture or individual.
Image. The reproduction of an appearance of something; 'imitation'.
Immotion. Movement.
Inscension. Climbing up, developing.
Intension. Climbing in.
Interaction. Reciprocal effects by two living beings (on nonliving things) on each other.
Interference. Exploitation and disturbance on a scale that destroys the ability of a organisms or their ecosystem to continue.
Industrialism. A form of social and economic organization characterized by large industry, mechanization, fossil-fuel use, and concentrations of workers.
Inversion. Turning in, entropy.
Kondratieff cycle. A long-term trade fluctuation named after the Russian economist.

Landscape. A heterogeneous land area composed of a cluster of interacting ecosystems that are repeated in similar form throughout. Landscapes vary in size, from bioregions to a few kilometers in diameter.
Law. A law is a stable pattern inherent in things; it has to be perceived or discovered (like gravity). Example of a cultural law: "Higher culture does not emerge in society until the basic needs of a segment of its members have been satisfied."
Limit. The word limit comes from the Latin word for boundary, border or frontier. It means boundary, restriction, or utmost extent, either largest or smallest.
Love. An intense feeling; mysterious, emotive force.
Matrix. The most extensive and connected landscape element type, which plays a significant role in landscape functioning. Or, that element surrounding a patch.
Maturity. Ability to move to a greater efficiency over time (in an ecological system).
Metalysis. The process of renewal by discrete units; 'loosening change'.
Metaphor. Based on the Greek words meaning "carrying beyond." A metaphor is a figure of speech that combines two frames of reference.
Metapolitics. Discussion of civilization as more than cities.
Model. A stylized or hypothetical representation of an object. A generalized description based on analogy or metaphor.
Morals. Customs or manners. The "way of going together" (from the Latin root).
Motion. Movement.
Nation. A people living in a territory united by a single government.
Nature. The wild environment of humanity, the self-making process of living. Nature is a feeling system, with mutual experience and history.
Net. An openwork fabric of woven threads forming a mesh in various sizes.
Noplace. Nowhere.
Optimum. What is an optimum? Must it have a less than maximum in every case? (See 'satisficing')
Order. A condition of comprehensible arrangement among elements of a group; a number

of successive differentiations.

Organicism. For L. Bertalanffy organicism was necessary to accomplish three specific jobs in biology: appreciation of wholeness (regulation), organization (hierarchy and level laws), and dynamics (process, behavior of open systems).

Overshoot is that growth beyond the carrying capacity of an ecosystem leading to collapse or die-off.

Panecology. The study of all forms of ecosystems, including human ideational.

Panethic. The human recognition of the importance of interrelations of all beings.

Participation. The state of being part of a system. Joining or sharing with others.

Pattern. Regularity in a system, from physical systems to cultural ones.

Perturbation. Change to a system beyond the normal range of variation.

Phenomenological spiral. A way of questioning and investigating "things as they are" by repeating under different circumstances, with new knowledge or perspectives.

Physiocryptology. The study of nature.

Place. Emotionally-invested space, living space.

Practice. Actual doing, habitual action, custom, repeated performance (from the Greek, *praktikos*, fit for action, from *prassein*, to do); the Greek *poieein*, means to make.

Principles. Fundamental rules that we can use to create images or models.

Process. A sequence of states, or a trajectory of states, describing as path. If it involves the emergence of new things it is described as an evolutionary process.

Productivity. The conversion of energy into flesh.

Properties. Qualities common to all members of a class. A property is an attribute proper to a thing or characteristic quality.

Quality. Refers to a basic nature or characteristic element. A characteristic of something; a feature; 'of what kind'

Quantity. An amount of something; the character of a proposition

Radical. Referring to root.

Remotion. Moving back, removing, departing.

Rescension. Climbing back.

Resilience. A measure of how fast variables return to equilibrium after a perturbation in a system.

Resistance. The degree to which a variable is changed after a perturbation.

Retension. Power for remaining.

Reverence. Schweitzer's 'honor-fear' a feeling of respect, awe or love.

Reversion. Turning back, or out, returning.

Rhythm. Durational quality; a regular movement

Rights. Claims in accordance with human law, rule or morality.

Ritual. A religious ceremony; the technique of constituting an order.

Rule. A social convention set up by the people in a culture.

Sacred. Worthy of reverence; something made holy.

Satisficium. An amount that leads to satisfaction, that may be more then the minimum, less than a maximum and different from an optimum.

Scension. Climbing. The universal process of emergence.

Science. A form of systematic knowledge derived from observation and experiment.

Sign. the suggestion of a presence of a quality; a gesture to convey meaning; 'seal.'

Signal. a sign serving as a means of communication.

Society. An interacting group of people sharing a common culture.
Stability. Ability to provide constant internal environment. Ability to maintain identity under the flow of external forces and disturbances.
Standard. A model of quality that can be repeated.
State. A condition in time. The state of an organism, for instance, depends on its history and its environment.
STEM. The universal field of space-time/energy-matter.
Sustainable. The ability of a system to maintain its structure and function indefinitely.
Symbol. Anything with a culturally-defined meaning.
System. A complex unit (in space-time) whose components keep structure and function stable despite changes and disturbances. A system as a complex object, every part of which is connected with other parts of the object in such a way that the whole possesses emergent properties that the parts lack.
Tao. Way; way of the universe (also uncapitalized).
Tention. Stretching.
Theory. A theory is a reasoned expectation, as opposed to practice. A scientific theory is a set of general explanatory statements about a natural process; the set is related as a model, which is a human linguistic construction, subject to human perspectives and limitations.
Tragedy. A dramatic work of struggle ending in ruin; 'goat song'.
Topolatry. Worship of place.
Topophilia. Generally, love of place. As coined by Y-F. Tuan: the "effective bond" between people and place.
Topopoetics. The art and science of place-making.
Trap. The use of resources by a people, where the replenishment rate is constant ands the rate of use exceeds it, resulting in ecosystem degradation that is less reversible. Agriculture is an energy trap, because it allows the concentration of energy, e.g., higher yields, but then it requires more energy put into the system to maintain it.
Trend. A temporary pattern in a human culture, such as human infertility or globalization of capital, which can be reversed.
Umwelt. A perceived world; von Uexkull's term for animal world.
Unity. A condition of accord; the state of being one.
Universe. All existing things; 'turning into one'.
Urban ecosystems. Specialized human ecosystems where production and living are concentrated. Most energy and materials are imported from outside the system.
Variables. Qualities with no fixed value, changeable.
Version. The universal process of turning (see ektropy, entropy, evolution).
Vitality. Activity, health.

War. The condition of open armed conflict between factions or countries. A formal nonmonotonic condition of destruction affecting all living beings.
Wealth. Having rare things. Having more than the minimal requirements to live. Having great quantities of things that are valuable as measured by price.
Wild. pre-reflective, undomesticated.
World. Our planet and its inhabitants; or the perspective of the inhabitants (from the German word for "man-image").

7.3. ***Bibliography***

Abrams, Charles. 1966. *Man's Struggle for Shelter in an Urbanizing World.* Cambridge: MIT Press.

Alexander, Christopher. 1977. *A Pattern Language.* New York: Oxford University Press.

Alexander, Christopher. 2002. *The Nature of Order. Book One: The Phenomenon of Life.* Berkeley: Center for Environmental Structure.

Anderson, E. N. 1996. *Ecologies of the Heart.* New York: Oxford University Press.

Andrewartha, H. G. and L. C. Birch. 1984. *The Ecological Web.* Chicago: University of Chicago Press.

Aristotle. 1952. *The Works of Aristotle.* Tr. by I. Bywater. W. Ross, ed. Oxford: Clarendon Press.

Arbib, Michael. 1972. *The Metaphorical Brain.* New York: John Wiley and Sons.

Bachelard, Gaston. 1969. *Poetics of Space.* trans. M. Jolas. Beacon Press, Boston, pp. 4-6.

Bacon, Edmund. 1967. *Design of Cities.* New York: Penguin Books.

Bacon, Francis. 1901. *Novum Organum.* J. Devey, ed. New York: P.F. Collier.

Baker, Richard St. Barbe. 1980. "A man of the trees: Ed Goldsmith interviewing Richard St. Barbe Baker." *Coevolution Quarterly* 25: 66-70.

Baldwin, Jr., A. D., J. de Luce, and Carl Pletsch, eds. 1994. *Beyond Preservation: Restoring and Inventing Landscapes.* Minneapolis: University of Minnesota Press.

Ball, Philip. 1999. *The Self-made Tapestry: Pattern Formation in Nature.* Oxford: Oxford University Press.

Barker, Roger J. 1968. *Ecological Psychology.* Stanford: Stanford University Press.

Barrett, William. 1978. *The Illusion of Technique*: A Search for Meaning in A Technological Civilization. New York: Doubleday.

Barton, Hugh, Editor. 2002. *Sustainable Communities*: The Potential for Eco-Neighborhoods. London: Earthscan.

Bates, Daniel G. 1998. *Human Adaptive Strategies: Ecology, Culture, and Politics.* Boston: Allyn and Bacon.

Bates, Marston. 1964. *Man in Nature.* Englewood Cliffs: Prentice-Hall.

Bateson, Gregory. 1987. *Steps to an Ecology of Mind.* Northvale, NJ: Jason Aronson Inc.

Bateson, Gregory. 1979. *Mind and Nature*: A Necessary Unity. New York: E.P. Dutton.

Becker, Ernest. 1973. *The Denial of Death.* New York: The Free Press.

Bellah, Robert N. et al. 1991. *The Good Society.* New York: Alfred A. Knopf.

Bell, J. S. 1964. On the Einstein Podolsky Rosen Paradox. *Physics* 1:195-200.

Benedict, Ruth. 1934. *Patterns of Culture.* New York: Houghton Mifflin.

Berger, John. 1973. *Ways of Seeing.* New York: Viking.

Bergonzi, B. 1969. *Great Short Works of Aldous Huxley.* New York: Harper and Row.

Bergstraesser, Arnold. 1962. *Goethe's Image of Man and Society.* Freiburg: Herder.

Berlyne, D.E. 1971. *Aesthetics and Psychobiology.* New York: Appleton Century Crofts.

Berry, Wendell. 1983. "People, Land, and Community," pp. 64-79 in *Standing by Words.* San Francisco: North Point Press

Bertalanffy, L. von. 1975. *Perspectives on General Systems Theory.* New York: G. Braziller.

Beston, Henry. 1971. *The Outermost House.* New York: Ballantine Books.

Black, C. A. 1968. *Soil Plant Relationships.* New York: Wiley.

Birch, Charles, and Cobb, Jr., John B. 1981. *The Liberation of Life.* Cambridge: Cambridge University Press.

Birch, Thomas H. 1982. "Man the Beneficiary?: A planetary perspective on the logic of wildland preservation." *International Dimensions of the Environmental Crisis.* R. Barrett, editor. Boulder: Westview Press.
Bly, Robert, ed. 1980. *News of the Universe*: Poems of the Twofold Consciousness. Sierra Club, San Francisco.
Bodley, John H. 1884. *Cultural Anthropology*. Mountain View, CA: Mayfield Publishing.
Boguslaw, Robert. 1965. *The New Utopians*. Englewood Cliffs: Prentice-Hall.
Bohm, D. 1980. *Wholeness and the Implicate Order*. London: Routledge and Kegan Paul.
Bonner, J.T. 1952. *Morphogenesis*: An Essay on Development. Princeton: Princeton University
Press. Bookchin, Murray. 1989. *Remaking Society*. New York: Black Rose Books.
Bookchin, Murray. 1989. *Toward an Ecological Society*. New York: Black Rose Books
Borgstrom, George. 1965. *The Hungry Planet*. New York: Macmillan Co.
Borgstrom, George. 1973. *Harvesting the Earth*. New York: Abelard-Schuman.
Botkin, Daniel B. 1990. *Discordant Harmonies*: A New Ecology for the Twenty-first Century. New York: Oxford University Press.
Boulding, Kenneth E. 1956. *The Image: Knowledge in Life and Society*. Ann Arbor: University of Michigan Press.
Boulding, Kenneth E. 1966. *Beyond Economics*. New York: Harper & Row.
Boulding, Kenneth E. 1969. "The economics of the coming spaceship earth," In *Population Evolution and Birth Control*. San Francisco: Freeman.
Brandeis, Louis D. 1935. *The Curse of Bigness*. New York: Viking Press.
Braungart, Michael, and William McDonough. 2002. *Cradle to Cradle*: Remaking the Way We Make Things. New York: North Point Press.
Bridgman, P. W. 1941. *The Nature of Thermodynamics*. Cambridge: Harvard University Press.
Bronowski, J. 1965. *Science and Human Values*. New York: Harper and Row.
Brown, G. S. 1979. *Laws of Form*. New York: The Julian Press.
Brown, Harrison. 1954. *The Challenge of Man's Future*. New York: Viking Press.
Brown, James H. and Brian A. Maurer. 1989. *Macroecology*: The Division of Food and Space Among Species on Continents. *Science* 243:1145-1150.
Brown, Lester. 1979. ``Crossing the threshold? Pressures on earth's biological systems." *Environment* 21(8):12-37.
Bryan, Frank and John McClaughry. 1991. *The Vermont papers*. Post Mills: Chelsea Green.
Bunge, Mario. 1959. *Causality. The place of the causal principle in modern science*. Cambridge: Harvard University Press.
Burns, Neal M.; Chambers, Randall M; and Hendler, Edwin, Eds. 1963. *Unusual Environments and Human Behavior: Physiological and Psychological Problems of Man in Space.* New York, The Free Press of Glencoe.
Burlingh, P. et al. 1975. *Computation of the Absolute Maximum Food Production of the World.* Wageningen, Netherlands: Agriculture University.
Callenbach, Ernest, et al. 1993. *Ecomanagement.* San Francisco: Berrett-Koehler.
Campbell, Joseph. 1969. *The Flight of the Wild Gander*. New York: Viking Press.
Campbell, Joseph. 1972. *The Masks of God: Primitive Mythology*. New York: Penguin.
Campbell, Joseph. 1972. *The Masks of God: Creative Mythology*. New York: Penguin.
Capra, Fritjof. 1975. *The Tao of Physics.* Berkeley: Shambhala.
Capra, Fritjof. 1982. *The Turning Point*. New York: Simon and Schuster.

Caratheodory, Alain M. *Wild Apples*. Wilmington: Mozart & Reason Wolfe, 1976.
Carson, Rachel. 1962. *Silent Spring*. Boston: Houghton Mifflin.
Casey, Edward S. 1996. How to get from Space to Place in a fairly short stretch of time: Phenomenological Prolegomena. In *Senses of Place*, ed. Steven Feld and Keith H. Basso. Santa Fe. pps. 13-52.
Catton, William R. 1982. *Overshoot*: The Ecological Basis of Revolutionary Change. Urbana: University of Illinois Press.
Cavalli-Sforza, Luca and Albert Ammerman. 1984. *The Neolithic Transition and the genetics of Populations in Europe*. Princeton: Princeton.
Cheng, T.C. 1970. *Symbiosis*. New York: Pegasus.
Chermayoff, Serge and Christopher Alexander. 1965. *Community and Privacy*. New York: Anchor Books.
Chew, G. F. 1970. Hadron bootstrap: Triumph or frustration. *Phys. Today* (October):23-28.
Churchman, C.W. 1968. *The Systems Approach*. New York: Delacorte Press.
Cicero, M.T. 1933. *De Natura Deorum*. Tr. by H. Rackham. London: Wm. Heinemann. Pps. ii, 154.
Clapp, Jennifer and Peter Dauvergne. 2005. *Paths to a Green World*. Cambridge: MIT Press.
Clark, Colin. 1958. ``World Population." *Nature* 181:1235-1236.
Clark, Colin. 1967. *Population Growth and Land Use*. London: Macmillan.
Clark, Grahame. 1977. *World Prehistory*. Cambridge: Cambridge University Press.
Clark, William C. 1991. "Managing Planet Earth," *Scientific American* September.
Cobb, John B. Jr. 1971. *Is It Too Late?* New York: Glencoe.
Cobb, John B. Jr. "Economism or Planetism: The Coming Choice," *Earth Ethics* 3(1):1-5.
Cobb, John B. Jr. 1992. *Sustainability: Economics, Ecology, and Justice*. Maryknoll: Orbis Books.
Coleridge, Samuel T. 2003. *Coleridge's Poetry and Prose*. New York: Norton.
Colinvaux, P. 1978. *Why Big Fierce Animals Are Rare*. Princeton: Princeton University Press.
Commoner, Barry 1971. *The Closing Circle*. New York: Knopf.
Coates, Gary J. and David Seamon. 1993. "Promoting a foundational ecology practically through Christopher Alexander's Pattern Language: The example of Meadowcreek," Ch. 14 in *Dwelling, Seeing, and Designing: Toward a Phenomenological Ecology*. Albany: State University of New York Press.
Conrad, Peter. 1999. *Modern Times, Modern Places*. New York: Knopf.
Costanza, Robert. Ed. 1991. *Ecological Economics*: The Science and Management of Sustainability. New York: Columbia University Press.
Costanza, R., B. G. Norton, and B. D. Haskell, eds. 1992. *Ecosystem Health*: New Goals for Environmental Management. Washington: Island Press.
Cottrell, Leonard. 1962. *The Horizon Book of Lost Worlds*. New York: Horizon.
Cramer, F. 1993. *Chaos and Order: The Complex Structure of Living Systems*. Weinheim: VCH. Cultural Survival Trust. 1992. "Veddhas say no to colonization plan," *Cultural Survival*, Spring 1992, Pp. 11-12.
Daly, Herman E. 1968. Economics as a life science. *Journal of Political Economy* 76:392-401.
Daly, Herman E. 1977. *Stead-State Economics*. San Francisco: Freeman.
Daly, Herman E. and John B. Cobb, Jr. 1989. *For the Common Good*. Boston: Beacon Press.
Daly, Herman E. 1993. The steady-state economy, In Herman Daly and Kenneth Townsend, eds., *Valuing the Earth: Economics, Ecology, Ethics*. Cambridge: MIT Press.

Daly, Mary 1978. *Gyn/Ecology* Boston: Beacon Press.
Dansereau, Pierre. 1957. *Biogeography: An Ecological perspective*. New York: Ronald Press.
Darling. F. Fraser and John P. Milton, eds. 1966. *Future Environments of North America*. Garden City: Natural History Press.
Darlington, P. J. 1957. *Zoogeography*. New York: Wiley.
Darwin, Charles. 1859. *The Origin of the Species by Means of Natural Selection*. London: Murray. Page 75.
Dasmann, Ray. 1972. *Environmental Conservation*. 3rd Ed. New York: Wiley.
Daubenmire, Rexford. 1970. *Steppe Vegetation of Washington*. Technical Bulletin Number 62. Pullman: WA Agricultural Experiment Station.
De Los Reyes, B.N. et al. 1965. A case study of the tractor and carabao-cultivated lowland rice farms in Laguna, crop year 1962-63. *Phil. Agric*. 49:75-94.
Diamond, J.M. 1975. The island dilemma: Lessons of modern biogeographic studies for the design of natural preserves. *Biol. Conserv*. 7:129-146.
Diamond, J.M.. 1997. *Guns, Germs, and Steel*: The Fate of Human Societies. New York: W. W. Norton.
Dietz, Thomas, and Eugene A. Rosa. 1994. "Rethinking the Environmental Impacts of Population, Affluence and Technology." Human Ecology Review, Summer/Autumn, 1.
Diole, P. 1974. *The Errant Ark*: Man's Relationship with Animals. New York: Putnam.
Doolittle, W. F. 1981. Is nature really motherly? *Coevolution Quarterly* 29:58-62.
Doxiades, C.A. 1975. *Building Entopia*. New York: Norton.
Doxiades, C.A. 1977. *Ecology and Ekistics*. Boulder: Westview Press.
Drengson, Alan. 1989. *Beyond Environmental Crisis*: From Technocrat to Planetary Person. New York: Peter Lang.
Drengson, Alan et al. 1994. "The ecoforester's way: An oath of ecological responsibility," *International Journal of Ecoforestry* 10(1):48.
Drucker, Peter. 1969. *The age of discontinuity*: Guidelines to our changing society. New York: Harper & Row.
Drucker, Peter. 1990. *The New Realities*. New York: Dutton
Dubos, Rene. 1967. *Man Adapting*. New Haven: Yale University Press.
Dubos, Rene. 1972. *A God Within*. New York: Charles Scribner's Sons.
Dubos, Rene. 1974. *Beast or Angel? Choices That Make Us Human*. New York: Charles Scribner's Sons.
Dubos, Rene. 1976. Symbiosis between the earth and humankind. *Science* 193:459-462
Dubos, Rene. 1980. *The Wooing of Earth*. New York: Charles Scribner's Son.
Dugatkin, L. A. 2000. *The Imitation Factor*. New York: Free Press.
Durham, William H. 1991. *Coevolution: Genes, Culture and Human Diversity*. Stanford: Stanford University Press.
Eckbo, Garrett. 1969. *The Landscape We See*. Berkeley: UC Press.
Eckholm, Eric P. 1976. *Losing Ground*: Environmental Stress and World Food Prospects. New York: W. W. Norton.
Eddington, Arthur. 1935. *Philosophy of Physical Science*. London: Textbook Publishers.
Edie, James. 1969. *New Essays in Phenomenology*. New York: Quadrangle.
Ehrlich, Paul. 1968. *The population bomb*. New York: Ballantine Books.
Ehrlich, Paul and Holdren, John P. 1971. Impact of Population Growth. *Science*. 171:1212-1217.

Ehrlich, Paul and A. Ehrlich. 1972. *Population, Resources, Environment.* San Francisco: Freeman.
Ehrlich, Paul. 1981. An ecologist standing up among seated social scientists. *Coevolution Quarterly* 31:24-35.
Eibl-Eibesfelt, I. 1970. *Ethology*: The Biology of Behavior. New York: Holt, Rinehart and Winston.
Eigen, Manfred, and P. Schuster. 1979. *The Hypercycle*: A Principle of Natural Self-Organization. Berlin: Springer-Verglag.
Eigen, Manfred, and P. Schuster. 1981. *Laws of the Game.* New York: Alfred A. Knopf.
Einstein, Albert and Leopold Infeld, 1960. *Evolution of Physics.* New York: Simon & Schuster.
Eisenberg, Evan. 1998. *The Ecology of Eden.* New York: Vintage.
Elton, C. 1966. *Animal Ecology.* New York: October House.
Evernden, Neil. 1981. *Out of Place* (unpublished manuscript).
Evernden, Neil. 1992. *The Social Creation of Nature.* Baltimore: Johns Hopkins.
Ewald, William R. Jr. 1968. *Environment and Change*: The Next Fifty years. Bloomington: Indiana University Press.
Eyre, Samuel. 1978. *The Real Wealth of Nations.* London: E. Arnold.
FAO. 1973. *Production Yearbook,* 1092. Vol. 26. Rome: FAO.
Feibleman, James K. *Mankind Behaving*: Human Needs and Material Culture.
Ferkiss, Victor C. 1969. *Technological Man*: The Myth and Reality. New York: Braziller.
Festinger, L. 1957. *A Theory of Cognitive Dissonance.* Stanford: Stanford University Press.
Finkelsten, D. 1972. The space-time code. *Phys Rev* 50, no. 12:2922.
Fitch, James M. 1961. *Architecture and the Esthetics of Plenty.* New York: Columbia University Press.
Foreman, Dave, ed. 1985. *Ecodefense: A Field Guide to Monkeywrenching* Tucson: Earth First! Books (p. 10).
Foreman, R.T.T. and M. Godron. 1986. *Landscape Ecology.* New York: John Wiley.
Forestry Commission, 1994. *Forest Landscape Design.* London: HMSO.
Forman, R. T. T. and Michel Godron. 1986. *Landscape Ecology.* New York: John Wiley.
Fowles, John. 1970. *The Aristos.* New York: The New American Library, Inc.
Fowles, John. 1979. Seeing Nature Whole. *Harper's.* 259:49-56
Fowles, John. 1980. Is nature necessary? *Harper's.*
Fox, M.W. 1974. *Concepts in Ethology: Animal and Human Behavior.* Minneapolis: U. of Minnesota Press.
Fox, M.W. 1976. *Between Animal and Man.* New York: Coward, McCann and Geoghehan Inc.
Fox, M.W. 1980. Returning to Eden: Animal Rights and Human Responsibility. New York: Viking Press.
Fox, M.W. 1997. *Concepts in Ethology*: Animal Behavior and Bioethics. Krieger.
Fox, M.W. 2001. *Bringing Life to Bioethics.* State University of New York Press.
Frankel, Otto. 1975. *Crop Genetic Resources for Today and Tomorrow.* O. Frankel and J. Hawkes, eds. New York: Cambridge University Press.
Frankel, O. H. and M. E. Soule. 1981. *Conservation and Evolution.* Cambridge: Cambridge University Press.
Fraser, J.T. 1975. *Of Time, Passion and Knowledge.* New York: George Braziller.

Fromm, Erich. 1956. *The Art of Loving*. New York: Harper.
Fromm. Erich. 1976. *To Have or To Be*. New York: Bantam Books.
Fuller, R. B. 1969. *Utopia or oblivion*: The prospects for humanity. New York: Bantam.
Fuller. R. B. 1970. *Operating Manual for Spaceship Earth*. New York Pocket Books.
Fuller, R. B. 1981. *Critical Path*. New York: St. Martin's Press.
Galbraith, John Kenneth. 1967. *The new industrial state*. New York: Houghton-Mifflin.
Gabel, Medard. 1979. *HO-PING: Food for Everyone*. Garden City, New York: Anchor Press/ Doubleday.
Galdston, Iago, ed. 1963. *Man's Image in Medicine and Anthropology*. New York: International Universities Press, Inc.
Gans, Herbert J. 1969. *People and Plans*. New York: Columbia University Press.
Gazzaniga, M.S. 1967. The split brain in man. *Sci. Am.* 217:24-29.
Georgescu-Roegen, Nicholas. 1971. *The Entropy Law and the Economic Process*. Cambridge: Harvard University Press.
Georgescu-Roegen, Nicholas. 1976. Bioeconomics: A new look at the nature of economic activity. In *The Political Economy of Food and Energy*, ed. L. Junker. Ann Arbor: The University of Michigan Press.
Gibbs, J. W. 1960. *Principles in Statistical Mechanics*. New York: Longmans, Green and Co.
Gibson, William. 1986. "Ecology and Justice", *Wilderness* (Summer): 52-56.
Gillard, E. Thomas. 1963. Evolution of Bowerbirds. *Sci. Am.* 209:38-46.
Ginsberg, Lee and Mark Colyvan. 2004. *Ecological orbits*. New York: Oxford University Press.
Globus, G., G. Maxwell, and I. Sarednik. 1976. *Consciousness and the Brain*. New York: Plenum Press.
Goldsmith, Edward. 1988. "The Way: An ecological worldview," Pp. 160-185 in *The Ecologist*, Vol. 18, No. 4/5.
Goldtooth, Tom B.K. 1995. Indigenous nations: Summary of sovereignty and its implications for environmental protection. In B. Bryant, ed., *Environmental Justice*. Washington: Island Press.
Golley, Frank B., K. Petrusewicz, and L. Ryszkowski, eds. 1975. *Small Mammals: Their Productivity and Population Dynamics*. New York: Cambridge University Press.
Goodman, Paul and Percival Goodman. 1947. *Communitas: Means of Livelihood and Ways of Life*. New York: Abrams.
Goodman, Paul. 1962. *Utopian Essays and Practical Proposals*. New York: Random House.
Goodman, Paul. 1970. *New reformation*: Notes of a Neolithic Conservative. New York: Random House.
Gould, S. J. 1977. *Ontogeny and Phylogeny*. Cambridge: Belknap Press.
Gould, S.J. 1977. *Ever Since Darwin*. New York: W.W. Norton.
Grande, John K. 2004. *Balance: Nature and Art*. New York: Black Rose.
Graves, Robert. 1948. *The White Goddess*. New York: Farrar Straus & Giroux. p. 260.
Gray, Russell D. 1988. Metaphors and methods. In Mae-Wan Ho and S. W. Fox, eds., *Evolutionary Processes and Metaphors*. New York: Wiley.
Gregg, Alan. 1955. A medical aspect of the population problem. *Science* 121 (3,50):681-2.
Greene, Patricia and Dean Apostol. 1994. Design for biodiversity. *Landscape Architecture* 85 (4):63-65.
Grene, M., ed. 1971. *Interpretations of Life and Mind*. New York: Humanities Press.

Gunderson, Lance H. and C. S. Holling. Eds. 2002. *Panarchy*. Washington: Island Press.
Gustavson, Carl G. 1976. *The Mansion of History*. New York: McGraw-Hill Book Company.
Gutkind, E. A. 1962. *The Twilight of Cities*. New York: Free Press of Glencoe.
Haldane, J.B.S. 1927. *Possible Worlds and Other Papers*. London: Chatto and Windus.
Hall, Edward T. 1969. *The Hidden Dimension*. Garden City, New York: Doubleday.
Hall, Edward T. 1976. *Beyond Culture*. Garden City, New York: Anchor Press.
Halprin, Lawrence. 1969. *The RSVP Cycles*: Creature Processes in the Human Environment. New York: George Braziller.
Hampden-Turner, C. 1981. *Maps of the Mind*. New York: Macmillan.
Harajan, 7.9.1935, p. 234. Gandhi.
Hardin, Garrett. "The tragedy of the commons," *Science* 162:1243-1248.
Hardin, Garrett. Ed. 1969. *Population, Evolution, and Birth Control*. San Francisco: Freeman.
Hardin, Garrett. 1977. *The Limits of Altruism: An Ecologist's View of Survival*. Bloomington: Indiana University Press.
Hardin, Garrett. 1980. "An ecolate view of the human predicament," in *Global Resources*. Baltimore: University Park Press.
Hardin, Garrett. 1985. *Filters Against Folly*. New York: Penguin Books. (Discounting the future or who can afford a forest? pp. 71-76. The effect of scale on values, pp. 128-140.)
Hare, Nathan. 1970. Black Ecology, *The Black Scholar* 1(April):2-8.
Harper, J. L. 1961. The Evolution and Ecology ... *Evolution* Vol. 15: 209-227.
Harrington, Michael. 1966. *The Accidental Century*. New York: Weidenfeld & Nicolson.
Hart, Richard. 1994. "Monitoring for ecosystem management," *International Journal of Ecoforestry* 10(2):74-75.
Hebb, D.O. 1958. Alice in Wonderland or psychology among the biological sciences. In: *The Biological and Biochemical Bases of Behavior*. H. Harlow and C. Woolsley, eds. Madison: University of Wisconsin Press.
Heidegger, M. 1960. *Being and Time*. 9th ed. Tubingen: Max Niemeyer Verlag. Heisenberg, W. 1958. Physics and Philosophy. New York: Harper Torch Books.
Heminger, Jr., S. K. 1974. *Touches of Sweet Harmony*. San Marino, California: Huntington Library.
Henberg, M. 1984. "Wilderness as playground." *Environmental Ethics* 6:253-263.
Henderson, Hazel. 1980. *Creating Alternative Futures*: The End of Economics. New York: Putnam.
Herber, Lewis. *Our Synthetic Environment*. (see M. Bookchin).
Heroditus. *The History Book* 1:22-23.
Hoffman, David. 1976. *Inuit land use on the barren grounds. Inuit Land Use and Development Project*, Vol. 2. Ottawa: Dept. of Indian and Northern Affairs.
Holling, C. S. 1973. Resilience and Stability of Ecological Systems. In: *Annual Review of Ecology and Systematics*, R. F. Johnston et al., eds., Vol. 4: 1-24.
Homer. 1956. *The Iliad*. Middlesex, England: Penguin Books.
Hoyle, Fred. 1977. *Astronomy and Cosmology*. San Francisco: Freeman.
Hughes, J. Donald. 1983. *American Indian Ecology*. El Paso: Texas Western Press.
Huizinga, Johan. 1955. *Homo Ludens*: A Study of the Play Element in Culture. London: Routledge & Kegan Paul.
Hulet, H.R. 1970. ``Optimum world population." *Bioscience* 20(3):160-161.

Humboldt, Alexander von. 1897. *Cosmos: A Sketch of a Physical Description of the Universe.* trans. E. Otte. London: H. G. Bohn.
Huxley, Aldous. 1945. *The Perennial Philosophy*. New York: Harper.
Huxley, Aldous. 1948. *Ape and Essence*. New York: Harper & Row.
Huxley, Aldous. 1956. Knowledge and Understanding. In: *Adonis and the Alphabet*. London: Chatto and Windus.
Huxley, Aldous. 1977. *The Human Situation*. P. Ferrucci, ed. New York: Harper & Row.
Huxley, Julian. 1964. *The Human Crisis*. Seattle: University of Washington Press.
Huxley, Julian, and Huxley, T. H. 1947. *Touchstone of Ethics*. New York: Harpers.
Huxley, T. H. 1856. "On natural history as knowledge, discipline and power."
Hyde, Richard. 2000. *Climate Responsive Design*. London: E & FN Spon.
Illich, Ivan. 1973. *Tools for Conviviality*. New York: Harper & Row.
Illich, Ivan. 1978. *Towards a History of Needs*. New York: Pantheon.
Imanishi, Kinji. 1952. *Man* (Japanese language). Tokyo: Mainichi-Shinbunsha.
International Union for Conservation of Nature. 1984. *World Conservation Strategy in Action*. Gland, Switzerland.
Jackson, Wes, W. Berry, and B. Colman. 1984. *Meeting the Expectations of the Land: Essays in Sustainable Agriculture*. San Francisco: North Point.
Jacobs, Jane. 1961. *The Death and Life of Great American Cities*. New York: Random House.
Jammer, Max. 1954. *Concepts of Space*. Cambridge: Harvard University Press.
Jantsch, E. 1975. *Design for Evolution*. New York: Braziller.
Jantsch, E. 1980. *The Self-Organizing Universe*. New York: Pergamon Press.
Jonas, H. 1974. *Philosophical Essays*: From Ancient Creed to Technological Man. Englewood Cliffs: Prentice-Hall Inc.
Kant, Immanuel. 1998. *Critique of Judgment*. New York: Cambridge University Press.
Kaplan, Rachel. 1983. "The role of nature in the urban context," in I. Altman and J. F. Wohlwill, eds. *Behavior and the Natural Environment*. New York: Plenum Press.
Kaplan, Richard. 1983. The role of nature in the urban context. Irwin Altman and Joachim Wohlwill, eds. *Behavior and the Natural Environment*. Plenum Press, New York, pp. 127159.
Kaplan, Robert. 2000. *The Coming Anarchy.* New York: Vintage Books.
Kaplan, S. 1973. "Cognitive maps; human needs and the designed environment." In W. F. E. Preiser, ed., *Environmental Design Research*. Stroudsberg: Dowden, Hutchinson & Ross.
Kellert, S.R. 1983. Affective, cognitive, and evaluative perceptions of animals. Irwin Altman and Joachim Wohlwill, eds. *Behavior and the Natural Environment*. Plenum Press, New York, pp. 241-265.
Kepes, Gyorgy. 1965. *Structure in Art and Science*. New York: G. Braziller.
Kieffer, George H. 1979. *Bioethics*: A Textbook of Issues. Reading, MA: Addison-Wesley Publishing Co.
Kirk, G. S. and J. E. Raven. 1957. *The Pre-Socratic Philosophers*. Cambridge University Press.
Klein, David R. 1970. *IUCN Publ. New Series* No. 16:209-242.
Klein, David. 1972. Toward an ecophilosophy. Tomte Symposium on Ecology and Land Use, Steinsgard, Norway.
Klein, David R. and R. G. White, eds. 1978. *Parameters of Caribou Population Ecology in Alaska*. Fairbanks: Biol. Papers Univ. AK, Special Report No. 2.

Klein, David. 1976. "Wilderness Part 1. Evolution of the Concept." *Landscape* 20: 36-41.
Klein, David R. 1981. Alternate species for northern animal production. *Can. J. Anim. Sci.* 61:7-15.
Klopfer, P. H. 1962. *Behavioral Aspects of Ecology*. Englewood Cliffs, NJ: Prentice-Hall. Kockelmans, J., ed. 1972. On Heidegger and Language. Evanston, IL: Northwestern University Press.
Koestler, A. and J.R. Smythies, eds. 1969. *Beyond Reductionism*: New Perspectives in the Life Sciences. London: Hutchinson.
Koestler, A. 1978. *Janus*: A Summing Up. New York: Random House.
Kohler, I. 1962. Experiment with goggles. *Sci. Am.* 206:62-86.
Kohler, W. 1947. *Gestalt Psychology*. 2nd ed. New York: Liveright Publishing Corp.
Kohr, Leopold. 1957. *The Breakdown of Nations*. New York: E.P. Dutton.
Kohr, Leopold, 1973. *Development Without Aid*. New York: Schocken Books.
Kohr, Leopold. 1977. *The Overdeveloped Nations*: Diseconomies of Scale. New York: Schocken Books.
Kozlovsky, Daniel G. 1974. *An Ecological and Evolutionary Ethic*. New York: Prentice-Hall.
Krebs, Charles J. 1985. *Ecology*. 3rd ed. New York: Harper & Row.
Kropotkin, P.A. 1972. *Mutual Aid: A Factor in Evolution*. New York: New York University Press.
Krutch, Joseph W. 1970. *The Best Nature Writing of Joseph Wood Krutch*. New York: Pocket Books
Krutch, Joseph W. *Human Nature and the Human Condition*. NC: NP.
Kuhn, Thomas. 1970. *The Structure of Scientific Revolutions*. Chicago: University of Chicago Press.
Kuhns, William. 1969. *Environmental Man*. New York: Harper and Row.
Lacan, J. 1968. *The Language of the Self*: The Function of Language in Psychoanalysis. Baltimore: Johns Hopkins Press.
Lackner, S. 1984. *Peaceable Nature*. New York: Harper and Row
Laing, R. D., H. Phillipson, and A. R. Lee. *Interpersonal perception*. NC: NP.
Lamarck, Jean. 1963. *Zoological Philosophy*. trans. H. Elliot. New York: Hafner Publishing Co (1809. Philosophie Zoologique).
Landers, Richard R. 1966. *Man's Place in the Dybosphere*. Englewood Cliffs: Prentice-Hall.
Lao Tse. 1963. *Tao Te Ching*. Translated by D. C. Lau. Middlesex, England: Penguin Books.
Lappe, Frances Moore and Joseph Collins. 1979. Food First: Beyond the Myth of Scarcity. New York: Ballantine.
Laszlo, Ervin. 1972. *Introduction to Systems Philosophy*: Toward a New Paradigm of Contemporary Thought. New York: Harper Torch Books
Laszlo, Ervin et al. 1977. *Goals for Mankind*. New York: E. P. Dutton.
Laszlo, Ervin. 1987. *Evolution: The Grand Synthesis*. Boston: New Science Library
Le Corbusier and Francois de Pierrefeu. The Home of Man.
Lieth, Helmut F. H., ed. *Patterns of Primary Production in the Biosphere*. Stroudsburg, PA: Dowden, Hutchinson & Ross, Inc.
Lieth, Helmut F H. and Robert Whittaker, eds. 1975. *Primary Productivity of the Biosphere*. New York: Springer-Verlag.
Leonard, George. 1978. *The Silent Pulse*. New York: E. P. Dutton
Leopold, Aldo. 1949. *A Sand County Almanac; And Sketches of Here and There*. New York:

Oxford University Press.
Levi-Strauss, C. 1963. *Structural Anthropology*. New York: Basic Books.
Levy-Bruhl, Lucien. 1975. *The Notebooks on Primitive Mentality*. Translated by Peter Leenhardt. Oxford: Basil Blackwell.
Lewin, K. 1951. *Field Theory in Social Science*. D. Cartwright, ed. New York: Harper and Row.
Likens, Gene E., ed. 1989. *Long-Term Studies in Ecology*. New York: Springer-Verlag.
Lincicome, D.R. 1969. The Goodness of Parasitism: A New Hypothesis. Thomas C. Cheng, ed. *Aspects of the Biology of Symbiosis*. Baltimore: University Park Press.
Lorenz, Konrad. 1952. *King Solomon's Ring: New Light on Animal Ways*. trans. M. K. Wilson. New York: Crowell.
Lorenz, Konrad. 1974. *Civilized Man's Eight Deadly Sins*. trans. M. K. Wilson. New York: Harcourt, Brace, Javonovich.
Lovejoy, Arthur O. 1964. *The Great Chain of Being: A Study of the History of an Idea*. Cambridge:Harvard University Press.
Lovejoy. Thomas and D. C. Oren. 1981. "The minimum critical size of ecosystems," in W.D. Billings et al., eds., *Forest Island Dynamics in Man-dominated Landscapes*. New York: Springer-Verlag.
Lovejoy, Thomas and Richard Bierregaard. In Soule, Michael. 1986. *Conservation Biology: The Science of Scarcity and Diversity*. Sunderland: Sinauer Associates.
Lovelock, James E. 1979. *Gaia*: A New Look at Life on Earth. Oxford: Oxford University Press.
Lovelock, James E. 1988. *The Ages of Gaia*: A Biography of Our Living Earth. New York: W.W. Norton.
Lovelock, James E. 1991. *Healing Gaia: Practical Medicine for the Planet*. New York: Harmony Books.
Lowenthal, David. *Environmental Perception and Behavior*.
Lucas, Oliver. 1990. *The Design of Forest Landscapes*. New York: Oxford University Press.
Lynch, Kevin. 1960. *The Image of the City*. NC: NP.
MacArthur, Robert H. 1972. *Geographical Ecology: Patterns in the Distribution of Species*. New York: Harper & Row.
MacArthur, R.H. and E.O. Wilson. 1967. *The Theory of Island Biogeography*. Princeton: Princeton University Press.
MacKaye, Benton. *From Geography to Geotechnics*. NC: NP.
Malinowski, Bronislaw. 1944. *A Scientific Theory of Culture and Other Essays*. Chapel Hill: University of North Carolina Press.
Mandelbrot, B. B. 1982. *The Fractal Geometry of Nature*. San Francisco: W.H. Freeman.
Mander, Jerry. 1991. *In the Absence of the Sacred*. San Francisco: Sierra Club Books.
Mander, Jerry and Edward Goldsmith, eds. 1996. *The Case Against the Global Economy*. San Francisco: Sierra Club Books.
Mandeville, Bernard de (see Lewis Mumford, 1922).
Mann, John. 1965. *Changing Human Behavior*. New York: Scribner.
Margalef, R. 1968. *Perspectives in Ecological Theory*. Chicago: University of Chicago Press.
Margulis, L. 1974. Five kingdoms—classification and the origin and evolution of cells. *Evol Biol* 7:45-48.
Margulis, Lynn and Dorion Sagan. 1986. *Microcosmos*. New York: Summit Books. (pp. 169-

174.)
Margulis, Lynn. 1991. Big trouble in biology: Physiological autopoiesis versus mechanistic neo-Darwinism. In John Brockman, ed., *Doing Science.* New York: Prentice Hall Press.
Marsh, G.P. 1964. *Man and Nature,* The Earth as Modified by Human Action. St Clair, MI: Scholarly Press.
Maruyama, Magorah. 1979. Transepistemological Understanding: Wisdom beyond theories. Maruyama, Magorah, ed. *Cultures of the Future.* The Hague: Mouton.
Maruyama, Magorah. 1980. Toward Cultural Symbiosis. In: *Evolution and Consciousness: Human Systems in Transition.* E. Jantsch and C. H. Waddington, eds. Reading: Addison-Wesley Publishing Co.
Marx, Leo. 1964. *The machine in the garden: Technology and the pastoral ideal in America.* Oxford University Press.
Maser, Chris. 1988. *The Redesigned Forest.* San Diego: R and E Miles.
Maser, Chris. 1990. *The Forest Primeval.* San Francisco: Sierra Books.
Maslow, A. H. 1968. *Toward a Psychology of Being.* 2nd ed. New York: Van Nostrand.
Maslow, A. H. 1971. *The Farther Reaches of Human Nature.* New York: Viking Press.
Maturana, H.R. and Varela, F: 1987. *Tree of Knowledge.* Boston: Shambhala.
Mayr, Ernest. 1942. *Systematics and the Origin of the Species.* New York: Columbia University Press.
Mazrui, Ali Al Amin. 1976. *A World Federation of Cultures: An African Perspective.* New York: Free Press.
McArthur, Robert H. 1972. *Geographical Ecology*: Patterns in the Distribution of Species. New York: Harper & Row.
McCarry, James. 1972. *The Quality of the Environment.* New York: Free Press.
McCay, Bonnie J. and James Acheson, eds. 1987. *The Question of the Commons.* Tucson: University of Arizona Press.
McCulloch, Warren. 1965. A Heterarchy of Values Determined by the Topology of Nervous Nets. *Journal of General Physiology* 43:6.
McDonough, William and Michael Braungart. 2002. *Cradle to Cradle.* New York: North Point Press.
McHale, John. 1969. *The Future of the Future.* New York: G. Braziller.
McHarg, Ian. 1969. *Design with Nature.* Garden City: Natural History Press.
McKeon, Richard. 1973. *Introduction to Aristotle.* Chicago: University of Chicago Press.
McLean, G.F., ed. 1978. *Man and Nature.* Calcutta: Oxford University Press.
McLuhan, Marshall. 1964. *Understanding Media.* New York: McGraw Hill.
McLuhan, Marshall and Quentin Fiore. *War and Peace in the Global Village.*
McLuhan, Marshall and Quentin Fiore. 1967. *The Medium is the Massage.* New York: Bantam.
Meadows, Donna et al. 1972. *The Limits to Growth.* A report for the Club of Rome's project on the Predicament of Mankind. New York: Universe books.
Meadows, Dennis. 1982. "Fallacies in resource planning," in Charles Hewett, T. Hamilton, and I. Anderson, eds., *Forests in Demand.* Boston: Auburn House.
Meadows, D. H., D. L. Meadows, and J. Randers. 1992. *Beyond the Limits.* Post Mills: Chelsea Green. (Forest distribution and deforestation globally, pp. 57-64.)
Mech, L. David. 1984. *The Wolf.* Minneapolis: University of Minnesota Press.
Meeker, Joseph. 1974. *The Comedy of Survival.* New York: Charles Scribner's Sons.

Mehra, Jagdish. 1973. The Physicist's Conception of Nature. Boston: D. Reidel Publishing Co.
Merchant, Carolyn. 1980. *The Death of Nature*: Women, Ecology, and the Scientific Revolution. San Francisco: Harper & Row.
Merleau-Ponty, M. 1964. *The Primacy of Perception.* J. Edie, ed. Evanston, IL: Northwestern University Press.
Merleau-Ponty, Maurice. 1968. *The Visible and the Invisible.* Translated by A. Lingis. Evanston, Illinois: Northwestern University Press.
Merriam, Thomas. 1977. The Disenchantment of the World. *The Ecologist,* Vol 7, Pp. 22-29.
Midgley, Mary. 1989. *Wisdom Information & Wonder*. New York: Routledge.
Mill, John S. 1963. Collected Works. 5 Vols. Toronto: University of Toronto Press.
Miller, George. 1956. The magic number seven plus or minus two. *Psych. Rev.* 63:81-97.
Miller, Jr. G. Tyler. 1992. *Living in the Environment.* Belmont, CA: Wadsworth Publishing Co.
Mollison, Bill. 1988. *Permaculture: A Designers' Manual.* Tyalgum, Aus.: Tagari Pubs.
Montaigne, Michel de. 1958. *Complete Essays.* Stanford: Stanford University Press.
Montesquieu. 1949. De l'Esprit des Lois (The Spirit of the Laws).
Moran, Emilio F., ed. 1990. *The Ecosystem Approach to Anthropology.* Ann Arbor, MI: University of Michigan Press.
More, Thomas. 1982. *Utopia.* London: Penguin Books.
Morgan, Conway Lloyd. 1925. *Emergent Evolution.* New York: Henry Holt and Co.
Morris, Richard and Michael W. Fox. 1978. *Animal Rights and Human Ethics.* Washington: Acropolis Books.
Mourelatos, A. P. D., ed. 1974. *The Pre-Socratics.* Garden City, New York: Anchor Press/Doubleday.
Mumford, Lewis. 1922. *The Story of Utopias.* New York: Boni and Liveright.
Mumford, Lewis. 1956. *The Transformation of Man.* New York: Harper and Row.
Mumford, Lewis. 1961. *The City in History: Its Origins, Its Transformations, and Its Prospects.* New York: Harcourt Brace and World.
Mumford, Lewis. 1966. *The Myth of the Machine.* Technics and human development. New York: Harcourt, Brace & World, 1966.
Mumford, Lewis. 1973. *Interpretations and Forecasts*: 1922-1972. New York: Harcourt, Brace, Jovanovitch.
Munson, R., ed. 1971. Man and nature: Philosophy. In: *Issues in Biology.* New York: Dell Publishing Co.
Neihardt, J.G. 1959. *Black Elk Speaks.* New York: Pocket Books.
Myers, Norman. 1984. *The Primary Source.* New York: Norton. Myers, Norman. 1984. *Gaia: An Atlas of Planet Management.* Doubleday and Company, Garden City, New York.
Naess, Arne. 1972. The shallow and the deep, long-range ecology movement. A summary. *Inquiry,* 16: 95-100
Naess. Arne. 1974. *Gandhi and Group Conflict* Oslo: Universitets forlaget.
Naess, Arne. 1987. Okologi, Samfunn og Livsstil (Norwegian version, later published as *Ecology Community and Lifestyle.* New York: Cambridge University Press).
Needleman, Jacob. 2003. *Sense of the Cosmos.* New York: Monkfish.
Nettle, Daniel and Suzanne Romaine. 2000. *Vanishing Voices: The Extinction of the World's Languages.* New York: Oxford University Press.

Nicolis, G., and Prigogine, I. 1977. *Self-organization in Non-equilibrium Structures*. New York: Wiley.
Norberg-Schulz, C. 1971. *Existence Space and Architecture*. New York: Praeger.
Norberg-Schulz, C. 1980. *Genius Loci*. London: Academy Editions.
Northrup, F. S. (in Morris and Fox).
Odum, Eugene P. 1970. ``Optimum population and environment: A Georgian microcosm." *Current History* 58:355-366.
Odum, Eugene P. 1971. *Fundamentals of Ecology*. 3rd Edition. Philadelphia: Wm. B. Saunders.
Odum, Howard T. and Elisabeth C. Odum. 1981. *Energy Basis for Man and Nature*. New York: McGraw Hill.
Odum, William. 1988. Predicting ecosystem development following creation and restoration of wetlands. In J. Zelazny and J. S. Feierabend, eds., *Increasing Our Wetland Resources*. Washington: National Wildlife Federation Proceedings.
Oliver, Paul. 1997. *Encyclopedia of Vernacular Architecture of the World*. New York: Cambridge University Press.
Olson, Steve. 2003. *Mapping Human History*. Boston: Mariner Books.
Ong, Walter J. 1967. *In the Human Grain*. New York: MacMillan Co.
Ophuls, William. 1977. Ecology and the Politics of Scarcity. San Francisco: W. H. Freeman & Co.
Ornstein, R., ed. 1973. *The Nature of Human Consciousness*. San Francisco: Freeman.
Orr, David W. 2004. *The Nature of Design*. New York: Oxford University Press.
Otten, C., ed. 1971. *Anthropology and Art*. Garden City, New York: American Museum of Natural History.
Owings, N. A. 1969. *The American Aesthetic*. New York: Harper & Row. P
Pagels, H. 1982. *The Cosmic Code*: Quantum Physics as the Language of Nature. New York: Simon and Schuster.
Palumbi, Steven R. 2001. *The Evolution Explosion*. New York: Norton.
Partridge, Eric. 1983. *Origins*. New York: Greenwich House
Passmore, J. 1974. *Man's Responsibility for Nature*: Ecological Problems and Western Tradition. NC: Duckworth.
Peate, I. C., ed. 1930. *Studies in Regional Consciousness and Environment*. Oxford: Oxford University Press.
Peirce, C. S. 1955. *Selected Writings of Peirce*. Edited by J. Buchler. New York: Dover Publishing Company.
Pepper, Stephen C. 1958. *Sources of Value*. Berkeley: University of California Press.
Pepper, S. 1961. *World Hypotheses*. Berkeley: University of California Press.
Perry, D. A., T. Bell, and M. P. Amaranthus. 1992. "Mycorrhizal fungi in mixed species forests and other tales of positive feedback, redundancy, and stability," in M. Cannell, D. Malcom, and P. Robertson, eds. *The Ecology of Mixed-species Stands of Trees*. Oxford: Blackwell Scientific.
Perry, D. A. 1995. *Forest Ecosystems*. Baltimore: Johns Hopkins University Press.
Pfeffer, R. 1972. *Nietzsche: Disciple of Dionysius*. Lewisburg: Bucknell University Press.
Piaget, J. 1968. *Logical Thinking in Children*. I. Sigel, comp. New York: Holt, Rinehart and Winston. Page 6.
Pimm, Stuart L. 1991. *The Balance of Nature*. Chicago: University of Chicago Press.

Pirie, N. W. 1976. *Food Resources*. London: Pelican Books.
Planck, Max. 1959. *The New Science*. New York: Norton.
Plato. 1961. The Collected Dialogues of Plato. E. Hamilton and H. Cairns, eds. trans. L. Cooper et al. New York: Pantheon Books.
Poe, Edgar Allan. 1909. *The Works of Edgar Allan Poe*. 10 Vols. New York: The Century Co.
Poincare, Henri. 1905. *Science and Hypothesis*. trans. W. S. G. London: The Walter Scott Publishing Co.
Polunin, Nicholas, ed. 1980. *Growth without Ecodisasters*? New York: Wiley.
Polunin, Nicholas and John H. Burnett. Eds. *Maintenance of the Biosphere*: Proceedings of the Third International Conference on Environmental Future. New York: St. Martin's.
Ponge, Francis. 1972. *The Voice of Things*. B. Archer, ed. New York: McGraw-Hill Book Company.
Popper, K.R. and J.C. Eccles. 1977. *The Self and Its Brain*. Berlin: Springer-Verlag.
Popper, K. R. 1982. The place of mind in nature. R.Q. Elvee, ed. *Mind in Nature*. San Francisco: Harper & Row
Portmann, Adolf. 1964. *New Paths in Biology*. New York: Harper & Row.
Prabhu, R. and U. Rao, eds. 1946. *The Mind of Mahatma Gandhi* Madras: Oxford University Press.
Prehoda, Robert W. 1967. *Designing the Future*. Philadelphia: Chilton.
Pribram, Karl. 1977. Problems Concerning... In: *Consciousness and the Brain*, Globus et al., eds. New York: Plenum Press.
Price, David H. 1990. *Atlas of World Cultures*. London: Sage Publishers.
Prigogine, Ilya. 1980. *From Being to Becoming*. San Francisco: Freeman.
Radcliffe-Brown, A.R. 1952. *Structure and Function in Primitive Society*. London: Cohen and West.
Ramsay, William and Claude Anderson. 1972. "Economic analysis: medication for ecosystems," *Managing the Environment: An Economic Primer*. New York: Basic.
Rappoport, Amos. 1969. *House Form and Culture*. Englewood Cliffs: Prentice-Hall.
Rapport, DJ 1995. "Ecosystem health: An emerging integrative science." In Rapport, DJ, CL Gaudet, and P. Calow, eds. *Evaluating and Monitoring the health of large scale ecosystems*. pp. 5-34. Heidelberg: Springer.
Rapport, DJ, C Thorpe, and HA Regier, 1979. Ecosystem medicine. *Bul Ecol Soc Am* 60:180-182.
Reichel-Dolmatoff, G. 1977. Cosmology as Ecological Analysis: A view from the Rain Forest. *The Ecologist*, Vol 7, Pp. 4-11.
Reinheimer, Herman. 1913. *Evolution by Co-operation*: A Study in Bio-economics. London: Kegan Paul.
Relph, Edward. 1976. *Place and Placelessness*. London: Pion.
Riedl, Rupert. 1978. *Order in Living Organisms*. Translated by R. P. Jafferies. New York: J. Wiley & Sons.
Rifkin, J. 1982. *Algeny*: The Last Magic (prepublication copy).
Rodin, L.E., N.I. Bazilevich, and N.N. Rozov. 1975. ``Productivity of the world's main ecosystems." IN: *Productivity of World Ecosystems*. Washington, D.C.: National Academy of Sciences.
Rodman, John. 1977. "The Liberation of Nature?" *Inquiry* 20:83-145
Rodman, John. 1977. Theory and practice in the environmental movement: Notes toward

an ecology of experience. In: *The Search for Absolute Values in a Changing World.* Tarrytown, New York: International Cultural Foundation.
Rolston, III, H. 1983. "Values Gone Wild." *Inquiry* 26:181-207.
Rorty, Richard. 1982. Mind as ineffable. In *Mind in Nature*, ed. R. Elvee. Harper and Row, San Francisco, p. 88.
Ross, W., ed. 1952. *The Works of Aristotle*. Oxford: Clarendon.
Roszak, Theodore. 1972. *Where the Wasteland Ends*. New York: Harper and Row.
Roszak, Theodore. 1979. *Person/Planet*. New York: Harper and Row.
Rothenberg J. and G. Quasha, eds. 1974. *America A Prophecy*. New York: Vintage Books.
Rudolfsky, Bernard. 1965. *Streets for People*. New York: Doubleday.
Ruesch, Jergen and Weldon Kees. 1956. *Nonverbal Communication*. Berkeley: University of California Press.
Sahlins, Marshall. 1968. *Tribesmen*. Englewood Cliffs: Prentice Hall
Sahlins, Marshall. 1972. *Stone Age Economics*. Chicago: Aldine Publishing.
Salk, Jonas. ND. *Survival of the Wisest*. New York: Harper and Row.
Sapir, E. 1949. *Selected Writings of Edward Sapir in Language, Culture and Personality*. D. Mandelbaum, ed. Berkeley: University of California Press.
Sarfatti, J., and B. Toben. 1975. *Space-Time and Beyond*. New York: Dutton.
Schaller, G.B. 1972. *The Serengeti Lion*. Chicago: University of Chicago Press.
Scheler, M.F. 1954. *The Nature of Sympathy*. Tr. by P. Heath. London: Routledge and Kegan Paul, Ltd.
Schleiden, M. and Schwann, T. (see Maynard Smith).
Schrodinger, Erwin. 1946. *What is Life?* The Physical Aspect of the Living Cell. New York: Macmillan.
Schumacher, E.F. 1973. *Small Is Beautiful*. New York: Harper and Row.
Schweitzer, Albert. 1949. *Out of My Life and Thought*. New York: Henry Holt and Co.
Schweitzer, Albert. 1957. *The Philosophy of Civilization*. Translated by C. T. Campion. New York: Macmillan Co.
Searles, H. 1962. The role of the nonhuman environment. *Landscape* (Winter 1961-1962):31-34.
Sears, Paul. 1957. *The Ecology of Man*. Eugene: Oregon State System.
Segall, Marshall H, D. T. Campbell, and M. Herkovits. *The Influence of Culture on Human perception*.
Sharp, Henry. Comparative ethnology of the wolf and the Chipewyan. In *Man and Wolf*. H. Frank, Ed. Dordrecht: Dr. W. Junk.
Sheldrake, Rupert. 1981. *A New Science of Life*: The Hypothesis of Formative Causation. Los Angeles: J. P. Tarcher, Inc.
Shepard, Paul. 1967. *Man in the Landscape*. New York: Alfred Knopf.
Shepard, Paul and D. McKinley, eds. 1969. *The Subversive Science*. Boston: Houghton Mifflin.
Shepard, Paul. 1974. Animal rights and human rites. *The North American Review* Winter, p.35.
Shepard, Paul. 1978. *Thinking Animals*. New York: Viking Press.
Shepard, Paul. 1982. *Nature and Madness*. San Francisco: Sierra Club Books.
Short, John Rennie. 2001. *Global Dimensions*. London: Reaktion Books
Simmons, I. G. 1989. *Changing the Face of the Earth*. New York: Basil Blackwell.

Simon, Herbert A. 1969. *The Sciences of the Artificial.* Third Ed. Cambridge: MIT Press.
Simpson, G.G. 1944. *Tempo and Mode in Evolution*. New York: Hafner Publishing Co.
Singer, Peter. 1981. *The Expanding Circle: Ethics and Sociobiology*. New York: Farrar, Strauss & Giroux. (Page 62).
Singer, S. Fred, ed. 1971. *Is There an Optimum Level of Population?* New York: McGraw-Hill.
Skolimowski, Henryk. 1981. *Ecophilosophy*. Boston: Marion Boyars
Slater, Phillip. 1974. *Earthwalk*. New York: Bantam Books.
Smith, Maynard J. 1968. *Mathematical Ideas Biology*. Cambridge: Cambridge University Press.
Smithsonian Editors. Annual II. *The Fitness of Man's Environment*.
Smuts, J. 1926. *Holism and Evolution*. Ann Arbor, MI: University Microfilms.
Snyder, Gary. 1969. *Earth House Hold*. New York: New Directions. P. 105.
Snyder, Gary. 1995. *A Place in Space*. Washington: Counterpoint.
Soleri, Paolo. 1969. *Arcology: The City in the Image of Man*. Cambridge: The MIT Press.
Soleri, Paolo. 1978. A response to "Fields of Danger". *The North American Review*, Spring: pp. 71-72
Soleri, Paolo. 1983. *The Food Chain*: A Celebration. Scottsdale, Arizona. Sorokin, Pitirim. *Social and Cultural Dynamics*.
Soule, Michael and Wilcox, B. A., eds. 1980. *Conservation Biology: An Evolutionary-Ecological Perspective*. Sunderland, MA: Sinauer Associates.
Soule, Michael. 1986. *Conservation Biology: The Science of Scarcity and Diversity*. Sunderland: Sinauer Associates, Inc.
Speck, W. A. 1975. Mandeville and the Eutopia seated in the brain. In Primer, Irwin, ed. *Mandeville Studies* (1670-1733). The Hague: Martinus Nijhoff.
Spencer, Herbert. 1969 (1876-1896). *Principles of Sociology*. London: Macmillan.
Stanley, Steven. 1981. *The New Evolutionary Timetable*: Fossils, Genes and the Origin of Species. New York: Basic Books.
Stevens, P. S. 1974. *Patterns in Nature*. Boston: Little Brown Co.
Stevens, Wallace. 1974. *The Collected Poems of Wallace Stevens*. New York: Alfred A. Knopf.
Stock, Doroty and Herbert Thelen. *Emotional Dynamics and Group Culture*.
Stokes, Kenneth M. 1994. *Man and the Biosphere*: Toward a Coevolutionary Political Economy Armonk New York: ME Sharpe.
Stone, Christopher D. 1974. *Should Trees Have Standing?* New York: Avon Books.
Stulman, Julius and Ervin Laszlo. 1973. Emergent Man. New York: Gordon and Breach.
Stratton, G.M. 1896. Some preliminary experiments on vision without inversion of the retinal image. *Psych. Rev.* 3:611-617.
Susser, Bernard. 1981. *Existence and Utopia*: The Social and Political Thought of Martin Buber. Rutherford: Fairleigh Dickinson University Press.
Szekielda, Karl-Heinz. 1988. *Satellite Monitoring of the Earth*. New York: John Wiley.
Tainter, Joseph A. 1988. *The Collapse of Complex Societies*. Cambridge: Cambridge University Press.
Tansley, A.G. 1935. The use and abuse of vegetational concepts and terms. *Ecology* 16:284-307.
Tax, Sol. Ed. 1960. *Evolution After Darwin*. Chicago: University of Chicago Press.
Tax, Sol, ed. *Horizons of Anthropology*. NC: NP.
Taylor, Gordon Rattray. 1968. *The Biological Time Bomb*. New York: The World Publishing

Company.
Thakur, S.C. 1978. A Touch of Animism. In *Man and Nature*. G.F. McLean, ed. Calcutta: Oxford University Press.
Theobald, Robert. 1968. *An alternative future for America*. Chicago: Swallow Press, 1968.
Tinbergen, Jan, Coordinator. 1976. *Reshaping the International Order*, Report of the Club of Rome. New York: Dutton.
Thines, George. 1977. *Phenomenology and the Science of Behavior*. London: George Allen & Unwin.
Thom, Rene. 1975. *Structural Stability and Morphogenesis:* An Outline of a General Theory of Models. trans. D.C. Fowler. Reading: W.A. Benjamin.
Thomas, David H. 1979. *Archaeology*. New York: Holt, Rinehart and Winston.
Thomas, Keith. 1983. *Man and the Natural World*. New York: Pantheon
Thomas, Lewis. 1975. *Lives of a Cell*. New York: Bantam.
Thompson, W. I. 1971. *At the Edge of History*. Harper & Row.
Thompson, W. I. 1974. *Passages About Earth*. New York: Harper and Row.
Thompson, W. I. 1976. *Evil and World Order*. New York: Harper and Row.
Thompson, W. I. 1981. *The Time Falling Bodies Take to Light*. New York: Harper and Row.
Tibbs, B. C. 1992. Industrial Ecology: An environmental agenda for industry, *Whole Earth Review* 77:4-19.
Todd, John. 1977. Towards a sacred ecology. In *Earth's Answer*. pp. 170-183. M. Katz et al., eds. New York: Harper & Row
Todd, N. J. and J. 1994. *From Eco-Cities to Living Machines*: Principles of Ecological Design. Berkeley: North Atlantic Books.
Toffler, Alvin. 1970. *Future shock*. New York: Random House.
Toynbee, Arthur J. *Change and Habit*. NC: NP.
Tuan, Yi-Fu. 1974. *Topophilia*: A Study of Environmental Perception, Attitudes, and Values. Englewood Cliffs: Prentice-Hall.
Tuan, Yi-Fu. 1995. *Passing Strange and Wonderful*. New York: Kodansha International.
Tucker, William. 1982. Is Nature Too Good for Us? *Harper's* (March).
Turnbull C.M. 1961. *The Forest People*: A Study of the Pygmies of the Congo. New York: Simon and Schuster.
Tylor, Edward B. 1871. *Primitive Culture*. London: Murray.
Tylor, E.B. 1958. *The Origins of Culture*. New York: Harper Torchbooks.
Uexkull, J. von. 1957. A Stroll Through the World of Animals and Men. Schiller, Claire, ed. 1957. IN: *Instinctive Behavior*. New York: International Universities Press Inc.
Ulanowicz, Robert E. 1986. *Growth and Development: Ecosystems Phenomenology*. New York: Springer-Verlag.
United Nations. 1993. "Combating Deforestation," in *Earth Summit 92*. Washington: United Nations.
Vaihinger, Hans. 1961. *The Philosophy of As If*. C. K. Ogden, trans,. Bloomington: Indiana University Press.
Varela, Francisco et al. 1974. Autopoiesis: The organization of living systems. *Biosystems* 5:187-196.
Varela, Francisco. 1978. *Principles of Biological Autonomy*. New York: North Holland.
Varela, Francisco. Laying down a path in walking. In *Gaia: A Way of Knowing*.
Vayda, Andrew P., ed. 1977. *Environment and Cultural Behavior: Ecological Studies in Cultural*

Anthropology. Austin: Texas Press.
Vitousek, Paul M. 1992. Global environmental change. *Ann Rev. Ecol and Syst*. 23:1-14.
Wackernagel, M. and W. Rees. 1996. *Our Ecological Footprint*: Reducing Human Impact on the Earth. Gabriola Island, BC: New Society.
Waddington, C.H. 1960. *The Nature of Life*. New York: Atheneum.
Waddington, C.H., ed. 1969. *Towards a Theoretical Biology*. Chicago: Aldine Publishing Co.
Waddington, C.H. 1975.*The Evolution of an Evolutionist* Ithaca: Cornell University Press.
Wagner, Philip L. 1960. *The Human Use of the Earth*. Glencoe: Free Press.
Walsh, R. 1981. *Towards an Ecology of Brain*. Jamaica, New York: Spectrum Publishers.
Waltner-Toews D. and E. Wall. 1997. Emergent perplexity: in search of post-normal questions for community and agroecosystem health. *Soc Sci Med*, 45(11):1741-9.
Warner, A. W., D. Morse, and T. E. Cooney. The Environment of Change.
Watson, Richard A. and Philip M. Smith. 1970. "The Limit: 500 Million." *FOCUS/Midwest* 7(49):40-42.
Weil, Simone. 1955. *The Need for Roots*. Boston: Beacon Press.
Weiss, Gerald. 1975. *Campa Cosmology*: The World of a Forest Tribe in South America. Anthropological papers, Vol. 52, Part 5, Pp. 217-588. New York: The American Museum of Natural History.
Weiss, P. A. 1973. *The Science of Life*. Mount Cisco, New York: Futura.
Weiss, P. 1967. One Plus One Does Not Equal two. In *The Neurosciences*: A Study Program. G.C. Quarton et al., eds. New York: Rockefeller University Press.
Weiss, P. 1969. The Living System: Determinism Stratified. In *Beyond Reductionism: New Perspectives in the Life Sciences*. A. Koestler and J. Smythies, eds. New York: The Macmillan Co.
Welch, Holmes. 1966. *Taoism: The Parting of the Way*. Revised ed. Boston: Beacon Press
Weltfish, Gene. 1965. *The Lost Universe*. New York: Basic Books.
Westermarck, Edward. 1912. *The Origin and Development of the Moral Ideas. Vol. 1*. London: Macmillan.
Westing, Arthur H. 1981. ``A world in balance." *Environmental Conservation* 8(3):177-183.
Wheeler, J., W. C. Misner, and K. Thorne. 1973. *Gravitation*. San Francisco: Freeman.
Wheelwright, P. 1962. *Metaphor and Reality*. Bloomington: Indiana University Press.
White, Leslie A. 1959. *The Evolution of Culture*. New York: McGraw Hill.
Whitehead, Alfred N. 1933. *Adventures of Ideas*. New York: Macmillan.
Whitehead, Alfred N. 1938. *Modes of Thought*. New York: Macmillan.
Whitehead, Alfred N. 1967. *Science and the Modern World*. New York: Free Press.
Whitehead, Alfred N. 1969. *Process and Reality*. New York: Free Press.
Whitehead, Alfred N. 1978. *Process and Reality* (corrected edition). New York: Free Press.
Whittaker, R.H., F.H. Bormann, G.E. Likens, and T.G. Siccama. 1974. ``The Hubbard Brook ecosystem study: Forest biomass and production." *Ecological Monographs* 44:233-252.
Whyte, L.L. 1965. *Internal Factors in Evolution*. New York: Braziller.
Wiener, Norbert. 1962. *Cybernetics*. Cambridge: MIT Press.
Wigner, E. 1963. The problem of measurement. *Am J. Phys*. 31: np.
Willard, B. E. et al. 1977. "Ethics of Biospheral Survival: A dialogue." In *Growth Without Ecodisasters?* pp. 505-535. N. Polunin, ed. New York: John Wiley & Sons
Wilson, E.O. 1975. *Sociobiology*: The New Synthesis. Cambridge: Belknap Press.

Wilson, E.O. 1984. *Biophilia.* Cambridge: Harvard University Press.
Wingo, Lowdon. ed. *Cities and Space: The Future Use of Urban Land.* NC: NP.
Wittbecker, A.E. 1970. *Eutopias: A Poetic Commonwealth of Earth.* (Printed Bound Essay). Newark: Shamrock Press.
Wittbecker, A.E. 1976. "The psychology of catastrophe: Environmental deterioration and rapid social change." *Proc. Marsh Inst.* 1:1-17.
Wittbecker, A.E. 1983. "Human populations related to ecosystem productivities." Contributed paper, Ecol. Soc. Am. annual meeting, Grand Forks.
Wittbecker, A.E. 1983. "Ecology, mythology, and holopoetic culture." Montreal: Proceedings XVII World Congress Phil.
Wittbecker, A.E. 1983. "Quantitative determinations of minimum wilderness areas for the planet as a whole." Contributed paper, 3rd World Wilderness Congress, Findhorn.
Wittbecker, A.E. 1986. "Palouse Ecosystem Restoration." 4th International Congress of Ecology, Syracuse.
Wittbecker, A.E. 1986. "The role of deep ecology in wilderness preservation." Contributed paper, *4th International Congress of Ecology*, Syracuse.
Wittbecker, A.E. 1986. "The place of human society in wilderness." *The Trumpeter.* 3(3):34-38.
Wittbecker, A.E. 1989. "Nature as self." *The Trumpeter* 6(3):77-81.
Wittbecker, A.E. 1991. "An empowered United Nations: Proposals for cooperation and survival." *Common Voice*: 1(1):1-8.
Wittbecker, A.E. 1992. "Drawings & Discussion of a Proposed Palouse Arcology." *Proc. Marsh Inst.* 17:44-61.
Wittbecker, A.E. 1992. "Setting community limits for long-term stability." *International Forum for Biophilosophy*, Budapest.
Wittbecker, A.E. 1995. "Saving common places: The Palouse," *Wild Earth* 5(1):54-58.
Wittbecker, A.E. 1995. "Limits and Goals for communities on the North Slope of Alaska," *Pan Ecology* 10(2):1-14.
Wittbecker, A.E. 1999. "Forestry as Ecosystem Medicine and Poetic Activity," Closing Address, *Forests for the Future*, Vancouver Island.
Wittbecker, A.E. 2001. "Ecological Thought Experiments" *Sofia Echo* Vol. 5, Issue 31, Aug 3-9, p. 12 (1st of 12-part series).
Woodwell, George M. and Robert Whittaker. 1968. ``Primary production in terrestrial ecosystems." *American Zoologist* 8:19-30.
Woodwell, George M. and Robert Whittaker. 1975. *Primary Productivity of the Biosphere.* New York: Springer-Verlag.
Wright, Robert. *Nonzero*: The Logic of Human Destiny. 2000. New York: Vintage.
Wynne-Edwards, V. C. Animal Dispersion in Relation to Social Behavior.
Young, L. B. *Population in Perspective.*
Zadeh, L. A. 1965. "Fuzzy Sets" *Information and Control* Vol. 8:338-353.
Zajonc, Robert. 1980. Feeling and thinking: Preferences need no inferences. *American Psychologist* 35:151-175.
Zeleny, Milan, ed. 1980. *Autopoiesis: A Theory of Living Organisms.* New York: North Holland.
Zonneveld, I. S. and RTT Forman, eds. 1990. Changing *Landscapes*: An Ecological perspective. New York: Springer-Verlag.

7.4. *Index* (Exploded Contents)

2.3.1. World as Field
2.3.1.1. Characteristics of Field
2.3.1.2. Operation of the Field
2.3.1.2.1. Motion
2.3.1.2.2. Version
2.3.1.2.3. Scension
2.3.1.2.4. Tension
2.3.1.2.5. Formation
2.3.1.2.6. Novation
2.3.1.3. Metaphysics in Action
2.3.2. World as Source
2.3.2.1. Materials (Things Plants & Animals)
2.3.2.1.1. Cultural use
2.3.2.1.2. Accumulation
2.3.2.2. Energy
2.3.2.2.1. Cultural Use
2.3.2.2.2. Quantity & Efficiency
2.3.2.3. Historical Mixture
2.3.2.3.1. Cycles
2.3.2.3.2. Scale
2.3.3. World as Sink
2.3.3.1. Structure of Place
2.3.3.1.1. Characteristics of Place
2.3.3.1.2. Connection & Isolation in Place
2.3.3.1.3. Limits of Place
2.3.3.1.3.1. Physical Limits & Locality
2.3.3.1.3.2. Limits to Freedom & Connections
2.3.3.1.3.3. Biological Limits & Carrying Capacity
2.3.3.1.3.4. Human Limits
2.3.3.1.3.4.1. Physical & Psychological Limits
2.3.3.1.3.4.2. Cultural Limits & Human Cosmologies
2.3.3.1.3.4.3. Economic Limits & Growth
2.3.3.1.3.4.4. Political Limits & Security
2.3.3.1.3.4.5. Necessity of Limits
2.3.3.2. Culture in Place
2.3.3.2.1. Dwellings & Territories
2.3.3.2.2. Patterns on the Land
2.3.3.2.2.1. Herding Paths
2.3.3.2.2.2. Agriculture Patches
2.3.3.2.2.3. Urban Paths & Patches
2.3.4. World as Flesh
2.3.4.1. Primacy of Perception
2.3.4.2. Primacy of Imagination
2.3.4.3. Primacy of Expression
2.4. *Connecting Dreams—Outopias or Eutopias*
3.0. Utopia: No Place
3.0.1. Images Cast Shadows
3.0.2. One World Through Reason—The Place of the UN (Idea & History)
3.1. *The Lure of Nowhere*
3.1.1. Getting Nowhere
3.1.1.1. The Gift of Bigness (& Corporate Bigness)

4.2.2.2.1.3.3.3.2.3.2. Institutional Slavery
4.2.2.2.1.3.3.3.2.3.3. Comprehensive Slavery
4.2.2.2.1.3.3.4. Disadvantages of Agriculture
4.2.2.2.1.3.3.5. Systems of Agriculture
4.2.2.2.1.3.4. Promise of Regenerative Agriculture
4.2.2.2.1.3.4.1. Natural Farming
4.2.2.2.1.3.4.2. Regenerative Permaculture
4.2.2.2.1.3.4.3. Mimic Agriculture (Land Institute)
4.2.2.2.1.3.5. Refitting Agriculture
4.2.2.2.2. Humanity as Agents of Change Forces of History
4.2.2.2.2.1. Humanity as Biological Agents
4.2.2.2.2.2. Humanity as Ecological Agents
4.2.2.2.2.3. Humanity as Geological or Climactic Force
4.2.2.3. Human Culture
4.2.2.3.1. Copying & Animal Culture
4.2.2.3.2. Imitation & Human Culture
4.2.2.3.2.1. Why Culture?
4.2.2.3.2.2. Definitions of Culture
4.2.2.3.2.3. Requirements for Culture
4.2.2.3.2.4. Characteristics of Culture
4.2.2.3.2.5. Components of Culture
4.2.2.3.2.5.1. Material Component of Culture
4.2.2.3.2.5.2. Nonmaterial Component of Culture
4.2.2.3.2.6. Functions of Culture
4.2.2.3.3. Effectiveness of Culture
4.2.2.3.3.1. Strengths of Culture
4.2.2.3.3.2. Weaknesses of Culture
4.2.2.3.4. Diversity & Universals of Cultures
4.2.2.3.5. Changes & Cultural Transformations
4.2.2.3.5.1. Patterns & Renewal
4.2.2.3.5.2. Challenges & Traps
4.2.2.3.5.3. Conflict Collapse & Ruin
4.2.2.3.5.3.1. Conflict & Complexity
4.2.2.3.5.3.2. Collapse
4.2.2.3.5.4. Intensification & Civilization
4.2.2.3.5.5. Cultural Conversion of Places
4.2.2.3.5.6. Global Culture
4.2.2.3.5.7. The Future in Balance: Cultural Exchange Health
4.2.3. Domiture & Pan Ecology
4.2.3.0.1. Human Needs
4.2.3.0.2. Cultural needs
4.2.3.1. Imagining Place: Human Cosmologies
4.2.3.1.1. Imagination & Metabolism
4.2.3.1.2. Holocosmological Framework
4.2.3.2. Fitting in Place — Image & Place
4.2.3.2.1. Living Field
4.2.3.2.2. Living Together
4.2.3.2.3. Making Things & Homes
4.2.3.3. Attaching to Place (Loving a Place)
4.2.3.3.1. Topophilia
4.2.3.3.2. Biophilia

4.4. *Managing the Place: Economics*
  4.4.1. History of Making a Living
  4.4.2. Styles of Economics
    4.4.2.1. Subsistence Reciprocity & Traditional Economics
      4.4.2.1.1. Households Reciprocity (Bands)
      4.4.2.1.2. Distribution (Tribes) Exchange & Accumulation
      4.4.2.1.3. Redistribution
        4.4.2.1.3.1. Chiefdoms
        4.4.2.1.3.2. Command Economies
          4.4.2.1.3.2.1. Tributary
          4.4.2.1.3.2.2. Socialism
        4.4.2.1.3.3. Market Economies
          4.4.2.1.3.3.1. Invisible Hands
          4.4.2.1.3.3.2. Controlling Minds
    4.4.2.2. Modern Model: Abstraction & Accumulation Economy
      4.4.2.2.1. Corporations Distribution (Independence without Limits)
      4.4.2.2.2. Ownership Capitalism & Scale
      4.4.2.2.3. Use of Commons
      4.4.2.2.4. Capitalism & Growth
      4.4.2.2.5. Growth of Industrialism
      4.4.2.2.6. Industrialism & Accidental Globalization
      4.4.2.2.7. Global Capital Conversion & Erasure
        4.4.2.2.7.1. Positive & Negative Effects of Global Economics
        4.4.2.2.7.2. Globalism & Localism
      4.4.2.2.8. Myths of Modern Economics
  4.4.3. Functions of Economics
  4.4.4. Holeconomics
    4.4.4.1. Ecological Economics
      4.4.4.1.1. Ownership & Responsibility
        4.4.4.1.1.1. Personal Ownership & Responsibility
        4.4.4.1.1.2. Ecological Responsibility of Corporations.
        4.4.4.1.1.3. Community or State Ownership
      4.4.4.1.2. Holistic Accounting
        4.4.4.1.2.1. Full Costing Accounting
        4.4.4.1.2.2. Work Tracking
      4.4.4.1.3. Steady State Community Economics
    4.4.4.2. Wealth
      4.4.4.2.1. Wealth of the Earth
        4.4.4.2.1.1. Resources
        4.4.4.2.1.2. Net Primary Production
      4.4.4.2.2. Wealth of Human Contributions to Place
        4.4.4.2.2.1. Labor as Wealth
        4.4.4.2.2.2. Production & Imagination as Wealth
        4.4.4.2.2.3. Holeconomic Wealth
      4.4.4.2.3. Poverty of Humanity & the Earth
    4.4.4.3. Holoeconomic Goals
      4.4.4.3.1. Expanding capital
      4.4.4.3.2. Diversifying Institutions
      4.4.4.3.3. Emphasizing Development
      4.4.4.3.4. Accommodating Limits
4.5. *Managing the Inhabitants of Places: Ecocybernics* (or Politics)

5.2.1.2.4. To Be Peaceful: Gandhian Nonviolence
5.2.1.2.5. To Share in the Governing Process
5.2.1.3. Responsibilities of Individuals in Groups
5.2.1.3.1. Kinds & Importance (Families, Community, County, State)
5.2.1.3.2. Responsibilities of Groups of Individuals
5.2.1.3.2.1. To Educate Individuals
5.2.1.3.2.2. To Ensure Health & Safety
5.2.1.3.2.3. To Ensure Resources & Food
5.2.1.3.2.4. To Provide Ways of Movement & Transport
5.2.2. Being Nations in Good Places
5.2.2.1. The Import of Recognizing Nations
5.2.2.1.1. Regional Differences between Nations
5.2.2.1.2. Regional Politics of Nations
5.2.2.2. Responsibilities of Nations
5.2.2.2.1. To Conserve Ecosystems: Boundaries
5.2.2.2.2. To Manage Resources: Distribution
5.2.2.2.3. To Maintain People: Population
5.2.2.2.4. To Provide Paths of Power for People
5.2.2.2.5. To Provide Opportunities for Needs of People
5.2.2.2.6. To Provide Paths for Resolution of Conflicts Rights & Justice
5.2.3. Being in an International Framework in Good Places
5.2.3.1. Necessity of the Framework (the UN Currently)
5.2.3.1.1. Structure of the UN Main Organs
5.2.3.1.1.1. General Assembly
5.2.3.1.1.2. Security Council
5.2.3.1.1.3. Economic & Social Council
5.2.3.1.1.4. Trusteeship Council
5.2.3.1.1.5. Secretariat
5.2.3.1.1.6. International Court of Justice
5.2.3.1.1.7. Special Agencies: IMF WB WHO
5.2.3.1.2. Purposes Goals & Actions of the UN
5.2.3.1.2.1.To Maintain Peace & Security
5.2.3.1.2.1.1. Disarmament
5.2.3.1.2.1.2. Peacemaking
5.2.3.1.2.1.3. Peacekeeping
5.2.3.1.2.2.To Develop Friendly Relations between Nations
5.2.3.1.2.3.To Cooperate in Solving Problems
5.2.3.1.2.3.1. Promote Rights
5.2.3.1.2.3.2. Promote Health
5.2.3.1.2.3.3. Promote Education
5.2.3.1.2.3.4. Development
5.2.3.1.2.3.5. Poverty Protect Vulnerable Africa
5.2.3.1.2.3.6. Human Assistance Emergency Services
5.2.3.1.2.4.To Harmonize the Actions of All Nations
5.2.3.1.2.4.1. Governance
5.2.3.1.2.4.2. Environment
5.2.3.1.3. Activities of the UN
5.2.3.1.3.1.Conferences
5.2.3.1.3.2.Years
5.2.3.1.3.3.Treaties & International law
5.2.3.1.4. Inadequacies of the UN

- 5.3.4.1.2. Group/Community Steps
  - 5.3.4.1.2.1. Defining Itself as a Community
    - 5.3.4.1.2.1.1. Setting Goals of the Community
    - 5.3.4.1.2.1.2. Implementing Ecological Goals
    - 5.3.4.1.2.1.3. Convening a Constitutional Meeting
  - 5.3.4.1.2.2. Balancing Budgets of the Community
    - 5.3.4.1.2.2.1. Income for the Community
      - 5.3.4.1.2.2.1.1. Taxes for the Community
          - 5.3.4.1.2.2.1.1.0.1. Traditional Old Kinds of Taxes
          - 5.3.4.1.2.2.1.1.0.2. Effects of Traditional Taxes
          - 5.3.4.1.2.2.1.1.0.3. New Taxes for Community or State
        - 5.3.4.1.2.2.1.1.1. Use Taxes for a Community
          - 5.3.4.1.2.2.1.1.1.1. Air Use Tax
          - 5.3.4.1.2.2.1.1.1.2. Water Use Tax
          - 5.3.4.1.2.2.1.1.1.3. Land Use Tax
          - 5.3.4.1.2.2.1.1.1.4. Elements Use tax
          - 5.3.4.1.2.2.1.1.1.5. Species Use Tax
        - 5.3.4.1.2.2.1.1.2. Loss Taxes for a Community
          - 5.3.4.1.2.2.1.1.2.1. Land Loss tax
          - 5.3.4.1.2.2.1.1.2.2. Nonrenewables Loss tax
          - 5.3.4.1.2.2.1.1.2.3. Slow Renewables Loss Tax
          - 5.3.4.1.2.2.1.1.2.4. Fast Renewables Loss Tax
          - 5.3.4.1.2.2.1.1.2.5. Waste Loss tax
        - 5.3.4.1.2.2.1.1.3. Adjustment Taxes for a Community
          - 5.3.4.1.2.2.1.1.3.1. Sin Adjustment Taxes
            - 5.3.4.1.2.1.4.1.1.3.1.1. Cigarette Tax
            - 5.3.4.1.2.1.4.1.1.3.1.2. Alcohol Tax
            - 5.3.4.1.2.1.4.1.1.3.1.3. Drug Tax
          - 5.3.4.1.2.2.1.1.3.2. Pollution Adjustment Tax
            - 5.3.4.1.2.2.1.1.3.2.1. Industrial
            - 5.3.4.1.2.2.1.1.3.2.2. Agricultural
            - 5.3.4.1.2.2.1.1.3.2.3. Personal
          - 5.3.4.1.2.2.1.1.3.3. Sale of Heritage Items Adjust Tax
          - 5.3.4.1.2.2.1.1.3.4. Financial Speculation Adjust Tax
        - 5.3.4.1.2.2.1.1.4. Distribution Taxes for a Community
          - 5.3.4.1.2.2.1.1.4.1. Heroic Luxuries Distribution Tax
          - 5.3.4.1.2.2.1.1.4.2. Heroic Income Distribution Tax
          - 5.3.4.1.2.2.1.1.4.3. Heroic Inheritance Dist. Tax
          - 5.3.4.1.2.2.1.1.4.4. Heroic Profit Distribution Tax
        - 5.3.4.1.2.2.1.1.5. Discussion of Missing Taxes: Corporate
      - 5.3.4.1.2.2.1.2. License Privileges for a Community
        - 5.3.4.1.2.2.1.2.1. Marriage License for a Community
        - 5.3.4.1.2.2.1.2.2. Reproduction License
        - 5.3.4.1.2.2.1.2.3. Driving & Vehicles Licenses
        - 5.3.4.1.2.2.1.2.4. Voting License
        - 5.3.4.1.2.2.1.2.5. Airwaves Spectrum License
        - 5.3.4.1.2.2.1.2.6. Business License
        - 5.3.4.1.2.2.1.2.7. Collecting License
        - 5.3.4.1.2.2.1.2.8. Weapons License
      - 5.3.4.1.2.2.1.3. Fees & Tolls for a Community
      - 5.3.4.1.2.2.1.4. Labor Use for Community

- 5.3.4.1.2.2.2. Payout for Community
  - 5.3.4.1.2.2.2.1. Issue vouchers from Community
    - 5.3.4.1.2.2.2.1.1. Basic Income Voucher
    - 5.3.4.1.2.2.2.1.2. Medical Voucher
    - 5.3.4.1.2.2.2.1.3. Productivity Resource Vouchers
      - 5.3.4.1.2.2.2.1.3.1. Carbon
      - 5.3.4.1.2.2.2.1.3.2. Others
    - 5.3.4.1.2.2.2.1.4. Education Voucher
    - 5.3.4.1.2.2.2.1.5. Children Voucher
    - 5.3.4.1.2.2.2.1.6. Other Vouchers
  - 5.3.4.1.2.2.2.2. Community Costs for Health & Maintenance
    - 5.3.4.1.2.2.2.2.1. Promote Self-reliance
    - 5.3.4.1.2.2.2.2.2. Operate Community Government
    - 5.3.4.1.2.2.2.2.3. Encourage New Business
- 5.3.4.1.2.3. Assess Community Performance
- 5.3.4.1.3. Future & Leadership in the Community
- 5.3.4.2. Contributing to a Nation
  - 5.3.4.2.1. Steps for a Nation & Catastrophic Steps
    - 5.3.4.2.1.1. Changing Nation to Larger Self in Environment
      - 5.3.4.2.1.1.1. Defining Secure Borders & Lines of a Nation
      - 5.3.4.2.1.1.2. Balancing Isolation & Connections of a Nation
      - 5.3.4.2.1.1.3. Balancing Immigration & Emigration of a Nation
      - 5.3.4.2.1.1.4. Securing Treaties with Neighboring Nations
      - 5.3.4.2.1.1.5. Setting Standards for Nation
        - 5.3.4.2.1.1.5.1. Ecosystem Standards for Nation
        - 5.3.4.2.1.1.5.2. Social Standards for Nation
        - 5.3.4.2.1.1.5.3. Economic Standards for Nation
        - 5.3.4.2.1.1.5.4. Technological Standards for Nation
      - 5.3.4.2.1.1.6. Establishing Rights in Nation
        - 5.3.4.2.1.1.6.1. Rights for Nature in Nation
          - 5.3.4.2.1.1.6.1.1. Opportunity to Flourish in Nation
          - 5.3.4.2.1.1.6.1.2. Freedom from Suffering & Death
        - 5.3.4.2.1.1.6.2. Rights for People of Nation
          - 5.3.4.2.1.1.6.2.1. Basic Rights in Nation
            - 5.3.4.2.1.1.6.2.1.1. Right to Healthy Environment
            - 5.3.4.2.1.1.6.2.1.2. Right to Be Secure
            - 5.3.4.2.1.1.6.2.1.3. Right to Opportunity for Home
            - 5.3.4.2.1.1.6.2.1.4. Right to Opportunity to Work
          - 5.3.4.2.1.1.6.2.2. Tradable Rights
            - 5.3.4.2.1.1.6.2.2.1. Right to Have Children
            - 5.3.4.2.1.1.6.2.2.2. Right to Share in Luxury
            - 5.3.4.2.1.1.6.2.2.3. Other Rights
        - 5.3.4.2.1.1.6.3. Work to Establish Justice
      - 5.3.4.2.1.1.7. Representing All People of Nation
        - 5.3.4.2.1.1.7.1. Obligation to Protect Rights & Privileges
        - 5.3.4.2.1.1.7.2. Obligation to Meet Basic Needs of Nation
        - 5.3.4.2.1.1.7.3. Integrate Everyone into Society
    - 5.3.4.2.1.2. Amend Constitutions to Control Corporations & Groups
      - 5.3.4.2.1.2.1. Encourage Small Businesses in Nation
      - 5.3.4.2.1.2.2. Protect Citizenship of Nation
    - 5.3.4.2.1.3. Create Long-term Ecological Planning for Nation

5.3.4.3.1.1.2.5. Being Fallible
5.3.4.3.1.1.2.6. Being Naïve
5.3.4.3.1.2. Starting Catastrophic Measures
5.3.4.3.1.2.1. Transfer Powers
5.3.4.3.1.2.2. Disarm Nations
5.3.4.3.1.2.2.1. Take Nuclear or Large-scale Weapons
5.3.4.3.1.2.2.2. Allow Personal Weapons
5.3.4.3.1.2.3. Start Year of Consideration
5.3.4.3.1.2.4. Start Equity Measures
5.3.4.3.1.2.5. Implement Ecological Goals
5.3.4.3.2. Managing Nations
5.3.4.3.2.1. Describe Functions, Charter & Constitution
5.3.4.3.2.2. Create New Structure & Describe Branches
5.3.4.3.2.2.1. General Assembly (Legislative) with All Nations
5.3.4.3.2.2.1.0.1. Define Limits of Global Law
5.3.4.3.2.2.1.0.2. Make Laws
5.3.4.3.2.2.1.1. Office of Normalization
5.3.4.3.2.2.1.2. Office of Standards & Rates
5.3.4.3.2.2.1.3. Office of Budget (Secure & Balance Budget)
5.3.4.3.2.2.1.3.1. Income for Global Association
5.3.4.3.2.2.1.3.1.1. Lease (Territories & Oceans)
5.3.4.3.2.2.1.3.1.2. Membership Dues
5.3.4.3.2.2.1.3.1.3. Taxes on Global Things
5.3.4.3.2.2.1.3.1.3.1. Use (Air, water cycles)
5.3.4.3.2.2.1.3.1.3.2. Loss Renewables, waste)
5.3.4.3.2.2.1.3.1.3.3. Adjustment (Sin pollution)
5.3.4.3.2.2.1.3.1.3.4. Distribution (luxuries profit)
5.3.4.3.2.2.1.3.1.4. Licensing of Global Activities
5.3.4.3.2.2.1.3.1.4.1. International Corporations
5.3.4.3.2.2.1.3.1.4.2. Satellites/air
5.3.4.3.2.2.1.3.1.4.3. Collect (Plants & Animals)
5.3.4.3.2.2.1.3.1.4.4. National Weapons
5.3.4.3.2.2.1.3.1.5. Fees & Tolls for Global Activities
5.3.4.3.2.2.1.3.1.5.1. Common Wealth & Art
5.3.4.3.2.2.1.3.1.5.2. Wilderness
5.3.4.3.2.2.1.3.1.6. Labor for Global Projects
5.3.4.3.2.2.1.3.2. Payout for Global association
5.3.4.3.2.2.1.3.2.1. Operate Organization & Agencies
5.3.4.3.2.2.1.3.2.1.1. Health Supplement
5.3.4.3.2.2.1.3.2.1.2. Education Supplement
5.3.4.3.2.2.1.3.2.1.3. Global Planning
5.3.4.3.2.2.1.3.2.1.4. Global Resource Monitoring
5.3.4.3.2.2.1.3.2.2. Restore Global Cycles or Places
5.3.4.3.2.2.1.3.2.3. Create Set-asides for Catastrophes
5.3.4.3.2.2.2. Executive Branch of Global Association
5.3.4.3.2.2.2.0.1. Oversee National Enforcement of Laws
5.3.4.3.2.2.2.0.2. Enforce Global Laws
5.3.4.3.2.2.2.1. Coordinators of Global Association
5.3.4.3.2.2.2.2. Security Council (Safety)
5.3.4.3.2.2.2.2.1. Assess Threats (Internal & External)
5.3.4.3.2.2.2.2.1.1. International Conflict

5.3.4.3.2.2.2.2.1.2. Global Threats
5.3.4.3.2.2.2.2.1.2.1. Geological
5.3.4.3.2.2.2.2.1.2.2. Solar System
5.3.4.3.2.2.2.2.2. Provide Security with Police Force
5.3.4.3.2.2.2.2.2.1. Expand Police Force
5.3.4.3.2.2.2.2.2.2. Use Police Force
5.3.4.3.2.2.2.2.2.2.1. Deal with Conflicts
5.3.4.3.2.2.2.2.2.2.2. Protect Against Disasters
5.3.4.3.2.2.2.3. Agencies of the Global Association
5.3.4.3.2.2.2.3.1. Food & Shelter Agency
5.3.4.3.2.2.2.3.2. All Energy Agency
5.3.4.3.2.2.2.3.3. Transportation Agency
5.3.4.3.2.2.2.3.4. Economies Agency
5.3.4.3.2.2.2.3.5. Education Support Agency
5.3.4.3.2.2.2.3.6. Health Agency
5.3.4.3.2.2.2.3.6.1. Health
5.3.4.3.2.2.2.3.6.2. Monitor Global Diseases & Threats
5.3.4.3.2.2.2.3.6.3. Chemicals & Drugs
5.3.4.3.2.2.2.3.7. Communications & Technology Agency
5.3.4.3.2.2.2.3.8. Research Agency
5.3.4.3.2.2.2.3.9. Business & Corporation Agency
5.3.4.3.2.2.2.3.9.1. Incorporation of Earth
5.3.4.3.2.2.2.3.9.2. Recharter Internat'l Corporations
5.3.4.3.2.2.2.3.10. Office of Personnel & Service
5.3.4.3.2.2.2.3.10.1. Permanent & Part-time
5.3.4.3.2.2.2.3.10.2. Office of Volunteers service
5.3.4.3.2.2.2.3.11. Office of Self-assessment & Future
5.3.4.3.2.2.3. Judicial Branch International Court of the AN
5.3.4.3.2.2.3.0.1. Monitor Other Branches & All Laws
5.3.4.3.2.2.3.0.2. Interpret Laws
5.3.4.3.2.2.3.1. Conflict Court
5.3.4.3.2.2.3.2. Environment Court
5.3.4.3.2.2.3.3. Office of Rights (Human & Nonhuman)
5.3.4.3.2.2.3.4. Tax Court
5.3.4.3.2.2.3.5. Police Enforcement Court
5.3.4.3.2.2.3.6. Office of International Trade & Equalization
5.3.4.3.2.2.4. Commission on the Earth
5.3.4.3.2.2.4.1. Environmental Surveys & Monitoring
5.3.4.3.2.2.4.1.1. Create Inventories
5.3.4.3.2.2.4.1.2. Create Monitoring of Global Territories
5.3.4.3.2.2.4.1.2.1. Atmosphere Monitoring
5.3.4.3.2.2.4.1.2.2. Oceans Monitoring
5.3.4.3.2.2.4.1.2.3. Deep Continents Monitoring
5.3.4.3.2.2.4.1.2.4. Bioregions Watersheds Monitoring
5.3.4.3.2.2.4.1.2.5. Antarctica & Space Monitoring
5.3.4.3.2.2.4.2. Ecodesign & Planning for Commission on Earth
5.3.4.3.2.2.4.2.1. Create Long-term Ecological Plans
5.3.4.3.2.2.4.2.2. Protect Planetary Hotspots
5.3.4.3.2.2.4.2.3. Keep Critical Areas Intact
5.3.4.3.2.2.4.2.4. Restore Large Areas
5.3.4.3.2.2.4.2.5. Anticipate Climate Change

7.5. **Name Index**

7.6. Biography

During a brief career in astrophysics and astronomy at the University of Arizona, where he worked on mathematical models of stars and on spectrometric analysis, Alan Wittbecker spent his daylight hours climbing trees and trying to track mountain lions; his companions were a mouse and squirrel, who shared his trailer near the observatory on Mount Lemon.

Encouraged by research budget cuts to pursue a different direction, Wittbecker went to graduate school in psychology, anthropology, philosophy, and ecology. As a graduate student in 1970, he was a cofounder of the G. P. Marsh Institute for Research in Ecology, where he worked for 22 years, including three separate years as Director, as the position rotated annually. He worked on a wide variety of projects, from forest monitoring and ecosystem restoration to country-wide wolf monitoring, in many countries, including Bulgaria, Canada, Mexico, Norway, and Russia. When funding was in short supply, he worked in other occupations, including librarian, systems engineer, editor, graphic artist, typesetter, housepainter, television repairman, cook, swimming coach, carpenter, clinical psychologist for a drug abuse clinic, Austin Healy auto mechanic, tree-planter, and college instructor.

In 1976, with three partners, Wittbecker cofounded Nieman Ryan Community Designs, specializing in private and urban local landscape design—but, also designing books, posters, journals, packages, landscapes, and buildings. In 1983, he finished his doctorate at International College, working with Michael W. Fox, John B. Cobb, Jr., Paolo Soleri, David Klein, Henryk Skolimowski, Neil Evernden, Paul Shepard, and Buckminster Fuller. He continued his postgraduate education in landscape ecology, forestry, conservation biology, zoology, and genetics.

In 1991, Wittbecker founded SynGeo ArchiGraph, a firm specializing in global and regional designs; he created designs for several bioregions, as well as international frameworks. A year later he set up the educational program for the new Ecoforestry Institute, becoming an Instructor in 1994, journal Editor in 1995, and Director from 1997 to 2006. He has worked on public and private forests from British Columbia to California, and on wildlife projects, from Siberia to Norway.

He is the author of eight books, including *REviewing REthinking REturning*, which won an Eppie award for best nonfiction on the web, *Fragments*, a finalist with the National Poetry Series, and *Good Theories Good Practices*, which won an award for best essay, for "The Health of Forests," and over 100 articles. He has also written series in ecology and forestry for newspapers and journals.

A veteran of the U.S. Air Force, Wittbecker is also a returned Peace Corps Volunteer from Bulgaria, where he monitored wolves in the Central Balkan Mountains. He has used his education and interests to explore a spectrum of ecological applications, from research on forest pests—larch casebearers, cedar powderworms, coyotes, and bears—to the political implications of the protection of species and habitats.

When not engaged in preservation activities, he enjoys walking, swimming, reading, and drawing, at the Altazor forest in western Idaho. To discuss any of these essays with him, contact him at home@eutopias.net.

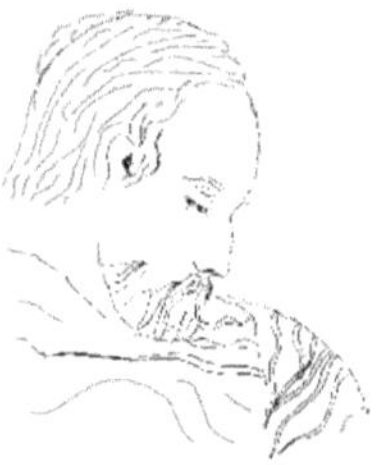

7.7. **Author's Note**

To make up for the loss of trees and their services, as a result of my use of paper in these books, I have planted over nine thousand trees, during a period of twenty years, at the Altazor Forest in Idaho. More plantings are planned in Oregon forests and Virginia forest farms.

Colophon

Type: Garamond
Display Type: Garamond
Book Design: Rian Garcia Calusa Designs
Cover Design: Rian Garcia Calusa (based on the 1934 cover page of Thomas More's *Utopia* by the Limited Editions Club)
Graphics: Alan Wittbecker
Author Drawing: Merissa DePasse, 1994
Editing: J. Garcia B. of Rian Garcia Calusa
Hardware: Macintosh G5
Software: Adobe InDesign & Acrobat
Furious Charge & Entertainment: Pippi Frog
Spiritual & Material Support: Precious Woulfe

www.ingramcontent.com/pod-product-compliance
Lightning Source LLC
LaVergne TN
LVHW061216100826
845148LV00004B/775

* 9 7 8 0 9 1 1 3 8 5 4 3 4 *